Pollution Prevention and Control Technology for Plating Operations

—Second Edition—

George C. Cushnie Jr.
CAI Resources, Inc.

ncms

A project sponsored by the National Center for Manufacturing Sciences

Library of Congress Cataloging-in-Publication Data

Cushnie, George C., Jr
Pollution Prevention and Control Technology for Plating Operations.
Includes bibliographical references.
1. Electroplating – Metal recovery.
2. Waste minimization – Handbooks, manuals, etc.
2008943337
CIP
ISBN 978-0-615266-39-8

Disclaimer
This book was prepared by CAI Resources as contractor to the National Center for Manufacturing Sciences, Inc. (NCMS). It is protected both under the U.S. Copyright Act and applicable state trade secret laws and may not be reproduced or disseminated without the express written permission of the NCMS. (NCMS members may reproduce and disseminate in accordance with their membership contracts with the NCMS and the NCMS bylaws and policies and procedures.) Neither NCMS, members of NCMS, nor any person or entity acting on behalf of them:

- **Makes any warranty or representation with respect to the accuracy, completeness, or usefulness of the information contained in this report.**
- **Makes any warranty or representation that the use of information, apparatus, method, or process disclosed in this report will not infringe any proprietary rights of a third party.**
- **Makes any warranty or representation that the information contained in this report is statistically reliable or has been prepared based on generally accepted analytical data collection procedures or has been verified in any manner.**
- **Assumes any liability with respect to the use of, or for damages resulting from the use of, any information, apparatus, methods, or process disclosed in this report.**

The information and any express or implied opinions in this book do not represent the opinions or views of the NCMS, members of NCMS, or any person or company acting on behalf of the NCMS. This book is intended only to be a factual report of responses submitted by a cross section of manufacturers and vendors in the electroplating industry. Reliance on the information contained herein is at the user's own risk.

Neither the NCMS, its members, or any person or entity acting on behalf of the NCMS makes any warranties whatsoever as to the contents hereof, and all warranties of merchantability and fitness for purpose are hereby disclaimed.

National Center for Manufacturing Sciences
3025 Boardwalk Drive
Ann Arbor, MI 48108-3266

PRINTED IN THE UNITED STATES OF AMERICA

Foreword

A lot has happened since 1994, back when ...

- The North American Free Trade Agreement (NAFTA) went into effect on New Years Day of that year, but the "giant sucking sound" pulling the coating industry south seemed somewhat quieter than some were expecting. (It got a lot louder a few years later, but by then it seemed to be coming from across the Pacific, without any help from NAFTA.)
- The electoral success of the Contract With America put the United States firmly on the track of small government. (I know, but just think how big government's footprint would be now if it hadn't.)
- Hexavalent chrome plating was clearly on its way out, due to safety and environmental concerns. (It's still on its way out, and will no doubt continue to be for some time to come.)
- A new book, officially called Pollution Prevention and Control Technology for Plating Operations, but known in the surface finishing industry as the "Bluebook", gave platers the first comprehensive overview of how the industry was responding to environmental concerns, and how to decide among the available options.

It's not entirely clear how much politicians have learned in the meantime, but platers have come a long way. Some technologies that were cutting edge in 1994 are now standard practice; others have given way to better ideas. But much technology has stood the test of time, and is still in routine use.

This new version of the Bluebook retains the detail contained in the original, but contains an extensive section bringing the old information up to date, covering both technology and best practices. It includes industry data and suggestions for improving environmental performance that have been gathered through the Strategic Goals Program, and through surveys conducted every year by the National Metal Finishing Resource Center. This level of detail was simply not available in 1994.

The business climate was challenging for the plating industry in 1994, and it is even more challenging today. At the same time, environmental regulation may be entering a new era, with the development of formerly obscure concerns like climate change. Customers are starting to demand information on the environmental impacts of all the processes that go into the products they buy, extending all the way up the supply chain – and definitely including platers and coaters. More than just the regulators are watching – the retailer selling the product you coated may want to know more about how you operate than you ever imagined you would be telling them.

The creators of the Bluebook offer this information to give you the fullest possible picture of what your options are as you face these new challenges, in the hope that you, your customers, and the planet, will be better off.

Paul Chalmer
Project Manager, Environmentally Conscious Manufacturing
National Center for Manufacturing Sciences
January 2009

Acknowledgement

This second edition "Bluebook" includes an update of techniques and technologies used by the metal finishing industry to comply with regulations, lower operating costs and thereby improve sustainability. Information for this update came from a number of sources, private and public. But perhaps the most valuable sources of information have been the U.S. metal finishing companies that participate in the Strategic Goals Program (www.strategicgoals.org). Every year during this now ten-year-old program, these companies have faithfully submitted data used to track their environmental progress, identify industry-wide trends, and establish performance benchmarks. The ongoing efforts of these companies and their generous cooperation have greatly enhanced the value of this second edition for the entire industry.

Table of Contents

SECTION 5 – Chemical Solution Maintenance

SECTION 6 – Substitute Technologies

SECTION 7 – Wastewater Treatment

SECTION 8 – Off-Site Metals Recycling

List of Exhibits

Section 1. P2 Technology for Common Metal Finishing Processes

Section 2. Overview of Project Results

Section 3. General Waste Reduction Practices

Section 4. Chemical Recovery

Section 5. Chemical Solution Maintenance

Section 1
P2 Technology for Common Metal Finishing Processes

1.1 INTRODUCTION

1.1.1 How the information is arranged

Section 1 contains an update to the original Bluebook, which is now presented in its entirety in Sections 2 through 8.

In the original Bluebook, technical information is arranged in sections covering specific topic areas, such as waste reduction, chemical recovery, and bath maintenance. Each section includes discussions of all of the technologies relevant to the topic, such as ion exchange, evaporation, or reverse osmosis. Information on each of the major plating baths or processes (nickel, copper, hard chrome, anodizing etc.) is dispersed throughout the technology discussions.

In Section 1, the order has been rearranged. Each plating bath is now discussed fully in a sub-section of its own. Grouping the discussion topics according to bath type has two main advantages:

- It enables users to develop a comprehensive strategy for upgrading each of the processes carried out in their facilities.
- It helps users appreciate which of the different variants of each pollution control technology are appropriate for the specific process they are examining.

References cited in Section 1 are indentified in footnotes. References for Sections 2 through 8 are found in the Reference section at the end of the book.

1.1.2 Information sources

The information presented in Section 1 is derived from various sources, including:

- the original Bluebook
- technical journals
- EPA and state environmental agency reports
- supplier's literature
- data collected from 1997 through 2008 from metal finishing facilities throughout the U.S. in conjunction with:
    - the Strategic Goals Program (SGP, 1997 – 2001)[1]
    - the Benchmarking Metal Finishing study (conducted in 1999, updated in 2003)[2]
    - National Metal Finishing Resource Center (NMFRC) surveys (2002 – 2008)[3]

The Benchmarking Metal Finishing study gathered information from 160 participants, who described process-specific pollution prevention practices that have worked in their shops. Most of these suggestions are practical and very cost effective to implement and many of them have been included in this update. This information is further supported by SGP data that were submitted by many of the survey participants and other facilities from 1997 through 2008.

In many cases, new technologies or variations of existing technologies have emerged since the publication of the Bluebook in 1994. A number of demonstration projects conducted since then have also provided very useful information for evaluating both technical capabilities and costs. This edition incorporates insights gleaned from the most recent sources.

[1] http://www.strategicgoals.org
[2] http://products.ncms.org/DEPT_RPT/RPT170408.htm
[3] http://www.nmfrc.org/ see also http://www.strategicgoals.org

1.1.3 Section format

Each process section has a similar format. Following a general overview, a set of basic pollution prevention objectives are presented in the left-hand column of a summary table, with each objective accompanied by a recommended strategy for attaining it. In general, the basic objectives include:

- Reducing dragout and water use
- Improving material utilization
- Reducing human exposure

To the right of the "Objectives and Strategy" column, a "Potential Practices" column lists specific actions that companies may take to implement each strategy. These actions are divided into two categories, basic and advanced. Actions in the advanced category typically involve somewhat greater effort and expense than do the actions in the basic category. We invite companies seeking to improve their environmental performance to use this information to identify practices and process improvements that can work in their operations, and to map out a comprehensive pollution prevention strategy.

1.2 RINSING

Product quality and operating costs are both highly dependent on effective rinsing. Inadequate rinsing causes dragout carryover which contaminates downstream process solutions and can cause mis-plated parts. Excessive and/or inefficient use of rinse water increases water use costs, including the costs of the raw water and chemical waste treatment. With any plating operation, there is an optimal range of rinse water use that can be attained by using the right equipment, configuration and procedures.

A significant portion of the original Bluebook is devoted to best practices for dragout reduction and rinsing (see Section 3). All of that information is still valid. Presented here is:

- status report on water use trends,
- good practices reminder,
- refocus on total dissolved solids as a control parameter, and
- update on rinse controller technology,

Exhibit 1-1. Water Use for SGP Facilities

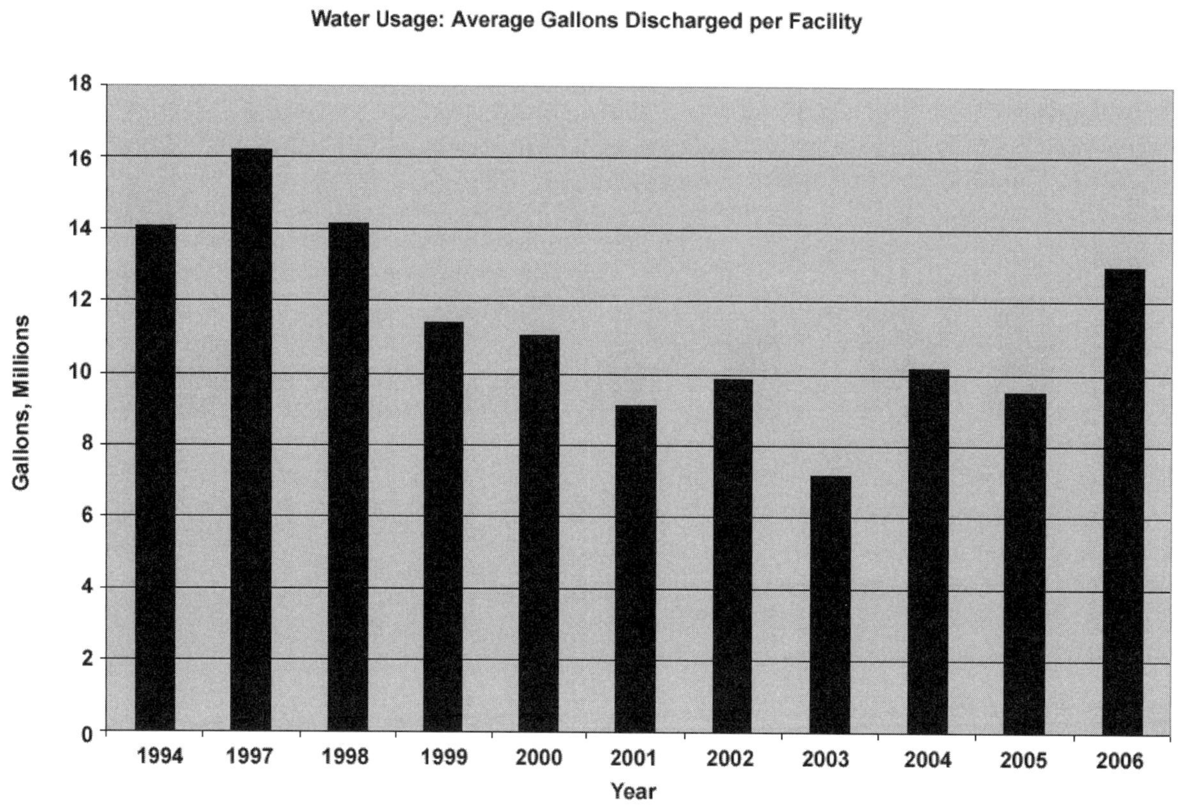

1.2.1 Water Use Trends

Water use by U.S. metal finishing shops has decreased significantly over the past 30 years. In the late 1970s, EPA conducted wastewater sampling at electroplating facilities as part of the rulemaking process for wastewater standards ("effluent guidelines"). Data from the supporting EPA document[4] indicate that the average discharge for facilities in their sample population (25 facilities) was 109,000 gallons per day (or 28,340,000 gallons per year).

New federal wastewater limits, went into effect in 1984. Most U.S. plating shops reduced water use as part of their overall strategy to meet the new limits By reducing water use, facilities were able to save significantly on both capital and operating costs for wastewater treatment, since both aspects of costs are closely related to the volume of wastewater flow. These reductions were relatively easy to attain. Since there had previously been little or no regulatory or economic motive to lower water use, most facilities were using more water than necessary. Therefore, basic measures, such as installing ***flow restrictors*** and turning off free-running rinses during non-production times, cut water use by 50% or more for many shops. Some facilities had already installed ***counterflow rinse*** systems, but according to the EPA development document, in 1979 the majority of plants relied on single rinse tanks.

The Bluebook survey (1993) asked facilities which year they installed their wastewater treatment system. 70% of the respondents indicated that they had a system in place by 1985, and 95% had one installed by 1990. By 1994, the baseline year for the Strategic Goals Program (SGP)[5], most plating shops had made decent progress on water use reduction, again mainly to reduce capital and operating costs for waste treatment. Exhibit 1-1 is a graph showing the average flow for SGP facilities for 1994 and 1997 through 2006. A comparison of these flow rates to those found in the EPA development document suggests that the industry significantly cut water use by 1994, and made further gains through about 2003. The values here are not normalized and therefore don't account for changes in production. However, Exhibit 1-2 shows normalized values for approximately the same time period,

[4] Development Document for Existing Source Pretreatment Standards for the Electroplating Point Source Category, August 1979 (EPA 440/1-79/003).

[5] More information on the SGP can be found at www.strategicgoals.org.

Exhibit 1-2. Water Use for SGP Facilities

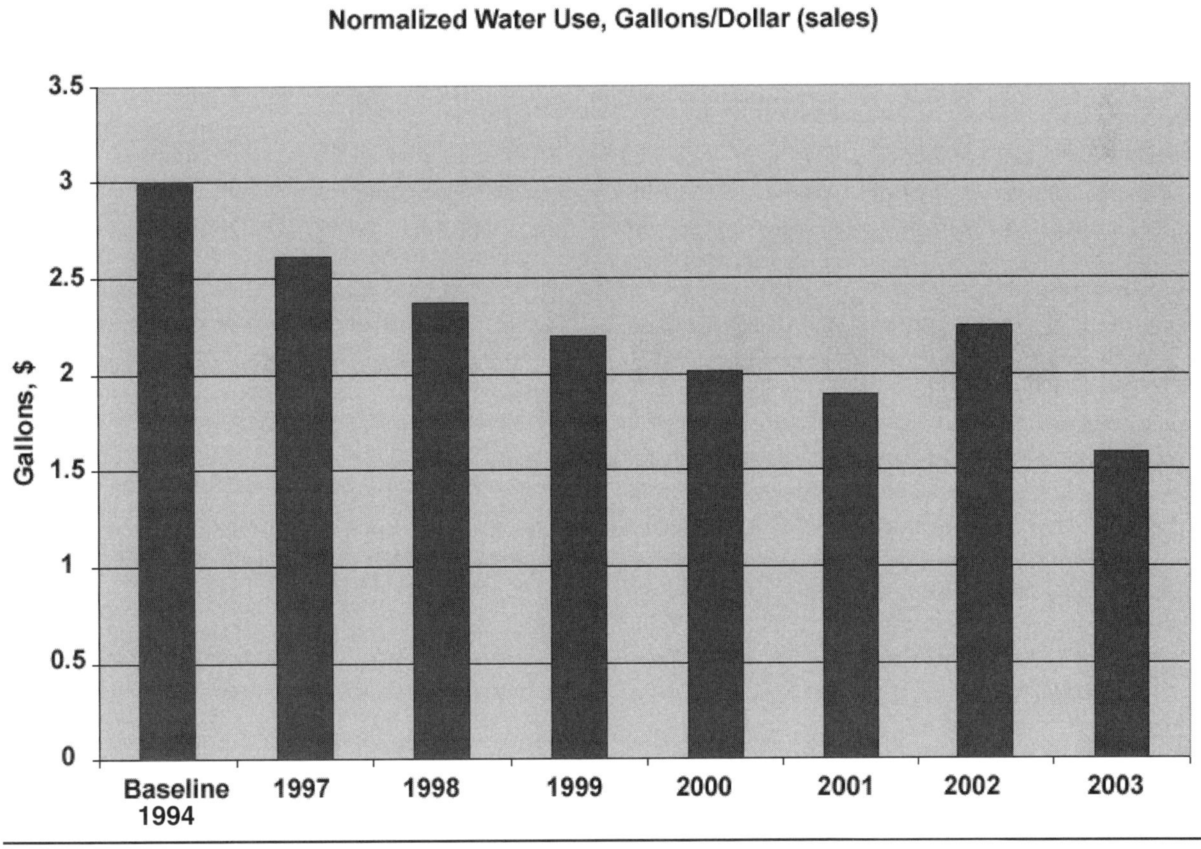

and it suggests approximately the same trend (2004 to 2006 were not calculated because a much smaller number of facilities submitted SGP data in those years).

How and why did facilities reduce water use from 1994 through 2003? The "how" is pretty easy to answer. Efficient rinsing systems, consisting of ***multiple rinse tanks*** and ***control mechanisms*** can reduce water use by a factor of several times, compared to the practices that were common in 1979. Although industry made significant progress by 1994, there was still plenty of low hanging fruit. Presumably, more and more facilities took advantage of this situation.

The "why" is more difficult to answer. There are most likely a number of reasons for this improvement. Cost savings was the most likely motivator. Water and sewer charges increased during those 10 years, as publicly owned treatment works (POTWs) upgraded their facilities and passed along price increases to industry. This trend is reflected in the SGP data, which show a 24% increase in the median water/sewer price from 2000 to 2006 (from $3.64 to $4.50 per 1,000 gal).

Another potential stimulus for water use reduction is the "green" movement that has impacted nearly all sectors of U.S. industry. Metal finishers were sympathetic to the goals of the green movement, as evidenced by their widespread participation in the SGP (over 400 SGP companies). Another factor is the improvement in information access made possible by the Internet. Industry now has quick access to free pollution prevention and compliance assistance information such as the ***National Metal Finishing Resource Center (NMFRC)***[6]. Also, the Internet makes it easier to locate and communicate with consultants and vendors that can assist with pollution prevention.

So, what happened following 2003 (see Exhibit 1-1)? It is difficult to say with certainty, because we are talking about a relatively short time period, and it may just be an anomaly of the data (fewer companies submitted SGP data after 2003 and perhaps a disproportionate percentage of low discharge facilities dropped out of the program). On the other hand, although it may be just coincidental, 2003 was the year that EPA decided not to impose the additional wastewater regulations on the electroplating industry that had been proposed under the ***Metal Products and Machinery Category***.[7] After a close call with some very stringent standards, some facilities may have eased up a bit. At any rate, the recent uptrend in water use will probably be erased over the next several years as industry is forced to squeeze down further on operating costs in order to remain competitive.

1.2.2 Overview of Best Practices for Rinsing

The ***optimal rinse system*** will vary from facility to facility due to variables such as part size/shape, level of automation and other site-specific factors. However, in each situation, if each of the following elements is present, rinse water use will be close to optimal:

- **Multiple rinse tanks.** Multiple rinse tanks plumbed in a ***counterflow configuration*** can reduce water use by 90% or more. A simple calculator, available free on-line, can be used to evaluate the impact of various rinsing strategies (more on this below).
- **Water control**. Any methodology that closely matches water use and workload will increase the efficiency of your rinse system. ***Flow restrictors*** as a standalone method are not sufficient, except on automated lines that are in continuous use. Examples of viable control methods include ***timer and conductivity controllers*** and ***optical sensors***, which are used on conveyor lines (more on controllers below).
- **Water circulation**. If you have poor mixing in your rinses, you are wasting water. Proper plumbing of the tanks will minimize short circuiting. ***Air or mechanical agitation*** (e.g., pumping) will eliminate the problem altogether.
- **Standard procedures**. Skipping stages in multiple tank arrangements is the most common rinse crime. Other forms of malpractice include sabotaging controllers and using supplementary water sources, such as a hose. Rules for rinsing should be established and enforced by management. Rinse tanks should be checked on a scheduled basis for total dissolved solids (TDS) concentration (more on TDS below).

Each of these factors is covered in detail in the original Bluebook. Some additional details on several of these items are presented in this update.

---

[6] An EPA-sponsored compliance assistance center (www.nmfrc.org).

[7] The Metal Products and Machinery (MP&M) regulation establishes national technology-based limits on pollutants in wastewater discharges from facilities that manufacture, rebuild or maintain metal parts, products, or machines. The proposed rule (Jan. 2001) included electroplating operations, however in 2003 this sector was dropped from the rule (see http://www.epa.gov/guide/mpm).

On July 1, 2008, EPA published the ***National Emission Standards for Hazardous Air Pollutants for Plating and Polishing Operations***.[8] This rule impacts many metal finishing operations (EPA estimates 2,900 facilities). It applies to facilities with plating, polishing or thermal spray processes that contain **cadmium, nickel, lead, manganese and/or chromium.** (The rule will not apply to chromium anodizing and electroplating, which are coverered under a previous rule.)

With this regulation EPA did not establish emission limits for plating and polishing operations as they did with chromium plating, but instead required plating and polishing facilities to implement specific pollution prevention (P2) and control practices that are related to air emissions, dragout and rinsing:

- minimize bath agitation when removing tank objects;
- maximize dripping of bath solution back into the tank by extending drip time when removing the tank objects and using drain boards (i.e., drip shields);
- optimize the design of barrels, racks, and parts to minimize dragout of bath solution, and
- use a wetting agent or fume suppressant in non-cyanide electroplating baths.

Therefore, affected facilities should also incorporate theses practices into their dragout control and rinsing systems.

1.2.3 Water Use Benchmarks

How does your water use compare with that of other shops? One way to find out is to compare your data with ***industry benchmarks***. In 2000, NCMS published a benchmarking report[9] (the benchmarks were updated in 2003) that covered various metrics, including water use, metals discharged, sludge generation, and energy use.

A unique aspect of the study results is that they account for the types of plating processes employed. For example, there are separate benchmarks for hard chromium plating, decorative chromium plating, nickel plating, zinc rack plating zinc barrel plating, anodizing, etc. The benchmarks also account for differences between certain sectors such as the auto manufacturing and fastener plating sectors. Therefore, the benchmarks can be tailored to any plating operation, even when multiple processes are present.

Water use benchmarks normalized to sales for common metal finishing processes are shown in Exhibit 1-3. The target benchmarks are the average water use for companies in the lower quartile (i.e. the best 25%).

The ***Strategic Goals Program (SGP)*** incorporated the benchmarks from this study into its program. Companies that participate in SGP automatically receive benchmark reports that show how well they measure up. These reports are particularly useful for identifying areas where significant progress can be made. For example the reports indicated how much money could be saved annually by meeting the benchmarks. You can view benchmarking reports at ***www.strategicgoals.org*** (select facility reports from the menu, and then choose "Benchmarking Analysis" on the subsequent page).

1.2.4 Using Total Dissolved Solids (TDS) as a Control Parameter for Water Use

In order to optimize rinse water use, finishers need data. Specifically, finishers should know:

[8]National Emission Standards for Hazardous Air Pollutants: Area Source Standards for Plating and Polishing Operations; Final Rule, Federal Register July 1, 2008.

[9] National Center for Manufacturing Sciences, Benchmarking Metal Finishing, June 2000. Available at www.nmfrc.org.

Exhibit 1-3. Water Use Benchmarks -- Gallons of Water Discharged Per Dollar of Sales

Process	Industry Average gal/$ sales	Target Benchmark, gal/$ sales
Zinc plating	4.79	2.5
Nickel plating	1.99	0.5
Decorative chromium plating	2.27	0.6
Electroless nickel plating	1.42	0.3
Anodizing	1.96	0.6
Hard chromium plating	0.2	0.0

Data source: National Center for Manufacturing Sciences, Benchmarking Metal Finishing, June 2000.

- the quality of their raw water supply
- the quantity of dragout generated by various parts and fixturing used
- target levels for rinse water quality and associated part cleanliness

A relatively easy and inexpensive means for generating these data involves measurement of ***total dissolved solids*** (TDS). In a matter of hours a finisher can generate data that are as useful as results generated by days of expensive consulting and laboratory work.

TDS is present in all natural waters. Typically, tap water has a TDS content of 100 to 400 mg/l, most of which is hardness (principally calcium and magnesium and associated anions). When water is used for rinsing, the TDS concentration increases significantly due to the presence of the plating chemicals that are washed off the parts. Therefore, rinse tanks must be flushed with "clean" water during processing. Good rinsing is necessary to prevent staining and other plating problems and to prevent cross contamination of process tanks. The "cleanliness" of rinse water can be measured by analyzing for TDS. The acceptable level of TDS varies depending on factors such as intermediate vs. final rinsing, type of finish, type of parts, etc. For example, rinsing following an acid dip may be adequate at 1,000 mg/l TDS, whereas rinsing following bright plating may require a TDS of 50 mg/l or lower. Some tap waters may not be sufficiently low in TDS for rinsing after bright plating applications, thus requiring processing (e.g. by ion exchange or reverse osmosis) of the tap water prior to use.

TDS measurements are usually made in a laboratory by filtering and then evaporating water and weighing the residue. A rapid TDS measurement can be made using specific-conductance measurements (a hand-held or ***"pocket" TDS instrument*** can be purchased for under $50). Such measurements indicate the capacity of a sample to carry an electrical current, which is in turn related to the concentration of ionized substances in the water, in units of milligrams per liter (mg/l). Most dissolved inorganic substances are in the ionized form and therefore contribute to specific conductance. Although this method provides only approximate TDS values, if it is properly used, it can serve as a guide to rinse water quality.

Exhibit 1-4. Rinse Calculator Data Input Form

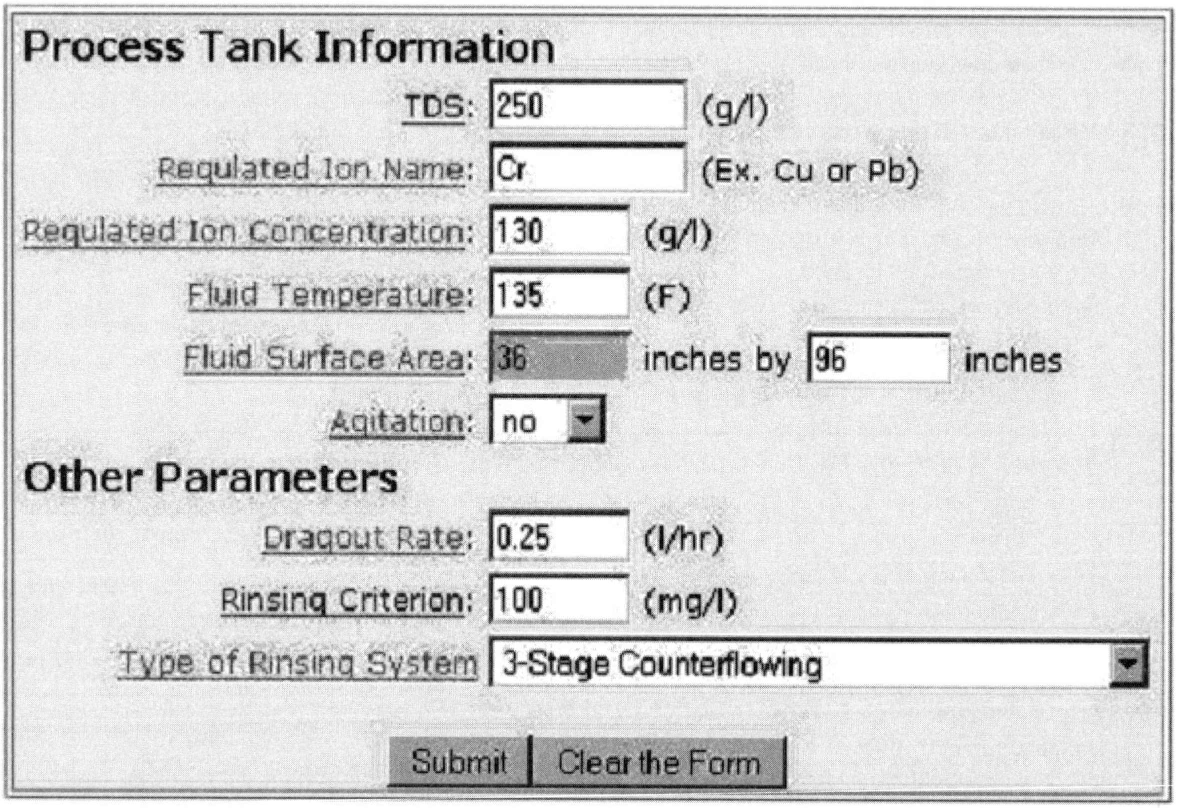

More sophisticated measuring devices can also be used to generate rinse water data with actual specific conductance readings, in units of microsiemens per cm (μS/cm) or the equivalent micromhos/cm (μmhos/cm). For practical purposes, you can convert μS/cm (or μmhos/cm) to mg/l of TDS by dividing by two (alternatively, multiple mg/l of TDS by 2 to arrive at μS/cm of specific conductance). For example, 1,000 μS/cm is approximately equal to 500 mg/l TDS.

A methodology for measuring dragout rates and rinse water quality can be found in Section 3.5.3. Once you have generated your own TDS data, you can use an on-line calculator to evaluate your rinse system and potential alternative configurations. The online tool is located at ***(www.nmfrc.org/subs/rinsecal.cfm)***. To use the tool, enter the process tank information and other parameters, and then select a type of rinse system. A diagram of the selected rinse system is then displayed. As an example, see the form and diagram in Exhibits 1-4 and 1-5, respectively.

1.2.5 Rinse Controller Update

Conductivity controllers. Conductivity controllers have been in use by the metal finishing industry for more than 25 years. A control system consists of three components:

- a probe or sensor located in the rinse tank that senses the conductivity of the rinse water;
- a transformer box that houses the solid state circuitry that controls the system; and
- a solenoid valve that opens and closes in response to signals from the circuitry.

Conductivity controllers work by responding to the actual conditions in the rinse tank. When dragout is introduced to the rinse tank, the probe senses a rise in conductivity. When the conductivity level rises above a set-point, the signal from the probe causes the circuitry to open the solenoid water valve. The value remains open until the probe senses a drop in conductivity below another set-point. Both of the set-points are operator-adjustable, to permit use over a range of desired water qualities.

Two types of conductivity sensors are available, conventional and electrodeless.

Conventional conductivity sensors consist of two electrodes that contact the water with a low-level electrical potential between them. The electrodes are sized and spaced to provide a known cell constant, which corresponds to a specific operating range that must match the conductivity range of the rinse water. Since the electrodes in a conventional sensor attract charged particles, they eventually become encrusted or fouled, causing the unit to malfunction. To ensure accurate conductivity measurements, sensors must be cleaned as frequently as every two weeks, and calibration checks must be performed monthly.

Exhibit 1-5. Rinse Calculator Output Diagram

3-Stage Counterflow

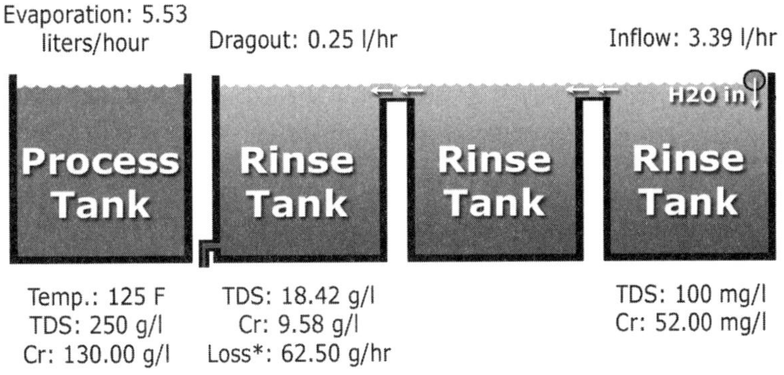

This system may support zero discharge by returning the rinsewater to the process tank, replacing evaporative loss. This arrangement often requires bath maintenance technologies.

* This is the grams per hour of chemistry dragged out of the process tank and discharged from the rinsing system.

Exhibit 1-6. Merit Partnership Project Conductivity Controller Costs

Cost Parameter	Conventional Type[a]	Electrodless Type[b]
Capital	$290	$1,140
Additional hardware	$100	$250
Installation	$400	$600
Total (per system)	$790	$1,990

[a] Conventional sensor, analyzer with no display, and analog set point.
[b] Electrodless sensor, analyzer with digital display, and programmable set point.

Electrodeless sensors, also referred to as ***inductive-type***, eliminate the fouling problems associated with conventional units. The electrodeless sensor consists of two toroids, or wire loops, sealed within a nonconductive casing. The toroids use magnetic fields to interact with the rinse water without contacting it directly.

- The first toroid creates a magnetic field that induces an electrical current in the water. The amount of current induced is proportional to the conductivity of the solution
- The current in the rinse water gives rise to its own magnetic field, which in turn induces a current in the second toroid. Thus the magnitude of the current in the second toroid is proportional to the conductivity of the rinse water.

Electrodeless sensors are easier to operate and maintain because they require less frequent cleaning (since the sensors do not directly attract ions, fouling is unlikely). An additional attribute of electrodeless sensors is their wide conductivity range as compared to conventional units.

Since the publication of the Bluebook (1994), there have been at least two good demonstrations projects that involved installation and monitoring of conductivity controllers.[10,11] Information from those demonstrations is discussed here.

Merit Partnership. The Merit Partnership project involved the installation and monitoring of both ***conventional conductivity*** and ***electrodless controllers***. A total of nine units were installed, three each from three different manufacturers. Six of the units were electrodless and three were conventional. The units were installed on rinses following acid activation, copper cyanide plating, nickel plating and decorative chrome plating. Eight of the nine units were wired to a common control panel, where operators could read conductivity measurements and adjust set-points. The other unit was mounted on a wall near the associated rinse tank.

Costs for the conductivity control units, including installation are shown in Exhibit 1-6.

The Merit Partnership demonstration was conducted over a three month time period. The water use and cost saving realized during the project are summarized in Exhibit 1-7. The study results indicate that the majority of savings are derived from reduced use of wastewater treatment chemicals. The study results noted that no adverse effects on product quality were observed during the demonstration.

Waste Management and Research Center. The Waste Management and Research Center (Champaign, IL) installed and monitored two electrodless (inductive loop-type) ***conductivity controllers*** at a plating shop that sought assistance in reducing their water use. The controllers were initially set at 1,000 μS and increased in 400 μS increments twice a week. This allowed the system to stabilize and gave production workers time to verify that there were no quality problems as a result of the increase in TDS. The final conductivity set point was 3,600 μS. Over time, no product quality problems were observed and the facility achieved an average water savings of 87.5% per rinse system and an annual water use reduction of 2,808,000 gal ($10,883 saved annually on water and sewer costs). The cost of the controllers was $1,370 each. Additional material and installation costs brought the installed cost to about $2,000 for each unit.

[10] Waste Management and Research Center. Pollution Prevention Case Study: Conductivity Control System Technology, March 2000.
[11] Merit Partnership. Reducing Rinse Water Use with Conductivity Control Systems, December 1996. [The Merit Partnership is a joint venture between U.S. Environmental Protection Agency (EPA) Region 9, state and local regulatory agencies, private sector industries, and community representatives. The partnership was created to promote pollution prevention (P2), identify P2 technology needs, and accelerate P2 technology transfer within various industries in southern California.]

Exhibit 1-7. Merit Partnership Project Water Use Savings

Metric	Per Month		Monthly Savings
	Before	**After**	
Rinse water use	516,000 gal	296,000 gal	$280
Wastewater discharged	516,000 gal.	296,000 gal	$110
Wastewater treatment chemical use	$4,000	$3,200	$800
Total Savings			$1,190

Timer Controls. Timer rinse controls consist of a push-button switch, a timer mechanism, and a solenoid valve. These units operate in a manner similar to conductivity controllers. However, instead of regulating rinse water flow on the basis of rinse tank water quality, the timer controls simply turn water on and off based on a pre-set time period.

In operation, a finisher lowers parts into the rinse tank and pushes a button (alternatively, a signal could be sent automatically by a momentary switch that is activated whenever a rack or barrel is lowered). The button or switch activates a timer and opens the solenoid valve for a preset time period. After that time period has expired the solenoid valve automatically closes.

The timer setting is selected through trial and error. It is best to select a time period that provides consistently clean rinse water, without excessive waste. Water quality can be periodically monitored using an inexpensive TDS tester, as discussed above. Once set, the time period is not changed unless the general trend of production changes.

Exhibit 1-8. Timer Rinse Control (Courtesy of Remco Engineering, Santa Barbara, CA)

The capital cost of ***timer rinse controls*** is similar to the price of conventional conductivity controllers. However, these units require less maintenance and generally have a longer life-span.

Timer controls continue to be an underutilized technology, in part because there are few commercial units on the market. One currently available unit is shown in Exhibit 1-8. The timer mechanism is built into a NEMA 6 waterproof epoxy box with an external raised pushbutton that energizes the timer. The timers consist of a complete unit with continuous adjustment from 6 seconds to 10 minutes. An alternative range is from 1 minute to 100 minutes.

1.3 ALKALINE CLEANING

1.3.1 Overview

As recently as 1994 (the baseline year for the Strategic Goals Program), metal finishers often used halogenated solvents to remove oils and other contaminants from work pieces prior to plating. But over the next few years, the industry largely abandoned solvent cleaning, except for a few specific applications.

The handwriting was already on the wall in 1994. The ***Montreal Protocol*** (an international agreement limiting the use of substances that destroy the Earth's protective stratospheric ozone layer) required participating nations, including the U.S., to phase out most of the halogenated solvents then in use, including the widely used trichloroethane (TCA). In addition a ***National Emission Standard for Hazardous Air Pollutants*** (NESHAP[12]), covering vapor degreasing and immersion (cold) cleaning, was enacted that year. Companies operating existing equipment had until December 1997 to comply with the requirements.

[12] The NESHAP was revised in May 2007.

Much of the cleaning that previously had been done with solvent is now accomplished using alkaline cleaning, the subject of this section. Because it avoids the air emissions and hazardous waste problems associated with solvents, alkaline cleaning can be considered a pollution prevention technology.

Of the solvent cleaning that remains in use today, most is carried out by the aerospace industry, by military depots, by facilities whose applications require precision cleaning operations, and (to a lesser extent) by some facilities involved in general finishing. Even these applications have undergone some changes: most of these facilities use new solvents that have been designed for environmental compliance. The cleaning/degreasing equipment has also changed significantly. It is often downsized, and is generally much more efficient and less polluting (for example, being operated as totally enclosed systems, or being equipped with carbon adsorption capture devices).

Once the 1994 NESHAP was in place, there was an immediate and major reduction in use of solvent cleaning over the next three years, followed by a less dramatic, but still steady decline. Exhibit 1-9 shows data from the ***Metal Finishing Strategic Goals Program***. These values are for quantities of solvent released to the air on a sales normalized basis.[13]

With the declining use of solvent cleaning, alkaline cleaning is now typically the initial operation performed on the plating line. Typically, the process is carried out in two stages using (1) a soak clean followed by (2) an electroclean and rinse. The cleaning stage is important, both for the quality of the finished product and for the overall environmental impact of the plating operation. If cleaning is not properly performed, the result is sure to be

- an increase in rejects (and consequent rework), and
- contamination of process solutions,

two very significant factors that increase waste generation.

Although alkaline cleaning can help prevent pollution, the cleaning baths and associated rinses can themselves be significant sources of pollution. Cleaning baths are the most frequently discarded process solutions on most finishing lines. Spent cleaners and rinse waters, typically treated on site, generate significant quantities of sludge. In addition, certain components of cleaners interfere with

[13] The SGP started collecting data in 1997, although most companies also submitted data for 1994, which served as a baseline. SGP data were collected by the NMFRC as recently as 2008, however there are insufficient data to look reliably past 2005.

Exhibit 1-9. Strategic Goals Program (SGP) Data -- Pounds of Toxic Organic Chemical Emissions per $1 Million Sales

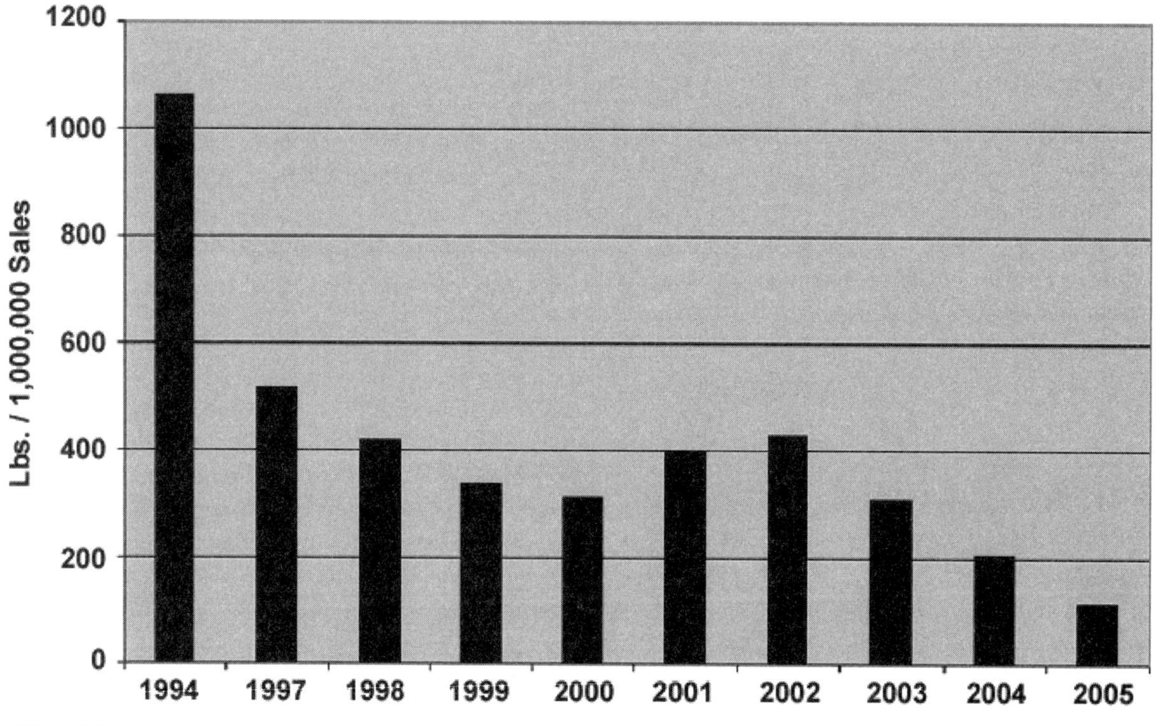

Exhibit 1-10. Summary for Pollution Prevention Strategy for Alkaline Cleaning

Objectives and Strategy	Potential Practices
Reduce Drag-Out and Water Use Strategy: Minimize drag-out to the maximum extent possible, control rinse water use and implement reactive rinsing.	Basic: • Reduce drag-out to decrease water use. • Install two-stage counterflow rinse tanks. Advanced: • Reuse acid rinse water for rinsing after alkaline cleaning. • Install conductivity, pH or timer controllers. • Implement closed-loop rinsing with drag-out tanks and a microfiltration system.
Improve Material Utilization Strategy: Select cleaners that don't cause treatment problems. Monitor the bath and make timely chemical additions. Maximize bath life/minimize solution disposal by removing bath contamination and using available maintenance technology. Avoid commingling alkaline cleaner wastes with other waste streams.	Basic: • Minimize use of cleaners that may pass through, interfere or upset the wastewater treatment processes (e.g., highly chelated, cyanide-containing). • Monitor bath for alkalinity (and/or other components as instructed by the supplier) and make timely additions. • Employ filtration and oil removal technology (oil separator/skimmer). • Recycle WWT sludge off-site. • Segregate cleaner waste streams from other wastes to improve wastewater treatment results. Advanced: • Monitor bath for individual bath components and make timely additions. • Use advanced bath maintenance technology (e.g., microfiltration) to keep contaminants at low levels and recycle drag-out with a closed-loop rinse system. • Consider use of bio-cleaners as an alternative to conventional cleaners.
Reduce Human Exposure Strategy: Eliminate use of solvents and cyanide.	Basic: • Convert solvent cleaning operations to alkaline cleaning. • Eliminate use of cyanide-bearing cleaners. Advanced: • None identified.

the precipitation of metals and can affect compliance with discharge regulations.

Exhibit 1-10 summarizes the pollution prevention and control recommendations for alkaline cleaning. The focus is on improved rinsing methods, bath maintenance, and strategies for on-site treatment. The remainder of this subsection provides supporting information for the recommended strategies, including new technology information not found in the original Bluebook.

1.3.2 DRAGOUT & WATER USE REDUCTION

Excessive dragout is costly because:

- it increases water use
- it results in the need for frequent additions of cleaner
- it increases wastewater treatment costs

1.3.2.1 Dragout

As with any metal finishing process, it is desirable to minimize dragout rates from cleaning tanks to the greatest possible extent. Consider implementing the following process changes, which can be effective methods for reducing dragout from cleaners:

- Modify rack or barrel design to improve drainage
- Improve part orientation to minimize solution carryover
- Operate cleaner bath at a higher temperature to reduce viscosity
- Increase dwell times over tank
- Rotate barrels over process tank
- Use spray rinsing as parts emerge from the cleaner

1.3.2.2 Rinsing

Counterflow rinsing provides the cleanest parts with the lowest water use. In counterflow rinsing, the rinse process is carried out in several stages. Clean water is introduced at the final stage, so the last water that the work piece contacts is the cleanest available. Water from the final rinse then moves upstream for use in the next-to-last stage, and so on.

It is generally advantageous to use a ***counterflow rinse*** following alkaline cleaning. A two-stage configuration is typically adequate, although a three-stage system can provide even greater reductions in water use. The rinse system should include some kind of automatic monitoring or control (such as a conductivity monitor or a timer) to ensure that operator inattention does not result in excess clean water consumption.

Benchmarking study participants reported several other successful rinsing practices, including:

- **Recovery rinsing. (See Section 3.4.2.3)** Recovery rinsing is applicable to alkaline cleaning (although seldom used for this purpose), especially where bath maintenance (e.g., microfiltration) is practiced. In general, when recovery rinsing is used, softened or deionized water should be used for rinsing. One benchmarking survey participant achieved a closed-loop cleaning operation by using three dragout tanks in a counterflow arrangement and a microfiltration system that processes both the bath and recovery rinse. Closing the loop has an additional advantage. The cleaning solution chemistry includes chelating compounds such as EDTA that can interfere with metals precipitation. By closing the loop, finishers can minimize treatment problems caused by the chelating compounds.
- **Reactive rinsing. (See Section 3.5.3.4)** During reactive rinsing, the rinse following acid dip is reused in the rinse tank following the alkaline cleaner. The acidic rinse helps to remove the alkaline cleaner film better than plain water. However, some facilities warn that use of reactive rinsing sometimes results in precipitated solids in the rinse water, especially with silicated cleaners.

1.3.3 IMPROVED MATERIAL UTILIZATION

During the benchmarking survey several respondents provided suggestions with regard to material utilization and these are summarized below:

- One company indicated that it reuses two-thirds of its used ***electrocleaner*** as makeup for its soak cleaner and, therefore, only uses fresh chemistry in the electrocleaner tank. Electrocleaner baths are more concentrated than soak cleaners due to the need for electrical conductivity. When spent, this company's electrocleaner has a sufficient concentration of components to be reusable as a soak clean. However, some fresh solution must also be added. It should be noted that, as one survey respondent pointed out, some electrocleaners do not work well as soak cleaners due to different chemistries.
- Several survey respondents indicated that they have implemented programs with more frequent bath analyses and that they analyze and subsequently adjust for individual bath components. One respondent analyzes cleaners once every 8-hour shift. A different respondent indicated that it has also increased analytical work to include contaminants such as oil and grease. This company uses these results to trigger bath disposal instead of routinely disposing of the solution based on a time schedule or amount of surface area processed.
- One survey respondent suggested switching from powdered cleaners to ***liquid cleaners***. Another respondent also recommends the switch because powdered cleaners may not completely dissolve in the bath, causing chemical waste and carryover of chemicals to the treatment system or sewer.
- Several survey respondents have substituted ***biological cleaners*** for conventional cleaning baths. This technology uses a specially formulated cleaner chemistry (slightly alkaline pH), which

permits the buildup of a microbial population that consumes oil and grease. One respondent indicated that they implemented use a microbial cleaner on one production line and due to the results plan to use it in all similar processes. A discussion of bio-cleaners is presented later in this section.

- Cyanide is used in the formulation of some alkaline cleaners; however, its use has declined significantly over the past 20 years. Several survey respondents indicated that they had recently eliminated use of cyanide cleaners as a P2 measure. One of these companies indicated they use bead blasters to supplement cleaning and although this is less efficient than their old process, it has eliminated use of cyanide-containing cleaners.

1.3.4 BATH MAINTENANCE TECHNOLOGY UPDATE

Bath maintenance is an important aspect of both P2 and cost management. A well maintained bath improves the performance of the alkaline cleaning operation, which in turn reduces reject and rework, and extends the life of the bath. The result is better parts at lower cost, *and* less waste.

Maintenance includes chemical monitoring and timely additions of fresh chemicals, as well as the use of various technologies to remove contaminants. Knowing what to look for is an important aspect of monitoring. Most finishers limit the analysis of their alkaline cleaner bath to a single component, such as alkalinity or conductivity. When the alkalinity drops below the recommended operating level, they add cleaner solution. This practice fails to take into account that specific components of the cleaning bath are usually degraded or consumed at different rates. A single test will not provide sufficient information.

Alkaline cleaning solutions are a mixture of chemicals, including surfactants, alkali salts, caustic soda, phosphates, and complexing agents. These baths build up concentrations of oil and solids during use. At some point, the cleaning efficiency of the bath is impaired and the solution is discarded, despite the fact that most of the bath's constituents are still usable. Alkaline cleaner bath maintenance technologies are primarily used to remove these suspended solids and oil. In many cases, heavy-duty cleaners must be replaced once per week.

Oil can be partially removed using an oil separator/skimmer or coalescer. Free oils can be removed by simple skimming. These methods work best with cleaner chemistries that are formulated to "split" oils. However, emulsified oils and colloidal solids are not affected by these devices. One benchmarking survey respondent reported adding a reagent to used cleaners that helps split emulsified oil. The oil layer is subsequently skimmed and the cleaner is returned to service.

Most solids can be removed by settling and/or cartridge filtration. Large solids are removed by simple filtration Colloidal solids and oil can be removed by ***microfiltration***. Several benchmarking survey respondents employ this technology. When implementing microfiltration technology it is sometimes necessary to change cleaning chemistry, since microfiltration works best with non-silicated emulsifying cleaners.

Many commercial microfiltration systems used for this application employ ceramic filter membranes in a crossflow filtration configuration. These membranes are a relatively new development that permits application of microfiltration to solutions and emulsions that are both heated and corrosive. The ceramic membranes are produced in a range of pore sizes that selectively permit a large percentage of the surfactants to pass through the membrane. Typical pore sizes are from 0.2 μ (microns) to 0.8 μ. Crossflow filtration, as opposed to barrier or "dead-end" filtration, permits the application of this technology to high solids-feed streams. With dead-end filtration, all of the feed solution is forced through filtration media by an applied pressure. With a high solids-feed stream, the pores of a deadend filtration device plug. With crossflow filtration, the fluid to be filtered is pumped across the membrane, parallel to its surface. By maintaining a high velocity across the membrane, the retained material is swept off the membrane surface.

EPA Testing of Microfiltration. Under the EPA Environmental Technology Verification (ETV) program (www.epa.gov/etv), an alkaline cleaner ***microfiltration*** technology was evaluated under actual production conditions. The verification test evaluated the ability of the Silverback™ Model 900 unit to remove oils and recover the alkaline cleaning chemistry.[14]

[14] Cushnie, George, Evaluation of Silverback Model 900 Alkaline Cleaner Recycling Systems, April 4, 2000. (http://www.epa.gov/etv/pubs/06_tp_silverback.pdf).

The Silverback™ Model 900 is shown in Exhibit 1-11 and an operational diagram is shown in Exhibit 1-12. The Membralox® Silverback™ Model 900 is an alkaline cleaner recycling system. This unit contains a Membralox® 7P19-40 module made up of seven alpha alumina elements, each with 19 lumens (channels) that are 4.0 mm in diameter. The inside of each channel is lined with an alpha alumina or zirconia membrane having a total surface area of 1.68 m^2 (18.1 ft^2), and a pore size of 0.2 μ. The manufacturer indicates that the membrane is impervious to nearly all chemicals except for phosphoric and hydrofluoric acids, and can tolerate temperatures up to 200°F.

The alkaline cleaner used at the test facility is CLEAN-R-120GR, which is composed primarily of sodium hydroxide, phosphate, glycol ether, octylphenol, and an amine. In use, alkaline cleaner contaminated with oil enters a two-compartment stainless steel tank through a prefilter that removes large particulate material from the feed stream. Free oil accumulates in the initial compartment and can be removed on a periodic basis through a drain port located on the upper part of the tank. The liquid then moves to a second tank compartment through a sub-surface passage, thereby leaving the floating oils in the first compartment. The liquid in the second compartment (referred to as the recirculation tank) is pumped through a microfiltration ceramic filter (0.2 μ). The filter reject returns to the recirculation tank and the recovered alkaline cleaner flows back to the cleaning process. At the demonstration site, the Silverback™ unit recovers 1.0 gpm of alkaline cleaner.

Two test runs of five days each were conducted during the ETV project. Test results show that the unit recovered greater than 95% of the cleaner chemistry, while average contaminant removal was 82% for total suspended solids (TSS) and 90% for oil.

Implementation of the technology reduced the disposal frequency of the alkaline cleaning solution at this facility from 15 times per year to two times per year. The overall volume of concentrated waste generated from alkaline cleaning was reduced by 67.5% and the weight of total solids in the waste products has been reduced by 58.9%.

Operating and maintenance (O&M) labor requirements were monitored during testing. The O&M labor requirement for the equipment was observed to be 3.75 hrs/wk. O&M tasks performed during the verification test include daily inspections of the unit and weekly cleaning of the tank and membrane.

A cost analysis of the Silverback™ Model 900 was performed using current operating costs and historical records from the demonstration site. The installed capital cost (1999) of the unit was $43,000 (includes $36,000 for the unit, $5,000 for storage tanks, and $2,000 for installation costs). The annual cost savings associated with the unit is $32,064. The projected payback period is 1.3 years.

The test results show that the microfiltration technology provides an environmental benefit by extending the bath life of the alkaline cleaner, thereby reducing the amount of liquid and solid wastes produced by the cleaning operation without removing the cleaning constituents of the bath. The economic benefit associated with this technology is low operating and maintenance labor and a payback period of approximately 1.3 years. As with any technology selection, the end user must select appropriate cleaning equipment and chemistry for a process that can meet their associated environmental restrictions, productivity, and cleaning requirement.

Biological Cleaners. A unique approach to bath maintenance that may work for some facilities is ***biological cleaners***, which are an alternative to conventional alkaline cleaners. Typically, these are mild alkaline emulsifying solutions that use biological digestion to remove residual oils from baths that operate at relatively

Exhibit 1-11. Diagram of the Silverback™ Model 900

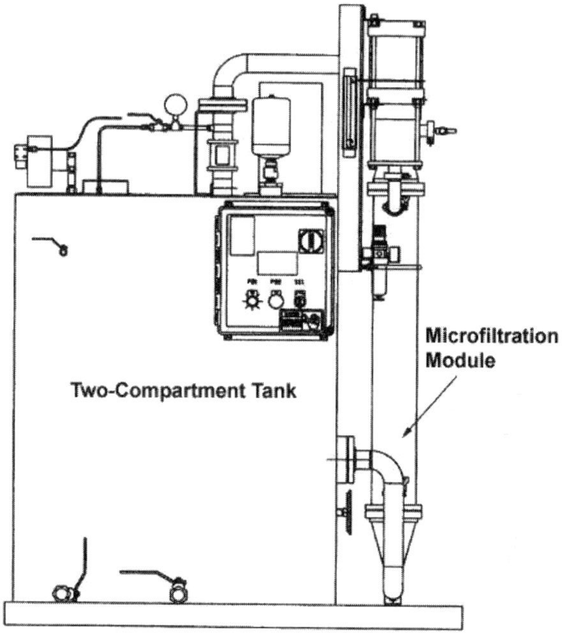

low temperatures. The system is essentially self-regulating, since the microbial activity will adjust itself to the amount of removed oil and grease present in the system. Bio-cleaner systems operate at temperatures between 104°F and 131°F, with a pH range of 8.8 to 9.2, conditions which provide a viable environment for microorganisms. The microbes present in the cleaner system are those normally found in industrial oils and greases.

The cleaning process takes place in two separate operations:

- When parts come in contact with the solution, the oil and impurities are emulsified into micro-particulates
- The micro-particulates are then consumed by microorganisms which are present in the bath

Although the bio-cleaner technology has been has been tested (see results below), it is a relatively new technology and should be approached with prudence. These cleaners introduce a biological process into the shop – a type of process that is otherwise foreign to most finishers. Successful use of this technology requires a different approach to bath operation and maintenance from that used with more common cleaner formulations. If the industry's use of this technology expands, implementation will become easier for each facility as operating experience is communicated through journal publications and reports.

EPA Testing of a Bio-Cleaner. An EPA ETV project evaluated the effectiveness of one particular bio-cleaner, the BioClean™ system.[15] The facility selected for the test was a zinc plating shop. At this facility, the cleaning solutions from four separate cleaning baths are pumped continuously into a holding tank that feeds the BioClean system (see Exhibit 1-13). After BioClean treatment the cleaning solution is returned, by gravity, into the holding tank, and is then pumped back into the cleaner tanks. This operation is run in a continuous mode, with level guards on the cleaner tanks that prevent overfilling.

The demonstration project resulted in the following conclusions:

- Implementing the BioClean System reduced the annual amount of alkaline cleaner waste requiring treatment from 34,400 gallons to 6,940 gallons.
- The O&M labor requirement for the equipment was observed to be 2 hrs/wk.
- The installed capital cost (1998) of the unit was $47,569 (includes $27,625 for the BioClean unit, plus $19,944 for installation to four work-centers). The annual cost savings associated with the BioClean System was $86,192. The projected payback period was 0.6 yrs.

[15] Eskamani, Gus, Environmental Technology Verification Report: Evaluation of BioCleanUSA, LLC Biological Degreasing System for the Recycling of Alkaline Cleaners ETV (http://www.epa.gov/etv/pubs/06_vr_bioclean.pdf)

Exhibit 1-12. Diagram of Alkaline Cleaner Recovery Operation

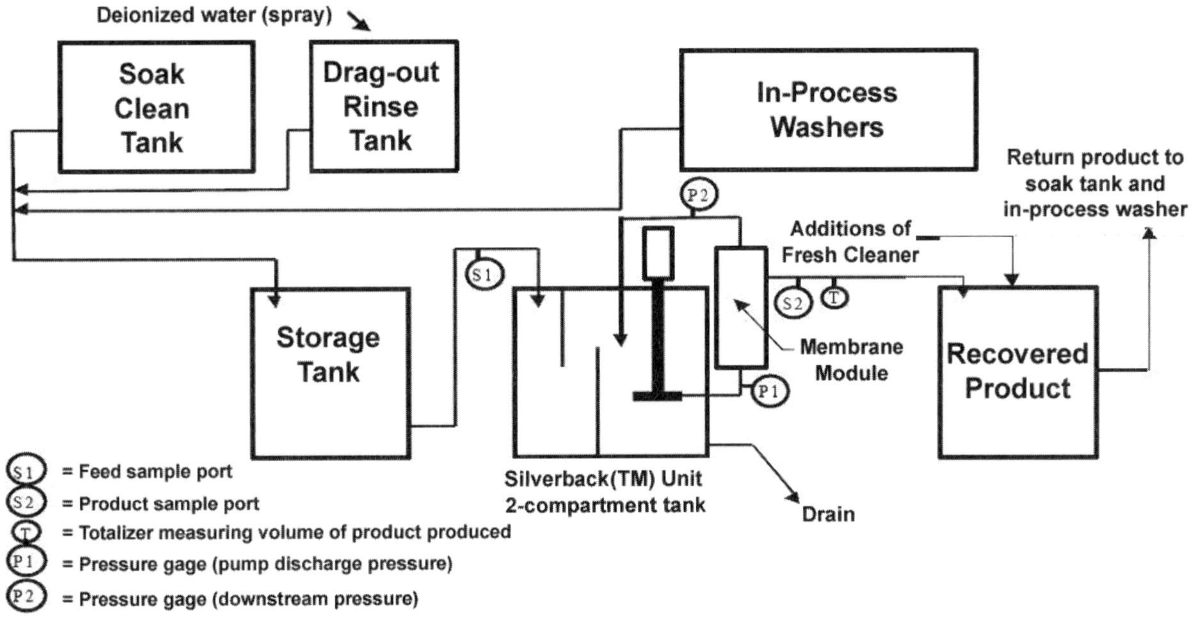

1.3.5 TREATMENT OF SPENT CLEANERS

Eventually, as the active ingredients are depleted, all alkaline cleaning solutions need to be discarded. Dumping a cleaner bath directly into a wastewater treatment system will usually upset the treatment process, and may result in a discharge violation for one or more metals. To avoid such problems, consider implementing one of these methods:

- **Bleed into General Flow**. To implement this option a shop would typically install a holding tank in the waste treatment area that is large enough to hold one or more cleaner tank dumps. Drain valves on the cleaner tanks would be piped to this holding tank. It would be necessary to schedule tank dumps to adequately space out the expected volume of spent cleaner. Once a cleaner dump is transferred to the holding tank, it would be bled at a predetermined rate (e.g., 1%) to the treatment process using a chemical metering pump.
- **Pretreatment of Spent Cleaner Before Combining with General Flow**. To implement this option, a shop would typically install a batch treatment system, rather than the holding tank discussed in the previous option. A conical bottom tank equipped with a mixer and chemical feed systems is recommended for this purpose. The spent cleaner is chemically treated, solids are removed through a valve in the bottom of the tank, and the treated liquid is metered into the general wastewater flow.
- **Hauling Spent Cleaner to a Treatment/ Disposal Site**. To implement this option, a facility would transfer spent cleaner from the cleaning tanks to drums suitable for holding or transporting hazardous waste (e.g., 55-gal. DOT-approved drums). Spent aqueous cleaning solutions may or may not be hazardous, depending its pH and toxic metal content.[16] The drums would be stored on-site and transported by a licensed transporter to an appropriate treatment or disposal site. If the spent cleaner were determined to be a hazardous waste, it would be handled in compliance with EPA's hazardous waste manifest system.

Exhibit 1-13. Diagram of Bio-Cleaner Operation

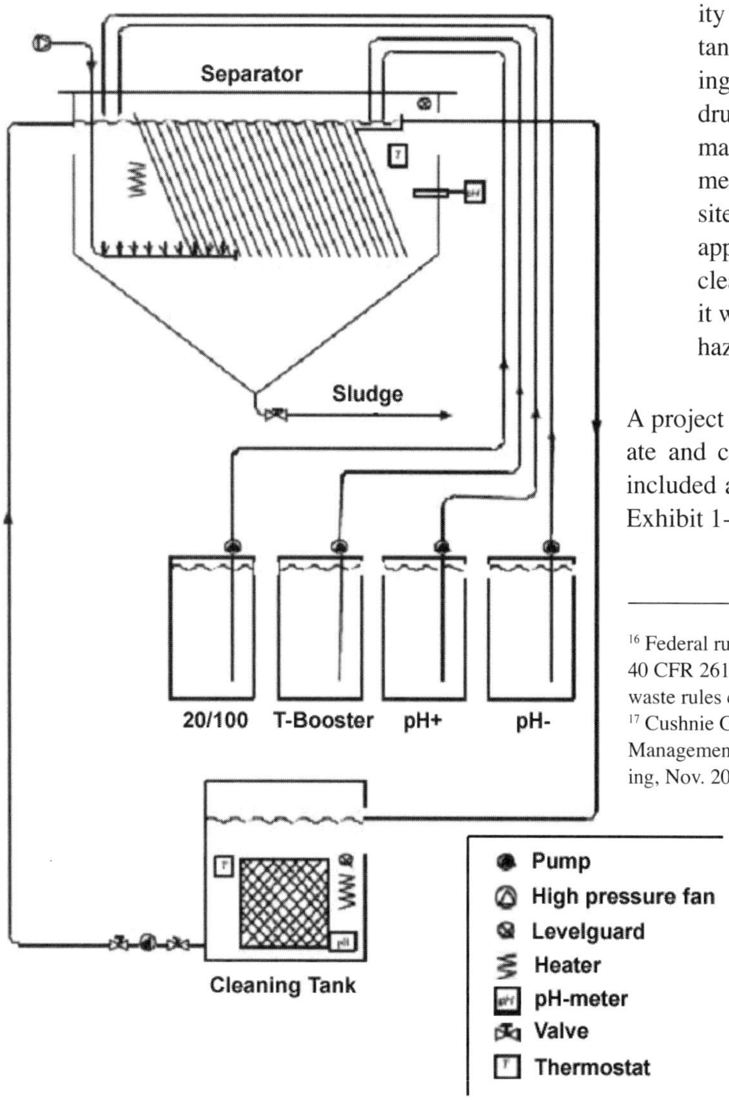

A project was conducted at a zinc plating shop to evaluate and compare these options.[17] The evaluation also included a ***microfiltration*** system for bath maintenance. Exhibit 1-14 shows a cost comparison of the four options

---

[16] Federal rules for determining if a waste is hazardous can be found at 40 CFR 261. State rules may supersede federal rules (state hazardous waste rules can be downloaded at: http://www.envcap.org/hwrl).

[17] Cushnie George, Chalmer, Paul, Marsh, Greg and Ferguson, David, Management of Spent Alkaline Cleaners, Plating and Surface Finishing, Nov. 2005 (www.nmfrc.org/pdf/psf2005/110522.pdf)

Exhibit 1-14. Cost Comparison of Alkaline Cleaner Management Options for Various Total Cleaner Capacities

Option	Sum of Annualized Capital and Operating Costs for Volume of Spent Cleaner Indicated				
	1,500 Gal./Yr	3,000 Gal./Yr	6,000 Gal./Yr	12,000 Gal./Yr	30,000 Gal./Yr
1. Bleed into general flow at 1% rate	$1,351	$2,503	$4,905	$9,610	$23,525
2. Batch Treatment	$1,991	$3,233	$5,965	$11,130	$26,325
3. Hauling	$2,720	$5,240	$10,580	$21,160	$51,400
4. Bath Maintenance*	$4,300	$4,550	$6,300	$10,050	$19,225

* Assumed that only 25% of volume indicated is dumped.

for five sizes of cleaning operations, ranging from 1,500 gal./yr. to 30,0000 gal/yr. of spent cleaner generated. For labor calculations it was assumed for options 1, 2 and 3, that cleaning baths are replaced every two months and that the cleaner is not replaced with option 4.[18] The costs shown in Exhibit 1-13 are the sum of annualized capital costs and operating costs. Exhibit 1-15 shows an example of a breakdown of these costs for one of the five cleaning operation sizes (12,000 gal./yr.).

The cost analysis in Exhibit 1-14 shows that for all but the largest cleaning capacity evaluated, bleeding spent cleaner into the wastewater is the most economical option. For the largest capacity, bath maintenance is the more economical option. Although the bleed method is usually less expensive to implement, in many cases, this option is not practical or feasible because wastewater flow is insufficient (and/or too variable) to dilute the spent cleaner to a point where it no longer interferes with the treatment processes. (Based on experimentation, a

[18] Therefore, the cleaning tank capacities (could be one or more tanks) for the five sizes of operations are 250 gal. (1,500/6 = 250), 500 gal., 1,000 gal., 2,000 gal., and 5,000 gal.

Exhibit 1-15. Breakdown of Annualized Capital and Operating Costs for 12,000 gal./yr.

Option	Annualized Capital Cost	Operating Costs				Sum of Annualized Capital & Operating Costs
		Cleaner Replacement	Treatment Chemicals	Disposal Fees	Labor	
1. Bleed into general flow at 1% rate	$400	$4,800	$1,800	$1,410	$1,200	$9,610
2. Batch Treatment	$1,200	$4,800	$1,800	$1,410	$1,920	$11,130
3. Hauling	$1,000	$4,800	$0	$12,000	$3,360	$21,160
4. Bath Maintenance	$4,300	$1,200	$450	$350	$3,750	$10,050

Assumptions:
10 year straight line depreciation for capital.
Labor cost: $20.00/hr. Includes labor for solution disposal, solution treatment, formulating new bath, waste treatment, waste management, and technology operation, where applicable.
On-site treatment cost for cleaner: $0.15/gal.
Replacement cleaner: $0.40 /gal.(after dilution to a 5% solution).
Sludge generation form on-site treatment of spent cleaners: 0.17 lbs dry solids/gal. of cleaner treated.
Sludge dewatered to 35% solids before disposal.
Sludge disposal cost: $0.25/lb.
Haul/treat spent cleaner: $1.00/gal

Exhibit 1-16. Test Results

Sample (500 ml)	Average Zinc Concentration Before Treatment	Ferric Chloride Dosage*	Final pH	Average Zinc Concentration After Treatment	% Zn Removal
100% wastewater	5.17 mg/l	0.4 ml	9.3	0.43 mg/l	91.6%
1% spent cleaner, 99% wastewater	5.16 mg/l	0.8 mg/l	9.4	0.56 mg/l	89.1%
3% spent cleaner, 97% wastewater	5.13 mg/l	1.6 ml	9.3	2.58 mg/l	49.7%

*Note: The required ferric chloride dosage was determined by preliminary testing, which showed that 0.8 ml of ferric chloride were needed for good settling and zinc removal for the 100% wastewater sample and that an additional 0.08 ml were needed for each ml of spent cleaner added.

maximum 1% bleed rate was identified; see Exhibit 1-16). It is also important to keep in mind that intentional dilution is unlawful, so adding clean water to the wastewater flow to achieve a treatable spent cleaner dilution rate is not an option.

Getting back to the cost analysis in Exhibit 1-14, if bleeding spent cleaner at a sufficiently low rate is infeasible, impractical or just too risky, then either batch treatment or bath maintenance are the most cost effective options. Batch treatment is more economical with small volumes of spent cleaner. With larger cleaning tank volumes, bath maintenance becomes more economical: when the generation rate of spent cleaner is between 6,000 and 12,000 gallons, the replacement costs for cleaner begin to outweigh the annualized cost of bath maintenance equipment. With a 30,000 gal. cleaner capacity, bath maintenance is clearly more economical than batch treatment.[19] In every case analyzed, hauling the spent cleaner was not economical, primarily due to the high labor cost associated with drumming and managing the waste.

For this analysis, microfiltration was used to establish the costs and benefits of bath maintenance. The underlying assumptions used for cost calculations were based on the previously discussed EPA ETV demonstration project.

Conclusions. Bleeding spent cleaner into the general wastewater flow is risky business, considering the low dilution rate that is required and the normal variability of wastewater flows. At the selected facility, a single cleaner dump has a volume of 1,500 gal. and would require 260,000 gal. of wastewater in order to dilute the spent cleaner to 1%. With an average daily wastewater flow of only 13,000 gpd, it would take 20 days to bleed the spent cleaner into the waste treatment system. Facility management considers such a slow bleed rate to be a non-practical solution and further, because of the risk of non-compliance associated with the bleed method, they have chosen to employ batch treatment. Their decision is corroborated by the results of the cost analysis shown in Exhibit 1-14.

Batch treatment of spent cleaners is an economical alternative for small operations. However, for facilities that use a sufficient volume of cleaner, bath maintenance is a more attractive alternative. The breakeven point appears to occur approximately at 12,000 gal. of spent cleaner generated per year. At this point, the sum of cleaner replacement cost savings and other savings begin to outweigh the annualized capital cost of bath maintenance equipment and other associated costs.

1.4 ACID DIPPING AND PICKLING

Acid ***pickling*** refers to processes aimed at removal of scale. (Scale is a surface oxide that is formed when a metal, such as steel, is cooled during the transformation from molten into solid form.) Acid ***dipping*** generally refers to processes that remove the alkaline film carried from the previous cleaning step, and that activate the surface of parts prior to plating. The most common acids used for this purpose are sulfuric, hydrochloric, nitric and phosphoric acids. The most significant waste products generated from acid dipping and pickling are:

[19] This cost analysis does not account for differences in cleaning chemistry that exist at most facilities (e.g., different cleaners for different alloys, soak clean vs. electroclean, etc.). However, some facilities have been able to implement bath maintenance by first reducing the number of different cleaners used. By doing so, they are able to use the same bath maintenance equipment for multiple cleaning tanks.

- Spent solutions
- Rinse water
- Wastewater treatment sludge
- Acid fumes
- Scrubber water blow-down

Pollution prevention and control efforts for these processes should be aimed at process changes that reduce pollution and worker hazards, rinsing considerations, bath maintenance/recovery methods, and other waste reduction opportunities. Exhibit 1-17 presents a pollution prevention strategy for acid dipping and pickling.

1.4.1 Water Use Reduction

Excess water introduced by drag-in can dilute acid baths, especially if they are operated at ambient temperatures. Excessive dragout is costly because it results in increased rinse water use, more frequent acid replacement, and increased wastewater treatment costs. To reduce drag-in and dragout, efforts should focus on:

- Rack design
- Part orientation
- Dwell times

Conductivity controls or pH controllers can regulate rinse water use. For final rinses following acid dips, a pH

Exhibit 1-17. Summary of Pollution Prevention Strategy for Acid Dipping and Pickling

Objectives and Strategy	Potential Practices
Reduce Dragout and Water Use Strategy: Minimize drag-in and dragout to the maximum extent possible, control rinse water use and implement reactive rinsing.	Basic: • Reduce drag-in to avoid diluting acid bath. • Reduce dragout to decrease water use. • Install two-stage counterflow rinse tanks Advanced: • Reuse acid rinse water for rinsing after alkaline cleaning. • Install conductivity, pH or timer controllers. • Raise bath temperature and lower bath concentration to reduce chemical loss.
Improve Material Utilization Strategy: Avoid solution disposal by minimizing bath contamination and using available maintenance technology. Recycle wastewater treatment (WWT) sludge.	Basic: • Avoid bath contamination (clean parts thoroughly before plating; minimize corrosion of busing, racks, fixtures and parts; and retrieve parts/fixtures which fall into the bath). • Use the lowest possible acid concentration and temperature to reduce the attack on base metals. • Use inhibitors in pickling baths to reduce the attack of acid on base metals. • Recycle WWT sludge off-site. Advanced: • Filter acid solution to remove suspended contaminants. • Use bath maintenance technology (i.e., acid sorption or diffusion dialysis technology) to keep metallic contaminants at low levels.
Reduce Human Exposure Strategy: Minimize acid air emissions to workplace and community. Use less-toxic, substitute technologies, where applicable.	Basic: • Employ adequate air speeds and ventilation rates to assure removal of fumes or mists from inhalation zones. • Monitor inhalation zones to evaluate exposure and immediately correct any inadequacies. • Lower bath temperature and acid concentration to reduce hazards to employees. Advanced: • Consider use of alternative chemicals such as acid salts.

range of pH 5 to 6, or a conductivity of 3,000 to 5,000 μmho, is generally considered acceptable. ***Timer rinse controls*** can also be used, with a timer setting adjusted so that it produces similar results.

A two-stage counterflow rinsing configuration following acid dipping or pickling will typically suffice. Some benchmarking survey respondents use ***reactive rinsing***, where the rinse following acid dip is reused in the rinse tank following the alkaline cleaner. The acidic rinse helps to remove the alkaline cleaner better than plain water. However, two survey respondents warned that use of reactive rinsing could result in precipitated solids in the rinse water. One survey respondent employs drag-in/dragout recovery rinsing and reports that it conserves acid.

1.4.2 Bath Maintenance

Bath monitoring and timely additions will improve performance and reduce waste generation. The efficiency of baths can be measured by a weight loss test. Acid baths are commonly contaminated with soils and dissolved metals, both of which reduce performance. Filtration is commonly used for removing particles. Dissolved metals can be removed by ***acid sorption*** and ***diffusion dialysis*** (see discussion below), although these technologies are generally not cost effective for small operations. ***Electrowinning*** can be used for removing copper and zinc from dilute sulfuric baths (10% by volume).

Acid Sorption. Acid sorption is described in the original Bluebook (see Section 5.5), where a diagram shows how the technology can be applied to an acid dip process. Since the publication of the Bluebook (1994) the U.S. Department of Defense has investigated use of this technology for their metal finishing depot activities (e.g., aircraft overhaul).[20] The DoD demonstration tests were conducted using a 1-inch diameter by 12-inch length, packed ion exchange resin column. A set of 10 experimental runs was carried out using concentrated sulfuric, hydrochloric, nitric, hydrofluoric, and phosphoric acids in combination with a variety of metals, including iron, nickel, chromium, zinc, and aluminum. The tests represented typical conditions from pickling, cleaning, activation, anodizing, and stripping procedures. Acid recovery averaged over 80 percent for all the acids that were tested. The process achieved recovery efficiencies in excess of 95 percent for the acids of greatest interest, such as hydrochloric, nitric, and sulfuric acids. Metal removal efficiencies were not reported.

The DoD demonstration offered the following economic analysis, which may be useful to those considering this technology.

The purchase of an acid purification system for the Jacksonville Naval Aviation Depot was compared to off-site disposal of spent acid. A summary of annual operating costs for the two options is shown in Exhibit 1-18. The following assumptions were used:

- Throughput: 175 lb/hr of acid
- Operation: 50 days/yr (400 hrs/yr)
- Acid recovery efficiency: 60-95%
- Average waste density: 15.0 lb/gal
- Labor (burdened): $79/hr

Economic Analysis Summary:

- The total installed cost is $29,153, including adsorption equipment ($19,100), mobile kit ($1,557), multi-tank manifold ($796), resin ($2,690), and set-up ($5,000).
- Operating costs for an adsorption unit is $38,461 vs. $92,500 for disposal.
- The calculated payback period for investment in the equipment/process: 7 months, using a 15-year analysis, 10% discount rate, and straight line depreciation over 10 years.
- The calculated annual savings is $54,039 in disposal costs.

Diffusion Dialysis. Diffusion dialysis is described in the original Bluebook (see Section 5.8). A diagram of an operational system for acid dip bath recovery is shown in Exhibit 1-19. This diagram illustrates the use of both ***continuous*** and ***batch*** operation. In the continuous mode, acid is withdrawn directly from an acid bath, and the recovered acid stream is returned directly to the bath. The stream containing the contaminants and unrecovered acid is sent to an industrial wastewater treatment system (IWTP). An advantage of operating in the continuous mode is that labor requirements are minimal. Once the operation has been established, the unit runs with only occasional oversight. Another advantage is that continuous processing of the bath acid results in a bath with consistent composition. The primary disadvantage of the continuous mode of operation is that the required processing rate is generally about three times the normal acid disposal rate for the bath, and the waste generation rate is correspondingly larger. In general, continuous operations are only applicable to acid baths with inventories greater

[20] Adsorption Treatment Systems to Recover Mineral Acid Solutions

Exhibit 1-18. Annual Operating Costs Comparison for Disposal and Adsorption Unit

Operational Cost Category	Dollars per Year	
	Disposal	Acid Sorption
Material:	$4,200	$1,004
Labor and Clean-up:	$0	$31,600
Laboratory:	$0	$5,000
Utilities:	$0	$145
Secondary Treatment (@$53/1,000 gal):	$0	$12
Disposal (@$0.81/lb):	$56,700	$0
Regulatory Compliance:	$31,600	$700
Total Operational Costs:	$92,500	$38,461
Net Annual Savings:		$54,039

than several thousand gallons. At smaller volumes, a batch operation, as shown, is more practical.

In 1999, the Environmental Protection Agency Region I supported research to test a diffusion dialysis acid recovery system. The resulting report documented eight applications of diffusion dialysis at four companies; the results are summarized in Exhibit 1-20.

The sizing of a dialysis system can be based on the volume of spent acid previously produced; the "rule of thumb" requires that, at a minimum, the volume of spent acid that was previously discarded should be recycled once through the dialysis unit over the same period of time it took to generate the spent acid. For example, if a 300-gallon acid bath is being disposed of once per month, then a 10 gallon per day (gpd) diffusion dialysis system

Exhibit 1-19. Diagram of Diffusion Dialysis Unit and Operating Scheme[21]

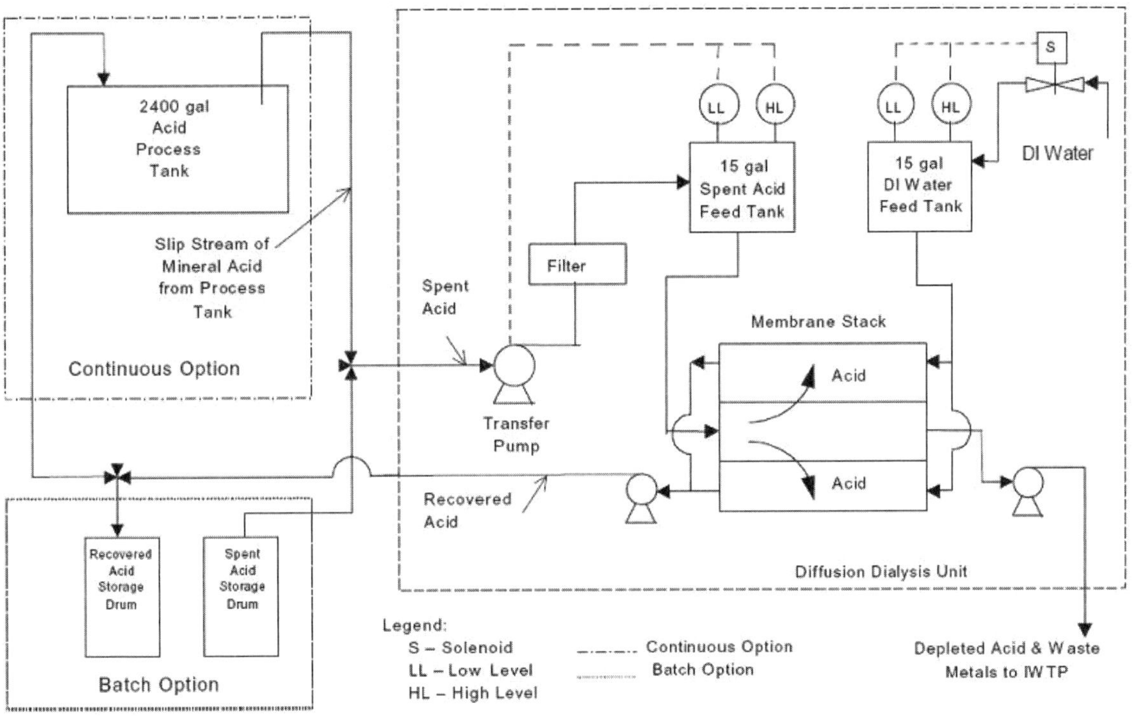

[21] U.S. Department of Defense, Environmental Security Technology Certification Program, ESTCP Cost and Performance Report (PP-9705), September, 1999.

Exhibit 1-20. Results from EPA Supported Testing of Diffusion Dialysis[22]

Acid	Acid Recovery, %	Contaminant	Contaminant Reduction, %
Nitric	96	nickel, iron, copper	66-86
Nitric	80	copper	88
Hydrochloric	90	chromium, iron	72-80
Hydrochloric	76	copper, iron	78-89
Sulfuric	94	copper	91
Sulfuric	89	nickel, iron, chromium	76-91
Sulfuric	90	aluminum, copper	89-94

would be recommended. The standard processing rate for diffusion dialysis systems is 0.025 gal/hr/sq ft (approximately 1 liter/hr/sq meter) of available anion exchange membrane surface area. To obtain the membrane area needed to process large volumes, the membranes are stacked between gasketed hydraulic flow spacers. These membrane stacks are usually standardized over a range of differing processing capacities. [23]

Comparison of Diffusion Dialysis and Acid Sorption. Diffusion dialysis and acid sorptiont compete directly for acid recovery applications. Based on vendor literature and journal articles it appears that there are approximately equal numbers of diffusion dialysis and acid sorption units in commercial use (about 500 to 750 of each).

Acid sorption is the older of the two technologies, having been first used around 1963. Early applications of acid sorption used available ion exchange resins and equipment. In 1975 a novel ion exchange technique called reciprocating flow ion exchange was applied to acid recovery, which essentially commercialized the acid sorption process. This method of ion exchange is characterized by

- fine particle size resins
- countercurrent elution
- short column heights
- fixed, over-packed resin beds
- short cycles

and a number of other features. The process is ideal for treating small volumes of concentrated solutions with a minimum of dilution or fluid intermixing in the resin column. Reciprocating flow ion exchange has seen wide application in the metal finishing industry for the recovery of a wide variety of metals. By employing this method, the performance of the acid sorption process was significantly improved.[24] There is only one major North American manufacturer of this technology (Eco-Tec, Ontario, Canada).

Diffusion dialysis was developed in the early 1980's in Japan and commercialized around 1990. There are at least three U.S. manufacturers of this technology.

Diffusion dialysis and acid sorption were objectively evaluated by Concurrent Technologies Corporation (Johnstown, PA) using bench scale equipment to compare their ability to remove metal contaminants and recover free acid from a contaminated mineral acid bath.[25] This comparison was based on one specific application: recovery of hydrochloric acid (HCl) from iron (Fe) contaminated hydrochloric acid solutions.

For the test, solutions were prepared in a laboratory to closely approximate actual spent hydrochloric acid activation and cleaning baths. The results of testing indicated:

- Diffusion dialysis achieved acid recoveries averaging 97% and iron rejection averaging 86%.
- Acid sorption achieved acid recovery and iron removal efficiencies of approximately 96% and 45%, respectively.

---

[20] Adsorption Treatment Systems to Recover Mineral Acid Solutions

[22] Bonner, Francis J., and Donatelli, Alfred A., Diffusion Dialysis and Acid Recovery in Metal Operations, The Toxics Use Reduction Institute, University of Massachusetts Lowell, 2001.

[23] Delaney, Sean and Bailey, Daniel, Acid Recycling at Plating for Electronics, Products Finishing Magazine.

[24] Pajun, Paul and Harrison, J., Anodizing Acid Purification using Resin Sorption Technology at Pioneer Metal Finishing, AESF Compliance Week, January 1997.

[25] Kinnel, Wayne A. and Roberts, David S., Mineral Acid Purification Opportunities, Concurrent Technologies Corp., May 1996.

The tests show that for this application, the two technologies achieve almost identical HCl recovery efficiencies; however, diffusion dialysis achieves greater percent rejection of the contaminant metal in the feed and therefore is the more effective technology. It should be noted that for other applications, different results may be obtained.

In terms of capital cost, the two technologies are similar; however, replacement costs for the diffusion dialysis membrane will exceed replacement costs for acid sorption resin over the lifetime of the equipment.

1.4.3 Process Conditions

Higher concentrations of acid and higher operating temperatures dissolve more of the substrate metal. Since dissolved metal will cause the bath to lose effectiveness, lower acid concentrations and lower bath temperatures are preferred from a pollution prevention (P2) standpoint. The less metal that is dissolved into the bath, the longer the life span of the solution. In addition, higher operating temperatures require use of energy, which is also a source of pollution. However, lower is not always better. Lower temperatures and acid concentrations increase the time necessary for pickling and oxide removal, and the finisher must consider the time constraints of the process. There is usually an operating range in which both time constraints and P2 objectives can be met.

1.4.4 Inhibitors

Inhibitors can be added to pickling baths to retard or stop the etching caused by the acid solution, thus providing a P2 benefit. ***Synthetic inhibitors*** made up of organic compounds are in common use today with pickling operations. As may be expected, the use of inhibitors increases the time needed to remove scale. When the concentration of inhibitor is increased, less base metal is dissolved and the required processing time increases, especially when low-concentration acids are employed. Yet P2 favors lower acid concentrations. The metal finisher must be prepared to deal with this trade-off.

1.4.5 Chemical Substitution

Mineral acids are effective at removing scale and activating metal surfaces. Because of acid fumes (e.g., hydrochloric and nitric acids) and safety concerns, however, substitutes are sometimes sought. Chemical substitution was implemented by several benchmarking survey respondents, including the use of ***acid salt substitutes*** for nitric, hydrochloric and sulfuric acids. Examples of acid salts used for descaling (pickling) include sodium fluoride, ammonium persulfate, and ferric sulfate. The acid salts can be mixed with other chemicals to give a more effective cleaning action. For example, hydrogen peroxide is sometimes added to acid salt solution to improve surface activation[26]

1.4.6 Uses for Spent Baths

Spent acid baths can be used as a wastewater treatment reagent for neutralizing highly alkaline wastes. Spent sulfuric baths containing iron can be used with chromium wastewater as a reducing agent. This practice may result in higher sludge production rates as compared to the use of conventional reagents such as sodium bisulfite.

1.5 ANODIZING

The most common electrolytes used for anodizing are:

- Chromic acid (Type I anodizing)
- Sulfuric acid (Type II anodizing)
- Sulfuric and oxalic acids (Type III anodizing)

The use of chromic acid raises the most significant environmental issues. One of the most important changes in anodizing that has taken place over the past 25 years is a significant reduction in the use of the chromic acid process. This change has occurred primarily because of environmental and health concerns, and the consequent regulations. The most common alternative chosen to replace chromic acid anodizing has been the sulfuric acid process. Chromic acid, however, is still the method of choice for certain applications. One example in particular where chromic acid anodizing remains popular is coating aircraft parts, especially those with recesses (e.g., spot welded assemblies and rolled seams), which trap electrolyte. Lingering sulfuric acid in recesses will cause corrosion and possible part failure. One benefit of the chromic acid process is that residual chromic acid does not appreciably attack aluminum alloys. In addition, the chromic acid process has a less deleterious effect on fatigue life overall than does sulfuric acid anodizing. Nevertheless, aircraft manufacturers continue to seek alternatives to chromic acid. During the past 15 years, the aerospace industry has used a substitute process with a ***sulfuric-boric acid (SBAA) electrolyte*** to further reduce the use of the chromic acid process.

[26] Tan, A. C., Tin and Solder Plating in the Semiconductor Industry: A Technical Guide, Springer Semiconductors, 1993.

P2 efforts associated with sulfuric acid anodizing have capitalized on opportunities related to rinsing, anode use, ***bath maintenance***, substitution and energy use. Exhibit 1-21 summarizes the following discussion regarding pollution prevention strategies for anodizing.

The anodizing process involves several additional operations, including pretreatment (e.g., etch) and post-treatment (e.g., sealers and dyes). Pollution prevention opportunities also exist for these ancillary processes.

1.5.1 Dragout and Water Use Reduction

A good rinsing configuration for this process is a two-stage ***counterflow rinsing*** arrangement. As with all processes, efforts should be directed at reducing ***dragout***, which has a direct bearing on rinse water use. Rinse water control can be accomplished using pH controllers, ***conductivity controls,*** or a ***timer rinse control*** system. For final rinses following anodizing, a commonly accepted conductivity range is 2,000 to 4,000 μmho.

Some ancillary processes performed in anodizing shops are covered under the Plating and Polishing air pollution regulation (e.g., nickel acetate and chromate seals).[27] For the affected processes (those that use cadmium, chromium, lead, manganese and/or nickel), shops must implement the required P2 practices; among which the following are related to dragout control and rinsing:

- minimize bath agitation when removing tank objects
- maximize dripping of bath solution back into tank by extending drip time when removing the tank objects and using drain boards (i.e., drip shields)
- optimize the design of barrels, racks, and parts to minimize dragout of bath solution

1.5.2 Sulfuric-Boric Acid Anodizing

The Boeing Company developed and qualified the boric sulfuric acid anodize (SBAA) process in 1990 as a direct replacement for the chromic acid anodizing[28] (CAA) process used on aluminum production pieces. The SBAA process consists of a sulfuric-boric acid anodizing bath and a dilute chromate sealer bath (about 70 mg/l chromic acid). SBAA is a commercially available process that provides a protective coating meeting all military and industrial specifications that apply to the chromic acid anodizing process.

The SBAA process was tested by the Naval Air Systems Command at NADEP North Island, and has been approved by the military as an alternative to the chromic acid anodizing process. The SBAA process has since been implemented at various military depots. Mil-C-8625C (the Mil-Spec that governs all anodized coatings) was revised to include the SBAA process as an option for aluminum anodizing.[29]

The following are processing characteristic of BSAA as compared to chromic acid anodizing (Boeing):

- Paint adhesion -- As good or better than chromic acid.
- Processing time -- 20 minute process, 30-60% less time than chromic acid.
- Energy Requirements -- Requires lower temperature and voltage, more energy efficient than the chrome-based process.
- Facilities Requirements -- Equipment may require material upgrade.
- Sealing -- Dilute chromate seal must be used. De-ionized water seal is not acceptable BSAA will take longer to reach the same extent of hydration as Chromic Acid Bath Performance.
- Long term bath performance is more reliable
- Licensing -- BSAA requires patent license. No fee if all parts are for Boeing.
- Other -- In dusty environments, addition of sodium benzoate or benzoic acid to BSAA prevents fungus growth.

1.5.3 Cathode Use

Sulfuric acid anodizing can be carried out using either lead or aluminum alloy cathodes. One benchmarking survey respondent reported substituting aluminum cathode rods for lead cathodes as a P2 measure. In addition to eliminating lead, the advantage of aluminum cathodes is that they require less current and less cooling, due to the greater current carrying capacity of aluminum as compared to lead. Aluminum alloy cathodes will also outlast lead cathodes, which will slowly dissolve unless an electrical charge is maintained on the cathodes when the tank is not in operation. [30]

---

[27] National Emission Standards for Hazardous Air Pollutants: Area Source Standards for Plating and Polishing Operations; Proposed Rule, Federal Register March 14, 2008 (pp. 14126 – 14151).

[28] Boeing Corporation (http://www.boeing.com/companyoffices/doingbiz/environmental/chromic.html).

[29] Joint Services Pollution Prevention and Sustainability Technical Library, Sulfuric/Boric Acid Anodizing, April, 2003.

[30] Schaedel, Fred C, Try a Holistic Approach to Improve Anodizing Processes, Products Finishing Magazine, Dec., 2006.

Exhibit 1-21. Summary of Pollution Prevention Strategy for Anodizing

Objectives and Strategy	Potential Practices
Reduce Dragout and Water Use Strategy: Minimize drag-in and dragout to the maximum extent possible, control rinse water use.	Basic: • Reduce drag-in to avoid diluting anodizing bath. • Reduce dragout to decrease water use. • Install two-stage counterflow rinse tanks Advanced: • Install conductivity, pH or timer controllers. • Spray rinsing using low volume.
Improve Material Utilization Strategy: Avoid solution disposal by minimizing bath contamination and using available maintenance technology. Recycle WWT sludge.	Basic: • Avoid bath contamination (clean parts thoroughly before plating; minimize corrosion of busing, racks, fixtures and parts; and retrieve parts/fixtures which fall into the bath). • Recycle WWT sludge off-site. • Use electrolysis to hel maintain chromic acid solutions. Advanced: • Use advance technology to maintain the bath. Use acid sorption or diffusion dialysis for sulfuric acid baths and membrane electrolysis or ion exchange for chromic acid baths. • Use bath additive to sulfuric acid bath to permit higher operating temperature and lower electrical consumption.
Reduce Human Exposure Strategy: Minimize acid air emissions to workplace and community. Use less-toxic, substitute technologies, where applicable.	Basic: • Employ adequate air speeds and ventilation rates to assure removal of fumes or mists from inhalation zones. • Monitor inhalation zones to evaluate exposure and immediately correct any inadequacies. • Convert from chromic acid anodizing to a non-chromium process whenever possible. Advanced: • Use bath additive to reduce surface tension of chromic acid baths.

1.5.4 Process Control

The anodizing process is referenced by more than 75 major aerospace, military and commercial specifications, some of which have overly generous ranges for acceptable chemical concentrations, operating parameters, (e.g., temperature) and impurity levels. Operating the anodizing process within narrow, more desirable parameter ranges will reduce pollution by minimizing rejects and rework, extending the life of baths, and lowering energy use. For example, free sulfuric acid concentration ranges for the various specifications range from 5.0% to 28.4%, although the ideal range is 10 to 11% for Type II and 11.0 to 15.5% for Type III (Schaedel).

Correct use of additives and modifiers can significantly boost efficiency and energy savings. One benchmarking survey participant reduced energy use with Type II anodizing by using a bath additive that allowed the bath to operate at a higher temperature. This reduced the electrical requirements for cooling the anodizing solution. Much of industry operates Type II baths at 70 °F. When a good additive and modifier are used, the bath can be run as high as 85 °F (Schaedel). The same survey respondent indicated that use of a lower-frequency pulse rectifier reduced energy consumption for their hard anodizing process. This same equipment is also applicable to Type II anodizing. Special secondary half-wave rectifiers can also provide an energy savings with Types II and III anodizing (Schaedel).

Exhibit 1-22. Process Control Savings at a Small Anodizing Shop

Cost Parameter	Monthly		
	Before	After	Annual Savings
Nickel acetate chemicals	26 gal	6.8 gal.	$4,140
Treatment chemicals	$380	$100	$3,360
Treatment labor	9 hours	2.4 hours	$3,960
Sludge generation	150 lbs.	39 lbs.	$670
Totals			$12,130

Bath additives can also extend the life of nickel acetate seals. One such additive adds wetting and dispersing agents to improve seal quality and prevent smut and white powder formation on parts. It also provides a hydrophobic anodic coating that helps water to run off parts, thus minimizing spotting.[31]

One small anodize shop (23 employees) documented saving of $12,130 per year through improved process solution monitoring and control (Merit Partnership). A breakdown of their cost savings is shown in Exhibit 1-22. In addition to implementing various control measures, their improvements included installing a filter system on the nickel acetate bath.

Sealers and Dyes. The Merit Partnership (a joint venture including U.S. EPA Region 9, state and local regulatory agencies, private-sector industries, and community representatives) successfully demonstrated the use of ***reverse osmosis*** (RO) for recovery of nickel acetate and black dye.[32] In each operation (see Exhibit 1-23), the overflow from the first-stage rinse tank is pressurized by a high-pressure feed pump. Particulates are removed from the feed solution by two 1-micron cartridge filters. The feed solution then flows in series through spiral wound modules containing thin-film composite RO membranes (two modules for nickel acetate and four modules for black dye operations). After separation, a portion of the concentrate stream is returned to the process bath in order to recover process chemicals. The remaining portion of the concentrate stream is conveyed to the recirculation tank, where it is temporarily stored before being recirculated through the RO unit. Permeate is conveyed to the second rinse tank and is reused as rinse water. A small amount of fresh deionized water from an outside source is added to the process baths in order to make up for evaporative water losses.

[31] Merit Partnership, Extending Metal Finishing Bath Life, Dec. 1997.
[32] Merit Partnership, Reverse Osmosis Applications for Metal Finishing Operations, Jan. 2002 revision.

Exhibit 1-23. Reverse Osmosis Recovery of Nickel Acetate Diagram

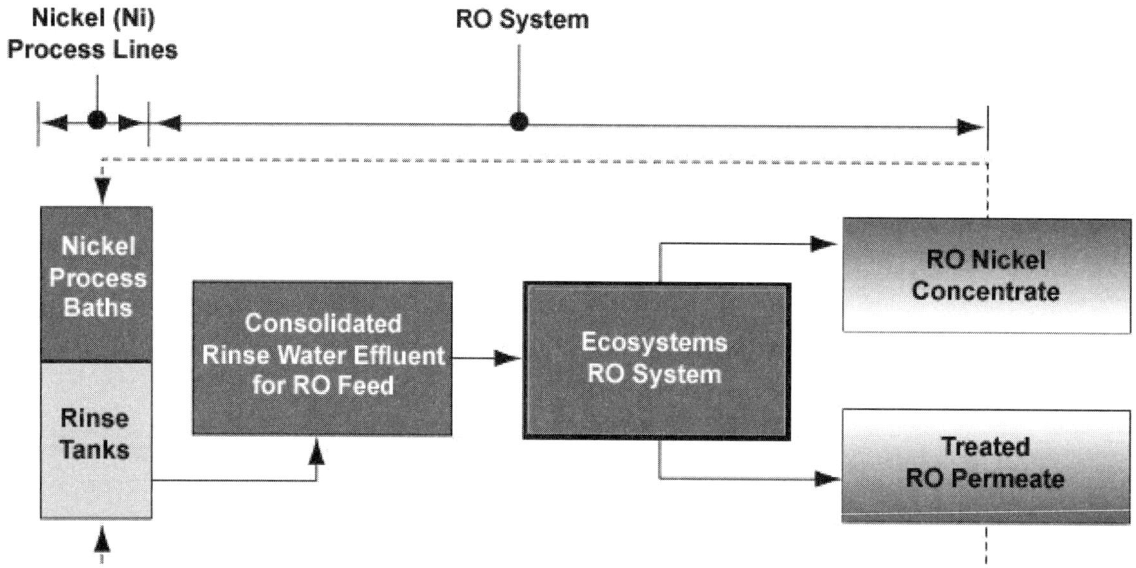

Exhibit 1-24. Monthly Cost Savings for RO/Nickel Acetate Installation

	Monthly Comparison		
	Before	**After**	**Savings**
Chemical additions	200 gal	40 gal	$1,939
Water use	200,000	2,000	$345
Discharge savings	200,000	0	$50
Total monthly savings			$2,334
Less O&M			$541
Net Savings			$1,792

The installed cost of the nickel acetate RO unit was $45,000. Annual operations and maintenance costs are $541 (mainly electricity and RO membrane cleaning). Cost savings for this installation are shown in Exhibit 1-24. The payback period for the RO installation was approximately 2 years, not including depreciation of the RO unit (i.e., tax savings).

1.5.5 Special Practices for Chromic Acid

The operation of chromic acid anodizing is affected both by environmental and by safety and health regulations directed toward the control of airborne emissions from the process:

- National Emissions Standards for Hazardous Air Pollutants (NESHAP) (40 CFR. §63.6) for chromium regulates air emissions to the environment.
- OSHA's chromium permissible exposure limit (PEL) regulates the amount of chromium in the air within the shop (29 CFR Parts 1910, 1915, et al.).

P2 practices can help meet these two regulations simultaneously. Chromium is released to the air from anodizing operations during the process, as gases created during electrolysis at the anode and cathode rise through the solution, break at the surface and release an aerosol to the air above the tank. A common P2 practice involves the use of ***fume suppressants*** that lower surface tension and/or create a foam blanket that reduces the amount of chromic acid mist released to the air.

There are two basic types of fume suppressants: ***wetting agents*** and ***foam blankets***. The difference between foam blankets and wetting agents is the way in which they reduce emissions:[33]

Exhibit 1-25. Diagram of a Process Tank with EED Cover

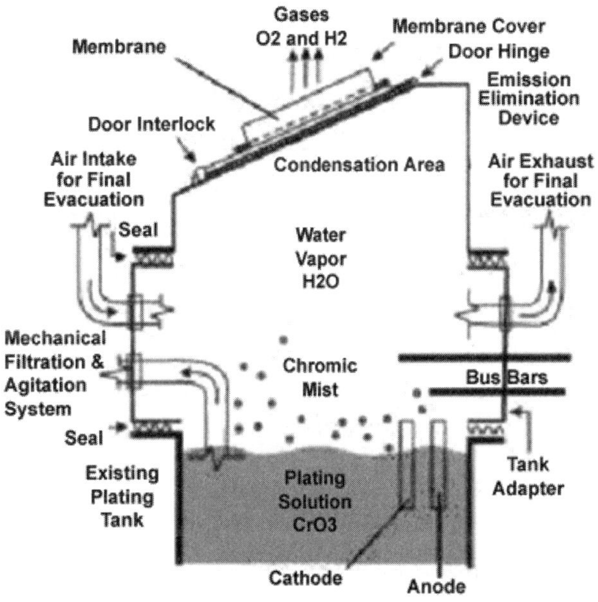

- A foam blanket fume suppressant generates a layer of foam across the surface of a solution when current is applied to that solution. The foam layer physically suppresses the mists produced on the surface of plating baths. Foam blankets do not prevent the formation of chromic acid mist – instead, they trap the mist under the blanket.
- Wetting agents lower the surface chemistry of plating baths. This causes gases to escape at the surface of the solution with a diminished "bursting" effect, reducing mist formation.

Another approach to preventing chromium losses to the air is to use ***tank covers*** that enclose the top of the anodizing tank. One commercial unit (Emissions Elimination Device, EED) incorporates membranes that allow gases to pass freely, while effectively blocking the escape of water vapor and chemical mists (see Exhibit 1-25). Prior to opening the tank cover, an evacuation system removes chromium-laden air, filters it, and returns the air to the plating room. Tank covers of this design eliminate the need for fume suppressants and a scrubber system.[34, 35]

[33] Riordan, Barrett J., Karamchandanl, Rohit T. Zitko, Larry J., and Cushnie George C., Capsule Report Hard Chrome Fume Suppressants and Control Technologies, United States Environmental Protection Agency Technology Transfer Office of Research and Development, Washington, DC 20460, EPA/625/R-981002 December 1998.

[34] Graves, Beverly, Encapsulating Hard-Chrome Plating Tanks, Products Finishing Magazine, Aug., 1999.

[35] Products Finishing Magazine, December, 1997.

1.5.6 Bath Maintenance

Chromic acid anodizing. Common bath contaminants associated with chromic acid anodizing include:

- Chloride – causes burning or pitting of parts: maintain below 0.3 g/l.
- Sulfate – gives rise to opaque appearance of coating: maintain below 0.5 g/l.
- Trivalent chromium – forms complexes that are ineffective for anodizing: maintain below 4 g/l.
- Aluminum – forms complexes that are ineffective for anodizing: maintain below 3 g/l.

Chloride and sulfate are often introduced via water that is added for evaporative makeup, and for rinsing parts. Using deionized water for these purposes can minimize chloride and sulfate introduction. Precipitation using silver oxide will remove sulfate if it builds up beyond tolerable limits. Limiting the total cathode surface to between one-fifth and one-tenth of the total anode area will minimize the production of trivalent chromium. This can be accomplished by shielding part of the cathode area with wire-reinforced glass plates. When trivalent chromium levels exceed tolerable limits, high current density ***electrolysis*** can be used to convert trivalent chromium to the required hexavalent form.

Various technologies have been marketed for removing dissolved aluminum, including ***ion exchange, ion transfer*** (i.e., non-ion-specific membranes including ceramic porous pots and units with polyfluorocarbon membranes), and ***membrane electrolysis*** (i.e., ion-specific membrane). Each of these technologies is discussed in Section 5. Of these three technologies, only ion exchange is presently a viable method for removing dissolved aluminum at a rate that prevents buildup of aluminum in the bath. Membrane electrolysis does an effective job; however only one company produced a suitable unit and, although they are still servicing existing units, they are no longer selling new equipment.[36] Ion transfer has only a marginal effect on dissolved aluminum concentration and it cannot maintain a chromic acid anodizing bath below tolerable limits. An EPA Environmental Technology Verification (ETV) test of an ion transfer unit was conducted at a chromic acid anodizing aerospace facility in 2000.[37] The results of the test showed that the unit does remove some dissolved aluminum from the bath, although the rate of removal was insufficient to prevent the bath from having to be discarded. During the ten week test, the unit was only able to remove about 25% of the aluminum that dissolved into solution during the test period.

Although ion exchange is an effective technology for maintaining chromic acid anodizing baths, it generates a significant volume of waste during resin regeneration. The waste stream is typically treated on-site using neutralization and precipitation, which generates a hazardous waste sludge (F019).

Additional information on chromic acid purification can be found in the hard chromium plating section.

Sulfuric Acid Anodizing. Bath maintenance is an important aspect of sulfuric acid anodizing. The anodizing process causes an increase in the concentration of dissolved aluminum. In conventional practice, the bath, or a portion of the bath, is replaced with fresh electrolyte. The spent solution is then typically neutralized and treated using a hydroxide precipitation process, a practice that creates sludge.

A fairly significant number of anodizers have implemented bath maintenance as an alternative to solution dumping. The most widely used technologies for this purpose are ***acid sorption and diffusion dialysis.*** Both technologies are discussed in the Acid Dipping process section.

Although they work by very different mechanisms, acid sorption and diffusion dialysis produce approximately the same results. Exhibit 1-26 shows operating performance data for an acid sorption unit applied to sulfuric acid anodizing baths.[38] For this installation, 96% of all free sulfuric fed to the unit is recovered for reuse. For every pound of aluminum removed by the unit, 2.5 pounds of sulfuric acid is lost to waste. This translates to a 70% reduction in sulfuric acid purchases and caustic soda neutralization costs when compared to a dump-decant scenario.

A benchmarking survey participant operates a hard coat anodize tank with a maximum aluminum content of 3.75 g/l. When that limit is exceeded, they use the spent hard coat bath as replacement for a sulfuric acid anodize bath,

[36] Personal communication with Dan Vaughan of Ion Sep Corporation, Rockland, DE.

[37] Start, Chris, Environmental Technology Verification Report, Evaluation of US Filter Corporation's RETEC® Model SCP-6 Separated Cell Purification System for Chromic Acid Anodize Bath Solution, VR-P2MF-10-02, September 2001.

[38] Jan, Paul Pajunen, P.E., Jim Harrison; Anodizing Acid Purification Using Resin Sorption Technology at Pioneer Metal Finishing, 18th AESF/EPA Conference, 1997.

Exhibit 1-26. Acid Purification Unit Operating Conditions

Stream	Sulfuric Acid, g/l	Aluminum, g/l	Flow l/hr.
Feed	200	7.0	432
Product	192	3.9	432
Byproduct	13.7	5.4	253
Water	-	-	253

a larger bath that requires a lower acid concentration and has a maximum allowable aluminum concentration of 18 g/l. When the concentrated hard coat bath is used, water is also added to dilute the acid, and the aluminum concentration is proportionally diluted to about 3 g/l. With periodic dumping of the sulfuric acid tank to a treatment system and replacing the sulfuric acid bath with the used hard coat solution, the aluminum sulfuric acid bath never exceeds the 18 g/l limit. The hard coat solution contains additives that are not needed in the sulfuric acid process, but these additives have not caused any operational problems with the sulfuric acid anodizing process. This practice of reusing the hard anodize bath has been in effect for about 15 years.

1.6 CADMIUM PLATING

The number of cadmium plating operations in the U.S. and worldwide has declined significantly during the past 20 years because of the considerable environmental and health concerns associated with cadmium metal. In some parts of the world, such as Europe, cadmium-plated parts cannot be imported or sold. Most cadmium electroplating has been replaced by zinc electroplating. Other popular alternatives include vacuum deposition of aluminum, and zinc and tin alloy plating.

Nevertheless, cadmium provides several distinct advantages over zinc; therefore, its use continues for certain applications. Cadmium:

- provides superior corrosion protection in marine environments
- is less apt to cause hydrogen embrittlement
- has a dense and adherent oxide film that does not form bulky corrosion products
- provides a conductive coating with good lubricity, suitable for fasteners (particularly in aerospace applications)

Seventeen benchmarking survey respondents (12.8%) perform cadmium plating. Two indicated that they are in the process of discontinuing cadmium plating. In an effort to retain customers while eliminating cadmium, one job shop offers to perform R&D for customers to find alternative coatings. Cadmium plating is currently 6.4% of this shop's total metal finishing sales ($9.5 million).

Exhibit 1-27 summarizes the following discussion regarding pollution prevention strategies for cadmium plating.

1.6.1 Water Use Reduction

Rinsing following cadmium plating presents challenges to the metal finisher. Due to the relatively low temperature of the cadmium bath, the opportunity to use dragout recovery rinsing is limited. However, some companies from the benchmarking survey do use dragout recovery rinsing or ***spray rinsing*** to capture and return ***dragout*** to the process tank. A company using spray rinsing is reportedly able to recover all dragout by spraying over a dead rinse and returning that solution. This company, which plates zinc die castings (rack and barrel), has experienced a buildup of zinc in its cadmium plating tank, which is attributed to spray rinsing of the barrels. The zinc contamination problem may force the company to make design changes.

The ***drag-in/dragout tank*** arrangement is applicable to cadmium plating. Generally, a good strategy is to focus on dragout minimization and to implement ***counterflow rinsing*** and good water use control. Dragout reduction can be achieved by:

- Increasing dwell time
- Rotating racked parts for better drainage
- Redesigning barrels to reduce the volume of trapped solution

Operating the cadmium plating bath at the lowest practical metal level will also lower dragout losses. Water use control can be accomplished by using ***flow restrictors***, ***conductivity controls***, shut-off valves, and ***timer rinse controls***. For final rinses following cadmium plating,

Exhibit 1-27. Summary of Pollution Prevention Strategy for Cadmium Plating

Objectives and Strategy	Potential Practices
Reduce Dragout & Water Use Strategy: Minimize dragout, employ counterflow rinsing, and exercise good rinse water use control.	Basic: • Reduce dragout (e.g., thoroughly drain parts over process tank, rotate barrels above tank using, use of insoluble anodes). • Operate the bath at the lowest practical cadmium concentration. • Install drip guards and return solution to process tank. • Install counterflow rinsing. Advanced: • Use conductivity controlled rinsing. • Implement recovery rinsing using a dragout tank and/or a spray rinse. • Install drag-in/dragout rinsing arrangement.
Improve Material Utilization Strategy: Avoid solution disposal by minimizing bath contamination and performing bath maintenance. Recycle WWT sludge.	Basic: • Avoid bath contamination (clean parts thoroughly before plating; minimize introduction of metallic impurities; use DI water for bath additions, and retrieve parts/fixtures which fall into the bath). • Mix insoluble and soluble anodes to avoid metal buildup in bath. • Use bath maintenance methods to remove impurities. • Recycle WWT sludge off-site. Advanced: • None identified.
Reduce Human Exposure Strategy: Eliminate cadmium electroplating.	Basic: • Convert cadmium electroplating processes to zinc electroplating. Advanced: • Substitute aluminum vapor deposition, zinc or tin alloy or other coating for cadmium electroplating processes.

a commonly accepted conductivity range is 500 to 700 μmho.

The Merit Partnership investigated ways of improving rinsing efficiency and reducing chemical losses for cadmium plating at a job shop.[39] The original and modified rinse systems for cadmium plating and chromate conversion coating from their report are shown in Exhibit 1-28. Overall, this system reduced the mass of metal and wastewater discharged to treatment by about 50%. The calculated payback of the modifications was 1.7 years.

On July 1, 2008 EPA published the ***National Emission Standards for Hazardous Air Pollutants for Plating and Polishing Operations.***[40] This rule impacts many metal finishing operations (EPA estimates 2,900 facilities). It applies to facilities with plating, polishing or thermal spray processes that contain cadmium, nickel, lead, manganese and/or chromium. (The rule will not apply to chromium anodizing and electroplating, which are covered under a previous rule.)

With this regulation EPA did not establish emission limits for plating and polishing operations as they did with chromium plating, but instead required plating and polishing facilities to implement specific pollution prevention (P2) and control practices that are related to air emissions, dragout and rinsing:

- minimize bath agitation when removing tank objects
- maximize dripping of bath solution back into tank by extending drip time when removing the tank objects and using drain boards (i.e., drip shields)
- optimize the design of barrels, racks, and parts to minimize dragout of bath solution
- use a wetting agent or fume suppressant in non-cyanide electroplating baths

[39] Merit Partnership, Modifying Tank Layouts to Improve Process Efficiency, Oct. 1996.

[40] National Emission Standards for Hazardous Air Pollutants: Area Source Standards for Plating and Polishing Operations; Final Rule, Federal Register July 1, 2008.

Exhibit 1-28. Cadmium Plating Rinse Systems

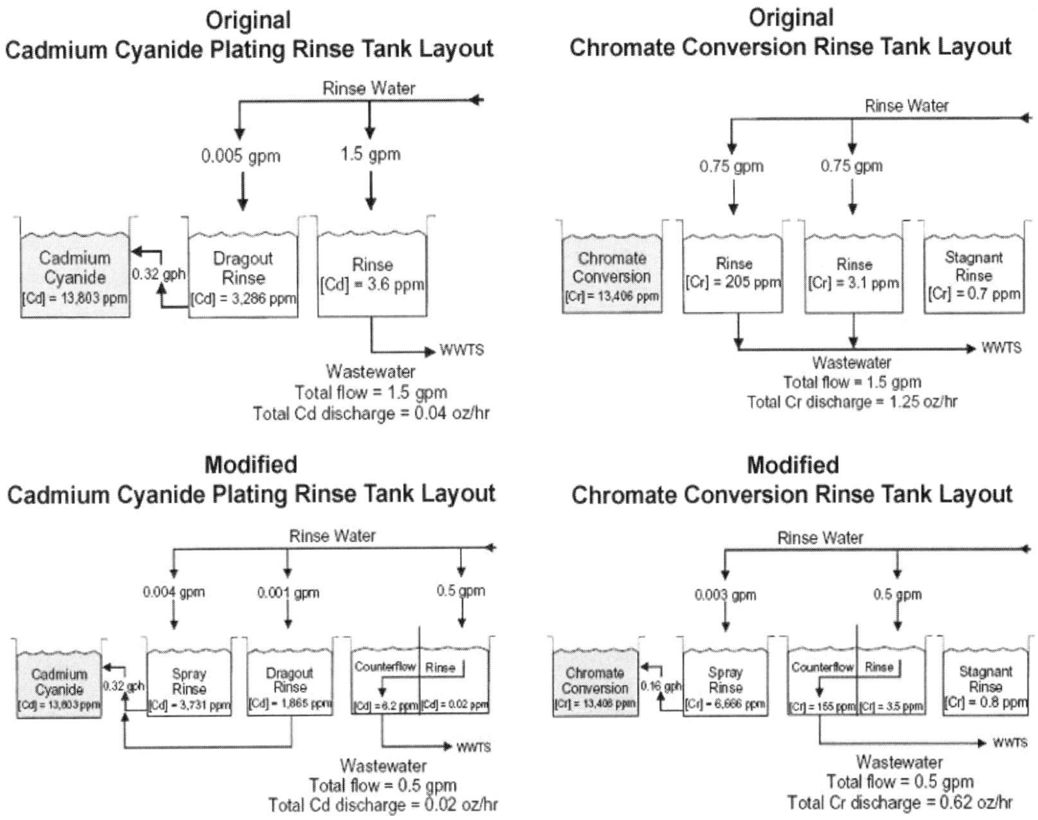

1.6.2 Bath Maintenance

Cadmium baths tend to increase in metal concentration over time, due to the difference between anode and cathode plating efficiency. Some platers compensate for this increase by adding other bath constituents to maintain recommended ratios. This procedure results in significantly higher levels of cadmium in the dragout. An alternative method is to control the cadmium metal concentration in the bath using a combination of insoluble anodes (carburized steel balls) together with soluble anodes. Other accepted ***bath maintenance*** methods include ***filtration*** and carbonate control (using chilling system).

1.6.3 Chemical Recovery

Various chemical recovery technologies have been used with cadmium electroplating. The most frequently used technologies are ***electrowinning, ion exchange*** and ***evaporative recovery*** (vacuum type). Electrowinning and ion exchange applications were demonstrated by EPA on a full scale installation.[41] Exhibit 1-29 shows the cadmium recovery system configuration. Water from the first rinse tank is passed through a filter (8 gpm) to remove suspended solids that would otherwise clog the resin bed. An anionic resin captures the cadmium-cyanide complex, and the water is returned to the second rinse. Samples of the rinse water collected before and after reovery were analyzed. The feed into the exchange unit contained about 2 to 8 mg/l of cadmium, 1 to 35 mg/l of cyanide, and 200 mg/l total dissolved solids (TDS). The flow out of the ion exchange unit was generally below 0.01 mg/l of cadmium and cyanide, and contained about 150 mg/l TDS. The pH of the rinse water was around 11.5, both before and after ion exchange. The resin is regenerated, using 15 to 20% sodium hydroxide, when the capacity of the resin bed is reached, and the regenerant (about 660 gallons per year) is processed in an electrowinning unit. The cadmium metal recovered by the electrowinning unit (about 70 pounds per year) is reused in the plating tank as anode material. Operating costs for this installation are shown in Exhibit 1-30. The capital cost of the system was $15,725 and the payback period for the system was less than one year.

[41] Gavaskar, Arun R.; Olfenbuttel, Robert F.; and Jones, Jody A., Cadmium and Chromium Recovery from Electroplating Rinsewaters, EPA Risk Reduction Engineering Laboratory, EPA/600/SR-94/050, April 1994.

Exhibit 1-29. Ion Exchange Recovery of Cadmium from Plating Rinse Water

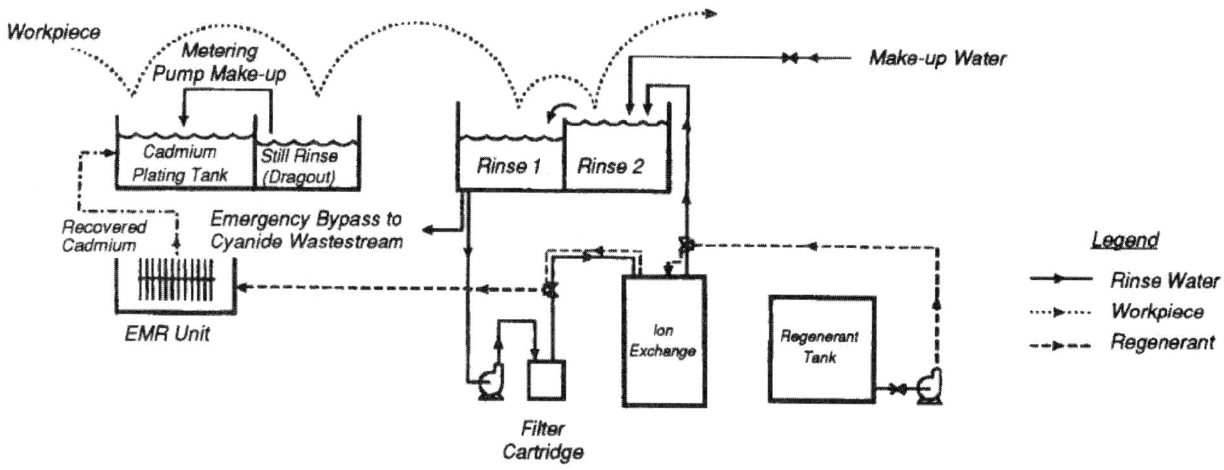

A second demonstration of electrowinning for cadmium was conducted by the Merit Partnership using a high surface area electrowinning unit (Exhibit 1-31). A heater was installed in the second rinse tank (ideally it would have been installed in rinse tank 1, but that was a plastic tank) and it was operated at 115 °F. The heater evaporates water, which concentrates the cadmium, and thereby improves the efficiency of electrowinning. After two months of operation, the unit had recovered 4.4 pounds of cadmium. The installed cost of the electrowinning system (including heater) was $9,010. The projected annual savings for the installation was $1,564, mainly due to reduced treatment costs. The estimated payback period was 8.7 years.

1.6.4 Non-Cyanide Cadmium Plating

Concerns over occupational health and safety, waste treatment costs, regulatory compliance requirements, and potential liability have encouraged the metal finishing industry to investigate ***non-cyanide plating technologies***. The earliest and most complete cyanide substitution that has taken place in the industry is the conversion from zinc cyanide to zinc chloride or zinc alkaline. Cadmium plating solutions that do not use cyanide are less common, but are commercially available. These include cadmium neutral or acid sulfate and cadmium acid fluoroborate plating solutions. Successful conversion to non-cyanide cadmium plating has been documented.[42] At this installation, the facility indicated that many of the anticipated difficulties or perceived shortcomings of a non-cyanide plating electrolyte did not materialize. Reported positive aspects of the non-cyanide bath included:

- Excellent throwing power -- internal anodes used with the cyanide bath for certain deep parts were not needed.
- Faster plating rate than cyanide and brightness is achieved more quickly -- the same current density and dwell time yields 0.2 mil more cadmium.
- Bath maintenance is simple and easy -- wet analysis of acid and metal is required once a week. Both components are stable, needing only small adjustments.
- The bath has shown a tendency to increase slowly in Cd concentration -- it is not necessary to remove the anodes from the tank during off hours.
- Brightener additions are based on amp-hr readings -- the consumption rate has been in line with the vendor's guideline of one gallon per 10,000 amp hr.

The reported problems with the non-cyanide bath include:

- Periodic ***carbon filtration*** treatment was necessary to eliminate staining on the inside diameter of some parts, and to restore brightener response in the Hull cell.
- When plating directly, machined C1215 steel parts developed roughness. Preliminary testing indicated that with enough cleaning and stripping cycles the parts would eventually accept a smooth acid-cadmium deposit. A nickel or copper strike is also a possibility if the application allows its use.

[42] Products Finishing Staff, Elimination of Cadmium Cyanide Waste Treatment, Products Finishing, Oct. 1997.

Exhibit 1-30. Operating Cost Comparison for Cadmium Recovery System

Item	Amount Used Per Year	Unit Cost	Total Annual Cost
Before recovery			
Freshwater	1,920,000 gal	$0.70/1,000 gal	$1,344
Wastewater treatment	1,920,000 gal	$22/1,000 gal	$42,240
Total			**$43,584**
With recovery system			
Freshwater	12,500	$0.70/1,000 gal	$9
Chemicals (NaOH)	96	$1.50/gal	$144
Energy	1,564 kWh	$0.075/kWh	$117
Labor	173	$25/hr	$4,325
Filter cartridges	6	$5	$30
EW anode plates	1	$30	$30
EW cathode plates	12	$30	$360
Labor	24 hr	$25/hr	$600
Regenerant disposal	660 gal	$22/1,000 gal.	$15
Filter disposal	6	$400/36 units	$67
Total			**$5,697**

- The effect of operating temperature was underestimated. Although the temperature never exceeded the maximum recommended in the operating guide, significant changes in the deposit brightness, uniformity, coverage and color (staining) occurred above 80°F. Also, brightener consumption tripled, which led to the installation of a cooling coil.
- There are a total of four additives for this process, making it somewhat difficult to interpret Hull cells and make necessary additions.
- The initial cost of equipment for this process can be high, due to the need for zirconium anode baskets.

1.6.5 Non-Chromate Conversion Coatings

Chromate conversion coating is a type of post-treatment applied to passivate cadmium and other metal deposits, to inhibit corrosion, and to promote coating adhesion. The traditional process uses toxic hexavalent chromium compounds. Recently, the industry has developed less toxic alternatives in order to comply with substance restriction legislation such as RoHS. One alternative is trivalent chromate conversion, which under testing by the Navy has shown equal or better effectiveness compared to hexavalent chromium coatings in a 1,000 hour neutral salt fog test.[43] Work by the Navy is continuing and includes evaluation of non-chromium conversion coatings.

1.7 COPPER PLATING

Operating environmentally efficient copper electroplating processes begins with the selection of the plating bath. Twenty years ago, most copper plating was carried out using cyanide baths. However, acid copper baths have become the predominant choice for meeting most metal finishers' needs.

Unfortunately, acid copper baths cannot be used to plate copper directly onto steel or zinc (both of which are commonly plated). In these cases, a copper cyanide strike (thin deposit) can be used, followed by an acid copper solution. This strategy greatly reduces the quantity of cyanide needed. Alternatively, copper pyrophosphate strike solutions are employed by some shops to avoid the cyanide bath when plating steel substrates. In addition, proprietary alkaline, non-cyanide copper plating processes may allow the use of a single solution to replace the two-step copper strike-plate process. However, the

[43] Matzdorf, Craig, Trivalent Chromium-based Post Treatment for Sacrificial Coatings, AESF/EPA Conference for Environmental and Process Excellence, Feb. 2003.

Exhibit 1-31. Cadmium Electrowinning Application

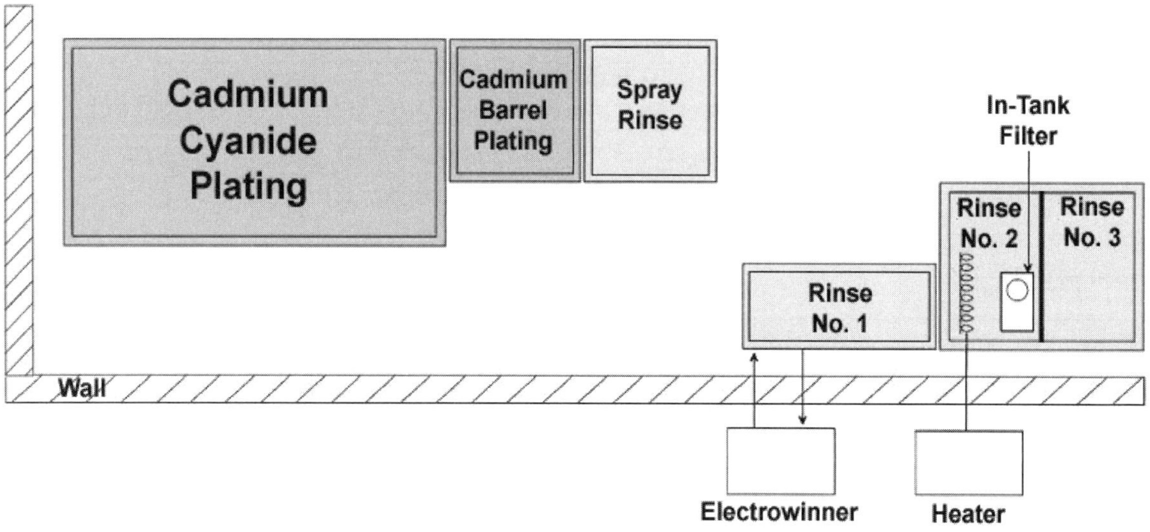

use of such solutions is not widespread. Exhibit 1-32 summarizes the following discussion regarding pollution prevention strategies for copper plating.

1.7.1 Water Use Reduction

Minimizing ***dragout***, using multiple rinse tanks, and controlling water use can achieve water use reduction. To reduce dragout, efforts should focus on:

- Rack design
- Part orientation
- Dwell times

To meet rinsing needs and lower water use, a two-stage ***counterflow rinsing*** configuration will typically suffice following copper plating.

Although ***dragout tanks*** for recovery are typically not employed with copper plating, they can be used with baths operated above 100°F. One benchmarking survey respondent reported using ***recovery rinsing*** on copper plating. On continuous production lines, where rinse water needs are relatively stable, flow can be controlled easily by periodically monitoring rinse water conductivity and manually adjusting flow, using a valve. If workflow is sporadic, a ***conductivity control*** or ***timer rinse control*** should be considered. These devices help to match rinse water use and production flow. For final rinses following copper plating, a commonly accepted conductivity range is 200 to 500 μmho.

1.7.2 Bath Maintenance and Control

Copper electroplating baths are rarely discarded. Common ***bath maintenance*** methods include ***filtration*** (removes suspended solids), carbon treatment (removes organic contaminants), low current density ***electrolysis*** (removes dissolved contaminants), and carbonate freezing or carbonate precipitation (removes excessive carbonate from cyanide baths).

Organic residues from buffing compounds, cleaners, tank and rack materials and other organics can contaminate the bath. Continuously filtering the plating solution through a filter chamber packed with approximately 0.7 grams of activated plating-grade carbon per liter of plating solution (0.1 oz per gallon) will remove small amounts of some organics.

Low concentrations of hexavalent chromium, which can be dragged into the bath on the racks, will produce dull and spotty deposits. Chromium can be reduced quickly to its trivalent form and precipitated by the addition of proprietary reducing agents directly to the plating bath. Hexavalent chromium also can be converted to the inactive trivalent state by using high-current-density electrolysis, or by adding sodium hydrosulfite or sodium stannite. There is disputed evidence that trivalent chromium left in the plating bath can be reoxidized by air back to hexavalent chromium. The best procedure is to precipitate the chromium and remove it by filtering.

Exhibit 1-32. Summary of Pollution Prevention Strategy for Copper Plating

Objectives and Strategy	Potential Practices
Reduce Dragout & Water Use Strategy: Minimize dragout, employ counterflow rinsing, and exercise good rinse water use control.	Basic: • Reduce dragout • Install counterflow rinsing. Advanced: • Use conductivity controlled rinsing. • Employ recovery rinsing for baths with elevated temperature.
Improve Material Utilization Strategy: Avoid solution disposal by minimizing buildup of zinc and metallic bath contamination. Recycle WWT sludge.	Basic: • Avoid bath contamination (clean parts thoroughly before plating; minimize introduction of metallic impurities; use DI water for bath additions, and retrieve parts/fixtures which fall into the bath). • Use common bath maintenance methods to remove impurities. • Use copper ball or nugget anodes in favor of copper bar. • Recycle WWT sludge off-site. Advanced: • Use equipment for automatic additions of certain bath components.
Reduce Human Exposure Strategy: Use non-cyanide copper electroplating processes.	Basic: • Convert copper cyanide electroplating processes to non-cyanide processes, wherever possible. Advanced: • Eliminate cyanide strike process, whenever possible.

Zinc contaminants will plate out simultaneously with the copper, causing brittle and brassy looking deposits. Zinc can be removed by dummying the bath at 0.2–0.4 A/sq dm. Dummying also removes many other metallic contaminants.

Alkaline cyanide solutions are relatively more difficult to control than acid sulfate solutions. Cyanide also involves special handling and treatment procedures because of its acute health hazards and waste disposal requirements. Recommendations for solution control include:[44]

- Use high-purity, oxide-free anodes for high-efficiency decorative copper plating. Lower quality anodes can be used in strike and medium-efficiency Rochelle-salt cyanide-copper plating baths.
- Use copper balls or nuggets in steel baskets, or slab anodes. Intermingle plain steel anodes with the copper anodes to control the buildup of copper in the bath.
- Bag the anodes (PP, nylon) in order to contain insoluble particles coming from the anodes.
- Keep the anode/cathode ratio between 1:1 and 2:1.

Acid copper sulfate baths are easier to maintain:

- Use quantitative analysis to control copper, sulfuric acid and chloride.
- Add proprietary addition agents, which control brightness, ductility, and leveling, on the basis of amp-hr. Also regulate their addition by using Hull cells and a copper analysis.

1.7.3 Material Usage

Converting from copper bar anodes to baskets of copper nugget or ball anodes can improve plating efficiency and material usage rates. Because the copper surface is maintained when nuggets or balls are used, consistently high plating speeds can be maintained. Bars typically cost about 10% more than nuggets, and not all of the bar material is consumed in the plating process. Although the work may be distributed evenly, the plating process does not deplete the copper evenly along the length of the rod. There is typically less erosion in the vicinity of the hook than at the opposite end. In time, this produces a spear

[44] Snyder, Dondald L., Choosing and Troubleshooting Your Copper Electroplating Process, Products Finishing Magazine,

shape that needs replacing. Typically, bars are replaced when about 15% of the copper material remains. Bars also hang from hooks in such a way that not all of the copper material is fully immersed in the plating bath. Inevitably, there will be unused material at the top. In basket operations, copper material is fully submerged, and all of the material is consumed. As the copper dissolves, the baskets are replenished, leaving no scrap material. Selling the unused bar material back to a copper distributor as scrap recovers some of the value of the original metal, but the unused copper still represents a sizable cost. A typical scrap return of $2.00 per lb still produces a scrap cost of $1.00 per lb, not taking into account the costs related to handling and transportation.[45]

1.7.4 Chemical Recovery

Various chemical recovery technologies have been used with copper electroplating. The most frequently used technologies are ***ion exchange, electrowinning,*** and ***evaporative recovery***. The most common approach to copper recovery is a combined system using both ion exchange and electrowinning.

An efficient IX/EW configuration is shown in Exhibit 1-33. A dragout rinse is used to recover the bulk of the copper and therefore allows for installation of smaller and less expensive IX and EW recovery equipment. The ion exchange unit, which is best applied to dilute rinse waters, maintains a low metal concentration in the final rinse. The ion exchange unit concentrates the metal into a regenerant stream, and the electrowinning unit removes the metal.

1.8 DECORATIVE CHROMIUM PLATING

Although some of the same strategies used for hard chromium plating are applied to decorative chromium plating, several key factors have caused companies to take some approaches that are specific to decorative chrome. These factors include:

- Lower plating bath temperature (<125°F)
- Higher average dragout rates
- Widespread use of a substitute plating solution (trivalent chromium).

The advantages to using the trivalent solution include the following:

- Trivalent chromium baths reduce human exposure to toxic hexavalent chromium compounds.
- During operation, the trivalent bath does not generate mist, as the hexavalent bath does.
- Trivalent baths also have a significantly lower chromium concentration and lower viscosity, both of which reduce the quantity of chromium lost due to dragout.

45 Panagiotakis, Gregory, The Case for Anode Baskets, Products Finishing Magazine, Nov. 1995.

Exhibit 1-33. Combined Ion Exchange and Electrowinning System

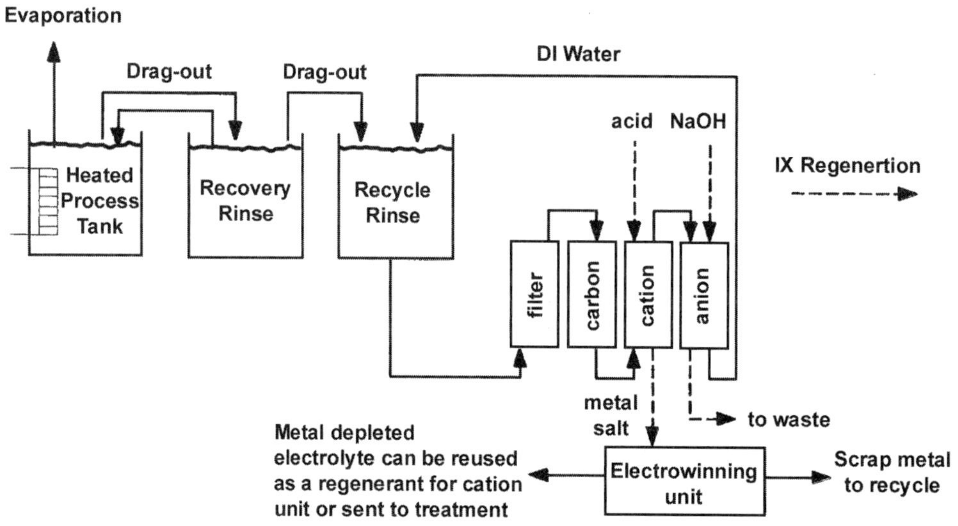

Some benchmarking survey respondents have successfully implemented trivalent chromium plating. But decorative chromium also presents some environmental challenges. For example, the lower bath temperature and higher dragout rate make implementing ***closed-loop rinsing*** more difficult than with hexavalent chromium plating. Exhibit 1-34 summarizes the following discussion regarding pollution prevention strategies for decorative chromium plating.

1.8.1 Water Use Reduction

A good formula for environmental success with decorative chromium plating includes dragout minimization, direct recovery, and ***counterflow rinsing*** with good water use control. ***Dragout*** can be minimized by orienting parts on racks such that they drain completely, slowly withdrawing parts from the process solution, suspending the parts over the process tank for drainage (see ***draining/rinsing over the plating tank***), and using a dragout recovery rinse followed by a multiple-stage counterflow rinse.

Another approach to dragout minimization is to operate the bath at a lower, but still acceptable, chromium concentration. A number of survey respondents have implemented most or all of these P2 elements.

Water use control is often achieved simply with ***flow restrictors*** due to the continuous nature of workloads. However, for decorative chromium lines that are not operated continuously, additional controls are needed, such as ***conductivity controls*** or ***timer rinse controls***. For final rinses following decorative chromium plating, a commonly accepted conductivity range is 100 to 300 μmho. Due to the lower bath temperature and higher dragout rates, most companies do not achieve ***closed loop*** rinsing (although two exceptions were identified by the benchmarking survey). As a result, companies frequently install recovery equipment, the most successful of which is the ***atmospheric evaporator***.

One company that has closed-loop rinsing uses a drag-in tank and three final rinses following plating. The tanks are organized in a counterflow arrangement. This company also has a separate "holding tank." Deionized water is fed to the last rinse and counterflows to the first rinse. The solution in the first rinse flows to a holding tank. The plating solution is circulated continuously to the holding tank. The holding tank provides flow equalization by balancing the incoming water flow rate and the evaporation rate in the plating tank. ***Bath maintenance*** (***porous pot***) is also performed in the holding tank, which saves plating tank space. The drag-in solution is circulated to the first rinse tank. A "bubble pump" (no moving parts) is used to circulate the drag-in tank solution (at 3.5 gpm). This arrangement is referred to as a ***drag-in/dragout recovery rinsing*** scheme. It contributes to closing the loop on rinsing.

Survey respondents listed additional P2 methods that relate to rinsing. These include using ***wetting agents*** to reduce viscosity, and ***spray rinsing*** parts over the process tank

1.8.2 Bath Maintenance

Returning dragout to the process tank may result in a buildup of contaminants in the plating bath. This buildup is a concern especially with zero wastewater operations. Contaminants, contributed from various sources, can lead to slower plating rates and poor deposits. The first line of defense against contaminant buildup is prevention. Examples of good practices include:

- Thoroughly cleaning parts before plating
- Minimizing corrosion of busing, racks, fixtures and parts
- Using deionized water for rinsing
- Avoiding drag-in of sulfate and chlorides
- Avoiding exclusively plating inner diameters
- Retrieving parts or fixtures that fall into the bath.

To deal with contaminant buildup, which occurs even when prevention measures are taken, companies can install bath maintenance technologies. Methods used include filtering, porous pots, and more advanced technologies such as ***membrane electrolysis*** and ***ion exchange***. Of the available technologies, ion exchange is the most effective and reliable. However, ion exchange creates a significant waste stream as a result of resin regeneration. This stream is typically treated on-site using neutralization and precipitation, and results in the formation of hazardous sludge (i.e., F006). Therefore, the value of preventing bath contamination cannot be overemphasized.

Use of ion exchange for removing metal impurities is particularly well-established for hexavalent chromium plating baths. In these situations, cationic impurities (e.g., iron, copper, and nickel) are being separated from the desired anionic plating species (i.e., hexavalent chromium). This can be accomplished using a strong acid cationic resin. Ion exchange purification using a trivalent chromium bath is a more significant challenge because

Exhibit 1-34. Summary of Pollution Prevention Strategy for Decorative Chromium Plating

Objectives and Strategy	Potential Practices
Reduce Dragout & Water Use Strategy: Minimize dragout and, to the maximum extent possible, return residual dragout to the process tank, and minimize rinse water use.	Basic: • Reduce dragout (e.g., good racking, slow withdraw from process solution, thoroughly drain parts over process tank, use wetting agent, and operate bath at lowest possible chromium concentration). • Install drip guards and return solution to process tank. • Perform initial rinsing (spray) over the process, if feasible. • Use dragout rinse tank with return of solution to process tank, followed by counterflow rinse. Advanced: • Install drag-in/dragout rinsing arrangement. • Install closed loop rinsing with multiple dragout tanks (2-4) connected in a counterflow arrangement. • Where dragout rate is high and surface evaporation is minimal, install auxiliary evaporation (e.g., atmospheric evaporator). • Recycle water from process tank heat exchangers.
Improve Material Utilization Strategy: Avoid solution disposal by minimizing bath contamination and using available maintenance technology. Recycle WWT sludge.	Basic: • Avoid bath contamination (clean parts thoroughly before plating; minimize corrosion of busing, racks, fixtures and parts; use DI water for rinsing; and retrieve parts/fixtures which fall into the bath). • Recycle WWT sludge off-site. Advanced: • Recycle air scrubber/mesh pad wash-down back to process tank. • Filter stripping and plating solution to remove suspended contaminants. • Use bath maintenance technology (e.g., ion exchange or membrane technology) to keep Cr+3 and metallic contaminants at low levels. This also reduces energy requirements for solution heating and plating.
Reduce Human Exposure Strategy: Minimize Cr air emissions to workplace and community. Use less-toxic, substitute chemistry.	Basic: • Employ adequate air speeds and ventilation rates to assure remove of mists from inhalation zones. • Monitor inhalation zones to evaluate exposure and immediately correct any inadequacies. Advanced: • Employ mesh pad mist eliminators to remove and recover chromium from ventilated air. • Control surface tension of plating bath. • Use trivalent chromium plating chemistry.

both the metal contaminants and the desired bath species are cationic. Trivalent baths are sensitive even to low concentrations of metal contaminants, including copper, zinc, and nickel (copper and zinc have been reported to cause discoloration and poor coverage at concentrations as low as 20 mg/l).[46] For trivalent chromium baths, purification can be achieved by using a chelating ion-exchange resin having a very high affinity for the cationic metal ion contaminants compared to the affinity for trivalent chromium. Commercial systems for this application that operate on a continuous basis (i.e., the plating bath is not taken out of operation) are available.

[46] Woodle, Roger and Hosea, j. Michael, Ion Exchange for Trivalent Chromium Plating Baths, Products Finishing Magazine, Nov. 1997.

1.8.3 Air Quality

Decorative hexavalent chromium plating causes the evolution of chromic acid mist. The quantity of misting is generally much less than with hard chromium plating. ***Fume suppressants*** can be used to lower the surface tension of the bath and thereby reduce the generation of fumes. One benchmarking survey respondent suggested that maintaining the surface tension below 45 dynes/cm greatly reduces the amount of mist being generated and reduces dragout.

Well-designed and operated ventilation systems prevent the chromium fumes from entering the workplace. In turn, the ventilated air is processed by ***scrubbers*** and/or ***mist eliminators*** to prevent emissions of chromic acid to the atmosphere.

Additional information on air quality issues for chromic acid baths is found in the chromic acid anodizing section.

1.9 ELECTROLESS NICKEL PLATING

Compared with solutions used with electroplating processes, electroless nickel (EN) plating baths have a very limited life span. Because these baths are discarded regularly, electroless nickel plating generates a significant quantity of waste. Exhibit 1-35 summarizes the following discussion regarding pollution prevention strategies for electroless nickel plating.

1.9.1 Bath Maintenance

Operating environmentally efficient electroless nickel plating processes begins with efforts to maintain the bath in a usable condition for an extended time. Minimizing outside contamination and maintaining the bath's constituents at proper concentrations can help.

In addition, technology is available to regenerate baths and keep them operating for an extended time. The overall electroless nickel deposition reaction can be generally written as:

$$3NaH_2PO_2 + 3H_2O + NiSO_4 = 3NaH_2PO_3 + H_2SO_4 + 2H_2 + Ni^0.$$

Sodium hypophosphite (NaH_2PO_2) reduces nickel sulfate to metallic nickel, and is oxidized to sodium orthophosphite (NaH_2PO_3). As the deposition of nickel proceeds, the orthophosphite concentration increases. Nickel salts and other constituents are added to the bath to make up for the chemicals consumed by the plating process. However, the process cannot continue indefinitely because of the buildup of by-products.

The quality of the nickel deposit diminishes over time and the nickel bath must usually be discarded after 7 to 12 ***metal turnovers*** (MTO). (One turnover is said to occur when the cumulative amount of nickel added is equivalent to the original quantity in the bath.) The deterioration of the bath is caused mainly by an excessive buildup of orthophosphite.

In recent years, ***electrodialysis*** (ED) has become a popular method to extend the life of electroless nickel baths. Originally, the ED process was used on a batch basis to rejuvenate the EN bath after it was partially contaminated, or after it had completely reached the end of its useful life.[47] Batch treatment, although relatively successful, had the drawbacks of having to remove the solution from production, and resulting in a "jig-saw" curve of bath contaminants, which leads to unpredictable performance results. Continuous processing was not originally used because of technical hurdles, including lack of available membranes that could tolerate the operating temperature of the EN bath. New developments have apparently overcome the technical issues. One promising approach is use of monopermselective ion exchange membranes and a modified arrangement of anionic and cationic ion exchange membranes in a system referred to as Electro-Dialysis of Electroless Nickel (EDEN).[48]

A flow diagram of continuously regenerated EN using the EDEN approach is shown in Exhibit 1-35. The electrodialysis stack is supplied with high volume pumps from two reservoir tanks for the diluate and the concentrate, respectively. The diluate tank is continuously supplied with a low volume flow of approximately 10 gph (37.5 l/hr) from the electroless nickel-plating tank. The diluate is cooled by a counterflow heat exchanger to limit the solution temperature in the stack to maximum 104°F (40°C). The current density supplied to the electrodialysis stack is controlled by a specific gravity sensor in the diluate reservoir, and therefore adjusts the stack performance to the actual plating bath loading or operating time. The second solution circuit of the concentrate comprises an automatic pH-control and automatic water

[47] The Merit Partnership, Extending Electroless Nickel Bath Life Using Electrodialysis, U.S. EPA Region 9, July 1998.

[48] Longfield, Peter and Orgill, Graham, The EDEN System for Electroless Nickel Bath Life Extension and Consistent Quality, AESF SUR/FIN 2005 Proceedings.

Exhibit 1-35. Summary of Pollution Prevention Strategy for Electroless Nickel Plating

Objectives and Strategy	Potential Practices
Reduce Dragout & Water Use Strategy: Reduce dragout, employ counterflow rinsing, and exercise good rinse water use control.	Basic: • Reduce dragout (e.g., good rack design, operator training, drain parts over process tank, and spray rinsing over tank). • Install drip guards and return solution to process tank. • Install counterflow rinsing. Advanced: • Install spray rinsing over the process tank. • Use conductivity controlled rinsing.
Improve Material Utilization Strategy: Minimize solution disposal (increase turnovers) by reducing introduction of contaminants and controlling bath constituents. Recycle baths/nickel.	Basic: • Avoid bath contamination (clean parts thoroughly before plating and use DI water for bath additions). • Employ adequate filtration. • Avoid overheating bath. • Cool bath when idle to reduce breakdown of constituents. • Strip tank walls to prevent excessive plate out. • Recycle baths (or plated-out nickel metal) off-site. Advanced: • Install electrolytic recovery of nickel from rinse water. • Use equipment for automatic analyses and additions of certain bath components. • Install electrodialysis equipment to prolong bath life.
Reduce Human Exposure Strategy: Avoid worker exposure to toxic chemicals and other hazards.	Basic: • Always enforce use of appropriate PPE. • Provide adequate ventilation. Advanced: • Change type of chemistry employed to avoid ammonium hydroxide.

dosing to keep the concentrations of sodium, sulfate and orthophosphite in the optimum range. The typical bleed stream is approximately 4 gph (15 l/h). The regeneration unit also contains an automatic controller for the nickel content in the plating bath (photometric analysis of Ni) and the pH value to allow for fully automated addition of the replenishment chemicals to the electroless plating bath (Longfield).

Under production conditions, the EDEN system has proven to have a bath life of 2,150 MTO.

1.9.2 Disposal of Spent Baths

When baths deteriorate and require disposal, companies may process them on-site to recover the nickel (plating the metal out onto steel wool), or may send the baths off-site for recycling. Some chemical suppliers offer bath recycling. Recycling avoids treating the spent electroless nickel bath, which is a difficult task due to the presence of complexing agents such as EDTA.

1.9.3 Dragout and Water Use Reduction

Benchmarking survey respondents employ common methods for ***dragout*** reduction such as good rack design and good racking practices, operator training, draining parts over the process tank (see ***draining/rinsing over the plating tank***), and water use control methods such as ***flow restrictors***. Rinsing following electroless nickel plating typically consists of two to three counterflow rinses. Some companies use ***recovery rinsing***, but there is concern that it may lead to faster deterioration of the bath. Also, some companies use ***spray rinsing*** over the bath.

On July 1, 2008 EPA published the ***National Emission Standards for Hazardous Air Pollutants for Plating and Polishing Operations.***[49] This rule impacts many metal finishing operations (EPA estimates 2,900 facilities). It applies to facilities with plating, polishing or thermal spray processes that contain cadmium, nickel, lead, manganese and/or chromium. (The rule will not apply to chromium anodizing and electroplating, which are covered under a previous rule.)

With this regulation EPA did not establish emission limits for plating and polishing operations as they did with chromium plating, but instead required plating and polishing facilities to implement specific pollution prevention (P2) and control practices that are related to air emissions, dragout and rinsing:

- minimize bath agitation when removing tank objects
- maximize dripping of bath solution back into tank by extending drip time when removing the tank objects and using drain boards (i.e., drip shields)
- optimize the design of barrels, racks, and parts to minimize dragout of bath solution

1.9.4 Air Quality

Electroless nickel processes should be ventilated to avoid emissions within the workplace. One survey respondent indicated it uses a ***scrubber*** to remove contaminants from the ventilated air stream. Two survey respondents switched from using a plating solution containing ammonium hydroxide for pH control to one with potassium carbonate in order to reduce employee exposure.

1.9.5 RoHS & ELV Compliant Electroless Nickel

Over the past few years, a number of new environmental directives have come out of Europe and Asia, encompassing mainly the automotive and electronics industries. The purpose of these directives is to reduce the amount of toxic materials entering the environment by limiting the amount of certain toxic substances allowed in a manufactured product and providing for the recyclability of the manufactured product. The three major directives affecting the metal finishing industry are:

- End of Life Vehicles (ELV, Directive 2000/53/EC) – Aims to reduce the amount of heavy metals contained in an automobile and provide for the recyclability of automobile components.
- Restriction of the Use of certain Hazardous Substances in Electrical and Electronic Equipment (RoHS, Directive 2002/95/EC) -- Restricts use of hazardous substances in electrical and electronic equipment.
- Waste Electrical and Electronic Equipment (WEEE, Directive 2002/96/EC) -- Imposes the responsibility for the disposal of waste electrical and electronic equipment (WEEE) on the manufacturers of such equipment.

The primary heavy metals addressed in these regulations are cadmium, lead, hexavalent chromium, and mercury. In electroless nickel plating, cadmium and lead are the major concerns; lead and cadmium are added to electroless nickel systems for stability and brightness. The ELV and RoHS Directives specify the limits for cadmium and lead in an electroless nickel deposit to less than 100 and 1,000 ppm, respectively. To meet this challenge, electroless nickel formulators have developed ELV and RoHS compliant baths. These systems can provide stable, simple-to-use processes capable of meeting any of the current electroless nickel specifications. The current downside to switching to these processes is that there will be a learning curve by both platers and suppliers as the new technology evolves and improves.[50]

1.10 HARD CHROMIUM PLATING

The key to environmental success with hard chromium plating is direct recovery. The hard chromium bath is operated at an elevated temperature (130 – 140°F). This results in surface evaporation rates of 0.05 to 0.07 gal/hr/ft^2. A 48-in. x 96-in. process tank operated at 140°F will evaporate 2.24 gal/hr. or about 54 gal over the course of 24 hours, which provides an opportunity to recover an equal volume of solution. As facilities approach 100% recovery of dragout, they must grapple with the buildup of bath impurities that will inevitably occur. Exhibit 1-37 summarizes the following discussion regarding pollution prevention strategies for hard chromium plating.

[49]National Emission Standards for Hazardous Air Pollutants: Area Source Standards for Plating and Polishing Operations; Final Rule, Federal Register July 1, 2008.

[50] Bellemare, Rich, RoHS and ELV Compliant Electroless Nickel, Products Finishing Magazine, Oct. 2005.

Exhibit 1-36. Flow Diagram Showing Continuous Regeneration of EN Solution

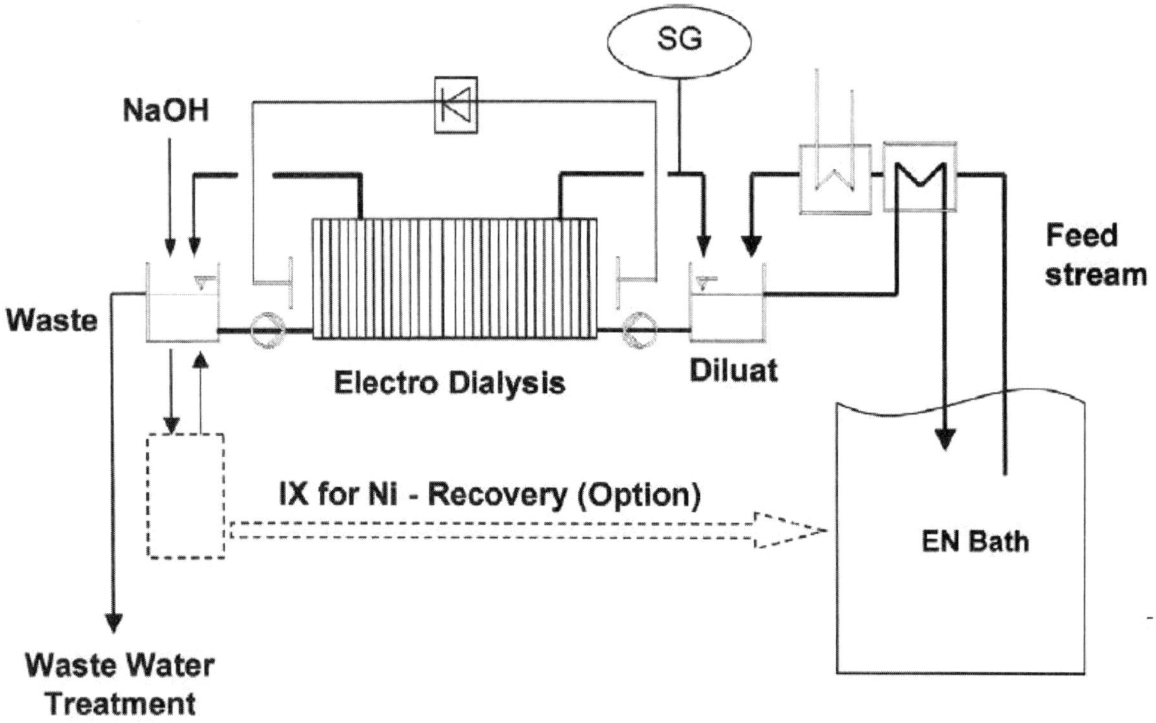

1.10.1 Dragout Recovery

Companies can maximize use of this recovery opportunity by returning concentrated solution to the process tank. Solution sources include

- drips from parts exiting the process tank
- solution caught by splash guards or trays
- ***scrubber*** water or mesh pad wash-down
- solution from rinsing

Because rinsing can generate a large volume of dilute chromium-bearing water, it is important to use a well-engineered, multiple-tank configuration that concentrates the ***dragout*** into the smallest possible volume.

In most cases, a three-stage ***counterflow rinsing*** arrangement will generate a sufficiently low flow that the entire volume of rinse water can be returned to the bath. In situations where there is insufficient surface evaporation to achieve zero wastewater discharge, an ***atmospheric evaporator*** can be installed to increase the overall evaporation rate.[51] Benchmarking survey respondents listed additional P2 methods that relate to rinsing. These include:

- use of ***drag-in/dragout recovery rinsing***
- splash guards on all tanks
- using ***wetting agents*** to reduce viscosity
- suspending racks over the process tank to drain
- ***spray rinsing*** parts over the process tank
- unracking over the process tank

1.10.2 Bath Maintenance

Returning dragout to the process tank may result in a buildup of contaminants in the plating bath. This buildup is of particular concern with zero wastewater operations. Contaminants, contributed from various sources, can lead to slower plating rates, poor deposits and higher energy bills. The initial line of defense against contaminant buildup is prevention. Examples of good practices include:

- Installing a ***dedicated etch tank*** for reverse etching
- Thoroughly cleaning parts before plating
- Minimizing corrosion of busing, racks, fixtures and parts
- Using ***deionized water*** for rinsing
- Avoiding drag-in of sulfate and chlorides
- Avoiding exclusively plating inner diameters
- Retrieving parts or fixtures that fall into the bath

[51] Mason, Sam, Case Study: Zero Sewer Discharge for a Hard Chrome Plating Facility; 20th AESF/EPA Conference, February, 1999.

Exhibit 1-37. Summary of Pollution Prevention Strategy for Hard Chromium Plating

Objectives and Strategy	Potential Practices
Reduce Dragout & Water Use Strategy: Minimize dragout and, to the maximum extent possible, return residual dragout to the process tank.	Basic: • Reduce dragout (e.g., use masking that does not trap excess solution, thoroughly drain parts over process tank, use wetting agent and/or operate plating bath at high end of allowable temperature range, and operate bath at lowest possible chromium concentration). • Install dragout racks over tanks or drip guards and return solution to process tank. • Perform initial rinsing (spray) over the process, if feasible. • Use dragout rinse tank with return of solution to process tank, followed by counterflow rinse. Advanced: • Install drag-in/dragout rinsing arrangement. • Install closed loop rinsing with multiple dragout tanks (2-4) connected in a counterflow arrangement. • Where dragout rate is high and surface evaporation is minimal, install auxiliary evaporation (e.g., atmospheric evaporator). • Recycle water from process tank heat exchangers.
Improve Material Utilization Strategy: Avoid solution disposal by minimizing bath contamination and using available maintenance technology. Recycle WWT sludge.	Basic: • Unrack parts over process tank to catch drips and rinse off masking in rinse tank prior to disposal. • Avoid bath contamination (clean parts thoroughly before plating; minimize corrosion of busing, racks, fixtures and parts; use DI water for rinsing, avoid drag-in of sulfate and chlorides; avoid exclusively plating Ids; and retrieve parts/fixtures which fall into the bath). • Recycle WWT sludge off-site. Advanced: • Recycle air scrubber/mesh pad wash-down back to process tank. • Filter stripping and plating solution to remove suspended contaminants. • Use bath maintenance technology (e.g., ion exchange or membrane technology) to keep Cr+3 and metallic contaminants at low levels. This also reduces energy requirements for solution heating and plating.
Reduce Human Exposure Strategy: Minimize Cr air emissions to workplace and community. Use less-toxic, substitute technologies, where applicable.	Basic: • Use mist suppressants or poly-balls to reduce the generation of mists from the chromium plating process. • Employ adequate air speeds and ventilation rates to assure remove of mists from inhalation zones. • Monitor inhalation zones to evaluate exposure and immediately correct any inadequacies. Advanced: • Employ mesh pad mist eliminators to remove and recover chromium from ventilated air. • Consider use of alternative coatings (e.g., electroless nickel, nickel alloys, spray coatings, vacuum coatings).

To deal with contaminant buildup, which occurs to a certain extent even when prevention is employed, companies can install ***bath maintenance*** technologies. Solution filtering is an effective method of removing particulate contaminants from the bath; however, it has no effect on dissolved contaminants, such as iron and other tramp metals. Methods used for dissolved metal removal include porous pots, membrane electrolysis, and ion exchange.

Various technologies have been marketed for removal of dissolved tramp metals (e.g., Fe, Cu, Ni), including ***ion exchange, ion transfer*** (i.e., non-ion-specific membranes including ceramic porous pots and units with polyfluorocarbon membranes), and ***membrane electrolysis*** (i.e., ion-specific membrane). Each of these technologies is discussed in Section 5. Of these three technologies, only ion exchange is presently a viable method for removing dissolved metals at a rate that prevents a buildup in the plating bath. Membrane electrolysis does an effective job; however only one company produced a suitable unit and although they are still servicing existing units, they are no longer selling new equipment.[52] Ion transfer has only a marginal effect on dissolved tramp metals and it cannot maintain a hard chrome plating bath below tolerable limits.

Two types of ***ion transfer*** units were tested by the Navy on hard chrome baths; an external ion transfer unit and an in-tank porous pot.[53] The external unit is the same technology tested by the EPA ETV program for purification of chromic acid anodizing solution (see anodizing section). Results of the testing showed that very little dissolved metals impurities are transferred through these membranes. The actual transport mechanism is the physical transfer of plating solution (bucketing or bailing) into the catholyte compartment. Once there, some "precipitation" of tramp metals occurs and they are removed as a sludge. Both systems are effective for conversion of trivalent chromium to hexavalent chromium. However, for this purpose, they are no more effective than simple "dummy plating," which can be accomplished using materials present in most chrome shops.

A diagram of the ion transfer system is shown in Exhibit 1-38. Testing was performed by attempting to purify 90 gal. (341 liters) of plating solution (process tank). The initial catholyte solution was made by adding chrome plating solution to tap water until a pH of 1.8 was achieved. During the test, the pH of the catholyte solution increased and it was necessary to transfer 6.6 gal/day (25 l/day) of plating solution from the "process tank" to the catholyte tank in order to maintain a pH of 1.8 to 2.0. This physical transfer of anolyte is the mechanism for moving the dissolved tramp metals into the catholyte tank; electrotransport of metals was "minimal."

Testing of the ***porous pot*** showed similar results. The Navy concluded that "The performance of ion transfer systems appears to be quite low for tramp metal removal". Further, the Navy stated that "Relatively high O&M labor requirements were experienced with both ion transfer systems."

[52] Personal communication with Dan Vaughan of IonSep Corporation, Rockland, DE.

[53] Koff, Jennie and Stencel, Nicholas, Evaluation of Performance Capabilities & Operational Requirements of Chromium Purification Technologies, Naval Facilities Engineering Service Center ; 16th AESF/EPA Pollution Prevention and Control Conf. (Feb. 1995).

Exhibit 1-38 Diagram of Ion Transfer System Used in Navy Testing

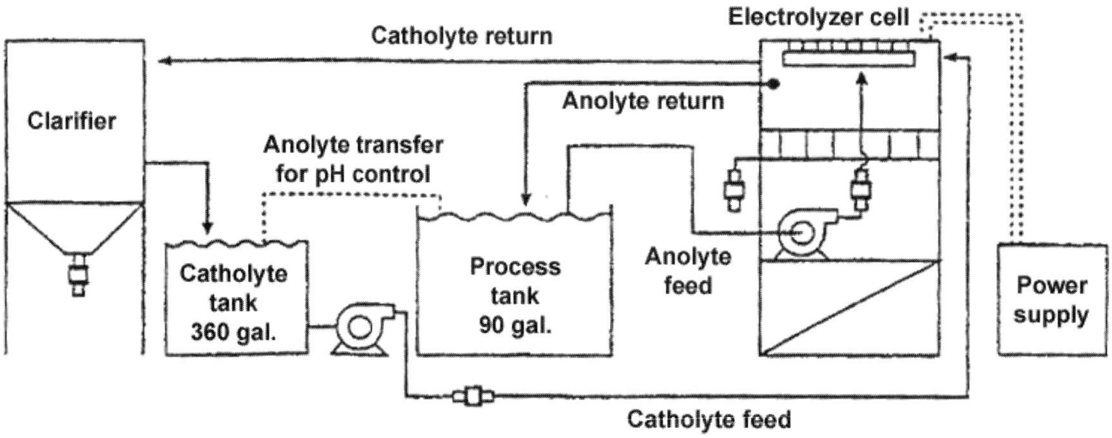

Although ion exchange is an effective technology for chromic acid anodizing bath maintenance, it has some limitations and drawbacks:

- IX generates a significant volume of chromium-bearing waste during resin regeneration, which is typically treated on-site using neutralization and precipitation, and results in production of a hazardous waste sludge (F006).
- Since resin is a plastic material, it can degrade if left in contact with a full strength chrome bath for extended periods of time.
- The concentrated nature of the chromic acid limits the exchange capacity that resin has for metals.

One commercial product for this application that overcomes most of the limitations is a ***reciprocating flow ion exchange technology***.[54] This system includes appropriately selected ion exchange resin and employs short cycle times that reduce the ion exchange resin/chromic acid contact time. In operation, a portion of the plating solution is transferred to a feed tank to allow cooling to minimize oxidative effects. The solution is then pumped through the short column of fine mesh strong acid cation resin at 60% of its normal bath strength. The resulting purified solution, with up to 90% of the metallic impurities removed in a single pass, is collected in a product tank for return back to the plating bath. Extra volume created by the dilution effect is not a problem at most shops, as normal evaporation losses from the chrome bath itself creates a hydraulic balance (although it could reduce the amount of recovery accomplished by dragout recovery rinsing). Regeneration is undertaken by using diluted sulfuric acid. Typical process specifications for this technology (supplied by the manufacturer) are shown in Exhibit 1-39. This same technology is applicable to decorative chromium plating and chromic acid anodizing.

Over the past 5 to 10 years there has been a flurry of research activity in the U.S. and abroad aimed at developing a better method of chromium bath purification. Among others, this includes:

- Laboratory investigations of porous pot and membrane electrolysis[55]
- Laboratory testing of a high pH treatment chemical process[56]
- Research on a method using a ceramic microfilter[57]
- Research on modifications to membrane electrolysis using two and three-compartment cells[58]
- Laboratory testing of a fuel cell cathode ion exchange membrane process[59]
- Laboratory testing of membrane electrolysis using a sulfuric acid catholyte[60]

Although some of this research shows promise, as yet no commercial products have resulted from this work. As such, there remains a significant opportunity for an effective chrome bath purification technology that can compete with ion exchange.

1.10.3 Air Quality

The operation of hard chrome plating is affected both by environmental and by safety and health regulations covering airborne emissions from the process:

- National Emissions Standards for Hazardous Air Pollutants (NESHAP) (40 CFR. §63.6) for chromium regulates air emissions to the environment.
- OSHA's chromium permissible exposure limit (PEL) regulates the amount of chromium in the air within the shop (29 CFR Parts 1910, 1915, et al.).

P2 practices can help meet these two regulations simultaneously. Chromium is released to the air from plating operations during the process, as gases created during electrolysis at the anode and cathode rise through the solution, break at the surface, and release an aerosol to the air above the tank. A common P2 practice involves the use of ***fume suppressants*** that lower surface tension

[54] Pajunen, Paul, and Sheedy, Michael, Metal Removal and Recovery from Process Solutions Using Short Bed Ion Exchange, SM Technical Paper 181, Eco-Tec Corporation, Nov. 2006.
[55] Donepudi, V.S., Khalili, N.R., Kizilel R. and Selman, J.R., Purification of Hard Chromium Plating Baths-Options and Challenges, Plating and Surface Finishing, April, 2002.
[56] Benaben, P. and Popakul, J., Purification & Recycling Process For Hexavalent Chromium Solutions Using a Low-cost Chemical Method, Plating and Surface Finishing; December, 2000.
[57] Stewart, David, R. Thoen, Paul, and Grove, Tom, In Tank Removal of Hard Chromium Plating Contaminates, Aerospace/Airline Plating & Metal Finishing Forum, March 2001.
[58] Kruglikov, Sergey S., Turaev, Dmitri Yu, The Use of Anionic Membranes for the Removal of Cationic Impurities from Chromium Plating Solutions, SUR/FIN 2004, June 2004.
[59] Ahmed, M. I., Khalili, N. R., Donepudi, V. S., Smotkin, E. S., Selman, J. R. and Holsen, T.M., In-Process Contaminant Removal from Hard Chromium Plating Baths Using a Fuel Cell Electrode Ion-Exchange Membrane Process, 22nd AESF/EPA Conference; January, 2001.
[60] Miller, P.O., Taylor, E.J., Zhou, C.D., and Renz, R.P., In-process Purification of Plating Baths, 21st AESF/EPA Conference; January, 2000.

Exhibit 1-39. Typical Operating Parameters for the Reciprocating Ion Exchange System

System Element	Composition, g/l				
	Chromic Acid Typical Range		H_2SO_4	Dissolved Metals	
Feed	220	270	-	5	15
Recovered Acid	130	160	-	0.6	1.8
Waste	<1	<1	20-24	2.0	2.0

and/or create a foam blanket that reduces the quantity of chromic acid mist released to the air.

There are two basic types of fume suppressants: ***wetting agents*** and ***foam blankets***. The difference between foam blankets and wetting agents is the way in which they reduce emissions:[61]

- A foam blanket fume suppressant generates a layer of foam across the surface of a solution when current is applied to that solution. The foam layer physically suppresses the mists produced on the surface of plating baths. Foam blankets do not prevent the formation of chromic acid mist – instead, they trap the mist under the blanket.
- Wetting agents lower the surface chemistry of plating baths. This causes gases to escape at the surface of the solution with a diminished "bursting" effect, reducing mist formation.

Another approach to preventing chromium losses to the air is to use ***tank covers*** that enclose the top of the anodizing tank. These are discussed in the chromic acid anodizing section.

Mesh pad mist eliminators can be used instead of standard wet scrubber systems to recover chromic acid that is exhausted from the surface of a tank. A well designed system includes an automatic spray wash down device that keeps the mesh pads clean while returning the chromium rinse back to the plating tank. The return of the wash-down solution will compete with recovery rising for available tank space generated by evaporation, and in some cases will require that supplemental evaporation be provided (e.g., installing an atmospheric evaporator, or keeping chrome tanks at operating temperature during idle hours).

1.10.4 Energy Conservation

Electricity is a significant operating cost for hard chrome plating, as shown by the Benchmarking Survey results. Fortunately, a number of P2 methods exist that can significantly reduce these energy costs:

- Keep total metallic impurities (e.g., Fe, Cu) below 3,500 mg/l. A conventional chromium plating solution containing 7,500 mg/l (1 oz/gal.) of total metallic impurities has a potential plating efficiency of 9 to 10%, while the same solution containing 3,500 mg/l (0.5 oz/gal) has an efficiency of 12 to 14% efficiency. This may translate into a 20 to 40% reduction in energy consumption and a corresponding reduction in mist generation. Maintain trivalent chromium concentration below 3.5% (3,500 mg/l).[62]
- Reduce anode to cathode distance by using ***conforming anodes*** instead of stick anodes. Stick anodes are inefficient because they cannot be placed sufficiently close to the part. Much of the electricity used in the process ends up heating the solution rather than depositing chromium. ***Conforming anodes*** can use 40% less electricity than stick anodes placed at five inches or more from the cathode.[63]
- Use eductor systems for bath agitation instead of compressed air.
- Install and maintain an energy efficient ventilation system. Check ventilation rates on a regular basis to be certain they are within the accepted standards (i.e., ANSI Z91).
- Insulate plating tank walls to prevent heat loss. Tank wall insulation will reduce solution heating costs by 80% or more.[64] Cover tanks when not in use.

---

[61] Riordan, Barrett J., Karamchandanl, Rohit T. Zitko, Larry J., and Cushnie George C., Capsule Report Hard Chrome Fume Suppressants and Control Technologies, United States Environmental Protection Agency Technology Transfer Office of Research and Development, Washington, DC 20460, EPA/625/R-981002 December 1998.

[62] Altmayer, Frank, Energy Conservation In Hard Chromium Plating, Plating and Surface Finishing, January, 2003.

[63] Advanced Tooling Corporation, Hard Chrome Plating Improvement Implementation No-Mask Anode Fixtures for Warner Robins AFB, Final Report, March 2005.

[64] Altmayer.

- Use enclosed chrome plating tanks to eliminate tank ventilation.[65]
- Purchase and maintain energy efficient air compressors. Compressed air accounts for up to 15% of the total energy used in a hard chrome shop.[66]
- Select rectifiers based on workload. Rectifiers are most efficient when they are used as close to maximum output voltage as possible. Convection-cooled and water-cooled rectifiers use less energy than forced air units for removing excess heat. Maintain rectifiers so that they are at peak performance.

1.10.5 Hard Chrome Plating Alternatives

Considerable effort has been made over the past 20 years to develop alternative processes that can match the characteristics, performance, and economics of hard chromium plating. However, no single alternative has been accepted as a replacement, and the hard chromium process is not likely to be superseded in the foreseeable future for many critical applications. Examples of alternative technologies that have been investigated and/or demonstrated include:

- thermal spray
- electroless nickel
- trivalent chromium
- electrodeposited nanocrystalline coatings
- plated nickel-cobalt alloys
- vacuum techniques

The most promising of the alternatives are thermal metal sprays, including the ***high-velocity oxygen fuel (HVOF)*** process. However, HVOF and other alternatives cannot provide a complete hard chromium replacement (e.g., HVOF is limited to line-of-sight applications), which leads to installation of substitute technologies on top of continued use of the hard chrome process. This situation makes it difficult for any substitute to be economically competitive, since the cost of O&M for multiple processes, including hard chrome, would have to be compared to the baseline of hard chrome alone. In addition, research has shown that hard chromium deposits have a number of unique properties that have not been duplicated by substitute processes; e.g., recent research shows that hard chrome contains ribbon-like inclusions (i.e., microribbons) which are unique to this process and positively influence its corrosion and wear resistance properties.[67] Further, the substitutes have their own operational disadvantages (e.g., HVOF: extreme noise levels, use of pressurized gases, difficulty of stripping deposits, hazardous waste, and toxic air emissions). Some DoD activities, including the Corpus Christi Army Depot, are hesitant to switch from hard chromium plating to HVOF or other alternatives.[68]

Some hard chrome substitute researchers have compared their technologies to outdated hard chrome methods and data in order to promote their substitutes. For example, one website[69] uses hard chrome plating rates from the 1930's for comparison, and appears to conclude that HVOF is a competitive technology. It is more realistic to use modern hard chrome plating methods, including use of ***conforming anodes*** and ***no-mask tooling*** to make such comparisons. Modern methods can compete successfully with, or even outperform HVOF in many respects, including production rates (based on parts per square foot of operating space, parts per unit cost, or other measurements of production rate). Coupled with pollution prevention methods such as covered tanks, closed-loop rinsing, and dedicated etch tanks, the hard chrome process can be operated with acceptable environmental, health and safety impacts.

1.11 NICKEL PLATING

Nickel plating pollution prevention practices include a wide range of activities and technologies relating to ***dragout*** prevention and recovery, rinse water minimization, plating ***bath maintenance***, and in-process recovery.

Exhibit 1-40 summarizes the following discussion regarding pollution prevention strategies for nickel plating.

1.11.1 Dragout and Water Use Reduction

Rinsing following nickel electroplating should include dragout recovery and counterflow rinsing. Some survey respondents have achieved zero water discharge using three to four recovery tanks connected in a counterflow arrangement. Nickel tanks that run at a low temperature may not evaporate sufficiently for the dragout to be com-

---

[65] Graves, Beverly, Encapsulating Hard-Chrome Plating Tanks, Products Finishing Magazine, Aug., 1999.

[66] Altmayer.

[67] Jones, Allen and Bishop, Craig, *Microribbons in Hard Chromium Deposits*, Journal of Applied Surface Finishing, 1(3), 249-262 (2006).

[68] National Center for Manufacturing Sciences (NCMS), Implementation of Inovative Hard Chromium Plating Tooling at Corpus Christi Army Depot Phase II, Commercial Technologies for Maintenance Activities (CTMA), June 2007.

[69] Hard Chrome Alternatives Team (www.hcat.org).

pletely fed back. In this case there are several methods (e.g., reverse osmosis, and evaporation) that can concentrate the rinse water before return. ***Recovery rinsing*** will accelerate the buildup of impurities, which must be dealt with (see bath maintenance); otherwise, bath disposal or recovery will be necessary. A drag-in/dragout configuration can also be employed; note however, an important warning when considering this approach. Dilute nickel solutions have been known to cause passivity of components, particularly when stray DC current is possible. To avoid this problem, use a non-metallic drag-in tank.[70]

When workloads are continuous, water use control can be achieved simply with ***flow restrictors*** and by closing water valves when plating lines are idle. However, for nickel lines that are operated intermittently, additional controls are needed, such as ***conductivity controls*** or ***timer rinse controls***. For final rinses following nickel plating, a commonly accepted conductivity range is 300 to 600 μmho.

One benchmarking survey company that has ***closed-loop*** rinsing employs three final rinses following nickel plating that are organized in a counterflow arrangement. This company also has a separate holding tank. Deionized water is fed to the last rinse and counterflows to the first rinse. The solution in the first rinse flows to a holding tank. The holding tank provides flow equalization for the deionized water flow rate and the evaporation rate in the plating tank. During the work week, the quantity of solution in the holding tank increases. Over the weekend, the idle nickel plating tank is kept at its normal operating temperature so that evaporation continues. During this time, a significant percentage of the solution in the holding tank is transferred to the plating tank. The return flow is pumped through a carbon filter, thus providing solution maintenance.

On July 1, 2008 EPA published the ***National Emission Standards for Hazardous Air Pollutants for Plating and Polishing Operations.***[71] This rule impacts many metal finishing operations (EPA estimates 2,900 facilities). It applies to facilities with plating, polishing or thermal spray processes that contain cadmium, nickel, lead, manganese and/or chromium. (The rule will not apply to chromium anodizing and electroplating, which are covered under a previous rule.)

With this regulation EPA did not establish emission limits for plating and polishing operations as they did with chromium plating, but instead required plating and polishing facilities to implement specific pollution prevention (P2) and control practices that are related to air emissions, dragout and rinsing:

- minimize bath agitation when removing tank objects
- maximize dripping of bath solution back into tank by extending drip time when removing the tank objects and using drain boards (i.e., drip shields)
- optimize the design of barrels, racks, and parts to minimize dragout of bath solution
- use a wetting agent or fume suppressant in the nickel electroplating bath

1.11.2 Chemical Recovery

One benchmarking survey respondent uses an ***ion exchange*** system for recovering nickel. This company's plating tank is followed by three rinses, connected in a counterflow configuration. A conductivity probe is present in the first rinse, nearest the plating tank (the most concentrated rinse). When the conductivity reaches a set point, solution in the first rinse is transferred to the ion exchange holding tank (220 gal). This flow is balanced by the addition of deionized water to the last rinse. The solution in the ion exchange holding tank is pumped through dual filters and the ion exchange column, which contains a cation bed that retains the nickel. The ion exchange unit is regenerated with sulfuric acid, and nickel sulfate is eluted. A small volume of hydrochloric acid is added to the recovered solution to destroy organics. The resulting nickel sulfate is tested using a Hull cell, and is subsequently used to replenish the plating bath. The water processed through the ion exchange unit is treated by a second ion exchange step to remove any residual nickel, and is then discharged to sewer, without the need for conventional treatment. Regenerant from the second ion exchange unit is recycled off-site.

In-process recovery (i.e., connected to the rinse system) is often used with nickel electroplating. The most common technology employed is the ***atmospheric evaporator.*** This technology is typically used in conjunction with recovery rinsing. Other applicable technologies include ***vacuum evaporators***, ***ion exchange***, ***electrowinning***,

[70] Fisher, Bryan, Avoid nickel plating losses – Protect the environment and improve profitability (extended version), Products Finishing Magazine, Sept. 2005.

[71] National Emission Standards for Hazardous Air Pollutants: Area Source Standards for Plating and Polishing Operations; Final Rule, Federal Register July 1, 2008.

Exhibit 1-40. Summary of Pollution Prevention Strategy for Nickel Plating

Objectives and Strategy	Potential Practices
Reduce Dragout & Water Use Strategy: Minimize dragout and, to the maximum extent possible, return residual dragout to the process tank, and minimize rinse water use.	Basic: • Reduce dragout (e.g., good racking, slow withdraw from process solution, thoroughly drain parts over process tank, use wetting agent, and operate bath at lowest possible nickel concentration). • Install drip guards and return solution to process tank. • Use dragout rinse tank with return of solution to process tank, followed by counterflow rinse. Advanced: • Install closed loop rinsing with multiple dragout tanks (3-4) connected in a counterflow arrangement. • Where dragout rate is high and surface evaporation is minimal, install auxiliary evaporation (e.g., atmospheric evaporator).
Improve Material Utilization Strategy: Avoid solution disposal by minimizing bath contamination and using available maintenance technology. Recycle WWT sludge.	Basic: • Avoid bath contamination (clean parts thoroughly before plating; minimize corrosion of busing, racks, fixtures and parts; use DI water for rinsing; and retrieve parts/fixtures which fall into the bath). • Use bath maintenance technology (e.g., filtration, carbon treatment, high pH) to keep metallic and organic contaminants at acceptable levels. • Recycle WWT sludge off-site. Advanced: • Install in-process recovery technology.
Reduce Human Exposure Strategy: Minimize Ni air emissions to workplace and community. Use less-toxic, substitute chemistry.	Basic: • Employ adequate air speeds and ventilation rates to assure remove of plating fumes from inhalation zones. • Monitor inhalation zones to evaluate exposure and immediately correct any inadequacies. Advanced: • Use less hazardous stripping solutions (avoid cyanide and nitric acid)

reverse osmosis, and ***electrodialysis***. Each of these technologies has both advantages and disadvantages:

- **Evaporation**. Atmospheric evaporations are relatively inexpensive and simple to operate. They are capable of recovering over 90% of nickel dragout and concentrating the nickel to 40 g/l Ni or more. Evaporators do not generate a waste stream like most other recovery technologies. One drawback is that all dissolved materials (including degraded brighteners) are recovered, which increases the need for bath maintenance. In addition, vacuum evaporators are expensive and are relatively complicated to operate and service.
- **Ion Exchange**. IX is capable of 97% recovery of nickel and it yields a relatively high concentration product stream (up to 40 g/l Ni). The IX resins must be periodically regenerated with acid (sulfuric or hydrochloric, depending on nickel bath chemistry), which creates a wastestream that must be treated. Some areas of the country have IX services that will regenerate columns off-site.
- **Electrowinning**. EW is relatively easy to operate and maintain. This technology has been applied to nickel, although it is better suited for recovery of other metals such as copper and precious metals. EW is not capable of removing nickel to low concentration levels, and the recovered nickel is difficult to harvest and reuse.

- **Reverse osmosis**. RO is capable of recovering over 90% of nickel while rejecting almost all dissolved materials, including brighteners (both usable and degraded). RO produces a recovery stream containing 10 to 20 g/l of nickel. The process could require an evaporation step if there is insufficient tank evaporation. The main weaknesses of RO are limited membrane life, the generation of a significant reject stream that contains some nickel and must be treated, and the loss of expensive brightener chemistry.
- **Electrodialysis**. ED can achieve 95% recovery of nickel. It concentrates the nickel salts sufficiently that in most cases the product stream can be returned directly to the plating tank. It has approximately the same cost as RO, and membrane replacement is a concern.

1.11.3 Additives

Anti-pitting agents are used to lessen pitting in nickel electroplating. The agent promotes the release of gas bubbles from the surface of the part. Anti-pitting agents also reduce surface tension of the solution, which, in turn, will help solution drain from the part, minimizing dragout.

1.11.4 Bath Maintenance

The key aspects of P2 for nickel electroplating operations are preventing bath contamination and maintaining the bath. Some important considerations and alternatives are discussed below.

During electroplating, the nickel anode dissolution efficiency is 100%. The cathode efficiency of nickel plating is typically 95–98%. When a moderate to high concentration of bath contaminants is present, the cathode efficiency can drop below 90%. This situation may cause the nickel content of a bath to increase over time, especially when recovery rinsing is used. To prevent the wasteful buildup of nickel metal, it is important to eliminate all sources of contaminants in the nickel plating bath. Even when contaminants are kept to a minimum, ***closed-loop nickel systems*** will generate an increase in nickel concentration. The holding tank concept used by one respondent (described under Water Use Reduction) may help in dealing with this problem. Copper anode bus bars can be a source of copper contamination, which is quite detrimental in nickel plating. A number of options are available that reduce the introduction of copper contamination in nickel plating:

- Wrap bus bars with a plating grade of nonadhesive vinyl tape, except for contact areas
- Plate copper bars with nickel
- Install PVC drain boards over the anode bars.

Parts fallen into the tank are a significant source of contamination and should be removed quickly. An inexpensive tank magnet can be used for retrieving ferrous parts. A siphon hose or makeshift shovel can be used to remove non-ferrous materials.

During electroplating, the organic additives of nickel baths produce decomposition products that, if allowed to accumulate, can result in unacceptable plating. To help reduce the occurrence of these problems, finishers treat the solution using ***activated carbon***, alone or in combination with hydrogen peroxide or potassium permanganate. If excessive carbon treatment is needed on a regular basis, the plating procedures should be reviewed to identify sources of organic contamination (e.g., drag-in or poor cleaning), and these should be eliminated.

Not all activated carbon products are suitable for use with nickel plating solutions. Suppliers of proprietary additives can recommend a suitable grade. The quantity of carbon required will depend on the degree of contamination.[72]

Although effective, carbon treatment increases downtime, creates liquid and solid waste, and increases employee contact with toxic materials. In addition, the treatment process can result in solution losses up to 20%.[73] There are alternatives to using activated carbon for organics removal. At least one supplier offers a polymeric resin system that provides continuous nickel solution maintenance.[74] The resin is regenerated with mild alkaline solution and is reused. Another alternative technology recently investigated is use of catalyzed hydrogen peroxide.[75] During long-term testing of this process, it significantly outperformed conventional hydrogen

[72] Bless, Diana R., A Review of Nickel Bath Life Extension, Nickel Recovery and Copper Recovery from Nickel Baths, Plating and Surface Finishing, April 2000.

[73] Fisher, Bryan, Avoid nickel plating losses, *Guidelines to help prevent the loss of nickel and its salts from the finishing shop, Products Finishing Magazine, Sept. 2005.*

[74] Atotech, Nikotect Adsorber System: Advanced Maintenance and Recycling Technology for Bright Nickel Plating Processes, www.atotech.com.

[75] Forth, Peter, Vibert, Art, Busch, Madeline and Stevenson, Sarah, Treatment and Control of Nickel Electroplating Baths Using Catalyzed Hydrogen Peroxide, 23nd AESF/EPA Conference, Jan. 2002.

peroxide/carbon treatment, especially with regard to detrimental organic breakdown products.[76]

Electrolytic treatment (dummying) can be effectively used to remove copper, zinc, iron, and excesses of certain organic brightening agents. A corrugated cathode is used in this treatment, and the bath is electrolyzed at 2 to 5 amps per square foot (asf). One survey respondent indicated that their facility has implemented continuous electrolytic treatment by adding a holding tank with a separate rectifier. Solution from the bath is transferred continuously to the holding tank and returned to the bath. The circulation loop includes ***carbon filtration***.

Metal precipitation or "high pH" treatment can be used for removal of trivalent chromium, aluminum, copper, iron, and silicates. Removal is accomplished by transferring the solution to a batch treatment tank, raising the temperature (145–150°F), and adding a slurry of plating-grade nickel carbonate (raises pH to 5.2). Hydrogen peroxide can be added to help oxidize the iron. The solution is filtered and returned to the plating tank. The solution may then be electrolyzed at low current density until deposit quality is acceptable.[77] One survey respondent pointed out that high pH treatment results in solution loss, sludge creation, and holding tank cleaning. Instead of performing this type of bath maintenance, this company periodically decants a portion of the bath and reuses it on-site or recycles it off-site.

1.11.5 Additional P2 Methods

Benchmarking survey respondents listed additional P2 methods for nickel electroplating. These include:

- Converting from a nitric acid strip to a less hazardous and regenerable sulfuric acid/ peroxide process
- Replacing cyanide stripper with non-cyanide stripper
- Employing equipment rotators to improve drainage of dragout
- Using a closed-loop chiller system to control bath temperature

1.11.6 Air Emissions

A cold, still, nickel plating bath most likely does not generate any detectable nickel-bearing mist. Also, given that the deposition efficiency of a nickel electrolyte is in the order of 96-98%, (and therefore there is very little gassing from electrolysis), the application of direct current for deposition is unlikely, by itself, to increase the presence of nickel in the immediate atmosphere by any significant quantity (Fisher). However, heating the solution and applying air agitation can cause significant nickel losses to the atmosphere. To reduce losses, consider operating the bath at the lowest practical temperature, using mechanical agitation (i.e., solution pumping or cathode bar movement). In addition, although not specifically used for nickel plating as yet, ***tank covers*** could be used to assure compliance with any new air pollution control rule (see the anodizing section for discussion of EED covers).

1.11.7 Nickel Strip Acid Recovery

The most common nickel stripping solution is nitric acid. During use, acid is consumed and the rate of stripping declines. Eventually the solution must be removed from service and disposed of (off-site disposal can run as high as $5.00 per gallon). At this point, some active nitric acid is still present, although it is diluted by the unusable portion. One method of maximizing the life of the nitric acid strip is to employ ***diffusion dialysis***, a membrane separation process commonly used for maintaining sulfuric acid anodizing baths.

When recovering acids using diffusion dialysis, an anion exchange membrane acts as a semi-permeable barrier placed between a flowing water stream and a flowing acid solution containing dissolved metals. The anion exchange membrane has fixed positive charges located on its surface. These positive charge centers attract the negatively charged anions in solution that come into close contact with the anion exchange membrane surface.

With nitric acid baths, the overwhelmingly predominant anion is the nitrate ion (NO_3^{-1}). As the nitrate ions in the acid solution are attracted to the membrane, they are also driven by the concentration difference to diffuse across the membrane to the water side. Simultaneously, the Law of Electroneutrality (in solution total charge must balance to zero) requires that the transfer of every nitrate ion (negative) into solution is accompanied by the transfer of one unit of positive charge. Positively charged ions such as Ni^{+2} or other metal ions, are strongly inhibited from crossing the positively charged membrane because of the

[76] Forth, Peter, No More Activated Carbon Treatments Required, Products Finishing Magazine, May, 2005.
[77] Dibari, George A., Nickel Plating, Elsevier Metal Finishing Guidebook and Directory, 2008.

repulsion between like charges. The hydrogen ion, present in the acid solution as H_3O^+ ions, is also positively charged, but is able to cross the membrane with very little hindrance. This is due to the high concentration of hydrogen ions in the acid solution, the small size of the hydrogen ion, and the ability of the hydrogen ion to transfer rapidly between one water molecule and another. The net effect is that both components of the nitric acid molecule, the hydrogen ion and the nitrate ion, diffuse across the membrane faster than the dissolved nickel. Typically, 80-90% of the acid is recovered, with 70-90% of the dissolved metals removed. [78]

1.12 TIN AND TIN–LEAD PLATING

There are several pollution prevention opportunities for tin and tin–lead plating operations. The most common focus on alternative processes for tin–lead, reducing water use, plating solution control and maintenance, and recovery. Exhibit 1-41 summarizes the following discussion regarding pollution prevention strategies for tin and tin–lead plating.

1.12.1 Alternative Processes

Tin–lead is commonly used by the printed wiring board (PWB) industry and for other electronics applications. In PWB manufacturing, tin is essentially just as effective as tin–lead as an etch resist and can replace tin–lead plating when solder-mask-over-bare-copper (SMOBC) is employed. Tin-only baths can be used as a substitute in many other cases, although not for all current tin–lead applications. Emerging substitute technologies include tin–bismuth alloys, conductive adhesives, and nickel–gold or nickel–palladium electroplating.

1.12.2 Dragout and Water Use Reduction

Metal finishers can achieve water use reduction in tin and tin–lead plating by minimizing ***dragout***, using multiple rinse tanks, and controlling water use. To reduce dragout, efforts should focus on rack design, part orientation, and dwell times. With continuous processes, such as reel-to-reel plating for electronics, an air knife can be used to further reduce dragout losses.

To reduce water use while still meeting rinsing needs following tin or tin–lead plating, a two-stage ***counterflow rinsing*** configuration will typically suffice. One survey respondent indicated that they use three-stage rinsing after tin plating. A ***dragout tank*** for recovery can be used with baths operated above 100°F (e.g., potassium stannate or sodium stannate).

On continuous production lines, where rinse water needs are relatively stable, flow can be controlled easily by periodically monitoring rinse water conductivity and manually adjusting flow, using a valve. If workflow is sporadic, a ***conductivity control*** or ***timer rinse control*** should be considered. These devices help to match rinse water use and production flow. For final rinses following tin and tin-lead plating, commonly accepted conductivity ranges are 70 to 500 μmho (final tin alkaline rinses should be operated below 340 μmho).

On July 1, 2008 EPA published the ***National Emission Standards for Hazardous Air Pollutants for Plating and Polishing Operations.***[79] This rule impacts many metal finishing operations (EPA estimates 2,900 facilities). It applies to facilities with plating, polishing or thermal spray processes that contain cadmium, nickel, lead, manganese and/or chromium. (The rule will not apply to chromium anodizing and electroplating, which are covered under a previous rule.)

With this regulation EPA did not establish emission limits for plating and polishing operations as they did with chromium plating, but instead required plating and polishing facilities to implement specific pollution prevention (P2) and control practices that are related to air emissions, dragout and rinsing:

- minimize bath agitation when removing tank objects
- maximize dripping of bath solution back into tank by extending drip time when removing the tank objects and using drain boards (i.e., drip shields)
- optimize the design of barrels, racks, and parts to minimize dragout of bath solution
- use a wetting agent or fume suppressant in the tin/nickel electroplating bath

1.12.3 Bath Maintenance

Bath maintenance can reduce the need to discard process solutions. The primary source of tin and tin–lead bath contamination is metal impurities from anodes. These

[78] Capalbo, Ralph, New Method for Recycling Acids at New Method Plating, Products Finishing Magazine, May 2004.

[79] National Emission Standards for Hazardous Air Pollutants: Area Source Standards for Plating and Polishing Operations; Final Rule, Federal Register July 1, 2008.

Exhibit 1-41. Summary of Pollution Prevention Strategy for Tin and Tin–Lead Plating

Objectives and Strategy	Potential Practices
Reduce Dragout & Water Use Strategy: Minimize dragout, employ counterflow rinsing, and exercise good rinse water use control.	Basic: • Reduce dragout • Operate the bath at the lowest practical metal concentration. • Install drip guards and return solution to process tank. • Install counterflow rinsing. Advanced: • Use conductivity controlled rinsing. • Install dragout rinsing arrangement with baths above 100°F.
Improve Material Utilization Strategy: Avoid solution disposal by minimizing bath contamination and performing bath maintenance. Recycle WWT sludge.	Basic: • Avoid bath contamination (clean parts thoroughly before plating; minimize introduction of metallic impurities by using high purity anodes; use DI water for bath additions, and retrieve parts/fixtures which fall into the bath). • Use bath maintenance methods to remove impurities. • Recycle WWT sludge off-site. Advanced: • None identified.
Reduce Human Exposure Strategy: Eliminate lead from electroplating processes	Basic: • Convert tin-lead processes to tin electroplating. Advanced: • Substitute tin-bismuth alloys, conductive adhesives, and nickel-gold or nickel-palladium for tin-lead electroplating processes.

contaminants can be partially avoided by purchasing high-quality, low-impurity anodes.

Dissolved metal impurities can be removed using low current density ***electrolysis*** (dummying). Stannic ion is an unavoidable contaminant that must be maintained at relatively low levels by performing continuous ***filtration*** and chemical treatment. Fluoborate baths are particularly subject to buildups of sludge. Therefore, continuous in-tank or external filtration during operation is recommended with these solutions. Organic bath additives are decomposed during plating and will concentrate in the bath. A carbon treatment can be performed on tin–acid and tin–lead baths to eliminate all organics. Fresh organic additive is then reintroduced.

Stannous tin (stannite) is a contaminant in a stannate solution. It can break down to produce sludge by forming colloidal tin. The presence of stannite may also produce rough, spongy deposits. To convert stannite back to stannate, dilute a small amount of hydrogen peroxide with water to make a 10% solution. Add this to the plating tank with a lot of agitation. After addition of the peroxide, the plating solution will need to be dummied until the excess peroxide is converted to water.[80]

Process control for methanesulfonic acid systems using insoluble anodes, where tin is not automatically replaced, often involves bath bailouts to maintain a sufficiently high tin concentration (due to the fact that all of the tin that is plated comes from the tin methanesulfonic acid). The bailouts result in high production of F006 wastewater treatment sludge. A novel approach to the problem using a ***membrane technology*** was developed at a reel-to-reel shop that uses a tin replacement cell hydraulically connected to the plating tank. The system is shown in Figure 12. In the plating tank, oxygen is liberated at the insoluble anode and tin is plated on the cathode, resulting in a lower tin concentration in solution and an increase in the free acid concentration. The solution is continuously recirculated from the plating cell to the anolyte chamber of the tin replenishment cell via the lower plating sump (not shown). In the tin replenishment cell, tin is electrochemically dissolved into the anolyte at the soluble tin anode, and it combines with the free acid produced in the plating cell, liberating hydrogen ions. The hydrogen ions pass through the membrane into the catholyte chamber, which

[80] Altmayer, Frank, Advice and Council: Tin Basics, Plating and Surface Finishing, May 2005.

is filled with methanesulfonic acid at approximately the same concentration as the plating bath. The ions replace those lost to the evolution of hydrogen gas at the tin replacement cell cathode. Some tin ions pass through the membrane (approximately 5%), and are deposited at the tin replacement cell. Bath bailouts were eliminated and the process control of the solution was simplified. Only small adds of tin methanesulfonic acid were required to make up for drag out losses. Waste treatment and sludge disposal costs were reduced, due to elimination of bailouts. The net annual savings after energy and membrane replacement costs was in excess of $40,000. The capital cost of the equipment was less than $30,000.[81]

1.12.4 Chemical Recovery

Various chemical recovery technologies have been used with tin and tin–lead electroplating. The most frequently used technologies are ***ion exchange***, ***electrowinning*** and evaporative recovery.

1.12.5 Disposal of Spent Baths and Sludge

Tin and tin–lead sludges do not have significant economic value and are by themselves usually not recycled off-site. However, one survey respondent indicated that their facility treats tin-bearing rinse waters and spent baths, and sends the resultant hydroxide sludge to a reclamation site.

1.12.6 Tin Strip Recovery

Off-site disposal or on-site treatment of spent nitric acid tin stripping solutions can be expensive. Diffusion dialysis has been used for these baths, but with limited success.[82] A novel approach to this problem was developed where a caustic stripper was substituted for the conventional nitric bath and electrowinning was used to capture the metal.[83] Using this process, tin concentrations were reduced from about 150 g/l to 30 g/l.

1.13 ZINC PLATING

Operating environmentally efficient zinc electroplating processes begins with the selection of the plating bath. Twenty years ago, most zinc plating was carried out with cyanide baths. Today, low-cyanide and non-cyanide plating solutions are available that meet most metal finishers' needs.

Many of the survey respondents identified conversion from cyanide to non-cyanide as a major P2 improvement. Several of the respondents indicated that complete conversion to non-cyanide was not possible due to special work requirements. In such cases, companies have converted one or more tanks to non-cyanide and maintain a single zinc cyanide tank, or have switched to a low-cyanide chemistry.

Zinc cyanide plating baths are significantly less affected by poor cleaning than alkaline non-cyanide baths. One respondent noted that it is possible to operate a zinc cyanide bath with floating oil. In contrast, alkaline non-cyanide zinc plating requires extensive cleaning, similar to that needed for nickel plating directly onto steel. Exhibit 1-42 summarizes the following discussion regarding pollution prevention strategies for zinc plating.

1.13.1 Bath Maintenance

Even with non-cyanide baths, solution chemistry control must be exercised to minimize pollution. Several survey respondents indicated that zinc plating bath control was an integral part of their P2 approach. Several survey respondents indicated that the use of automatic chemical bath additions improves control. One survey respondent controls the use of brightener by storing the chemical in totes, which platers cannot access without permission. This restriction has reduced brightener usage by 30%. Important aspects of solution control include controlling metal buildup, making proper additions of bath components, and contamination control. Metal buildup can occur when zinc anodes dissolve during idle times. This situation can be avoided by removing the anodes from the solution when it is not in use. This practice is more effective for alkaline baths than for zinc chloride baths.

Zinc metal concentration can also be controlled in cyanide and alkaline non-cyanide plating solutions by using inert anodes (i.e., steel or carbon). Using deionized water for evaporative makeup, retrieving fallen parts or racks, ***filtration***, and removing metallic impurities can improve contamination control. One respondent uses permanent media filters rather than disposable filters to reduce the quantity of waste being discarded. ***Electrolysis*** is used to remove lead and cadmium. Various chemical treatments followed by filtration are also used to remove metallic impurities, including:

---

[81] William Copping; SUR/FIN 2003; 2003 Jun Novel Technique for Replenishing Tin & Controlling Tin Baths

[82] Ford, Christopher and Coletta, Ted, Reclamation of Nitric Acid from Solder Strip, The Massachusetts Toxic Use Reduction Institute, 1995.

[83] McKesson, Doug, Miracles in Tin Stripping, PC Fab Magazine, June 1999.

- Sulfide reagents (lead, cadmium removal)
- Sodium hydrosulfite (chromate removal)
- Zinc dust (cadmium and lead removal)
- Hydrogen peroxide (iron removal).

Activated carbon treatment is employed to remove organic impurities (cleaner components, plasticizers, and decomposed additives). One survey respondent completely removes plating solution from their zinc plating tank when it is not in production mode. This company indicated that this practice prevents buildup of organics on the anode baskets, extends the service life of anode baskets, prevents zinc metal buildup in the plating bath, and provides a safeguard against tank failure.

1.13.2 Dragout and Water Use Reduction

Rinsing following zinc plating presents challenges to the metal finisher. Due to the relatively low temperature of the zinc bath, there is limited opportunity for use of ***drag-out recovery rinsing***. However, some proprietary zinc chloride baths can be operated as high as 130°F, which is sufficiently high to make ***recovery rinsing*** feasible and effective. One respondent uses an ***atmospheric evaporator*** to increase evaporation and recover plating solution. Although several survey respondents indicated they employ recovery rinsing, most rely more on several other aspects of rinsing: ***dragout*** minimization, ***counter-flow rinsing***, and water use control. Survey respondents have implemented dragout reduction by:

- Increasing dwell time
- Rotating racked parts for better drainage
- Redesigning barrels to reduce the volume of trapped solution
- Rotating barrels above the plating tank after they exit the solution (see ***draining/rinsing over the plating tank***)
- Training operators

Operating the zinc plating bath at the lowest practical metal level or using excess ***wetting agent*** will also decrease dragout losses. Numerous respondents cited water use control as a key aspect of pollution prevention on the zinc plating line. The equipment employed includes ***flow restrictors***, ***conductivity controls***, shut-off valves, and ***timer rinse controls***. For final rinses following zinc plating, commonly accepted conductivity ranges are less than 630 μmho for zinc acid and 300 to 1,200 μmho for cyanide zinc. Although the traditional dragout recovery rinse is not used by many of the benchmarking survey respondents, at least two companies employ drag-in/ dragout recovery rinsing tanks to recover a portion of the dragout.

1.13.3 Chemical Recovery

Various chemical recovery technologies have been used with zinc plating, although there are fewer total applications than with most other plating metals. The most frequently used technologies are evaporative recovery and ***electrowinning***. ***Ion exchange*** has also been used, but with less success than the other two recovery technologies.

Zinc is the least recovered of the common electroplated metals due to:

- technology issues -- recovery technologies used by metal finishers are less efficient and effective with zinc than other metal applications
- lack of incentive – the relatively low value of zinc and the relatively high zinc effluent discharge standards (especially in comparison with local standards for metals such as chromium, cadmium, and lead).

Off-site recovery, although not within the true definition of pollution prevention, is highly applicable to zinc wastewater treatment sludges, particularly when they are not combined with wastes of other metals. There are two reasons that make off-site recycle an attractive option:

- When wastewater treatment sludge is generated from segregated zinc cleaning, plating (on carbon steel) and/or zinc stripping processes, that sludge is exempt from the scope of RCRA hazardous waste F006 (40 CFR 261.31). To obtain this status, any chromating wastes, if present, would have to be handled on a segregated basis. Therefore, unless the sludge is a characteristic hazardous waste (e.g. corrosive), then it is non-hazardous, and the rules that apply to management, transportation and disposal are relaxed.
- There are a number of licensed facilities around the U.S and Mexico that recycle zinc wastes. Zinc sludges can be easily recycled into usable industrial products, including construction material.

Recycling of zinc wastes is carried out by large zinc waste processing facilities in the U.S. and Mexico that employ

Exhibit 1-42. Summary of Pollution Prevention Strategy for Zinc Plating

Objectives and Strategy	Potential Practices
Reduce Dragout & Water Use Strategy: Minimize dragout, employ counterflow rinsing, and exercise good rinse water use control.	Basic: • Reduce dragout (e.g., thoroughly drain parts over process tank, rotate barrels above tank using, use wetting agent). • Operate the bath at the lowest practical zinc concentration. • Install drip guards and return solution to process tank. • Install counterflow rinsing. Advanced: • Use conductivity controlled rinsing. • Install drag-in/dragout rinsing arrangement.
Improve Material Utilization Strategy: Avoid solution disposal by minimizing buildup of zinc and metallic bath contamination. Recycle WWT sludge.	Basic: • Avoid bath contamination (clean parts thoroughly before plating; minimize introduction of metallic impurities; use DI water for bath additions, and retrieve parts/fixtures which fall into the bath). • Recycle WWT sludge off-site. • Remove anodes from tank when not in use. Advanced: • Use inert anodes. • Use equipment for automatic additions of certain bath components. • Use bath maintenance methods to remove impurities.
Reduce Human Exposure Strategy: Use non-cyanide zinc plating processes.	Basic: • Convert zinc cyanide processes to non-cyanide processes, wherever possible. Advanced: • Mechanical plating can be used as an alternative coatings for some applications.

rotary kilns or other technology. The most common waste processed by these facilities is electric arc furnace dust generated by steel making/recycling processes. Zinc electroplating sludges are an ideal secondary feedstock for the rotary kilns. The kilns utilize a high temperature metal recovery (HTMR) process for zinc recycle. At temperatures over 2000 °F, valuable metals are volatilized and captured for further processing into refined metal products, which are sold for industrial uses. The slag from the kiln produces an iron-rich material (IRM) which can be used as construction material or as filter media in wastewater treatment applications.

1.14 OTHER METAL FINISHING PROCESSES

This section contains benchmarking survey respondents comments related to processes other than those specifically covered in previous sections. Readers may find some of these comments useful when evaluating their own practices and identifying new opportunities for pollution prevention. The comments are essentially unedited except for spelling and other minor corrections to improve readability. Square brackets denote paraphrasing of respondent comments, or editorial comments.

Readers should fully investigate and evaluate any options before implementing them.

1.14.1 Phosphating

The following comments are pollution prevention methods benchmarking survey respondents used to address phosphating.

- Automatic additions to phosphate baths— resulting in less human exposure.
- Changed to a chemistry that generates only one quarter of the sludge previously produced. Iron control system used on zinc phosphate bath.

- Installed motor-controlled hoists to reduce dragout. Use counter flow rinsing. Conducted operator training. Installed bulk storage system and eliminated use of waste drums.
- Filter phosphate solution so we don't have to decant periodically. [One respondent indicated that this is not a very effective practice.]
- Converted existing rinse tanks to counterflow.
- Use water restrictors. Increased dwell time. Installed drip pans. Use vacuum evaporator for recovery. Installed oil skimmer and microfilter on cleaner. Reduced concentration of chromium in sealers (magnesium phosphate) and use non-chromium sealers (zinc phosphate.
- We have tried non-chromium sealers and will move into one shortly.
- Increased analysis of phosphate solution.
- Added acid filter. Increased analysis and bath maintenance.
- Use counterflow rinses. Use acid wetters.
- Use manual titration for prep and phosphating baths. Substituted water-soluble oil whenever customer requirements permit. Implemented operator training to minimize dragout and water use and to extend bath life.
- Use counterflow rinsing.
- Installed counterflow rinses to reduce rinse water use, installed ***flow meters*** at all rinse tanks to reduce water use, installed oil separator on the cleaner system. Installed heat exchangers to heat zinc phosphating bath in place of live steam, saving zinc phosphate bath from displacement overflow.

1.14.2 Passivation

The following comments are pollution prevention methods survey respondents used to address passivation.

- We are pilot testing of an acid sorption unit.
- Extended dwell time over process tank.
- Use deionized water for bath makeup. Deionized water used throughout passivation line.
- Maintain racking orientation to achieve best drainage; allow sufficient drainage over process tanks.
- Replaced nitric acid process with a proprietary citric-acid-based process. The new process is less toxic and dangerous; however it is more expensive to operate. [Another survey respondent is testing a citric acid process.]
- Installed oil skip process to reduce the oil flowing into the system, thereby reducing chemical usage.
- Changed acids and reduced the quantity of acid on site.

1.14.3 Black Oxide

The following comments are pollution prevention methods survey respondents used to address black oxide.

- Use dead rinses. Reduced acid concentration in pickle. Use flow restrictors.
- Use deionized water for rinsing and replenishment.
- Use counterflow rinsing.
- More efficient burners purchased (natural gas reduction).
- Flow restrictors on water rinses. Plumbed existing rinse tanks to counterflow.
- Use oil recovery drip trays and recover oil from spin dryers.
- Water restrictors, dwell time, drip pans, caustic cleaner skimming and microfiltration, elimination of solvent-based cleaners, lower-VOC water-based oils, cold evaporation of water from waste production oils (a 65% reduction).
- Automatic additions to the bath resulting in less human exposure. Better method for disposing of the sludge generated in the bath through waste treatment.
- New line installed—less hydroxide loss and better process control.
- Basic P2 methods. Equipment rotators for better drainage, cleaner recycling.

1.14.4 Electropolish

The following comments are pollution prevention methods survey respondents used to address electropolishing.

- Save rinse; closed-looped rinse system with evaporator used to maintain temperature of process tanks.
- Found end users for used electropolishing solution. Closed manufacturing loop to purify acidic rinse water. Sludge from clarifier is Class V nonhazardous.
- Rinse water is reused for different rinses up the line.
- Cover tanks when not in use to prevent adsorption of moisture from the air.

Section 2
Overview of Project Results

2.1 DESCRIPTION OF PROJECT

2.1.1 General

The National Center for Manufacturing Sciences (NCMS) sponsored a project to assess pollution prevention and control technology available to the plating industry. Project activities included extensive surveys of the plating industry and vendors of technologies and an in-depth literature review. The plating industry survey was performed in cooperation with the National Association of Metal Finishers (NAMF). This book is the primary product of the project.

The project results are useful for all those associated with pollution prevention and control in the plating industry. The results show which treatment, recovery and bath maintenance technologies have been most successful for different plating processes and the costs for purchasing and operating these technologies. The project results also cover trends in chemical substitution, the identification of compliance-problem pollutants, sludge generation rates, off-site sludge recovery and disposal options, and many other pertinent topics. Section 2 provides an overview of the approach taken by the project team, as well as an outline of key results.

2.1.2 Relationship of Report and Database

This text and the associated database (no longer available) are a unique and dynamic method of transferring information among users of technology. The book organizes information and data from various sources, while the database gives readers a mechanism to investigate the basis of any conclusions and, if they chose, to conduct their own analyses. For example, using the database, the reader has the opportunity to further explore specific facts found in the book concerning the technologies and the plating shops that use them or to perform statistical analyses on categories of data that may answer the reader's own unique technical, regulatory or economic questions. The book and database are dynamic in the sense that an updated version of each will be prepared once these results are disseminated and feedback is received. This will include additional data that will be accepted from new respondents to the Users Survey and Vendors Survey (see definitions of the surveys in Section 2.1.3).

2.1.3 Project Scope

A diagram showing the various stages of the project is provided in Exhibit 2-1. The assessment was conducted using three major sources of information and data:

(1) A mailed questionnaire-survey sent to approximately 2,000 electroplating and metal finishing plants (Users Survey).

(2) A mailed questionnaire-survey of approximately 60 vendors of pollution prevention and control equipment and off-site metals recycling services (Vendors Survey).

(3) A literature search that gathered approximately 520 articles, reports, conference papers and other sources of relevant information.

Exhibit 2-1. Overview of Project Activities

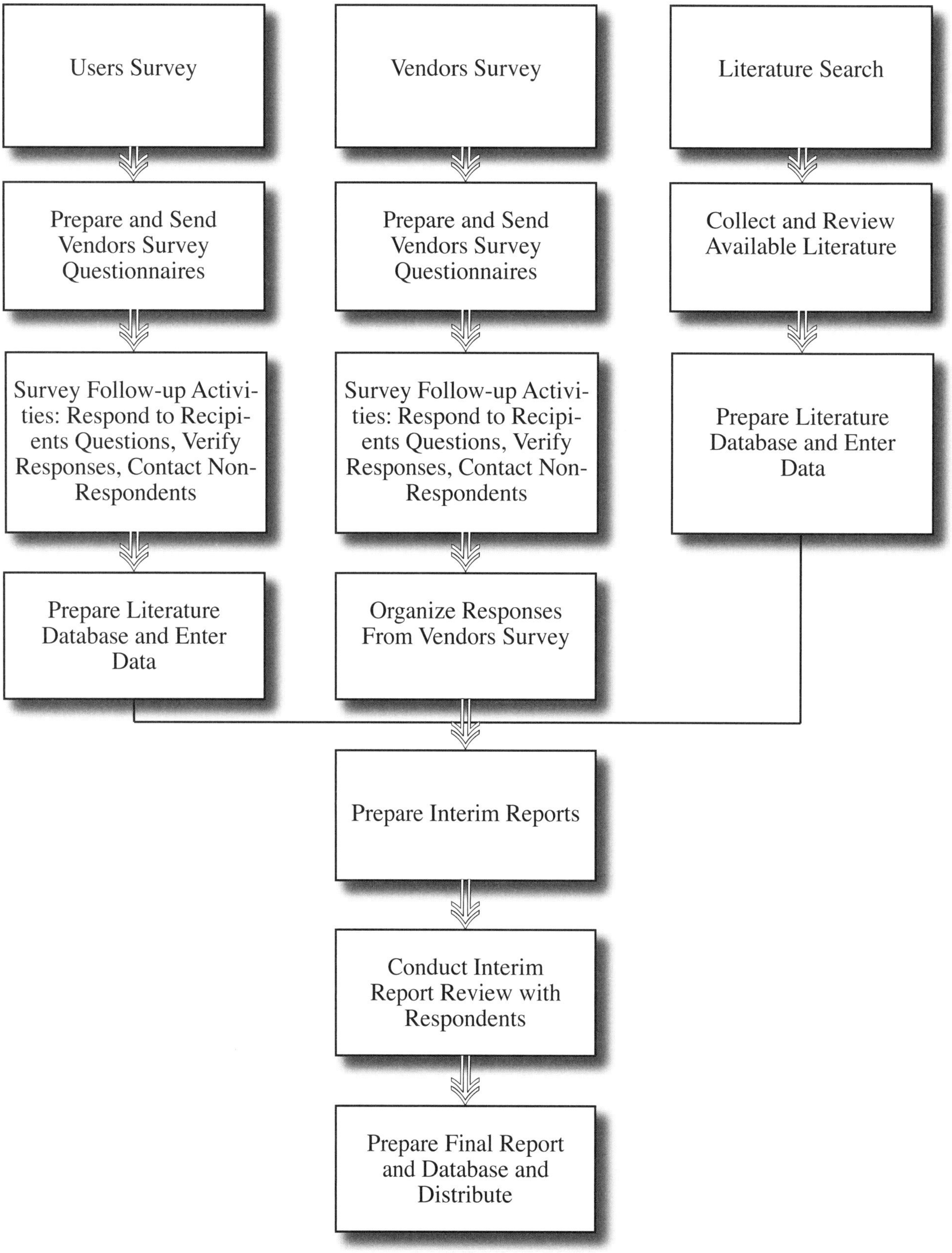

Using the collected materials, information and data, a set of seven Interim Reports was written covering the following topics:

- Interim Report No. 1: Overview of Project Results
- Interim Report No. 2: General Waste Reduction Practices
- Interim Report No. 3: Chemical Recovery
- Interim Report No. 4: Chemical Solution Maintenance
- Interim Report No. 5: Process Substitution
- Interim Report No. 6: Wastewater Treatment
- Interim Report No. 7: Off-Site Metals Recycling

The Interim Reports were distributed to the respondents of the Users and Vendors Surveys and other interested parties. Based on comments received during the review process, the Interim Reports were revised and subsequently consolidated into this text. The text and associated database will be updated once these results are disseminated and feedback is received. This will include additional data that will be accepted from new respondents to the Users Survey and Vendors Survey.

The most important element of the project was the Users Survey. During this project phase, a questionnaire was developed and sent to approximately 2,000 electroplating and metal finishing plants. The questionnaire development process included the preparation of several drafts of the survey form that were tested on representative plating shops. Also, during the development of the questionnaire, input was received from the National Association of Metal Finishers (NAMF), American Electroplaters and Surface Finishers (AESF), other industry personnel and the U.S. Environmental Protection Agency (EPA).

The Users Survey questionnaire covered seven major areas:

- **Facility Identification and Characterization Data** (company name, point of contact, number of employees, age of plant, general types of plating equipment used, primary business areas, and production rates).

- **Wastewater Data** (type of discharge: direct, indirect, zero; plating and other industrial wastewater flow rates; effluent limitations; and compliance difficulties).

- **Process Identification** (metal cleaning including solvent use, plating, stripping, aluminum finishing, other surface finishing processes, number of process tanks of each type and total volumes of specific process solutions).

- **Pollution Prevention and Recycle Technology Identification** (frequency of use and level of success for: good operating practices, rinse water and drag-out reduction, chemical recovery technologies, bath maintenance technologies, product changes, bath chemistry changes, and off-site recycling).

- **Technology Data Sheets** (details for advanced chemical recovery/recycling and bath maintenance technologies including: manufacturer, date of purchase, capital and operating costs/savings, skill requirements and man-hour requirements for operation and maintenance, satisfaction ratings for vendors and the technology itself, performance and operational and maintenance details, residuals generation, and future plans of the technology user).

- **End-of-Pipe Treatment** (details of processes employed, ratings of manufacturers, chemical reagent requirements and costs, labor requirements, residuals generation, and associated compliance problems).

- **Identification of Facility Problems and Needs** (environmental problems, technical information needs, technical resources that are utilized, ideas on dissemination and use of the project results).

The mailing list for the Users Survey included all members of the National Association of Metal Finishers, participating member companies of the National Center for Manufacturing Sciences that operate plating shops, military plating shops and captive plating shops that were identified by Elsevier Science Publishing Co. (*Metal Finishing* magazine).

To store, organize and utilize the data from the Users Survey, a database was prepared.[1] Because the database would be made publicly available, respondents were assigned a code number which was entered into the database in place of a company's or individual's name. Database control software was developed using a commercial program (Clarion™). The database was then utilized to provide statistical results from the Users Survey.

The commercial software that was selected is a modern, professional, relational database language that allows the production of a free-standing program to control and use the database without any royalty, user fee, or distribution requirement or cost. This is unlike many other languages that are subject to distribution fees or requirements to buy "runtime" software to operate the programs. Furthermore, the resulting program is user-friendly and menu-driven and can be used without any knowledge of database language or specific expertise in software. The hardware requirements include an IBM AT, PS/2 or 100% compatible computer, IBM DOS 3.1 or higher operating system, 640 KB ram, EGA or VAG color monitor, one 5-1/4" or 3-1/2" diskette drive, a hard drive and a parallel laser printer with full IBM-US character set.

Although for most users the original version of the database is expected to be sufficient, other users may want a different format that is more familiar to them. As such, converted formats (e.g., dbase™) are available.

The database (no longer available), which like this report is publicly available, contains most of the data entered into the Users Survey questionnaires (most comments or experiences that did not fit into the database format have been included in this text). Although much of the Users Survey data has already been evaluated and summarized in this text, individuals or organizations may want to use the database to find out specific details concerning a given technology application or to prepare their own statistical analyses or summaries. Therefore, the availability of the database enhances the usefulness of the text. As an example, Section 7 covers technologies used for end-of-pipe treatment. A reader of the text may locate from data tables a shop using a specific combination of technologies that the reader is investigating. Using the database, the reader can quickly determine what plating processes that shop has in place and the volumes of solutions that are present, the effluent limitations that the shop is required to meet, any compliance problems they have reported, what methods of recovery/recycle or bath maintenance are also used that may impact the end-of-pipe process selection or other specific facts that may have influenced the choice of treatment technologies.

2.1.4 Project Organization and Participants

This project was sponsored by the National Center for Manufacturing Sciences. Partial funding for the project was provided by the Joyce Foundation. The project was managed by Paul D. Chalmer, Ph.D. of NCMS. Highly valued technical support and guidance during the project were provided by the following individuals (company and/or affiliation shown in parenthesis):

- Frank Altmayer (Scientific Control Laboratory and the American Electroplaters and Surface Finishers Society)
- Brenda Harrison (Texas Instruments)
- Dave Marsh (Marsh Plating and the National Association of Metal Finishers)
- Brian Manty (United Technologies, Pratt & Whitney)
- Brian Ramey (the former General Motors, Allison Gas Turbine Division)
- Bill Sonntag (National Association of Metal Finishers)

The key project participants in this study were the plating shop personnel that completed survey forms and provided other technical information. The members of the National Association of Metal Finishers provided approximately 1,800 man-hours of time collecting data and completing the forms. The time spent by other individuals from plating shops totaled approximately 200 man-hours.

Another key source of information and data for this project was provided by the manufacturers and vendors of pollution prevention and control equipment and operators of off-site metals recovery operations. These companies are identified in Section 2.1.6. Individuals from these companies that contributed most significantly are: Fred Reinhard (Kinetic Recovery Corporation); Serge Astaud (i^3); Lyle Kirman (Kinetico Engineered Systems, Inc.); Larry Zitko, (Chrometech, Inc.); Stratton Tragellis (Memtek Corporation); and Jon Youngson (Midwest Air Products, Co.).

The contractor for this project was CAI Resources. George Cushnie was the CAI project manager and the author of the surveys and this text. The Users Survey database development was performed by a consultant, George Ryder. Mark Eelman of CAI Resources provided extensive technical and economic input regarding chemical recovery and bath maintenance technologies. Peter Gallerani of Integrated Technologies developed cost curves from vendor data. George Pollin and Leslie Cushnie of CAI provided technical input and assisted in conducting the surveys and organizing the literature database. Leslie Cushnie also managed the publication of this text. Nova Ryder performed data input for the Users Survey. Kim Tomlinson prepared graphics and performed word processing. Tom Higgins and Jeff Enzminger of CH2M

Hill provided a portion of the technical literature used during the project.

2.1.5 Description of the Users Survey

A portion of the Users Survey results are presented in Section 2.2 and in various other sections of the text. Some key details of the Users Survey are presented here, so that readers may understand the methodology employed to gather and organize the data.

The survey form employed during the Users Survey was developed through an iterative process. Draft versions of the form were prepared by the contractor and reviewed by the National Center for Manufacturing Sciences Project Manager and Steering Group, the AESF, other industry members and EPA. Two draft versions of the form were sent to a portion of the NAMF membership to be completed and critiqued. The iterative survey development process employed these various sources to improve the survey form in four ways:

(1) improved the clarity of questions
(2) improved the format of the forms
(3) minimized technical and typographical errors
(4) expanded the technical coverage of the form to better meet the objectives of the project.

These improvements resulted in clearer and more in-depth responses from platers, an increase in the technical and regulatory scope of the questionnaire, and an improvement in the amenability of the data to inclusion in a database. The iterative process also permitted the contractor to improve the efficiency of mailing, non-response follow-up, and response handling techniques. The inclusion of EPA and AESF in the survey form review process expanded the technical base from which the survey was developed.

The Users Survey was sent to approximately 2,000 companies and government facilities that perform electroplating. A significant portion of these shops and the majority of the respondents were members of the National Association of Metal Finishers. Most of these NAMF shops are job or contract shops. This type of plating business processes parts for manufacturers of products, and is differentiated from a captive shop, which is an internal plating operation at a manufacturing plant. Other recipients included: members of the Connecticut Association of Metal Finishers (mostly job shops); NCMS members (mostly captive shops of Steering Group companies); captive shops randomly selected from the mailing list of the Metal Finishing magazine; and military plating shops. The number of plating shops included in the survey and the number of respondents from each group are shown in Exhibit 2-2.

Exhibit 2-2. Summary of Questionnare Recipients and Respondents

Affiliation	Recipients	Respondents
NAMF (mostly job shops):	781	294
Connecticut Association of Metal Finishers (mostly job shops):	70	5
Military (Air Force, Navy and Army plating shops):	8	0
Metal Finishing Mailing List (randomly selected captive shops):	1,000	16
Miscellaneous Captive & Job Shops:	112	3
Totals	1,971	318

2.1.6 DESCRIPTION OF THE VENDORS SURVEY

The Vendors Survey was conducted to supplement the data from the Users Survey. Whereas the focus of the Users Survey was on determining the performance and operating costs for technologies, the focus of the Vendors Survey was on identifying the types, capital costs and design features of commercially available equipment. Also, companies offering off-site metals recycling services were surveyed. Twelve specific questionnaires were prepared to solicit information on the following topics:

- Ion Exchange
- Electrowinning
- Electrodialysis
- Diffusion Dialysis
- Membrane Electrolysis
- Atmospheric Evaporators
- Vacuum Evaporators
- Reverse Osmosis
- Ultrafiltration/Microfiltration
- Acid Sorption (acid retardation)
- End-of-Pipe Treatment
- Off-Site Metals Recycling

Each of the technology survey forms requested respondents to provide a specific list of applications for their equipment (e.g., recovery of nickel from rinse water), the number of units sold for each application, available equipment sizes or capacities, materials of construction, optional equipment, expected equipment life-span and equipment prices.

Thirty-five manufacturers/vendors of these pollution prevention and control technologies and services completed survey forms or provided equivalent information. These companies are listed in Exhibit 2-3.

2.1.7 Description of the Literature Search

The literature search gathered information from a wide range of sources including: journals, books, conference proceedings, reports, and technical brochures. The results were cataloged using a database and organized by subject (e.g., waste reduction, end-of-pipe treatment, regulations), technology (e.g., reverse osmosis, filtration), and applications (e.g., chromium plating, aluminum finishing, printed circuit board manufacturing). More than 500 literature sources were identified, reviewed, and entered into the database for use on the project and they are listed in the reference section of this text.

2.2 OVERVIEW OF USERS SURVEY RESULTS

This section presents an overview of the Users Survey data. More in-depth summaries and discussions of the data are presented in the other sections.

2.2.1 Characterization of Respondents

Exhibit 2-4 provides some basic facility characterization data for the survey respondents. The identity code of each shop is shown in column 1. These codes are used throughout the text where they are expressed as PS 001 through PS 318 (PS stands for plating shop and is used in the text to differentiate the Users Survey data from the literature references, which are simply numbered 1 through 522). The location of the shop is shown in column 2 and is expressed as a region of the U.S. The regional codes are as follows:

- NE Northeast
- SE Southeast
- MW Midwest
- SC South Central
- W West
- FW Far West
- NW Northwest

Although each of these regions has been specifically defined in terms of which states they contain, that information is not presented in the database or within this text in order to further preserve the confidentiality of the responses. Several respondents sent in anonymous survey forms and their location was unknown. The region for these shops is designated as "X". The total number of employees for each shop is shown in column 3. Additional data on employees are contained in the database, including the number of equivalent full-time employees working in the plating shop and a breakdown by labor category (platers, rackers, skilled trades, process engineers, chemists, environmental engineers, and other). Column 4 shows the year that the shop commenced metal finishing operations at the location for which the survey form was completed. Column 5 indicates the percentage of the plating work that is contract work (job shop). Where the percentage is less than 100%, the remaining percentage is captive shop work. The next two columns indicate the type of plating equipment that is present; either rack or barrel. The database contains additional plating equipment data concerning the level of automation present (e.g., automatic with manual loading vs fully automatic). The last column presents annual production data in terms of square feet of plating, pounds of parts processed or amp-hours of rectifier usage. Since most job shops are unable to track their work volume using these units or other non-fiscal units, this column contains many blanks. Exhibits 2-5 (a) (b) and (c) graphically display some of the facility characterization data.

2.2.2 Metal Finishing Process Characterization

Exhibit 2-6 presents a portion of the process data. Column 1 shows the process name. Column 2 indicates the type of the process, either aluminum finishing, cleaning, other (e.g., conversion coating and passivation), plating, solvent cleaning, or stripping. The third column indicates the number of respondents that employ each process and the last column indicates the total volume of their process tanks. Collectively, the respondents to the Users Survey operate 167 different types of metal finishing processes. The 25 most frequently found processes are identified in Exhibit 2-7 (excludes pre-plating, post-plating and stripping processes).

Exhibit 2-3.
List of Manufacturers, Vendors, and Recycling Companies that Participated in the Project

Company	Location	Telephone Number
Alpha Omega	Long View, TX	903/297-7272
Bio-Recovery Systems Inc.	Las Cruces, NM	505/523-0405
Calfran International Inc.	Springfield, MA	413/732-3616
ChromeTech, Corp.	Windsor, OH	216/272-5195
CP Chemicals	Fort Lee, NJ	201/944-6020
Cyano Corp.	Detroit, MI	313/843-9243
Delta	Preston,WA	206/222-4544
Eco-Tec Inc.	Pickering, Ontario	416/831-3409
Eltech System Corp.	Fairport Harbor, OH	216/357-4080
Encycle	Corpus Christi, TX	512/289-0300
Hard Chrome Plating Consultants	Cleveland, OH	216/631-9090
Horsehead Resource Development Co.	Palmerton, PA	800/253-5579
i^3	Woodbridge, VA	703/491-1190
Inmetco	Ellwood City, PA	412/758-5515
Ionsep Corporation	Rockland, DE	302/798-7402
Jaynor, Inc.	Falls Church, VA	703/256-2122
JWI, Inc.	Holland, MI	616/772-9011
Kinetic Recovery Corporation	Edina, MI	612/941-0739
Kinetico Engineered Systems, Inc.	Newbury, OH	216/564-5379
Koch Membrane Systems	Imington, MA	508/657-4250
LICON, Inc	Pensacola, FL	904/434-5088
Memtek Corporation	Billerica, MA	508/667-2828
Midwest Air Products Co.	Owosso, MI	517/723-8881
NAPCO	Terryville, CT	203/589-7800
Netzsch, Inc.	Exton, PA	215/363-8010
Osmonics Inc.	Minnetonka, MN	612/933-0141
Parkson Corp.	Ft. Lauderdale, FL	305/974-6610
Penfield Liquid Treatment Systems	Plantsville, CT	203/621-9141
Poly Products Corporation	Henderson, NV	702/566-3000
Prosys Corp.	Chelmsford, MA	508/250-4940
QPS Inc	Bristol, CT	203/582-7518
Recontek	Newman, IL	217/837-2461
The GraverCo.	Union, NJ	908/964-2400
U.S. Filter	Warrendale, PA	412/772-0044
World Resources	McLean, VA	703/734-9800

2.2.3 Wastewater and Discharge Characterization

Exhibit 2-8 provides an overview of the wastewater data contained in the Users Survey database. Column 1 shows the shop code. Column 2 indicates the type of wastewater discharge (*direct*—to a stream or river, *indirect*—to a publicly owned treatment works or POTW, or *zero discharge*—no wastewater discharge). Column 3 indicates the average discharge rate from the plating operations. The database also contains data for the maximum plating discharge and the average and maximum total industrial discharge. Column 4 indicates the amount of flow reduction that has been achieved through the implementation of pollution prevention techniques and methods. The next two columns indicate the type of effluent regulations that apply to each shop. These are shown as either standard or non-standard (i.e., CFR 413 or CFR 433) and aquatic-toxicity-based.

The majority of the respondents to the Users Survey are indirect dischargers. The percentage of shops that are either indirect, direct, both indirect and direct, and zero discharge are shown in Exhibit 2-9. These data indicate that captive shops are more likely to be direct dischargers than are job shops. EPA estimates in 1984 indicated a similar trend (ref. 517).

The electroplating discharge rates (average daily flows) of the survey respondents vary from 0 gpd to 420,000 gpd (some higher discharge rates were reported for combined plating and non-plating industrial discharges). The average and median plating discharge rates for respondents were 34,600 gpd and 14,000 gpd, respectively (see Exhibit 2-10 for a graphical summary of discharge data). Many shops indicated that they have made drastic progress in reducing wastewater flow rates, the most significant of which are listed in Section 3.

Exhibit 2-4. Partial Summary of Facility Characterization Data

Shop Code	Location	Number of Employees	Year Shop Started Plating	% Job Shop	Plating Type Rack	Plating Type Barrel	Annual Production Rate
1	NE	35	1985	100	X		
2	MW	72	1945	100	X	X	
3	NE	75	1942	100		X	
4	NE	45	1956	100		X	
5	MW	31	1950	100	X	X	
6	NE	9	1947	100	X		
7	NE	65	1940	100	X		
8	MW	40	1945	100	X	X	
9	MW	75	1973	100	X	X	
10	FW	38	1985	100	X	X	
11	FW	20	1954	100	X	X	
12	NE	25	1940	100	X	X	
13	NE	45	UNK	100	X	X	
14	NE	34	1986	100	X		
15	NE	84	1984	100		X	
16	MW	40	1969	100	X	X	
17	MW	50	1986	100	X		
18	NE	20	1954	100	X		7,000,000 amp-hr
19	FW	34	1988	100	X		
20	NE	93	1961	100	X	X	
21	MW	24	1951	100	X		
22	MW	100	1964	100	X		
23	MW	47	1983	100	X	X	
24	MW	166	1984	100	X		1,000,000 sq ft
25	MW	62	1930	0	X	X	
26	MW	28	1968	100			
27	FW	70	1965	100	X		
28	MW	8	1968	100	X	X	
29	MW	23	1960	100	X		900,000 lbs
30	FW	6	1975	100		X	
31	NE	15	1981	100	X	X	
32	NE	23	UNK	100			
33	FW	15	1975	100	X	X	
34	FW	20	1967	100	X	X	
35	MW	17	1983	100	X	X	
36	FW	30	1978	100	X	X	
37	FW	12	1972	100			
38	NE	50	1949	100	X		
39	MW	40	1963	100	X	X	
40	SC	11	1961	100	X	X	
41	NE	35	1947	100			4,110,845 sq ft
42	MW	8	1962	100	X		
43	SE	28	1950	100	X	X	
44	FW	9	1964	100	X	X	
45	MW	130	1942	100	X		
46	NE	20	1965	100	X		
47	MW	12	1937	100	X		
48	MW	40	1951	100	X	X	
49	FW	45	1980	100	X		
50	MW	25	1946	100	X	X	
51	MW	35	1966	100	X	X	7,400,000 sq ft
52	NE	26	1940	100	X		
53	NE	18	1913	100	X	X	
54	MW	17	1967	100		X	
55	FW	20	1960	100	X	X	
56	NE	4	1983	100			
57	MW	24	1985	100		X	
58	NE	36	1953	65	X		
59	W	21	1961	100			
60	MW	16	1982	0	X	X	
61	FW	10	1971	100	X	X	
62	FW	10	1971	100	X		
63	MW	24	1973	0	X	X	
64	NE	27	1960	100	X	X	160,000 sq ft
65	NE	100	1959	100	X		4,531,000 sq ft
66	NE	30	1980	0	X		
67	MW	20	UNK	100	X		
68	FW	18	1987	100	X	X	
69	FW	58	1964	100	X		
70	W	25	1950	100	X	X	
71	MW	8	1954	0	X		
72	MW	44	UNK	100	X	X	
73	NE	9	1971	100			
74	SC	25	1965	100	X		
75	NE	3	1972	100			
76	MW	24	1979	100	X	X	
77	W	110	1959	0	X	X	
78	FW	15	1977	100	X		
79	MW	85	1964	100	X		
80	NE	12	1940	100	X		
81	MW	9	1962	100		X	
82	MW	250	1962	100	X		240,000,000 amp-hr
83	NE	5	1952	100	X	X	
84	SC	28	1978	100	X	X	6000,000 lbs
85	FW	40	1957	95	X		
86	NE	68	1931	100	X	X	
87	MW	220	1964	100	X	X	
88	MW	31	1968	100		X	
89	MW	75	1960	100	X		
90	MW	13	1984	100	X	X	325,000 sq ft
91	FW	43	1960	100	X	X	
92	FW	65	1976	100	X		
93	MW	41	1927	100	X	X	
94	MW	20	1976	100	X		
95	FW	60	1975	100	X	X	4,000,000 sq ft
96	MW	27	1991	100	X	X	
97	MW	13	1977	100	X	X	
98	MW	150	1973	100			
99	MW	21	1951	100	X	X	
100	SC	8	1983	50	X	X	
101	X	60	1978	100			
102	SE	135	1962	90	X	X	6,173,717 lbs
103	MW	26	1992	0	X	X	6,249,800 lbs
104	MW	17	1986	100	X	X	
105	MW	80	1973	100	X	X	
106	NW	38	1940	100	X	X	
107	FW	6	1986	100	X		
108	FW	25	1975	0	X	X	
109	SC	40	1962	100	X	X	
110	NE	50	1981	100	X		
111	FW	12	1958	100	X		
112	MW	9	1980	100	X	X	
113	W	50	1965	100	X		
114	SE	7	1983	100	X	X	
115	MW	15	1968	100			3,000,000 lbs
116	NE	34	1952	100	X		
117	MW	8	1956	100		X	
118	SC	40	1958	40	X	X	
119	FW	20	1970	100	X	X	
120	X	16	1987	100	X		
121	FW	31	1978	100	X	X	
122	MW	1	UNK	100		X	
123	FW	140	1958	100	X	X	
124	MW	144	1919	100	X	X	
125	MW	48	1976	100	X		12,000,000 sq ft
126	SC	43	1982	100	X	X	
127	FW	5	1972	0	X		
128	MW	52	1980	100		X	17,000,000 lbs
129	MW	25	1963	100	X		
130	SC	15	1978	60	X	X	
131	MW	80	1953	100	X		
132	MW	50	1991	100	X	X	
133	NE	10	1981	100	X	X	
134	MW	4	1982	10	X		
135	FW	19	1978	100	X		
136	MW	14	1979	100	X	X	
137	SE	170	1986	100	X		
138	FW	40	1947	100	X		
139	NE	27	1952	100		X	
140	NE	60	1970	100	X		
141	MW	28	1959	100	X	X	12,000,000 sq ft
142	SC	25	1961	100	X	X	
143	X	65	1950	0	X	X	
144	MW	10	1963	100			
145	FW	11	1959	100	X	X	
146	FW	85	1981	100	X	X	5,400,000 sq ft
147	MW	25	1964	100			
148	FW	34	1948	100	X		
149	MW	61	1966	100	X	X	8,006,325 amp-hr
150	MW	151	1969	100	X		
151	FW	6	1945	100	X	X	
152	MW	19	1984	100	X		
153	FW	14	UNK	10	X	X	10,000 sq ft
154	FW	20	1977	100	X		6,575 lbs
155	FW	40	1976	100	X		
156	MW	25	1972	100	X		
157	FW	2	1978	0	X		
158	FW	83	1968	100	X	X	
159	NE	35	1970	100	X		
160	SE	35	1961	100	X	X	

Exhibit 2-4 (continued). Partial Summary of Facility Characterization Data

Shop Code	Location	Number of Employees	Year Shop Started Plating	% Job Shop	Plating Type Rack	Plating Type Barrel	Annual Production Rate
161	MW	22	1966	100	X		
162	MW	35	1956	100	X	X	
163	W	30	1984	100	X	X	75,409,000 sq ft
164	MW	25	1972	100	X	X	
165	MW	57	1977	100	X		
166	FW	82	1952	100	X		
167	MW	40	1964	100	X		
168	FW	19	1978	100	X		900,000 sq ft
169	NE	110	1990	0	X	X	
170	NE	29	1955	100	X		
171	SC	16	1980	100	X	X	
172	MW	45	1945	100	X		
173	SE	65	UNK	100	X		
174	FW	104	1980	100	X		
175	SE	13	1990	100	X		
176	MW	130	1977	100	X	X	
177	FW	15	1961	95	X		1,000,000 sq ft
178	NE	32	1965	100	X	X	
179	MW	11	1983	100	X	X	
180	FW	120	1989	100	X	X	
181	NE	45	1953	100	X	X	
182	FW	34	1980	10	X	X	
183	FW	17	1974	100	X		
184	FW	88	1979	100	X	X	
185	FW	10	1983	0	X		
186	NE	15	1978	100	X	X	
187	NE	45	1978	100	X		
188	MW	7	1967	100	X	X	
189	FW	7	1983	100	X	X	
190	NE	17	UNK	100		X	
191	FW	33	1968	100	X		2,835,948 amp-hr
192	MW	50	1983	100	X	X	
193	SE	15	1972	100	X		
194	MW	12	1940	100			50,000,000 amp-hr
195	SC	7	1985	100	X		
196	NE	136	1980	100	X	X	
197	FW	111	1951	0	X		
198	MW	4	1955	100	X		
199	MW	62	1981	100		X	25,000,000 lbs
200	NE	16	1959	100	X		5,000 sq ft
201	FW	26	1962	100	X	X	
202	FW	23	1967	100			110,000,000 amp-hr
203	FW	95	1952	100	X	X	
204	MW	190	1947	100	X	X	
205	SC	9	1958	0	X		
206	FW	31	1958	100	X		209,000 amp-hr
207	NE	40	1937	100	X	X	868,000 sq ft
208	X	29	1988	100	X	X	
209	SE	47	1966	0	X	X	
210	FW	115	1960	100			
211	NE	5	UNK	100			
212	NE	11	1956	100	X		
213	NE	195	1965	0	X		2,871,200 lbs
214	MW	125	1947	100	X	X	80,000,000 lbs
215	MW	35	1947	100	X		
216	X	16	1929	100	X	X	
217	SE	19	1951	0			
218	FW	4	1975	100	X		
219	FW	27	1941	100	X	X	
220	FW	32	1978	100	X		
221	NE	7	1987	0	X		
222	MW	14	1979	100	X	X	
223	NE	327	1950	100	X	X	
224	MW	6	1986	100	X	X	
225	MW	39	1977	100		X	
226	FW	30	1958	80		X	
227	FW	9	1973	0			
228	MW	90	1940	100	X		2,600,000 sq ft
229	MW	18	1956	100	X	X	
230	FW	30	1973	100	X		
231	MW	65	1963	100	X	X	
232	NE	60	1920	100	X	X	
233	NE	19	1973	100	X		
234	SE	39	1972	100	X		
235	MW	57	1955	100	X	X	
236	MW	125	1951	100	X	X	
237	MW	83	1981	100	X		
238	FW	387	1911	100	X		
239	NE	27	1925	100	X	X	
240	SE	23	1982	100		X	8,000,000 lbs
241	NE	28	1956	100	X		
242	FW	21	1959	100	X	X	
243	MW	25	1980	100		X	18,000,000 lbs
244	FW	82	1971	100	X	X	
245	SC	45	1968	100	X	X	1,600,000 sq ft
246	MW	60	1935	100	X		
247	MW	52	1952	55	X	X	
248	W	37	1966	100	X	X	
249	FW	6	1983	100	X	X	
250	MW	30	1957	100	X	X	16,000,000 lbs
251	FW	65	1939	100	X		
252	MW	14	1948	100	X		
253	NE	80	1967	100	X	X	15,000,000 sq ft
254	MW	32	1952	0	X	X	
255	FW	20	1983	100	X		1,000,000 sq ft
256	NW	130	1987	0	X	X	6,000,000 lbs
257	SE	300	1986	0	X		
258	MW	95	UNK	100	X	X	
259	FW	4	1946	100	X		
260	MW	42	1971	100	X	X	
261	NE	7	UNK	0	X	X	
262	FW	35	1960	100	X	X	
263	MW	87	1987	100	X		9,000,000 sq ft
264	NE	70	1972	100	X	X	
265	FW	700	1987	100	X		108,000 sq ft
266	SC	34	1975	80	X		80,000 sq ft
267	MW	10	1963	100	X	X	408,900 sq ft
268	SE	110	1969	100	X	X	4,556,432 sq ft
269	MW	10	1937	100	X		
270	NE	25	1976	100	X		800,000 lbs
271	FW	15	1959	100	X		36,656,700 amp-hr
272	FW	63	1959	100	X	X	
273	MW	56	1942	0			
274	NE	35	1986	100	X		
275	MW	79	1986	0	X		
276	NE	40	1946	100	X	X	
277	NE	250	1957	0		X	
278	SC	300	1967	0	X		2,815,000 sq ft
279	SE	65	1986	100	X	X	
280	MW	50	1972	100	X		2,500,000 sq ft
281	MW	750	1930	0	X	X	
282	SE	29	1958	10			
283	NE	350	1987	0	X	X	
284	SE	8	1990	50	X		
285	NE	250	1981	0		X	300,000 lbs
286	NE	123	1867	0	X	X	
287	NE	100	1905	0	X	X	
288	NE	38	1947	100	X	X	133,056,000 amp-hr
289	SC	320	1967	0	X	X	1,000,000 lbs
290	SC	400	1987	0	X		
291	FW	400	1960	10	X		
292	MW	250	1960	0	X	X	400,000 lbs
293	FW	99	1970	100			
294	FW	30	1947	100	X		
295	NE	3,000	1949	0	X	X	
296	MW	490	UNK	0	X		40,000,000 lbs
297	NE	18	1959	100	X	X	
298	MW	35	1950	100	X		11,500,000 sq ft
299	MW	85	1958	0			20,000,000 sq ft
300	MW	110	1959	100	X	X	15,908,750 sq ft
301	SC	15	1968	100	X	X	
302	MW	31	1969	100	X	X	
303	FW	250	1983	50			
304	MW	13	1971	100	X	X	
305	NE	260	1950	0	X		
306	X	1	UNK	100	X		
307	MW	90	1972	100	X	X	
308	NE	12	1985	100	X	X	
309	NE	25	1983	100	X	X	
310	FW	7	1985	100	X		
311	MW	60	1972	100	X		3,000,000 sq ft
312	NE	20	1959	100			
313	X	40	1940	0	X		
314	SE	271	1962	0		X	16,500,000 lbs
315	MW	60	1948	100	X	X	
316	NE	95	1930	100	X	X	
317	NE	27	1944	100	X	X	
318	SC	15	1960	100	X		

Exhibit 2-5(a). Geographical Distribution of Respondents

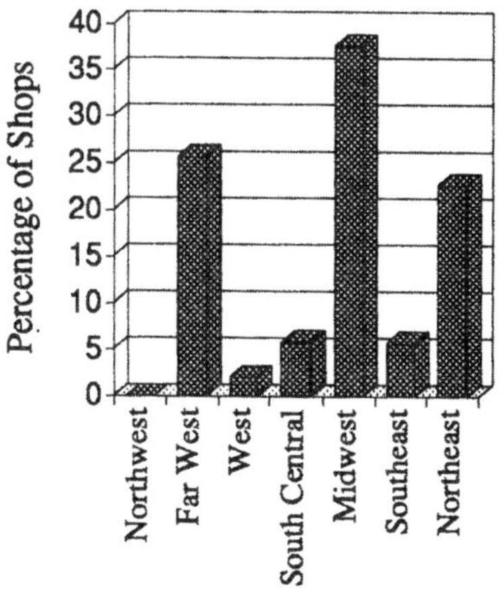

Exhibit 2-5(b). Shop Age Distribution (Year shop commenced metal finishing operations)

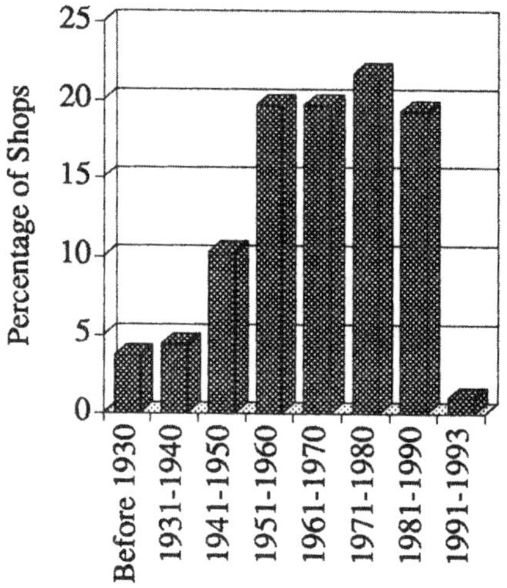

Exhibit 2-5(c). Company Size Distribution by Number of Employees

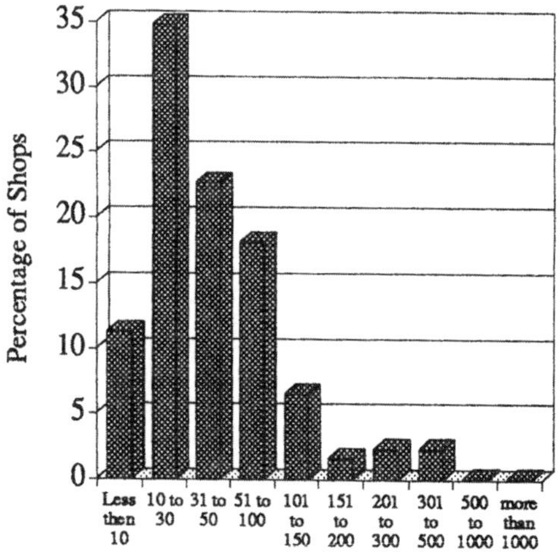

Exhibit 2-6. Distribution of Respondents' Processes

Process Name	Type of Process	Number of Respondents Using Process	Total Volume of Process Tanks, gal
Anodize, Boric/Sulfuric	Alum. Finish	5	8,501
Anodize, Chromic Acid	Alum. Finish	44	35,846
Anodize, Hardcoat	Alum. Finish	47	78,109
Anodize, Oxalic Acid	Alum. Finish	3	2,430
Anodize, Phosphoric Acid	Alum. Finish	4	446
Anodize, Sulfuric Acid	Alum. Finish	83	131,988
Bright Dip	Alum. Finish	28	12,668
Deoxidize	Alum. Finish	68	57,470
Desmut	Alum. Finish	64	51,482
Dye	Alum. Finish	69	92,006
Etch	Alum. Finish	82	81,165
Seal	Alum. Finish	54	43,924
Seal, Dichromate	Alum. Finish	37	31,340
Seal, Nickel	Alum. Finish	52	42,065
Seal, Sodium Bicarbonate	Alum. Finish	1	350
Abrasive Blasting	Cleaning	91	0
Acid Clean/Pickle, H2S04	Cleaning	131	117,025
Acid Clean/Pickle, H3P04	Cleaning	36	17,577
Acid Clean/Pickle, HCl	Cleaning	239	393,460
Acid Clean/Pickle, HF	Cleaning	38	14,222
Acid Clean/Pickle, HN03	Cleaning	51	41,784
Acid Clean/Pickle, Mixed	Cleaning	29	29,468
Acid Clean/Pickle,HFB03	Cleaning	1	48
Acid Salts	Cleaning	14	35,696
Alk., Elect.Clean	Cleaning	7	12,733
Alk., Elect.Clean(Non-CN)	Cleaning	232	530,060
Alkaline, Cyanide	Cleaning	26	19,608
Alkaline, Descaling	Cleaning	55	33,367
Alkaline, Soak(Non-Cyan)	Cleaning	265	634,158
Alkaline, Ultrasonic	Cleaning	2	777
Emulsion Cleaning	Cleaning	10	3,363
Salt Bath Descaling	Cleaning	3	200
Spray Washers	Cleaning	28	24,080
Tumbling	Cleaning	49	0
Alkaline Resist Develop	Other	1	120
Ammonium Bifluoride	Other	2	4,600
Black Oxide	Other	55	17,547
Bright Dip of Cadmium	Other	29	3,439
Bright Dip of Cu/Cu Alloy	Other	61	4,907
Bright Dip of Nickel	Other	2	350
Cadmium, Vacuum	Other	1	0
Chem Film	Other	1	1,051
Chem Milling of Al	Other	1	250
Chem Milling of Ti	Other	1	500
Chromate on Aluminum	Other	121	73,532
Chromate on Black Oxide	Other	1	150
Chromate on Cadmium Plate	Other	75	29,132
Chromate on Copper	Other	1	160
Chromate on Magnesium	Other	21	4,211
Chromate on Phosphate	Other	1	100
Chromate on Silver	Other	4	2,545
Chromate on Zinc	Other	19	11,715
Chromate on Zinc Plate	Other	114	200,395
Chromium Etch	Other	4	9,100
Electropolish	Other	24	20,829
Etch,Chromic Acid	Other	1	4,000
Flame Spray	Other	0	
Galvanizing	Other	2	6,000
Magnesium HAE	Other	2	2,246
Mass Finishing	Other	8	5,555
Mechanical Plating	Other	6	2,160
Nital Etch	Other	9	2,086
Passivation	Other	67	17,941
Passivation, Nitric	Other	58	12,420
Passivation, SIS	Other	2	650
Phosphate, Calcium	Other	1	0
Phosphate, Iron	Other	18	8,222
Phosphate, Manganese	Other	40	13,539
Phosphate, Titanium	Other	1	400
Phosphate, Zinc	Other	74	78,800
Tin Hot Dip	Other	1	30
Tin-Lead Solder Dip	Other	1	2,000
Titanium Anodize	Other	1	365
Brass	Plating	55	39,262
Bronze	Plating	16	8,646
Cadmium, Cyanide	Plating	94	127,092
Cadmium, Non-Cyanide	Plating	8	4,905
Chrome, Black	Plating	2	1,160
Chromium, Hard	Plating	81	322,227
Chromium,Decorative(Cr+3)	Plating	26	22,685
Chromium,Decorative(Cr+6)	Plating	91	148,982
Copper	Plating	4	860
Copper, Cyanide	Plating	120	102,341
Copper, Electroless	Plating	10	7,626
Copper, Fluoborate	Plating	3	510
Copper, Pyrophosphate	Plating	4	600
Copper, Strike(CN)	Plating	65	37504
Copper, Strike(Non-CN)	Plating	13	9116
Copper, Sulfate	Plating	61	107,460
Gold, Cyanide	Plating	64	10,629
Gold, Non-Cyanide	Plating	19	3,927
Indium	Plating	5	151
Iron	Plating	3	3,003
Lead	Plating	6	9,721
Lead-Tin	Plating	36	14,136
Nickel	Plating	18	18,995
Nickel, Black	Plating	2	230
Nickel, Brite	Plating	21	54,380
Nickel, Electroforming	Plating	3	2,191
Nickel, Electroless,Boron	Plating	4	401
Nickel, Electroless,Other	Plating	5	3,290
Nickel, Fluoborate	Plating	1	710
Nickel, Semi-Brite	Plating	3	32,420
Nickel, Sulfamate	Plating	82	55,879
Nickel, Sulfate	Plating	18	40,875
Nickel, Teflon	Plating	3	175
Nickel, Tin	Plating	1	1,000
Nickel, Watts	Plating	132	541,552
Nickel, Woods	Plating	32	20,732
Nickel,Electroless,Phosph	Plating	94	76,047
Palladium	Plating	2	565
Platinum	Plating	2	24
Rhodium	Plating	16	208
Silver, Cyanide	Plating	78	36,258
Silver, Non-Cyanide	Plating	3	205
Silver, Strike(CN)	Plating	2	630
Tin, Acid	Plating	86	49,644
Tin, Alkaline	Plating	30	17,130
Tin-Lead (PC Boards)	Plating	8	1,840
Zinc, Acid	Plating	33	212,188
Zinc, chloride	Plating	1	100
Zinc, Cyanide	Plating	48	202,632
Zinc, Non-Cyanide	Plating	90	441,934
Zinc/Cobalt	Plating	5	4,225
Zinc/Iron	Plating	4	3,515
Zinc/Nickel	Plating	3	1,264
Zincate	Plating	9	3,302
1,1,1-Trichloroethane	Solvent Clean	92	785
Acetone	Solvent Clean	4	0
Chlorofluorocarbons	Solvent Clean	11	150
Freon	Solvent Clean	1	0
Methyl Ethyl Keytone	Solvent Clean	15	0
Methylene Chloride	Solvent Clean	13	0
Perchloroethylene	Solvent Clean	36	50
Stoddard	Solvent Clean	1	0
Trichloroethylene	Solvent Clean	53	1,100
Xylene	Solvent Clean	1	0
Aluminum	Stripping	1	2
Anodize	Stripping	49	22,888
Anodize, Caustic	Stripping	2	1,700
Anodize, Chrome-Phosphate	Stripping	1	50
Brass	Stripping	6	3,220
Bronze	Stripping	3	1,290
Cadmium	Stripping	53	6,642
Chromium	Stripping	77	51,915
Copper	Stripping	45	10,913
Gold	Stripping	9	766
Magnesium HAE,Strip	Stripping	1	450
Nickel	Stripping	138	50,487
Nickel,Electroless,Strip	Stripping	1	4,300
Plasma,Strip	Stripping	1	50
Silver	Stripping	31	2,056
Tin	Stripping	6	970
Tin-Lead	Stripping	32	3,338
Tungsten Carbide,Strip	Stripping	1	412
Zinc	Stripping	60	47,650
Zinc/Iron	Stripping	1	220

The respondents to the Users Survey are required to meet either CFR 413 (Electroplating Categorical Standards), CFR 433 (Metal Finishing Categorical Standards), or non-standard effluent limitations. Non-standard limitations are more stringent than the categorical standards for one or more pollutant parameters. Some of the non-standard limitations are written in terms of pollutant mass and flow rates (e.g., 0.37 lbs/day of chromium with a maximum flow of 40,000 gpd) rather than concentration limitations. The percentage of respondents that are required to meet each type of effluent limitation are as follows:

40 CFR 413: 28%
40 CFR 433: 8%
Non-Standard: 64%

In addition to concentration or pollutant mass discharge standards, 16% of the respondents indicated that they are also subject to aquatic-toxicity-based effluent standards. These limits require that an industrial wastewater be sufficiently treated such that certain percentages of organisms (typically fish and water fleas) are able to survive in the effluent for a given time period.

More detailed data on regulations and the compliance experience for each shop are contained in the database. For example, the Users Survey asked platers to list the pollutant parameters for which they have compliance difficulty. A summary of their responses is shown in Exhibit 2-11.

2.2.4 Pollution Prevention

Various terms are used to group or categorize the different methods employed by industry in their pollution control strategies, e.g., pollution prevention, source control, waste minimization, and in-plant changes. One popular term used extensively by EPA since 1988 is pollution prevention. This term is defined by EPA as the maximum feasible reduction of all wastes (wastewater, solid waste and air emissions) generated at production sites. It involves the judicious use of resources through source reduction, energy efficiency, reuse of input materials during production and reduced water consumption. Pollution prevention covers a broad range of pollution control methods employed by plating shops, such as drag-out and rinse water reduction, chemical recovery from rinse waters, bath maintenance, and material substitution. Each of these aspects of pollution prevention have been addressed by the Users Survey. An overview of the survey results are presented here and details are contained in Sections 3, 4, 5 and 6.

For platers, pollution prevention has emerged as an important set of tools used to help attain compliance and reduce operating costs. Widespread success has been achieved using simple methods and techniques that reduce drag-out losses and rinse water use. More than 90 percent of the shops indicated that they utilize these tools and have benefited from them. Although some shops have had great success with chemical recovery

Exhibit 2-7. The Tvrenty-Five Most Frequently Operated hIetal Finishing Processes

Process Name	Percent of Shops Using Process	Average Volumeof Solution per Shop (gal)*	Process Name	Percent of Shops Using Process	Average Volume of Solution per Shop (gal)*
1. Nickel (Watts) plating	42	4,102	14. Silver (CN) plating	25	46
2. Zinc (non-CN) plating	39	5,276	15. Chromate on cadmium	24	388
3. Chromate on zinc plate	39	1,594	16. Zinc phosphate	23	1,065
4. Chromate on aluminum	38	607	17. Tin-leadplating	21	256
5. Passivation (all types)	38	244	18. Gold (CN) plating	20	166
6.Copper(CN)plating	38	852	19. BrightdipofCu/alloy	19	80
7. Cadmium (CN) plating	30	1,360	20. Copper (sulfate) plating	19	1,761
8. Electroless nickel plating	30	809	21. Black oxide	17	319
9. Decorative Cr (+6) plate	29	1,637	22. Brass plating	17	713
10. Tin (acid) plating	27	581	23. Hard coat anodizing	15	1,661
11. Nickel (sulfamate) plate	26	681	24. Zinc (CN) plating	15	4,221
12. Sulfuricacidanodizing	26	1,590	25. Chromicacidanodizing	14	814
13. Hard chromium plating	25	3,978			

*Often involves a multiple number of tanks per shop containing the same solution.

Exhibit 2-8. Respondents' Wastewater and Discharge Characterization

Shop Code	Type of Discharge	Plating Discharge gpd	Flow Reduction gpd	Type of Standards	Aquatic?	Shop Code	Type of Discharge	Plating Discharge gpd	Flow Reduction gpd	Type of Standards	Acquatic?
001	INDIRECT	9,000	ND	NONSTD	NO	081	INDIRECT	6,000	38,487	NONSTD	NO
002	INDIRECT	80,000	ND	CFR4 3 3	NO	082	INDIRECT	100,000	0	NONSTD	NO
003	INDIRECT	225,000	ND	NONSTD	NO	083	INDIRECT	8,000	0	NONSTD	NO
004	INDIRECT	160,000	ND	CFR4 13	NO	084	INDIRECT	25,000	0	NONSTD	YES
005	INDIRECT	100,000	ND	CFR4 13	NO	085	INDIRECT	28,407	0	CFR413	NO
006	ZERO	0	ND	NONSTD	YES	086	INDIRECT	70,000	0	NONSTD	NO
007	DIRECT	20,000	ND	NONSTD	YES	087	INDIRECT	300,000	√	CFR413	NO
008	INDIRECT	58,000	ND	NONSTD	NO	088	INDIRECT	103,000	0	CFR413	NO
009	INDIRECT	190,000	ND	CFR4 13	NO	089	INDIRECT	12,500	5,000	CFR413	NO
010	INDIRECT	6,000	ND	NONSTD	YES	090	INDIRECT	4,334	700	CFR413	NO
011	INDIRECT	4,500	ND	CFR4 13	NO	091	INDIRECT	5,500	1,000	NONSTD	NO
012	INDIRECT	31,000	ND	CFR4 13	NO	092	INDIRECT	12,000	8,000	CFR413	NO
013	DIRECT	160,000	ND	NONSTD	YES	093	INDIRECT	38,500	20,000	CFR413	NO
014	INDIRECT	22,000	ND	NONSTD	NO	094	DIRECT	11,500	5,000	NONSTD	YES
015	BOTH	30,000	ND	NONSTD	YES	095	DIRECT	50,000	15,000	NONSTD	NO
016	INDIRECT	50,000	ND	CFR4 13	NO	096	INDIRECT	41,000	20,000	NONSTD	NO
017	INDIRECT	1,200	ND	NONSTD	NO	097	INDIRECT	9,416	√	NONSTD	NO
018	INDIRECT	8,500	ND	NONSTD	NO	098	INDIRECT	76,000	0	CFR413	NO
019	INDIRECT	7,200	ND	NONSTD	NO	099	INDIRECT	496	0	NONSTD	NO
020	INDIRECT	35,000	ND	CFR4 13	NO	100	ZERO	0	8,000	NONSTD	YES
021	ZERO	0	1,000	NONSTD	NO	101	INDIRECT	30,000	0	NONSTD	NO
022	INDIRECT	70,000	60,000	CFR413	NO	102	INDIRECT	50,000	25,000	NONSTD	NO
023	INDIRECT	42,000	12,000	NONSTD	NO	103	INDIRECT	26,000	0	NONSTD	NO
024	INDIRECT	20,000	0	CFR4 13	NO	104	INDIRECT	17,600	4,000	CFR433	NO
025	INDIRECT	55,000	0	NONSTD	YES	105	INDIRECT	22,000	2,500	CFR433	NO
026	INDIRECT	5,000	0	NONSTD	NO	106	INDIRECT	30,000	10,000	CFR413	NO
027	INDIRECT	29,000	0	CFR4 13	NO	107	INDIRECT	3,000	12,000	CFR433	NO
028	INDIRECT	8,000	4,000	NONSTD	NO	108	INDIRECT	ND	200	CFR413	NO
029	INDIRECT	10,000	5,000	CFR413	YES	109	DIRECT	38,000	5,000	NONSTD	NO
030	INDIRECT	500	100	CFR4 3 3	NO	110	INDIRECT	8,600	8,600	NONSTD	YES
031	INDIRECT	7,800	8,000	NONSTD	NO	111	INDIRECT	6,200	300	CFR413	NO
032	DIRECT	13,000	0	NONSTD	NO	112	DIRECT	9,500	3,000	CFR413	NO
033	INDIRECT	22,000	0	CFR4 13	YES	113	INDIRECT	15,000	10,000	CFR413	NO
034	INDIRECT	70,000	30,000	CFR413	NO	114	INDIRECT	1,300	0	NONSTD	NO
035	INDIRECT	900	4,100	CFR413	NO	115	DIRECT	7,500	5,000	NONSTD	NO
036	INDIRECT	2,704	50,000	CFR4 13	NO	116	DIRECT	10,000	35,000	NONSTD	YES
037	DIRECT	5,000	0	NONSTD	NO	117	INDIRECT	750	0	CFR413	YES
038	INDIRECT	1,400	1,000	NONSTD	NO	118	INDIRECT	42,630	190,000	CFR433	NO
039	INDIRECT	75,000	0	NONSTD	NO	119	INDIRECT	1,500	8,500	NONSTD	NO
040	INDIRECT	20,500	0	NONSTD	NO	120	INDIRECT	5,000	3,000	CFR413	NO
041	INDIRECT	18,000	12,000	NONSTD	NO	121	INDIRECT	6,000	1,000	NONSTD	NO
042	INDIRECT	5,000	4,000	NONSTD	NO	122	INDIRECT	175,000	0	NONSTD	NO
043	INDIRECT	50,000	5,000	NONSTD	NO	123	INDIRECT	26,500	√	CFR413	NO
044	INDIRECT	3,000	1,000	CFR413	NO	124	INDIRECT	354,956	25,000	CFR413	NO
045	INDIRECT	90,000	20,000	CFR413	NO	125	INDIRECT	230,000	15,000	NONSTD	NO
046	INDIRECT	6,500	5,500	NONSTD	YES	126	INDIRECT	12,000	10,000	CFR413	NO
047	INDIRECT	1,400	8,000	NONSTD NO	NO	127	INDIRECT	4,000	2,000	NONSTD	NO
048	INDIRECT	30,000	0	NONSTD NO	NO	128	INDIRECT	4,600	10,000	NONSTD	NO
049	INDIRECT	5,500	4,500	NONSTD NO	NO	129	INDIRECT	27,000	5,000	NONSTD	NO
050	INDIRECT	20,000	18,000	NONSTD NO	NO	130	INDIRECT	9,938	5,000	NONSTD	NO
051	INDIRECT	30,371	28,800	NONSTD	NO	131	INDIRECT	9,938	√	NONSTD	YES
052	ZERO	0	500	CFR413	NO	132	INDIRECT	90,000	√	CFR413	NO
053	INDIRECT	15,000	25,000	CFR413	NO	133	ZERO	0	3,000	NONSTD	NO
054	INDIRECT	44,872	5,000	CFR413	NO	134	INDIRECT	2,000	300	NONSTD	NO
055	INDIRECT	5,000	0	NONSTD	NO	135	INDIRECT	5,000	500	NONSTD	NO
056	INDIRECT	1,000	500	NONSTD	NO	136	INDIRECT	19,800	6,000	CFR413	NO
057	INDIRECT	8,000	0	CFR413	NO	137	BOTH	100,000	0	NONSTD	YES
058	INDIRECT	78,000	10,000	CFR413	NO	138	DIRECT	500	0	NONSTD	YES
059	DIRECT	10,000	80,000	NONSTD	NO	139	DIRECT	52,000	75,000	NONSTD	NO
060	INDIRECT	10,000	5,000	CFR413	NO	140	INDIRECT	10,000	15,000	CFR413	NO
061	INDIRECT	9,000	7,000	NONSTD	YES	141	INDIRECT	30,000	0	NONSTD	YES
062	INDIRECT	1,500	3,000	CFR413	NO	142	INDIRECT	11,500	11,500	CFR413	NO
063	INDIRECT	67,700	3,000	CFR433	NO	143	INDIRECT	40,000	20,000	NONSTD	NO
064	INDIRECT	10,759	10,000	CFR413	NO	144	INDIRECT	5,500	0	NONSTD	NO
065	INDIRECT	70,000	0	NONSTD	YES	145	INDIRECT	4,500	0	NONSTD	YES
066	INDIRECT	50,000	10,000	CFR433	NO	146	INDIRECT	7,200	2,000	NONSTD	NO
067	INDIRECT	2,200	200	NONSTD	NO	147	INDIRECT	600	0	NONSTD	NO
068	INDIRECT	4,300	9,000	NONSTD	NO	148	DIRECT	1,100	0	NONSTD	NO
069	INDIRECT	26,880	0	NONSTD	NO	149	INDIRECT	35,000	10,000	CFR433	NO
070	INDIRECT	32,000	0	NONSTD	NO	150	INDIRECT	100,000	400,000	CFR413	NO
071	ZERO	0	0	NONSTD	NO	151	DIRECT	10,000	0	NONSTD	NO
072	INDIRECT	115,000	√	NONSTD	NO	152	INDIRECT	32,500	0	NONSTD	YES
073	INDIRECT	6,000	0	CFR413	NO	153	INDIRECT	2,200	7,000	CFR433	YES
074	DIRECT	10,600	0	NONSTD	YES	154	INDIRECT	8,000	3,500	CFR413	YES
075	ZERO	0	100	NONSTD	YES	155	INDIRECT	10,000	14,000	NONSTD	NO
076	INDIRECT	50,000	0	NONSTD	NO	156	INDIRECT	16,800	5,000	NONSTD	NO
077	INDIRECT	55,000	0	NONSTD	NO	157	INDIRECT	3,000	1,000	CFR433	NO
078	ZERO	0	1,500	NONSTD	NO	158	INDIRECT	48,000	0	CFR413	NO
079	DIRECT	18,045	4,500	NONSTD	NO	159	INDIRECT	7,000	15,000	NONSTD	NO
080	ZERO	0	28,000	NONSTD	NO	160	INDIRECT	60,000	17,000	NONSTD	NO

Exhibit 2-8 (continued). Respondents' Wastewater and Discharge Characterization

Shop Code	Type of Discharge	Plating Discharge gpd	Flow Reduction gpd	Type of Standards	Aquatic?	Shop Code	Type of Discharge	Plating Discharge gpd	Flow Reduction gpd	Type of Standards	Acquatic?
161	INDIRECT	11,000	√	CFR413	NO	240	INDIRECT	1,000	0	CFR413	NO
162	DIRECT	44,000	0	CFR413	NO	241	INDIRECT	7,500	8,000	NONSTD	YES
163	INDIRECT	9,047	1,000	NONSTD	NO	242	INDIRECT	800	15,000	NONSTD	NO
164	INDIRECT	2,000	1,000	NONSTD	NO	243	INDIRECT	47,600	0	NONSTD	NO
165	INDIRECT	2,029	√	NONSTD	NO	244	INDIRECT	25,810	6,048	NONSTD	YES
166	INDIRECT	35,000	45,000	CFR413	NO	245	INDIRECT	24,400	41,000	CFR413	NO
167	INDIRECT	60,000	0	NONSTD	NO	246	INDIRECT	ND	√	CFR413	NO
168	INDIRECT	ND	40,000	NONSTD	NO	247	INDIRECT	109,000	0	CFR433	NO
169	INDIRECT	1,500	8,000	NONSTD	NO	248	DIRECT	75,000	0	NONSTD	NO
170	INDIRECT	12,500	√	NONSTD	NO	249	ZERO	0	2,000	NONSTD	NO
171	INDIRECT	12,000	3,500	NONSTD	NO	250	INDIRECT	11,000	80,000	NONSTD	NO
172	INDIRECT	70,000	80,000	CFR413	NO	251	DIRECT	40,000	√	CFR433	NO
173	INDIRECT	42,400	5,000	NONSTD	NO	252	INDIRECT	14,000	0	CFR413	NO
174	INDIRECT	26,000	15,000	NONSTD	YES	253	DIRECT	31,582	5,185	CFR433	NO
175	INDIRECT	8,000	40,000	NONSTD	NO	254	DIRECT	54,800	6,000	NONSTD	YES
176	INDIRECT	99,484	0	CFR413	NO	255	INDIRECT	3,000	2,000	NONSTD	NO
177	INDIRECT	9,163	3,000	NONSTD	NO	256	INDIRECT	140,000	36,000	NONSTD	NO
178	INDIRECT	12,000	0	NONSTD	NO	257	INDIRECT	2,000	10,000	NONSTD	NO
179	INDIRECT	7,930	14,000	NONSTD	NO	258	INDIRECT	170,000	0	CFR413	NO
180	INDIRECT	35,000	10,000	NONSTD	NO	259	INDIRECT	750	250	NONSTD	NO
181	INDIRECT	38,110	10,000	CFR413	NO	260	INDIRECT	6,589	0	NONSTD	YES
182	INDIRECT	20,000	20,000	CFR413	NO	261	DIRECT	20,000	√	NONSTD	NO
183	INDIRECT	9,000	500	NONSTD	NO	262	INDIRECT	30,000	20,000	NONSTD	NO
184	INDIRECT	11,000	110,000	CFR413	NO	263	INDIRECT	40,000	0	CFR413	NO
185	INDIRECT	700	5,000	NONSTD	NO	264	INDIRECT	77,900	20,000	CFR413	NO
186	INDIRECT	7,500	2,500	NONSTD	NO	265	INDIRECT	20,000	10,000	CFR413	NO
187	INDIRECT	3,000	25,000	CFR433	NO	266	INDIRECT	56,607	0	NONSTD	NO
188	INDIRECT	5,280	2,000	NONSTD	NO	267	INDIRECT	9,000	16,000	NONSTD	NO
189	INDIRECT	8,000	0	NONSTD	NO	268	INDIRECT	17,000	70,000	CFR413	NO
190	INDIRECT	18,000	0	NONSTD	NO	269	DIRECT	2,700	500	NONSTD	NO
191	INDIRECT	6,000	1,500	NONSTD	NO	270	ZERO	0	√	NONSTD	NO
192	INDIRECT	40,000	0	CFR413	NO	271	ZERO	0	0	NONSTD	NO
193	ZERO	0	√	NONSTD	NO	272	INDIRECT	35,000	√	NONSTD	NO
194	ZERO	0	√	NONSTD	NO	273	DIRECT	813,000	0	NONSTD	YES
195	ZERO	0	0	NONSTD	NO	274	INDIRECT	47,000	15,000	CFR433	NO
196	INDIRECT	90,800	20,000	CFR413	NO	275	INDIRECT	150,000	0	NONSTD	NO
197	INDIRECT	ND	√	CFR413	YES	276	INDIRECT	42,000	5,000	NONSTD	NO
198	ZERO	0	250	CFR413	NO	277	DIRECT	60,000	10,000	NONSTD	YES
199	INDIRECT	20,000	0	NONSTD	YES	278	INDIRECT	40,000	0	NONSTD	NO
200	INDIRECT	4,800	√	NONSTD	YES	279	INDIRECT	50,000	10,000	NONSTD	NO
201	INDIRECT	19,448	20,000	CFR413	NO	280	INDIRECT	142,000	960	NONSTD	NO
202	ZERO	0	1,100	NONSTD	NO	281	INDIRECT	73,000	30,000	NONSTD	NO
203	INDIRECT	22,368	14,000	CFR413	NO	282	ZERO	0	100	NONSTD	NO
204	INDIRECT	208,720	0	NONSTD	NO	283	INDIRECT	16,844	15,000	NONSTD	NO
205	INDIRECT	5,000	0	NONSTD	NO	284	DIRECT	600	0	NONSTD	YES
206	INDIRECT	9,000	5,000	NONSTD	NO	285	INDIRECT	8,000	0	CFR433	NO
207	INDIRECT	18,000	50,000	NONSTD	NO	286	INDIRECT	8,000	0	NONSTD	NO
208	ZERO	0	0	NONSTD	NO	287	DIRECT	50,000	0	NONSTD	YES
209	INDIRECT	11,500	20,000	NONSTD	NO	288	INDIRECT	15,000	10,000	NONSTD	NO
210	INDIRECT	20,000	0	NONSTD	NO	289	DIRECT	48,000	0	CFR433	YES
211	DIRECT	2,000	4,000	NONSTD	NO	290	DIRECT	700	0	NONSTD	NO
212	INDIRECT	5,400	√	CFR413	NO	291	DIRECT	350	150	NONSTD	YES
213	INDIRECT	50,000	80,000	CFR433	NO	292	INDIRECT	60,000	100,000	CFR433	NO
214	INDIRECT	72,000	0	NONSTD	NO	293	INDIRECT	70,000	1,100	NONSTD	YES
215	INDIRECT	49,000	20,000	CFR413	NO	294	ZERO	0	0	NONSTD	NO
216	INDIRECT	3,000	3,000	CFR413	NO	295	DIRECT	17,000	0	NONSTD	NO
217	DIRECT	130,000	0	CFR433	NO	296	BOTH	1,700,000	200,000	NONSTD	YES
218	INDIRECT	51,820	20,000	NONSTD	NO	297	ZERO	0	0	NONSTD	NO
219	INDIRECT	87,900	0	NONSTD	NO	298	INDIRECT	70,000	90,000	CFR413	NO
220	INDIRECT	17,420	0	NONSTD	NO	299	INDIRECT	1,400	0	CFR433	NO
221	INDIRECT	12,000	10,000	NONSTD	NO	300	INDIRECT	90,000	9,600	NONSTD	NO
222	INDIRECT	6,000	0	NONSTD	NO	301	DIRECT	50	0	NONSTD	NO
223	DIRECT	45,000	0	NONSTD	YES	302	INDIRECT	86,400	0	CFR413	YES
224	INDIRECT	6,000	3,000	NONSTD	NO	303	INDIRECT	37,000	15,000	CFR433	NO
225	INDIRECT	40,000	25	NONSTD	NO	304	INDIRECT	3,700	0	NONSTD	NO
226	INDIRECT	25,000	20,000	CFR413	NO	305	INDIRECT	11,000	0	CFR433	NO
227	INDIRECT	4,000	1,000	NONSTD	YES	306	INDIRECT	50,000	10,000	CFR413	NO
228	INDIRECT	40,000	3,000	CFR413	NO	307	DIRECT	58,606	0	CFR413	YES
229	DIRECT	45,000	20,000	CFR413	NO	308	INDIRECT	17,500	0	CFR433	NO
230	INDIRECT	15,000	7,500	CFR433	NO	309	INDIRECT	1,450	25,000	NONSTD	NO
231	INDIRECT	64,800	30,000	CFR413	YES	310	ZERO	0	5	NONSTD	NO
232	DIRECT	29,000	40,000	NONSTD	YES	311	INDIRECT	115,000	0	CFR413	NO
233	ZERO	0	600	NONSTD	NO	312	INDIRECT	1,200	200	CFR433	NO
234	INDIRECT	1,174	0	NONSTD	NO	313	INDIRECT	35,000	10	CFR413	NO
235	INDIRECT	10,650	40,000	CFR413	NO	314	DIRECT	40,000	12,000	NONSTD	NO
236	INDIRECT	56,000	0	NONSTD	YES	315	INDIRECT	75,000	0	CFR413	NO
237	INDIRECT	76,000	0	CFR413	NO	316	INDIRECT	360,000	40,000	CFR413	NO
238	INDIRECT	420,000	44,000	CFR413	NO	317	ZERO	0	0	CFR433	NO
239	INDIRECT	18,000	10,000	NONSTD	NO	318	INDIRECT	13,000	0	NONSTD	NO

ND = no data √ = flow reduction attained, but not quantified.

technologies, these have generally been much less frequently applied than drag-out and rinse water reduction efforts. The most successful of the chemical recovery technologies is atmospheric evaporation, which is also generally regarded as the least technically sophisticated recovery technology. Bath maintenance technologies are less frequently purchased than are chemical recovery technologies, but have generally been more successful. Exhibit 2-12 shows ratings given by the respondents for some common pollution prevention methods. The following subsections provide overviews of key pollution prevention topics.

2.2.4.1 Drag-Out and Rinse Water Reduction

For the typical electroplating job shop, the drag-out of process solutions and the subsequent contamination of rinse waters are the major pollution control problems. Section 3 contains an explanation of the basic principles of drag-out theory and explores the function and applicability of the various drag-out minimization techniques in use today. Because of the importance of drag-out and drag-out loss prevention, numerous questions in the Users Survey were related to this topic. The responses to these questions are statistically evaluated in Section 3 and summarized in this subsection.

The Users Survey asked respondents to indicate and rate the usefulness of the methods that they employ to reduce the formation or loss of drag-out and the usage rate of rinse water. A summary of their responses is presented in Sections 3.4 and 3.5. The most frequently used drag-out reduction methods are: allowing parts/racks to drip over process tanks; the use of drag-out rinses; reducing the speed of rack/part withdrawal; use of drip shields; and positioning the workpiece to minimize solution holdup. On the average, all of the drag-out reduction methods that are used by the respondents have been successfully applied. Some shops had specific problems with one or more methods (e.g., buildup of bath contaminants). These problems are discussed in Section 3 along with potential solutions.

The most frequently used methods of reducing water use involve the application of: flow restrictors; counterflow rinses; manually turning off water; and air agitation. As with drag-out methods, the rinse water reduction methods have been generally successful, with the highest success ratings given to the use of flow restrictors and counterflow rinsing.

2.2.4.2 Chemical Recovery Technologies

Chemical recovery technologies are used by platers to separate plating chemicals from rinse waters or other solutions or to concentrate them, thereby making them available for reuse/recycle. According the respondents of the Users Survey, chemical recovery technologies are most frequently purchased to (in order of frequency): help meet effluent regulations; reduce wastewater treatment costs; reduce plating chemical purchases; and reduce the quantity of waste shipped off-site.

The Users Survey requested platers to provide detailed technical, performance and operating cost data for chemical recovery technologies. Also, during their survey, vendors were requested to provide technology descriptions, operating data and capital cost data. As a result of obtaining data from these two sources, plus the information from the extensive literature review, this text contains a substantial quantity of information for the following chemical recovery technologies: electrodialysis, electrowinning, atmospheric evaporators, vacuum evaporators, ion exchange, reverse osmosis and meshpad mist eliminators. A separate subsection of the report is devoted to each of these technologies. Within each subsection, the following are provided: technology

Exhibit 2-9. Distribution of Respondents by Type of Discharge

Type of Discharge	Percentage of All Respondents with the Type of Discharge Indicated*	Percentage of Job Shop Respondents with the Type of Discharge Indicated*	Percentage of Captive Shop Respondents with the Type of Discharge Indicated*
Direct Discharge	12.6%	10.8%	22.0%
Indirect Discharge	78.9%	80.6%	70.0%
Both Direct andIndirect	0.9%	0.7%	2.0%
Zero Discharge	7.5%	7.8%	6.0%

*Based on 318 respondents, of which 268 are job shops and 50 are captive shops.

Exhibit 2-10. Average Plating Discharge Rate of Survey Respondents (gallons per day)

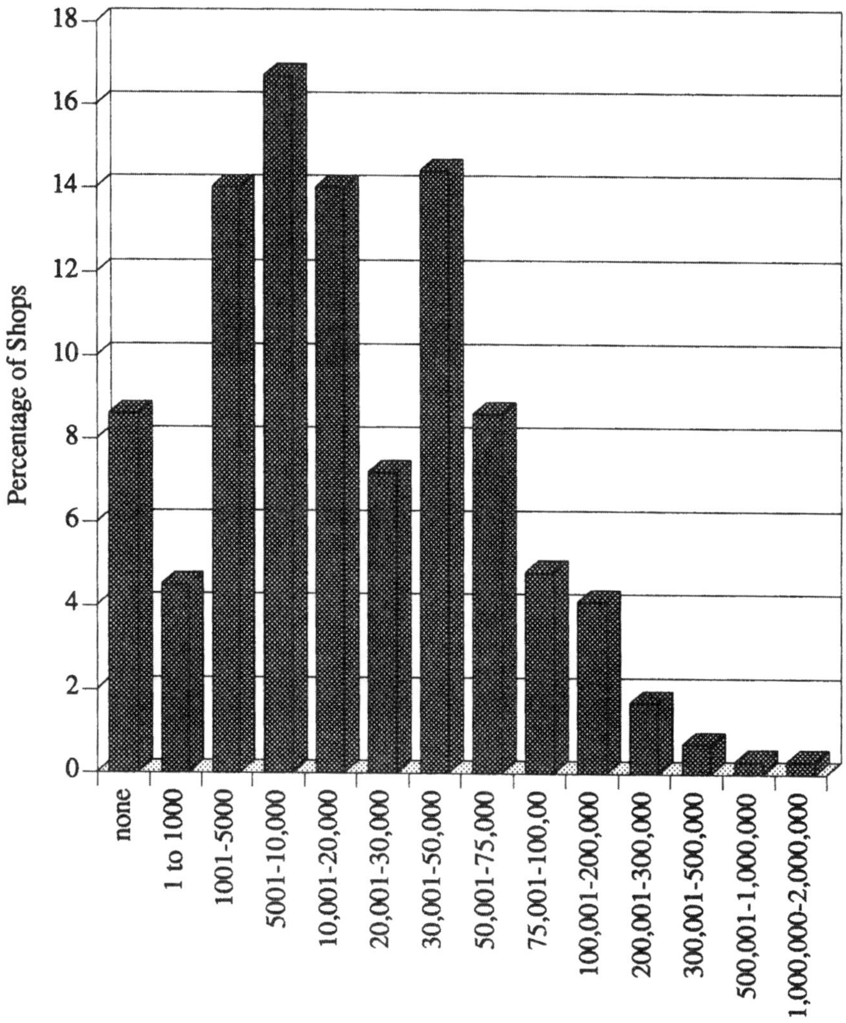

Exhibit 2-11.
Pollutant Parameters for Which Compliance Difficulty was Reported by Respondents

Pollutant Parameter	Percentage of Respondents Reporting Compliance Difficulty	Pollutant Parameter	Percentage of Respondents Reporting Compliance Diffculty
Nickel	19%	Lead	7%
Zinc	19%	Cyanide (amenable)	2%
Chromium (total)	17%	Chromium (+6)	2%
Copper	12%	Silver	2%
Cyanide (total)	12%	Total Toxic Organics	1%
Cadmium	1 0%		

Note: Some shops listed two or more parameters.

Exhibit 2-12. Ratings for Pollution Prevention Methods and Technologies

Method	Average Rating*
Good Operating Practices	3.9
Drag-Out Reduction (all methods)	3.5
Rinse Water Reduction (all methods)	3.8
Chemical Recovery:	
Atmospheric Evaporators	3.8
Vacuum Evaporators	3.5
Ion Exchange	3.2
Electrowinning (all)	3.1
Electrowinning (excluding high surface area)	3.4
Electrodialysis (one data point)	1.0
Reverse Osmosis	3.0
Weighted Avg. for Chemical Recovery	3.4
Bath Maintenance:	
Microfiltration	—
Ion Exchange	3.9
Acid Sorption	4.3
Ion Transfer	3.5
Membrane Electrolysis	3.0
Diffusion Dialysis	—
Weighted Avg. for Bath Maintenance	3.7

*Based on a scale of 1 to 5 with 5 being the most successful. The different methods are defined in Sections 3 through 5.

overview; development and commercialization; applications and restrictions (with diagrams showing different potential configurations); technology/equipment description; capital costs; operating costs; performance experience; and residuals generation.

Exhibit 2-13 presents a summary of the chemical recovery applications addressed by the Users Survey data.

Many installations of chemical recovery technologies and advanced bath maintenance (see Section 2.2.4.3) have not been successful (approximately 30 to 40 percent). The survey respondents indicate that failure is most frequently caused by: maintenance problems, misapplication of the technology (often due to ignorance on the part of manufacturers' representatives and/or the plating shop personnel), poor design, inability to purchase replacement parts (usually manufacturer went out of business), poor technical support by manufacturers, improper operation of technology by shop personnel, technically too complex for employees, chemical recovery caused a build-up of contaminants in plating bath, recovery process destroyed plating chemicals, recycled water was of insufficient quality, chemical product was insufficiently concentrated for return to plating bath, inadequate capacity, and high residuals generation.

Maintenance problems were the most frequent cause of system failure. The maintenance problems most often reported with advanced technologies are: low quality system components, mechanical problems with pumps and valves, damage to or fouling of components by plating chemicals, and excessive labor requirements for system cleaning.

Exhibit 2-14 indicates the operational status of the chemical recovery and bath maintenance technologies purchased by survey respondents.

2.2.4.3 Solution Maintenance Methods and Technologies

Metal finishing solutions are subjected to a variety of forces that cause them to become unusable. The key contributing factors are: (1) depletion of bath chemicals; (2) chemical break-down of process chemicals or chemical side reactions; (3) contamination from impurities in make-up water, chemicals or anodes; (4) anodic/cathodic etching of parts and inert electrodes; (5) corrosion of parts, racks, bussing, tanks, heating coils, etc.; (6) drag-in of non-compatible chemicals; (7) buildup of by-products (e.g., carbonates); (8) breakdown of maskant, fume suppressant and wetting agents; (9) errors in bath additions; and (10) airborne particles entering the tank.

Solution maintenance replaces the practices of: (1) using a fresh chemical solution until it is degraded and replacing it with fresh solution or (2) decanting a portion of a degraded solution and replacing it with fresh solution. In both cases, the spent solution is usually either treated on-site or transported to a treatment/disposal site. On-site treatment is not always possible because concentrated wastes may upset treatment facilities designed primarily for treating dilute rinse waters.

Two major categories of solution maintenance were identified during the project: preventative and corrective. Within this text, preventative solution maintenance refers to the practices that avoid bath contamination or involve monitoring and adjusting of solution chemistry. Corrective solution maintenance refers to the practice of removing contaminants from the bath, whether they are dissolved or particulate, organic or inorganic. Both preventative and corrective solution maintenance involve the use of methods, techniques and technologies. Meth-

Exhibit 2-13. Distribution and Ratings of Chemical Recovery Technologies

Process Name	ED[a] Rating (applications)	EW Rating (applications)	ATM EV Rating (applications)	VAC EV Rating (applications)	IX Rating (applications)	MP Rating (applications)	RO Rating (applications)
Anodize, Chromic Acid	-	-	-	-	ND (1)-	4.0 (2)	-
Anodize, Hardcoat	-	2.0 (1)	-	-	-	-	-
Brass	-	3.5 (2)	4.0 (2)	-	-	-	-
Cadmium, Cyanide	-	2.7 (29)	3.0 (4)	3.0 (3)	2.0 (4)	-	5.0 (1)
Cadmium, Non-cyanide	-	ND (1)	4.0 (1)	-	-	-	-
Chromate (Aluminum)	-	-	-	-	5.0 (1)	-	-
Chromate (Zinc)	-	-	-	-	ND (1)		
Chromium Etch	-	-	-	4.5 (2)	-	-	-
Chromium, Hard	-	-	4.3 (4)	-	2.0 (2)	4.1 (15)	-
Chromium, Decorative (Cr^{+6})	-	-	3.5 (11)	4.0 (5)	5.0 (3)	-	-
Chromium, Decorative (Cr^{+3})	-	-	5.0 (2)	-	4.0 (2)	4.0 (1)	-
Copper, Electroplating[b]	-	3.0 (15)	3.7 (4)	4 0 (1)	ND (3)	-	2.0 (1)
Gold, Electroplating[c]	-	4.0 (4)	-	-	4.0 (13)	-	-
Lead-Tin	-	-	4.0 (1)	-	-	-	-
Nickel, Electroplating[d]	1.0 (3)	4.0 (6)	4.1 (20)	4.0 (3)	3.8 (12)	-	1.0 (4)
Nickel, Electroless Plate	-	ND (1)	3.0 (2)	-	1.7 (3)	-	-
Seal, Nickel	-	-	-	-	-	-	4.0 (1)
Silver, Electroplate	-	4.4 (12)	2.0 (1)	-	-	-	-
Tin, Acid		ND (1)					
Zinc, Cyanide	-	1.7 (7)	3.0 (2)	1.7 (3)	ND (1)	-	1.0 (1)
Zinc, Non-cyanide[e]	-	2.0 (7)	3.7 (9)	-	2.6 (7)	-	5.0 (1)
Zincate	-	-	4.0 (1)	-	-	-	-

[a] Technology key: ED=electrodialysis, EW=electrowinning, ATM EV=atmospheric evaporation, VAC EV=vacuum evaporation, IX=ion exchange, MP=meshpad mist eliminator, RO=reverse osmosis.

Ratings are based on a scale of 1 to 5, with 5 being the most successful. Number of applications from the Users Survey is shown in parenthesis. "-" = no applications, ND = no data. Not all of the shops that indicated they have applications rated the technologies. Therefore, the ratings may be based on fewer data points than the number of applications given. There were more applications of atmospheric and vacuum evapora~ors than indicated. In some cases it could not be determined if the units were atmospheric or vacuum type units. The data were ignored for these shops (34 data points) in prepar~ng this table.

[b] Includes cyanide plate, cyanide strike and sulfate baths.

[c] Includes cyanide and non-cyanide baths.

[d] Includes bright, sulfamate, sulfate, Watts and Woods baths.

[e] Includes acid and alkaline baths.

Exhibit 2-14. Operational Status of the Chemical Recovery and Bath Maintenance Technologies Purchased by Survey Respondents

Technology	% of Units Purchased that are Still Operating	Average Age of Operating Units (years)	Average Age of Non-Operating Units (years)	Age of Oldest Operating Unit (years)
Recovery:				
Atm Evaporation	90%	5 4	6.1	18
Vacuum Evaporation	74%	8.3	13.8	15
Ion Exchange	61%	6.5	5.4	18
Electrowinning	59%	5.5*	6.9	12
Electrodialysis	0%	—	9.0	—
Reverse Osmosis	50%	4.0	9.3	6
Bath Maintenance:				
Microfiltration	50%	4.4	1.0	—
Ion Exchange	73%	9.0	1.0	18
Acid Sorption	100%	4.9	—	15
Ion Transfer	70%	5.0	7.7	7
Membrane Electrolysis	60%	—	4.0	9
Diffusion Dialysis	—	—	—	—

ods and techniques are typically procedural in nature or low capital items that can be implemented quickly and have an almost immediate payback. Technologies are generally equipment packages that have a moderate to high capital cost and payback periods of one year or greater. Most preventative measures are either methods or techniques. Corrective measures include both methods/techniques such as dummy plating and technologies such as microfiltration.

Within this text, the corrective technologies, which are generally less familiar to platers, are covered in detail. The methods of preventative and corrective solution maintenance that are commonly applied by plating shops (e.g., filtration) are more familiar to platers and therefore are covered less extensively. The bath maintenance technologies addressed by the project include: microfiltration, ion exchange, acid sorption, ion transfer, membrane electrolysis and diffusion dialysis. Exhibit 2-15 presents a summary of the corrective technology use by respondents to the Users Survey.

2.2.4.4 Substitute Technologies

The results of the Users Survey show that respondents have made significant strides in reducing or eliminating the use of chlorinated solvents, cadmium, cyanide and chromium. Sometimes referred to as the four Cs, these materials have been identified by EPA as key targets for control within the metal finishing industry. Approximately 60 percent of the respondents attempted material input changes that potentially reduce or eliminate the use of one or more of the four Cs or another pollutant problem. Based on the comments received from respondents, these changes were made in an effort to reduce the impacts of their processes on the environment and worker health, to help meet environmental regulations and to reduce operating costs.

Although most of the material input changes attempted by survey respondents have been successful, there have been some failures and in many cases, even with successful changes, there have been adverse production impacts. Section 6 summarizes the status of change in these areas and conveys the attitudes and concerns of the respondents.

2.2.5 Wastewater Treatment/Sludge Management

The Federal electroplating and metal finishing pretreatment wastewater standards were developed by EPA by identifying commonly used treatment practices, and determining their effectiveness by collecting effluent data from well-operated systems. Conventional treatment (a series of unit processes used extensively by industry that have provided reliable treatment for many electroplating operations) was selected by EPA as the standard system. Therefore, for most plating shops, use of conventional treatment will provide sufficient pollutant removal to meet discharge standards. There are two major exceptions to this rule. First, many plating shops are regulated by local discharge standards that are more stringent than the Federal standards and conventional treatment may be

Exhibit 2-15. Distribution and Ratings of Bath Maintenance Technologies

Process Name	MF[a]	IX	AS	IT	ME	DD
Acid Clean	-	-	ND (2)	-	-	-
Alkaline Clean	ND (1)	-	-	-	-	-
Anodize, Chromic Acid	-	3.0 (2)	-	-	-	-
Anodize, Hard coat	-	-	4.0 (1)	-	-	-
Anodize, Sulfuric Acid			4.5 (3)			
Chromium, Hard	-	4.3 (7)	-	3.4 (12)	3.0 (3)	-
Chromium, Decorative (Cr^{+6})	-	-	-	4.0 (3)	-	-
Chromium, Decorative (Cr^{+3})	-	5.0 (1)	-	-	ND(1)	-
Chromium Etch	-	-	-	-	ND (2)	-
Chromium Strip, (NaOH)				4.0 (1)		
Copper Strip, (CrO_3)		2.0 (1)				
Nickel, Electroplating[b]	-	ND (1)	-	-	-	-

[a] Technology key: MF=microfiltration, IX=ion exchange, AR=acid sorbtion, IT=ion transfer, ME=membrane electrolysis, and DD=diffusion dialysis. Ratings are based on a scale of I to 5, with 5 being the most successful. Number of applications from the Users Survey is shown in parenthesis. "-" = no applications, ND = no data.

[b] Includes bright, sulfamate, sulfate, Watts and Woods baths.

insufficient to meet these limitations. Second, the treatment systems selected by EPA for establishing the Federal standards were those systems that EPA determined to be "properly operating facilities." For example, EPA omitted facilities that: (1) did not have well operated treatment processes; (2) had complexing agents (e.g., non-segregated wastes from electroless plating); and (3) had dilution from non-plating wastewaters. As a result, some plating facilities may not meet the properly operated facility criteria used by EPA and may have difficulty meeting Federal standards using conventional treatment.

In cases where conventional treatment is insufficient to meet discharge limitations for a given facility, there are three basic choices for attaining compliance: (1) correct or upgrade the existing processes; (2) make internal changes (e.g., improve rinsing, add recovery, segregation of waste streams) to "normalize" the wastewater, (3) use conventional treatment plus additional treatment (i.e., polishing), and (4) use alternative treatment reagents and/or processes. Information on each of these methods is covered in this text.

Pretreatment standards for the electroplating industry were first established in 1974, but it was not until promulgation of 40 CFR 413 on September 7, 1979 that Electroplating Categorical Pretreatment Standards became a reality. Several years later, EPA promulgated the Metal Finishing Categorical Standards (40 CFR 433). Prior to the existence of Federal standards, plating shops were regulated locally (if at all), presumably with wide variation in effluent limitations and levels of enforcement. Most plating shops did not have treatment systems for cyanide destruction and metals removal. Approximately 12 percent of the surveyed plating shops that were in business in 1975 (excludes zero discharge shops) indicated that their initial treatment system was installed by that year. Exhibit 2-16 presents a breakdown of the data in five year increments. These data indicate that by 1985, after the compliance dates for Federal regulations, 70 percent of the surveyed plating shops had installed their initial treatment system (excludes zero discharge shops).

Most plating shops installed conventional treatment to meet Federal regulations. Although it is difficult to assess exactly the respondents' data concerning end-of-pipe technology, it appears that an early trend occurred during the late 1970's and early 1980's when a significant percentage of shops attempted to utilize advanced technology in place of conventional treatment. These early efforts generally resulted in failure and the shops later resorted to conventional systems. One prominent

Exhibit 2-16. Distribution of End-of-Pipe System Installation Dates

Initial Treatment System Installed By:	Percent of Shops*
1975	16%
1980	32%
1985	70%
1990	95%
1993	98%

*Percent of shops in business by the date indicated with operating treatment systems. Excludes zero discharge shops.

example of this trend is the implementation of high surface area electrowinning as an end-of-pipe technology. Between 1979 and 1983, approximately 4 percent of the shops (excludes zero discharge shops) in existence installed this technology at an average cost of $66,360. Only one of these systems is currently operating and that unit was extensively modified by its user. Early failures such as these appear to have had a negative impact on advanced technology. No single technology has since emerged as a significant replacement for conventional treatment. In fact, changes in end-of-pipe methods have tended toward simpler technologies. This conclusion is supported by the fact that the most significant technology change with respect to end-of-pipe treatment since 1975 is the use of sludge dehydration equipment (i.e., sludge dryers) to reduce the volume of sludge shipped off-site (29 percent of the respondents have installed this relatively simple technology with approximately 80% purchased between 1988 and 1993). Approximately 10% of all the shops surveyed presently rely on non-conventional treatment methods (includes zero discharge shops). The most popular non-conventional end-of-pipe treatment methods (ion exchange, evaporation, and membrane technology) are addressed in Section 7.

It should be noted that the majority of respondents to the Users Survey were job shops. More frequent use of advanced end-of-pipe technology may exist in other industry segments such as captive aerospace facilities.

Approximately 8% of the shops surveyed have attained zero discharge (24 shops). These shops are generally smaller and less diverse than the shops with discharges. The maximum, average and median number of employees at the zero discharge shops is 30, 16 and 15, respectively (for all shops responding to the survey, the employee figures are: maximum = 3,000, average = 67 and median = 35). Of the zero discharge shops, 58% are primarily hard chrome platers. The hard chrome process is one of the easiest to operate as a close-loop because of the high ratio of evaporation to drag-out (i.e., permits use of spray rinsing over the bath, drag-out recovery rinsing, etc.). The remaining zero discharge shops operate various metal finishing processes, including: cadmium, nickel and zinc plating; conversion coating; and aluminum finishing. Details of their metal finishing processes and pollution prevention and control technologies are contained in the NCMS database and summarized in this text.

One of the most frequent concerns of platers is the availability and cost of disposal for treatment process residuals (mainly F006 sludge). Of the respondents that generate F006 sludge, the average and median generation rates are 158,272 lbs/yr and 50,000 lbs/yr, respectively (excludes shops with zero generation rates and PS 273 which generates a large volume of dilute sludge). These shops spend an average of $0.52/lb for sludge disposal (includes, where applicable: transportation, stabilization, disposal or recycle, and taxes) (median cost is $0.25/lb). The unit costs for sludge disposal vary widely, depending mostly on the annual quantity of sludge generated. For example, shops that generate less than 10,000 lbs/yr pay an average rate of $1.58/lb, while shops generating more than 1 million pounds per year pay an average of $0.05/lb.

Discussions of sludge generation rates and costs are presented in Sections 7 and 8. Section 7 provides data from each respondent covering sludge generation rates, the location of their disposal site, the distance that sludges are hauled, the solids concentration of the sludge, and the disposal charges. Many platers (33% of the respondents) are using off-site metals recyclers as an alternative to land disposal of their treatment residuals and spent process solutions. Section 8 identifies the recycling companies used by the respondents, presents an overview of their recovery processes (provided by the recycling companies themselves), presents criteria for determining the applicability of off-site recycling, and compares the costs of recycling to land disposal.

2.2.6 Miscellaneous Survey Data

Exhibit 2-17 presents a summary of data that helps to identify the environmental problems and needs of the survey respondents. For each topic shown, the questionnaire listed potential answers that were checked-off by the respondents. The following observations were made from reviewing these data:

- The most pressing environmental problems cited by the respondents were the increasing costs of compliance, frequently changing regulations and meeting effluent discharge standards.

- Areas where the respondents indicated that they need additional technology transfer are more frequently process related (e.g., non-cyanide finishing) than technology related (e.g., electrodialysis).

- The most common types of environmentally related technical resources available to the respondents are an in-house process engineer or an in-house environmental engineer. Due to the fact that most of the respondents were job shops, as opposed to large manufacturing plants, very few of the responding shops had multiple technical personnel available.

- A broad range of technical resources are used by the respondents for obtaining information on pollution prevention. The most common resources are professional journals, vendors, conferences or conference proceedings, books and consultants.

- When asked to rate information dissemination methods for the results of this project that they would potentially use, the respondents rated the printed report the highest.

- The average number of man-hours that the respondents spent gathering data and completing the survey questionnaire was 5.8 hours.

The survey questionnaire also asked platers to list environmental technology needs that are most important to their company. No checklist was present in the survey form for this question and therefore a broader range of responses was received than for similar questions. All of the responses are contained in the database; the following is a list of the needs that at least two or more platers offered (the number of shops that listed each of these needs is shown in parenthesis).

- Alternatives to solvent cleaning/degreasing (37 shops)

- Better cyanide plating alternatives or controls (35 shops)

Exhibit 2-17. Identification of Facility Problems and Needs

Most Pressing Environmental Problems Facing Your Company

Problem	No. of Responses
Lack of disposal sites	57
Meeting eMuent discharge limits	122
Eliminating use of solvents	76
Frequently changing regulations	174
Meeting air emission standards	74
Increasing costs for compliance	231

Technology Transfer Insufficiency

Information Need	No. of Responses
Ion Exchange	41
Reverse Osmosis	32
Electrodialysis	31
Ultrafiltration	39
Diffusion Dialysis	27
Non-Chromium Finishing	62
Cadmium Plating Alternatives	53
Non-Cyanide Finishing	82
Non-chlorinated Solvent Cleaning	76
Evaporation	32

Available Technical Resources

Tvpe of Resource	No. of Responses
In-House Process Engineering Dept.	15
Corporate Process Engineering Dept.	3
In-House Process Engineer	126
In-House Environmental Engineering Dept.	5
Corporate Environmental Engineering Dept.	5
In-House Environmental Engineer	103
Other Technical Resource	5

Information Sources Most Frequently Used

Tvpe of Source	No. of Responses
In-House Engineer	64
In-House Chemist	72
In-House, Other	14
Consultant	164
Vendor	203
Books	163
Professional Journals	211
Other Literature	95
Attend Conferences or Read Proceedings	171
Trade Associations	21

Rating of Information Dissemination Results

Method	No 5's	No-4's	No-3's	No.2's	No.l's	No.0's	Avg.
Printed Report	236	30	15	6	9	21	4.6
On-Line System	10	14	38	40	96	119	2.0
On-Disk System	25	40	44	36	60	112	2.7
Workshop	42	71	52	42	29	81	3.2
Other	4	8	4	1	13	287	2.6

Based on a scale of 1 to 5, with 1 being the least likely method to be used and 5 being the most likely. Average Man-Hours to Complete NCMS/NAMF Survey 5.8 hrs.

- Improved methods for water reduction, closed-loop processing, source reduction, recycling or zero discharge (28 shops)
- Better cadmium plating alternatives or controls (24 shops)
- Improved methods or more affordable end-of-pipe treatment (21 shops)
- Alternative to chromium metal finishing (includes all uses such as anodizing, plating and conversion coating) (20 shops)
- Affordable options for F006 sludge recycle or disposal (8 shops)
- Alternatives to or recycling technologies for nickel (6 shops)
- Improvements to or alternatives for zinc plating (5 shops)
- Improved methods of bath maintenance (5 shops)
- Chromium recovery technologies (4 shops)
- Cost effective treatment or recovery for zinc (4 shops)
- Better and/or more affordable air emissions control technologies (4 shops)
- Affordable and/or small membrane filtration or chemical recovery systems (3 shops)
- Electroless nickel baths with long life span (2 shops)

Exhibit 2-18 provides a summary of the responses given by survey respondents when asked to list aspects of pollution prevention that the Users Survey failed to cover. Most frequently, their responses were directed at regulatory and enforcement issues. Others topics listed by respondents that were particularly interesting include: air emissions and controls, stormwater runoff, hazardous waste not related to wastewater, printed circuit board operations, painting operations, analytical and legal costs, pollution prevention effects on plating quality, and training requirements. Some of these topics will be included in updated versions of the Users Survey.

2.3 FUTURE SURVEY EFFORTS OF THE NATIONAL CENTER FOR MANUFACTURING SCIENCES

The initial NCMS project activities have established a benchmark assessment of pollution prevention and control technology for plating operations. The project products will assist platers in various ways, including: sharing ideas for drag-out and water use reduction; providing useful technology descriptions; providing an explanation of EPA's pollution prevention concept; summarizing cost and performance data from actual technology installations involving chemical recovery, bath maintenance and waste treatment; summarizing platers' experiences with alternative process chemicals that may reduce the hazardousness of the plating processes and the resultant wastes; and providing detailed data for off-site metals recycling options. This assessment has been made possible by the efforts of numerous platers, technology vendors and other interested parties. Hopefully, this industry will continue these efforts by participating in updates of the assessment. Each iteration of the assessment process will refine our technical knowledge and lead to more cost effective means of complying with environmental regulations.

For additional information about this program contact the NCMS Project Manager.[2]

[1] The database is a publicly available product of this project. For information contact CAI Resources (434/286-7781).
[2] Paul Chalmer: NCMS; 3025 Boardwalk Drive, Ann Arbor, MI: 48108-3266; 313/995-4911.

Exhibit 2-18. Aspects of Pollution Prevention and Control that the Survey Failed to Cover

I.D. Code	Section 7 Datafile Memorandum Notation
015	Evaluation of the commitment of management to pollution prevention and control.
019	Conflicts between state and Federal regulations
040	Unequal enforcement of regulations.
041	Negative impacts of regulatory enforcement drains resources and discourages pollution control improvements with risks.
042	Aspects of interaction with local authorities on reasonableness of pollution control (effluent) guidelines.
045	Does not account for differences in plating shops.

continued on the next page ...

Exhibit 2-18. Aspects of Pollution Prevention and Control that the Survey Failed to Cover

I.D. Code	Section 7 Datafile Memorandum Notation
055	Forced changes in treatment; real costs of compliance.
062	Detail on effect of training and good people; pollution effect of redoing reject parts; and product life.
064	Regulatory aspects-conflicting regulations-too many agencies, each with its own requirements and laws.
071	Disposal of materials other than liquids and sludges, i.e. tapes, plastics, rags, etc.
088	Wastewater treatment effluent analysis and cost; assistance from POTWs.
093	The devastating effects that lower discharge limits would have on companies.
061	The government regulations.
104	Analytical and legal costs
106	Monitoring methods and frequency by sanitary district.
113	The adversarial methods being employed by government agencies and the untrue environmental hype driving legislation.
116	Pollution prevention aimed at manufacturing and large business; metal removers can't use pollution prevention, and personal loads of regulation.
120	State does not enforce regulations on non-treat equip. shops, it only fines them. This puts compliance shops at disadvantage.
132	Cost associated with the operation of pretreatment systems.
146	Realistic cost/benefits analysis. Cost of systems versus value to the environment or the country in general.
149	Interest in scheduled publication informing plating shops of new or used technologies and rules for pollution prevention.
152	Pollution control problems with painting and solvents.
154	Amount and frequency of regulations and the agencies inability to provide necessary compliance information.
162	Runaway bureaucracy and too many government departments need too many reports containing same information.
164	Storm water regulations and cost. Also treatment system analytical costs.
186	Cost and effort to satisfy pollution control regulations.
188	Air quality and storm water runoff.
214	The amount of time needed to complete all necessary Federal and state reports and surveys.
222	Non-electroplating waste removal systems.
130	Air quality.
235	Training key personnel.
245	The competitive reality of inequities of regulatory compliance, Non-compliant facilities s/b eliminated.
250	Hazardous waste generation not related to water discharge.
257	Printed wire board processes such as etching, photochemical milling and resist processing solutions.
258	Chemical container disposal methods, i.e. drums, buckets and bags.
273	Air pollution control.
294	Air pollution prevention.
298	Availability of information regarding closed-loop recovery and waste disposal of various materials.
300	Enforcement of existing regs. and variable new entry requirments into industry.
316	Management commitment to pollution prevention

Section 3
General Waste Reduction Practices

3.1 INTRODUCTION

Section 3 covers general waste reduction practices. The implementation of these practices is considered the first step in an effective pollution control program. These methods are relatively inexpensive to implement and they reduce the need for more expensive recovery, recycle and treatment technologies. The effective implementation of waste reduction practices requires an organized and efficient approach; otherwise, the effects will be isolated within the shop and will have little impact on overall recovery or compliance goals. Within the past fifteen years, a significant effort has been made by EPA and industry to establish guidelines for the effective implementation of waste reduction practices. Some key elements of those guidelines are presented in Section 3.2. A large number of potential waste reduction practices have been identified from the literature search. These practices, which have been categorized into three groups, are reviewed in Sections 3.3, 3.4 and 3.5 and summary tables are presented at the end of this section. The three groups of waste reduction practices are:

- Good Operating Practices
- Drag-Out Loss Reduction
- Rinse Water Reduction

3.2 POLLUTION PREVENTION AND SOURCE CONTROL

Various terms are used to group or categorize the different methods employed by industry in their pollution control strategies, e.g., pollution prevention, source control, waste minimization, and in-plant changes. One popular term used extensively by EPA is pollution prevention. This term (see definition in Section 3.2.1) covers a broader range of pollution control methods than is presented in Section 3. Other aspects of pollution prevention such as chemical recovery, bath maintenance and process chemistry changes are covered in Sections 4 through 6. Pollution prevention is a concept that includes not only the technical methods for reducing waste generation, but also the organizational and administrative tools for implementation. To familiarize the reader with the pollution prevention approach, key aspects of this concept are presented in the following subsections.

3.2.1 Definitions

Pollution prevention, as defined by EPA, is the maximum feasible reduction of all wastes (wastewater, solid waste and air emissions) generated at production sites. It involves the judicious use of resources through source reduction, energy efficiency, reuse of input materials during production and reduced water consumption. Source reduction is the highest priority in the U.S. EPA's Environmental Management Options Hierarchy (Exhibit 3-1). It is defined as the use of materials, processes, or practices that reduce or eliminate the creation of pollutants or wastes at the source (ref. 26, 302).

There are two general methods of source reduction that can be used in a pollution prevention program: product changes and process changes. Product changes are changes in the composition or use of the intermediate or end products that are performed by the manufacturer with the purpose of reducing waste from manufacture, use, or ultimate disposal of the products. This method of pollution prevention is addressed in Section 6. Process changes are manufacturing modifications that impact

Exhibit 3-1. U.S. EPA's Environmental Management Options Hierarchy

Method	Example Activities	Example Applications
Source Reduction (Highest Priority)	• Environmentally friendly design of new products • Product changes • Source elimination	• Modify product to avoid solvent use • Modify product to extend coating life
Recycling	• Reuse • Reclamation	• Solvent recycling • Metal recovery from a spent plating bath • Volatile organic recovery
Treatment	• Stabilization • Neutralization • Precipitation • Evaporation • Incineration • Scrubbing	• Thermal destruction of organic solvent • Precipitation of heavy metal from a spent plating bath
Disposal	• Disposal at a permitted facility	• Land disposal

Source: ref 25

how the product is produced. These methods of pollution prevention are discussed in this section and in Section 6. Process changes include: input material changes, technology changes and improved operating practices. Typically, improved operating practices can be implemented more quickly and at less expense than input material and technology changes. Examples of these source reduction options are shown in Exhibit 3-2.

There are a number of pollution control measures that are applied only after wastes are generated. They are, therefore, not correctly categorized as pollution prevention. Exhibit 3-3 provides some examples of procedures that are waste handling, not source reduction measures. Because these procedures are classified as waste handling, their implementation may not provide pollution prevention credits under certain state or Federal statutes. However, their implementation may be beneficial both in terms of environmental protection and cost reduction and therefore should not be overlooked.

Facilities are required to have pollution prevention programs for wastes classified as hazardous. The applicable Federal regulations include:

- The Resource Conservation and Recovery Act (RCRA) requires that facilities with permits for treatment, storage, or disposal implement a program to reduce the volume or quantity and toxicity of these wastes.

- The Pollution Prevention Act (PPA), specifies that facilities required to report releases to the U.S. EPA for the Toxic Release Inventory (TRI)

Exhibit 3-2. Examples of Source Reduction Methods

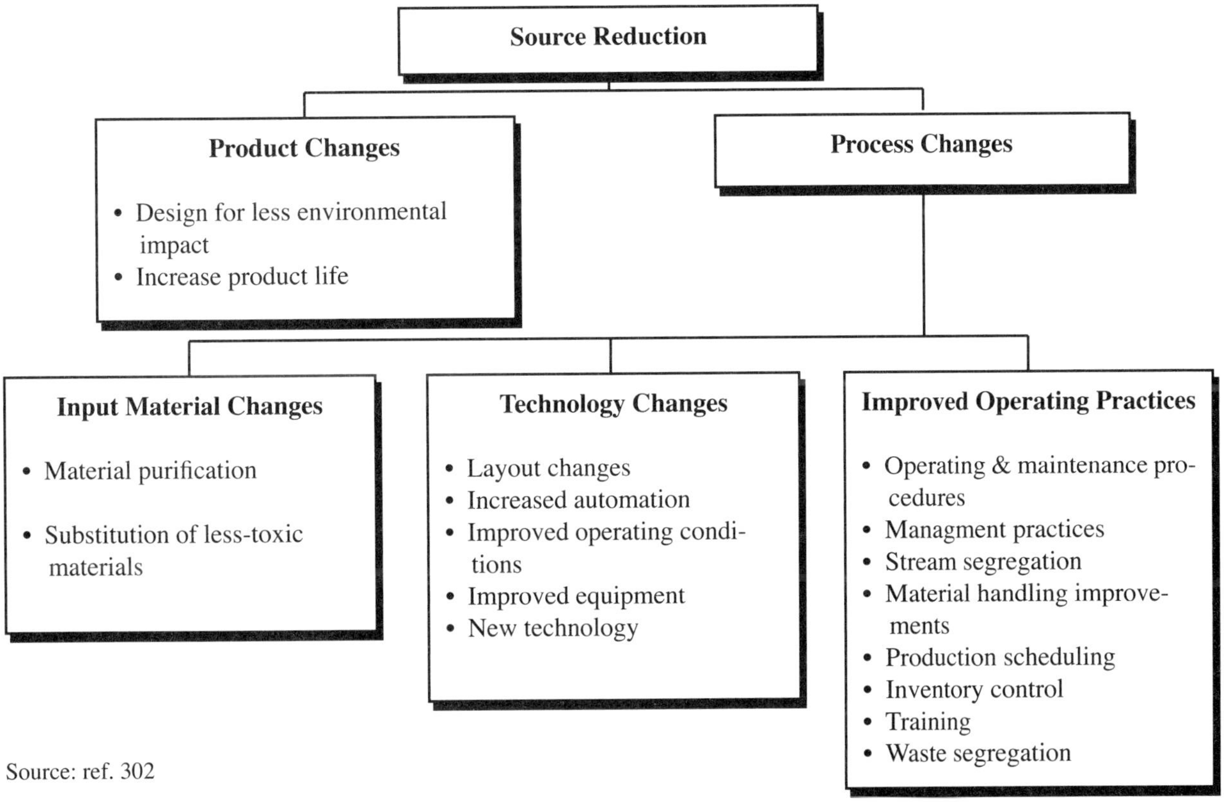

Source: ref. 302

Exhibit 3-3. Waste Handling Methods That Are Not Considered Pollution Prevention or Source Reductions Options by the EPA

The following are, according to EPA, not pollution prevention measures because they are taken after the waste is created:

- **Off-site recycling:**
  Off-site recyling (e.g., sending ion exchange conisters to centralized waste treatment facility or F006 sludge to a metals reclamation facility) is an excellent waste management option. However, it can create pollution during the recycling procedure.
- **Waste treatment:**
  Waste treatment involves changing the form or composition of a waste stream through controlled reactions to reduce or eliminate the amount of pollutant. Examples include: metals precipitation; cyanide oxidation; and stabilization and solidification or encapsulation of sludge.
- **Concentrating hazardous or toxic constituents to reduce volume:**
  Volume reduction operations, such as dewatering, are useful treatment approaches, but they do not prevent the creation of pollutants. For example, pressure filtration and drying of F006 sludge prior to disposal decreases the sludge water content and waste volume, but it does not decrease the mass of heavy metals in the sludge.
- **Diluting constituents to reduce hazard toxicity:**
  Dilution is applied to a waste stream after generation and does not reduce the absolute amount of hazardous constituents entering the environment.
- **Transferring hazardous or toxic constituents from one environmental medium to another:**
  Many waste management, treatment, and control practices used to date have simply collected pollutants and moved them from one environmental medium (air, water, or land) to another. An example is scrubbing to remove chromic acid from a ventilation system.

Source: adapted from ref. 302

Exhibit 3-4. State Legislation Promoting Pollution Prevention as of January 2003

State	Legislation
Alaska	Solid and Hazardous Waste Management Act
Arizona	Amendments to Arizona Hazardous Waste Management Statutes
California	Hazardous Waste Reduction and Management Review Act
Connecticut	Environmental Assistance to Business Act
Delaware	Waste Minimization/Pollution Prevention Act
Florida	Pollution Prevention Act
Georgia	Amendment to Hazardous Waste Management Act
Illinois	Toxic Pollution Prevention Act
Indiana	Amendment to Environmental Code
Iowa	Toxic Pollution Prevention Act
Kentucky	(no title)
Louisiana	Waste Reduction Law
Maine	Reduction of Toxics Use, Waste and Release Act
Massachusetts	Toxic Use Reduction Act
Minnesota	Toxic Pollution Prevention Act
Mississippi	Comprehensive Multimedia Waste Minimization Act
New Jersey	Pollution Prevention Act
New York	Hazardous Waste Management Act
North Carolina	Hazardous Waste Management Act
Oregon	Toxic Use Reduction and Hazardous Waste Reduction Act
Rhode Island	Hazardous Waste Facility Planning Act
Tennessee	Hazardous Waste Reduction Act
Texas	Waste Reduction Policy Act
Vermont	Hazardous Waste Management Act
Washington	Hazardous Waste and Substance Reduction Act
Wisconsin	Hazardous Substances, Toxic Pollutants, Hazardous Waste Use and Release Reduction

provide documentation of their procedures for preventing the release of these materials or for reusing these materials. The PPA encourages that the preferred method of preventing pollution is to reduce the volume of waste generated at the source and to perform reuse (closed-loop recycling) whenever possible.

- The Comprehensive Environmental Response, Compensation, and Liability Act (CERCLA), requires generators of hazardous waste to evaluate and document their operations.

In addition to Federal regulations covering this topic, many states have enacted legislation that requires pollution prevention. As of January 2003, a total of 26 states had passed such legislation. These states are listed in Exhibit 3-4.

Source reduction methods have been well documented and utilized within the metal finishing industry for more than 20 years to: (1) reduce input material use, waste generation, energy use, and production costs; (2) minimize worker exposure to hazardous chemicals; and (3) improve product quality (ref. 1).

EPA has developed a systematic procedure for identifying, evaluating and implementing source reduction and other pollution control methods (Exhibit 3-5), which is termed the Pollution Prevention Assessment Procedure. Published materials are available that specifically relate this methodology to the metal finishing industry (ref. 25, 378).

3.2.2 Pollution Prevention Programs

This section presents an approach for establishing a pollution prevention program and initiating the assessment and implementation of pollution prevention measures. The overall approach includes an organizational frame-

Exhibit 3-5. EPA's Pollution Prevention Assessment Procedure

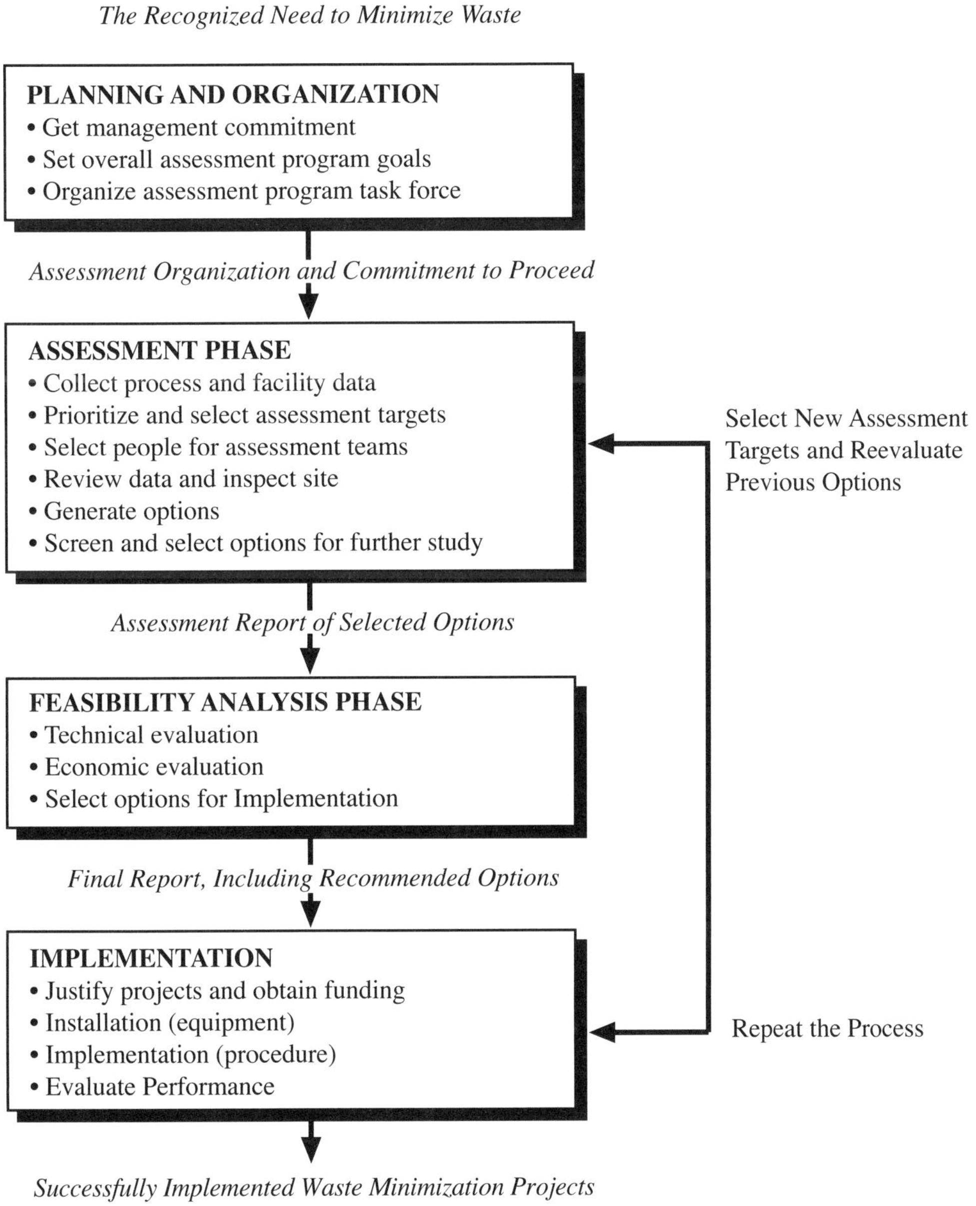

work needed for creating a pollution prevention program, guidelines for pollution prevention planning and goal setting, procedures for collecting and organizing information and data, techniques for evaluating and ranking pollution prevention options and methods for implementing, monitoring and assessing program elements.

Portions of the approach are extracted from EPA's waste minimization strategy (ref. 26), which is presented in Exhibit 3-5. Also included are successful elements of existing corporate pollution prevention programs and field proven techniques which have been identified from the NCMS Survey.

The User Survey (see description in Section 2), asked platers if their shops had established a formal pollution prevention program. A summary of responses to this question and others relating to pollution prevention programs and good operating practices is presented in Exhibit 3-6. Of the 318 plating shops responding to the survey, 161 (or 50.6%) indicated that they have established a formal pollution prevention program. The average success rating given by respondents for this pollution prevention method was 3.69. This rating is slightly lower than the average rating for all good operating practices. It should be noted that, although its rating is below the average for all methods, it still received a rating in the upper one-half of the rating scale (see description of rating scale on Exhibit 3-6), as did all of the methods.

3.2.2.1 Pollution Program Organization

The regulations and incentives which create a need for pollution prevention were identified in Section 3.2.1. To meet these needs effectively, small companies, corporations and Federal facilities must initially develop a well defined pollution prevention policy and establish a pollution prevention organization or task force with the committed fiscal and administrative authority. Without top management support and a well defined organization with functional and financial capabilities, a pollution prevention program will be short lived and will not have a substantial impact on waste generation. Some guidelines for pollution prevention policy development and task force organization are presented in the following subsections.

3.2.2.2 Pollution Prevention Policy

A pollution prevention policy is a statement which reflects the general goals of the program. The following example is a pollution prevention mission statement used by a major aerospace corporation (Source: confidential contribution).

> *"Our goal is to implement an effective pollution prevention program to eliminate liquid, solid and gaseous waste generation. Program priorities shall emphasize the protection of employees, community and the environment while maintaining product quality and business objectives."*

Both the pollution prevention policy and the task force should be established in a manner that cuts across the entire organization, involving all employees and especially the workers with direct contact with waste generation. Pollution prevention will not be effectively achieved without the acceptance and cooperation of the operators. Often, operators cause waste and they are in the best position to prevent it.

Policy statements can be written to communicate clearly that pollution prevention is an important measure of

Exhibit 3-6. Summary of Users Survey Data Relating to Good Operating Practices

Good Operating Practice Method	Number of Shops Using Method[1]	Percentage of Shops Using Method[1]	Success Rating[2]
1. Establish a formal policy for pollution prevention	158	49.7	3.58
2. Establish a formal pollution prevention program	161	50.6	3.69
3. Conduct employee education for pollution prevention	217	68.2	3.69
4. Strict chemical inventory control	207	65.1	3.83
5 Use specific personnel for chemical additions	273	85.8	4.16
6. Perform routine bath analyses	293	92.1	4.37
7. Maintain bath analyses/addition logs	273	85.8	4.35
8. Have written procedure for bath makeup and additions	229	72.0	4.19
9. Use process baths to max. extent (no dump schedule)	234	73.6	4.21
10. Remove cadmium and zinc anodes from bath when idle	78	24.5	3.58
11. Regularly retrieve fallen parts/racks from tanks	250	78.6	3.72
12. Establish preventive maintenance program for tanks	185	58.2	3.75
13. Install overflow alarms on process tanks	50	15.7	3.36
14. Install other spill/leak detection systems	47	14.8	3.64
15. Use control method to add water to process tanks	108	34.0	3.87
16. Perform regular maintenance of racks/barrels	207	65.1	3.80
17. Preinspect parts to prevent processing of rejects	177	55.7	3.77
Average			3.86

[1] Based on 318 survey responses.
[2] Respondents were asked to rate the success level of the method on a scale of 1 to 5, with 1 being the least successful and 5 being the most successful.

Exhibit 3-7. Data from the Users Survey Regarding the Categories of Labor Employed and Technical Resources Available

Technical Resource Description	Number of Responses
In-House Process Engineering Department	11
Corporate Process Engineering Department at another location	3
In-House Process Engineer (i.e., one or two persons rather than a full department)	121
In-House Environmental Engineering Department	2
Corporate Environmental Engineering Department at another location	4
In-House Environmental Engineer (i.e., one or two persons rather than a full department)	101
Other	2

employee performance, like production yield and quality. Waste generation tracking should be instituted at the most basic level possible to provide quantitative measures of performance.

Of the 318 plating shops responding to the Users Survey, 158 (or 49.7%) indicated that they have established a formal policy statement with regard to pollution prevention and control. The average success rating given by the respondents for this pollution prevention tool was 3.58 (see Exhibit 3-6).

3.2.2.3 Task Force Organization

The size of the program task force will depend heavily on the size of the company. At a minimum, the task force should include members of any group or department in the company that manages, generates, handles or in some other manner affects wastes. Some members, who work directly with waste generation, may be constantly involved in the program while others, who have less direct contact with the waste generating processes, may only attend major planning meetings. Potential departments and personnel that would be appropriately involved from a major manufacturing company include (ref. 26):

- Plant vice president
- Metal finishing department manager
- Process engineer responsible for the operation of metal finishing processes
- Facilities engineer responsible for the maintenance of metal finishing equipment
- Wastewater treatment department supervisor
- Staff environmental engineer
- Hazardous waste coordinator

Larger companies, with successful programs, typically have 10 to 20 people on the task force. Some small companies have successful programs with as few as 2 to 4 members. In any case, a single person should not be given the sole responsibility for the program. This will isolate the program from the employees and generate a narrow range of ideas.

The organization of the task force may be dictated by a company's structure. As with any group, a leader is needed to chair meetings, motivate members and maintain communications. In large companies, the leader typically comes from the environmental department. The leader should be an experienced individual with knowledge of both environmental and production issues. Besides a task force leader, the organizational structure may consist of committees, headed by committee leaders, with well defined responsibilities. These committees could be organized by production areas, specific pollution prevention projects or task force responsibilities such as data collection. Typically, the organization structure and size will change as the program matures through the assessment phases and into implementation. The scope of the program will determine whether full-time participation is required by the leader or other team members (ref. 26, 303).

Employees can become involved in the program through working groups and motivated with financial incentive programs. Larger companies with Total Quality Management can utilize this organizational structure to encourage the generation of pollution prevention ideas.

The Users Survey requested shops to provide information about categories of employees and technical resources available at their sites. Their responses provide some insight to the types of personnel that are available to participate on a task force. A partial summary of these data is shown in Exhibit 3-7. Additional data identifying labor categories (e.g., process engineers, chemists, skilled trades, etc.) employed by the survey respondents is contained in the database.

3.2.2.4 Pollution Prevention Goals

The first priority of the task force should be to establish goals that are consistent with the policy adopted by management (ref. 26). The general policy of the program establishes broad and long-term objectives, whereas program goals should be quantifiable to establish a clear guide as to the success of the program. The following are two examples of goals used by major companies.

- **Goal Example 1**
  A chemical company has adopted a corporate-wide goal of 5 percent waste reduction per year with waste specific goals for individual facilities (ref. 26).

- **Goal Example 2**
  An aerospace company has set the following waste reduction goals, using 1988 as a base year: 30 percent reduction by 1993, 50 percent reduction by 1995 and 80 percent reduction by 1997. In addition to general waste reduction goals, certain priority chemicals (e.g., chlorinated solvents) were identified for which specific reduction goals were established (ref. confidential contribution).

3.2.2.5 Pollution Prevention Planning

The task force should have the responsibility of developing a strategic pollution prevention plan and periodically updating the plan (e.g., annually). The plan should outline an approach to meeting program goals by: identifying specific action items; indicating a time frame for their planned completion; and indicating the responsible person or department. Early plan elements should focus on data collection and assessment and subsequent elements should focus on implementation. As the program progresses the plan will become more defined.

3.2.2.6 Assessment/Feasibility Analysis

The assessment/feasibility analysis is the engineering phase of the pollution prevention program (See Exhibit 3-5). It involves the collection of baseline data, data evaluation, identification of pollution prevention targets, and the identification, evaluation and selection of pollution prevention options. EPA has published assessment manuals (ref. 25, 26) that contain worksheets that can guide the user through these various steps. The problem with using the EPA manuals is the resultant mass of worksheets generated from performing an assessment and feasibility analysis for all but the smallest of operations. For an average size manufacturing plant, use of the EPA manual will generate 200 to 500 pages of forms. As an alternative methodology, firms can utilize the basic EPA approach and develop their own forms that are tailored to their production processes and treatment operations. Use of the manual in this manner is encouraged by EPA.[2]

3.3 GOOD OPERATING PRACTICES

Good operating practices, a subset of pollution prevention methods, are organizational and procedural activities that reduce the generation of waste. Generally, these are not equipment oriented methods of waste reduction, although the implementation of some good operating practices can result in significant capital expenditures when implemented plant-wide. Many of the good operating practices identified will provide product quality improvements and operating cost reductions in addition to reducing waste generation. Also, they will generally improve the working environment of a shop, including health and safety aspects.

3.3.1 Employee Awareness and Education

Employees are often the fundamental cause of the generation of waste and conversely, they are in the most advantageous position to employ pollution prevention and control. Without employee cooperation, even the best efforts of management will be ineffective or futile. As stated by one survey respondent: Operator input into pollution prevention is essential for their "buy-in." Further, most pollution prevention methods will fail unless the operators believe that the methods will work (PS 257).

Employee awareness and education begins with a clear company policy with regard to the environment and plans for pollution prevention and control (see Section 3.2). The policy must be conveyed to the employees and reinforced in various ways in order to create a sufficiently positive attitude toward meeting the company's environmental goals.

There are three stages to instilling a good pollution prevention attitude in employees and providing them with the needed knowledge to perform successfully. These are: prior to job assignment, during job training, and ongoing education throughout employment (ref. 303). This training should include the following elements:

- How, why and where waste is produced and how to minimize it (e.g., good rinsing practices).

- Preventive maintenance methods that reduce waste generation (e.g., tank/liner inspection and repair).

- Company rules for handling process chemicals and making process (tank) additions.

- Procedures for handling spills/leaks.

- How to operate pollution prevention and control technologies in their working area.

- Where to go for assistance with a non-routine problem.

- Capabilities and capacities of waste treatment processes.
- Environmental regulations and how they relate to the processes the employees operate.

- Why pollution prevention is important (cost, regulations, health and safety, improved working environment, improved environment).

- How related waste management operating costs (e.g., chemicals, water, waste treatment, hazardous waste) impact employee wages.

The advantages of establishing company rules and employee training will be reduced or eliminated unless the program has a method of measuring success and can deal with those who refuse to participate. Success should be quantified, whenever possible. This means companies should collect chemical use and waste generation data, maintain records and periodically evaluate the records. Data collection and record keeping are discussed in Section 3.3.2.

Bonuses, awards, plaques and other forms of recognition are often used to provide motivation, and to boost employee cooperation and participation. In some companies, meeting waste minimization goals is used as a measure for evaluating the job performance of managers and other employees (ref. 26).

Of the 318 plating shops responding to the Users Survey, 217 (or 68.2%) indicated that they conduct employee education for pollution prevention. The average success rating given by the respondents for this pollution prevention tool was 3.69 (see Exhibit 3-6.)

3.3.2 Chemical Tracking and Record Keeping

The major sources of pollution from plating operations are process chemicals. Process chemicals become pollution through both use and misuse, resulting in wastewater generation, spills/leaks, spent solutions, sludge and air emissions. To be fully effective, a pollution prevention and control program must track and record chemical purchases, chemical use, and waste generation. The following is a list of chemical and waste data that merit consideration for record keeping:

- Chemical purchases
- Chemical inventory
- Bath analyses (see Section 3.3.3)
- Process (tank) reformulations and chemical additions
- Partial tank discharges (i.e., decanting) and total tank discharges (i.e., batch dumps)
- Water use per rinse tank or plating line (see Section 3.5.2.6)
- Total wastewater flow
- Wastewater treatment chemical use
- Spent process solution analyses
- Waste treatment sludge analyses
- Waste location

To increase the utility of the chemical use and waste generation data, corresponding production data should also be collected and recorded. These data can be used to identify decreases or increases in chemical use and waste generation that are due to production changes rather than operational practices.

The chemical data can also be used to develop material balances for individual processes, a sequence of processes or an entire facility. A material balance, which is often shown as a block diagram, indicates the material inputs and outputs for the selected operation. Creating a material balance for a process such as chromium plating forces one to consider the various chemical inputs and losses such as chromic acid additions, emissions, drag-out, chromium plated onto parts, bath maintenance losses, etc. A material balance for chromium use at the Tinker AFB plating shop is shown in Exhibit 3-8.

Several questions in the Users Survey were related to chemical tracking and record keeping. These include

Exhibit 3-8. Material Balance for Chromium Use at Tinker AFB Plating Shop

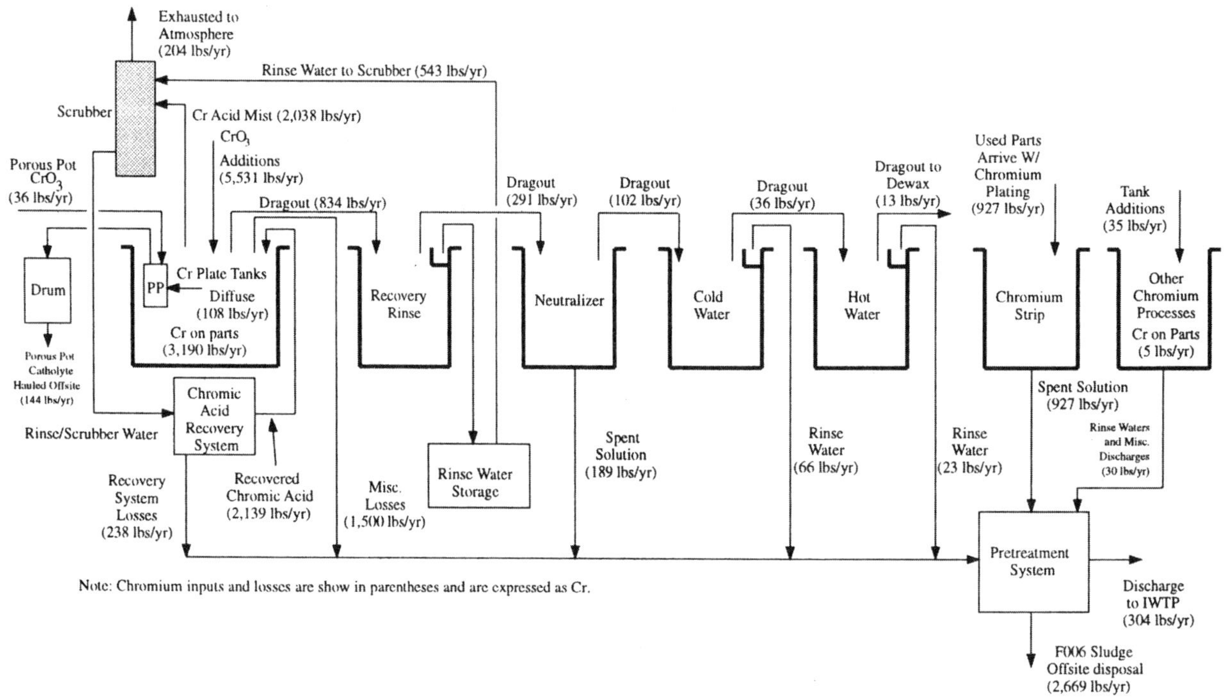

questions 4, 6 and 7 of Exhibit 3-6 and question 8 of Exhibit 3-16. The response to question 4 showed that 207 shops (or 65.1 %) have instituted strict chemical inventory control. The response to questions 6 and 7 indicate that 393 shops (or 92.1 %) perform routine bath analysis and 273 shops (or 85.8%) maintain bath analyses and/or tank addition logs. The response to question 4 (Exhibit 3-16) indicates that 37 shops (or 11.6%) use water flow meters and/or accumulators to track fresh water use at each tank or plating line. Of these various pollution prevention options, the use of routine bath analyses and maintaining logs had the highest success ratings (4.37 and 4.35 respectively).

3.3.3 Chemical Purchasing, Storage, Usage, and Handling

Proper purchasing, storage, usage, and handling of chemicals increases the percentage of raw materials that reach their intended process without spills, leaks or other types of losses that could result in waste generation. Some basic guidelines for good operating practices include (ref. 303):

- Purchasing:
  - Standardization of materials, i.e., using the minimum number of materials in all operations. Many times the decision to use one material over another is based on operator preference, rather than on a technical or economic requirement. Written specifications can improve purchasing and reduce waste (PS 257).
  - Avoid over-purchase of materials.
  - Avoid collecting free samples of process chemicals from vendors. Only accept amounts needed for testing purposes.

- Storage:
  - Utilize a dedicated/protected storage area.
  - Space containers in storage areas to facilitate inspection.
  - Label all containers.
  - Stack containers according to manufacturers' instructions to prevent cracking and tearing from improper weight distribution.
  - Separate incompatible materials in storage such as cyanides and acids.
  - Raise containers off the floor in the storage area to inhibit corrosion from "sweating" concrete.

- Handling/Use:
  - Establish written procedures for process (tank) formulation and additions.
  - Use specifically assigned personnel to formu-

late baths and make tank additions.
- Perform routine bath analyses and maintain bath analyses logs and tank formulation/ addition logs.
- Use process baths to the maximum extent possible (do not employ a dump schedule).
- Remove zinc anodes from the baths when they are idle to prevent a buildup of metal in the baths.
- Implement multiple use of certain materials (e.g., soak cleaner is reused as an electrocleaner by PS 068).
- Implement statistical process control (SPC) to improve the efficiency of chemical use (PS 257).

Survey questions that relate to chemical usage, handling and storage include questions 4, 5, 6, 7, 8, 9, and 10 (Exhibit 3-6). The responses to several of these questions were previously discussed in Section 3.3.3. The response to question 5 indicated that 273 shops (or 85.8%) use specifically assigned personnel for chemical additions and the response to question 8 indicated that 229 shops (or 72.0%) have written procedures for bath make-up and chemical additions. These two pollution prevention options received particularly high average success ratings. The response to question 9 indicates that 234 shops (or 73.6%) use process baths to the maximum extent possible rather than employ a dump schedule and the response to question 10 indicates that 78 shops (or 24.5% of all shops or approximately one-half of shops performing cadmium and/or zinc plating) remove cadmium or zinc anodes from baths when they are idle. The latter practice prevents a buildup of dissolved metal in the bath, particularly with zinc plating.

3.3.4 Preventive Maintenance

In general, preventive maintenance is an important element in operating and maintaining a plating facility. With regard to pollution prevention, preventive maintenance can minimize chemical losses due to leaks and can reduce the potential for a catastrophic loss due to a tank failure. Specific areas where preventive maintenance can reduce pollution generation include:

- Periodic inspection of tanks and tank liners with replacement or repair of damaged or corroded units.
- Regular replacement of seals on chemical pumps and filter systems.
- Inspection and repair of racks and barrels, particularly focusing on loose coatings that can hide dragout and nodules of plated metal that can dissolve in process tanks.
- Regularly checking tank bottoms for parts, fixtures and racks that have fallen and will corrode.

Of the 318 plating shops responding to the Users Survey, 185 (or 58.2%) indicated that they have established a preventive maintenance program for tanks, 207 (or 65.1%) indicated that they perform regular maintenance of racks and/or barrels and 250 (or 78.6%) indicated that they regularly retrieve fallen parts/racks from tanks (see questions 11, 12 and 16 in Exhibit 3-6). The success ratings for these three items fall within a narrow range (3.72 to 3.80).

3.3.5 Leak/Spill Prevention and Control

Chemical losses from leaks and spills can equal or outweigh the losses due to routine production operations. If small leaks from pumps, filters or tanks go unnoticed or ignored over a long time period, the overall loss can be very significant. Catastrophic losses, such as a tank failure caused by corrosion, will cause more immediate results. Several methods for reducing the potential of chemical losses from these sources were identified from the literature and Users Survey. These include:

- Conduct preventive maintenance of pumps, filters, tanks, etc., as discussed in Section 3.3.4.
- Employ a controlled method of adding make-up water to process tanks (do not permit use of unattended hoses).
- Install overflow alarms on all process tanks and especially on tanks that are heated and require regular evaporative replacement.
- Install double-walled tanks and for added protection install a sight tube that will indicate if a leak of the inner wall has occurred. Two shops indicated use of this method for cyanide baths (PS 124, PS 133)[3].
- Implement company rules for tank additions and other chemical transfers.

- Construct secondary containment with segregation that would permit reuse of spilled material. For example, install berms around process tanks, external filter systems (PS 149) and pumps (PS 257).

- Install pH, ORP, moisture sensors, and/or conductivity sensors with an associated alarm system in bermed areas (PS 146, PS 155), sumps (PS 089, PS 165, PS 172), drain lines, or around treatment tanks (PS 036). Example: shops PS 089 and PS 165 installed moisture sensing alarms in sumps that under normal operation are dry. Of the 318 plating shops responding to the Users Survey, 50 (or 15.7%) indicated that they have installed overflow alarms on process tanks and 47 (or 14.8%) have installed other spill or leak detection systems (see questions 13 and 14 in Exhibit 3-6).

3.4 DRAG-OUT REDUCTION

For the typical electroplating job shop, the drag-out of process solutions and the subsequent contamination of rinse waters are the major pollution control problems. This section explains the basic principles of drag-out theory and explores the function and applicability of the various drag-out minimization techniques in use today.

Because of the importance of drag-out and drag-out loss prevention, numerous questions in the Users Survey were related to this topic. The responses to these questions are discussed throughout the following subsections.

3.4.1 Drag-Out Reduction Principles

Electroplaters are well aware that drag-out varies considerably among the various parts plated at their shops. For example, the volume of drag-out in rack plating differs visibly from that in barrel plating. When a barrel emerges from a process tank, it usually carries with it over 10 times more solution than does a typical rack. In addition to the obvious effects of rack and barrel design and shape of parts, there are more subtle factors that affect the volume of drag-out. These parameters include viscosity and chemical concentration, surface tension, and temperature (ref. 1).

The viscosity of a plating process solution can be described as its resistance to flow or removal by another liquid (in this case, rinse water), caused by molecular attraction forces. The difference between high and low viscosity can be demonstrated with honey and water. A much thicker film will form on a knife dipped in honey than on one dipped in water. Honey, therefore, has the higher viscosity because of its cohesive and adhesive qualities. The same effect can be observed with plating solutions. If two identical surfaces are immersed in separate chromium baths with concentrations of 53 ounces per gallon (397 g/L) and 33 ounces per gallon (247 g/L), respectively, the lower concentration bath will produce 73 percent less volume of drag-out (ref. 1).

Surface tension is another physical phenomenon that has a significant effect in the plating shop. According to kinetic theory, molecules of a liquid attract each other. At the surface of a solution, such as a plating bath, the molecules are subjected to an unbalanced force because the molecules in the gaseous phase are so widely dispersed. As a result, the molecules at the surface are under tension and form a thin, skin-like layer that adjusts to create a minimum surface area. The property of surface tension causes liquid droplets to assume a spherical shape.

In the plating process, the volume of solution that clings to a workpiece surface depends partly on surface tension. The force of surface tension appears to be most significant at the bottom edge of the part as it passes through and leaves the process solution. This force and the resultant volume of drag-out appear to be greatly affected by the orientation of the part relative to the surface of the liquid (ref. 1).

The third factor that influences drag-out volume is the temperature of the process solution. Temperature is interrelated with viscosity and surface tension. As the temperature of a plating solution is increased, its viscosity, surface tension, and, therefore, drag-out volume are reduced. A possible exception is when a part is withdrawn too rapidly from a hot process solution, evaporation concentrates the film and impedes drainage. This problem, however, can be overcome by reducing withdrawal time and using a fog spray rinse on the parts as they emerge from the plating solution (ref. 306).

Estimations of drag-out have been presented in the literature (see Exhibit 3-9). However, these values should only be used for gross estimates. Actual measurements of drag-out are needed for accurate evaluations of pollution prevention options. Techniques for measuring drag-out are presented in various references (ref. 1, 20, 305). A simple and effective method for measuring drag-out is presented in Section 3.5.3.1.

3.4.2 Drag-out Reduction Techniques

Many devices and procedures are used successfully to reduce drag-out. These techniques usually are employed to alter viscosity, chemical concentration, surface tension, velocity of withdrawal, and temperature. Also used are drag-out tanks and similar equipment for capturing lost plating solution and for returning it to the bath.

Most drag-out reduction methods are inexpensive to implement and are repaid promptly through savings in plating and processing chemicals. An additional savings many times the cost of the changes is realized through decreased operating costs of a pollution control system. The reduced drag-out will decrease the need for treatment chemicals and, subsequently, the volume of sludge produced.

For some process solutions, return of drag-out may be impractical. In the case of processing baths which become steadily contaminated by use, the return of drag-out would simply increase the frequency of dumping (ref. 305). For example, PS 116 reported that drag-out recovery with electropolishing was "counterproductive."

The following subsections provide further detail and discuss the results of the Users Survey with respect to dragout loss prevention. A summary of the Users Survey data relating to drag-out loss prevention is presented in Exhibit 3-10. As with the good operating practices, there is a high usage rate of these pollution prevention methods among respondents and all of the methods have received a success rating above the mid-point of the rating scale.

Exhibit 3-9. Estimations of Drag-Out Generation Presented in the Literature

A. Average Drag-Out Losses (Soderberg)

Nature of Work Drainage	Drag-Out Rate (gal/1,000 ft^2)
VERTICAL	
Well drained	0.4
Poorly drained	2.0
Very poorly drained	4.0
HORIZONTAL	
Well drained	0.8
Very poorly drained	10.0
CUP SHAPES	
Well drained	8.0
Very poorly drained	24.0

B. Drag-Out Amounts Based on Hogaboom's Work

SolutionType	Drag-Out Rate (gal/1.000 ft^2) Flat Surfaces	Contoured Surfaces
Brass	0 95	3 3
Cadmium	1.00	3.1
Chromium (33 oz/gal)	1.18	3.0
Chromium (53 oz/gal)	4.53	11.9*
Copper Cyanide	0.91	3.2
Watts Nickel	1.00	3.8
Silver	1.20	3.2
Stannate Tin	0.83	1.6
Acid Zinc	1.30	3.5
Cyanide Zinc	1.20	3.8

It is of interest to note the effect of increased viscosity. There is roughly a threefold increase in drag-out volume with less than double the concentration increase.

Source: ref. 1

3.4.2.1 Minimizing Drag-Out Formation

The most obvious source of pollution in a metal finishing shop is the drag-out of various processing baths into subsequent rinses. The amount of pollutants contributed by drag-out is a function of factors such as the design of the racks or barrels carrying the parts to be plated, the shape of the parts, and plating procedures. As previously discussed, several interrelated parameters of the process solution, including the concentration of process chemicals, temperature, viscosity, and surface tension also impact pollution levels.

Many devices and procedures can be used successfully to reduce drag-out. These techniques are usually employed to alter those important and interrelated process solution parameters.

3.4.2.1.1 Controlling Plating Solutions

As a rule, as the chemical content of a solution is increased, its viscosity increases. Increased viscosity contributes not only to a large volume of drag-out, but also to a higher chemical concentration of drag-out. The consequent need for more rinse water creates additional pollution control problems. Plating baths can often be operated at significantly lower concentrations than those recommended by chemical manufacturers.

Of the 318 plating shops responding to the Users Survey, 110 (or 34.6%) indicated that they have lowered the concentration of a plating bath to reduce drag-out losses. The average success rating given by the respondents for this pollution prevention tool was 3.33 (see question 16 of Exhibit 3-10).

As mentioned in Section 3.4.1, temperature is interrelated with viscosity and surface tension. A total of 57 survey respondents (or 17.9%) indicated that they have tried increasing bath temperature to reduce viscosity and hence decrease drag-out formation. The average success rating for this method was 3.14 (see question 17 of Exhibit 3-10), which was the second to the lowest success rating of all pollution prevention options listed in the survey form.

For years wetting agents have been used in process solutions to aid in the plating process. These substances are used, for instance, in bright-nickel plating to promote disengagement of hydrogen bubbles at the cathode. They are also used as an aid to drag-out reduction. A wetting agent is a substance, usually a surfactant, that reduces the surface tension of a liquid, causing it to spread more readily on a solid surface. A typical plating bath solution has a surface tension close to that of pure water at room temperature or about 0.0050 lb/ft. The addition of very small amounts of surfactants can reduce surface tension considerably—to as little as 0.0017 to 0.0024 lb/ft (ref. 1, 39, 305). Further additions of the wetting agent will not lower the surface tension appreciably beyond this point (ref. 521).

Kushner (ref. 1) estimates that the use of wetting agents will reduce drag-out loss by as much as 50 percent, although no test data or other quantitative information are presented. He recommends the use of non-ionic wetting agents that are not harmed by electrolysis in the plating bath. Platers contemplating the use of a wetting agent for drag-out reduction should conduct experiments to determine their potential benefit before implementation. Also, platers should investigate the compatibility of a wetting agent with the bath chemistry before use. Some process baths (e.g., hard chromium) can only tolerate certain products (ref. 482).

In addition to reducing drag-out, wetting agents are used to reduce chromic acid mist formation during plating. The lower surface tension achieved with wetting agents reduces the effect that oxygen and hydrogen bubbles (generated at the tank's electrodes) have when they reach the surface of the bath (ref. 521).

Of the 318 plating shops responding to the Users Survey, 103 (or 32.4%) indicated that they use a wetting agent to reduce drag-out. The average success rating given by the respondents for this pollution prevention tool was 3.12 (lowest rated method).

One survey respondent that uses wetting agents made an interesting point, "I have no way of telling if wetting agents are effective." That shop rated this method at a success level of two. Shops that would like to quantify the effect of wetting agents on drag-out can perform simple drag-out tests like those described in the literature (ref. 1, 20, 305) using a sample of their standard bath and one to which a wetting agent has been added. Also, the surface tension of a bath can be measured with an inexpensive device called a stalagmometer.

Kushner further suggests keeping the concentration of all dissolved salts at the minimum needed for proper operation. To follow this recommendation, the plater should not permit substances to buildup in the plating bath, if it is possible to control and maintain them at the proper level. For example, cyanide baths are often permitted to

Exhibit 3-10. Summary of Users Survey Data Relating to Drag-Out Loss Prevention

	Drag-Out Reduction Method	Numer of Shops Using Method[1]	Percentage of Shops Using Method[1]	Sucess Rating[2]
1.	Drag-out rinse tanks; return chemicals to process bath (manual)[3]	194	61.0	3.77
2.	Drag-out rinse tanks; return chemicals to process bath (automatic)[3]	62	19.5	3.81
3.	Drip tanks; return chemicals to process bath (manual)	86	27.0	3.40
4.	Drip tanks; return chemicals to process bath (automatic)	21	6.6	3.24
5.	Reducing speed of rack/part withdraw (manual)[3]	121	38.1	3.23
6.	Reducing speed of rack/part withdraw (automatic)[3]	69	21.7	3.61
7.	Allow rack/part to drip over plating tank (manual)	192	60.4	3.44
8.	Allow rack/part to drip over plating tank (automatic)	82	25.8	3.79
9.	Using drag-in/drag-out arrangement (manual)	66	20.8	3.39
10.	Using drag-in/drag-out arrangement (automatic)	34	10.7	3.82
11.	Fog/spray rinses installed over process bath (manual)	60	18.9	3.67
12.	Fog/spray rinses installed over process bath (automatic)	34	10.7	3.35
13.	Air knives that blow off drag-out (manual)	7	2.2	3.14
14.	Air knives that blow off drag-out (automatic)	18	5.7	3.72
15.	Drip shields between tanks	181	56.9	3.68
16.	Lower bath concentrations	110	34.6	3.37
17.	Increasing solution temperatures	57	17.9	3.14
18.	Using a wetting agent to reduce surface tension	103	32.4	3.12
19.	Positioning workpiece to minimize solution holdup	165	51.9	3.75
	Average			3.50

[1] Based on 318 survey responses.
[2] Respondents were asked to rate the success level of the method on a scale of 1 to 5, with 1 being the least successful and 5 being the most successful.
[3] The terms manual vs automatic refer to the type of plating operation to which the drag-out reduction method is applied (e.g., manual hand line vs. automated plating machine).

buildup very high carbonate concentrations even though the concentration level could be controlled by treatment (e.g., carbonate freezing). Such a buildup could increase drag-out by as much as 50 percent (ref. 305).

3.4.2.1.2 Position on Rack

The metal finisher's primary consideration in the positioning of workpieces on a rack is proper exposure of the parts to the anodes for optimal coverage and uniform thickness of the electrodeposit. Drainage and rinsability are secondary, but very important considerations to pollution control. Also, damage to the workpiece surface can be caused by insufficient or inefficient rinsing, and succeeding process solutions can be contaminated by drag-in of unremoved chemicals from the previous solution (ref. 305).

Several rules apply to the position of work on plating racks for drag-out minimization. The basic principle, however, is that every object can be positioned in at least one way that will produce the minimum quantity of drag-out. This position could be determined by experiment, but unless a significant number of similar items are to be plated, it may be advisable to follow the suggestions of Kushner and Wallace (ref. 1, 305):

- Tilt all solid objects with plane or single-curved surfaces so that drainage is consolidated, that is, twist or turn the part so that the clinging fluid will flow together and off the part by the quickest route.
- If possible, avoid racking parts directly over one another to prevent lengthening the drainage path of the plating solution.
- Avoid table-like surfaces by tipping the part, but not at the expense of forming solution "pockets."
- Orient parts so that only a small surface area comes in contact with the liquid surface as it leaves the plating solution.

From the Users Survey, 165 (or 51.9%) shops indicated that they concern themselves with workpiece position in an effort to reduce drag-out formation. The average success rating given by the respondents for this pollution prevention method was 3.75 (see question 19 of Exhibit 3-10).

3.4.2.1.3 Workpiece Withdrawal

The velocity at which work is withdrawn from the process tank has a major effect on drag-out volume. The faster an item is pulled out of the tank, the thicker the drag-out layer will be because viscosity forces do not have a chance to operate and a much larger volume of liquid will cling to the surface (ref. 305). The effect is so dramatic that Kushner (ref. 1) suggests that most of the time allowed for withdrawing and draining the item should be used for withdrawal. An automatic machine that performs smooth, gradual withdrawal usually will drag out less solution per item racked than will manually operated equipment.

Questions 5 and 6 in Exhibit 3-10 indicated that reducing the speed of withdrawal is practiced by 69 respondents (or 21.7%) with automatic plating equipment and 121 respondents (or 38.1%) with manual lines. The success ratings were 3.61 and 3.23 respectively.

3.4.2.1.4 Design and Maintenance of Racks & Barrels

The transport of chemicals inside loose rack coatings from one process to another is not uncommon. For example, chromium may appear in rinse waters that are discharged into a plant's industrial sewer some distance from the chromium process tanks. The chromium often reaches these remote areas by way of loose rack coatings. Increased attention to rack maintenance not only will eliminate this potential problem but also will contribute to a welcomed reduction in the number of workpieces rejected because of poor contact (ref. 305).

Improving the design of racks, baskets, etc. to reduce the amount of solution they can trap will also reduce drag-out. For example, one shop reported great success by replacing old baskets with ones that had larger holes (PS 55). Many shops indicated that they have redesigned their racks to reduce pollution (e.g., PS 025, PS 058, PS 089, PS 159, PS 173, PS 192, PS 196, PS 229, PS 301, PS 316). Similarly, many barrel platers reduced drag-out by increasing the hole size of the barrels or making other design changes (e.g., PS 132, PS 214, PS 229).

As a group, the survey respondents indicated that regular rack/barrel maintenance was an important pollution prevention method. A total of 207 respondents (or 65.1%) indicated that they implement this method and the average success rating was 3.77 (see question 16 of Exhibit 3-6).

3.4.2.2 Direct Drag-Out Return

Commercially available equipment for the recovery of plating bath chemicals includes types that apply such principles as ion exchange, reverse osmosis, electrodialysis, and evaporation. These devices usually are applied to a single plating operation where they concentrate the salts in the rinse water, return them to the plating bath, and recycle the purified water to rinse tanks (ref. 305).

Although effective, these recovery technologies are capital intensive. Before the purchases of such equipment, metal finishing facilities usually implement simple methods of drag-out recovery that require much less capital and are simpler to operate. After using these methods and establishing new drag-out conditions, the plater can consider the applicability of additional recovery through commercially available units (ref. 305). This section describes methods that directly return the drag-out to the process tank. In the following section (Section 3.4.2.3) methods are described that recover the drag-out in tanks and then return it to the process bath.

3.4.2.2.1 Draining/Rinsing Over the Plating Tank

After a rack or barrel is removed from a process tank, the drag-out drains from the item and it returns directly to the bath as long as the item is held over the tank. This simple method of direct drag-out return can be maximized on a hand-line by installing a bar over the process line on which the operator can hang a rack or hook. On automatic machines, the unit can be programmed to increase dwell time above the process tank. For barrel operations, the barrel can be rotated over the process tank to help free the drag-out. Altmayer indicates that continuous barrel rotation is not always the best technique for freeing drag-out. He suggests that cup shaped parts should be rotated several times and then stopped and that free draining parts should not be rotated or rotated once and then stopped (ref. 482). As with many drag-out reduction methods, some experimentation will help to identify the optimal procedure.

Exhibit 3-11. Example of Spray Rinsing for Barrel Plating

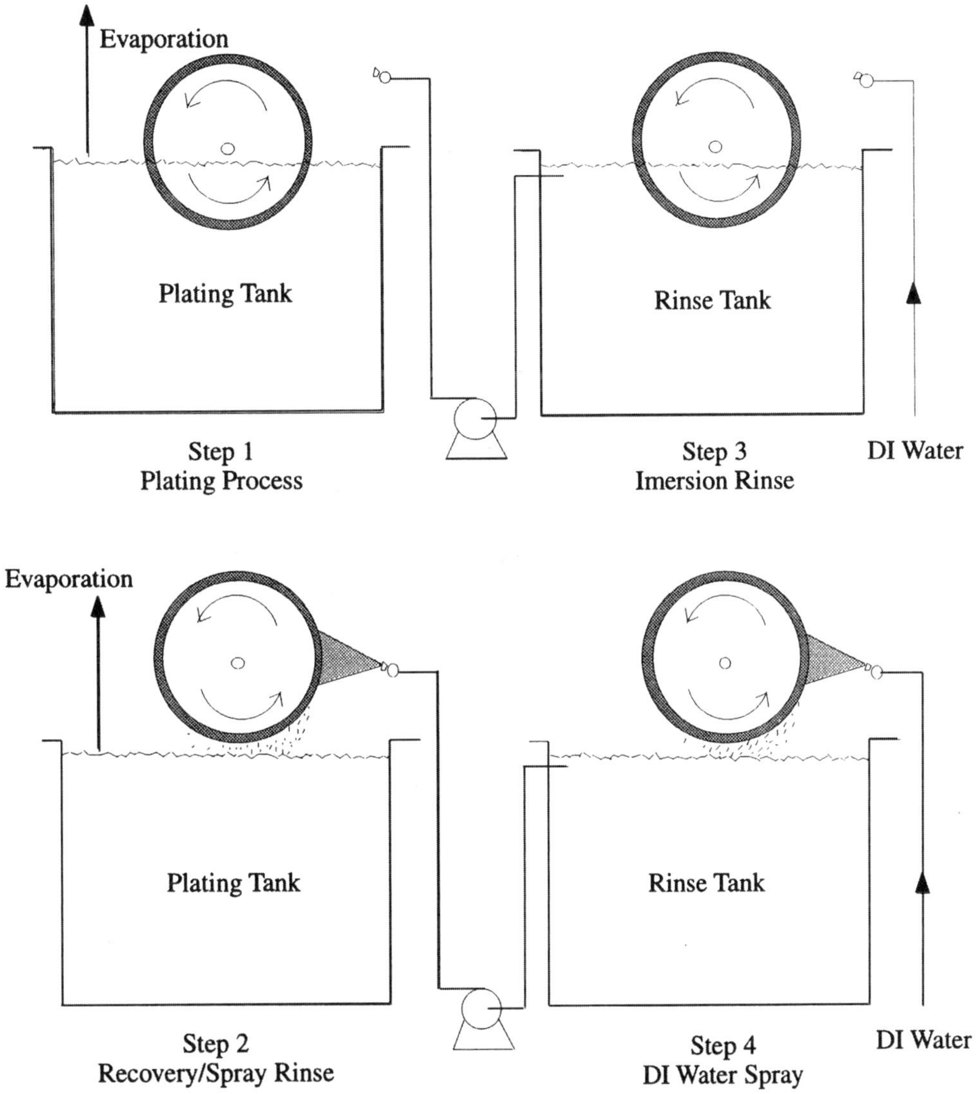

The quality of a metal finishing process can be diminished if the drag-out is permitted to dry on the part. This can cause staining, peeling (PS 183), passivation, or it may prevent complete rinsing. To increase the drag-out removal rate over the process tank, rinsing with small amounts of water can be employed. The amount of water that can be used will depend on the water balance for a given process tank. The water balance is affected mostly by evaporation. Process solutions operated at temperatures greater than 120°F often have sufficient surface evaporation such that rinsing can be performed over the tank. However, using this method may reduce or eliminate the potential benefit from other drag-out recovery methods (e.g., use of a drag-out tank).

Rinsing over the tank can be performed by flood rinsing (e.g., hose), spray rinsing, or fog rinsing. The use of flood rinsing is not practical except for very high temperature baths with high drag-out rates. Spray rinsing uses less water than flood rinsing. With the proper selection of spray nozzles this can be a very efficient method of direct drag-out return. Nozzle selection should consider: flow rate, spray velocity, and spray pattern. Air-assisted sprays are also utilized, which are generally more efficient than plain water sprays. Sprays can be hand-held or mounted on the tank rim. For automatic plating machines, the sprays are controlled to operate only when the part exits the bath. For example, PS 193 employs an automatic rinse spray that actuates during vertical lift for a period of 7 seconds.

Exhibit 3-12. Diagram of a Drag-Out Reduction Device Designed for Barrel Plating (Courtesy of Kinetic Recovery Corporation, Edina, MN)

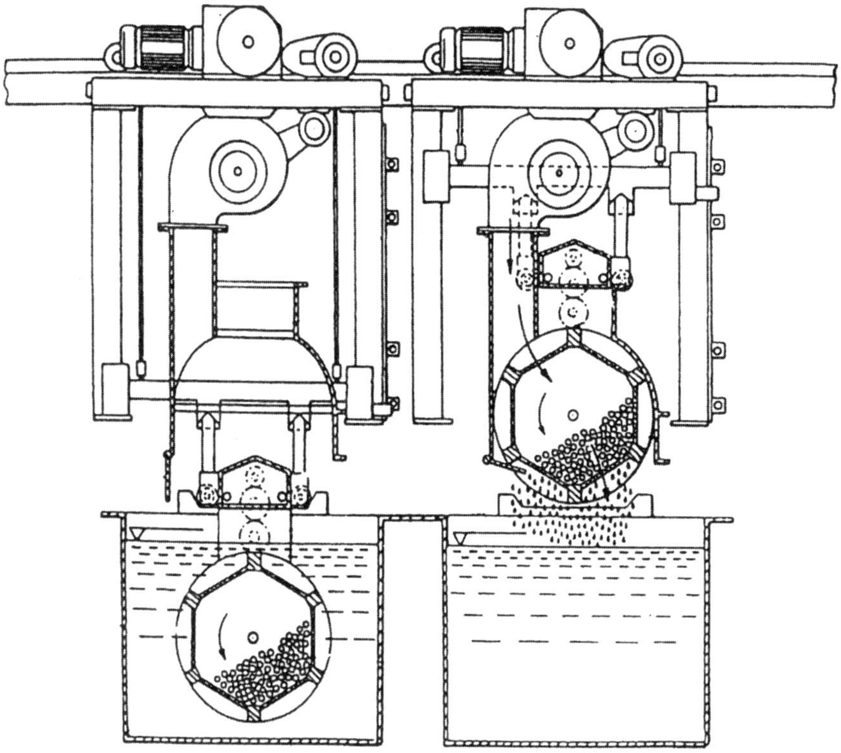

Operational Sequence:
1. The barrel hoist carries an attached hood, ventilation system, and sealing mechanism.
2. The rotating barrel is lifted by the hoist out of the process bath or rinse tank into the hood and sealing mechanism.
3. The vent creates an air current that pushes the liquid out of the rotating barrel, back into the tank.

The use of spray rinsing is not limited to racked parts. It can also be applied to barrel lines. Exhibit 3-11 shows how a barrel plating spray rinse can be employed.

A novel barrel draining method is shown in Exhibit 3-12. This system, which can be retrofit to an existing barrel plating line, makes use of an air purge to push solution out of a barrel. Vendor tests showed that simple rotation removes 37% of the drag-out after 20 seconds (no further dripping) while the new air purge system removes 63% of the drag-out (tests performed with brass plating solution).

Fog rinsing is used at exit stations of process tanks. A fine fog is sprayed on the work, diluting the drag-out film and causing a run-back into the process solution. Fog rinsing is applied when process operating temperatures, high enough to produce a high evaporation rate, allow replacement water to be added to the process in this manner. Fog rinsing prevents dry-on patterns by cooling the workpieces, but it may preclude the use of a drag-out tank as a recovery option. For fog rinsing to be effective, work must be withdrawn from the process tank at a slow rate (ref. 305).

As with any pollution prevention method, fog/spray rinsing over the bath has potential drawbacks and problems. When used in conjunction with a ventilated tank, the spray may increase the pollutant loading to a scrubber or be directly vented to the atmosphere. Fog/spray rinsing may be messy as indicated by one respondent (PS 124), or worse, may cause splashing on nearby workers. One shop indicated that the nozzles required frequent maintenance (i.e., unplugging) and must be occasionally repositioned to point in the correct direction (PS 124).

These methods of drag-out reduction generally work best for automated operations. Manual operations are too dependent on the behavior of the operator. For example, operators "rush the dwell time" when draining over process tanks (PS 086) and "doesn't work due to tired platers ... also operators want to increase production and do not take the time to hold the racks to allow extra drainage time" (PS 210).

Survey questions that relate to draining and fog or spray rinsing over process baths include questions 7, 8, 11 and 12 of Exhibit 3-10. The response indicates that 192 (or 60.4%) and 82 (or 25.8%) of the survey respondents implement draining, respectively for manual or automatic plating operations. The average success ratings were 3.44 (manual) and 3.79 (automatic). Fog and spray rinsing over the process tank is less frequently used by the respondents. As shown in Exhibit 3-10, 50 (or 18.9%) respondents use this method with manual operations and 34 (or 10.7%) use this method with automatic operations. The average success ratings were 3.67 (manual) and 3.35 (automatic).

3.4.2.2.2 Other Methods of Direct Drag-Out Return

The following are miscellaneous methods of direct drag-out return that are not discussed elsewhere in the report.

A drain board or drip shield is a tilted surface placed between process and rinse tanks that catches the drips from racks or barrels as they are transferred between tanks, thus preventing the drag-out from falling to the floor. The solution on the drain board returns to its original tank by gravity flow. The drain surface can be plastic or metal. For acid solutions, the best materials are vinyl chloride, polypropylene, polyethylene, and Teflon®-lined steel. Stainless steel should be used for hot alkaline solutions. It is important that the drain surface be positioned at an angle that allows the plating solution to return to the bath (i.e., rather than the subsequent rinse) (ref. 305).

The use of drain boards or drip shields was a frequently used method of drag-out loss prevention by survey respondents. As indicated in question 15 of Exhibit 3-10, 181 survey respondents (or 56.9%) use this method. The average success rating was 3.68.

Another direct drag-out return method, the air knife, is a device that blows an intensive air stream at a part/rack as it exits the bath causing the drag-out to be blown off. The use of air knives is limited due to: (1) the potential to dislodge parts from racks; and (2) the drying effect of the air stream which may cause staining, passivation, etc. Concerning the second limitation, Altmayer suggests that if the air is humidified to near saturation that drying will not occur (ref. 482). One respondent noted that the air supply must be oil-free (PS 257). One modern use of this device is with semi-aqueous cleaning systems, where the air knife is employed to minimize the quantity of solvent drag-out entering the water rinse system.

Air knives are not frequently used according to the survey respondents. Only 7 (or 2.2%) (manual) and 18 (or 5.7%) (automatic) of the shops surveyed employ this method of drag-out loss prevention. These shops rated the success level of air knives at 3.14 (manual) and 3.72 (automatic).

3.4.2.3 Drag-Out Recovery and Return

3.4.2.3.1 Drip Tank

A drip tank is an ordinary rinse tank that, instead of being filled with water, simply collects the drips from racked parts and barrels after plating and before rinsing. The drip tank is useful with work that involves continuous dripping over a period of time. Barrel plating, therefore, is a better candidate than rack plating for use of drip tanks. With barrel plating, the barrel should be rotated while it is suspended over the drip tank to ensure maximum drainage (see barrel rotation discussion in Section 3.4.2.2.1). When a sizable volume of solution has been collected in the drip tank, it can be returned to the plating bath. Because drag-out is not diluted with water when using a drip tank, this technique is especially applicable to lower temperature process solutions (ambient to 120°F).

Using a drip tank will restrict the use of an additional rinse tank, when floor space is limited. As will be discussed, an additional rinse tank, used as a drag-out tank or in a counterflow arrangement, is usually much more beneficial than a drip tank since a drip tank only recovers the drag-out that freely flows off the part/rack. The determining factors are the volume of drag-out, part configuration (i.e., drainability) and the evaporation rate in the process tank.

A total of 21 (or 6.6%) of the survey respondents use drip tanks with automated plating equipment and 86 (or 27.0%) with manual plating lines. The success ratings for drip tanks were 3.24 (manual) and 3.40 (automatic), which are lower than the success ratings of drag-out tanks.

3.4.2.3.2 Drag-Out Tank

The drag-out tank is a rinse tank that initially is filled with pure water. As the plating line is operated, the drag-out rinse tank remains stagnant and its chemical concentration increases as more work is processed. Air agitation is often used to aid the rinsing process because there is no water flow within the tank to cause turbulence. The presence of a wetting agent is also helpful, according to Kushner (ref. 1). After a period of operation, the solution in the drag-out tank can be used to replenish the losses to the plating bath. If sufficient evaporation has taken place, a portion of the drag-out tank solution can be added directly to the plating bath (e.g., using a transfer pump). Evaporation usually will be sufficient with baths, such as chromium and nickel plating solutions, that are operated at elevated temperatures. Low-temperature baths, such as cadmium or zinc plating solutions, have minimum surface evaporation and often their temperature cannot be significantly increased without degrading heat-sensitive additives. Reportedly, new additives, which are not as readily degraded by heat, have been developed for many of these plating baths. These additives might make operation of the plating bath possible at higher temperatures, facilitating drag-out recovery. Usually the value of the recovered chemicals is much greater than the increased energy cost associated with operating the bath at a higher temperature (ref. 305).

Diagrams showing the use of drag-out tanks and other rinsing configurations discussed in this section are shown in Exhibit 3-13. Drag-out tanks can be combined with counterflow rinsing to provide both chemical recovery and flow reduction. Combinations of rinse configurations are discussed in Section 3.5.3.6.

As a rough estimate, drag-out recovery, which is also referred to as recuperative rinsing (ref. 13, 14), will reduce drag-out losses by 50 percent or more. The efficiency of the drag-out tank arrangement can be increased significantly by adding a second drag-out tank. Use of a two-stage drag-out system usually reduces drag-out losses by 70 percent or more. In some cases, multiple drag-out tanks (e.g., three to five tanks) can be used to completely close the loop and return essentially 100 percent of drag-out (ref. 305).

The potential benefits (i.e., percent recovery) of drag-out rinse tanks can be estimated using either derived or empirical equations, a discussion of which follows. Actual results will depend on local factors (e.g., part configuration, operating practices, humidity).

The drag-out rate and evaporation rate are the key parameters that determine what percentage of the drag-out can be recycled back to the process tank. Various mathematical formula have been used to estimate the recovery rate (ref. 1, 39, 301). Exhibit 3-14 presents estimates for common conditions that can be used in lieu of the more complex equations.

Various methods can be employed for estimating or measuring drag-out, as discussed in Section 3.4.1. Evaporation rates of process tanks depend mostly on the operating temperature of the bath and to a lesser degree on the intensity of solution agitation. Where there is little evaporation in a process tank (e.g., baths operated below 110°F) there is little benefit from drag-out recovery tanks. PS 114 gave this method a rating of two for this reason. Various formulae and graphs are published for estimating evaporation (ref. 1, 39, 305). The following formulae are considered accurate, however, due to the relative ease of measuring evaporation rates, actual measurements should be used whenever possible.

Surface evaporation rates:

Still Tank: $\text{gal/hr/ft}^2 = e^{(0.03236T-7.20)}$

Agitated Tank: $\text{gal/hr/ft}^2 = e^{(0.02655T-5.95)}$

Where T = °F for process bath

Source: ref. 316

The transfer of solution between drag-out tanks and the plating tank and the addition of fresh make-up water to the system can be accomplished in several ways. Ryder (ref. 16) recommends that transfers to the plating tank be accomplished using a small pump (magnetic drive, seal-less types) which is activated by a "dead-man" switch. The dead-man switch only permits solution transfer while the switch is depressed. If the operator leaves, the solution transfer automatically stops, which prevents catastrophic tank overflows. For adding make-up water, Ryder suggests using a level controlled valve (local float controlled) in the first rinse. When the solution level in the first rinse is lowered (i.e., after solution is transferred to the plating bath) the float switch is activated and fresh water is added to the final rinse. Ryder further suggests the use of a water control valve on the inlet water line for shut-off during non-operating periods.

With multiple rinse tank arrangements, the transfer of solution from rinse tank to rinse tank can be accom-

Exhibit 3-13. Diagrams Showing Common Rinsing Configurations

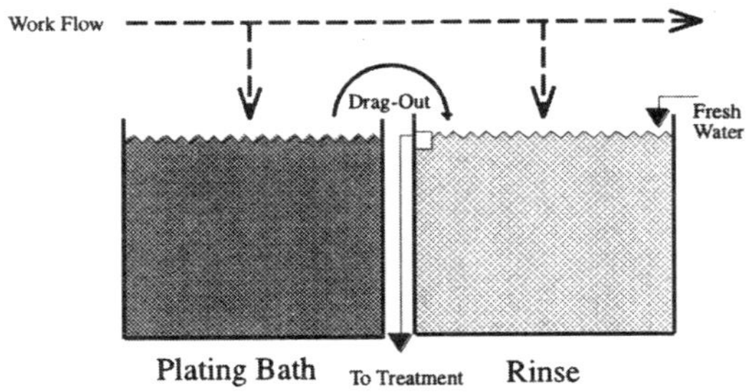

Rinse Type: Single Overflow
Applications: All Process Baths
Water Use Rate: Extremely High
Recovery Rate: None

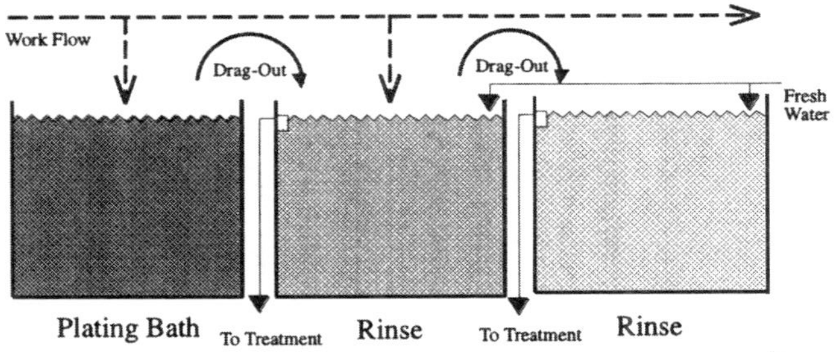

Rinse Type: Two Stage Parallel
Applications: All Process Baths
Water Use Rate: Moderate
Recovery Rate: None

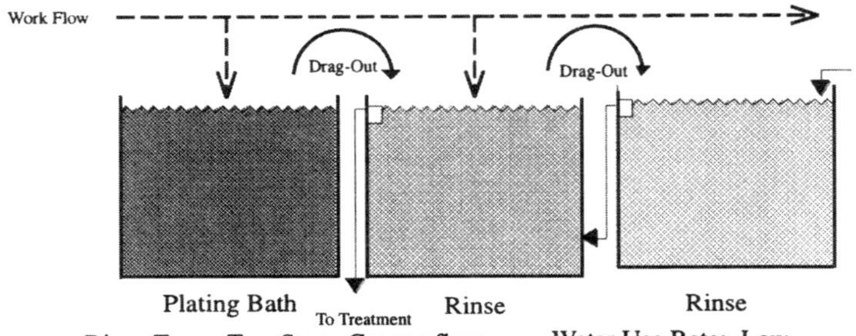

Rinse Type: Two Stage Counterflow
Applications: All Process Baths
Water Use Rate: Low
Recovery Rate: None

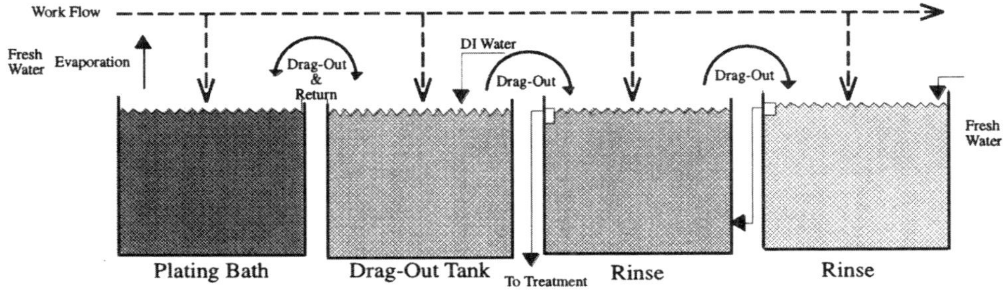

Rinse Type: Single Stage Recovery/Two Stage Counterflow Rinse Water Use Rate: Low
Applications: Most Heated Process Baths, Especially Plating Solutions
Recovery Rate*: Low (70 - 90 °F), Moderate (90 - 120 °F), or High (>120 °F)

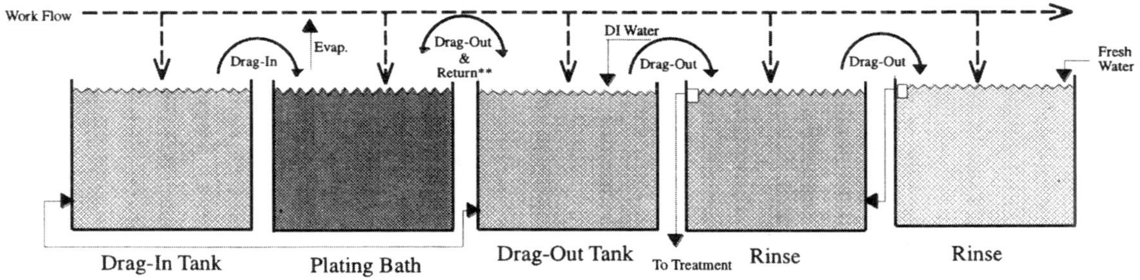

Rinse Type: Drag-In/Drag-Out Recovery/Two Stage Counterflow Rinse Water Use Rate: Low
Applications: Most Process Baths, Especially Ambient to Low Temp. Plating Solutions
Recovery Rate*: Moderate (70 - 90 °F), High (90 - 120 °F), or Higher (>120 °F)

*Depends on Process Tank Temperature

**Drag-Out Return Only Possible With Heated Process Tanks

Exhibit 3-14. Drag-Out Tank Recovery Rates for a Range of Common Conditions

Number of Drag-Out Tanks Installed	Drag-Out Recovery Rate, % — Evaporation to Drag-Out Ratio (1 to 5) E/DO = 1	E/DO = 2	E/DO = 3	E/DO = 4	E/DO = 5
1	50.0	66.7	75.0	80.0	83.3
2	66.7	85.7	92:3	95.2	96.8
3	75.0	93.3	97.5	98.8	99.4
4	80.0	96.8	99.2	99.7	99.9
5	83.3	98.4	99.7	99.9	100

Source: ref. 20

plished in the same manner as a flowing counterflow rinse system. These are discussed in Section 3.5.

The use of an automatic drag-out return system was described by Roy (ref. 4). In this system, chemical metering pumps were used to return drag-out from the rinse tank to the plating tank. The pumps were controlled by a level sensor in the plating tank. Roy also described a more complicated configuration where multiple drag-out tanks on different lines were connected to a "sump tank" using U-tubes. A level sensor in the sump tank controlled the water level in the drag-out tanks. As the drag-out return pumps drop the level in the sump tank, a level sensor turns on fresh water solenoids and the operating level is quickly restored.

It should be noted that although drag-out reduction can be a very effective means of pollution prevention, it may also present the plater with a new set of problems. In particular, by reducing drag-out, the plater reduces the purging of bath contaminants. The contaminants are contributed to process baths mainly by a breakdown of process chemicals and low concentration constituents in the fresh water (e.g., hardness). Other sources include: cross contamination due to transporting dripping racks over tanks, corrosion of bus bars, racks, anodes, tanks, etc., and airborne contaminants. Several shops identified contaminant buildup as a significant problem with drag-out reduction (PS 102, PS 124, PS 156). For example, PS 156 attempted to operate their shop on a near closed-loop basis. This facility, which now operates with an average flow of 16,800 gpd had reduced their flow, at one time, to 3,000 gpd. Their flow reduction efforts were partly discontinued after "contaminants entered the closed loop and caused plating problems." They also noted that their low volume discharge was contaminated with nickel and cyanide although these related processes were "closed loop." The presence of nickel and cyanide in their discharge may have resulted from cross contamination of baths caused by using the same racks for multiple lines. The presence of nickel and cyanide may also have been present when the flow rate was higher, but less noticeable due to dilution.

In some cases, drag-out reduction/recovery may result in serious degradation of the bath. To minimize the impact of contaminants, platers must do one or both of the following: (1) treat the raw rinse water prior to use with ion exchange and/or reverse osmosis technologies, (2) perform bath maintenance. Bath maintenance technologies are discussed in Section 4 rather than in this section. However, as a brief example of using bath maintenance with drag-out recovery, Exhibit 3-15 shows how one respondent has integrated electrolytic purification (dummying) into a closed-loop nickel rinsing configuration.

Another problem identified with drag-out tanks was the staining of parts by the drag-out rinse (PS 124 and PS 133). According to these shops, this was partly caused by drying of the solution on the part after drag-out rinsing.

The results of the Users Survey show that drag-out tanks are employed by 194 respondents (or 61.0% of all respondents) with manual plating operations and 62 (or 19.5% or all respondents) with automatic plating operations. The success ratings are 3.81 and 3.74, respectively.

3.4.2.3.3 Drag-In/Drag-Out Rinsing.

Drag-in/drag-out rinsing (also referred to as double-dipping) involves rinsing in the same solution before and after plating (see Exhibit 3-13). This can be achieved by using a single rinse tank or two hydraulically connected rinse tanks, usually located on opposite sides of the process tank. In the latter case, which is most applicable to

Exhibit 3-15. Examples of Rinsing Arragnements Used by Survery Respondent PS 118

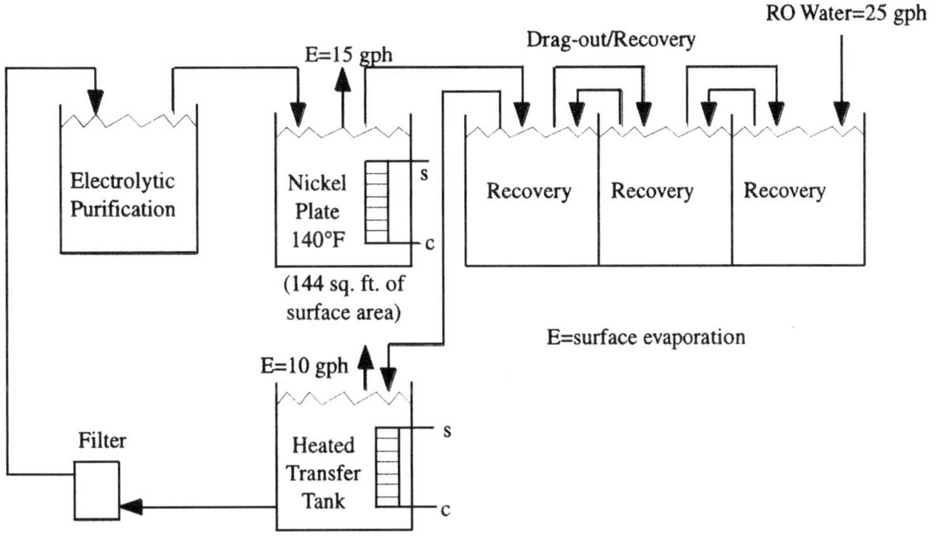

Figure A: Closed-Loop Nickel Plating With Elelctrolytic Purification

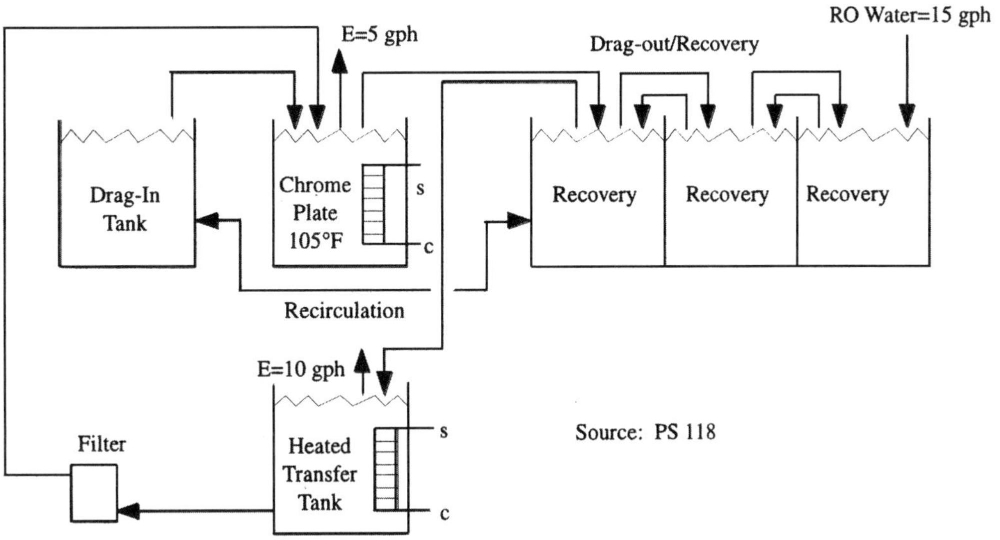

Figure B: Closed-Loop Chromium Plating With Drag-In/Drag-Out Arrangement

automatic plating machines, the rinse water is recirculated between the two rinse tanks using a transfer pump to maintain equal concentrations of chemicals in the tanks. The advantage of a drag-in/drag-out arrangement is that plating chemicals rather than pure rinse water are transferred into the process tank by incoming racks or barrels. This increases the recovery efficiency of the recovery rinse.

The drag-in/drag-out system finds application with plating baths that have a low to moderate evaporation rate and especially with baths that tend to increase in volume (i.e., equivalent to a negative evaporation rate). This condition, referred to as "solution growth" is common to zinc cyanide baths (e.g., PS 051), where the volume of drag-in (water from the preceding rinse) can be greater than the sum of drag-out and evaporation. The recycle ratio, which determines recovery efficiency, is calculated as the volume of recycled rinse plus the volume of drag-out divided by the volume of drag-out. The recycle ratio, therefore, is greater with a drag-in/drag-out system than a common recovery tank. If the evaporation rate is low, the difference between the recycle ratios for common recovery and drag-in/drag-out systems is significant. When evaporation ratios are high, the difference is less. Generally, the use of a drag-in/drag-out arrangement will increase the recovery rate by 25 to 40 percent (ref. 305). The use of atmospheric evaporators for eliminating solution growth, an alternative to the drag-in/drag-out system, is discussed in Section 4.2.

As with drag-out tanks, the drag-in/drag-out arrangement can result in bath contaminant buildup, as indicated by PS 124, who rated the success of this method as a "1."

As with drag-out tanks, the drag-in/drag-out arrangement can result in bath contaminant buildup. Also, the use of a drag-in tank adds an extra labor step to the recovery rinsing process. PS 173 indicated that although this method worked, only a few of their employees will take the time to use it.

Exhibit 3-15 shows a drag-in/drag-out arrangement used by PS 118. Rinse water is circulated between the drag-in and first drag-out tank by pumping from one to the other and having a gravity return by raising the height of one tank. They are able to operate on a closed-loop basis except when plating certain assembled tubular pieces that have excessive drag-out. A porous pot is used to prevent excessive contaminant buildup.

A substantial number of survey respondents use drag-in/drag-out rinsing configurations. A total of 66 (or 20.8%) of the respondents indicated that they use this method with manual operations and 34 (or 10.7%) use it with automatic operations. The success ratings for drag-in/drag-out rinsing are 3.39 (manual) and 3.82 (automatic).

3.5 RINSE WATER REDUCTION

The results of the Users Survey show that many plating shops have made significant strides in reducing water use. Sixty-eight percent of the respondents indicated that they have reduced flows by implementing pollution prevention. For shops that were able to quantify their achievements, the average flow reduction was approximately 20,000 gpd (or about 30%). Some shops were reluctant to make estimates of flow reduction because process and production changes that have occurred during the same time period make such estimates difficult (includes at least 19 shops). The following shops achieved the most dramatic flow reductions (all of these shops reduced their flows by 50,000 gpd or more or by 90% or more) (shown are the original and current flow rates and the year of the original flow rate data):

- **PS 022:** from 140,000 gpd to 70,000 gpd (50% reduction since 1980)
- **PS 036:** from 52,700 gpd to 2,700 gpd (95% reduction since 1978)
- **PS 059**: from 90,000 gpd to 10,000 gpd (89% reduction since 1977)
- **PS 118**: from 232,630 gpd to 42,630 gpd (82% reduction since 1983)
- **PS 139**: from 127,000 gpd to 52,000 gpd (59% reduction since 1986)
- **PS 150**: from 400,000 gpd to 100,000 gpd (75% reduction since 1986)
- **PS 172**: from 150,000 gpd to 70,000 gpd (53% reduction since 1975)
- **PS 184**: from 121,000 gpd to 11,000 gpd (91% reduction since 1982)
- **PS 207**: from 68,000 gpd to 18,000 gpd (74% reduction since 1986)
- **PS 213**: from 130,000 gpd to 50,000 gpd (62% reduction since 1985)
- **PS 250**: from 91,000 to 11,000 gpd (88% reduction, base year not given)
- **PS 268**: from 87,000 gpd to 17,000 gpd (80% reduction since 1987)
- **PS 292**: from 160,000 to 60,000 gpd (63% reduction since 1985)
- **PS 296**: from 1,900,000 gpd to 1,700,000 gpd (11% reduction, base year not given)
- **PS 298**: from 160,000 gpd to 90,000 gpd (70% reduction since 1990)
- **PS 309**: from 25,000 gpd to 1,450 gpd (94% reduction since 1984)

A summary of the Users Survey data relative to methods for reducing rinse water use is presented in Exhibit 3-16.

Although the plating industry as a whole has significantly reduced water use during the past 10 to 15 years, many plating operations can further reduce water use by improving the efficiency of their rinsing operations. The advantages of reducing water use include: (1) lowering operating costs by reducing the size of water bills; (2) reducing the quantity of treatment chemicals used (treatment chemical use is mostly dependent on the mass of contaminants, but a portion of treatment chemical use is

Exhibit 3-16. Summary of Users Survey Data Relating to Rinse Water Reduction

Rinse Water Reduction Method	Numberof Shops Using Method[1]	Percentage of Shops Using Method[1]	Success Rating[2]
1. Manually turning off rinse water when not in use	209	65.7	3.63
2. Conductivity or pH rinse controls	51	16.0	3.25
3. Timerrinsecontrols	36	11.3	3.78
4. Flow restrictors	222	69.8	4.10
5. Counter current rinses	217	68.2	4.21
6. Spray rinses	124	39.0	3.82
7. Air agitation in rinse tanks	185	58.2	3.71
8. Flow meters/accumulators to track rinse water	37	11.6	3.73
9. Reactive rinsing or cascade rinsing	76	23.9	3.79
Average			3.78

[1] Based on 318 survey responses.
[2] Respondents were asked to rate the success level of the method on a scale of 1 to 5, with 1 being the least successful and 5 being the most successful.

related to hydraulic loading (see Section 7 which covers end-of-pipe treatment); (3) potentially improving the removal efficiency of waste treatment systems; and (4) reducing the needed size of future end-of-pipe treatment systems and certain types of recovery technologies.
Central to the reduction of rinse water use is the required quality of water used in rinsing. Simply reducing the flow rate of water in a rinse system, without regard to water quality, may cause loss of plating quality or appearance or it may cause the contamination of the next tank in the plating sequence. Various rinse water quality criteria are presented in the literature. These are discussed in Section 3.5.3.1.

Various methods of water use reduction have been identified in the literature and throughout the Users Survey. These methods have been categorized into three groups: (1) optimizing the rinse tank design; (2) controlling the rate of rinse water use; and (3) using alternative rinsing configurations. The following is a discussion of each group.

3.5.1 Optimal Rinse Tank Design

The key objectives with regard to optimal rinse tank design are to attain fast removal of drag-out from the part and complete dispersion of the drag-out throughout the rinse tank. When these objectives are achieved, the time necessary for rinsing is reduced and the concentration of contaminants on the part when it leaves the rinse tank are minimized for a given rinse water flow rate. The following are rinse tank design elements that help to achieve fast drag-out removal and complete mixing. These methods can be combined to develop an optimal rinse tank design for a given workload. The optimal rinse tank design for different shaped parts, racks and barrels will usually result in the selection of a different combination of design elements. When parts with various configurations are rinsed in the same tanks, it may be necessary to compromise the design to provide adequate rinsing for all parts.

- Select the minimum size rinse tank in which the parts can be rinsed and use the same size for the entire plating line.
- Locate the water inlet and discharge points of the tank at opposite positions in the tank to avoid short-circuiting.
- Use a flow distributor/sprayer to feed the rinse water evenly.
- Use air agitation, mechanical mixing or other means of turbulence.
- Use spray rinsing for flat surfaced parts.
- Use ultrasonics, where applicable.

Air agitation was a relatively common element of rinse tank designs used by survey respondents. A total of 185 (or 58.2%) of the respondents reported its use. The average success rating was 3.71 (see Exhibit 3-16).

Spray rinsing is used by 124 (or 39.0%) of the survey respondents. The success rating for this method was 3.82.

3.5.2 Controlling the Flow Rate of Rinse Water Use

Regardless of the type of rinse tank arrangement employed (e.g., single overflow, counterflow), water use reduction can be achieved by coordinating water use and water use requirements. When these two factors are perfectly matched, the rinse water use for a given work load and tank arrangement is optimized. Four methods of coordinating water use and water requirements were identified during the Users Survey and literature search. Each of these methods is discussed in the following subsections. Some methods are applicable to a range of plating operations while others are more relevant to specific conditions (e.g., small manual operations, large automated machines). Some of the methods can be combined to optimize water use.

3.5.2.1 Flow Restrictors

Flow restrictors are inexpensive devices that are connected in-line with the tank's water inlet piping to regulate the flow of water through the pipe. They are typically an elastomer washer that flexes under pressure such that the higher the water pressure, the smaller the hole available for flow passage. Therefore, they maintain a relatively constant flow under variable water pressures. Flow restrictors are available in a wide range of sizes (0.1 gpm to more than 10 gpm). The smaller sized restrictors are most commonly used with multiple counterflow rinse tank arrangements and the larger ones are commonly used with single overflow rinses. Some restrictors aerate the water as it passes through, in a manner similar to a kitchen faucet (venturi effect).

Flow restrictors are applicable to nearly all rinse systems. A possible exception is a rinse tank equipped with a conductivity controller (see Section 3.5.2.3). With conductivity controllers, the instantaneous water flow rate is unimportant since the controller stops water flow based on the low conductivity set point of the controller and the conductivity of the water in the rinse tank. Therefore, restricting the flow will only reduce the time needed to dilute the rinse water to the conductivity set point and will not effect the total volume of water used.

Flow restrictors as a stand alone method of rinse water control are only effective with plating lines that have constant production rates, such as automatic plating machines. Even in such cases, to use water efficiently, the plater must have a means of stopping water flow during non-production periods. With variable production rates, flow restrictors alone will not provide the necessary coordination of rinse water need and use. One method for improving this coordination is to install a timer rinse control (see Section 3.5.2.4).

Generally, the size of a flow restrictor is selected to provide adequate rinsing for all parts. This means that the maximum rinse water flow requirement is the governing factor and that on the average, the flow will be higher than necessary for good rinsing. This fact is a sufficient reason for supplementing the control provided by a flow restrictor.

Flow restrictors are widely used by the respondents to the Users Survey. A total of 222 (or 69.8%) of the respondents employ this method of water use reduction. The average success level of this method is 4.10.

Only one survey respondent cited any problems with flow restrictors. That shop indicated that their platers usually remove the flow restrictors when they need to fill an empty rinse tank quickly and that the platers occasionally forget to replace them (PS 173).

3.5.2.2 Manual Control of Water Flow

Manual control of water flow simply refers to manually opening and closing water valves to adjust flow or to turn the water flow on or off. This method of control is obviously dependent on the operator and usually results in inconsistent water use. Several shops emphasized that this was a particular problem at their facility (PS 133, PS 176, PS 180, PS 265, and PS 276).

Combining manual control with flow restrictors reduces the variability of water flow; however, it does not address the problem of water use during idle production periods. Manual control can be improved by installing a main water valve for an entire plating line that stops water flow to all rinse tanks in that line.

A manual control method reported by one shop involved dumping and refilling all rinse tanks at one time, at the end of the day (PS 055). This shop is relatively small and has an average discharge of 5,500 gpd.

Of the 318 respondents, 209 (or 65.7%) use manual water control. The average success level for the respondents is 3.63.

3.5.2.3 Conductivity Controls

These units consist of three components: (1) probe or sensor located in the rinse tank that senses the conductivity of the rinse water; (2) transformer box that houses the solid state circuitry that controls the system; and (3) a solenoid valve that opens and closes in response to signals from the circuitry. In use, when drag-out is introduced to the rinse tank, the probe senses a rise in conductivity above a set-point which is picked up by the circuitry and the solenoid water valve is opened. The value remains open until the probe senses a drop in conductivity below a set-point. The set-points are operator-adjustable to permit use over a range of desired water qualities.

Conductivity rinse controls have been effectively used to reduce rinse water use (ref. 316). However, in some cases, they have been removed from service due to maintenance problems (ref. 21). The results of the Users Survey show a moderate level of usage for conductivity controllers (51 respondents or 16.0%) and the lowest success rating among rinse water reduction methods (3.25).

A report prepared for the military (ref. 379) listed three criticisms of conductivity controllers. First, the units are sometimes disabled or overridden by operators who object to the appearance of a controlled rinse, which may be less clear than a free-flowing rinse. Operators often override the units by placing the probe into a process tank or a bucket of process solution (i.e., causes solenoid to remain open). This problem can be controlled by shortening the length of the probe's cable or by running the cable through a PVC pipe (if using the later method, be certain the piping arrangement permits access to the probe for periodic cleaning). Second, the controllers do not sense non-ionic contaminants and rinse tanks may become contaminated with particulates such as dust. Third, the units require frequent preventive maintenance to remain operable. In particular, the probes become encrusted or fouled, especially in alkaline rinse waters. For example, one shop reported that they are "difficult to keep clean" (PS 086). Another shop indicated that they are a high maintenance item and susceptible to fouling (PS 114). Also, probes sometimes become entangled with plating racks and are damaged or disconnected.

One shop reported that they tried conductivity controllers and found them to cause high water use (PS 069). Presumably, the range of settings available on their units did not satisfy their rinse criteria. The shop discontinued the use of these devices and reverted to manual control and flow restrictors. A different problem was reported by another shop that indicated their controllers were not triggered by acid at a concentration that obviously impacted rinse quality (PS 124).

Altmayer suggests that conductivity controls using probes to measure the resistance (conductance) between two wires are very troublesome to operate, but that the controllers using inductive loop type probes, which are less susceptible to fouling, can be very successful. Altmayer also related to the experience of PS 069 and PS 124 (see above paragraph), saying that controllers must be specified to control within a certain range of conductivity and that choosing this range is the hardest part of the design. He stated that this problem is mainly caused by the lack of information and data available. Most guidance is based on total dissolved solids (TDS) as opposed to metal ion content, which is more familiar to the plater (ref. 482).

3.5.2.4 Solenoid Valve on Automated Plating Machines

One large job shop reported that they have installed solenoid valves on the water lines feeding each of their automated plating machines (i.e., semi-automated return-type). The valves automatically close when power to the plating machine is turned off.

3.5.2.5 Timer Rinse Controls

Timer rinse controls consist of a push-button switch and timer mechanism and a solenoid valve. These units operate in a manner similar to conductivity controllers; however, rather than regulating rinse water flow on the basis of rinse tank water quality, the timer controls simply turn water on and off based on a pre-set time period.

In operation, a plater lowers parts into the rinse tank and pushes a button (alternatively, a momentary switch could be used that is activated by lowering a rack or barrel). The button or switch activates a timer and opens the solenoid valve for a preset time period. After that time period has expired the solenoid valve automatically closes.

The timer setting is selected through trial and error. It is best to select a time period that provides consistently clean rinse water, without excessive waste. Once set, the time period is not changed unless the general trend of production changes.

Fewer of the Users Survey respondents have employed timer rinse controls than conductivity controllers. The number of respondents using timer rinse controls was 36 (or 11.3%). The average success rating was 3.78, which is significantly higher than for conductivity controllers. Several shops indicate that they have employed both. One such shop rated their success with the timer rinse control as "5", while rating their success with conductivity controllers as "1".

3.5.2.6 Flow Meters and Accumulators

These devices by themselves do not reduce water use. However, they make the metal finisher aware of water use rates and are useful in identifying excessive water use.

Flow meters and accumulators are most useful when installed on fresh water lines feeding individual rinse tanks or, at a minimum, on pipes feeding individual plating lines. Meter readings taken over an extended time period will show trends in water use. Using these data, shop management can identify specific locations where excessive water use occurs and can correct the problem before long-term wastage has resulted.

One shop that installed flow accumulators indicated that they were useful for specific studies, but in general were found to be a rather expensive way to generate usage numbers of "questionable validity" (PS 124).

3.5.3 Alternative Rinsing Configurations

3.5.3.1 Generating Rinsing Data

To evaluate rinsing alternatives effectively, the plater needs drag-out and water use data, rinse quality criteria and equations that utilize these data and criteria. Of these items, the drag-out data are the most difficult and expensive to obtain. The high cost is due to a need for expensive analytical work. A well proven method of generating drag-out data, that is easy and inexpensive to apply, is discussed in this subsection. This method, which utilizes conductivity measurements, was originally described by Mohler (ref. 2).

In order to utilize this methodology, a conductivity meter is required. Using a conductivity meter greatly reduces the analytical costs of generating rinsing data. In a matter of hours a plater can generate a quantity of data equal to that generated by days of laboratory work. Also, unlike using laboratory analytical methods, with the conductivity meter approach there is almost no lag time associated between the taking of samples and generating of results. Most plating shops have combination pH/conductivity meters in their shops that can be used for this purpose. Alternatively, a portable unit of sufficient quality can be purchased for under $100. Exhibits 3-17 and 3-18 present the conductivity values and rinsing criteria used in the following method.

1. Use conductivity measurements to determine the present concentration of a plating solution in a rinse tank (example: sulfuric acid pickle).

   - Measure the conductivity of your tap water (or DI water if used):
     a = 100 μmho (micromhos)

   - Measure the conductivity of the rinse water. If a multiple rinse tank arrangement is present (e.g., counterflow), make the conductivity measurement in the last rinse (i.e., least contaminated).
     b = 5,500 μmho

   - In a container, add one ml of the pickle solution to one liter of rinse water and measure the conductivity:
     c = 6,250 μmho

   - The increase in conductivity for one ml of pickle solution per one liter of rinse water is:
     c - b = 6,250 - 5,500 = 750

   - The concentration of pickle solution in the rinse water is:
     e = (b - a)/(c - b) = 5,400/750 = 7.2 ml/l

   - The equivalent sulfuric acid in the rinse is (see Exhibit 3-17, 6,300 μmho = 1,000 mg/l):
     1,000 (c - a)/6,300 = 6,150,000/6,300
     f = 976 mg/l equivalent H_2SO_4

   - Compare this concentration to the criteria in Exhibit 3-18 to determine if the proper rinse flow rate is being used. If necessary, adjust the rinse flow and repeat the measurements until the rinse criteria is met.

2. Use conductivity measurements to determine the drag-out rate.

   - First determine the rinse water flow rate. Record the number of racks processed per hour. Note the valve setting, turn off the flow, bail out six

Exhibit 3-17. Conductives of Typical Plating Solutions (Conductivity given in μmho)

Bath	40 mg/l	100 mg/l	150 mg/l	750 mg/l (0.1 oz/gal)	1,000 mg/l (0.13 oz/gal)	10,000 mg/l (1.3 oz/gal)
Brass (cyanide)		130	300	1,490	2,000	13,000
Bright Dip	185		700	3,500		
Bronze (cyanide)			240	1,200		
Cadmium (cyanide)		130	260	1,310	1,740	13,000
Chromic Acid	120		450	2 250	3,()00	25,000
Cleaner, Light Duty		110		1 150	1,540	13,000
Cleaner, Heavy Duty		170		2,250	2,960	27,000
Copper CN Strike		260		1,600	2,540	10,000
Copper, High Concentration				900		
Copper Rochelle				1,900		
Copper Sulfate		60		1,400	1,630	4,000
Dichromate Brightener		60	100	480	640	5,900
Gold (cyanide)			260	1,300		
Hydrochloric Acid		700		5,000	11,000	103,000
Iron Sulfate		90			870	6,450
Lead Fluoborate		10			590	6,400
Nickel		60		640	530	3,900
Nitric - Sulfuric Bright Dip		230			4,680	43.000
Platinum				1,200		
Rhodium			700	3.500		
Silver (cyanide)		80		680	910	8,500
Sodium Cyanide		190			1,870	17,000
Sodium Hydroxide		400		3,450	5,800	53,000
Sulfuric Acid		600		4,000	6,300	48,000
Tin (acid)				500		
Tin (alkaline)		10	70	340	390	4,750
Zinc (cyanide)		100	280	1.390	1,800	14,000
Zinc (acid)		50		630	470	3,300

Exhibit 3-18. Rinse Criteria for Permissible Levels of Contamination

Type of Rinse	Normal Range for Adequate Rinsing (mg/l)
Following cleaner/acid pickle	400 to 1,000
Following functional plating	100 to 700
Following bright plating	5 to 40

sources: ref 2, 3, 16, 20

or more inches of water, set the flow and record the rate of rise of the water in inches per minute. From the number of racks processed and the rate of rise of the water calculate the flow as liters per rack (alternatively flow can be expressed in l/ft^2 of wetted area, l/barrel, or another relevant unit of measure):

$$g = 8.0 \text{ l/rack}$$

- Second, measure the drag-in per rack. Turn off the flow, agitate the rinse, measure the conductivity, process a number of racks of work through the rinse, agitate the rinse, and measure the conductivity again. From the μmohs increase per rack calculate the drag-in ml/rack:

$$h = 50 \text{ ml/rack}$$

3. Use the equation presented in Section 3.5.3.2 to evaluate the effect of adding additional rinse tanks.

3.5.3.2 Counterflow Rinsing

Electroplaters have long reduced water use by employing several rinse tanks connected in series. Fresh water flows into the rinse tank located farthest from the process tank and overflows, in turn, to the rinse tanks closer to the process tank (see Exhibit 3-13). This technique is termed counterflow (or countercurrent) rinsing because the work piece and the rinse water move in opposite directions. Over time, the first rinse becomes contaminated with drag-out and reaches a stable concentration which is lower than the process solution. The second rinse stabilizes at an even lower concentration which enables less rinse water to be used then if only one rinse tank were in place. The more counterflow rinse tanks (three-stage, four-stage, etc.), the lower the rinse rate needed for adequate removal of the process solution.

Counterflow rinsing systems are not without drawbacks. The negative aspects of counterflow rinsing include: (1) cost of additional rinse tanks; (2) loss of valuable production space; and (3) an increase in production time/labor.

The rinse rate needed for adequate cleaning is governed by an exponential equation that depends on the concentration of plating chemicals in the drag-out, the concentration of plating chemicals that can be tolerated in the final rinse tank before poor plating results, and the number of counterflow rinse tanks. The mathematical rinsing models are based on complete rinsing (i.e., removal of all drag-out from the part/fixture) and complete mixing (i.e., homogeneous rinse water). These conditions are not achieved or even approached unless there is sufficient residence time and agitation in the rinse tank. More typically, each added rinse stage reduces rinse water use by 50 percent.

The most commonly applied counterflow rinsing model follows:

$$R_c = (C_t/C_r)^{1/n}$$

where,

R_c = rinsing ratio

C_t = concentration in the preceding tank

C_r = required concentration in the last rinse tank

n = number of rinse tanks

This model does not predict required rinse rates accurately when the value of R_c falls below 10. Also, complete rinsing will not be achieved unless there is sufficient residence time and agitation in the rinse tank.

The following example demonstrates the use of the counterflow rinsing equation. A typical Watts nickel plating solution contains 270,000 mg/l of total dissolved solids, and the selected final rinse concentration is 37 mg/l of dissolved solids. The ratio of C_t/C_r is 7,300 and approximately 7,300 gallons of rinse water are required for each gallon of process solution drag-in with a single-tank rinse system. By installing a two-stage rinse system, water requirements are reduced to 86 gallons of water per gallon of process solution drag-in (assumes 100% rinsing efficiency). The same degree of dilution is obtained in the final rinse, and the rinse water consumption is reduced by 99 percent. The mass flow of pollutants exiting the rinse system remains constant, however, the pollutants are much more concentrated with a two-stage rinse system than with a single-tank rinse system.

From the User Survey results, 217 respondents (or 68.2%) employ counterflow rinsing. These respondents gave this method of water use reduction a success rating of 4.21, the highest for any water use reduction method.

3.5.3.3 Cascade, Reactive, and Dual Purpose Rinsing

Cascade rinsing refers to the practice of reusing rinse water multiple times in different rinse tanks for succeeding less critical rinsing. Reactive rinsing is similar, but it refers to cases where a chemical reaction takes place as a result of using the rinse water for multiple purposes. An example is reusing the rinse water following acid cleaning as rinse water following alkaline cleaning. In this case, the acid rinse water helps to remove the viscous alkaline film remaining on a part after alkaline cleaning.

Dual purpose rinsing refers to the practice of using the same rinse tank for rinsing following more than one process tank. It provides essentially the same results as cascade and reactive rinsing but uses a fewer number of rinse tanks. Often, the employment of dual purpose rinsing means transporting a dripping rack/part over a considerable distance. This can result in dripping onto floors and/or the accidental contamination of other tanks. An exhaustive evaluation of dual purpose rinsing is presented by Mohler (ref. 3). Mohler presents methods and guidelines for ascertaining the accumulated concentration of chemicals in counterflow, dual-purpose rinses in order to determine the feasibility and economics of this technique.

Typically, cascade and reactive rinsing are used for automatic plating machines whereas dual purpose rinsing is used for manual operations.

Use of any of these methods must closely consider the combined chemistry in the rinse tank to prevent undesirable reactions that may impact worker safety (e.g., hydrogen cyanide formation) or work quality (e.g., precipitation of solids).

Reactive or cascade rinsing is used by 76 (or 23.9%) of the survey respondents to reduce water usage. These respondents gave this method an average success rating of 3.79.

3.5.3.4 Chemical Rinsing

The technique of chemical rinsing has been used by the metal finishing industry for many years. One of its earliest applications was to eliminate staining from chromium solution, which is notoriously difficult to rinse. By simply making the first rinse after chromium plate a stagnant rinse containing sodium bisulfite, the drag-in of hexavalent chromium was converted to trivalent chromium. The rinsability of the workpiece in the second rinse was improved considerably by changing the chemical nature of the film on the workpiece in the stagnant rinse and by reducing film concentrations before attempting to rinse by diffusion. The same principles are frequently employed in "neutralizing" dips.

The application of chemical rinsing to plant effluent treatment, known in the industry as integrated waste treatment, has been described by Lancy and Pinner (ref. 305). Aside from the environmental benefits, this type of rinsing also prevents the majority of heavy metal solids formed in the chemical rinse from reaching the succeeding water rinses by removing these materials in an external settling vessel. Removal of these solids is accomplished by flowing the chemical rinse solution to a treatment reservoir. The overflow from the reservoir is pumped back to the rinse tanks, forming a complete closed-loop system. Integrated treatment gained some popularity in the 1970's, but is believed to be in little use today, mostly due to high maintenance requirements.

3.5.3.5 Spray Rinsing

Spray rinsing is employed in various manners to reduce drag-out losses and rinse water use. Spray rinsing over process tanks (Section 3.4.2.2) provides direct recovery of drag-out. Spray rinse tanks can be used as drag-out tanks, single rinses, or multiple rinses.

A common use of spray rinsing is to substitute a spray rinse tank for an overflow rinse tank. Depending on the part configuration, spray rinsing generally uses from one-eighth to one-fourth the amount of water that would be used for equivalent dip rinsing (ref. 305). Spray rinsing is most effective for flat-surfaced parts and is less effective with recessed and hidden surfaces.

Although less commonly used, a spray rinse can substitute for a drag-out tank. Several shops indicated that they consider this a very successful pollution prevention measure (e.g., PS 177, PS 242, PS 275). Using this method, one shop has reduced their discharge from 15,000 gpd to an average flow of 800 gpd (PS 242). PS 275 employs a fog spray rather than a common spray rinse (success rating of "4").

Combined spray and dip rinse tank designs are employed where the bottom portion of a rinse tank acts as a dip tank and the upper portion a spray rinse. A weir is located at approximately the middle of the tank which maintains the solution level in the tank. In operation, the rack is lowered into the dip rinse, raised above the solution level, and sprayed with fresh water. This combination rinse can be nearly as effective as a counterflow rinse, but takes up the floor space of one tank.

The design of spray rinses must consider the size and shape of the part. Spray nozzles are available in many sizes and spray patterns, and should be selected appropriately. Usually, the pressure in the waterline is sufficient to operate an effective spray rinse, however, higher spray velocities can be obtained by pumping.

A special application of the spray rinse is a patented unit (ref. 6) that contains five to seven progressively cleaner rinse solutions in separate compartments. The solutions are successively pumped (up to 20 gpm) to a spray rinse tank and drain back to the unit. During each cycle, only the water use in the first spray is discarded or processed for recovery. The subsequent sprays are collected for reuse in the following cycles. The advantage of this unit is that it provides the effect of multiple counterflow rinsing with use of a single rinse tank. The floor space requirement of the unit is 7.5 square feet (five stage rinse unit) or 11.0 square feet (seven stage rinse unit).

Although it is not widely used, spray rinsing is applicable to barrel plating operations as previously shown in Exhibit 3-11.

One shop reported that their platers are mistrustful of the efficiency of spray rinses. Whenever plating quality problems arise, the platers target the spray nozzles as the cause (PS 176). Another shop complained that spray rinses often clog and must be cleaned or replaced (PS 230).

The Users Survey showed a moderate to high usage rate of spray rinsing. Most frequently it was used as a water use reduction method and to a lesser degree for drag-out loss prevention (fog or spray rinse over process tank). As a water use reduction method 124 (or 39.0%) of the shops employed spray rinsing. These respondents gave spray rinsing an average success rating of 3.82.

3.5.3.6 Combined Drag-Out Loss/Rinse Water Reduction Rinsing Arrangements

Many plating shops combine drag-out tanks and overflow rinsing in the same rinse systems (see Exhibit 3-13). For example, a four rinse system could consist of two drag-out tanks connected in series and two free-flowing rinses connected in series (counterflow). Alternatively, the system could consist of three drag-out tanks in series and a single overflow rinse or a drag-in/drag-out arrangement and two counterflow rinses. Various rinsing configurations can also be combined with chemical recovery technologies, as discussed in Section 3. The optimal rinse configuration will depend on numerous factors including: (1) the evaporation rate in the plating tank, (2) the drag-out rate, (3) the rinse water quality requirement (final rinse), (4) process chemical costs, (5) alternative technology recovery costs, (6) water costs, and (7) wastewater treatment/sludge disposal costs.

In general, when more of the available rinse tanks are used as drag-out tanks, the process chemical and wastewater treatment operating costs are lowered and water use costs are increased. The reverse is true when more tanks are used for counterflow rinsing than for drag-out recovery tanks.

The optimal configuration can be determined through mathematical means which must be supported by data collection (i.e., drag-out and evaporation measurements, production rates, etc.) for producing accurate results. Examples of such analyses are presented in ref. 305 for chromium and nickel plating. As an alternative to using the rinsing equations to perform the calculations, a modeling program can be employed. A commercially available software program (ref. 317) permits an analysis considering up to five rinse stations and the use of supplemental evaporative recovery. This program permits the user to add recovery rinses, change tank volumes, experiment with process chemistries, add evaporators, and change workload to find the combination that makes the most environmental and economic sense (ref. 398).

3.6 REUSE OF TREATED WASTEWATER

Some shops responding to the Users Survey indicated that they reuse treated wastewater in their rinse tanks. Most of the reported reuse of treated wastewater is practiced at shops with conventional treatment (i.e., hydroxide precipitation) (PS 054, PS 091, PS 109, PS 123, PS 128, PS 199). This reuse is generally done in non-critical rinses (e.g., PS 199 reuses effluent in rinses following alkaline soak cleaners and acid dips). Some shops filter the treated wastewater before reusing it (e.g., PS 123 pumps it through a sand filter). Two shops reused the filtrate from a filter press as rinse water (PS 109 and PS 168).

Those contemplating the reuse of chemically treated water should be cautioned that a hydroxide precipitation process generates an effluent that is high in dissolved solids (e.g., sodium) that will contaminate subsequent process tanks. More extensive treatment is needed to produce a suitable quality of water for most rinsing needs. For example, two shops responding to the Users Survey reuse treated water after: (1) hydroxide precipitation/ion exchange (PS 068) and (2) hydroxide precipitation/ultrafiltration/reverse osmosis (PS 233). Ion exchange is generally not a good choice of technology for recycling treated wastewater. The treated wastewater will contain a high concentration of anions that must be removed to produce a reasonable water quality. Applications of ion exchange recycling are discussed in Sections 4.4 and 7.5.2. Membrane filtration, a viable option for effluent recycling, is discussed in Section 7.5.3.

3.7 CLOSED-LOOP PROCESSING AND ZERO DISCHARGE

Closed-loop processing and zero discharge are two terms that are often used by platers, vendors, consultants and regulators (ref. 206, 223, 229, 258, 294, 397, 469, and 470). Various definitions are used for these terms, most of which recognize that 100 percent recovery/reuse of all materials (process chemicals, other chemicals, water, sludge, etc.) is not practical, economically feasible nor efficient from an energy standpoint.[4] More realistically, all metal finishing shops as well as individual metal finishing processes generate some form of residuals. The residuals are typically in one or more of four common forms: wastewater, spent process solutions, sludge, or air emissions. Often, some recovery/reuse can be implemented, but with every technology scheme or configuration there is some residual generated. Often there is a trade-off between the quantity and characteristics of two or more of the four common residuals. As a simple example, closed-loop rinsing after chromium plating will result in a build-up of contaminants in the bath. The rinse water discharge can be eliminated, but the bath will have to be either discarded or "purified." If the bath is purified, the purification process (e.g., porous pot, membrane electrolysis, or ion exchange) will result in a residual that must be properly discarded.

Of the 318 respondents to the Users Survey, 24 (or 7.5%) indicated that they have achieved zero discharge (these shops can be identified from Exhibit 2-8). Many other shops indicated that they are working towards this goal. Of the shops that have achieved zero discharge, 58% are primarily hard chrome platers. The hard chrome process is one the easiest to operate without a wastewater discharge because of the high ratio of evaporation to drag-out (i.e., permits use of recovery rinsing). Also, as indicated in Section 4.2.5, the respondents that have achieved zero discharge are generally smaller shops, which tend to be less complex and less automated. As a result, these shops generally have simple, manual processes that if given sufficient control can be operated without a wastewater discharge. The means implemented by these shops to achieve zero discharge is most frequently recovery rinsing and atmospheric evaporation. However, eight respondents have implemented advanced technology schemes in order to achieve zero discharge. The technologies implemented by these shops include vacuum evaporation, ion exchange, and membrane filtration. These systems are discussed in Section 7.

There are various economic benefits of operating closed-loop processes or a zero discharge shop, including: lower water and sewer charges, less effluent monitoring and paperwork, and reduced chance of compliance excursions/fines. Some firms are driven to zero discharge by specific regulatory conditions rather than economics, such as discharge permits with low numerical or mass loadings, TDS limits or aquatic-toxicity-based standards and low in-stream dilution factors. Furthermore, some facilities are practically isolated from sewers or streams (ref. 516). A negative aspect of zero discharge is the loss of flexibility that is provided by a wastewater treatment system/discharge. This loss can translate into higher costs for hauling spent process solutions to disposal sites, for example, if unexpected process solution contamination occurs. Also, one source indicates that closed-loop systems may pose regulatory conundrums, in that without a wastewater discharge permit, firms lose the RCRA treatment exemption and become subject to Part B permits (ref. 223).[5] That source suggests that firms intent on eliminating wastewater discharge should negotiate a "closed-loop" Clean Water Act permit (i.e., NPDES or POTW) with the regulatory authority. Legal requirements and options for companies that wish to operate zero discharge facilities are discussed in the literature (ref. 516).

Zero discharge is generally achieved in stages: (1) prepare a plan (e.g., as part of a pollution prevention program, see Section 3.2); (2) implement good operating practices (Section 3.3); (3) minimize drag-out which causes wastewater generation (Section 3.4); (4) modify rinsing practices to reduce/eliminate flows to treatment (Section 3.5); (5) implement bath maintenance[6] and chemical recovery (Sections 4 and 5); (6) continuously reassess flow rates, process chemical/contaminant concentrations in recovered streams, costs, and benefits. Additional suggestions for zero discharge programs can be found in ref. 223.

3.8 OTHER RINSING INFORMATION

This subsection contains information on methods applied for rinse water use reduction that are not discussed in detail elsewhere in the report.

One shop indicated that they use in-tank filtration in large rinse tanks (PS 228) and another shop indicated that they use external filtration for some rinses (PS 275). The filtration technology, which is more frequently applied to process baths than to rinse tanks, removes particulate matter from solutions. The source of particulates for these two shops was not indicated on their survey forms. However, it is known that suspended solids in rinses can be caused by precipitation (e.g., mixing of alkaline cleaner drag-out and metal-bearing drag-out) or airborne material.

Although it is not widely applied to rinsing, the use of ultrasonics can improve its effectiveness. Ultrasonics loosen soil particles that are not completely removed during the cleaning step. Although one would expect the use of ultrasonics to be limited to small rinse tanks and critical operations due to its high capital costs, one shop reported that it uses ultrasonics in a large rinse tank (12' x 30" x 40" deep) following electropolish (PS 022). Another respondent suggests that ultrasonics be used in rinse tanks whenever feasible (PS 068).

Most existing process lines suffer from a lack of rinsing. When use of additional counterflow rinse tanks is suggested, the most common complaint is that there is no space, thus rinsing suffers and flow rates remain high. To gain additional tankage, some shops partition large rinse tanks into two or more smaller sections (ref. 9).

3.9 SUMMARY OF GENERAL WASTE REDUCTION PRACTICES

Summary tables have been prepared based on the information and data from the Users Survey and literature search and the discussion presented in this section. Exhibit 3-19 presents a summary of improved operating practices that reduce waste generation. Exhibit 3-20 presents a summary of drag-out reduction methods and Exhibit 3-21 presents a summary of rinse water use reduction methods.

Section 3 Footnotes:

[1] Section 4 of the Users Survey addressed pollution prevention methods and techniques. The platers were requested to rank, on a scale of one to five, the success of each method or technique that they used (1 = unsuccessful and 5 = highly successful).

[2] Additional information regarding pollution prevention programs, including examples of assessments, is available from EPA (ref. 25, 302).

[3] One of these shops fabricated doubled walled tanks themselves and installed sight glasses (PS 133). During follow-up conversation, this shop warned that brass sight glasses, which are the most commonly used, are not applicable to cyanide containing tanks. They suggest using either stainless steel or PVC (PVC sight glass available from U.S. Plastic Corporation, Lima, OH). PS 133 also suggests placing the sight glass in a protected, front corner of the tank (i.e., should be readily visible, but protected from damage).

[4] The most frequently accepted definitions of the terms closed-loop processing and zero discharge refer to the elimination of a Clean Water Act regulated wastewater discharge rather than the complete elimination of wastes of all types; usually associated with partial recovery of plating chemicals.

[5] The Resource Conservation and Recovery Act requires firms to obtain a permit for the operation of hazardous waste treatment processes. Such systems are exempt from permitting if they treat wastewaters that are discharged under the Clean Water Act.

[6] Some form of bath maintenance must be implemented concurrently with drag-out recovery, especially when high contaminant loadings are present.

Exhibit 3-19. Improved Operating Practices That Reduce Waste Generation (1 of 3)

Method	Applications	Restrictions	Advantages/Benefits		Disadvantages	Extent of Use
			Pollution Prevention	Other		
Waste stream segregation of contact and non-contact (e.g., cooling) wastewater (i.e., segregate non-contact process waters and treat only process water).	All plating facilities with both contact and non-contact wastewaters.	Some non-contact waters (e.g., steam condensate from chromium plating heating coils) may be contaminated.	Eliminates dilution of process water prior to treatment which can increase treatment efficiency.	Reduces treatment reagent usage and operating costs.	High capital cost.	Cooling and other non-contact process waters (e.g., rectifier cooling) are routinely segregated (e.g., PS 197). PS 116 indicated that operating non-contact waters on a closed-loop was helpful.
Maintain physical integrity of all tanks conduct periodic inspection of tanks and liners. immediately repair, as required.	All process and rinse tanks.	None identified.	Prevents leaks from process rinse tanks that would contribute to waste generation.	Reduces potential for combining acid and cyanide solutions on floors that can result in HCN formation.	None identified.	59% of the survey respondents have established preventative maintenance program.
Establish written procedures for bath make-up and additions. Limit chemical handling to trained personnel. Keep tank addition logs.	All plating facilities.	None identified.	Prevents discarding process solutions due to incorrect formulation or contamination.	Improves plating solution and work quality consistency. Improves shop safety.	None identified.	71% of the survey respondents have written procedures.
Use a compatible wetting agent in preceding rinse to reduce drag-in of rinse water.	Cold processing tanks that tend to build-up volume.	Not needed with heated process tanks.	Minimizes decanting due to volume buildup.	Maintains bath near optimal operating concentration.	Adds a new variable to plating line's chemistry.	Not determined.
Avoid use of oil contaminated air for air agitation by adding air filtration unit or converting to low oil system (e.g., regenerative blower).	All plating facilities.	None identified.	Reduces generation of spend bath due to oil contamination.	May improve work quality and reduce reject rate.	Moderate capital cost.	Not determined.
Immediately retrieve fallen parts, racks, etc. from tanks (e.g., use magnet on a pole to retrieve fallen steel parts).	All plating facilities.	None identified.	Reduces bath dumps due to contamination from corroding parts, racks, etc.	Value of parts, racks, etc. saved.	None identified.	79% of the survey respondents regularly retrieve fallen parts.
Pre-inspect parts to prevent processing of obvious rejects; pre-cleaning of "clean parts; descaling of parts without scale: etc.	All plating facilities,	None identified.	Reduces number of parts processed which reduced waste generation.	Reduces overall operating costs. May improve product quality.	None identified.	56% of the survey respondents pre-inspect parts.
Eliminate obsolete processes and/or unused or infrequently used processes (e.g., using contract plater for infrequently used process).	All plating shops.	None identified.	Reduces risks associated with hazardous chemicals. Creates floor space to add countercurrent rinses or other pollution prevention methods.	Creates a safer and cleaner working environment.	None identified.	Not determined.
Install overflow alarms on 811 process tanks to prevent tank overflow when adding water to make up for evaporative losses.	Applicable to all process tanks, especially those operated at elevated temperatures.	None identified.	Minimizes potential for catastrophic loss of process solution via overflow.	Prevents loss of expensive chemicals.	Total reliance on alarm may result in overflow if system fails. Capital intensive, especially sophisticated systems.	16% of the survey respondents employ overflow alarms.

Exhibit 3-19. Improved Operating Practices That Reduce Waste Generation (2 of 3)

Method	Applications	Restrictions	Advantages/Benefits		Disadvantages	Extent of Use
			Pollution Prevention	Other		
Conductivity~and pH measurement instruments and alarm system for detecting significant chemical losses. Sophisticated systems automatically telephone response personnel if shop is unattended.	All plating facilities.	None identified.	Identified process solution overflows and leaks before total loss occurs. Alerts treatment operators to potential upset condition.	Reduces losses of expensive plating solutions.	Maintenance of instruments and alarm system required.	14% of the survey respondents have installed spill/leak detection systems.
Use dead-man switches on water feed lines for manually filled tanks. Do not use unmanned hoses for filling tanks.	All plating facilities for formulating bath or adding evaporative makeup to heated plating tanks.	None identified.	Reduces potential for overfilling process tanks.	None identified.	None identified.	35% of the survey respondents use a controlled method of water addition.
Proper rack and barrel design and maintenance to reduce dragout losses and cross-contamination of baths.	All plating facilities that use racks and barrels.	None identified.	Reduces dragout and resultant waste generation. Reduces frequency of bath dumps caused by contamination.	Reduces overall operating costs.	None identified.	65% of the survey respondents perform regular rack/barrel maintenance.
Combine the sensitization and actuation solution.	Electroless plating operations.	May be restricted by proprietary process chemistry.	Eliminates a rinsing step.	Provides additional floor space.	Possible reduction in the activity of the catalyst.	Not determined.
Regular stripping of copper or nickel from electroless tank wells prevents excessive buildup and loss of metal.	Electroless copper and nickel plating.	None identified.	Minimizing metal buildup on tank walls prevents the rapid buildup of metal and associated major stripping operation.	Reduces input chemical costs.	None identified.	Not determined.
Mask areas that should not be plated to avoid unnecessary etching and to limit corrosion.	Electroplating processes	Maskant must be compatible with bath chemistry and solution temperature.	Reduces buildup of dissolved metals in plating solutions that may result in bath dumps. Masking holes may reduce dragout.	Reduces corrosion of parts.	Masking/demasking may generate waste.	Not determined. Presumed extensive use in processes where thick metal deposits are applied (e.g., hard chrome).
Filter and recycle (closed-loop) deburring water (i.e., rinsewater recycled to deburr units and metal fines recovered).	Printed circuit board shops.	None identified.	Reduces quantity of wastewater generated.	Copper fines from PCB deburring water may be sold as scrap.	None identified.	Assumed to be widely used.
Control material purchases to minimize obsolete material disposal. Includes periodic review of future plating needs and chemical inventory.	All plating facilities.	None identified.	Reduces hazardous waste generation.	Reduces chemical purchases.	None identified.	65% of the survey respondents maintain strict chemical inventory control.
Use process baths to maximum extent possible before discarding. Eliminate dump schedules. Perform more frequent chemical analyses.	All plating facilities. Especially applicable to cleaning and stripping baths that are frequently discarded.	Requires analytical or statistical process control (SPC) capability.	Prevents discarding of solutions prematurely.	Reduces chemical costs. Chemical adjustment of baths will improve work quality.	Requires monitoring of baths.	74.3% of the survey respondents use process solutions to the maximum extent possible.

Exhibit 3-19. Improved Operating Practices That Reduce Waste Generation (3 of 3)

Method	Applications	Restrictions	Advantages/Benefits		Disadvantages	Extent of Use
			Pollution Prevention	Other		
Reduce bath dumps by using filtration to remove suspended solids contamination. Includes use of either external or in-tank filtration.	Cleaning, plating and other process solutions.	Some process solution (e.g., chemical milling) cannot be filtered due to high suspended solids content.	Extends bath life. Reusable filter cartridges reduce solid waste generation.	Improves bath performance.	External filtration system may leak. Most sites use non-reusable filters that are discarded as hazardous waste.	The most commonly used method of corrective bath maintenance (see interim Report No. 4).
Cascade use of process chemicals to obtain maximum use before discarding (e.g., use of used cleaning or stripping solution prior to use of fresher solution).	Cleaning and stripping solutions.	Critical cleaning operations. Many automated processes.	Reduces chemical usage and waste generation for applicable processes.	None identified.	Batch treatment causes production downtime. Most purification methods generate waste products.	Not determined.
Remove cadmium and zinc anodes from bath when it is idle. Anodes baskets can be placed on removable anode bars that are lifted from tank by an overhead hoist.	Cadmium and zinc cyanide baths.	None identified.	Eliminates cadmium/zinc buildup causing decanting of solution due to galvanic cell set up between steel anode basket and cadmium/zinc anodes.	Maintains bath within narrow Cd/Zn concentration providing more predictable plating results.	None identified.	25% of the survey respondents use this technique.
Use both insoluble and soluble anodes in the same bath to balance cathode and anode efficiencies.	Baths normally operated with only soluble anodes (e.g., Cd).	None identified.	Minimizes decanting due to metal concentration buildup.	Maintains bath near optimal operating metal concentration.	None identified.	Not determined.
Nickel plate copper bus bars to reduce the rate of corrosion and subsequent contamination of the bath.	Highly applicable to chromium plating because of attack of CrO_3 on Cu.	Not identified.	Reduces bath contamination.	May improve bath performance.	None identified.	Not determined. One shop indicated that they used this method (PS 108).
Employ reusable robbers and shields in place of disposal tapes.	Hard chrome plating and nickel buildup plating.	Single items or small production runs.	Reduces volume of hazardous waste (i.e., contaminated solids).	Reduces purchase of masking materials.	Requires labor for design/fabrication.	Not determined, only mentioned by one shop (PS 165).

Exhibit 2-20. Drag-Out Reduction Methods that Reduce Waste Generation

Method	Applications	Restrictions	Advantages/Benefits		Disadvantages	Extent of Use Based on Survey Results
			Pollution Prevention	Other		
Minimize the formation of drag-out by reducing plating bath viscosity with higher bath temperatures, lower bath chemical concentrations, and wetting agents.	Applicable to plating solutions that have a wide range for operating conditions (e.g., chromium and nickel plating).	Some baths are highly heat sensitive (e.g., most cyanide baths). Wetting agents are not compatible with all process solutions. PS 139 indicated that lowering their bath conc. reduced their production rate and that wetting agents caused foaming.	Reduces pollutant mass loading on treatment processes, treatment reagent usage, and resultant sludge generation. May improve treatment operation/removal efficiency.	Reduces chemical purchases and overall operating costs.	Returns both process chemicals and impurities. May require more frequent bath maintenance to remove returned impurities. Higher bath temperatures increase energy use.	Drag-out reduction methods have been widely publicized and are implemented to some extent at most plating shops.
Minimize the formation of drag-out by: redesigning parts and racks/barrels to avoid cup-shapes, etc., that hold solution; properly racking parts; and reducing rack/part withdraw speed.	Applicable to all plating operations, but especially captive shops. Rack/part withdraw speed is more easily controlled on automatic plating machines than with manual operations.	Job shops have little or no control over part design and generally use universal racks that may not optimize drag-out reduction.	Reduces pollutant mass loading on treatment processes, treatment reagent usage, and resultant sludge generation. May improve treatment operation/removal efficiency.	Reduces chemical purchases and overall operating costs.	Replacement of racks/barrels is capital intensive.	Parts redesign is mostly restricted to captive shops. Proper positioning of parts is practiced by 52% of the survey respondents. Reducing part withdraw speed is practiced by approximately 50% of the survey respondents.
Capture drag-out before rinsing and return it to the plating bath. Methods include: allowing rack to drip over plating tank/drip shields; drip tanks; and air knives.	Because plating solution is not diluted with rinse water these methods are applicable to both low and high temperature baths.	Some solutions (e.g., chrome) stain parts if permitted to drain until dry. in other cases (e.g., nickel strike), passivation may occur if the part dries.	Can recover a substantial percentage of drag-out, depending mostly on part configuration and "drainability."	Reduces chemical purchases and overall operating costs.	Returns both process chemicals and impurities. May require more frequent bath maintenance to remove returned impurities.	One or more methods used by nearly all plating shops.
Install fog rinses or sprays over process tanks to remove drag-out as rack/part exits bath.	Baths with high operating temperatures (>130°F) and/or low through-put operations (e.g., hard chrome plating).	Should not be used where chemical mists are drawn into air exhaust systems, increasing air pollution. Cannot be used with low temperature baths where fog or spray rinse rate exceeds bath's surface evaporation rate. Not effective for parts with recessed surfaces.	Can inexpensively recover a substantial portion of drag-out and does not require additional tankage.	Prevents drying of parts which can cause passivation or staining of surface.	Coordination of water use and evaporation rate must be maintained. Restricts use of drag-out tanks. May increase emission losses.	Used by approximately 25% to 30% of the survey respondents.
Use drag-out rinse tanks and return chemicals to the process bath. Multiple drag-out tanks (two to four) in series increase the recovery rate.	Plating baths operated at temperatures above 110°F. Can be used with lower temperature baths if evaporative equipment is also installed. Dragout tanks can utilize spray rinsing.	Not applicable to cleaning baths or other process solutions that are rapidly depleted or susceptible to self-contamination.	Can recover from 50% to nearly 100% of drag-out for applicable operations. Reduces subsequent rinsing requirements and water use	Reduces chemical purchases and overall operating costs.	Returns both process chemicals and impurities. May require more frequent bath maintenance to remove returned impurities.	Used by 70% to 80% of the survey respondents.

Exhibit 3-21. Rinse Water Reduction Methods that Reduce Waste Generation

Method	Applications	Restrictions	Advantages/Benefits		Disadvantages	Extent of Use
			Pollution Prevention	Other		
Install flow restrictors to control the flow rate of water.	Can be used as stand-alone method for continuous plating operations (e.g., automated plating machines).	Must be used in conjunction with other methods for plating operations with variable production rates.	Reduces water use and aids in reducing variability in wastewater flow.	Very inexpensive to purchase and install.	Eventually restrictors wear out and must be replaced.	71% of the survey respondents use flow restrictors.
Turn off rinse water when not in use.	Small plating operations where water flow is highly noticeable.	Most plating shops require an automatic control method to assure efficient water use.	Coordinates water use and production when properly implemented.	No capital expenditure associated with manual operation.	Effectiveness depends too heavily on plating personnel.	Widely used by smaller shops as primary method of flow reduction. 66%` of the survey respondents use this water control method.
Install conductivity or timer rinse controls to match rinse water needs with use.	Operations with variable production loads.	Unnecessary for continuous plating operations (e.g., automatic plating machines).	Coordinates water use and production when properly implemented.	Provides automatic control of water use.	Conductivity and pH control systems require frequent cleaning and maintenance of probes.	Used by approximately 25% of the survey respondents.
Use optimal rinse tank design that prevents short-circuiting of water flow. Includes use of air agitation to assure complete mixing.	Applicable to nearly all rinsing operations.	High air agitation rates may knock small parts off of racks.	Maximizes effectiveness of other water reduction methods.	Improves effectiveness of rinsing and cleanliness of parts.	Replacement of existing tanks may not be economically practical.	Newer plants generally have good rinse tank design. Air agitation is used by 58% of the survey respondents.
Use spray rinsing to mechanically remove drag-out from the surface of parts.	Can be used as initial and/or final rinse for parts with exposed surfaces.	Limited to line-of-site applications. Not applicable to parts with excessive hidden surfaces.	Significant water reduction for applicable parts.	Can be combined with other water reduction methods (e.g., counter-current rinsing).	Higher initial cost than overflow rinse.	38.7% of the survey respondents use spray rinses.
Use countercurrent rinse arrangement with two to four tanks in series, depending on drag-out rate.	Highly useful at all plating operations. A two-tank arrangement is usually sufficient for low production operations.	Three or four tanks may result in diminishing returns for low production operations, especially considering labor costs.	Major water reduction can be achieved.	High impact on water bills. May reduce the size of needed recovery/treatment equipment.	High floor space requirement. May increase processing time.	Heavily used throughout industry. Employed by 69% of the survey respondents.
Use cascade or reactive rinsing.	Primarily limited to cleaning operations.	Chemical reactions may result in precipitation and poor rinsing.	Moderate water reduction potential.	May improve cleaning process and work quality.	Requires transfer/routing of rinse water.	25% of the survey respondents use cascade or reactive rinsing.
Track water use with flow meters and accumulators. Keep logs on water use for individual operations.	Applicable to all plating operations. Water usage is best tracked for individual rinse tanks or plating lines.	None identified.	Identifies problem areas including inefficient processes or personnel.	Helps management to determine cost for individual plating processes.	Significant initial cost. Labor needed to collect and track data.	Only 12% of the survey respondents track water use with flow meters/accumulators.

Section 4
Chemical Recovery

4.1 INTRODUCTION

Various technologies are used by plating shops to separate plating chemicals from rinse waters and air emissions or to concentrate them, thereby making them available for reuse/recycle. The most common technologies are:

- Atmospheric Evaporation
- Vacuum Evaporation
- Ion Exchange
- Electrowinning
- Electrodialysis
- Reverse Osmosis
- Meshpad Mist Eliminator

Although many of the recovery technology applications identified by the respondents have been successful, there are also many unsuccessful applications. The success/failure rate reported by respondents to the Users Survey varied by technology and application. Exhibit 4-1 presents a summary of the technologies and applications identified during the Users Survey and it shows the average success rating given by the respondents. Technology success was measured by respondents on a scale of 1 to 5, with 1 being the least successful and 5 being the most successful.

When evaluating success and failure ratings of the Users Survey, one must consider the reasons why the respondents purchased the recovery technologies. A summary of these data is presented in Exhibit 4-2. In the survey form, the respondents were given a choice of five specific reasons why they may have purchased the technology, plus an opportunity to list "other" reasons. The question was worded and formatted in a way that multiple responses were possible. The results are somewhat surprising in that reducing chemical purchases (i.e., recovering the plating chemicals), which is the primary function of recovery technologies, was not the most frequently given response. Rather, meeting or helping to meet effluent regulations was cited much more frequently. Reducing plating chemical purchases, reducing the quantity of waste shipped off-site and reducing wastewater treatment costs were cited approximately the same number of times as reasons for purchasing the recovery technologies.

The remainder of this section presents the results of the Users Survey, Vendors Survey and literature search with regard to the chemical recovery technologies employed by the survey respondents. The Users Survey requested platers to provide detailed technical, performance and operating cost data for chemical recovery technologies. The vendors were requested to provide technology descriptions, operating data and capital cost data. As a result of obtaining data from these two sources, plus the information from the extensive literature review, this section contains a substantial quantity of information for the following chemical recovery technologies: electrodialysis, electrowinning, atmospheric evaporators, vacuum evaporators, ion exchange, reverse osmosis and mesh pad mist eliminators. A separate subsection of the report is devoted to each of these technologies. Within each subsection, the following are provided: technology overview; development and commercialization; applications and restrictions (with diagrams showing different potential configurations); technology/equipment description; capital costs; operating costs; performance experience; and residuals generation. The capital cost curves

Exhibit 4-1. Distribution and Ratings of Chemical Recovery Technologies

Process Name	Recovery Technologies[a] ED	EW	ATMEV	VAC EV	IX	ME	RO
Anodize, Chromic Acid	NA	NA	NA	NA	NA	4.0 (2)	NA
Anodize, Hardcoat	NA	2.0 (1)	NA	NA	NA	NA	NA
Brass	NA	3.5 (2)	4,0 (2)	NA	NA	NA	NA
Cadmium, Cyanide	NA	2.8 (26)	3.0 (4)	3.0 (3)	2.0 (4)	NA	5.0 (1)
Cadmium, Non-cyanide	NA	NA	4.0 (1)	NA	NA	NA	NA
Chromate (Aluminum)	NA	NA	NA	NA	5.0 (1)	NA	NA
Chromium Etch	NA	NA	NA	4.5 (2)	NA	NA	NA
Chromium, Hard	NA	NA	4.3 (4)	NA	2.0 (2)	4.1 (15)	NA
Chromium, Decorative (Cr^{+6})	NA	NA	3.4 (10)	4.0 (5)	5.0 (1)	NA	NA
Chromium, Decorative (Cr^{+3})	NA	NA	5.0 (2)	NA	NA	NA	NA
Copper, Electroplating[b]	NA	3.0 (14)	3.7 (3)	4.0 (1)	NA	NA	2.0 (1)
Gold, Electroplating[c]	NA	4.0 (4)	NA	NA	4.0 (5)	NA	NA
Lead-Tin	NA	NA	NA	4.0 (1)	NA	NA	NA
Nickel, Electroplating[d]	1.0 (2)	4.0 (5)	4.0 (18)	4.0 (3)	3.8 (10)	NA	1.0 (1)
Nickel, Electroless Plate	NA	ND (1)	NA	NA	3.0 (1)	NA	NA
Seal, Nickel	NA	NA	NA	NA	NA	NA	4.0 (1)
Silver, Electroplate	NA	4.4 (11)	2.0 (1)	NA	NA	NA	NA
Zinc, Cyanide	NA	1.3 (6)	3.0 (2)	1.7 (3)	NA	NA	1.0 (1)
Zinc, Non-cyanide[e]	NA	2.0 (5)	3.7 (8)	NA	2.6 (5)	NA	5.0 (l)
Zincate	NA	NA	4.0 (l)	NA	NA	NA	NA

[a] Technology key: ED=electrodialysis, EW=electrowinning, ATM EV=atmospheric evaporation, VAC EV=vacuum evaporation, IX=ion exchange, ME=meshpad mist eliminator, RO=reverse osmosis. Number of applications from the Users Survey is shown in parenthesis. NA= no applications. ND=no data.
[b] Includes cyanide plate, cyanide strike and sulfate baths.
[c] Includes cyanide and non-cyanide baths.
[d] Includes brite, sulfamate, sulfate, Watts and Woods baths.
[e] Includes acid and alkaline baths.

contained in Section 4 are based on data collected from the technology vendors and the operating cost curves are based mainly on data from platers. Both the capital and operating cost information are expressed in 1993 dollars. A labor cost of $25 per hour (includes overhead) and an electricity cost of $0.10/kWh have been used, where applicable, in calculating operating costs.

4.2 ATMOSPHERIC EVAPORATORS

4.2.1 Overview

Atmospheric evaporators are the most widely used method of chemical recovery in the plating industry. Of the 318 plating shops responding to the Users Survey, 71 (or 22.3%) have employed atmospheric evaporators for chemical recovery. Some shops have purchased or built two or more units for different recovery applications, resulting in a total number of 91 units used by survey respondents. The literature cites an extreme case, where a Virginia shop employs 14 atmospheric evaporators and has achieved zero discharge of wastewater effluent (ref. 33). Most of the atmospheric evaporators used by survey respondents were commercial units although approximately 5% were built in-house. By comparison, 26% of the electrowinning units were built in-house. The low percentage of home-made atmospheric evaporation units is most likely due to the low capital cost and successful track record of the commercial units.

Atmospheric evaporators are also used by some shops to concentrate liquid plating wastes prior to hauling them off-site for treatment/disposal, thereby reducing transportation costs and in some cases treatment/disposal costs. A total of 3 shops responding to the survey used atmospheric evaporators for this purpose.

An atmospheric evaporator is a device that evaporates water to the atmosphere. The commercial atmospheric evaporator used for recovery in the plating shop consists of a pump to move the solution, a blower to move the air, a heat source, an evaporation chamber in which the solution and air can be mixed, and a mist eliminator to remove any entrained liquid from the exit air stream (see Exhibit 4-3). The evaporation chamber is usually filled with packing material or finned panels to increase the air

Exhibit 4-2. Reasons Why Respondents Purchased Chemical Recovery Technologies

Why the Technology Was Purchased[b]	Recovery Technologies[a]							
	ED	EW	ATM EV	VAC EV	IX	ME	RO	Totals
To meet or help meet effluent regulations	1	38	42	14	18	4	4	**121**
To reduce plating chemical purchases	0	14	42	12	8	S	4	**85**
To reduce the quantity of waste shipped off-site	0	20	34	12	13	2	2	**83**
To reduce wastewater treatment costs	1	20	39	11	12	2	2	**87**
To improve product quality	0	1	9	0	S	1	3	**19**
Other	0	3	6	1	0	3	0	**13**

[a] Technology key: ED=electrodialysis, EW=electrowinning, ATM EV=atmospheric evaporation, VAC EV=vacuum evaporation, IX=ion exchange, ME=meshpad mist eliminator, RO=reverse osmosis.
[b] Multiple reasons for purchasing technologies were often given by a respondent for the same technology/application.

to water interface. In operation, the temperature of the solution being evaporated is elevated and the heated solution is introduced into the evaporation compartment. Air from the plating room is then blown through the compartment where it accepts the water vapor, and is then vented out of the chamber.

Commercial units are advertised to have evaporation rates of 10 to 90 gph, depending on the size of the unit and operating conditions (e.g. solution temperature). Often actual evaporation rates are considerably less because the atmospheric conditions within most plating shops do not match the ideal conditions under which the manufacturers rate their systems. To meet higher evaporative requirements, it is feasible to utilize multiple atmospheric evaporators in series. However, the use of atmospheric evaporators is generally limited by energy costs to applications where the required evaporation rate is 50 gph or less. Beyond this point, vacuum evaporators (Section 4.3) are more cost effective, considering both capital and operating costs.

The key attributes of atmospheric evaporators include: (1) low capital cost; (2) simple operation and low maintenance; (3) very high recovery rates can be achieved (usually 90% to nearly 100%); (4) no additional reagents are needed; and (5) no sludges or only small quantities of sludges are generated (when used in recovery application). The major negative aspects of this technology are: (1) high energy requirement (i.e., requires constant solution heating and during the winter months there is a heat loss due to venting the exhaust to the outside); (2) the air that is vented by these devices must be discharged to the outside (due to contaminants present and its high humidity) and may be a regulated source of air pollution; (3) because moisture is exhausted to the atmosphere it cannot be reused as rinse water as with vacuum evaporators[1]; (4) evaporators return contaminants to the bath and may reduce bath life; and (5) spray/fog rinsing over the bath or fume suppressants (floating chemical type or plastic balls) are not compatible with atmospheric evaporators since they reduce the head room in the plating tank and limit the return of rinse water/dragout.

Most commercial atmospheric evaporator units have the same principals of operation. To achieve chemical recovery, solution from a heated plating tank is fed to and concentrated by the evaporator and returned to the plating tank. This approach reduces the volume of solution in the plating tank, thereby "making room" for the recovery rinse water/dragout to be added to the plating bath. Often two or more recovery rinse stations are used to minimize the overall rinse water requirements of the process and increase the recovery rate of plating chemicals. Less frequently, atmospheric evaporation is applied to ambient or low temperature baths. In this case, the recovery rinse water may be fed to the evaporator from a heated transfer tank, which increases the overall evaporative capacity of the system. The latter application is often restricted by the maximum temperature that can be applied to the solution, since heat sensitive components of the bath could be destroyed.

[1] Some users of vacuum evaporators may view this as an advantage of the atmospheric evaporator because the distillate from the vacuum types is not always of sufficient quality for reuse and must sometimes be treated as wastewater.

Although some commercial atmospheric evaporators have their own heat source, nearly all units employed for plating chemical recovery in use today use the heat in the plating tank (and/or heated transfer tank) as the energy source for evaporation. Any solution above room temperature that is pumped to the evaporator will be returned at a lower temperature. The temperature difference is primarily due to the heat that goes into evaporation, which is (ref. 358):

- 540 calories/g of water evaporated, or
- 2,137 Btu/l (8,090 Btu/gal), or
- 626 watts/l (2,371 watts/gal), or
- 0.02137 "gas company" therms/l (0.08090 therms/gal), or
- 0.0637 boiler hp/gal

The heat taken from the plating bath must be replaced by the tank heating system (e.g., immersion heaters or steam system) in order to maintain the operating temperature of the bath. During winter months, when a plating shop is heated, the room's ambient air that is exhausted by the evaporator (typically 300 to 3,000 cfm) must be replaced. These two elements make up the bulk of the non-labor operating costs for atmospheric evaporation.

The humidity and temperature of a plating shop will significantly affect the evaporative capacity of an atmospheric evaporator, especially for low to moderate solution temperature applications. If the air in the plating shop is very humid (e.g., 90% relative humidity) before entering the evaporation chamber, it will hold a limited amount of additional water and the evaporation rate will be affected. From thermodynamic tables, it is known that warmer air holds more moisture than cooler air. For example, one pound of air will hold 0.015 pounds of water at 70°F. By comparison, that same pound of air will hold 0.220 pounds of water at 120°F. In all cases, the air entering the evaporation chamber is heated to a maximum temperature equal to a point less than the solution temperature (the closeness of the air and solution temperatures will depend on the evaporator design and resultant heat transfer efficiency). With high temperature solutions, the air temperature will reach 100°F or more. At this temperature level, the original moisture content of the air is relatively small in comparison to the new capacity of the air, hence relative humidity plays a lesser role.

The above example also indicates that preheating of the plating shop air before introducing it to the evaporation chamber would improve performance. Although air heat-

Exhibit 4-3. Common Commercial Atmospheric Evaporator Used for Chemical Recovery (Courtesy of Poly Products Corporation)

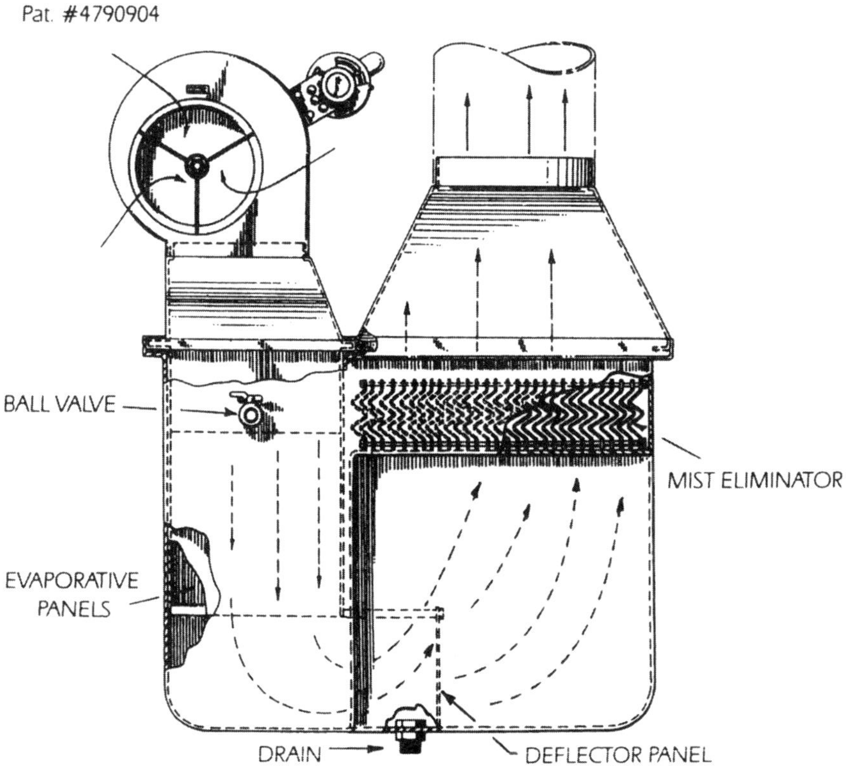

ing systems are discussed in the literature (e.g., ref. 299), no commercial units designed for plating shops were identified with this feature. The strategy of manufacturers of commercial units is to maximize air flow and increase the water/air contact area rather than increase air temperature.

Platers may be tempted to use outside air for make-up to their atmospheric evaporator, especially during winter months in colder climates, to reduce the loss of heated indoor air. This strategy generally does not work. Although the outside air may be dryer than the inside air, its low temperature will have an overriding impact on the evaporation process. The low temperature of the air will prevent it from reaching a sufficiently high temperature in the evaporation chamber to attain a reasonable water holding capacity.

In many cases, the types of commercial units used for concentrating wastes before off-site disposal are the same types of units used for recovery. However, there are also available specially designed waste concentration units. These devices usually have a direct heat source and operate at much higher temperatures than the recovery units. Higher temperatures can be used with wastes since there is no concern for the integrity of the chemical components. Because higher temperatures are used, the materials of construction for these units differ from the recovery units. Evaporation of water from wastes may be viewed by regulatory agencies as thermal treatment and they may require a RCRA permit for the operation of these units, depending on the interpretation of the application. Regulatory aspects of evaporators are not discussed in this report, but should be closely investigated before purchasing and operating evaporation equipment.

4.2.2 Development and Commercialization

Atmospheric evaporators are essentially adjuncts to plating tank evaporation. They increase the head room in the plating tank, thereby increasing the quantity of rinse water/dragout that can be recovered using recovery rinsing. The use of recovery rinsing in the plating industry was documented more than 50 years ago. The use of evaporators for chemical recovery extends back more than 40 years (see reference list in ref. 1). Most of the early evaporators used by the plating industry were the vacuum type, which was basically a technology transfer from the chemical processing industry. These were large units, often with capacities of 400 to 600 gph (1,500 to 2,300 l/hr). Large units were used because water conservation and pollution control were less important at that time, resulting in higher flow rates. Also, energy was much less expensive at that time and therefore there was less incentive to minimize flows prior to evaporation. In 1974, with the advent of rapidly rising energy costs, there began a movement to down-size recovery systems (ref. 300). Coupled with the Federal pollution control standards, first promulgated in 1979, plating shops turned to the small and inexpensive atmospheric evaporators for chemical recovery. Based on the results of the Users Survey, it is apparent that purchases of this technology have continued into the 1990's.

There are approximately 25 companies that manufacture evaporative recovery equipment applicable to the plating industry. This includes vacuum and atmospheric evaporators for both chemical recovery and waste concentration (ref. 421). The firms most often mentioned in the Users Survey were Poly Products Corporation, Techmatic Inc. and NAPCO.

4.2.3 Applications and Restrictions

Exhibit 4-4 shows the three basic configurations used for application of atmospheric evaporators for chemical recovery. Application AE-1 is used mostly for elevated process baths (>120°F), AE-2 for moderate temperature baths (100 to 120°F) and AE-3 for ambient or low temperature baths (ambient to 100°F). In each case, an important aspect to the implementation of this technology is the incorporation of recovery rinsing. The quantity of recovery rinse solution that can be added into a bath equals the sum of the evaporation from the tank's surface and the evaporation caused by the atmospheric evaporator (i.e., ignoring any differences in drag-in and dragout). For some processes, where the operating temperature is high and dragout is sufficiently low, a closed-loop configuration can be employed. Several of the respondents to the Users Survey indicated that they have achieved a closed-loop (e.g., PS 003 for nickel, PS 213 for nickel and chrome). Generally, these shops used either a three or four stage recovery rinse. Lower temperature processes and those with high dragout rates will usually require a free running rinse (or countercurrent arrangement) following the recovery rinse tank in order to maintain sufficiently clean water in the final rinse. The effects of rinsing configurations on recovery and plating quality are discussed in detail in Section 3.

The strategy with moderate temperature baths is to connect the atmospheric evaporator to a heated transfer tank with the idea that the solution in the transfer tank can be heated to a higher temperature than the maximum operat-

Exhibit 4-4. Three Primary Configurations for the Application of Atmospheric Evaporation to Chemical Recovery

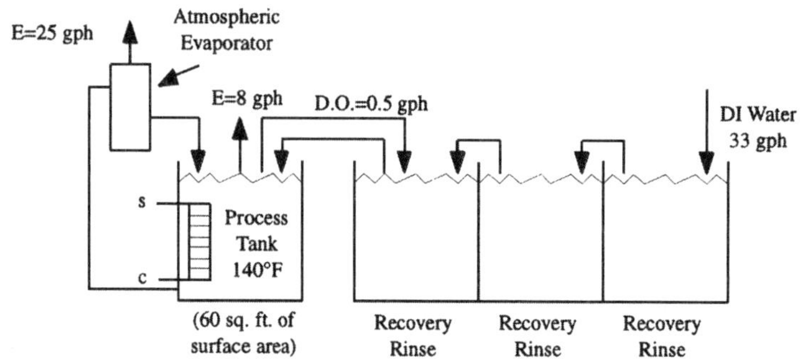

Configuration AE-1: High Temperature Process

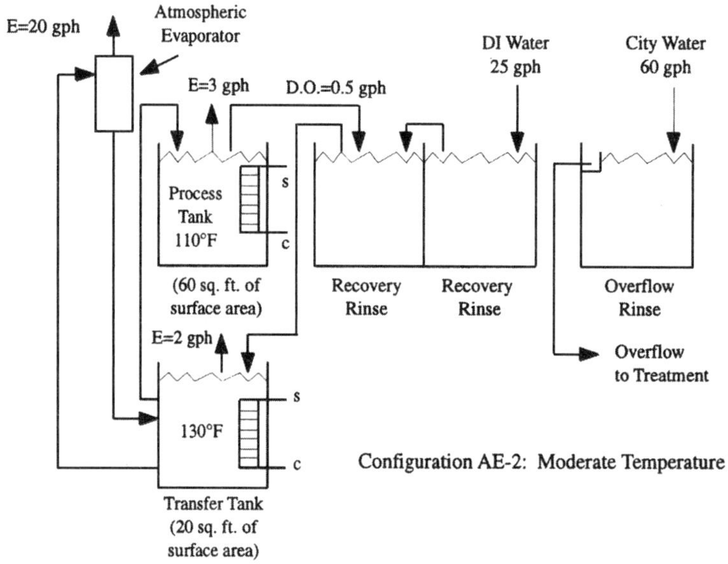

Configuration AE-2: Moderate Temperature Process

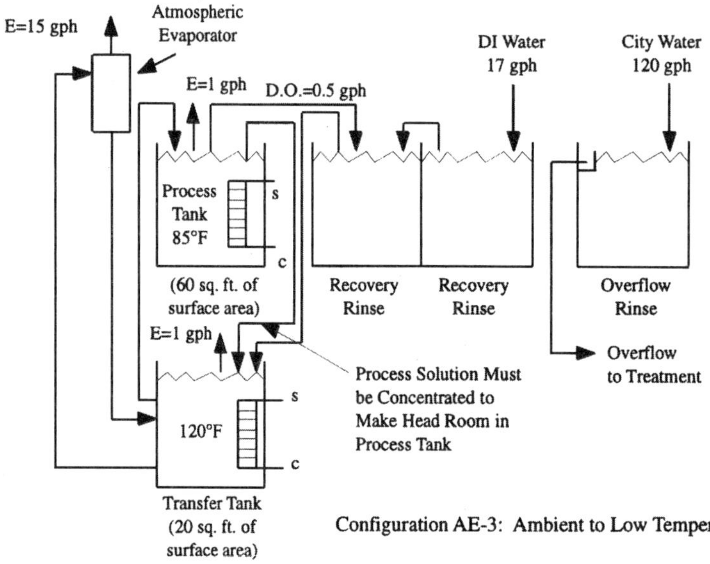

Configuration AE-3: Ambient to Low Temperature Process

ing temperature of the plating process. For example, a transfer tank used with acid zinc can be heated to 140°F, whereas the maximum operating temperature of the process bath is about 90°F. By connecting the evaporator to the transfer tank, the evaporation rate will be approximately tripled. A number of shops responding to the Users Survey indicated that they were using heated transfer tanks (e.g., PS 098, PS 252 and PS 278).

With low or ambient temperature baths, where there is no appreciable surface evaporation from the process bath, a bleed from the process bath to the transfer bath is used to make the needed headroom. Alternatively, a second evaporator could be used. For some ambient baths that contain wetting agents, drag-in may exceed dragout. The design capacity of the atmospheric evaporator must account for this difference plus the desired recovery rinse rate. In some cases, an atmospheric evaporator may be used simply to create the headroom needed to prevent discarding "extra" plating solution created by these conditions (ref. 355). One such application was found in the Users Survey (PS 214).

The Users Survey also showed that many shops were incorrectly using their atmospheric evaporators. Approximately 20% of the shops connected either an unheated recovery rinse or an unheated plating tank to the evaporator. Generally, these facilities experienced below average results. Performance experience is discussed in Section 4.2.6.

With all applications of atmospheric evaporators, the user should install water treatment in the form of ion exchange and/or reverse osmosis to remove the hardness and other impurities in the raw water that would otherwise accumulate in the process tank.

Recovery applications that were identified during the Users Survey are presented in Exhibit 4-5. The most common recovery application for atmospheric evaporators is nickel plating. On a combined basis, the nickel applications accounted for nearly 30% of all applications (includes bright nickel and electroless nickel applications).

When used in conjunction with hard chrome plating or other processes that have solution heating and cooling requirements and ventilation requirements, the atmospheric evaporator can serve as a recovery unit, cooling device and an air scrubber (ref. 299). Because the unit would be using waste heat from the tank for evaporating excess water, the evaporation process operates essentially free. The appropriate tank configuration is basically the same as shown in Exhibit 4-4, application AE-1. The plating tank's exhaust air would serve as the inlet air to the evaporator, which would remove its chromic acid mist. In such cases, the exhaust of the evaporator is usually connected to the existing ventilation system and the evaporator's blower is eliminated. Because of anticipated regulations for hard chrome plating air emissions, it is questionable that the evaporator's mist eliminator would adequately substitute for a future emission control

Exhibit 4-5. Atmospheric Evaporation/Process Matrix

Proeess Solution	Percentage of Atmospheric Evaporator Applications (%)	Percent of Process Solutions for which there is an Atmospheric Evaporator Application*
Brass	3.2	3.7
Cadmium, cyanide	6.6	3.3
Chromium, decorative (Cr+3)	3.2	8.7
Chromium, decorative (Cr+6)	18.0	11.8
Chromium, hard	6.6	5.1
Copper, cyanide	8.2	2.6
Nickel, not specified	3.2	1 1.7
Nickel, sulfate	3.2	1 1.8
Nickel, Watts	18.0	8.9
Zinc, acid	9.8	18.2
Zinc, cyanide	3.2	4.4
Zinc, non-cyanide	4.9	2.4
Other	1 1.5	—

*For example, there were 54 brass plating processes identified during the Users Survey. Of these, two processes had an associated atmospheric evaporation/chemical recovery unit (2/54 = 3.7%).

device. At this time there are insufficient data to evaluate this application.

Atmospheric evaporators are not applicable in cases where the solution temperature cannot be raised above approximately 85°F, either in the process tank itself or in a heated transfer tank. Generally, these are solutions that either contain highly heat sensitive components or fume excessively when heated or aerated. Also, atmospheric evaporators should not be applied to any process solution which cannot be maintained through use of methods and/or technologies that replenish active bath ingredients or remove the contaminants that build-up as a result of recovery rinsing. When applied in these cases, the atmospheric evaporator hastens the disposal rate of the bath resulting in essentially the same mass of chemicals discarded as if recovery were not practiced. Methods of bath maintenance for process solutions are covered in Section 5.

Atmospheric evaporators should not be applied to solutions that foam significantly when air agitated (e.g., high cyanide baths and still nickels). Such solutions will foam in the evaporation chamber and render the system inoperable. A simple jar test is recommended by one manufacturer to determine if foaming is a potential problem. This test is accomplished by placing a sample of the solution in a jar, vigorously shaking it and then observing to see if the foam quickly disappears (ref. Poly Products file).

4.2.4 Technology/Equipment Description

The equipment described in this section is manufactured by four firms that responded to the Vendors Survey. Three of these companies represent approximately 90% of the total number of atmospheric evaporators purchased by the respondents of the Users Survey (not counting home-made units or in cases where the manufacturer was not given).

The manufacture of atmospheric evaporators most often found in the Users Survey data is Poly Products Corporation. They manufacture four different models: ET-II Junior, ET-II, ET-III and ET-III-W. The ET-II Junior, ET-II, and ET-III are progressively larger units of the same design. The ET-III-W has a modified design, intended for use with solutions that have a high solids content. Each model has the same basic appearance and operating principles (refer back to Exhibit 4-3). During operation, pumped solution is circulated at approximately 10 to 45 gpm from the process tank or transfer tank to the unit and is gravity drained back to the tank. In the evaporator, solution is sprayed onto 700 to 1,000 ft^2 of evaporative panels to humidify the blower air that is forced through these panels. The air is provided by a blower sized from 1/2 to 1-1/2 hp, depending on the model. The humid air then passes through a chevron mist eliminator to remove entrapped chemical solution droplets and is ducted to the outdoors. Poly Products offers an optional mesh pad mist eliminator to further decrease chemical emissions. The ET units are manufactured of molded polyethylene. The largest ET unit occupies a floor space of 48 in. x 34 in. All models have maximum air temperature limits of 104°F and solution limits of 160°F (ref. Poly Products file).

Techmatic Inc. manufactures the MAX-EVAP™. There are four models available: MAX-EVAP, Super MAX-EVAP, MAX-EVAP Cr and Super MAX-EVAP Cr. The MAX-EVAP and Super MAX-EVAP are general purpose units with advertised evaporation rates of 50 to 60 gph and 80 to 90 gph, respectively. The bodies of these units are manufactured from polyethylene. The MAX-EVAP Cr and Super MAX-EVAP Cr are designed specifically for hexavalent chromium plating solutions. These two models are manufactured from steel and are lined with Koroseal (PVC). As with the operation of the Poly Products line, solution is circulated from the process tank or transfer tank to the unit and is gravity drained back to that tank. In the evaporator, solution is sprayed onto 5 to 10 ft^3 of polyethylene packing. Air is drawn through the unit by the shop's existing ventilation system or air movement is provided by an optional blower. The exhaust air passes through a "vertical extruded mist eliminator" (not defined by manufacturer) to remove chemical mists. The MAX-EVAP unit occupies a floor space of 63 in. x 21 in. and has a height of 31 in. (blower option measures 55 in. high) (ref. Techmatic file).

Kinetic Recovery manufactures a unit with a somewhat different design than the two previously discussed units. With the Kinetic Recovery unit (Exhibit 4-6) the process solution is pumped from its tank up to the liquid distributor of the evaporator which is located at the top of the cylinder. The liquid trickles over a tower packing down to the bottom drain of the cylinder and returns to the process tank. A blower, located at the bottom of the cylinder, blows air (300 cfm) upward through the packing to the top of the evaporator. When the air goes through the packing (pressure drop 0.3 in. of superfacial vapor velocity of 1.5 fps) it becomes saturated with vapor from the process solution. The cylindrical design improves air flow distribution by reducing dead space. On the top of the cylinder there is a vent transition which contains a mist eliminator (12 in. flexi-chevron). There are two view ports located on the unit, one at the liquid distributor and another below the packing. The basic Kinetic Recovery unit is manufactured from polypropylene and they offer construction in PVDF. The PVDF unit operates with a maximum solution temperature of 180°F.

NAPCO manufactures atmospheric evaporators with two basic designs: (1) cross-flow air pattern and (2) vertical air flow pattern. Both types of units are normally constructed of polyethylene, but a variety of plastics are also available, depending on the customers needs. The basic NAPCO unit consists of a blower, recirculation pump, high surface area packing and demister baffling. Numerous options are available, including heating coils, temperature controls, level controls and make-up solenoids and pumps. Their standard size units range from 10 gpm process feed/1,200 cfm air flow (NAPVAP Jr) to 30 gpm

Exhibit 4-6. Commercial Atmospheric Evaporator with Cylindrical Design (Courtesy of Kinetic Recovery Corporation)

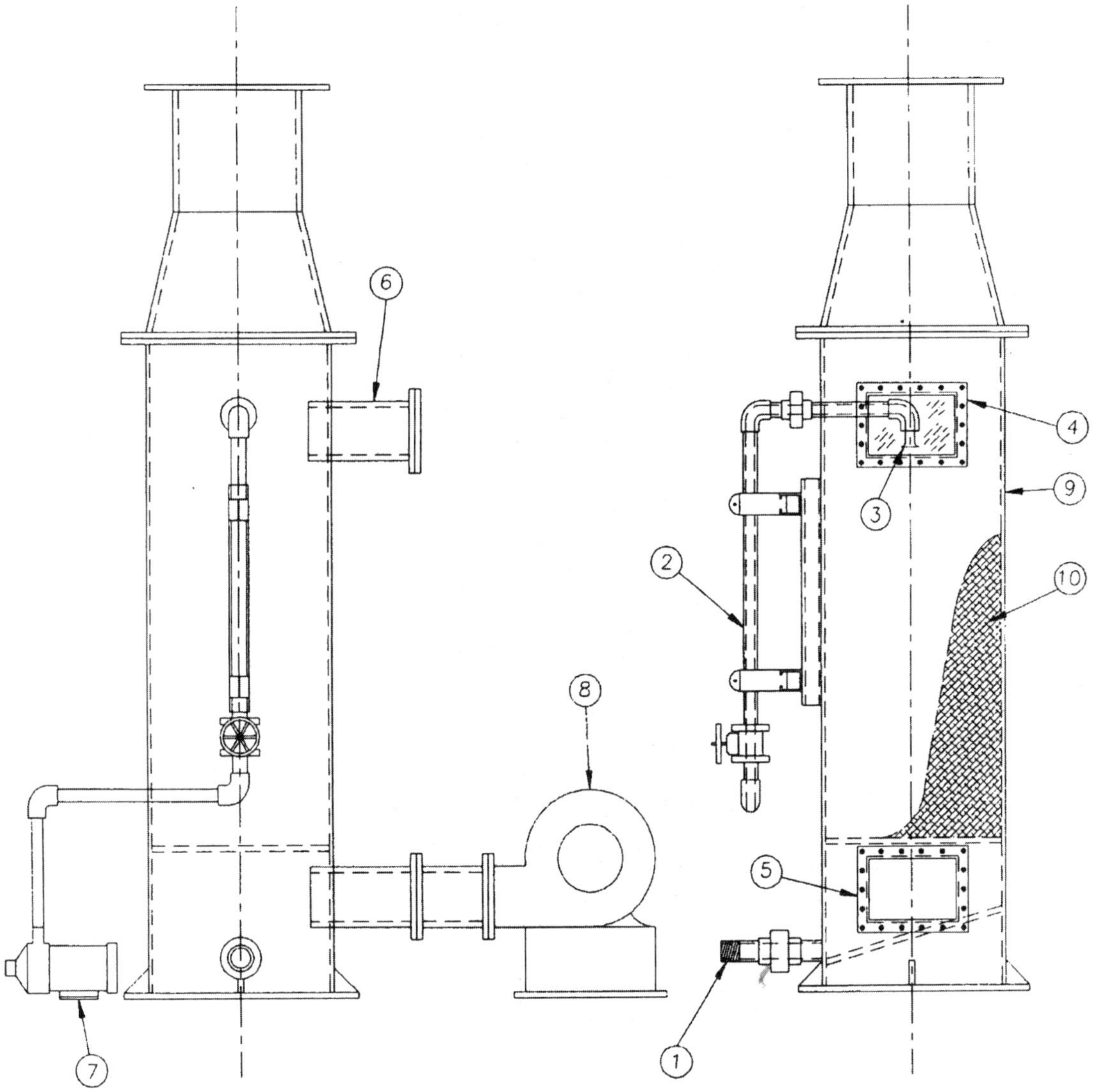

Description:

1 - Discharge
2 - Feed Pipe
3 - Spray Nozzle
4 - Sight Glass
5 - Duct Inlet
6 - View Port
7 - Pump
8 - Blower
9 - Main Cylinder
10 - Tower Packing

process feed/4,000 cfm air flow (NAPVAP Sr). NAPCO has been marketing their equipment to the plating industry since 1985 and have sold approximately 160 units to 100 different plating shops.

For each of these four commercial evaporators, the heat used for evaporation comes from the plating or transfer tank (unless a heating coil option is specified). Heat taken from the plating bath must be replaced by the tank heating system (e.g., immersion heaters or steam system) in order to maintain the operating temperature of the bath. Most tank heating systems are designed to provide a quick heat-up and are underutilized after the operating temperature is maintained. Therefore, auxiliary tank heating equipment is usually not necessary when an atmospheric evaporator is installed (unless a transfer tank is also installed). However, each potential installation should be evaluated before purchasing evaporative equipment. Nomographs and methods of calculating tank heating requirements are available in electroplating engineering references (e.g., ref. 341).

The atmospheric evaporators previously discussed are used primarily for chemical recovery, but are occasionally applied to waste concentration. Some evaporators are designed specifically for waste concentration. Two such units include the Technotreat Wastewater Evaporator and the Samsco Water Evaporator.

The Technotreat unit consists of an enclosed carbon steel (stainless steel available) tank equipped with electric immersion heaters (stainless steel or titanium). Liquid waste is fed into the evaporator by a feed pump from a drum or storage tank. The solution level in the evaporator is maintained by a level switch which activates the feed pump. The steam is vented by an assisting electric blower. As water is evaporated, the waste is concentrated. Oil floats to the top and is drawn off periodically. Sludge and viscous materials are removed through a bottom drain. The Technotreat unit comes in 100 and 200 gpd capacities. The range of costs is approximately $10,000 to $15,000 for the standard materials of construction (ref. 360).

The SAMCO Water Evaporator combines direct heating and air flow. The unit consists of an enclosed tank (carbon steel or 304 or 316 stainless steel) that contains a serpentine gas-fired heat exchanger and blower (530 cfm to 2,000 cfm) that draws air in through both the burner and an opening in the top of the tank. In operation, solution is fed to the tank in either a batch or continuous mode. The solution is heated to boiling (approximately 212°F) by the heat exchanger. The action of the blower draws air across the surface of the heated liquid, sweeping away water vapor as it breaks to the surface. The moisture-saturated air and the flue gases leave the tank via separate passages and are joined together at the blower entrance. The two air streams are mixed in the blower and released up a stack. Free oils that float on the surface of the liquid overflow a trough into an external waste receptacle. Solids that settle onto the sloping bottom of the tank are removed via a clean-out port. The SAMCO Water Evaporator comes in three capacities: 15 gph, 33 gph and 63 gph. The area dimensions of the largest unit are 65 in. x 68 in.

Installation of an atmospheric evaporator is a relatively easy task with the exception of the ventilation ducting. The basic units are shipped pre-piped and preassembled. For ventilation, ductwork is usually routed directly to the outdoors through a roof penetration. Alternatively, the exhaust of the evaporator can be connected by ductwork to an existing ventilation system. In this case, an evaporation system blower is unnecessary and the air would be drawn through the evaporation chamber rather than being pushed or blown through. As discussed in Section 4.2.3, this configuration is occasionally used with hard chromium plating.

Installation and operation of an atmospheric evaporator may require a new air permit or revisions to an existing air permit.

4.2.5 Costs

4.2.5.1 Capital Costs

The basic equipment cost for atmospheric evaporators is relatively low. However, all installations will experience some installation costs and most installations will require auxiliary equipment. The most common and significant installation cost is for exhausting the air exiting the evaporator. Ductwork must be run to either to an existing ventilation duct or, more frequently, through a roof penetration. Other installation work includes connecting power and water to the evaporator, rearranging of other equipment or tanks, installation of controls and installation of a transfer tank. Auxiliary equipment may include, for example, a transfer tank, additional recovery rinse tanks, an additional heat exchanger or a DI water system.

Capital cost estimates for atmospheric evaporators are shown in Exhibit 4-7. These costs are presented as a

Exhibit 4-7. Equipment Costs for Atmospheric Evaporators as a Function of Solution Temperature

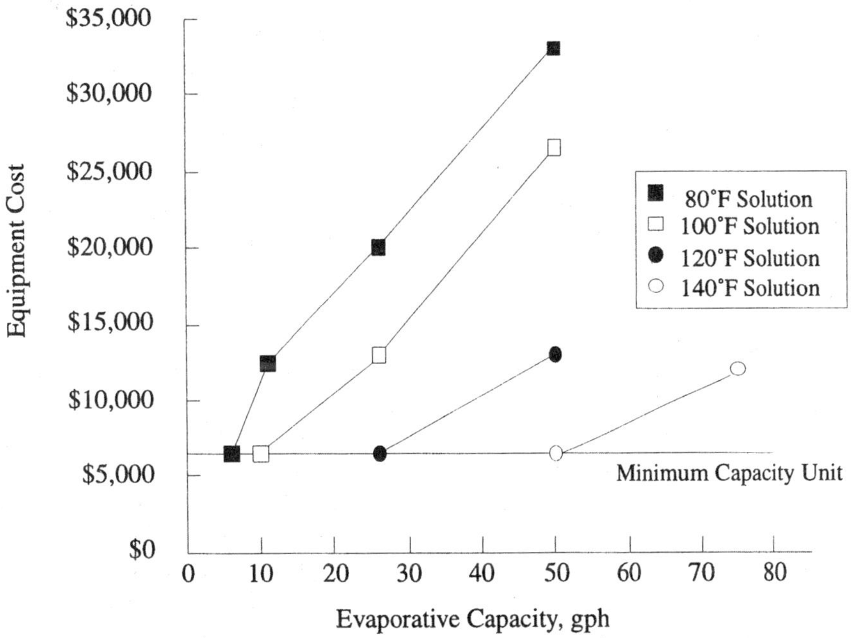

Note: Total installed costs (includes basic equipment, auxillary equipment and installation) are approximately 180% of equipment costs.

Exhibit 4-8. Operating and Maintenance Costs for Atmospheric Evaporators

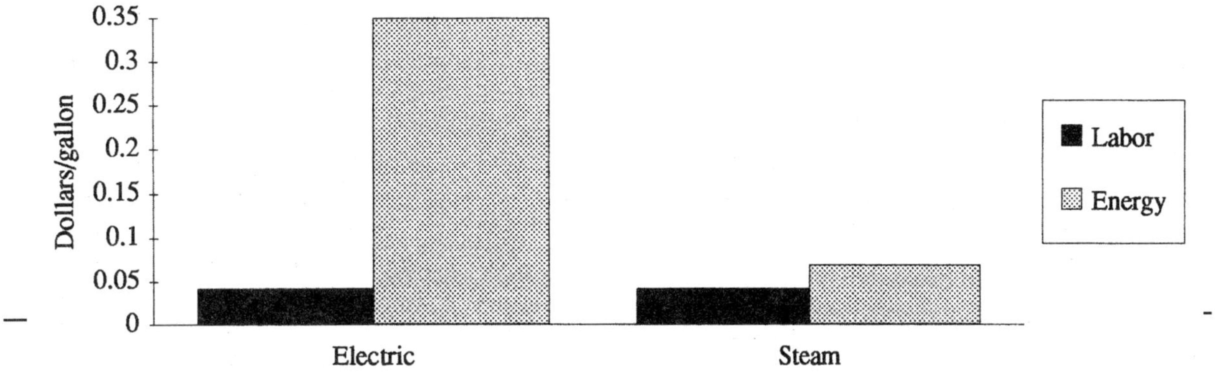

function of evaporative capacity (gph) over a range of process solution temperatures (either the process tank or transfer tank, whichever is fed to the evaporator). Exhibit 4-7 shows the basic equipment costs, which were derived by taking the median costs from the three vendor survey respondents (where a single evaporator is unable to provide the desired evaporation capacity, multiple units were assumed). The installed capital costs are approximately 180% of the basic equipment costs (based on Users Survey data).

4.2.5.2 Operating Costs

The major operating costs for atmospheric evaporators include O&M labor and energy. Estimates of these operating cost components are shown in Exhibit 4-8. From the Users Survey, the average O&M labor is 157 hrs/yr. In constructing the operating cost graph, it was assumed that this level of labor is adequate for a unit evaporating 15 gph, 24 hrs/day for 260 day/yr. The energy cost shown in Exhibit 4-8 is for replacement heat in the process tank and for operating a pump. The energy cost does not account for energy loss due to ventilation of shop air during winter months.

4.2.6 Performance Experience

A partial summary of the NCMS Users Survey data relative to atmospheric evaporators is presented in Exhibit 4-9. There are a number of observations that can be made from these data and other data contained in the database and literature:

- The majority of shops that have used atmospheric evaporators have found them to perform adequately. The average satisfaction level for this technology is 3.8 (on a scale of 1 to 5, with 5 being most satisfactory), which is the highest level rating for any of the chemical recovery technologies, except for meshpad mist eliminators. Also, 81 percent of the shops indicated that this technology satisfied the need for which it was purchased. The following is a breakdown of the reasons why shops purchased this technology:

  To meet or help meet effluent regulations: 46
  To reduce plating chemical purchases:.......................45
  To reduce the quantity of waste shipped off-site: 35
  To reduce wastewater treatment costs: 39
  To improve product quantity: 9
  Other: 6

- The use of atmospheric evaporators generally did not impact production quality or the rate of production. The following responses were provided:

	Product Quality	Production Rate
Improved	7	3
No Change	52	53
Decreased	6	9

- Where the product quality was impacted (PS 098, PS 118, PS 139 and PS 306), PS 098 and PS 139 indicated that contaminant buildup occurred in their chromium plating and nickel plating baths and PS 118 indicated that product quality is only impacted occasionally due to low rinse water flows.

- Where the production rate was impacted (PS 098, PS 139, PS 252, PS 262 and PS 306), PS 098 and PS 252 attributed work slowdowns to bath contamination.

- Most plating shops indicated, that based on their experience with this technology, they would purchase the same type of equipment from the same vendor. The following is a breakdown of their responses:

  Purchase the same technology from the same vendor: 73%
  Purchase the same technology from a different vendor: ... 12%
  Purchase a different technology: 12%
  Do nothing: 3%

- The reported savings from use of atmospheric evaporators was mostly due to reduced purchases of plating chemicals. The average annual savings per shop were slightly less than the sum of the average capital costs plus annual operating costs. This indicates, that on the average, shops experienced a payback period of approximately one year.

- Very few survey respondents reported the evaporation rate they were achieving with their atmospheric evaporator. Presumably, this is because such data are not routinely collected. The highest evaporation rate reported by a respondent was 60 to 75 gph (PS 183). Oddly, with this particular application, the shop connected the evaporator to a heated, 150°F cadmium cyanide transfer tank. (This is not a recommended application because heating of the cadmium solution will destroy cyanide and create carbonates.) PS 183 indicated in their survey form that they would have achieved a closed-loop "if it were not for the carbonate problem." For more traditional applications (nickel and chrome operated at 130 to 140°F), the highest reported evaporation rates were in the range of 20 to 50 gph (e.g., PS 228, PS 213).

- Some of the plating shops that gave atmospheric evaporation a lower rating may have misapplied the technology. The most frequent misapplication is the use of this technology for the concentration of ambient temperature process solutions or dragout. The commercial units do not have a direct heat source, but rather use heat from the process or transfer tank to evaporate water (see complete explanation in Section 4.2.1). When applied to an ambient solution, very little evaporation will take place. For example, PS 081 attains an evaporation rate of only 3 gph from a unit connected to a zinc cyanide bath that is operated at 85°F. PS 279 attains a rate of 2 to 5 gph for an ambient acid zinc

Exhibit 4-9. Summary of Users Survey Data for Atmospheric Evaporators

Shop			Year	Equipment Costs			Annual Operating Costs				Annual Savings					Use	Sat.	Future	Shop
Code	Application	Vendor	Purchased	Equip.	Other Cap.	Total	Non-Labor	hrs	Labor	Total	Bath Chem.	Treat. Chem.	Disposal	Other	Total	Code	Level	Decision	Code
003	Nickel, Watts	Poly Products Corp.	1988	$3,985	$5,000	$8,985	ND	ND	ND	ND	ND	ND	ND	ND	ND	1	3	1	003
015	Nickel, Watts	NAPCO	1988	$5,000	$500	$5,500	ND	ND	ND	ND	ND	ND	ND	ND	ND	1	5	1	015
015	Z inc, Non-Cyanide	NAPCO	1988	$5,000	$500	$5,500	ND	ND	ND	ND	ND	ND	ND	ND	ND	1	5	1	015
018	Cr+6, Decorative	In-House	1986	$500	$2,000	$2,500	$250	25	$375	$625	$10,000	$500	$500	$0	$11,000	1	0	1	018
018	Nickel, Sulfate	In-House	1986	$500	$2,000	$2,500	$1,000	100	$1,500	$2,500	$30,000	$1,500	$1,500	$0	$33,000	1	4	1	018
020	Nickel, Watts	Poly Products Corp.	1991	$5,000	$500	$5,500	$500	ND	ND	$500	$6,000	$0	$775	$1,000	$7,775	1	5	1	020
029	Chromate on Aluminum	Gumm Chemical	1989	$5,000	$1,000	$6,000	$1,000	500	$5,000	$6,000	$1,000	$1,000	$12,000	$1,000	$15,000	1	5	1	029
038	Nickel, Watts	Brucer	1985	$3,000	$2,000	$5,000	ND	20	$200	$200	ND	ND	ND	ND	ND	1	3	3	038
040	Cadmium, Cyanide	Poly Products Corp	1985	$5,000	$1,000	$6,000	$100	20	$120	$220	ND	ND	ND	ND	ND	3	1	4	040
045	Nickel, Watts	Max-Evap (2 units)	1989	$5,000	$1,200	$6,200	$400	ND	ND	$400	$20,000	$20,000	$0	$0	$40,000	1	4	2	045
053	Cadmium, Non-Cyanide	Poly Products Corp.	1985	$5,000	ND	$5,000	ND	ND	ND	ND	ND	ND	ND	ND	ND	2	4	1	053
053	Cr+6, Decorative	Poly Products Corp.	1985	$5,000	ND	$5,000	ND	ND	ND	ND	ND	ND	ND	ND	ND	1	4	1	053
053	Copper, Cyanide	Poly Products Corp.	1985	$5,000	ND	$5,000	ND	ND	ND	ND	ND	ND	ND	ND	ND	1	4	1	053
053	Nickel, Watts	Poly Products Corp.	1985	$5,000	ND	$5,000	ND	ND	ND	ND	ND	ND	ND	ND	ND	1	4	1	053
054	Zinc, Acid	Not Given	1989	$10,000	$2,500	$12,500	$100	1	$10	$110	$40,000	$0	$0	$0	$40,000	1	5	1	054
071	Chromium, Hard	In-House	1975	ND	ND	ND	ND	ND	ND	ND	$10,000	$0	$0	$0	$10,000	1	4	2	071
077	Chromium, Hard	Poly Products Corp.	1991	$2,500	$500	$3,000	ND	ND	ND	ND	ND	ND	ND	ND	ND	1	5	1	077
081	Z inc, Cyanide	Techmatic	1989	$4,300	$1,000	$5,300	$400	20	$200	$600	$1,100	$0	$0	$0	$1,100	1		1	081
089	Brass	Poly Products Corp.	1988	$4,300	$3,000	$7,300	$400	10	$150	$550	$0	$300	$0	$0	$300	1	4	2	089
098	Cr+6, Decorative	Techmatic Inc	1985	$3,000	$6,500	$9,500	$1,000	50	$500	$1,500	$0	$0	$0	$0	$0	1	2	3	098
101	Silver, Cyanide	Not Given	1987	$3,500	$350	$3,850	$2,400	260	$3,900	$6,300	$0	$0	$0	$0	$0	2	2	3	101
105	Nickel	Corrotec (OH)	1992	$6,000	$2,700	$8,700	$1,200	250	$11,875	$13,075	$1,400	$1,800	$700	$1,500	$5,400	1	5	1	105
118	Nickel, Watts	In-House	1989	$110,000	$30,000	$140,000	$15,000	ND	ND	$15,000	$20,000	$2,715	$6,000	$1,300	$30,015	1	5	1	118
139	Nickel, Watts	Napco Inc.	1987	$10,000	$3,500	$13,500	ND	ND	ND	ND	$5,000	$0	$0	$0	$5,000	2	1	4	139
160	Zinc, Acid	Not Given	1988	$4,000	$1,000	$5,000	ND	100	ND	ND	$0	$0	$0	$0	ND	1	5	1	160
172	Cr+6, Decorative	Poly Products Corp	1989	$1,800	$3,000	$4,800	ND	ND	ND	ND	ND	ND	ND	ND	ND	1	3	1	172
175	Nickel, Watts	Poly Products Corp.	1990	$4,500	$2,500	$7,000	ND	ND	ND	ND	ND	ND	ND	ND	ND	1	4	1	175
181	Chromium, Hard	Kontek Ecology Sys.	1992	$3,200	$1,050	$4,250	ND	ND	$576	$576	$0	$0	$0	$0	ND	1		1	181
183	Copper, Cyanide	Poly Products Corp.	1986	$5,000	$5,500	$10,500	$6,000	ND	ND	$6,000	$4,800	$6,000	$3,000	$2,400	$16,200	1	4	1	183
191	Cr+3, Decorative	Poly Products Corp.	1985	$2,830	$2,494	$5,324	$150	50	$1,000	$1,150	ND	ND	ND	ND	ND	1	5	1	191
193	Cr+3, Decorative	Techmatic Inc.	1982	$12,000	$5,700	$17,700	$3,500	150	$2,250	$5,750	$28,500	$10,000	$0	$16,500	$55,000	1	5	1	193
193	Nickel, Watts	Techmatic Inc.	1982	$12,000	$8,600	$20,600	$3,500	150	$2,250	$5,750	$28,500	$10,000	$0	$16,500	$55,000	1	5	1	193
209	Z inc, Acid	Maxi Vap	1991	$4,000	$500	$4,500	$800	100	ND	$800	$2,600	$2,600	$500	$633	$6,333	1	4	1	209
213	Cr+6, Decorative	Napco Inc.	1988	$7,400	ND	$7,400	ND	ND	ND	ND	ND	ND	ND	ND	ND	1	4	2	213
213	Nickel, Brite	Napco Inc.	1988	$6,500	ND	$6,500	ND	12	$300	$300	$0	$0	$0	$0	ND	1	5	1	213
214	Cadmium, Cyanide	Not Given	1988	$5,000	$2,000	$7,000	ND	ND	ND	ND	ND	ND	ND	ND	ND	2	2	3	214
214	Z inc, Acid	Not Given	1988	$5,000	$2,000	$7,000	ND	ND	ND	ND	ND	ND	ND	ND	ND	1	2	3	214
224	Nickel, Woods	Maxi-Vap	1986	$4,500	$1,500	$6,000	ND	ND	ND	ND	ND	ND	ND	ND	ND	1	4	1	224
228	Nickel, Watts	Techmatic	1990	$3,500	$2,000	$5,500	$1,000	ND	ND	$1,000	$30,000	$5,000	$7,500	$0	$42,500	1	4	1	228
239	Cadmium, Cyanide	Not Given	1989	$3,000	$500	$3,500	ND	400	$4,000	$4,000	$2,500	$2,500	$2,500	$2,500	$10,000	1	5	1	239
241	Cr+6, Decorative	Techmatic	1988	$4,000	ND	$4,000	ND	ND	ND	ND	$5,000	$0	$0	$0	$5,000	1	4	1	241
245	Chromium, Hard	In-House	1984	$2,000	$6,000	$8,000	$2,000	15	$195	$2,195	$2,000	$2,000	$2,000	$0	$6,000	1	4	1	245

Exhibit 4-9 (continued). Summary of Users Survey Data for Atmospheric Evaporators

Shop			Year	Equipment Costs			Annual Operating Costs				Annual Savings					Use	Sat.	Future	Shop
Code	Application	Vendor	Purchased	Equip.	Other Cap.	Total	Non-Labor	hrs	Labor	Total	Bath Chem	Treat. Chem.	Disposal	Other	Total	Code	Level	Decision	Code
247	Zinc, Acid	Poly Products Corp.	1991	$4,700	$2,000	$6,700	ND	ND	ND	ND	ND	ND	ND	ND	ND	1	3	1	247
252	Cr+6, Decorative	Not Given	1987	$7,000	$5,500	$12,500	$3,000	200	$2,000	$5,000	$2,500	$0	$0	$350	$2,850	1	3	2	252
252	Copper, Cyanide	Techmatic Inc.	1987	$7,000	$5,500	$12,500	$3,000	200	$2,000	$5,000	$2,000	$0	$0	$300	$2,300	1	3	2	252
252	Nickel, Watts	Techmatic Inc.	1987	$7,000	$5,500	$12,500	$3,000	200	$2,000	$5,000	$2,000	$0	$0	$350	$2,350	1	3	2	252
254	Z inc, Acid	Not Given	1984	$1,400	$1,000	$2,400	ND	ND	ND	ND	$3,200	$1,600	$5,600	$0	$10,400	1	4	1	254
255	Brass	Poly Products Corp.	1990	$4,600	$1,000	$5,600	$2,000	100	$1,000	$3,000	$2,000	$5,000	$1,000	$0	$8,000	1	4	1	255
255	Cr+6, Decorative	Poly Products Corp	1991	$5,000	$1,000	$6,000	$2,000	100	$1,000	$3,000	$4,000	$5,000	$2,000	$0	$11,000	1	4	1	255
258	Cadmium, Cyanide	Poly Products Corp.	1989	$5,467	$1,200	$6,667	$1,600	12	$180	$1,780	ND	ND	ND	ND	ND	1	4	1	258
261	Alkaline, Ultrasonic	Poly Products Corp.	1992	$6,685	$500	$7,185	$1,000	ND	ND	$1,000	ND	ND	ND	ND	ND	1	4	3	261
262	Cr+6, Decorative	Techmatic Inc.	1985	$5,000	$1,200	$6,200	$500	20	$300	$800	$500	$1,000	$1,000	$0	$2,500	1	3	3	262
278	Dye	Poly Products Corp.	1988	$9,000	$7,000	$16,000	$2,000	120	$1,200	$3,200	ND	ND	ND	ND	ND	3		1	278
279	Zinc, Non-Cyanide	Technic Inc.	1991	$3,500	$1,000	$4,500	ND	ND	ND	ND	$2,500	$0	$500	$0	$3,000	1	3	1	279
282	Cr+6, Decorative	Water Technology	1989	$13,000	$10,000	$23,000	$12,000	1,000	$10,000	$22,000	$0	$0	$0	$0	$0	1	1	3	282
286	Nickel	Napco Inc.	1988	$14,000	$11,600	$25,600	$2,000	5	$250	$2,250	$5,000	$2,000	$750	$750	$8,500	1	4	1	286
293	Zincate	Not Given	1985	$3,000	$500	$3,500	ND	ND	ND	ND	ND	ND	ND	ND	ND	1	4	1	293
298	Zinc, Cyanide	Poly Products Corp.	1992	$5,000	$4,000	$9,000	$500	520	$7,800	$8,300	ND	ND	ND	ND	ND	1	3	1	298
300	Cr+6, Decorative	Poly Products Corp.	1990	$4,985	$2,000	$6,985	$250	ND	ND	$250	ND	ND	ND	ND	ND	1	4	1	300
300	Nickel, Sulfate	Techmatic Inc.	1988	$17,680	$7,600	$25,280	$1,000	ND	ND	$1,000	ND	ND	ND	ND	ND	2	4	2	300
306	Cr+6, Decorative	Not Given	1985	$5,000	$50,000	$55,000	ND	200	$4,000	$4,000	$30,000	$0	$0	$5,000	$35,000	1	5	1	306
306	Nickel, Brite	Not Given	1985	$5,000	$50,000	$55,000	ND	200	$4,000	$4,000	$40,000	$0	$0	$5,000	$45,000	1	5	1	306
306	Nickel, Semi-Brite	Not Given	1985	$5,000	$50,000	$55,000	ND	200	$4,000	$4,000	$40,000	$0	$0	$5,000	$45,000	1	5	1	306
308	Copper, Cyanide	Not Given	unk	ND	ND	ND	ND	ND	ND	ND	ND	ND	ND	ND	ND	1		1	308
308	Ni, Electroless,Other	Not Given	unk	ND	ND	ND	ND	ND	ND	ND	ND	ND	ND	ND	ND	1		1	308
314	Zinc, Non-Cyanide	Not Given	unk	ND	ND	ND	ND	ND	ND	ND	$25,000	$0	$0	$0	$25,000	1		1	314
317	Ni,Electroless,Phosph	Napco	1989	$5,000	$4,000	$9,000	$1,000	24	$240	$1,240	$0	$0	$5,000	$0	$5,000	1	3	1	317
	Averages (5)		1987	$7,077	$5,932	$13,009	$2,159	157	$2,254	$4,412	$10,178	$2,013	$1,321	$1,502	$16,230		3.8		

Notes:

(1) Other capital costs include installation and ancillary equipment.

(2) Use Codes: 1-currently operating; 2-not currently operating and no intension for future use; 3-not currently in use, but intend to use in future.

(3) Satisfaction levels (manufacturer and technology) 1 to 5, with 1=lowest and 5=highest.

(4) Future decision codes: 1-purchase the same technology from the same vendor; 2-purchase the same technology from a different vendor; 3-purchase a different technology; 4-do nothing.

(5) Average total equipment cost is the sum of average equipment and other capital costs. Average annual operating cost is the sum of average non-labor and labor costs. Average savings is simply the average of reported savings. This method of calculation probable understates the average annual savings of the users.

(6) ND = no data

application. Other plating shops that may have misapplied the technology include (based on a review of supplied schematics): PS 040, PS 045, PS 101, and PS 143.

- The high incidence of misapplication for this technology may be due in part to the fact that 54% of the commercial atmospheric evaporators purchased by survey respondents were sold to them by manufacturer's representatives rather than direct purchases. The manufacturer's representatives may not have had sufficient knowledge of the technology to recommend proper installation.

- In some cases, performance was hampered by operational and maintenance problems. These are discussed in Section 4.2.7.

4.2.7 Operational and Maintenance Experience

The following summarizes the respondents O&M experiences and provides operating labor information relative to atmospheric evaporators.

- The quantity of labor required for the operation of this technology is relatively low compared to other recovery technologies. For shops providing data, the average number of annual operating hours per evaporator were: 157 hours per year. The skill requirement commonly needed for operating this technology is a trained technician, a wastewater treatment system operator, a plumber/pipe fitter or common labor. The following is a breakdown of the responses for skill requirements:

Environmental Engineer:	0
Process/Chemical Engineer:	3
Chemist:	5
Consultant:	0
Plumber/Pipe Fitter:	15
Electrician:	9
Vendor:	I
Senior-Level Plater:	8
Junior Level Plater:	11
Trained Technician:	24
Wastewater Treatment Plant Operator:	16
Common Labor:	16

- Approximately 90 percent of the atmospheric evaporator installations identified in the Users Survey were in operation at the time of the survey. The following is a breakdown of the responses for current operating status:

Currently in use:	89.6%
Not currently in use & have no intention of future use:	7.5%
Not currently in use, but intend to use in the future:	3.0%

- The average percent of downtime experienced with this technology was 7%. Only 15 percent of the respondents indicated that their downtime was greater than 5% (those with greater than 5% downtime included: PS 045, PS 101, PS 105, PS 139, PS 213, PS 239, PS 282, PS 298, PS 300 and PS 317).

- Generally, this technology is free of complex mechanical operational and maintenance problems because of the simple design and limited number of moving parts. There are however, some common maintenance requirements, most of which relate to cleaning. A list of the most frequently reported maintenance requirements follows (the percentage of all respondents identifying the problem is given in parenthesis):

Cleaning of evaporation chamber packing or evaporative panels:	11%
Cleaning of nozzles that spray solution over packing or evaporative panels:	4.7%
Maintaining pumps:	4.7%
Unplugging of pipes:	1.6%
Maintaining timers:	1.6%

- Fifty-seven percent of the plating shops that reported cleaning as a maintenance item have purchased the units equipped with evaporative panels rather than the plastic packing. The company that manufactures the panel-type unit has recently introduced a model that is designed to better handle high solids conditions (see 4.2.4). In some cases, the cleaning requirement was blamed on carbonates that precipitated from aerating cyanide containing solutions in the evaporator (PS 089, PS 101, PS 239, PS 258). In two other cases, crystals that formed from the drying of plating chemicals were blamed (PS 045, PS 172).

- According to the literature, plating chemical crystals can be removed from the packing and nozzles by a weekly recirculation of hot water through the evaporator with the fan shutoff. Salts that have formed on the nozzles and packing will dissolve (ref. 355). Carbonate deposits present a more significant problem and most likely will require mechanical (e.g., scraping) removal.

- One shop indicated that operation of their atmospheric evaporator for chromium plating solution recovery resulted in degrading of the unit and piping and that the manufacturer has replaced their evaporator twice in the past five years (PS 252). This shop also operates units for cyanide copper and nickel plating and has not experienced any problems with these other units. A leaking problem around welds was reported by another shop (PS 300). Eventually, PS 300 discontinued use of three of their atmospheric evaporation units because of leaking.

- Many shops indicated that use of an atmospheric evaporator has caused a build-up of contaminants in the plating bath. The most frequently identified problem was carbonate build-up in cyanide containing baths such as zinc, cadmium and copper (ref. PS 081, PS 089, PS 183). Respondents also indicated problems with contaminant build-up in chromium baths (PS 089, PS 172, PS 252 and PS 255), and nickel baths (PS 038, PS 105 and PS 139). PS 139 reported a build-up of brightener and iron in their nickel bath. Twice PS 139 has tried evaporative recovery for nickel and abandoned their efforts.

- Other bath problems attributed to use of an atmospheric evaporator include a breakdown of nickel bath chemicals (PS 039) and increased bath chemistry maintenance for trivalent chromium plating (PS 191).

- Several shops reported that climatic conditions effect the evaporative capacity of their atmospheric evaporator (PS 160, PS 183 and PS 224). These three shops are located in different regions (survey regions SE, FW and MW).

4.2.8 Residuals Generation

Most shops using atmospheric evaporators did not report the generation of any residuals. Presumably, this is because most residuals that would form in the evaporator (e.g., carbonates) will, for the most part, gravity drain back to the process tank. Two shops that reported residuals from the evaporator were PS 089 and PS 113. PS 089 uses a home-made evaporator for chromium solution recovery and generates a "minimal" quantity of sludge. PS 089 generates 30 gal/mth of sludge from evaporative recovery of cadmium solution.

Several other shops listed the carbonates removed from their cyanide containing plating solutions as a residual generated by evaporation. Presumably, they feel that the evaporator created this waste product. The most significant quantity is generated by PS 183 (80 lbs/week).

4.3 VACUUM EVAPORATORS

4.3.1 Overview

Vacuum evaporators are one of the earliest technologies used in the plating industry for chemical recovery. However, vacuum evaporators are currently used less frequently than some other recovery technologies, such as atmospheric evaporators (see Section 4.2). This is primarily due to the fact that the average vacuum evaporation unit costs approximately ten times more than the average atmospheric unit. Also, the vacuum units have more sophisticated and expensive operational and maintenance requirements. Of the 318 plating shops responding to the Users Survey, 23 shops (or 7.2%) have employed vacuum evaporators (30 total units of which approximately 80% were still in operation at the time of the survey) for chemical recovery, whereas, 71 shops (or 22.3%) have used atmospheric evaporators (86 total units). Another six shops (or 1.9%) use vacuum evaporators as end-of-pipe technologies to concentrate their wastes prior to off-site hauling and disposal. This section discusses recovery applications of this technology and Section 7.4.4 addresses end-of-pipe applications.

A vacuum evaporator is a distilling device that vaporizes water at low temperatures when placed under a vacuum. The following explanation of the kinetic theory of liquids and in particular vapor pressure helps in understanding this phenomena.

Liquids as well as gases are in constant motion in varying degrees, depending upon the chemical composition of that matter and the temperature and pressure applied to it. Molecules near the surface have a tendency to escape into the surrounding atmosphere. In open systems, most of these molecules do not return to the liquid and the substance is said to vaporize. In a closed system, molecules return to the liquid in proportion to their concentration in the gaseous phase. Eventually a steady state is reached where the quantities of molecules leaving and returning to the liquid are equal. The vapor is then said to be saturated and the pressure exerted by these escaping molecules is referred to as vapor pressure (ref. 361). Since the kinetic energy of all molecules increases with increasing temperature, so does the vapor pressure. When a liquid reaches the temperature at which its vapor pressure becomes equal to that of the atmosphere above it, boiling occurs. This is the rapid evaporation from all parts of the liquid mass, with bubbles of vapor forming in the interior and rising to the surface. Liquids with appreciable vapor pressure may be caused to boil over a wide range of temperatures by decreasing or increasing the pressure of the atmosphere above it (ref. 362). For example, water boils at 212°F at sea level, but will boil at room temperature if the pressure above it is reduced to about 0.4 psi (ref. 361).

Vacuum evaporators depend on the fact that water, when introduced into a vacuum, tends to boil off, or vaporize. The rate of vaporization is directly related to the level of the vacuum and the temperature of the solution. In operation, heated solution is introduced into the vacuum chamber, the boiling point of the solution is reduced by the vacuum and the resultant vapor (distilled water) is removed from the chamber. The vapor can be either discharged or can be condensed for return to the process (e.g., as rinse water).

Vacuum evaporation systems are relatively complex and are therefore more expensive to construct and maintain than the more simple atmospheric systems discussed in Section 4.2. There are several types of vacuum evaporators used in the plating industry: rising film, flash type, and submerged tube. Generally, each consists of a boiling chamber which is under a vacuum, a liquid/vapor separator and a condensing system. Site-specific conditions and the mode of operation influence the selection of one system over another.

Two techniques have been applied successfully to reduce steam demand for evaporation; both involve reusing the heat value contained in the vapor from the separator. The most common technique is to use a multiple-effect evaporator. Essentially, these are vacuum evaporators in series with different boiling points, made possible by varying the pressure between effects (subsequent effects have lower pressures). The driving force of a multiple effect system is the pressure drop from the first to the last effect. The solution to be concentrated is fed into the boiling chamber of the first effect and external heat is introduced to volatilize the water. The water vapor is then condensed at a different vacuum level and the energy is used to heat the subsequent vacuum chamber. Therefore, the same energy is used several times in multiple stages.

The second technique is to use a mechanical compressor. With this equipment, the water vapor from the separator enters the suction of the compressor where its temperature and pressure are increased. The vapor is then desuperheated and enters the reboiler. Thus the latent heat of evaporation, normally lost to the condenser is recycled by the compressor, providing a temperature difference across the heat exchanger. The needed energy then comprises only the power for the pressure increase to provide the temperature difference.

There are a number of advantages accruing to vacuum systems. Among them are the fact that they are essentially independent of the requirement to heat and move large volumes of air, thus reducing the air pollution problem, at least when compared to atmospheric systems. Further, they are operated at relatively low temperatures, which could be of considerable importance in systems that handle temperature-sensitive products. Additionally, vacuum systems are advantageous with alkaline cyanide solutions which would build up carbonates more rapidly with atmospheric evaporators because the latter type aerates the solution.

4.3.2 Development and Commercialization

Evaporation has been used for centuries for food and beverage processing. Modern vacuum evaporator design for industrial use dates back to the early 1900's with the development of the rising film evaporator. Plating applications for vacuum evaporators began in approximately 1949, when systems were used to recover chromic acid (ref. 300). The early plating applications were purchased primarily to reduce operating costs through chemical conservation. Subsequently, evaporation was sporadically implemented as a pollution control method in response to local discharge standards. In some cases, this was a sufficient method of meeting the local standards for targeted pollutants such as cyanide and chromium. The early evaporators used for plating applications were the same types of units used by other industry segments (e.g., chemical processing, dairy, food and beverage industries). These units had large capacities, due in part because water conservation and pollution control were less important at the time, resulting in higher flow rates. Also, energy was much less expensive. In 1974, with the advent of rapidly rising energy costs, there began a movement to down-size recovery systems. Coupled with the Federal pollution control standards, first promulgated in 1979, plating shops turned to smaller and more energy efficient vacuum evaporators and the less expensive atmospheric evaporators for chemical recovery. Energy efficiency was achieved by the employment of multi-effect vacuum units and mechanical vapor compression. During the 1980's and early 1990's firms have sought methods of low or zero effluent discharge to reduce their regulatory requirements. As a result, the newer energy efficient vacuum evaporators were applied as end-of-pipe technologies (ref. 300, 375).

There are approximately 25 companies that manufacture and/or supply evaporative recovery equipment applicable to the plating industry. This includes vacuum and atmospheric evaporators for both chemical recovery and waste concentration (ref. 421). Of these companies, three firms have responded to the vendors survey (LICON, Inc., Calfran Int., and QPS Inc.).

4.3.3 Applications and Restrictions

Two common configurations for the application of vacuum evaporators to chemical recovery are shown in Exhibit 4-10. In configuration VE-1, a closed-loop process is achieved using a three stage rinse system with the feed from the first rinse being concentrated by the evaporator and directed to the plating bath. A solution purification step is shown, which removes bath contaminants that would buildup in the bath due to the closed-loop process. Typical solution purification technologies used for this purpose include ion exchange and/or carbon filtration. PS 125 employs this configuration using a cation exchange unit to remove contaminants from its decorative chromium plating rinse water/dragout. PS 124 has a similar arrangement. PS 082 installed a cation exchange unit and electrolytic purification unit that are connected to a storage tank. The use of multiple-stage rinsing is nearly always required with evaporator applications in order to minimize the quantity of water to be evaporated. The survey respondents used a minimum of two and a maximum of four rinsing stations. The second configuration (VE-2) shows an open process, where a small portion of the dragout is not recovered. Also shown in VE-2, is a direct bleed from the bath to the evaporator. This may be required for ambient or low temperature baths, where there is a limited surface evaporation rate and insufficient "head-room" in the plating tank to return the concentrated dragout/rinse water.

Vacuum evaporators are applied to the recovery of a wide range of plating solutions. They are especially applicable in situations where atmospheric evaporators are either technically or economically impractical. This includes: (1) the recovery of heat sensitive chemicals (e.g., cyanide plating baths); (2) the recovery of chemicals that are sensitive to air oxidation (e.g., cyanide plating baths or the stannous tin bath); (3) low or ambient temperature plating solutions where there is no appreciable surface evaporation; (4) the recovery of solutions that contain volatile components; and/or (5) where high evaporation rates (e.g., >20 to 40 gph) are necessary to achieve recovery and atmospheric evaporators become too expensive (i.e., energy cost) to operate (ref. 299).

The results of the Users Survey and Vendors Survey showed that vacuum evaporators are applied to a range of plating and finishing solutions. These identified applications are shown in Exhibit 4-11.

Although vacuum evaporators may provide an energy savings over the atmospheric types of evaporators, neither one is economically practical to purchase or operate where large volumes of low concentration solutions are involved. In those cases, ion exchange or reverse osmosis are the more cost effective methods of recovery (ref. 299).

4.3.4 Technology/Equipment Description

4.3.4.1 General

This subsection discusses commercially available vacuum evaporation equipment that is manufactured and/or sold by vendor survey respondents. This is intended to provide the reader with information and data on a cross section of available equipment. Mention of trade names or commercial products is not intended to constitute endorsement for use.

There is a wide range in design of vacuum evaporators, although the majority of these devices work on the princi-

Exhibit 4-10. Two Configurations for the Application of Vacuum Evaporators for Chemical Recovery

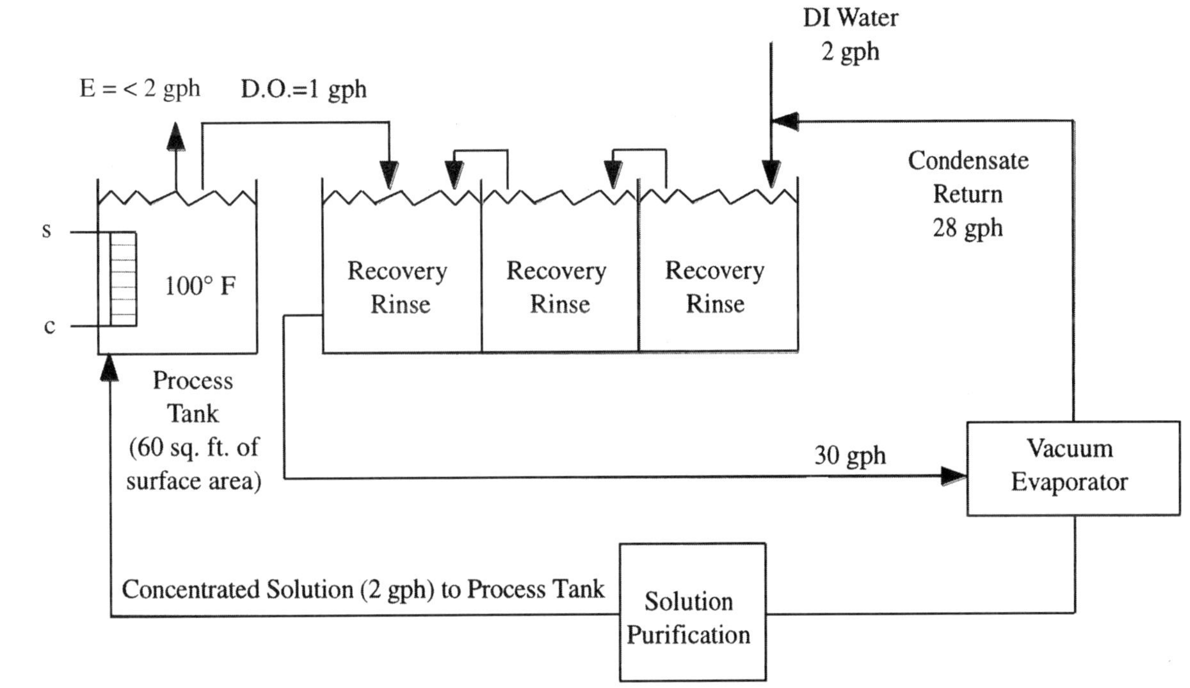

Vacuum Evaporation Configuration VE-1: Closed-Loop System

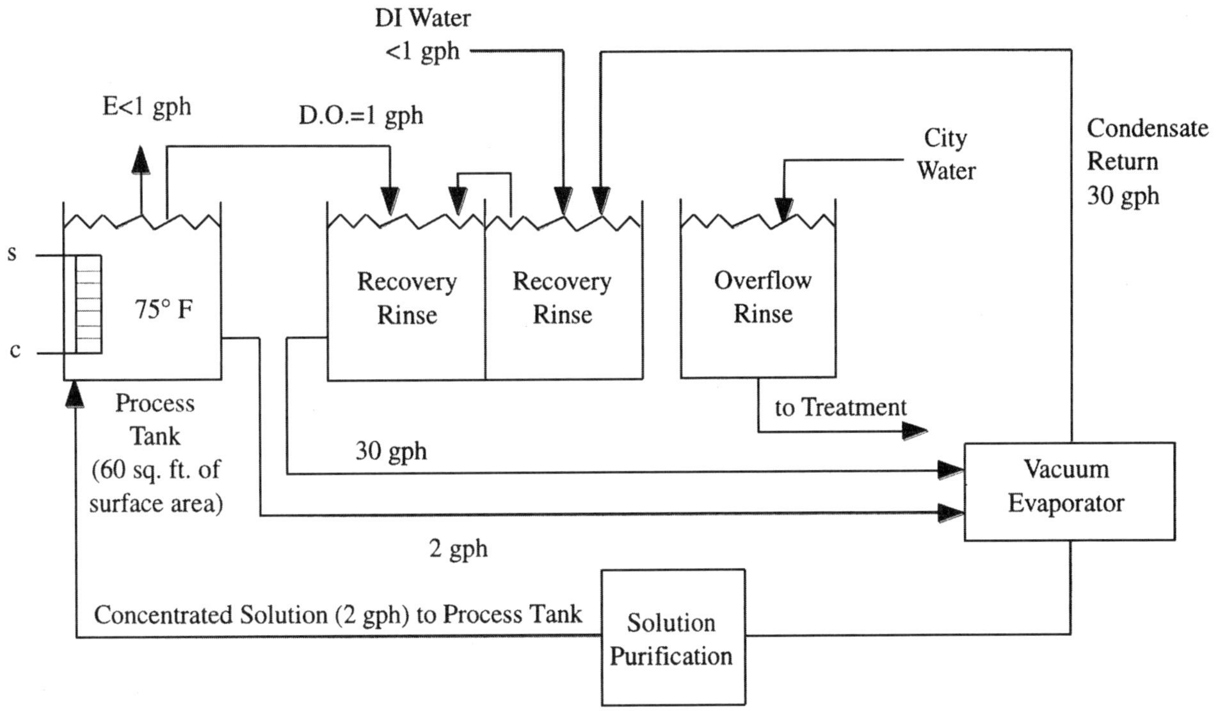

Vacuum Evaporation Configuration VE-2: Open Rinse System

Exhibit 4-11. Applications of Vacuum Evaporation Identified During the Users and Vendors Surveys

	Number of Users Survey Applications[1]	Number of Vendors Survey Applications
Alkaline Cleaners	0	30
Brass, cyanide	0	25
Bronze	0	2
Cadmium, cyanide	3	8
Chromium, decorative (Cr^{+3})	1	0
Chromium, all Cr^{+6}	7	51
Copper, acid	0	11
Copper, cyanide	1	0
Gold	0	20
Lead-tin	1	20
Nickel, electroplating	3	58
Nickel, electroless	0	3
Phosphate	0	8
Silver, cyanide	0	10
Sulfuric acid	0	6
Zinc, acid	0	3
Zinc, alkaline (non-CN)	0	1
Zinc, cyanide	3	65
End-of-pipe concentration	5	80

[1] Not all uses of vacumn evaporators by the Users Survey respondents were determined.

ples described in Section 4.3.1. Vacuum evaporators are built by various manufacturers for different applications. Exhibit 4-12 classifies vacuum evaporators according to the way water is vaporized. This design element helps to differentiate between some of the commercial equipment available to the electroplater. It should be noted that not all manufacturers of plating evaporation equipment are represented in this exhibit.

As with any technology group, the vacuum evaporation industry has developed their own terminology for their equipment and its components. Some of their commonly used terms are defined in Exhibit 4-13.

The following subsections describe the types of vacuum evaporators that are applicable to the plating industry. Where information is available, specific commercial units are briefly described.

4.3.4.2 Submerged Tube Evaporators

The submerged tube evaporators, which includes the short and long tube vertical (LTV) types and the horizontal tube type, are termed natural circulation evaporators, because no pump or other recirculation device is employed. These units, which are older, but still widely used types of vacuum evaporators, are sometimes referred to as calandria type evaporators. With the basic design (short tube type), a vertical tube bundle is placed inside a vertical cylindrical evaporator shell. The tubes or tube sheets, usually two to six feet in length, span the body diameter. The liquid level in the body is typically maintained such that 50% of the tube sheets are immersed. Liquid circulates through the tubes at a rate many times greater than the feed rate. The liquor travels up through the tubes and down a central pipe called a "downcomer." Steam or water vapor condenses on the outside surface of the tubes and the liquor is heated and boiled inside the tubes. The circulation of the liquid is achieved because of the difference in specific gravity between the liquor and vapor in the tubes plus a vapor lift effect. This combined phenomenon is known as the thermosyphon effect and it is the design basis for all natural circulation evaporators, which includes the falling film types (ref. 376, 422, 423). In some cases, an agitator, located inside or beneath the downcomer, is used to increase circulation in salting-type applications.

Generally, the submerged tube evaporators are less expensive to purchase than rinsing film or flash units of equal capacity. Steam or thermal demand is the same as for rising film (ref. 376). The evaporators find applica-

Exhibit 4-12. Classification of Commercially Available Evaporators for Plating Applications According to the Way Water is Vaporized

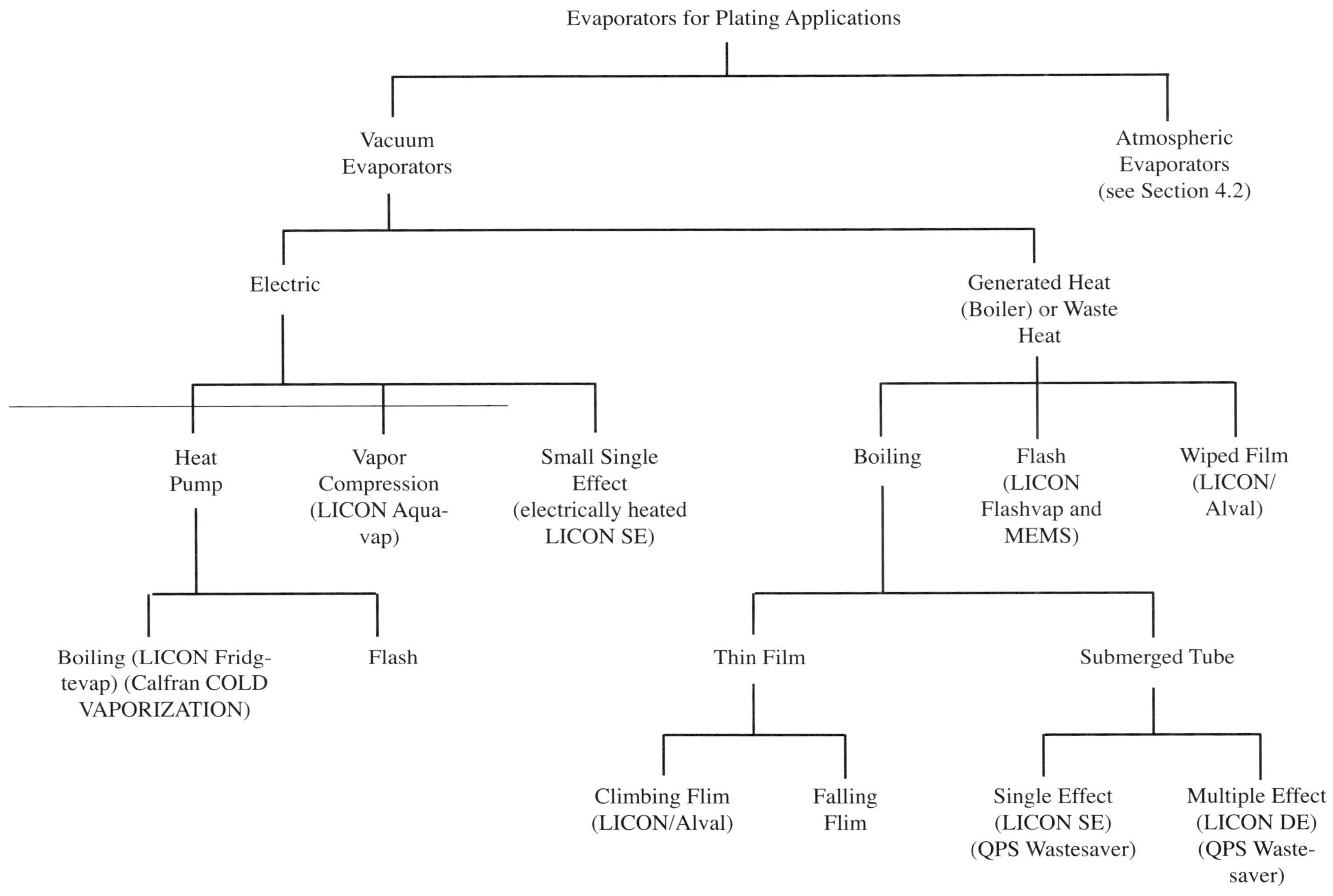

Exhibit 4-13. Common Vacuum Evaporator Terminology

Bodies	Also called vapor heads or flash chambers; the unit where vapor-liquid separation takes place. Also a label given to the basic module of an evaporator, comprising one heating element and one flash chamber.
Calandria	The unit in which heat transfer takes place.
Compression Evaporation	A process in which evaporated vapor is compressed to a higher pressure level and then condensed. The compressed vapor provides heat required for evaporation. Energy economy obtained by multi-effect evaporation can often be equalled in a compression evaporation system.
Condenser	Evaporator equipment used to condense the vapor from the last effect or used as an intercondenser in multi-stage vacuum producing systems.
Cooling Water	Water used in the condenser of an evaporator to condense the vapor phase that exists in the separator. Cooling water can come from cooling towers or reservoirs, or once-through process water can used.
Direct Contact Condenser	Condenses vapors as they are contacted with the cooling medium. There are no heat transfer surfaces and therefore the vapor pressure drop is low. Due to contact with the process solution, the cooling water will require waste treatment.
Effect	One or more bodies boiling at the same time.
Evaporator	The entire system of effects, not necessarily one body or one effect.
Fouling	Formation of deposits other than salt or scale.
Heat Pump	A device in which a refrigerant fluid is continuously circulated through a closed cycle. The gaseous refrigerant is first compressed and then allowed to condense by giving out heat to a cooler environment. The liquid refrigerant is then throttled to low pressure, by means of an expansion valve and allowed to evaporate in another heat exchanger, drawing in heat. The vapor is then circulated to the compressor to begin another cycle.
Intercondenser	A condenser within a multistage system that condenses some of the steam between stages and reduces steam consumption.
Mechanical Compression	Uses a compressor driven by a mechanical drive to compress all the overhead vapors.
Mechanical Pumps	One of two devices (see steam jet ejectors) used to produce a vacuum. More energy efficient than steam jet ejectors but less reliable due to the number of moving parts involved.
Precondenser	A condenser that removes process vapors and permits the use of smaller vacuum pumps.
Refrigeration	Systems that absorb heat not wanted or needed and reject it elsewhere. Heat is removed from the system by evaporation of a liquid refrigerant and is rejected by condensation of the refrigerant vapor.
Scaling	Growth or deposition on heating surfaces of a material that is either insoluble or has a solubility that decreases with an increasing temperature.
Staging	Two or more sections of a single effect evaporator operating at different concentrations.
Steam Jet Ejectors	One of two devices (see mechanical pump) used to produce a vacuum. Jet ejectors usually have lower initial costs, lower maintenance costs (fewer moving parts) but higher operating costs (less energy efficient) than mechanical systems. Multistage jet ejectors (connected in series) improve thermodynamic efficiency.
Surface Condenser	A condenser with a heat transfer surface on which condensing occurs. The heat transfer surface is subject to fouling, corrosion and plugging. Generally more expensive than direct contact condensers. Cooling water is not contaminated with the process solution.
Thermal Compression	Uses a steam jet to compress a fraction of the overhead vapors with high pressure steam.
Thin Film	Thin films of solution are created in evaporators to maximize the liquid surface area per unit volume of flow and therefore facilitate rapid evaporation.

Sources: ref. 373, 374, 423.

tion for processing mildly scaling liquors and relatively viscous solutions (ref. 422).

LICON Inc., a manufacturer of electroplating evaporation equipment and a Vendors Survey respondent, manufactures single effect and double effect submerged tube evaporators. These devices have been applied to the concentration of chromium (Cr^{+3} and Cr^{+6}), zinc chloride, nickel chloride, nitric acid, and sulfuric acid bearing waters as well as mixed waste streams (ref. LICON file).

QPS manufactures the Wastesaver® submerged tube evaporator, which is available with either single, double or triple effects. Their newer units have a pumpless liquid transfer system that reportedly eliminates problems commonly associated with mechanical liquid transfer equipment (e.g., pumps, seals, impellers, etc.). These units are manufactured with capacities ranging from 25 gph to 1,500 gph. The basic units are manufactured from stainless steel with titanium offered as an option.

4.3.4.3 Rising Film (Climbing Film)

The basic rising film evaporator consists of an evaporator body, separator and condenser. The evaporator body is a shell-and-tube heat exchanger. Liquid feed enters the bottom of the heat exchanger, it is preheated until it reaches the boiling point and it then moves up the tube. The vapor generated occupies the center of the tube and the liquid is forced to the tube wall. As the fluid travels up the tube, more vapor is formed resulting in a higher central core velocity. The upward velocity of the vapor forces any remaining liquid to the tube wall and continues to provide an upward motion. As the process continues, the higher vapor velocities result in thinner and more rapidly moving liquid films. This design provides a high heat transfer coefficient and relatively short residence time (ref. 373, 375, 376).

Evaporation is typically accomplished at pressures of 1.3 to 7.5 psia (67 to 388 mm Hg absolute), thereby lowering the boiling point to 110° to 180°F (43° to 82°C). The wastewater leaves the body and enters the separator where the water vapor is separated from the heavier plating solution. The plating solution is either returned directly to the bath or held in an integral reservoir. The vapor leaving the separator is condensed in a shell-and-tube heat exchanger and the distillate is directed to the rinse tanks (ref. 376).

Commercially available rinsing film evaporators used in the plating industry are manufactured by LICON/Aval and Corning. Several existing plating applications of rinsing film evaporators identified in the Users Survey were manufactured by the Pfaudler Company.

4.3.4.4 Falling Film

Liquid enters the top of the evaporator and a liquid film is formed by gravity, which then flows down the heat transfer surface. During evaporation, vapor fills the center of the channel and as the momentum of the vapor accelerates, the film becomes thinner. Also, the solution accelerates in velocity as it descends inside the tubes because of gravity and the drag of the vapor. Since the vapor is working with gravity, a falling film evaporator produces thinner films than a rising film evaporator for any given set of conditions. This gives rise to shorter residence times and a further improvement over the rising film types in heat transfer. With these devices, liquid is usually separated from the vapor in the bottom liquid chamber of the body.

The falling-film evaporator is particularly useful in applications involving heat sensitive chemical solutions. This is due to a low "driving force" or temperature difference between the heat-transfer medium and the liquid (ΔT's less than 15°F compared to 25°F or more for the rising film) (ref. 375, 377).

No commercial electroplating applications of the falling film evaporator were identified during the Users or Vendors Surveys, although they presumably exist due to the widespread commercialization of these devices (ref. 373, 377).

4.3.4.5 Wiped Film Evaporator

Feed is introduced at the top of the evaporator and is spread by wiper blades on to the vertical cylindrical surface inside the unit. Evaporation takes place as the thin film moves down the evaporator wall. The heating medium is usually high pressure steam. Use of the wiped film evaporator is limited primarily to highly viscous liquids and the stripping of solvents. The high number of moving parts, such as the rotor and wiper blades, may result in higher maintenance costs than other types of evaporators (ref. 375).

LICON Inc. manufactures a wiped film evaporator (Stratavap) with capacities from 5 to 700 gph, but no plating shop applications were identified in the Users or Vendors Surveys for this device or other wiped film evaporators.

4.3.4.6 Flash Evaporators

Unlike with thin film types (e.g., falling film or wiped film), with flash evaporators, vaporization does not occur on the heat exchanger surfaces. Instead, liquor flashes as it enters a separator, crystallization takes place, and a suspended slurry results. Since evaporation does not take place on a heat transfer surface, the tendency for scale to deposit is significantly reduced. The flash evaporation system can be used in single or multiple effects.

The LICON Inc. Flashvap is sold as an end-of-pipe industrial waste concentrator.

4.3.4.7 Thermal Compressor Evaporators

The thermal compressor evaporators are not, by themselves, a separate category of evaporator. Rather, they are evaporators, such as a rising film type, that uses a steam jet ejector or thermocompressor in order to increase steam economy. They can be designed with either single or multiple effects, although the thermocompressor is normally used on a single effect evaporator or only on the first effect of a multiple effect evaporator. Typically, the addition of a thermocompressor will provide an improved steam economy equal to the addition of another effect, but at lower cost. They should be considered only when high pressure steam is available. Because of their smaller size in comparison to an additional effect, they are favored in applications where space limitations exist. A disadvantage of these units is that the condensate is sometimes contaminated with product traces and may have to be treated, rather than reused as rinse water.

No applications of thermal compression were identified during the Users or Vendors Survey.

4.3.4.8 Heat Pump Evaporator

A heat pump is a device that upgrades a heat source to a higher temperature, thus rendering it more useful. With conventional evaporator/heat pump operation, a refrigerant, upon boiling, absorbs the heat that would otherwise be rejected in a condenser. The refrigerant vapor is compressed to a pressure adequate to permit the vapor to be condensed in the calandria, thereby providing the heat needed for evaporation. The condensate from the calandria is flashed into the condenser, thereby completing the cycle (ref. 373). The heat pump eliminates the waste of single and double effect designs, but does cost electrical power to operate the heat pump. Therefore, it is not applicable to plating shops where waste heat is available. Also, it is generally confined to small flows (≤ 100 gph) due to the range of heat pumps available.

LICON Inc. manufactures the Fridgevap (3 to 100 gph) heat pump evaporator, in which the solution is evaporated at around 100°F (40°C). This unit finds application where heat sensitive chemicals are involved.

Calfran, Int. manufactures a line of heat pump evaporators that they term COLD VAPORIZATION™. These include the PTU series (immersion coil design) and STU series (reaction vessel type) for applications of 1,000 gpd or less and greater than 1,000 gpd, respectively and the VTU series designed for low solids feed streams (75 to 1,000 gpd). Their basic materials of construction include 316 stainless steel heat exchangers and PVC shells. Their units are also available in all stainless steel design and heat exchangers are available in titanium and Hastelloy.

4.3.4.9 Mechanical Vapor Recompression (MVR)

The MVR evaporator is the highest priced evaporator type used in the electroplating industry and it is also the most energy efficient. The MVR evaporator is similar to a conventional single-effect evaporator, except the vapor released from the boiling solution is compressed (adds energy) in a mechanical compressor. This compressed water vapor condenses and gives up its latent heat, which is used to vaporize more water from the liquid that is being concentrated. The following example from the literature shows the potential operating cost savings from using the MVR evaporator (ref. 375).

Exhibit 4-14 shows an evaporator with a liquid boiling point of 212°F (atmospheric pressure). All of the water vapor that is boiled off passes to a compressor. In order to keep the energy input to the system as low as possible, the pressure boost across the compressor is limited. In the majority of cases, this pressure boost will correspond to a saturated temperature rise in the region of 15°F or less. In this example, there is a pressure boost of 4.5 psi across the compressor. Assuming that there is a pressure loss of 0.5 psi in the system, the effective pressure on the steam side of the evaporator is 18.7 psi. This compressed water vapor condenses and gives up its latent heat, which is used to vaporize more water from the liquid that is being concentrated. The latent heat of vaporization of water at atmospheric pressure is 970 Btu/lb. Note that it only requires a theoretical energy input of 18 Btu/lb to raise the water vapor from 14.7 to 19.2 psia. The theoretical steam economy, therefore, is 970/18 = 54. When compressor efficiency is taken into account, this figure is brought down to between 32 and 35 which is another way of saying that the MVR system is equivalent to an evaporator with 32-35 effects (see definitions in Exhibit 4-13). However, when the electricity cost for the compressor drive is taken into account, the MVR system then becomes the economic equivalent of just under a 19 effect evaporator.

The MVR has another definite advantage over steam. The condensate is available at high temperature and is ideal for evaporator feed preheating, particularly if the condensate rate is as high as 90% of the feed rate, i.e., a 10:1 concentration ratio within the evaporator. There are many such evaporators in operation where the sole energy input to the system is through the compressor with steam requirements limited to approximately 15 minutes during start up (ref. 375).

An example of a commercial MVR evaporator used by the plating industry is the LICON Inc. Aquavap. This evaporator has an auxiliary flash stage and is capable of achieving concentrations of 500,000 mg/l or more. Evaporative capacities for the Aquavap range from 50 to 600 gph. Existing plating applications include: concentrations of zinc phosphate rinses (multiple units totaling 1,800 gph), concentration of RO reject (300 gph), and end-of-pipe wastewater concentration (50 to 600 gph) (ref. LICON Inc. file).

Exhibit 4-14. Mechanical Vapor Recompression Evaporator/Condensor Schematic

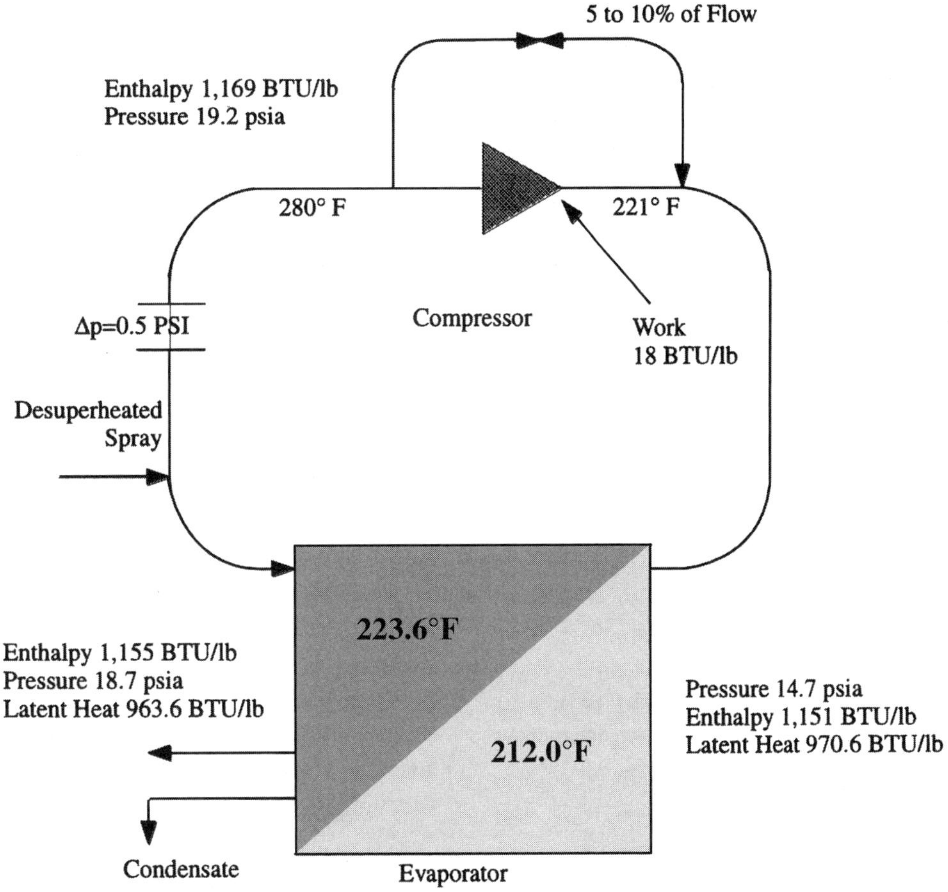

Effective Steam Economy:	970 ÷18 = 54
Efficiency of Drive & Compressor:	65 ---> 72%
Inefficiency due to By-Pass:	5 ---> 10%
Actual Effective Steam Economy:	32 ---> 35 Effects

Source: ref. 375

Exhibit 4-15. Utility Requirements for Single-Effect Evaporators

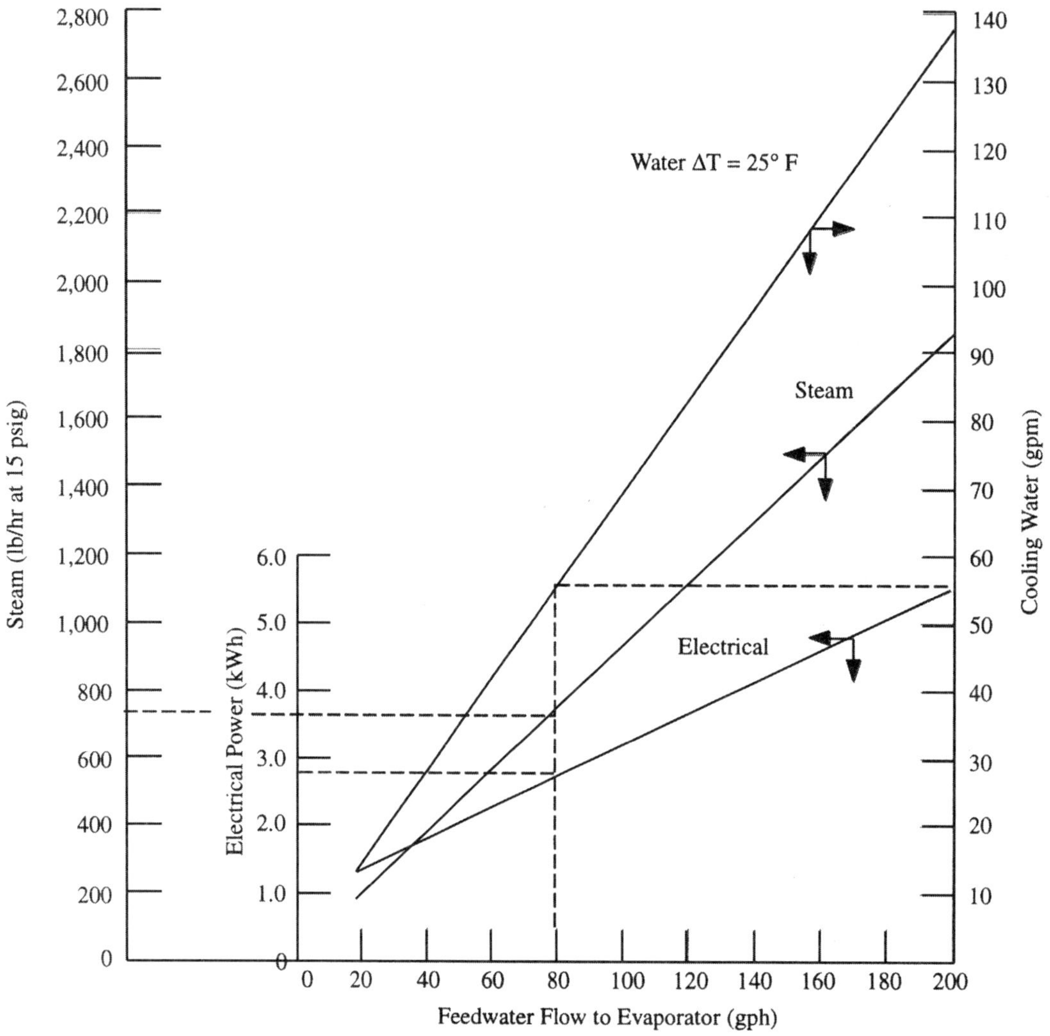

Notes: Evaporation at 1.3 psia. Wastewater feed at 100°F. Cooing water supply at 70°F. Steam at 15 psig. Rising film or submerged tube.
Example: A single effect evaporator boiling 80 gph of wastewater at 1.3 psia pressure has the following requirments - steam (15 psig), 730 lb/hr; electric power, 2.9 kWh; cooling water at ΔT = 25°F, 56 gpm.

4.3.4.10 Multiple Effect Evaporators

Multiple effect evaporators are not a specific type of evaporator, but rather a design element employed to improve the energy efficiency of the evaporation process.

Most evaporators used in the plating industry are single-effect units. Single-effect evaporators operate with one boiler or evaporator section. The water vapor is condensed or exhausted to the atmosphere. Approximately 1.1 pounds (0.5 kg) of steam is consumed in evaporating each pound of water from the plating solution (ref. 376).

Exhibit 4-15 shows the utility requirements for single-effect evaporators as a function of liquid flow rates to the evaporator. The electrical demand is associated with power requirements of the vacuum pump, recirculation pump, and feed pump. As a rule, the cooling water rates are based on a temperature rise of 25°F (14°C) across the condenser (ref. 376). For example, from Exhibit 4-14, if the wastewater flow rate to the evaporator is 80 gal/hr (303 l/hr), the steam rate is 730 lb/hr (331 kg/hr) for 15 lb/in^2 gauge (1,536 mm Hg absolute) steam. The electrical demand is 2.9 kWh and the cooling water rate is 56 gal/min (212 l/min). For atmospheric evaporators where no cooling water is used, the steam rate would be at least 20 percent higher (ref. 376).

A general application of a double-effect evaporator, is shown in Exhibit 4-16. The basic principle is to use the heat given up by condensation in one effect to provide the reboiler heat for another effect. In the system shown in Exhibit 4-16, approximately 50 percent of the wastewater is concentrated in the first effect using steam. The vapor from the separator of the first effect enters the second-effect reboiler and condenses to provide the thermal energy required to reach the final concentration of the plating solution.

The steam and cooling water rates for the double-effect unit in Exhibit 4-17 are approximately 50 percent of those required for the single-effect unit.

Some platers using double-effect units achieve an additional benefit by recovering two different plating baths simultaneously. However, care should be taken in employing this arrangement however, because there is a possibility of cross-contaminating baths (ref. 376).

Multiple effect evaporation, when used in the plating industry, is most often applied to submerged tube evaporators, rinsing film and flash types (ref. 376). An alternative method for reusing the heat value contained in the vapor from the separator is to employ a mechanical compressor (see Section 4.4.2.9).

4.3.5 Costs

4.3.5.1 Capital Costs

The basic equipment costs and installed costs for vacuum evaporators are indicated in Exhibits 4-18 and 4-19 for heat generated (steam) and electric types. Equipment costs will vary depending on the materials of construction; costs shown are for the basic materials offered by the manufacturer. Evaporators are currently marketed with a wide range of construction materials to resist the corrosiveness of various plating chemicals. The more popular materials include titanium, tantalum, borosilicate glass, stainless steel and carbon steel. Most evaporators are supplied as package units and only require the hook-up of utilities before start-up. However, some ancillary equipment is required (e.g., tanks), which are reflected in the installed costs. The installed cost estimate (140% of basic equipment cost) is based on Users Survey data.

Due to the capital intensive nature of this technology, it is prudent that the buyer make every effort to reduce the flow rate of the feed stream by employing pollution prevention measures. Methods of flow reduction are discussed in Section 3.

When selecting a vacuum evaporator, the plater should consider, in addition to costs, the following: (1) availability, quantity and quality of steam, hot water or waste heat (i.e., if unavailable or insufficient, then choose one of the electric units); (2) cooling water requirements; (3) electrical power requirements; (4) maximum temperature that can be applied to the feed stream (i.e., concern for heat sensitive chemicals); (5) expected feed rate; (6) required solids concentration of product (i.e., how concentrated must the plating solution be before it can be returned to the bath); (7) anticipated use of distillate; (8) materials of construction (depends on both the type and maximum concentration of chemicals); (9) controls (most units have microprocessor controls for automatic operation and manual override); (10) auxiliary equipment requirements (e.g., bath maintenance technologies for removal of contaminants that will be returned to the bath by the evaporator); and (11) O & M requirements (level of expertise required and number of man-hours per year).

4.3.5.2 Operating Costs

The primary operating costs for vacuum evaporators are labor, energy and cooling water. Energy and operating labor costs per gallon evaporated are shown in Exhibit 4-20. In this exhibit, a distinction is made between operating costs for segregated recovery and the concentration of mixed waste streams (e.g., end-of-pipe). Higher O&M costs can be expected for end-of-pipe applications because the solutions are evaporated to higher solids levels, which increases fouling and scaling.

4.3.6 Performance Experience

A partial summary of the user data relative to vacuum evaporation is presented in Exhibit 4-21. There are a number of observations that can be made from these data and other data contained in the database and literature:

- The average satisfaction level for chemical recovery applications is 3.5 (on a scale of 1 to 5, with

Exhibit 4-16. Double-Effect Evaporation for Chemical Recovery

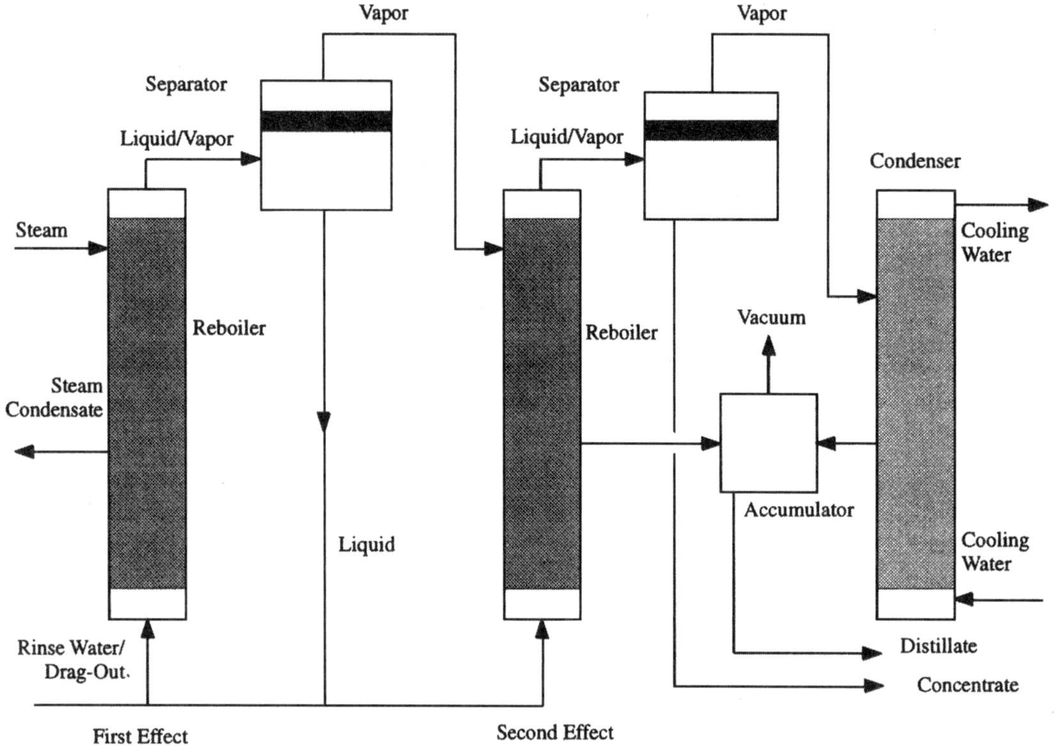

Source: ref. 376

Exhibit 4-17. Utility Requirements for a Double-Effect Evaporator

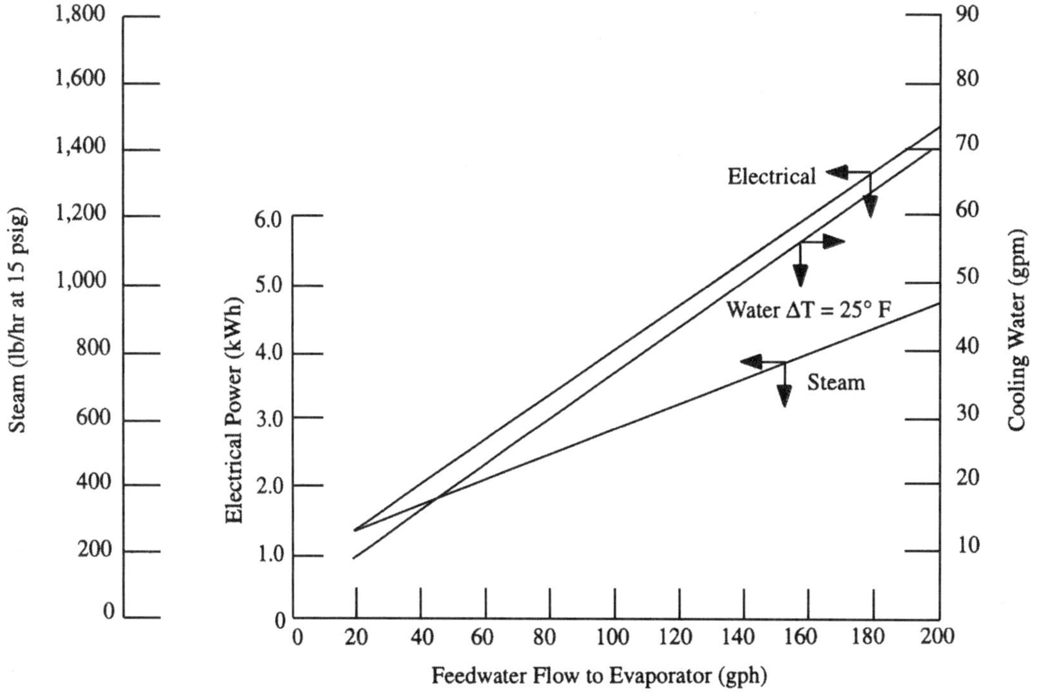

Notes: Wastewater feed at 100°F; cooling water supply at 70°F; and steam supply at 15 psig.

Source: ref. 376

Exhibit 4-18. Equipment Costs for Steam Type Vacuum Evaporators

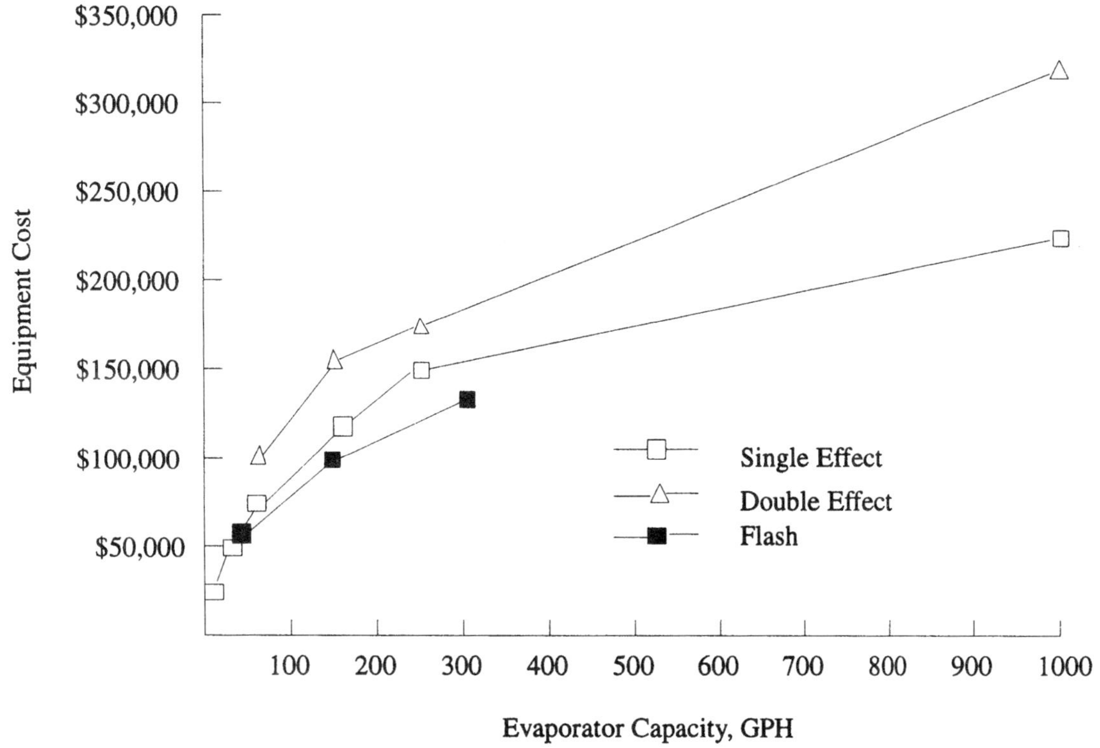

Note: Total installed costs (includes basic equipment, auxillary equipment and installation costs) are approximately 140% of equipment costs.

Exhibit 4-19. Equipment Costs for Electric Type Vacuum Evaporators

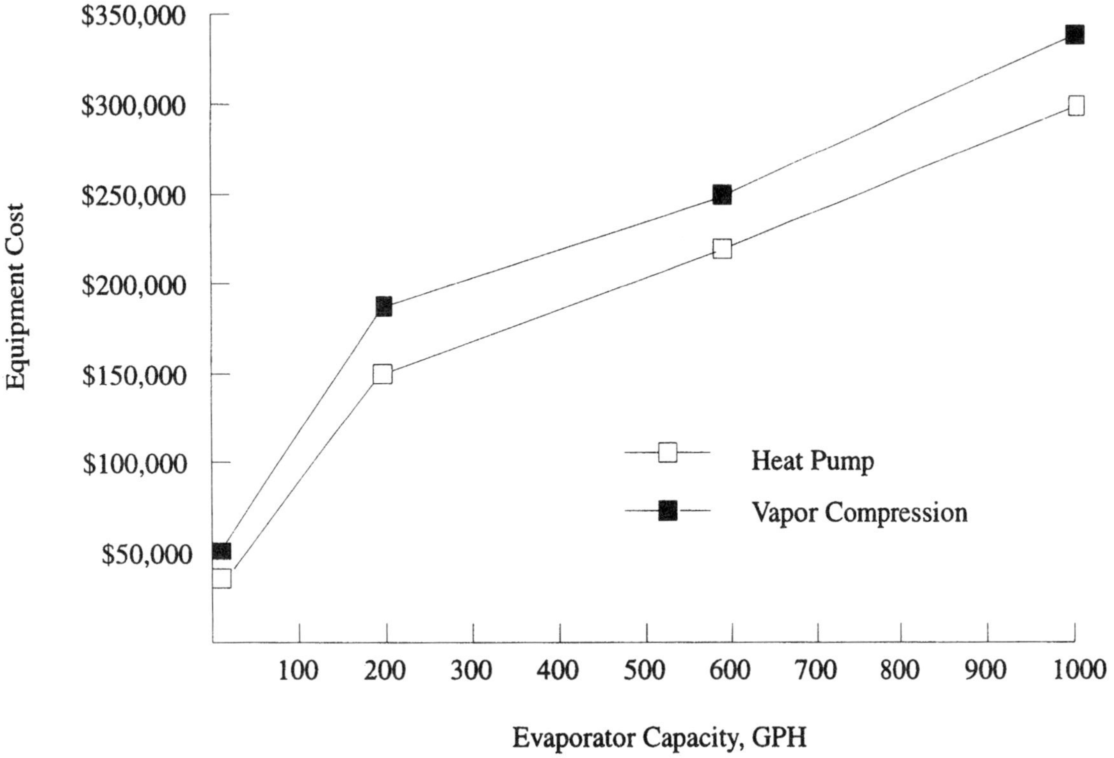

Note: Total installed costs (includes basic equipment, auxillary equipment and installation costs) are approximately 140% of equipment costs.

Exhibit 4-20(a). Operating and Maintenance Costs for Vacuum Evaporators (Segregated Recovery)

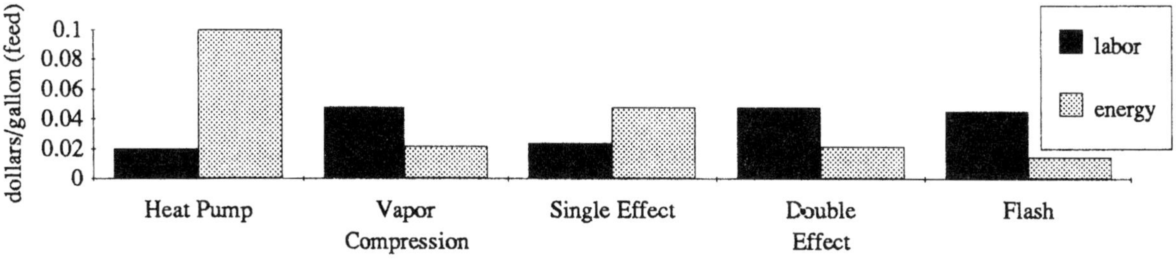

Exhibit 4-20(b). Operating and Maintenance Costs for Vacuum Evaporators (Mixed Waste Concentration)

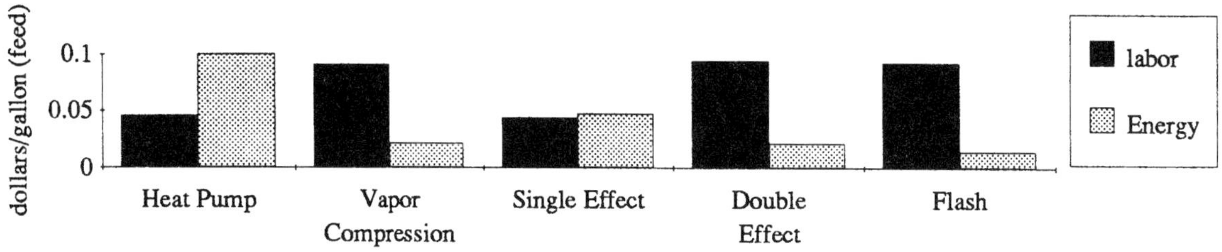

5 being most satisfactory), which higher than the average rating for all recovery technologies. Seventy-five percent of the shops using vacuum evaporation for chemical recovery indicated that this technology satisfied the need for which it was purchased. The following is a breakdown of the reasons why shops purchased this technology:

To meet of help meet effluent regulations: 14
To reduce plating chemical purchases: 12
To reduce the quantity of waste shipped off-site: 12
To reduce wastewater treatment costs: 11
To improve product quantity: 0
To close-loop a particular process: 1

- Vacuum evaporators were successful for most applications identified in the Users Survey except for zinc-cyanide plating solution recovery. The average annual savings from using vacuum evaporators exceeded the sum of the average capital cost plus the average annual operating cost. The most significant savings were derived from reductions of bath chemical and treatment chemical usage.

- The use of vacuum evaporation as a recovery technology generally did not impact production quality or the rate of production for the survey respondents. The following responses were provided:

	Product Quality	Production Rate
Improved	1	1
No Change	13	13
Decreased	2	0

- PS 298 indicated that use of their evaporator decreases product quality because their distillate is contaminated and not adequate for good rinsing. PS 102 also indicated that their unit decreases product quality.

- The respondents indicated, that based on their experience with this technology and, if given the opportunity, they would:

Purchase the same technology from the same vendor: 12
Purchase the same technology from a different vendor: 4
Purchase a different technology: 3
Do nothing: 0

- Two of the respondents indicated that their vacuum evaporation system was the cause of an effluent compliance excursion (PS 039 and PS 088). PS 280 did not respond to the question. All other respondents indicated that their vacuum evaporation system was not the cause of an effluent compliance excursion.

Exhibit 4-21. Summary of Users Survey Data for Vacuum Evaporators

Shop Code	Application	Vendor	Year Purchased	Equipment Costs: Equip.	Equipment Costs: Other Cap.	Equipment Costs: Total	Annual Operating Costs: Non-Labor	Annual Operating Costs: Hrs	Annual Operating Costs: Labor	Annual Operating Costs: Total	Annual Savings: Bath Chem.	Annual Savings: Treat. Chem.	Annual Savings: Disposal	Annual Savings: Other	Annual Savings: Total	Use Code	Sat. Level	Future Decision	Shop Code
034	Zinc, Cyanide	McDermit	1980	$35,000	$8,000	$43,000	ND	250	$5,000	$5,000	$500	$200	$1,000	$2,500	$4,200	3	2	2	034
039	Zinc, Cyanide	Waste Saver	1975	$50,000	$4,000	$54,000	$9,000	400	$2,400	$11,400	$5,000	$0	$0	$0	$5,000	2	1	2	039
082	Cr+6, Decorative	Pfaudler	ND	$150,000	$25,000	$175,000	$50,000	1,000	$15,000	$65,000	$130,000	$115,000	$30,000	$0	$275,000	1	4	1	082
088	Cadmium, Cyanide	Pfaudler	1982	$23,000	$2,050	$25,050	$10,000	500	$6,000	$16,000	$9,900	$15,000	$5,000	$0	$29,900	1	4	1	088
102	Copper, Cyanide	McDermid	1972	ND	ND	ND	ND	ND	ND	ND	ND	ND	ND	ND	ND	2	4	3	102
123	Lead-Tin	Calfran	1991	$60,000	$2,000	$62,000	$6,912	2,080	$14,560	$21,472	$214,700	$0	$34,260	$0	$248,960	1	4	1	123
124	Cr+6, Decorative	Corning	1980	$108,210	$0	$108,210	ND	ND	ND	ND	ND	ND	ND	ND	ND	1	4	1	124
125	Cr+6, Decorative	In-House	1985	$35,000	$10,000	$45,000	ND	200	$2,600	$2,600	$18,500	$0	$0	$0	$18,500	1	4	1	125
125	Nickel, Watts	In-House	1985	$70,000	$210,000	$280,000	$15,000	200	$2,600	$17,600	$38,000	$0	$0	$0	$38,000	1	4	1	125
132	Cadmium, Cyanide	Corning	1981	ND	ND	ND	ND	ND	ND	ND	ND	ND	ND	ND	ND	2	1	3	132
143	Cadmium, Cyanide	Water Vap	1992	$10,000	$500	$10,500	$3,000	ND	ND	$3,000	$0	$0	$500	$1,000	$1,500	1	4	1	143
196	Cr+3, Decorative	Corning	ND	$240,000	$25,000	$265,000	$10,000	200	$2,000	$12,000	$450,000	$15,000	$12,000	$18,000	$495,000	1	4	1	196
197	Cr+6, Decorative	Macdermid Inc.	1979	$44,646	$56,890	$101,536	$28,565	2,920	ND	$28,565	ND	ND	ND	ND	ND	1	4	1	197
197	Nickel, Watts	Macdermid Inc.	1978	$41,521	$56,950	$98,471	$40,017	2,920	ND	$40,017	ND	ND	ND	ND	ND	1	4	1	197
197	Nickel, Watts	Macdermid Inc.	1979	$54,172	$56,960	$111,132	$40,016	2,920	ND	$40,016	ND	ND	ND	ND	ND	1	4	1	197
213	Chromium Etch	Corning	1988	$80,557	ND	$80,557	ND	ND	ND	ND	ND	ND	ND	ND	ND	2	5	1	213
275	Cr+6, Decorative	Corning	1984	$75,000	$6,500	$81,500	$10,000	250	$4,250	$14,250	$200,000	$200,000	$38,000	$0	$438,000	1	4	2	275
280	Chromium Etch	Corning	1990	$90,000	$8,000	$98,000	ND	ND	ND	ND	$16,400	$80,700	$10,000	$0	$107,100	1	4	3	280
298	Zinc, Cyanide	Wastesaver Corp	1991	$43,000	$29,000	$72,000	$15,600	780	$7,800	$23,400	$31,000	$13,000	$0	$3,000	$47,000	1	2	2	298
	Average		1983	$71,183	$31,303	$100,644	$19,843	1,125	$6,221	$21,451	$92,833	$36,575	$10,897	$2,042	$142,347		3.5		

PS 082 purchased their unit used.
PS 088 purchased their unit second hand, but unused.

Notes:

(1) Other capital costs include installation and ancillary equipment.
(2) Use Codes: 1-currently operating; 2-not currently operating and no intension for future use; 3-not currently in use, but intend to use in future.
(3) Satisfaction levels (manufacturer and technology) 1 to 5, with 1=lowest and 5=highest.
(4) Future decision codes: 1-purchase the same technology from the same vendor; 2-purchase the same technology from a different vendor; 3-purchase a different technology; 4-do nothing.
(5) "√" indicates that a savings was realized, but not quantified.
(6) ND = no data

- Several respondents provided the following quantitative performance data:
  - PS 082 indicated that the supplier stated capacity of their unit is 300 gph and that the actual capacity is 175 gph.
  - PS 123 has an evaporation rate of only 6 gph. It is used to make head-room in their tin-lead plating tank so that recovery rinsing can be used. The feed to their unit (tin-lead plating solution), has a concentration of 16 to 18 oz/gal and the concentrated return has a concentration of 32 to 36 oz/gal.
  - PS 088 indicated that the capacity of their unit is 60 gph and that they are able to operate their cadmium plating process on a closed-loop basis. Their dragout rate is 1.5 gph and they have a three stage counterflow rinse system feeding the evaporator. The cadmium bath is operated at 75°F and there is essentially no surface evaporation.
  - PS 124 indicated that their unit has a capacity of 90 gph.
  - PS 125 indicated that their unit has a capacity of 100 gph.
  - PS 196 indicated that both the supplier stated capacities and actual capacities of their units (3) were 90, 75 and 50 gph.
  - PS 213 indicated that both the supplier stated capacity and actual capacity of their unit was 75 gph.
  - PS 298 indicated that their "unit does not meet levels stated in promotional" and that the "quality of distilled water is poor." The supplier stated capacity of their unit is 100 gph and the actual capacity is 70 to 80 gph.
- PS 132 indicated that their evaporator "never performed as sold." No details of their problems were provided.
- PS 034 expressed their feelings about their unit as follows: "Poor design, good technology."

4.3.7 Operational and Maintenance Experience

The following summarizes the respondent's O&M experiences and provides operating labor information relative to vacuum evaporators.

- The average number of annual man-hours spent for operating and maintaining a vacuum evaporation unit were: 657 hrs/yr. The skill requirement commonly needed for operating this technology is trained technician or a wastewater treatment plant operator. The following is a breakdown of the responses for skill requirements:

Environmental Engineer:	1
Process/Chemical Engineer:	1
Chemist:	1
Consultant:	1
Plumber/Pipe Fitter:	4
Electrician:	4
Vendor:	1
Senior-Level Plater:	4
Junior Level Plater:	1
Wastewater Treatment Plant Operator:	8
Trained Technician:	9
Common Labor:	1
Other:	0

- The most frequent and significant operational and maintenance problems identified with vacuum evaporation include: (1) mechanical problems with pumps; (2) damage to components by aggressive plating chemicals; and (3) contamination build-up in the plating bath.
- Approximately 26% of the total number of vacuum evaporation units reported in the survey forms are no longer in use. On the average, these units were purchased 16 years ago. Of those units still in use, the average age is 6 years. The oldest working unit was 14 years old.
- PS 034 indicated that they have weekly problems with their pump and vacuum system.

They attribute the pump problems to improper design. PS 039 also indicated that they problems with the vacuum pump. On a second unit, PS 039 had problems with the eductors. They have abandoned use of both of these units.

- Although they have installed both ion exchange and electrolytic purification, PS 082 indicated that they have trouble keeping their chromium bath free of contaminants because of the closed-loop recovery process. PS 102 used their unit for 4 years and then abandoned its use because of plating bath (copper, cyanide) contamination. PS 125 is experiencing a build-up of sodium and chloride in their nickel bath.

- PS 088 reported that their cooling water was too warm in the summer to effectively condense the vapors and operate their system.

- PS 088, which employs a four stage counterflow rinse prior to evaporation, indicated that users of this technology should concentrate on reducing rinse water flow and the resultant feed to the evaporator.

- PS 124 indicated that they need to clean the condenser of their unit approximately twice per year.

- PS 124 indicated that the maximum feed concentration to their unit is 1 oz/gal CrO_3 (presumably because higher concentrations will etch the glass of their Corning evaporation unit.) PS 196 reported some etching of their glass unit. PS 280, which operates a Corning unit with a fluoride bath, indicated that this application results in a shorter than average equipment life-span and higher maintenance costs. Their unit is three years old.

- PS 125 indicated that they must operate one of their units at <150°F because of nickel brightener considerations (nickel plating) and the other unit at ≤140°F due to fluoride considerations (decorative chromium plating).

- PS 196 indicated that they must operate their unit at >150°F to boil the chromium solution.

- PS 298 indicated that their distillate is some times contaminated beyond the limit for good rinsing. When this occurs, they recycle the distillate to the first rinse rather than the final rinse of their 3 stage counterflow rinse system. This results in a need to blow down rinse water to the treatment system.

4.3.8 Residuals Generation

Vacuum evaporators used for the recovery of plating chemicals create three streams: (1) the concentrate, which is reused in the plating bath; (2) the distillate, which is reused in the rinse system; and (3) cooling water, which is recycled using a cooling tower, reused as rinse water or discharged to the sewer. Cooling water requirements were previously described in Exhibit 4-14 and 4-17.

Only two respondents indicated that they generate any residual wastes other than the three streams mentioned. PS 143 indicated that they generate approximately 30 gal/mth of sludge, which they mix with their other F006 sludge and sent to off-site recycle. This sludge is generated from the treatment of the distillate (cyanide oxidation, precipitation, filter press and sludge dryer), which they do not send to the rinse tanks. PS 298 reported that carbonates collect at the bottom of the evaporation chamber forcing periodic removal. They remove approximately 100 lbs/mth of this material and treat it on-site.

4.4 ION EXCHANGE

4.4.1 Overview

Ion exchange is a technology that is very familiar to the plating industry. The five most common applications of this technology by the plating industry are: (1) treatment of raw water (e.g., city water) to produce high quality rinse water (includes both softening and deionization); (2) chemical recovery from rinse water; (3) treatment of plating baths to remove contaminants (e.g., tramp metals); (4) as a primary end-of-pipe treatment process; and (5) as a polishing end-of-pipe treatment process to comply with stringent effluent limitations. Chemical recovery applications of ion exchange are covered in this section. Bath maintenance and end-of-pipe treatment applications are covered in Sections 5 and 7, respectively. Raw water treatment is not covered by this project, although some respondents indicated in their survey responses that they use ion exchange for raw water treatment (e.g., PS 127, PS 196, PS 203).

Ion exchange is a chemical reaction wherein an ion from solution is exchanged for a similarly charged ion attached to an immobile solid particle (i.e., ion exchange resin). Ion exchange reactions are stoichiometric (i.e., predictable based on chemical relationships) and reversible. The resins are normally contained in vessels referred to as columns. Solutions are passed through the columns and the exchange occurs. Subsequently, when the capacity of the resins is reached, the ions of interest, which are attached to the resin, are removed during a regeneration step where a strong solution containing the ions originally attached to the resin is passed over the bed.

The strategy employed in using this technology is to exchange somewhat harmless ions (e.g., hydrogen and hydroxyl ions), located on the resin, for ions of interest in the solution (e.g., plating chemicals). In the most basic sense, ion exchange materials are classified as either cationic or anionic. Cation resins exchange hydrogen ions for positively charged ions such as nickel, copper and sodium. Anion resins exchange hydroxyl ions for negatively charged ions such as chromates, sulfates and cyanide.

Ion exchange resins are usually contained in vessels referred to as columns. The basic column consists of a resin bed which is retained in the column with inlet and outlet screens, and service and regeneration flow distributors. Piping and valves are required to direct flow and instrumentation is required to control regeneration timing. The systems are typically operated in cycles consisting of the following steps (ref. 39, 348):

1. **Service (exhaustion)** - Water solution containing ions is passed through the ion exchange column or bed until the exchange sites are exhausted.
2. **Backwash** - The bed is washed (generally with water) in the reverse direction of the service cycle in order to expand and resettle the resin bed.
3. **Regeneration** - The exchanger is regenerated by passing a concentrated solution of the ion originally associated with it through the resin bed; usually a strong mineral acid or base.
4. **Rinse** - Excess regenerant is removed from the exchanger; usually by passing water through it.

4.4.2 Development and Commercialization

Ion exchange is a versatile separation process with potential for broad application in the metal finishing industry. The ion exchange process has been commercially available for many years, but early use was primarily for water deionization or softening. Widespread interest in the process for metal finishing pollution prevention and control is a more recent application that has grown rapidly over the past 10 to 20 years.

Although the ion exchange process was first recognized in 1850, and defined as a reversible process in 1858, the first practical application of the process did not occur until 1905 when the inorganic sodium aluminosilicate cation exchanger was synthesized and used to soften water. In the early 1930's a cation material was developed from sulfonated coals, which extended the operable pH range of the process. In 1935, condensate polymers were first used as structural substrates for various functional groups, thereby creating both anion and cation exchange materials. The early anion product was not the strong base anion exchanger available today. Instead, it acted as an acid adsorbent capable of removing only free acids from the process stream. The cation and anion forms were used in series for the first time and deionization was accomplished; however it was limited in application. Strong base anion exchange materials were developed in the 1940's that extended the use of the process. In 1945 synthetic organic polymers were developed using styrene and divinylbenzene as the monomers. These were later prepared in bead form (earlier, only granular products were available). This gave rise to the modern "gel" type ion exchange materials. Improvements during the 1950's focused on reducing the fouling of resins by organic substances (ref. 383). One of the earliest applications of ion exchange to metal finishing mentioned in the literature was a hard chromium bath maintenance system operated by the Rock Island Arsenal in 1952 (ref. 368). The use of ion exchange as a pollution control and chemical recovery tool progressed steadily as pollution control regulations became more widespread and more stringent. As with many technologies, ion exchange use for pollution control in the metal finishing industry increased very rapidly in the late 1970's and early 1980's with the implementation of the Federal effluent guidelines. The most recent developments have focused on expanding the range of applications (e.g., cyanide plating and electroless plating baths) with new resins, equipment and techniques.

4.4.3 Applications and Restrictions

Ion exchange is used for a variety of purposes in the metal finishing shop, including: treatment of raw water; recovery of plating chemicals from rinse water; purification of plating solutions; wastewater treatment; and wastewater polishing. The following discussion of applications and restrictions focuses on use of this technology for chemical recovery from rinse water. The other applications, with the exception of raw water purification, are discussed in other interim reports.

Ion exchange is a useful technology for recovering plating chemicals from dilute rinse waters. Two common configurations are shown in Exhibit 4-22, application IX-1 is referred to as metal scavenging. It uses only one type of ion exchange resin, either anion or cation, depending on the charge of metal or metal complex being recovered. Because this system does not have both cation and anion resins, the rinse water will not be fully "deionized" and cannot be reused as rinse water for common rinsing purposes. In AE-2, both anion and cation resins are employed and the rinse water can be recirculated in a closed loop. With both of these configurations, rinse water containing a dilute concentration of plating chemicals is passed through an anion and/or cation column (or dual columns of the same type) and the metals are removed from the rinse water and held by the ion exchange resin. When the capacity of the unit is reached, the resin is regenerated and the metals are concentrated into a manageable volume of solution. Depending on the chemical nature of the process, the regenerant (eluate) solution can be returned directly to the plating tank for reuse, further processed and returned, or the metals can be recovered by another technology such as electrowinning (see Section 4.5). The most common applications for these configurations of ion exchange are with the recovery of copper, nickel, and precious metals.

Application IX-3 (Exhibit 4-23) is a bath maintenance configuration that is only applicable to chromic acid solutions. The chromium bearing rinse water is passed through a cation column to remove trivalent chromium, and tramp metals such as iron, nickel and aluminum and it is then used as make-up for evaporation in the bath. This application prevents the build-up of contaminants in the bath, a process that is hastened by recovery rinsing. Since hexavalent chromium is an anion, it is not removed by the cation resin.

Dragout recovery tanks (see Section 3) are used with ion exchange systems whenever feasible. In operation, the dragout tanks return the bulk of the plating chemicals to the plating bath and an ion exchange column captures only the residual quantities of chemicals. The needed size of the ion exchange unit and its regeneration frequency are therefore reduced.

Some of the respondents to the Users Survey misapplied the ion exchange technology for chemical recovery. For example, PS 261, operates a Watts nickel plating line with a 140°F bath and a four stage counter flow rinse. This shop feeds the ion exchange system from the first rinse, which is the most concentrated, and returns the treated water to the last rinse. Instead, this shop should utilize the configuration shown in Exhibit 4-22. Application IX-1. The use of a recovery rinse will greatly reduce the nickel load on the ion exchange columns.

As a recovery technology, ion exchange should be applied to dilute rinse waters. It is not applicable to concentrated dragout solutions or plating baths (although it can be used as a bath maintenance technology to remove tramp metals, see Section 5). A major limitation of this process is that many plating baths are more concentrated than the ion exchange regenerant. Therefore, it should not be used in a "bleed and feed" system, where spent bath is bled to the rinse water. The result in these cases is that the chemicals are diluted in the rinse water, collected by ion exchange, regenerated (using costly chemicals), and recovered in a lower concentration than they started (ref. 40).

Ion exchange is applicable to a wide range of plating processes. Exhibit 4-24 shows the applications identified by the Users and Vendors Surveys.

Ion exchange is used much more frequently for metal recovery from non-cyanide solutions than for cyanide solutions (e.g., cadmium cyanide, copper cyanide, zinc cyanide), with the exception of gold cyanide. When applied to cyanide solutions, the task of recovery is more difficult due to the nature of the cyanide complex. As explained by Reinhard (ref. 342), when a cation followed by an anion column arrangement is used, the cyanide complex in the rinse water is decomposed into free metal cations and cyanide anions and these cations and anions are exchanged for H^+ and OH^- ions in the corresponding resin beds. When the cation resin bed becomes exhausted there is insufficient acidity left to decompose the cyanide complex which is an anion and it will be exchanged for an OH^- ion in the anion resin bed. This presents a significant problem since the cyanide complex will accumulate in the anion resin and because of its high affinity to the resin matrix, regeneration will not remove it. If this occurs, the resin may require replacement. This problem is not a concern with gold cyanide applications because in most cases the resins are incinerated during the recovery process, rather than regenerated. Incineration of spent gold resin is an economically acceptable practice because of the relatively small quantity of resin used and its low cost in comparison to the gold contained on the resin.

Exhibit 4-22. Common Configuration for Application of Ion Exchange for Chemical Recovery

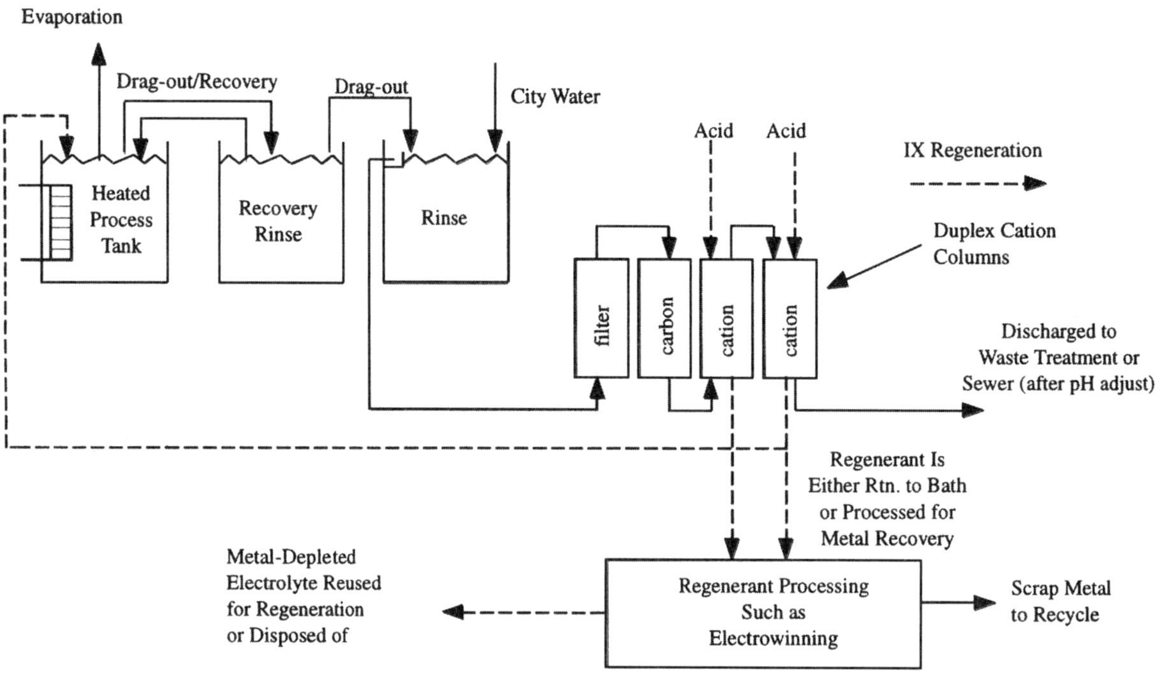

Application IX-1: Ion Exchange Recovery-Metal Scavenging Configuration

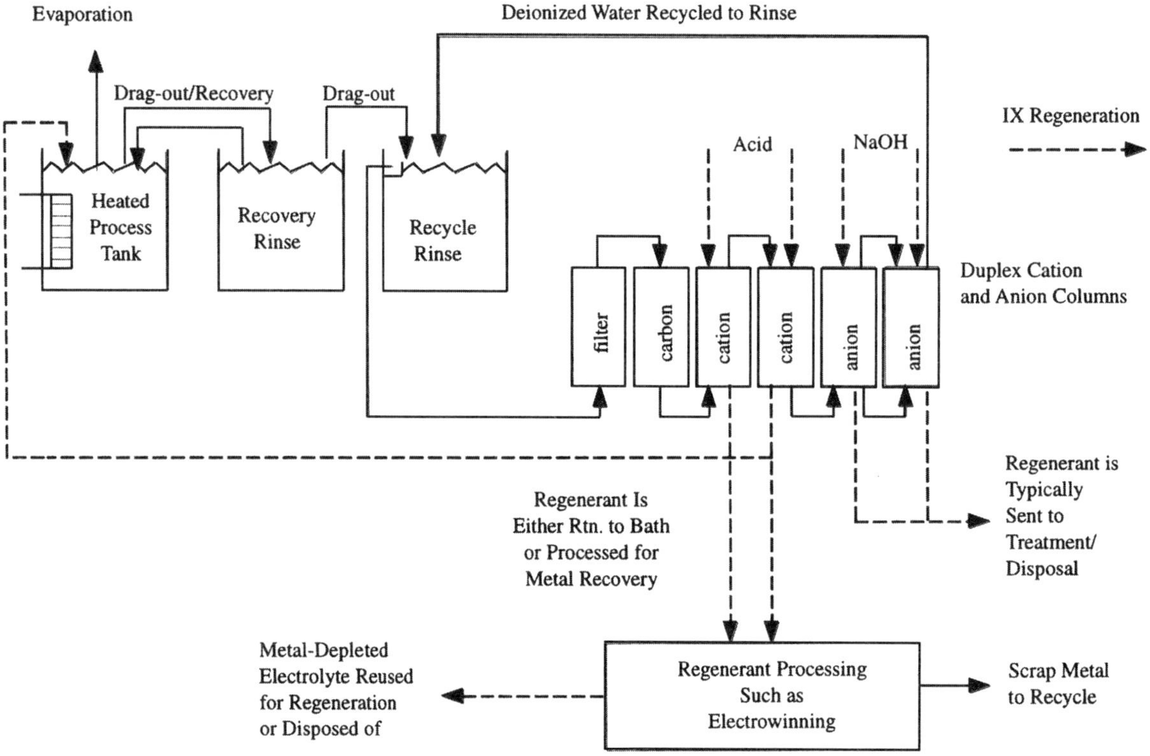

Application IX-2: Ion Exchange Recovery-Metal Recovery/Deionized Water Recycle

Exhibit 4-23. Additional Configuration for the Application of Ion Exchange for Chemical Recovery

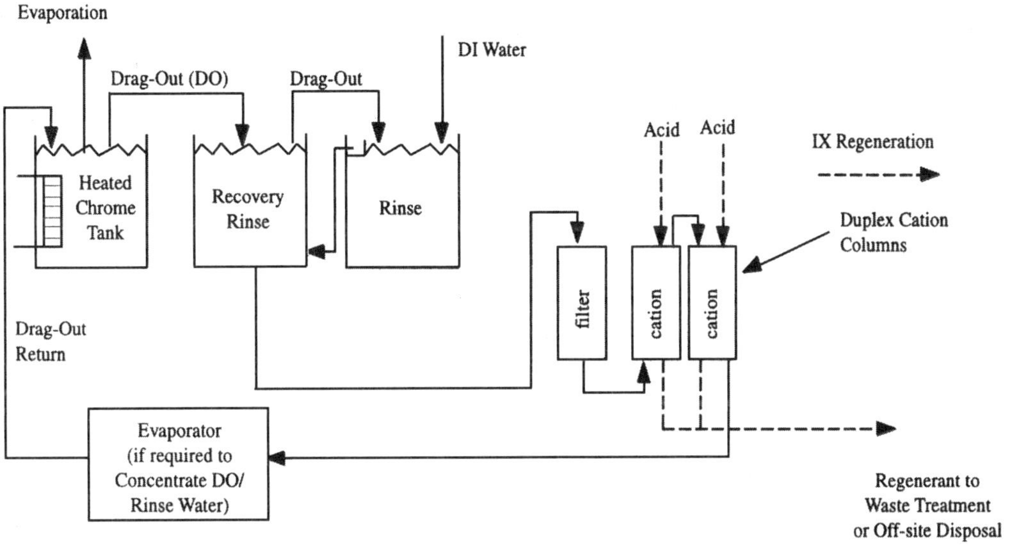

Application IX-3: Ion Exchange Recovery-Chromium Drag-Out Return

Exhibit 4-24. Listing of Ion Exchange Applications Identified from the Users and Vendors Surveys

Application/Process Solution	**Number of Applications Identified During Surveys[1]** Users Survey	**Vendors Survey**
Recovery Applications:		
Brass Plating, cyanide	0	1
Copper Plating, not specified	0	40
CopperPlating, acid	0	6
Copper Plating, cyanide	0	1
Copper Plating, electroless	0	2
Chromium, Cr^{+3}	0	1
Chromium, Cr^{+6}	5	6
Cadmium, cyanide	3	0
Gold, Cyanide	9	0
Gold, non-cyanide	1	0
Lead Plating	0	20
Nickel Plating, not specified	1	1
Nickel Plating, Watts	9	5
Nickel Plating, sulfamate	2	0
Nickel Plating, electroless	1	2
Tin Plating	0	10
Tin-Lead Plating	0	2
ZincPlating, cyanide	0	0
Zinc, non-cyanide	3	1
Metal Scavenging, not specified	0	45
Solution Pruification[2]	9	25
Rinse Recirculation, no recovery[3]	10	75
End-of-PipeTreatment, primary[3]	8	0
End-of-Pipe Treatment, polishing[3]	1	21

Notes:
[1] This exhibit compares the number of ion exchange applications identified during the Users and Vendors Surveys. The Users Survey data includes the responses from 318 plating shops. The Vendors Survey data includes responses from Kinetico Systems Engineering, Kinetic Recovery Corporation and Memtek Corporation.
[2] Details contained in Section 5.
[3] Details contained in Section 7.

Another operational problem with cyanide solutions can be caused by excessive free cyanide in the rinse water. If this condition exists, the free cyanide will remove the heavy metal from the cation resin and form a new metal-cyanide complex.

4.4.4 Technology/Equipment Description

The initial part of this section describes some of the more important design elements of ion exchange systems and the latter part presents a description of commercially available equipment.

4.4.4.1 Ion Exchange Resins and Columns

A wide range of ion exchange resins are manufactured, the choice of which depends mainly on the type of metal being recovered and the chemical composition and characteristics of the solution being treated. Properly matching the ion exchange resin and the process chemistry should result in efficient operation, quality byproducts and lower operating costs. Inappropriate selection of resin can result in total system failure.

Resins can be broadly classified as strong or weak acid cation exchangers or strong or weak base anion exchangers. Strong acid resins are so named because their chemical behavior is similar to that of a strong acid. The resins are highly ionized in both the acid and salt form. In a weak acid resin, the ionizable group is a carboxylic acid (COOH) rather than the sulfonic acid group (SO_3H) used in strong acid resins. These resins behave similarly to weak organic acids that are weakly dissociated. Because weak acid resins have an affinity for hydrogen ions, they have a limited exchange capacity when used to treat solutions with a pH below 6.0 (ref. 39).

Like strong acid resins, strong base resins are highly ionized and can be used over the entire pH range. Weak base resins are like weak acid resins, in that the degree of ionization is strongly influenced by pH. Consequently, weak base resins exhibit minimum exchange capacity above a pH of 7.0 (ref. 39).

A major advantage of both weak acid and weak base resins is that they can be regenerated much more efficiently than the strong acid and strong base resins. The weakly ionized resins can be regenerated using slightly greater than the stoichiometric reagent requirements, whereas the strongly ionized resins require significantly more regenerant (ref. 39).

Many specialty resins, such as chelating resins, are also in commercial use. Chelating resins that exhibit a high selectivity for heavy metal cations over other cations in solution have been commonly used in metal finishing, especially in the past ten years. Because of their selectivity, they are especially useful for end-of-pipe polishing following hydroxide precipitation. Chelating resins are also used in recovery with electroless copper and electroless nickel plating solutions. Generally, chelating resins cannot be used at low pH (<4) and a pH adjustment step is typically needed before the ion exchange process.

Exhibit 4-25 provides some general guidance on the types and capacities of resins used for common metal finishing chemical recovery applications, the chemicals used for regeneration and the method of recovery.

Most industrial applications of ion exchange used fixed-bed column systems, the basic component of which is the ion exchange column. The column must:

- Contain and support the ion exchange resin
- Uniformly distribute the service and regeneration flow through the resin bed
- Provide space to fluidize the resin during backwash
- Include the piping, valves, and instruments needed to regulate flow of feed, regenerant, and backwash solutions

After the feed solution is processed to the extent that the resin becomes exhausted and cannot accomplish any further ion exchange, the resin must be regenerated. Resin capacity is usually expressed in terms of equivalents per liter (eq/l) of resin. An equivalent is the molecular weight in grams of the compound divided by its electrical charge or valence. For example, a resin with an exchange capacity of 1 eq/l could remove 37.5 gram of divalent zinc (Zn^{+2}, molecular weight of 65) from solution (ref. 39).

The hydraulic loading of resins will vary considerably from application to application, depending on: column design; type of resin employed; concentration of metal in

Exhibit 4-25. Information and Data for Ion Exchange Resins Used for Metal Finishing Applications

Plating Solution	Resin Type	Capacity lbs. Metal/ft^3	Regenerant	Typical Recovery Method(s)
Watts Nickel	strong acid cat.ion	1.8	H_2SO_4 or HCl	electrowin Ni or use nickel sulfamate in bath
Sulfamate Nickel	strong acid cation	1.8	sulfamic acid	nickel suffamate used in bath
Acid Copper	strong acid cation	2.0	H_2SO_4 or HCl	electrowinning
Acid Zinc	strong acid cation	1.8	HCI	zinc chloride used in bath
Cyanide Copper[1]	strong acidcation	1.5	H_2SO_4	electrowinning
Cyanide Zinc[1]	strong acid cation	1.2	H_2SO_4 or HCl	electrowinning
Cyanide Cadmium[1]	strong acid cation	1.3	H_2SO_4	electrowinning
Cyanide[1]	medium base anion	1.6	NaOH	treatment
Tin/Lead Fluoroborate	strong acid cation	6.5	methyl-sulfamic acid	electrowinning
Chrome Plating Cation	strong acid cation	1.7	H_2SO_4 or HCl	treatment
Chrome Plating Anion	medium base anion	3.1	NaOH	membrane electrolysis, or pass through cation resin, then use chromic acid in bath
Final Polishing	selective cation (chelating)	2.7	H_2SO_4 or HCI/NaOH	treatment

[1] Cyanide metal solutions are treated first by a strong acid cation resin (cyanide complex destruction and metal recovery) followed by a medium base anion resin (cyanide recovery). This process is discussed in Section 4.4.3.

Source: Based on data provided by Dr Fred Reinhard, Kinetic Kecovery Corporation.

solution; other chemical characteristics of the feed solution (e.g., pH); and the allowable concentration of metal in the column effluent. Typical hydraulic loadings range from 2 to 3 gpm of rinse water per cubic foot of resin.

4.4.4.2 Integrated vs Modular Designs

An integrated ion exchange system design is one in which the various components needed to perform the ion exchange recovery and regeneration functions are connected within the one unit. Such systems may also have attached electrowinning units and/or chemical treatment systems for processing the regenerant. The modular or point source design separates the ion exchange column from the regeneration and regenerant processing equipment. With the modular design, the columns are transported to a central station for regeneration (in some cases the modules are hard piped). The regeneration station can be either in the plating shop or at an off-site location (i.e., centralized waste treatment facility). The modular ion exchange strategy can reduce capital costs for small to medium-sized applications where low to moderate regeneration frequency is required. Also, the modular units are considerably smaller and therefore do not occupy as much production area floor space as integrated units (i.e., if the regeneration station is remotely located to a non-production area). However, operating costs are usually higher for modular systems due to the labor needed for transporting the modules and connecting them to the regeneration station (or for changing operating modes and valve positions for hard piped modular systems) and initiating regeneration. Some commercial ion exchange modules have the appearance of large cans and are referred to as ion exchange canisters. With this type of unit, the canisters can be stacked upon one another to combine anion and cation types or to increase the resin bed volume. Standard column designs are also available.

4.4.4.3 Single vs Duplex Column Operation

Duplex column ion exchange systems are used in many chemical recovery operations, especially where a continuous feed flow is expected. Dual column configurations avoid downtime during regeneration. Two different duplex column arrangements are commonly used. In one arrangement, which is referred to as parallel/standby, the feed stream flows through either one column or the other, but never both. The off-line column is regenerated and then is held in reserve until the other column is ready for regeneration. This is a somewhat inefficient use of the two columns since column switching must take place before breakthrough occurs, which happens before the resin is completely loaded with ions of interest. In the second case, which is referred to as lead/lag, the two columns are placed in series flow. During operation, the majority of metal removal is accomplished in the first column (lead column) until it approaches capacity. The process can continue until the first column is essentially loaded to full capacity with ions of interest, since the second column (lag column) will remove the breakthrough of the first column. After breakthrough is reached, the first column is taken off-line for regeneration and then put back into service as the lag column. The roles of the two columns continue to be reversed following each regeneration. The switching of the columns, initiating of regeneration and other functions of modern ion exchange equipment is usually controlled by a microprocessor.

A special single column design that is widely used is referred to as reciprocating flow ion exchange (RFIE). A commercial application of this design is the Recoflo® (ref. 39, 349, 364, 365). The Recoflo® system is characterized by its short resin beds (6 to 24 in. as compared to 40 to 60 in. minimum for standard counterflow design) containing fine particle size resin beads. The Recoflo® process operates in a counterflow mode (see Section 4.4.1.4) with approximately a five minute cycle. During the cycle, the feed stream is fed for approximately 2.5 minutes, then regeneration occurs for approximately 2.5 minutes. The cycles are controlled by a microprocessor and are continuously repeated. The short cycle time, which is made possible by the short resin bed, emulates a continuous flow process.

4.4.4.4 Counterflow vs Cocurrent Flow/ Regeneration

One method of categorizing the operation of different ion exchange systems is by the direction of the service flow (e.g., rinse water) vs the direction of the regeneration flow. The two configurations that can be used are shown schematically in Exhibit 4-26. With cocurrent operation, the service flow and the regeneration cycle flow in the same direction and with countercurrent flow, they flow in opposite directions (as shown, service flow can be either downward or upward). Countercurrent flow is considered by most sources to be the more efficient method (e.g., ref. 42).

With cocurrent flow the hydrogen ions displace metal ions from the top to the lower portion of the bed. Complete removal of these ions can only be accomplished by the use of excessive levels of acid regenerant. With normal regenerant usage, there is a "heel" left at the exit end of the column (i.e., undisplaced metal ions). On the following service cycle, the desired exchange reaction occurs in the upper portion of the bed. However, as the hydrogen ion concentration increases toward the lower section of the bed, some reexchange with previously undisplaced metal ions leads to metal ion "leakage."

Exhibit 4-26. Comparison of Cocurrent and Counterflow Regeneration

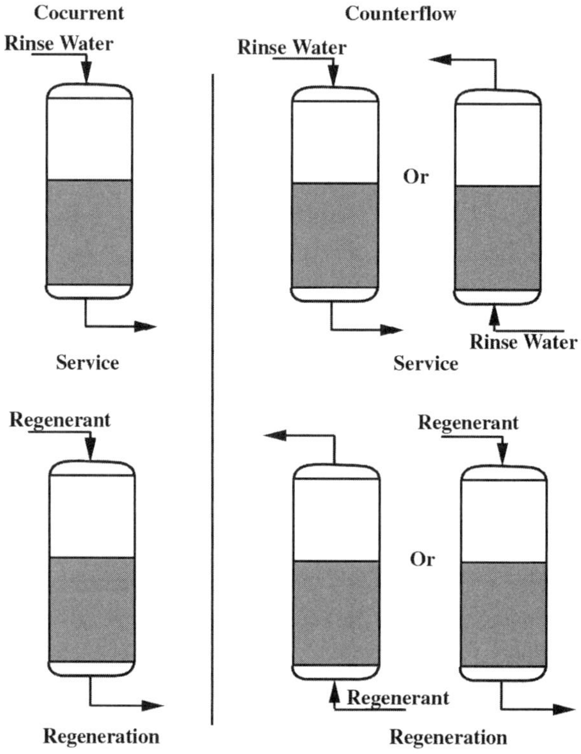

Source: ref. 43

After regeneration of the counterflow system, the residual ions are in the top of the bed, with the bottom being fully converted to hydrogen. Thus, there are no residual metal ions present at the bottom of the bed to permit the leakage reaction to occur on the subsequent service cycle (ref. 43).

In addition to reduced ion leakage, counterflow regeneration can increase operating capacities, decrease the need for waste stream pH adjustment, and reduce water rinsing requirements (ref. 43).

4.4.4.5 Other Equipment/Design Considerations

In addition to the basic ion exchange column, auxiliary equipment is employed for various purposes, among which include: resin bed channeling and fouling prevention; pH adjustment of the feed stream; solution pump and flow control; need for regeneration identification; and regeneration cycle control.

Pretreatment of the feed stream is usually performed. Filtration is a basic requirement for nearly all ion exchange applications. If solids are permitted to enter the ion exchange bed, they will often create an uneven film on the top of the bed that acts as a plug. The solids will impede flow and cause channeling through the bed. Channeling of the feed solution will result in incomplete usage of the bed and inefficient processing (ref. 349). Most commonly, cartridge filtration is used for this purpose. Multimedia filters are sometimes used in high flow applications, where changing of the cartridge filters would be too time consuming (ref. 348). Other types of pretreatment include pH adjustment and carbon filtration. The adjustment of pH is used for certain applications where resin capacity can be enhanced by increasing or lowering the pH. Carbon filtration is used to remove certain organics such as oils that can become irreversibly sorbed by ion exchange resins and oxidants such as peroxide that can oxidize and ruin the resins (ref. 348).

The means for identifying the point at which regeneration should be initiated varies among commercially available equipment. The methods employed depend on the overall design of the system (e.g., a lead/lag unit may be able to tolerate some ion leakage from the first column whereas a single column may not), the tolerable leakage concentration, the variability of the feedstream, the ion(s) of concern, and the solution chemistry. Some equipment uses direct measurement methods to identify breakthrough. Examples of applicable methods and instruments include: conductivity meters, sometimes used in conjunction with pH meters; specific ion probes; and colorimetric analytical methods. A different strategy is to regenerate a column based on elapsed time or flow. These latter methods are applicable to feed streams with relatively constant parameters.

Other design features of ion exchange systems, such as controls, are discussed in Section 4.4.1.6.

4.4.4.6 Commercial Equipment

This subsection contains a description of commercially available ion exchange equipment that is manufactured and/or sold by vendor survey respondents. This is intended to provide the reader with information and data on a cross section of available equipment. Mention of trade names or commercial products is not intended to constitute endorsement for use.

Kinetico Engineered Systems, Inc. (Kinetico) manufactures various ion exchange equipment including that applicable to chemical recovery, bath maintenance, raw water treatment, and end-of-pipe polishing. Since 1980, they have sold more than 300 units. The Kinetico Complete Ion Exchange (CIX) series is applicable to metals recovery and water recycling with configurations similar to that shown in Exhibit 4-22 (applications IX-1 and IX-2). The CIX series are integrated packaged units with standard sizes ranging from 30 to 500 gpm. All units are dual column design for continuous operation (lead/lag configuration) with counterflow regeneration. Regeneration is initiated automatically by a patented control valve that monitors flow or by a manual override (push button). The control valve, which is non-electric, also directs the flow of solution for the regeneration cycle. Power for the valve comes from the flow and pressure of the water.

Kinetic Recovery Corporation offers modular ion exchange and regeneration systems for metals recovery and water recycling (see Exhibit 4-22, applications IX-1 and IX-2) as well as bath maintenance and end-of-pipe polishing. The modular ion exchange systems, which are dual column, are packaged with a cartridge filter system, flow meter, level sensor, feed pump, valve assembly, conductivity monitoring system and pump controls. The separate regeneration system (Exhibit 4-27) is capable of

Exhibit 4-27. Modular Ion Exchange Equipment
(Courtesy of Kinetic Recovery Corporation)

Cation/Anion Systems, including conductivity controls

Model	D200C	D300C
Resin Column	2 ft^3	3 ft^3
Column Dimensions:		
Height	50"	73"
Flange Diameter	14"	14"
Base Tank Dimensions:		
Length	40"	40"
Width	28"	28"
Height	23"	23"
Capacity	34"	34"

Total Height of System	73"	96"
Operating Flow Rate	6-8 gpm	8-10 gpm
Pump Motor	0.75 hp	2.0 hp
Maximum Operating Pressure	25 psi	25 psi
Maximum Operating Temperature	110°F	110°F

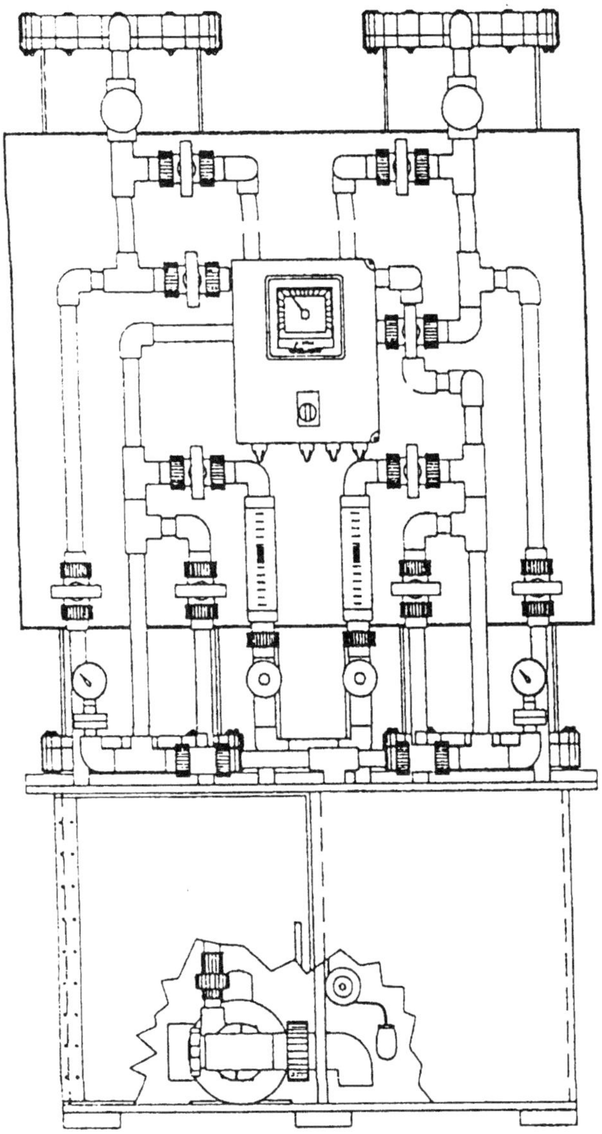

Modular Ion Exchange Unit

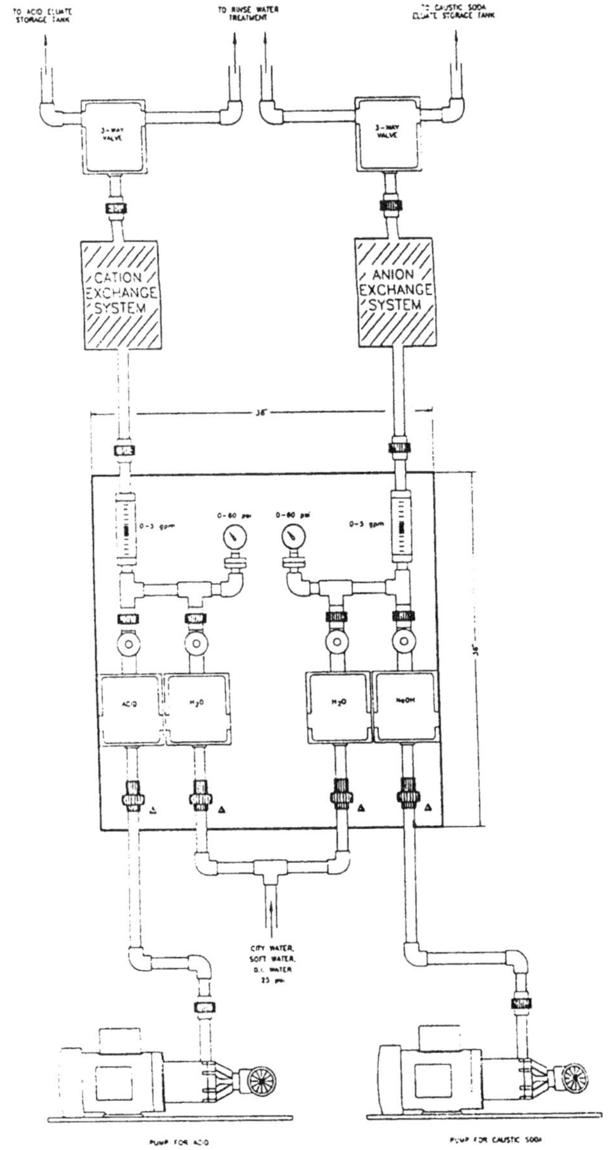

Dual Regeneration System

Exhibit 4-28. Ion Exchange Equipment for Lead-Bearing Rinse Waters (Courtesy of MemtekCorporation)

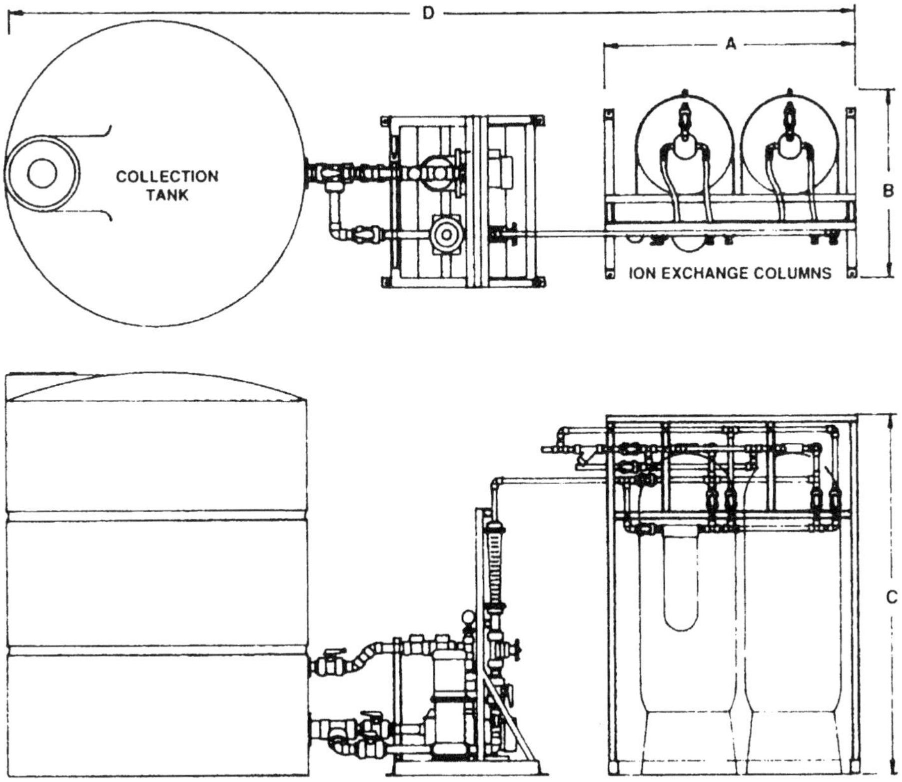

Pre-Conditioning Section   Ion Exchange Section

Specifications (Typical)*

System Designation		**PBX-10**	**PBX-12**	**PBX-16**	**PBX-24**
Flow (maximum)	gpm	2	4	7-10	15-20
Lead Capacity	lbs./column	3	6	10-16	22-30
Metal Removed (design)	lbs./week	0.75	1.5	2.5-4.0	5.5-7.5
Dimensions					
A. Length		55"	55"	61"	61"
B. Width		34"	34"	36"	44"
C. Height		65"	70"	83"	83"
D. Overall Length		12'	13'	13'	15'
Collection Tank Volume	gallons	220	500	750	1,000
Power Rating	volts/amps	460/5	460/5	460/7.5	460/10

Standard system components are pre-plumbed and pre-wired in the factory for easy on-site installation and include the following items:

- One (1) pre-conditioning HDPE tank
- One (1) pH monitor/controller
- One (1) level control probe with high level alarm
- Two (2) chemical feed pumps
- One (1) mixing pump
- One (1) lot Memtek proprietary ion exchange resin (PBX-R7)
- One (1) process pump
- One (1) cartridge filter
- Two (2) fiberglass ion exchange columns
- One (1) console for automatic pumps & pH control

*The specifications shown illustrate typical sizes, dimensions, and power requirements.

regenerating cation and anion exchange resins simultaneously. It consists of two chemical feed pumps, a valve assembly/control board with electrically actuated ball valves, flow meter and globe valve and a regeneration controller with a PLC and an operator interface. The valve assembly board allows the adjustment of the ion exchange into several modes of operation (parallel, series) and regeneration. With the operator interface unit, the regeneration mode can be selected and all operating conditions (mode, time elapsed, alarm condition) are displayed.

Memtek Corporation manufactures integrated ion exchange equipment for metal recovery and water recycling and end-of-pipe treatment. Their metals recovery/ water recycle series (Rinse water Maintenance System or RMS™) is a dual column (lead/lag) packaged system with prefiltration, conductivity monitors and alarms, regenerant make-up tanks with level controls, and microprocessor controller. Memtek markets specific ion exchange systems for electroless copper and lead applications (see Exhibit 4-28). These two system include a preconditioning pH adjustment step.

Ionics International Ltd. (i^3) offers both a semi-automatic integrated ion exchange system and a modular system with separate ion exchange and regeneration stations. Both the integrated and modular ion exchange systems are dual column (lead/lag) systems used for metals recovery and water recycle. The integrated systems typically include two anion and two cation columns containing approximately 6 ft^3 of resin per column. Regeneration is manually initiated and automatically controlled. The modular systems are hard piped to the regeneration station rather than transported. The modular units typically permit a service flow rate of 10 gpm.

4.4.5 Costs

4.4.5.1 Capital Costs

Capital costs for modular and integrated ion exchange systems for metal recovery and water recycle (anion and cation columns) are presented in Exhibit 4-29. The modular ion exchange system costs are based on two responses to the Vendors Survey . The graphs show the costs for one and two modules of 2 ft^3 and 3 ft^3 capacities each. A single regeneration station is included in the cost for both the one and two module cases. Typically, one regeneration station can serve up to ten modules. Therefore, capital costs would be expected to increase in the step-wise manner shown in Exhibit 4-29 up to 20 ft^3/30 ft^3 resin capacities. The installed costs were estimated based on the projected costs for electrical and piping, including two components of electrical costs (service module and regeneration station) and three components of the piping costs (service module, regeneration module and service/regeneration module interface). The Vendor 1 system is hard piped to the regeneration station and the Vendor 2 system is transported (e.g., hand truck) to the regeneration station. Therefore, the vendor 2 installed cost includes only the first two piping components.

The integrated system capital costs include automated and semi-automated systems. The fundamental difference between the two types is that the regeneration cycle of the automated system is automatically initiated and the unit goes back into service automatically, whereas the semi-automatic system requires an operator to initiate regeneration and to put it back into service. Both systems operate automatically after regeneration or service is initiated. The installed cost for each system is the same, based on estimates for electrical and piping costs.

For small, manual applications, capacity is usually expressed in terms of resin volume, to which capital costs are directly related. Larger system capacities are more often expressed in terms of flow rate, but direct vendor-to-vendor pricing comparisons based only on capacity units such as gallons-per-minute can be quite misleading. Based only on flow rate specifications of four vendors, Exhibit 4-29 does not account for several cost factors including the amount of customization, the precise level of automation (which can vary considerably within the categories of "automatic" and "semi-automatic"), the type and quality of metering and monitoring instrumentation, and the general design strategies and criteria pursued by the manufacturer. These factors, in part, explain the wide range of prices quoted for similar flow rates.

Manual systems are often sized to provide an acceptable service period. Larger columns offer the benefit of fewer regenerations or replacements, less downtime and less labor expense. Automatic systems, on the other hand, are sized to handle the expected flow rate. While still a major design consideration, the service period has less impact on the user in terms of planning and labor. Thus, for highly automated systems, more frequent regenerations of smaller columns is a viable design strategy. An extreme example of this strategy is the reciprocation flow ion exchange, described in Section 4.4.4.3.

Exhibit 4-29(a). Capital Costs for Fully Automatic, Water Recycle Ion Exchange Equipment

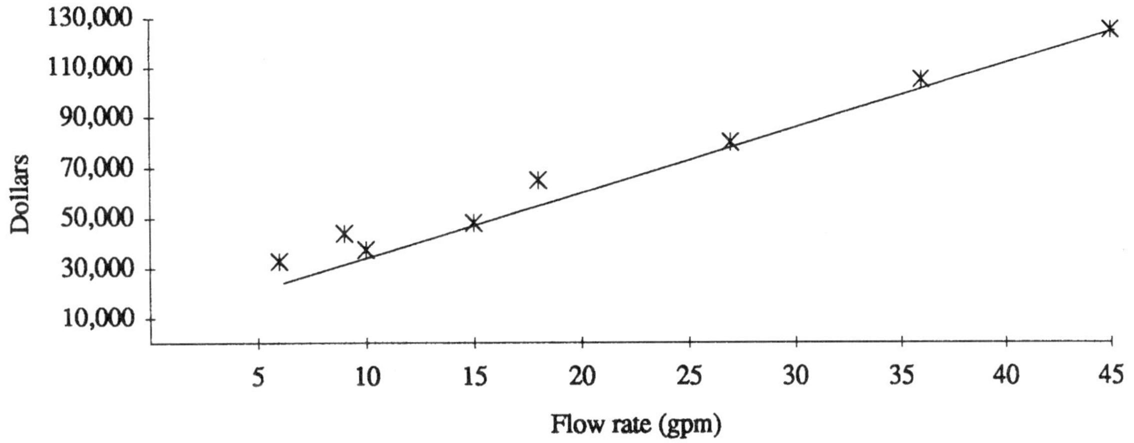

Exhibit 4-29(b). Capital Costs for Fully Automatic, Metal Scavenging Ion Exchange Equipment

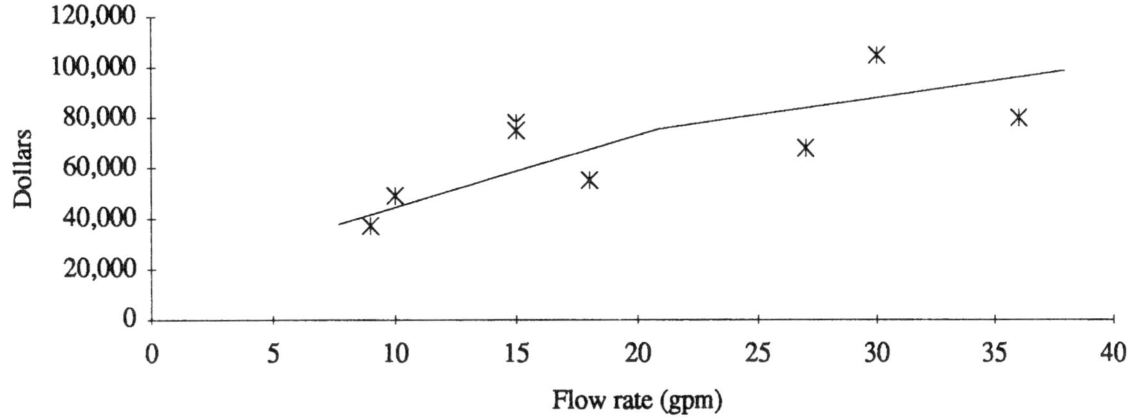

Water recycling systems are, in general, more expensive than metal-scavenging units. These systems require both anion and cation columns, which alone roughly doubles the regenerations required, and the cation resin must remove all cations, including non-regulated common cations such as calcium, sodium and potassium. In most cases, these factors make automation highly desirable or an outright requirement.

Installation expenses are site-specific but can be significant typically 5 to 40 percent or more of basic equipment costs). Shops currently employing trenches may require extensive plumbing to segregate the ion exchange stream from other wastewater. On the other hand, modular systems with off-site regeneration may incur no significant installation costs at all. In general, water-recycle systems that service several sources will require the largest installation outlay.

4.4.5.2 Operating Costs

Labor, regeneration chemistry, resin replacement, and energy are the major operating cost categories. Exhibit 4-30 presents operating costs for various operating modes of the ion exchange technology. This operating cost graph is based on a water-recycle application handling copper sulfate rinse water and is not necessarily representative of a wide range of applications.

Labor costs are significantly affected by the automation level of the system and automation capital costs are often quickly returned. Undersized or mis-applied equipment can greatly impact labor and other costs (see Section 4.4.7).

Resin life is usually measured in years, but can be shortened by misuse and improper application. Resin fouling,

Exhibit 4-30. Operating and Maintenance Costs for Ion Exchange Systems

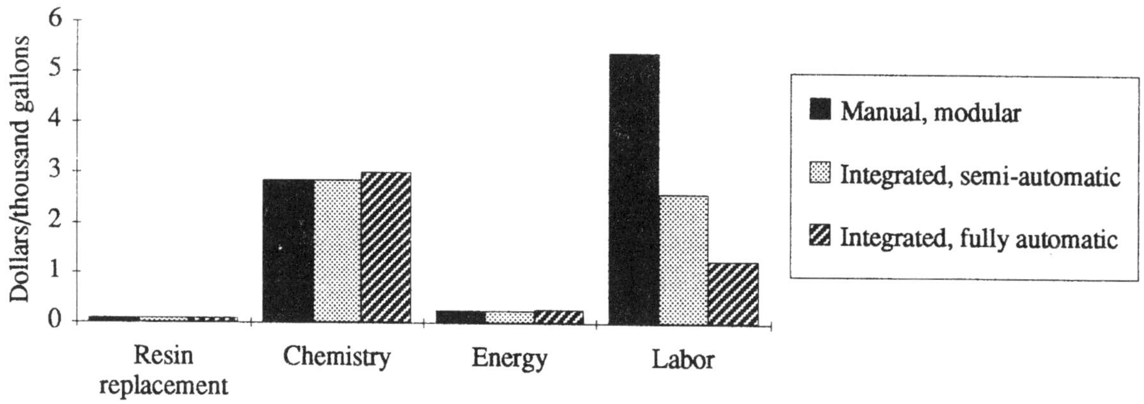

mentioned by several respondents (see Section 4.4.7), is usually a result of a marginal application, misuse, or insufficient upstream filtration or pre-treatment. Instrumentation designed to halt ion exchange system operation when harmful levels of chemistries enter the feed stream can be cost-effective where spills and accidental dumps are possible.

The operating costs estimates are based on the following assumptions:

Feed Characteristics (rinse water)/Resin Capacity

- Copper sulfate plating process generating rinse water containing 50 mg/l Cu^{++}.
- Resin capacity of 38 eq. Cu^{++}/ft^3 (=12,900 gal between regenerations for 2 ft^3 column).
- Assume two anion regenerations for each cation regeneration (=1.79 days between regenerations for 2 ft^3 column).

Energy

- 1 hp-hr/300 gal
- $.10/kWh (= $0.25/1,000 gal)

Regeneration Chemicals

- Assume 4 bed volumes for cocurrent (Vendor 1-Modular) and 2 bed volumes for counterflow (all others).
- H_2SO_4 (Modular 1): 12 gal (conc.) @ $2/gal per 12,900 gal flow (= $1.9/1,000 gal)
- NaOH (Modular 1): 24 gal (conc.) @ $2/gal per 12,900 gal flow (= $3.8/1,000 gal)

Resin Replacement

- Assume 5 year life with 3% mechanical loss per year.
- Cation: $200/$ft^3$ (= $0.03/1,000 gal)
- Anion: $400/$ft^3$ (= $0.06/1,000 gal)

Labor

- $25.00/hr
- Modular 1: 1.0 hr/day (= $2.71/1,000 gal)
- Modular 2: 2.0 hr/day (requires transport) (= $5.43/1,000 gal)
- Semi-automatic: 1.0 hr/day (= $2.71/1,000 gal)
- Automatic: 0.5 hr/day (= $1.36/1,000 gal)

4.4.6 Performance Experience

A summary of the Users Survey data for ion exchange recovery applications is presented in Exhibit 4-31. Within this exhibit, the response data has been grouped by types of plating solution. These include: nickel, chromium, non-cyanide zinc, cadmium cyanide, and gold cyanide. A general discussion of the ion exchange data is presented) followed by specific information regarding each of the different types of applications.

The following information and data summarizes the performance experience of the survey respondents.

- The average satisfaction level for ion exchange recovery is 3.2 (on a scale of 1 to 5, with 5 being most satisfactory), which is lower than the average level rating for chemical recovery in general (weighted average for all chemical recovery applications is 3.4). Fifty-four percent of the shops indicated that this technology satisfied the need for which it was purchased and another 11% indicated that it partially satisfied their need. Thirty-five

Exhibit 4-31. Partial Summary of Users Survey Data for Ion Exchange Metals Recovery Applications

				Capital Costs			Operating Costs (Yr.)		Annual Savings								Satisfaction Levels (4)		
Shop Code	Application	Manf.	Year	Equipment Cost	Other Capital	Total	Non-labor	Labor Hrs.	Water	Process Chemicals	Treatment Chemicals	Sludge Disposal	Other	Total	Use Code (3)	Down Time %	Manf. Support	Tech-nology	(5) Future Decision
015	Nickel, Watts	NAPCO	1990	$50,000	$0	$50,000								ND	1			3	1
063	Nickel, Watts	Inviroserve Co.	1989	$150,000	$50,000	$200,000		1,200	√		√	√		√	1	5	4	4	1
105	Nickel, Sulfamate	Kinetico	1993	$12,000	$7,000	$19,000								ND	3			4	1
124	Nickel, Watts	Eco-Tec Inc.	1977	$60,849	$0	$60,849								ND	2	100		3	3
139	Nickel, Watts	Engineer/Consultant	1993	$60,000	$30,000	$90,000								ND	3				1
161	Nickel, Sulfamate	Dayton Water Systems	1991	$5,200	$5,000	$10,200	$7,500		$2,000	$1,000		$200		$3,200	1	3		5	1
163	Nickel	Bartix Eng. Systems	1987	$15,000	$0	$15,000	$2,477	1,000				$10,000		$10,000	1	1	5	5	1
196	Nickel, Electroless	Dayton Water Systems	ND	$100,000	$3,000	$103,000	$3,000	150			$1,000	$1,000		$2,000	1	2		3	3
212	Nickel, Watts	Technology, Inc.	1989	$12,000	$3,000	$15,000								ND	2				3
261	Nickel, Watts	Eco-Tec Inc.	1986	$60,974	$0	$60,974	$2,000	70						ND	1	1		4	1
263	Nickel	Kinetico	1988	ND	$500	$500	$500	250	$5,000					$5,000	2	10		2	4
317	Nickel, Electroless	Millipore	1989	$10,000	$2,500	$12,500	ND	ND	$0	$0	$0	$0	$0	ND	2	ND	1	1	ND
318	Nickel, Electroless	Tiree Ind. (DMR 20)	1985	$2,000	$700	$2,700	ND	100						ND	2	20	3	1	3
	Average-Nickel		*1988*	*$52,602*	*$8,955*	*$61,557*	*$3,095*	*534*								*17*		*3.7*	
														ND					
052	Chromium, Hard	Ionsep	1988	$56,000	$0	$56,000	$500	200						ND	3	2		3	1
080	Chromium, Hard	Burt Process Tech.	1984	$3,025	$500	$3,525	$1,000	250						ND	2	99		1	3
161	Cr, Decorative (+6)	Dayton Water Systems	1992	$500	$2,000	$2,500	$5,000	100	$1,000	$1,300		$200		$2,500	1	5	4	5	1
163	Chromate, Aluminum	Bartix Eng. Systems	1987	$15,000	$0	$15,000	$3,064	1,000				$10,000		$10,000	1	1	5	5	1
305	Cr, Decorative (+6)	Haz. Matls. Mangt. Team	1992	$22,000	$0	$22,000	$6,000	50	√	√			√	√	1	5	5	5	1
308	Cr, Decorative (+3)	Haz. Matls. Mangt. Team	1990	$17,500	$7,000	$24,500	$4,000	50	$10,000	$20,000				$30,000	1	ND	4	4	1
	Average-Chromium		*1988*	*$18,631*	*$625*	*$19,256*	*$2,391*	*388*								*27*		*3.5*	
														ND					
061	Zinc, Acid	Bio Recovery	1989	$123,000	$0	$123,000	$20,000	1,040						ND	2	75		1	3
126	Zinc, Acid		1992	$40,000	$0	$40,000								ND	1		3	4	1
130	Zinc, Non-Cyanide	Bio Recovery	1990	$100,000	$0	$100,000								ND	3	50		3	3
245	Zinc, Non-Cyanide	Memphis Mobile	1984	ND	ND	ND								ND				1	3
289	Zinc, Non-Cyanide	Quin-Tec	1992	ND	ND	ND								ND	1			4	1
	Average-Zinc		*1989*	*$87,667*	*$0*	*$87,667*	*$20,000*	*1,040*								*63*		*2.6*	
														ND					
025	Cadmium, Cyanide	Ebbco	1991	$4,000	$1,200	$5,200	$3,400	300						ND	1	1		3	1
229	Cadmium, Cyanide	Unk	1984	$400,000	$350,000	$750,000	$100,000	2,080	$10,000	$2,000	$10,000	$30,000		$52,000	1	2		3	3
245	Cadmium, Cyanide	Memphis Mobile	1984	ND	ND	ND								ND				1	
254	Cadmium, Cyanide	Illinois Water Treatment	1985	$54,000	$2,000	$56,000	$8,000							ND	2	20		1	4
	Average-Cd-CN		*1986*	*$152,667*	*$117,733*	*$270,400*	*$37,133*	*1,190*								*8*		*2.0*	
														ND					
064	Gold, Cyanide	In-House				$0	$1,300							ND	1	5		4	1
123	Gold, Cyanide		1975	$2,000	$1,500	$3,500	$20,000			$1,344,000				$1,344,000	1	10		5	1
179	Gold, Cyanide	Technic	1983	$2,000	ND	$2,000	$200			$2,000				$2,000	1	1		5	1
283	Gold, Cyanide	RFE Industries	1979	ND	ND	ND				$1,320				$1,320	1	10		3	1
287	Gold, Cyanide	Pennfield	1979	ND	ND	ND								ND	1	1		3	1
	Average-Gold		*1979*	*$2,000*	*$1,500*	*$3,500*	*$7,167*							ND		*5*		*4.0*	
														ND					
	Average-All		*1979*	*$44,050*	*$4,714*	*$48,764*	*$11,529*	*553*						$11,802		*20*	*4.3*	*3.2*	

(1) MRS is the Metropolitian Recovery Systems, a CWT facility in Minnesota.

(2) Other capital costs include installation and ancillary equipment.

(3) Use Codes: 1-currently operating; 2-not currently operating and no intention for future use; 3-not currently in use, but intend to use in future.

(4) Satisfaction levels (manufacturer and technology) 1 to 5, with 1=lowest and 5=highest.

(5) Future decision codes: 1-purchase the same technology from the same vendor; 2-purchase the same technology from a different vendor; 3-purchase a different technology; 4-do nothing.

(6) "√" indicates that a savings was realized, but not quantified.

(7) Average total equipment cost is the sum of average equipment and other capital costs, excluding data from PS 229. Average annual operating cost is the sum of average non-labor and labor costs. Average savings is simply the average of reported savings, excluding PS 123. This method of calculation probable understates the average annual savings of the users.

(8) ND = no data

percent indicated that it did not satisfy the need for which it was purchased. The following is a breakdown of the reasons why shops purchased this technology (not all shops responded and some shops gave multiple reasons):

To meet or help meet effluent regulations: 21
To reduce plating chemical purchases: 8
(includes all four gold applications)
To reduce the quantity of waste shipped off-site: 14
To reduce wastewater treatment costs: 13
To improve product quantity: 6
Other: 0

- The use of ion exchange generally did not impact production quality or the rate of production. The following responses were provided:

	Product Quality	Production Rate
Improved	5	1
No Change	16	18
Decreased	3	4

- The majority of plating shops indicated, that based on their experience with this technology, if given the chance they would purchase the same type of equipment from the same vendor. The following is a breakdown of their responses:

Purchase the same technology from the same vendor: .18
Purchase the same technology from a different vendor: 0
Purchase a different technology: 8
Do nothing: 2

- The major savings from the operation of ion exchange for non-gold applications was due to reduced water use and reduced sludge generation. For gold applications, the major savings was due to the recovered gold.

4.4.6.1 Nickel Plating Performance Experience

Thirteen respondents to the User Survey provided detailed data regarding their experience applying ion exchange to nickel rinse waters. Two of these responses cover systems that were being installed at the time of the survey and therefore, no performance data are available from these shops (PS 105, PS 139). Three shops used the process for a short time period, but abandoned their efforts due to disappointing results (PS 212, PS 317, PS 318). None of the remaining eight shops currently operate the traditional recovery configurations (Exhibit 4-22), where the cation regenerant is subsequently processed by electrowinning to recover the nickel or nickel sulfate is returned to the bath. At one time, PS 124 recycled nickel sulfate regenerant directly to the bath. However, this shop discontinued use of the process due to "continuous and repeated" equipment failure. PS 015 purchased an electrowinning unit for nickel recovery, but the unit did not work and therefore, they selected to treat the regenerant instead. Also, one of the two new installations plans to operate with metal recovery (PS 139). Three shops use off-site recycling for regeneration of their columns (PS 161, PS 196 and PS 063) and another shop regenerates on-site and sends the regenerant to an offsite metals recycling firm (Inmetco) (PS 261).

Although no on-site recovery is presently practiced, the configurations used by the respondents involve the recycle of water. These applications have been included in this recovery section because they process a segregated nickel rinse water with ion exchange. As such, that they most likely experienced many of the same performance and O&M problems as recovery operations.

The following performance details were provided by survey respondents:

- The two most successful installations among survey respondents are PS 161 and PS 261. PS 161 sends their used ion exchange columns to off-site services for regeneration, most recently using Dayton Water Systems for regeneration of both nickel and chromium columns. This shop has used this form of off-site recycling since 1972. Originally, their equipment was purchased from Pollutronics and the columns were regenerated in Cleveland (company name not specified). They used Culligan's service in the late 1970's until 1991, when they switched to Dayton Water Systems "because of their recycling ability." After contacting Dayton Water Systems, it was learned that they do not currently provide a regeneration service for heavy metals, but may resume this service in the near future. The ion exchange columns used by PS 161 contain 3.2 ft^3 of mixed (anion and cation) resin. During the past year, PS 161 sent 18 nickel columns and 8 chromium columns to Dayton Water Systems. The cost of the service was $336 per column for regeneration plus a $70 per month rental fee for the equipment (therefore, the total annual cost for most recent year is $6,888 for nickel and $3,528 for chromium). According to

PS 161, Dayton Water Systems sent the regenerant to Inmetco for recovery of the metal (see Section 8 for information on off-site recycling).

- Only one of the shops indicated that an ion exchange unit applied to nickel plating was the cause of an effluent compliance excursion (PS 196). This shop stated: "Excursions happen when the mixed bed of resin is saturated with copper and nickel and will no longer polish the water." This shop uses an off-site service for regeneration and is charged by the number of columns used. Therefore, the shop may tend to load the resin beds to their maximum or beyond their breakthrough point.

- PS 261 reported a satisfaction level of 4 for ion exchange applied to nickel rinse waters. Their configuration is unusual in that the ion exchange unit is connected to the first rinse tank (i.e., most concentrated). Rinse water from the first rinse is pumped through the ion exchange column and is recycled to the last rinse (four-stage counterflow system). With this configuration, the ion exchange column receives the most concentrated rinse water. The normal configuration (see Exhibit 4-22) makes use of dragout recovery to reduce the chemical load on the ion exchange system and recycles the final rinse. Using their configuration, PS 261 probably experiences the need for frequent regeneration. Their regenerant is sent to Inmetco for off-site recycle.

- PS 317 and PS 318 applied ion exchange to rinse waters from electroless nickel plating. Both shops eventually abandoned their efforts to make the technology work. PS 317 described their use of the technology as a poor experience and indicated that their unit "required more water to restore the resin beads than water being processed." PS 318 indicated that the "process did not meet expectations or effluent guidelines."

4.4.6.2 Chromium Plating Performance Experience

Five respondents to the Users Survey provided some detailed data on their experience using ion exchange for chromium recovery. One of these shops uses an off-site regeneration/recycle service (PS 161) for regenerating columns applied to decorative chromium rinses (chromic acid, fluoride, barium carbonate). This shop also uses the off-site service for nickel recovery, details of which are provided in Section 4.4.6.1. Another respondent (PS 305) regenerates their unit on-site and sends the regenerant (sodium dichromate) to an off-site recycle company (Inmetco). One of the respondents recovers chromium from combined chromic acid anodizing and chromate conversion coating rinse waters and recycles the anion regenerant (dichromate) to the dichromate seal tank (PS 001). This system includes both anion and cation columns. The cation column removes contaminants such as aluminum and the anion column removes hexavalent chromium. Rinse water from the first rinse is filtered and passes through the columns and is returned to the final rinse. The cation column regenerant is treated and the anion column regenerant is transferred to the dichromate tank. Two of the four respondent's systems are not currently in use, but were previously used for chromium recovery from hard chrome plating rinse waters. One of these respondents intends to put their system back into use if the need arises (PS 052). Presently they have zero discharge, which is achieved through dragout recovery rinsing. Bath contaminants are removed using membrane electrolysis and a porous pot. The other shop tried the ion exchange process for a short time period in 1984 and does not intend to reuse it in the future (PS 080).

The following information and data summarize the performance experience of the five survey respondents.

- In one case, performance was hampered by operational and maintenance problems. PS 080 indicated that: "Initial attempts to use ion exchange for removal of impurities from chromic acid rinse water failed. Resins became fouled and could not be regenerated and maintain plating production throughput requirements." It should be noted that this system was relatively inexpensive ($3,025) as compared to the other chromium recovery units. The respondent indicated that the supplier stated capacity was 200 gpd and that the actual capacity was 0 gpd. The respondent provided a diagram in the survey that seems to indicate they were processing rinse water from dragout tanks. This solution may have been overly concentrated with chromic acid for the apparently small ion exchange unit, causing almost immediate total loading of the resin. Also, their low capital expenditure may indicate that prefiltration was not included with the purchased ion exchange equipment.

- PS 001 indicated that the supplier stated capacity and the actual capacity of their ion exchange system was 6 gpm at a feed concentration of 100 mg/l CrO_4. At this flow rate the system recovered 98% of the chromium.

- None of the shops indicated that an ion exchange unit applied to chromium recovery was the cause of an effluent compliance excursion.

4.4.6.3 Non-Cyanide Zinc Performance Experience

Five respondents to the Users Survey provided data on their experience with ion exchange used with non-cyanide zinc plating. Only two of these respondents provided sufficient detail to be considered useful for this report (PS 061 and PS 130). Both of these systems were manufactured by the same company and were purchased just one month apart. One ion exchange system was applied to acid zinc and the other to alkaline zinc. As will be discussed, the performance of the two systems was reportedly very different. The more successful of the two systems (PS 130) is a combination ion exchange and electrowinning system, similar in configuration to that shown in Exhibit 4-22. The other shop (PS 061) used ion exchange as an end-of-pipe treatment for "zinc bearing rinses and selected bath dumps," with the ion exchange regenerant going to an evaporator for concentration and the concentrated solution being hauled to an offsite recycle firm. The ion exchange processed wastewater was pH adjusted and discharged to a city sewer. This shop does not have a conventional hydroxide precipitation system following the ion exchange process. Originally this shop planned to return the zinc chloride concentrate from the evaporator to the zinc plating tank. As discussed later in this section, this plan did not work.

The following information and data summarize the performance experience of the two survey respondents.

- One of the two shops providing detailed data on zinc applications indicated that their ion exchange unit was not the cause of an effluent compliance excursion (PS 130). The other shop indicated that their overall treatment system, which included zinc and chromium ion exchange, evaporation and pH adjustment, caused an effluent limitation excursion. Their treatment system was put into operation in December, 1989. On July 4, 1990, they received a cease and desist order from their control authority.

- The performance of the ion exchange system operated by PS 061 was hampered by operational and maintenance problems. These are discussed in Section 4.4.6.3.2. The supplier stated capacity of this application was 18 gpm and the actual capacity was 10 to 12 gpm.

- PS 061 originally planned to recover the zinc chloride regenerant by concentrating it with evaporation and then reusing it in the zinc plating tank. This was not possible due to: (1) Chromium was present in the regenerant, and to remove the chromium would have required redesigning the ion exchange system. (2) The amount of zinc chloride generated by "false regenerations (a constant problem)" was much more volume than could be used in the plating tank, even after evaporation. (3) The low pH of the regenerant kept destroying the heaters in the evaporation unit and the steam from the system was so acidic that it corroded a hole in the shop roof. This shop also tried to recover the zinc using electrowinning. They found that it was too difficult to maintain a chemical balance during the process and that chlorine gas was liberated. Both problems made the process ineffective and the electrowinning unit was never put into full operation.

- Although PS 130 gave their system a satisfaction level rating of 3, they indicated that its downtime due to O&M problems was 50%.

- Both of the shops indicated, that based on their experience with this technology, they would purchase a different technology if given the opportunity to repeat the process.

4.4.6.4 Cadmium Cyanide Performance Experience

Four respondents indicated that they have used ion exchange for chemical recovery with a cyanide plating solution other than gold cyanide. These four shops included four applications to cadmium cyanide rinse waters and one for zinc cyanide rinse waters (one shop did both).

The shop using both applications (PS 229) installed their equipment in stages during the early to mid-1980's and it was integrated with an HSA electrolytic recovery system. The shop provided insufficient details to draw any conclusions, however, they offered the following as a summary of their experience with the combined system: "Costs in 11 years far outweigh benefits. Would have been much more practical to install conventional precipitation technology."

The other three respondents also provided sketchy information for their applications. One of the units was a pilot system purchased in 1991 that is still being used in 1993 (PS 025). The response from that shop indicate that they are relatively satisfied with the system. PS 245 used a rented ion exchange column for cadmium recovery and found that the resins "loaded very quickly and did not appear to be the right application." They intend to use the equipment following conventional precipitation for water reuse. Since these two applications require different types of resins, it appears that this respondent has not received sufficient technical support to utilize this technology. PS 254 purchased ion exchange equipment in 1985 for cadmium recovery. They later eliminated the ion exchange process because it was recycling ferrocyanides to the plating bath.

The following information and data summarizes the performance experience of the survey respondents.

- The major savings from the operation of ion exchange for this application was the result of water use reduction, treatment chemical reduction and sludge volume reduction.

- PS 229 reported that the capacities of their ion exchange columns were 50 percent of the supplier stated capacity (2 lb/column vs 4 lb/column). They also indicated that they were unable to reuse either their zinc chloride or cadmium chloride regenerant due to an excess volume problem in the plating tank (i.e., insufficient surface evaporation from the low temperature baths). They evaporate water from the zinc chloride regenerant (500 gpy) and send the concentrated solution to a landfill. They electrowin cadmium from that regenerant (1,000 gpy) and then send the metal-depleted regenerant to off-site disposal. A discussion of cyanide plating applications is presented in Section 4.4.3. Low cation capacity may be due to excess free cyanide. Methods for concentrating the solution and making headroom in the plating tank using evaporation are discussed in Sections 4.2 and 4.3. Electrowinning of the cadmium regenerant is an acceptable method of recovery. The recovered metal can often be used as anode material. It is possible to increase the recovery rate by reusing the cation regenerant following electrowinning (see Exhibit 4-22). PS 229 also indicated that the fact that their equipment manufacturer went out of business within two years of installing their ion exchange equipment hindered the performance of their application.

- None of the shops completing the survey indicated that an ion exchange unit applied to cyanide plating rinse waters was the cause of an effluent compliance excursion.

4.4.6.5 Gold Cyanide Performance Experience

Five user survey respondents provided some detailed data on their experience with ion exchange used for gold recovery. All but one of these shops send their used resins to their gold solution supplier for credit toward bath chemicals. The other shop regenerates the resin on-site and uses electrowinning to recover the metal (PS 123). The following information and data summarizes the performance experience of the survey respondents.

- The average satisfaction level for ion exchange applied to gold cyanide is 4.0 (on a scale of 1 to 5, with S being most satisfactory), which is higher than the average level rating for ion exchange chemical recovery in general. Also, all of the shops indicated that this technology satisfied the need for which it was purchased. In each case the need was identified as reducing plating chemical purchases or other (i.e., reclaim gold). The range of savings from gold recovery for these shops was $1,320 (PS 283) to $134,400 (PS 123) per year.

- Only one of the shops provided any capacity data. PS 179 indicated that they experience an average resin loading of 4 troy ounces per cubic foot of resin. They typically use 1 ft^3 of resin per year.

- In some cases, performance was hampered by operational and maintenance problems. These are discussed in Section 4.4.7.

4.4.7 Operational and Maintenance Experience

Of the 27 shops providing data, 17 (or 63%) were still operating their ion exchange equipment at the time of the survey. The average age of the operating systems was 7.0 years (only 4.2 years excluding gold plating applications). The average percentage of downtime experienced by the respondents was 20% (only 8.7% excluding PS 124, PS 080 and PS 061).

The following summarizes the respondents operating labor information. Details on O&M experiences are discussed for each type of application in Sections 4.4.7.1 through 4.4.7.5

- Fourteen shops provided operating labor data. For these shops, the average number of annual operating hours per ion exchange system were: 553 hrs/yr. The skill requirement commonly needed for operating this technology is a trained technician or a wastewater treatment plant operator. The following is a breakdown of the responses for skill requirements:

Environmental Engineer:	5
Process/Chemical Engineer:	5
Chemist:	4
Consultant:	3
Plumber/Pipe Fitter:	4
Electrician:	4
Vendor:	1
Senior-Level Plater:	6
Junior Level Plater:	4
Trained Technician:	14
Wastewater Treatment Plant Operator:	13
Common Labor:	3
Other:	0

4.4.7.1 Nickel Plating O&M Experiences

Reported O&M problems for nickel plating ion exchange applications relate to equipment failure and process chemistry concerns. Of the 11 shops providing data, six (or 63%) were still operating their ion exchange equipment at the time of the survey. The average age of the operating systems was 4.4 years. The average percentage of downtime experienced by the respondents was 17% (only 4% excluding PS 124).

The following O&M experiences were reported by respondents:

- PS 124 was the earliest installation (1977). Considering the cost ($6O,849 total installation cost), this was apparently a major recycling project for the shop. The shop provided the following responses concerning the equipment and the process: "Continuous and repeated failure of cheap valves used in design caused unit to plug up". "Return of calcium sulfate to plating tank clogged pipes and anode bags."
- PS 212 reported that: "(The ion exchange system) was used for a short period of time. It never did the job it was intended to do."
- PS 263 reported that their equipment, which they refer to as "prototype", "worked well, but created plating problems." This was a 4 gpm unit that recycled rinse water from the final rinse of a muti-stage counterflow rinse system. The shop believes that their plating problem was caused by the water in the final rinse (i.e., that which is recycled to the ion exchange column) because it was too clean and resulted in passivation of the nickel plated parts before chrome plating (i.e., the subsequent process).
- PS 261 was the only shop that reported that they had no significant O&M problems.
- PS 063 reported O&M problems with probe indicators and the balancing of chemicals.

4.4.7.2 Chromium Solutions O&M Experiences

Reported O&M problems for chromium ion exchange applications relate to resin fouling and equipment failure. Of the four respondents providing data for this application, two were still in operation at the time of the survey. The average age of operating systems was 3.5 years. The percentage of downtime experienced by the respondents was 27% (only 3% excluding PS 080).

The following summarizes the respondent's O&M experlences:

- PS 080 indicated that their ion exchange system failed due to resin fouling. A discussion of this shop's system is presented in Section 4.4.6.2.1.

- PS 052 indicated that their ion exchange regeneration process was very time consuming. They also indicated that only dilute rinse waters can be effectively processed by the unit and that their rinse waters are occasionally too concentrated for the unit.

- PS 001 indicated that their ion exchange system was also sensitive to the chromium concentration of the rinse water (i.e., feed to the ion exchange unit) and that to prevent overloading of the resin they initially rinse over the plating bath before using the rinse tanks.

- PS 305 indicated that they cannot process the contents of their dragout tank with ion exchange, without oxidizing the resin. (Note: The recovery of chemicals from concentrated solutions is not a good application for ion exchange. This technology is best applied to dilute solutions.)

4.4.7.3 Non-Cyanide Zinc Plating O&M Experiences

Reported O&M problems for non-cyanide zinc plating ion exchange applications relate to equipment failure and resin contamination with oil. Of the two respondents providing data for this application, none were still in operation at the time of the survey. One was purchased in December, 1989 (PS 061) and the other in January, 1990 (PS 130). PS 130 indicated that they intend to use their system in the future, although no specific plans were discussed in their survey form.

The following summarizes the respondents O&M experiences:

- PS 061 provided the following information that relates to O&M problems: "Never believe an engineer again! This technology should be never used to treat zinc wastewater from a job shop. Way too touchy-system, must be simple-discontinued July 4, 1990." "Pump failure, resin fouled, computer failure, float switch failure, heater failure, and on and on and on-" "Nothing worked for this project." This same shop offered the following regarding technical restrictions of the technology that they have encountered: "Chemical concentration & flow rates are critical! Failure of any one part can shut down the whole system without warning-[sic]"

- PS 130 indicated that the reason they are no longer using their zinc ion exchange system was a problem with oil in the rinse water feed to the unit that fouled the resin. They indicated that they plan on using the system in the future, after making changes to their cleaning process. This shop indicated the ion exchange system was not working 50% of the time as a result of the oil problem.

4.4.7.4 Cadmium Cyanide Plating O&M Experiences

Reported O&M problems for cadmium cyanide plating ion exchange applications relate to equipment failure, frequency of regeneration, excess regenerant production and bath contamination. Of the four shops providing data, two systems were operational at the time of the survey. The average age of the operating systems was 5.5 years. The average percentage of downtime reported by the respondents was 8%.

The following summarizes the respondents O&M experiences:

- PS 229 reported that their ion exchange units were overloaded due to high dragout rates and improper rinsing practices by their operators (discussed in Section 4.4.6.4.1). They also indicated that they felt their O&M problems were related to the manufacturer going out of business.

- PS 245 reported that their ion exchange columns loaded quickly and that the frequency of regeneration was very high.

- PS 254 eliminated their ion exchange process because it was returning ferrocyanides to the plating bath.

4.4.7.5 Gold Cyanide Plating O&M Experiences

Generally, the ion exchange units used for this purpose are small (< 0.5 ft^3 of resin) and are free of complex regeneration and control systems. As a result, the operational and maintenance problems are minimal and the units that are purchased remain in use. In fact, all of the gold cyanide ion exchange installations identified in the Users Survey were in operation at the time of the survey. The average age of the units was 14 years. The only specific problem identified by a respondent was flow blockage caused by algae growth. The average percentage of downtime experienced by the respondents was 5%.

4.4.8 Residuals Generation

The primary residuals from ion exchange recovery processes are the regenerants (eluates) and backwash solutions. The regenerants are concentrated wastes and the backwash is dilute. Both solutions are either caustic or acidic, depending on the resin type and application. High metal bearing regenerates (typically cation resin) are: (1) sometimes reused directly in the bath; (2) further processed to recover the metal (e.g., electrowinning); (3) waste treated; or (4) sent to an off-site recovery facility. Low metal bearing regenerants (typically anion resin) and backwash solutions are typically treated on-site. Waste treatment processes generate sludge that is an EPA listed hazardous waste (F006).

The volume of regenerant produced will depend on the regeneration requirement (e.g., lbs of acid per ft^3 of resin) and the concentration of acid used (typically 1 to 5%). The regeneration requirement will depend on the resin type, application (metal or complex being recovered) and the configuration (cocurrent vs counterflow). Typical volumes of regenerant are 20 to 50 gal/ft^3 of resin. The volume of regenerant waste is sometimes reduced by reusing the last portion of the regenerant, which will be less contaminated with metal and contains free acid. Backwash volumes depend mostly on the equipment design and the application. Typically, backwashing generates 25 to 75 gal/ft^3. The backwash is partly reused by some equipment vendors as make-up water for regenerant, in an effort to reduce the total waste volume generated. Because backwash contains only dilute concentrations of pollutants it is typically not a major concern and is treated on-site and discharged. However, for shops working toward zero discharge, the backwash volume could present a significant problem. Both backwash and regenerant can be processed by evaporation to reduce the volume requiring disposal. However, this increases the capital and operating costs of the system. Also, evaporation of hazardous wastes is sometimes regulated as a RCRA technology and may require a permit to operate.

4.5 ELECTROWINNING

4.5.1 Overview

Electrowinning is one of the two most widely used methods of metal recovery in the plating industry, the other being atmospheric evaporation (Section 4.2). Of the 318 plating shops responding to the Users Survey, 61 (0r 19%) have employed this technology. Some shops have purchased or fabricated two or more units for differrent applications, resulting in a total number of 80 electrowinning units used by the 318 survey respondents.

Electrowinning is most frequently used to: (1) reduce the mass of inexpensive regulated metals (e.g., zinc, copper, lead) and cyanide being discharged to treatment and thereby reduce the quantity of treatment reagents used and the quantity of sludge generated and/or (2) recover expensive common metals (e.g., nickel and cadmium) or precious metals (e.g., silver and gold) for recovery/recycle and thereby reduce overall operating costs. In either case, electrowinning is most often applied for gross metal recovery from concentrated solutions such as dragout rinses or ion exchange regenerant. Used in this manner, it is not sufficient as a stand-alone technology to meet discharge standards. Reticulate cathode, high surface area (HSA) or high mass transfer (HMT) cathode designs also make this technology applicable to some dilute metal bearing solutions (e.g., overflow rinses). The reticulate cathode units have been proven to be effective in maintaining the metal concentration of recirculated rinses to less than 1 mg/l. The HSA units have been advertised as a method of compliance (ref. 98) and in the late 1970's and early 1980's attempts were made to use the electrowinning technology in this manner. However, none of Users Survey respondents are currently discharging an effluent from an HSA or HMT unit without further treatment for metals removal. Some non-continuous discharges of batch-treated solutions are found. However, for these cases, the volume of the discharge is insignificant compared to total wastewater flow.

The basic unit of the electrowinning technology is the electrolytic cell: two electrodes (anode and cathode) are placed in a solution containing ions, where there occurs a movement of ions toward the charged electrodes. Dissolved metals in the electrolyte are reduced and deposited on the cathode. The deposited metal is removed by mechanical (e.g. scraping) or chemical means and either reused as anode material or sent off-site for refining/reuse or disposal.

The types of cathodes used in electrowinning can be grouped into three categories. These include, in order of increasing surface area: (1) flat plate, (2) expanded metal, wire mesh or reticulate plate, and (3) porous or woven carbon and graphite types. The flat plate cathodes are used for applications of gross metal recovery from concentrated solutions (e.g., >1 g/l of metal). The expanded metal, wire mesh, or reticulate plate and the porous or woven types are used for recovering metals from solutions with lower metal concentrations, with the latter group effective in some cases in the low mg/l range. Reticulate cathodes, which permit flow-through of the electrolyte, have an effective surface area of approximately 10 times the face or apparent area of the cathode. Porous or woven cathodes have internal pores that also permit flow-through of the electrolyte and have a surface area up to 13,000 times greater than the apparent area.

There are several common terms used in describing the equipment and processes relative to electrowinning. The basic electrolytic cell is composed of two electrodes, one anode (positive charge) and one cathode (negative charge). The chemical reactions that take place at the anode are oxidations and the reactions at the cathode are reductions. The solution is referred to as an electrolyte. When a direct current (D.C.) is applied to the cell, the anions present in the electrolyte migrate toward the anode and the cations migrate toward the cathode. An important controlling factor in the process is the amount of current flowing through the cell. The level of current is measured in amperes per unit area of electrode (typically, amperes per square foot) and is referred to as the current density. Current density affects the nature of the electroplated deposit, the distribution of the deposit, the current efficiency, and to some extent whether a deposit forms at all. In electrowinning, the theoretical quantity of metal that is deposited onto the cathode is described by Faraday's Law. This law states that the amount of chemical change produced by an electric current is proportional to the quantity of electricity used (ref. 350). Some of the electric current is used for reactions other than metal deposit. Electroplaters refer to the ratio of desired chemical change (deposit) to the total chemical change as the current efficiency, usually expressed as a percentage of current applied.

As indicated previously, the current density has a substantial impact on the rate of metal deposit. It is desirable to operate electrowinning processes at the maximum current density where good deposition still takes place. The current density should, however, not exceed that which deposits metal faster than ions can diffuse through the electrolyte. When the thin film of electrolyte surrounding the cathode is depleted of metal ions, a condition referred to as concentration polarization occurs. This results in an adverse effect on the current efficiency as well as the quality of the deposit due to excessive hydrogen evolution at the cathode and oxygen evolution at the anode. The allowable or critical current density is determined by the concentration of metal ions in the electrolyte and the thickness of the film surrounding the cathode. Innovations in the design of electrowinning devices have generally focused on extending the operating range of the process by: (1) increasing the surface of the cathode (i.e., high surface area), or (2) reducing the thickness of the film using agitation or heating (ref. 349, 351).

For most applications, the primary use of electrowinning is the recovery of metal. However, when performed with an electrolyte containing cyanide, the process also oxidizes some of the cyanide at the anode (alternatively CN can be oxidized with hypochlorite ions which result from the electrochemical oxidation of chloride ion in a basic medium). Although the anodic reactions are given less consideration in most applications, they can play an important role in the economic viability of the process by reducing the treatment reagent requirements for end-of-pipe treatment. Anodic reactions including cyanide destruction and organic complexing agent destruction (e.g., treatment of an electroless copper bath) were examined in detail by Waiux and Nguyen (ref. 123).

4.5.2 Development and Commercialization

Electrowinning is presumed to be one of the earliest methods of metal recovery used in the plating industry, although no specific reference to its use prior to the 1960's was found (ref. 128). One of the reasons for its presumed early development and current widespread use is the fact that this process emulates the electroplating process. As such, it is readily accepted and understood by the plating industry. Further, for these same reasons, there are fewer system failures caused by misapplications or operational errors than with other recovery technologies such as ion exchange or membrane technologies. The electrowinning technology also has technical roots and contemporary applications in other industries, including electrorefining of copper, extraction of metallic aluminum from bauxite, and recovery of silver from photographic film manufacturing and developing operations.

Electrowinning can be performed using very simple equipment. As such, many plating shops have constructed units in-house. The effectiveness of these home-made units varies from shop to shop. Commercial units for plating applications are manufactured/sold by at least 40 companies (ref. 421). Many of these units, like the home-made models are also relatively simple in design. As discussed in Section 4.5.1, the efficiency of the electrowinning process is impacted by several electrochemical factors. Some commercial units, through the incorporation of more sophisticated design elements, minimize the impact of the electrochemical forces that reduce plating efficiency. Typically, these design elements are limited to the commercial units and are not found to any significant extent with the home-made versions.

4.5.3 Applications and Restrictions

Exhibit 4-32 shows the three basic configurations in which electrowinning was successfully applied by the shops responding to the survey.

The most common configuration (EW-1) employs an electrowinning unit connected directly to a dragout tank. Alternatively, the solution from the dragout tank can be periodically transferred to a holding tank that is connected to the electrowinning unit. Either of these arrangements can be used with flat plate or reticulate cathode units. The reticulate cathode types will maintain the rinse system at a lower metal concentration (in some cases below 1 mg/l) but, because the cathodes are not reusable, the operating costs will be higher. The operation of the flat plate cathode types are more significantly affected by fluctuations in the metal concentration of the electrolyte. Therefore, if the plating operation causes sharp fluctuations in the dragout tank concentration, the user should consider the use of a side tank or a reticulate type of cathode. The HSA cathode units should be directly connected to the dragout tank. This will permit them to maintain a low steady state concentration of metal in the dragout tank.

Electrowinning removes metal from the dragout solution, but does not remove all dissolved solids. For this reason, the dragout solution must be occasionally discarded or purged to prevent the build-up of dissolved solids (e.g., acid). When this occurs, any residual metal in the dragout solution will be lost.

The metal recovery efficiency (i.e., the percentage of metal recovered from dragout) of the first configuration depends on two key factors: (1) the average concentration of metal in the dragout tank and (2) the mass of metal in the purge. The concentration of metal in the dragout tank is important because it determines the mass of metal that will be carried over by dragout to the next rinse, which is treated. This factor points out the weakness of the flat plate cathode types. These units operate efficiently only when the metal concentration is high (usually 1 to 5 g/l of metal). Therefore, the dragout tank must be operated until this level is achieved, which in turn increases the loss of metal to the free running rinse. The higher surface area of the reticulate and HSA units allow the user to operate the dragout tank at a lower metal concentration and therefore reduce metal losses. Further, these types of electrowinning units generate a purge with a lower metal concentration.

The second configuration (EW-2) is a combination of ion exchange and electrowinning. This configuration potentially has a much higher metal recovery efficiency than the first configuration. It addresses both of the factors that impact metal recovery efficiency. The ion exchange unit maintains a low metal concentration in the final rinse, thereby almost eliminating dragout losses. The ion exchange unit concentrates the metal into a regenerant stream and the electrowinning unit removes the metal. Residual metal in the regenerant is of less concern than the first configuration since it can be reconcentrated by the ion exchange unit. For the same reason, a flat plate cathode will suffice for this second configuration.

In some cases, the reticulate cathode units can be substituted for the second type of configuration. When such a unit maintains the dragout rinse in the low mg/l range, the metal recovery efficiency of the process would approach that of the ion exchange/electrowinning combination. Some recent operating data for a copper recovery application using this configuration are presented in Section 4.5.4.3.

The third configuration shows the recovery of metal from a spent process solution. Either the flat plate or reticulate cathode type of unit can be used in this configuration. The reticulate cathode type will provide greater metal recovery efficiency because it can lower the metal concentration of the spent bath below that of the flat plate. Because the reticulate cathodes are not reusable, its higher recovery efficiency comes at an increased operating cost.

Electrowinning is applied to a wide variety of chemical solutions in the electroplating industry. The literature indicates that the metals that are most commonly recovered by electrolytic treatment are gold, silver, copper, cadmium, and zinc. The metal recovery applications identified from the Users Survey are shown in Exhibit 4-33. This exhibit indicates the number of survey respon-

Exhibit 4-32. Three Primary Configurations for the Application of Electrowinning for Metal Recovery

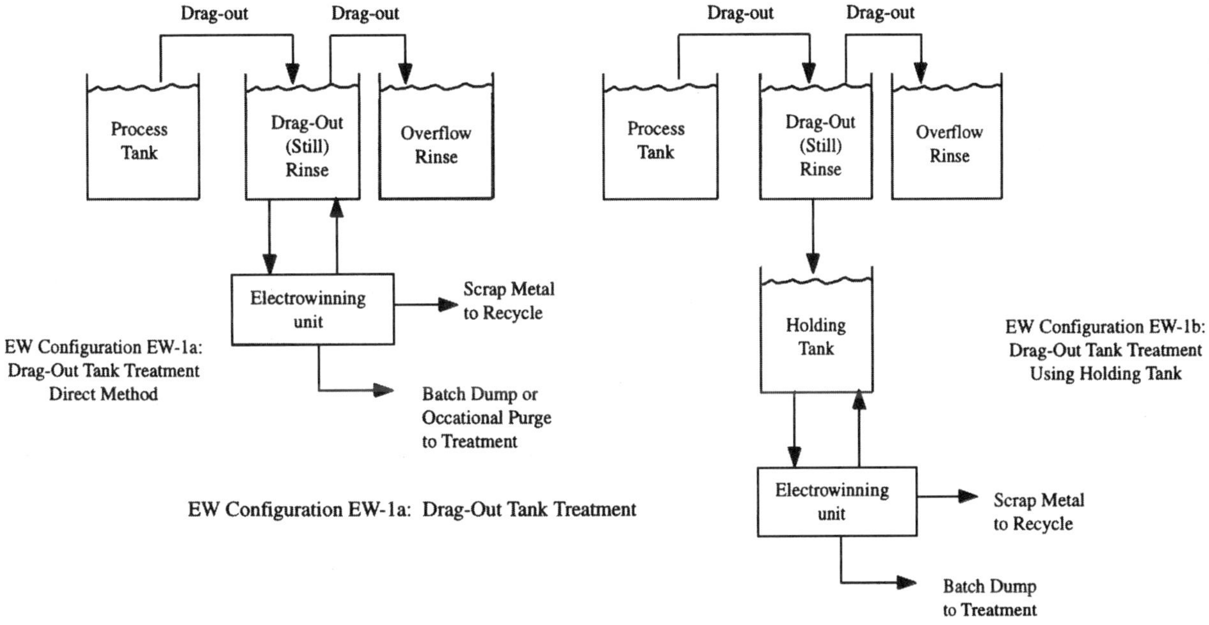

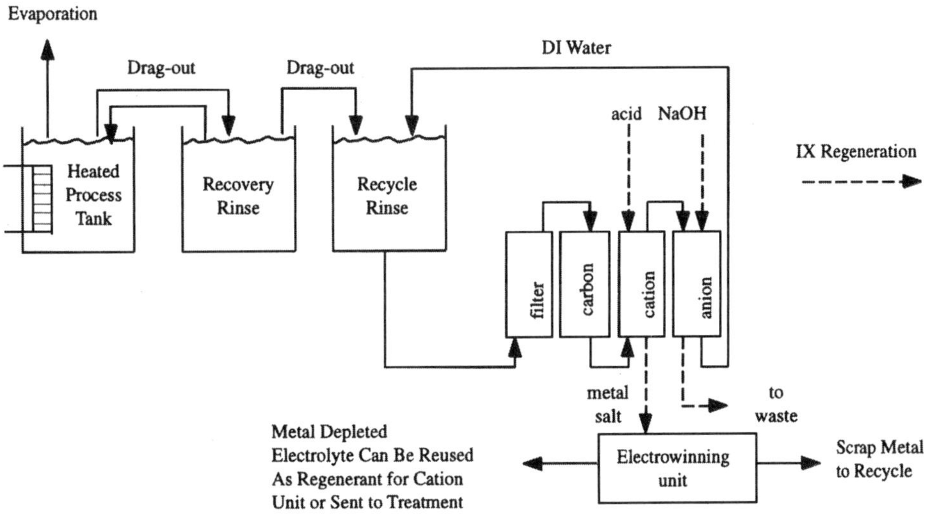

EW Configuration EW-2: Ion Exchange/Electrowinning Combination

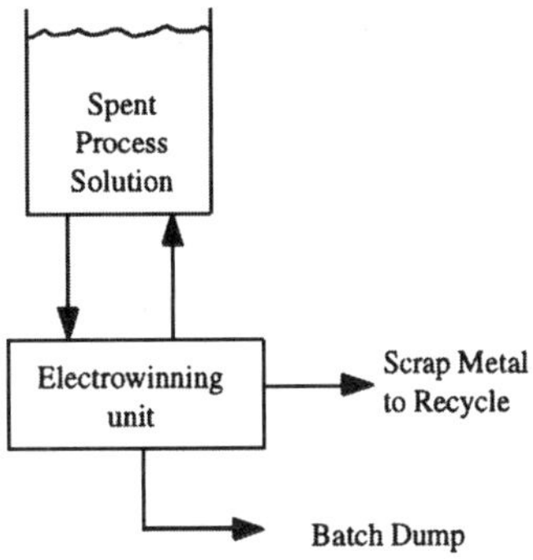

EW Configuration EW-3: Spent Process Solution

dents that applied electrowinning to each of the processes and the average satisfaction level of the technology for that application, based on a scale of 1 to 5 (1 equals the lowest satisfaction level and 5 equals the highest).

For practical purposes, the degree to which a metal can be recovered by electrowinning can be determined by its position in the Electromotive Series (see Exhibit 4-34). Metals that have more positive standard electrode potentials plate more easily than the ones with less positive potentials. As an illustration, the more noble metals, such as silver and gold, can be removed from solution to less than 1 mg/l using flat plate cathodes whereas with copper and tin, a concentration in the range of 0.5 to 1 g/l or more is required for a homogeneous metal deposit. Equations for accurately estimating the potential for a given application were presented by Brown (ref. 349) and Bailey and Chan (ref. 128).

It is interesting to compare the satisfaction levels in Exhibit 4-33 to the position of the metal in the electromotive series. The satisfaction levels for silver, copper, cadmium and zinc cyanide plating (the most common applications of the respondents) fall into nearly the exact order as the metal's position in the electromotive series.

Although copper, cadmium and zinc have a lower position in the electromotive series than precious metals and they received only moderate to low satisfaction levels from survey respondents, this is not to say that these applications cannot be successfully performed. With the application of proper engineering and good equipment selection these electrowinning applications can be highly successful, as indicated by some of the respondents. For additional data, Exhibit 4-35 groups potential electrowinning applications based on their frequency of use and success in industry and the general difficulty of the application. These rankings are based on input from electrowinning vendors and information from the literature. Included in this exhibit are a much broader range of metals than those identified in the Users Survey.

Although there are limitations for electrowinning nearly every metal, chromium is the only commonly electroplated metal that is not recoverable using electrowinning. Nickel recovery is possible, but it requires close control of pH and therefore is less frequently performed than, for example, cadmium or copper. Also, Altmayer suggests that nickel recovery is hampered by the absence of inexpensive suitable inert anodes that do not give off chlorine gas and disintegrate (ref. 482).

Solutions such as electroless plating solutions containing chelated metals, reducing agents and stabilizers are more

Exhibit 4-33. Electrowinning Applications Identified from the Users Survey

Process Solution	Number of Applications	Average Satisfaction Level
Anodize, hardcoat	1	2.0
Brass plate, cyanide	2	3.5
Cadmium, strip	1	ND
Cadmium plate, cyanide	26	2.8
Copper plate, cyanide	8	2.8
Copper plate, sulfate	5	3.0
Copper strike, cyanide	1	4.0
Gold plate, cyanide	2	ND
Gold plate, non-cyanide	2	4.0
Nickel strip	1	4.0
Nickel plate, sulfamate	2	4.0
Nickel plate, watts	2	ND
Nickel plate, woods	1	ND
Nickel plate, electroless	1	ND
Passivation, nitric	1	1.0
Silver strip	1	4.0
Silverplate, cyanide	11	4.4
Zinc plate, cyanide	6	1.3
Zinc plate, non-cyanide	5	2.0
Total/Average	**57**	**3.1**

ND = no data

Exhibit 4-34. Electromotive Series

Electrode Reaction			E^0, Volts at 25°C
$S_2O_8^{=} + 2e^-$	---------->	$2SO_4^{=}$	+2.01
$H_2O_2 + 2H+ + 2e^-$	---------->	$2H_2O$	+1.78
$Au^+ + e^-$	---------->	Au_m	+1.69
$2Cl + 2e^-$	---------->	$2Cl^-$	+ 1.36
$^1/_2 O_2 + 2H^+ + 2e^-$	---------->	H_2O	+1.23
$Pd^{++} + 2e^-$	---------->	Pd	+0.99
$Ag^+ + e^-$	---------->	Ag	+0.80
$Fe^{+++} + e^-$	---------->	Fe^{++}	+0.77
$Cu^+ + e^-$	---------->	Cu	+0.52
$Cu^{++} + 2e^-$	---------->	Cu	+0.34
$2H^+ + 2e^-$	---------->	H_2	+0.00
$Pb^{++} + 2e^-$	---------->	Pb	-0.13
$Sn^{++} + 2e^-$	---------->	Sn	-0.14
$Mo^{++} + 2e^-$	---------->	Mo	-0.20
$Ni^{++} + 2e^-$	---------->	Ni	-0.25
$Co^{++} + 2e^-$	---------->	Co	-0.28
$Cd^{++} + 2e^-$	---------->	Cd	-0.40
$Fe^{++} + 2e^-$	---------->	Fe	-0.44
$Zn^{++} + 2e^-$	---------->	Zn	-0.76
$Mn^{++} + 2e^-$	---------->	Mn	-1.19
$Al^{++} + 3e^-$	---------->	Al	-1.66
$Mg^{++} + 2e^-$	---------->	Mg	-2.36
$Na^{++} + e^-$	---------->	Na	-2.71

Source: ref. 349

Exhibit 4-35. Potential Applications for Electrowinning of Metals Identified from the Vendors Survey and Literature

Group	Metals (electrolyte)
Group 1 Includes metals with a high potential for successful application. All metals listed in this group are commonly recovered using electrowinning.	brass (cyanide), cadmium (cyanide), copper (acid, cyanide), gold (cyanide), silver (cyanide), zinc (cyanide)
Group 2 Includes metals with a high potential for successful application, however, metals listed in this group are less commonly recovered using electrowinning than those in Group 1.	antimony, cadmium (ammonium sultate), iridium, lead (acid), palladium, ruthenium, rhodium, selenium, tin (acid, alkaline)
Group 3 Includes metals with a moderate potential for successful application. May require chemical adjustment of the electrolyte or special equipment (e.g., unusual anodes).	cobalt, copper (electroless, strong acid, ammononical etches), gold (strip), indium, lead (fluoborate), nickel (Watts, woods, sulfamate, electroless), silver (thiosulfate), tin-lead (fluoborate), zinc (acid)
Group 4 Low potential for success. No known instances of use.	aluminum, barium, beryillium, boron, calcium, cadmium (strip), chromium, iron, magniesium, manganese, mercury, molybdenum, silicon, tantalum, titanium, tungsten, vandium

Sources: ref. 99, vendor files

difficult for the direct application of electrolytic recovery. However, there was one survey respondent that indicated they were successfully electrowinning nickel from a spent electroless solution (PS 188). Another shop (PS 164) is in the process of starting up a unit for the same purpose. One vendor (ref. 349) indicated that these baths can be processed by electrowinning after undergoing pretreatment (e.g., selective ion exchange) to break the metal-chelate bond. Another reference suggests that reducing and oxidizing agents can be combined to neutralize their effects; e.g., a printed circuit board shop can mix spent micro-etch and electroless copper baths and with proper pH adjustment create a solution that can be treated by electrowinning (ref. I3 file). Another reference indicates that electroless copper can be processed using electrowinning, but that anode life will be short (ref. 99).

Fluoborate solutions (e.g., tin, tin-lead) are not commonly treated using electrowinning due to their attack upon anode materials including iridium oxide coated titanium and niobium. However, one source (ref. 287) indicates that titania ceramic anodes coated with iridium can provide a successful application. This material and its application have been recently commercialized (ref. Kinetico file).

Certain corrosive solutions (e.g., certain etchants) may also pose problems for electrowinning because metal that is plated on the cathode may be etched off as quickly as it is plated (ref. 348). One reference suggests that increasing the current density will partially overcome the etching action of ammonomical etches when electrowinning copper from these solutions, but that complete removal is difficult to achieve (ref. 99).

4.5.4 Technology/Equipment Description

This subsection contains names and/or descriptions of commercially available electrowinning equipment that is manufactured and/or sold by vendor survey respondents or discussed in the literature. This is intended to provide the reader with information and data on a cross section of available equipment. Mention of trade names or commercial products is not intended to constitute endorsement for use.

4.5.4.1 General

The typical electrowinning system consists of a tank that holds the electrolyte, sets of anodes and cathodes, a pump for transferring solutions from a feed tank to the electrolyte tank, rectifier, and controls.

Most electrolyte tanks are manufactured from polypropylene, although one of the surveyed manufacturers (ref. Eco-Tec file) also used lined steel tanks. The tanks range in size from approximately 10 to 1,500 gal. Rectifier output amperage ranges from approximately 25 to 5,000 amps (ref. vendor files), with the smallest units used primarily for precious metals (e.g., Au, Ag, Rh) recovery (ref. vendor files and 111).

The three most common types of electrolytic metal recovery equipment use either: (1) parallel flat plate cathodes, (2) reticulate cathodes; and (3) fibrous or high surface area cathodes. Generally, the parallel flat plate cathode units are used with concentrated metal solutions, the reticulate cathode units work over a wide range of concentrations, and high surface area cathode units are used exclusively with solutions containing dilute metal concentrations.

Various materials are used in the fabrication of anodes and cathodes. Until the 1960's, graphite and lead alloys were the most preferred anode materials. However, their high overpotential requirement and degradable nature presented significant drawbacks. More recently, anodes are commonly being manufactured of titanium and niobium and coated using the solid phase roll bonding method with precious metals, metallic oxides and/or their alloys and fluoride resistant metal composites (ref. 128). These types of electrodes are generically referred to as dimensionally stable anodes. The advantages of the newer anodes over the lead alloy anodes include: (1) produce higher purity product (deposit); (2) low oxygen overpotential increases current efficiency; and (3) corrosion resistance provides higher durability and stability.

Most commonly, flat plate, wire mesh and expanded metal cathodes are fabricated from stainless steel, reticulate cathodes are metal coated foam and high surface area cathodes are fabricated from carbon fibers. Additional details of cathode design are discussed in Sections 4.5.4.2 through 4.5.4.4.

Exhibit 4-36 indicates the different materials used for electrode fabrication by the four manufacturers that responded to the Vendors Survey for their most common applications. It also includes other types of materials identified in the literature and known to be in common use.

4.5.4.2 Flat Plate Cathode Units

The flat plate design is often referred to as the conventional method of electrowinning because of its long stand-

Exhibit 4-36. Materials of Construction for Electrodes Used for Common Electrowinning Applications

Electrolyte	Cathode Material	Anode Material
Cadmium Cyanide	2, 3, 4	3, 6, 9
Cadmium Sulfate	3, 4	6, 9
Copper Cyanide	4	10
Copper Persulfate Microetch	14	14
Copper Peroxide Microetch	14	14
Copper Pyrophosphate	14	14
Copper Sulfate	2, 3, 4	1, 7, 10, 11
Gold Cyanide	2, 3, 4	7, 11
Lead Fluoborate	4	12
Lead Sulfamate	4	7
Nickel Sulfate	1, 2, 3, 5	7, 10, 11
Silver Cyanide	2, 3, 4	3, 7, 11
Tin Alkaline	2, 3	8, 11
Tin-Lead Fluoborate	2, 3	8, 11
Zinc Alkaline	13	7
Zinc Cyanide	1	10
Zinc Unspecified	1	10

Legend

1. Proprietary
2. Stainless steel mesh
3. Stainless steel sheet
4. Reticulated copper coated
5. Reticulated nickel
6. Carbon fiber
7. Dimensionally stable anodes (DSA®)
8. Ebonex®
9. Ebonex® iridium oxide or platinum coated
10. Lead alloy
11. Platinum clad titanium, niobium, tantalum
12. Graphite
13. Zinc mesh
14. Unknown

Sources: Data provided by Eltech International Corporation, Kinetico Engineered Systems, Eco-Tec Inc., and Memtek Corporation. These companies manufacture electrowinning equipment that employ either flat plate or reticulate cathodes. Also, the data are based on common material usage as identified from the literature.

ing role in the plating industry as well as other industries. Conventional electrowinning equipment is found in a variety of configurations. The basic design consists of a tank containing alternating flat sheets of cathodes and anodes. Commercially available electrolytic recovery units, used for waste treatment and recovery, have total cathode surface areas ranging from 1 ft2 to 200 ft2. Such units are extremely small in comparison to those used for primary copper production. An average copper refinery producing 500 tons per day of copper utilizes approximately 2.6 million square feet of total electrode area (ref. 349). A packaged recovery unit generally is supplied with a reactor tank or cell, copper bussing, cathodes, anodes, rectifier, current controller, recirculation pump, internal piping, and valves.

With the parallel flat plate electrode units, the recovered metal is removed in strips or slabs and can be sold to a refiner or used in-house by electroplaters as an anode material. Several variations of the conventional electrowinning process are used. Variations in design are typically aimed at overcoming electrode polarization and low ion diffusion rates which reduce recovery rates in low concentration solutions. This is typically achieved by reducing the thickness of the diffusion layer through agitation of the solution or movement of the cathode.

Flat plate electrowinning units are usually operated on a batch basis, although continuous configurations are also in use. With a batch operation, a solution containing metal ions is added to the electrowinning cell tank or continuously circulated from a side tank and a D.C. electrical current is applied. As the recovery process proceeds, metal ions are plated onto the cathode and the solution becomes depleted. Typically this process is halted when the deposition rate drops below a given point or when the metal deposit thickness reaches approximately 1/4 to 3/8 inch. The plated metal sheets can then be pealed from the cathode and reused or sold. It is possible for the plated deposit to envelop the cathode, making removal nearly impossible. This problem can be overcome by employing a technique termed current shadowing that gradually reduces the current density at the outer edges of the cathode plate. Another method is to use non-conductive edge strips. However this may result in the production of dendrites at the juncture of the edge-strips (ref. 349).

4.5.4.3 Wire Mesh, Expanded Metal and Reticulate Cathode Units

The wire mesh, expanded metal and reticulate cathode designs are aimed at increasing the surface area of the cathode. The wire mesh and expanded metal (appearance of floor grating) types are usually fabricated from stainless steel. Reticulate is a term used by at least one manufacturer of electrowinning units to describe their cathodes (ref. Eltech file). The term reticulate, which means having veins arranged like the threads of a net, accurately describes the appearance of this type of the cathode. The manufacturer also describes the reticulate cathode as a "foam metal cathode" (ref. 105).

The metalized surface of the reticulate cathode is rough and therefore has a greater actual surface area than its geometric surface. The manufacturer indicates that the surface area is 10 times greater than the apparent area. The higher surface area permits use of the units at lower metal concentrations than possible with conventional flat plate cathodes of the same size. One user of this technology (PS 196) indicated that it treats cadmium to below 5 mg/l, but that a significant concentration of residual cyanide remains. A diagram of a reticulate cathode electrowinning system used by nine survey respondents is shown in Exhibit 4-37.

The wire mesh and expanded metal types are used as anodes in a plating bath after they have been plated with metal in the electrowinning unit (ref. 130). The reticulate cathodes are not reusable. When they are fully coated with metal, they are either sent off-site for sale as scrap or are discarded, depending on the type and purity of the deposit and the ability of a reclaim site to deal with the non-metallic core of the cathode. Operations where the cathodes are discarded are referred to as extractive methods of electrowinning (ref. 421).

Vendor provided data for the electrowinning treatment of a copper cyanide bath using a reticulate cathode design is shown in Exhibit 4-38. Operating data for a reticulate cathode unit provided by the Naval Facilities Engineering Service Center (Port Hueneme, CA) are graphically displayed in Exhibit 4-39. These tests were performed on a printed circuit board line over a time period of 432 hours (18 days). The electrowinning unit is a Retec Model 6 (21 ft^2 of reticulate cathode surface area). Exhibit 4-39 (a) shows the copper concentration in the dragout rinse tank (56 gal) during the test period (same set-up as EW configuration EW-1a, Exhibit 4-32). The highest concentra-

Exhibit 4-37. Diagram of an Electrowinning Unit Employing Reticulate Cathodes (Courtesy of ELTECH Systems Corporation)

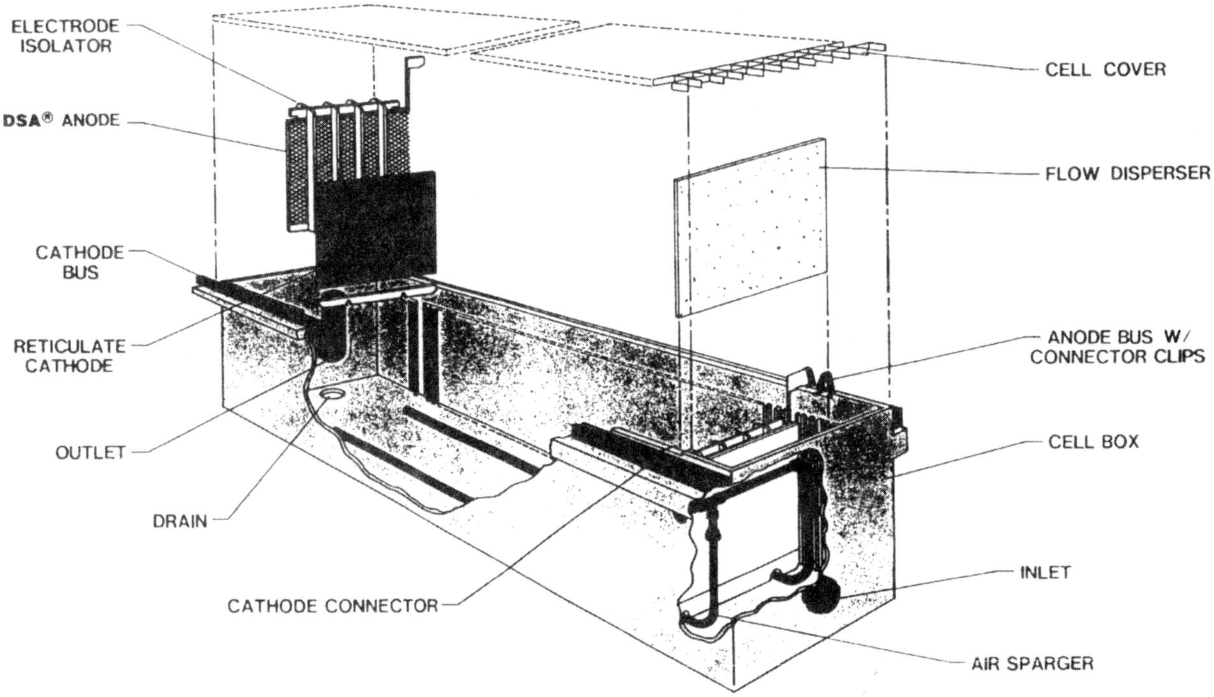

Exhibit 4-38. Example of the Performance of a Reticulate Cathode Electrowinning Unit Treating a Copper Cyanide Bath

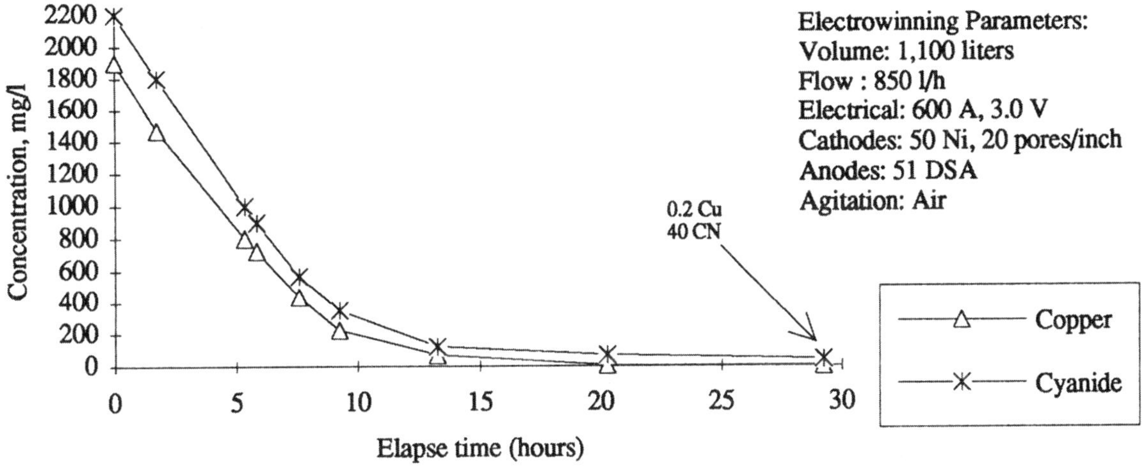

Source: ref. 103

Exhibit 4-39(a). Example of the Electrowinning Performance on Copper Sulfate Static Rinse

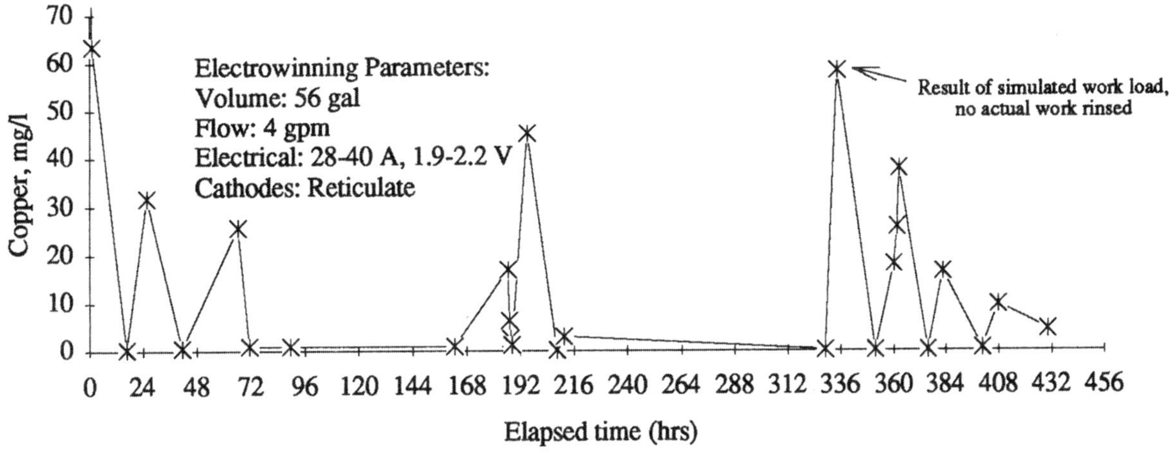

Exhibit 4-39(b). Work Load Through Copper Sulfate Static Rinse

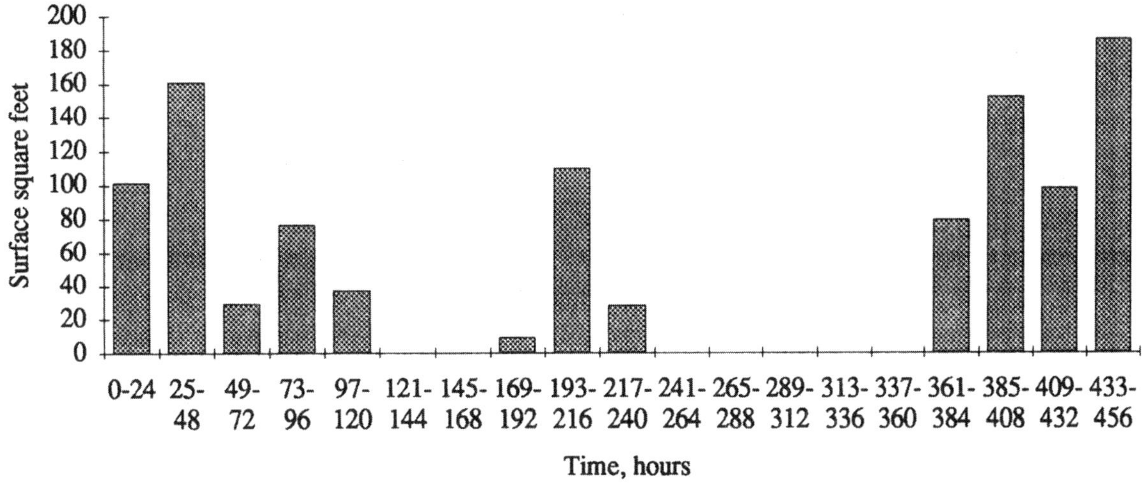

Data provided by Jennie Koff, Naval Facilities Engineering Service Center, Port Hueneme, CA.

tion measured in the dragout rinse during the test was 64 mg/l Cu. The copper concentration invariably fell to less than 1 mg/l overnight and during any idle period of a few hours duration. During one segment of the test, the copper concentration fell from 16 mg/l to 1.5 mg/l in 2 hours; during another segment, the copper concentration fell from 25.7 mg/l to less then 1 mg/l in 5 hours. The data suggest that for the conditions present at this facility, the copper concentrations will generally remain below 60 mg/l in the dragout rinse and will reach 1 mg/l or less within approximately 5 hours or less after plating has ceased.

A reticulate and disposable cathode is often used for gold electrowinning. A small commercial unit, operating with only 25 amps output, is shown in Exhibit 4-40. The cathode of this unit is placed directly into a small dragout tank. This unit is applicable to the recovery of most precious metals. It was used by four respondents to the Users Survey. The metal deposited onto the cathode is recovered chemically and/or thermally (dissolved in acid from cathode, precipitated, then melted or simply melted from the cathode) (ref. Gold Bug File).

Exhibit 4-40. Diagram of a Precious Metal Electrowinning In-Tank Unit (Courtesy of Precious Metals Processing Consultants, Inc.)

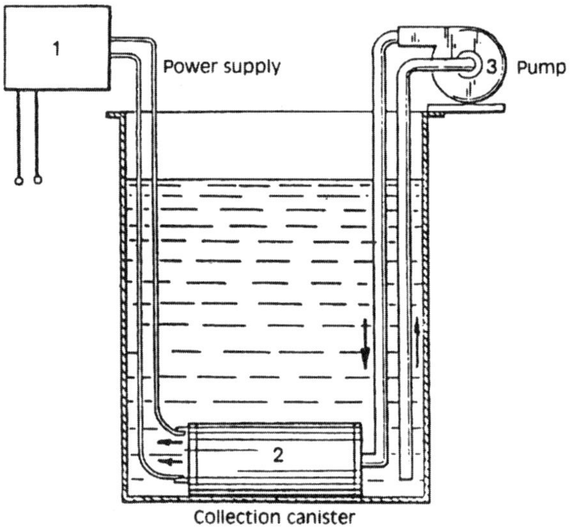

4.5.4.4 High Surface Area Cathode Units

High surface area units are used in rinsing operations, where low concentrations of metals are desired. The advantage of maintaining a lower equilibrium concentration is two fold; first, the percentage of material recovered is increased and second, the free rinse after the recovery rinse may be sufficiently dilute to be sewered without treatment. High surface area units extract the metal onto cathodes made of fibrous material such as carbon. The high surface area allows for metal removal at solution concentrations much lower than flat plate cathode types and even the reticulate types. The fiber cathode is regenerated by passage of a strip solution through the unit and reversal of the current. Plating solutions can sometimes be used as the strip solution and returned to the bath for reuse. More commonly, the concentrated metals in the strip solution are removed by a second electrolytic unit, employing conventional electrowinning.

One commercially available carbon-fiber cathode system employs a three dimensional flow-through type assembly, consisting of carbon fibers woven into layers of fabric secured to the electrical distribution feeder sheets in a plastic coated frame (ref. 128 and Baker Brothers file). The high surface area units have been mostly applied to recovery of metals from the rinses of cyanide based plating processes (e.g., cadmium, copper, zinc, gold and silver). These units remove metal ions to low concentrations and also oxidize the cyanide in the rinse water. Other applications noted in the literature include: copper etch, electroless copper, acid gold, acid silver, tin-lead fluoborate and tin-lead sulfate solutions.

Cyanide oxidation with HSA units can be performed with the addition of sodium chloride electrolyte to the rinse, although the practicality of the process is not widely accepted. With this method, the chloride ions are oxidized to chlorine at the anode and react with cyanide in the rinse (ref. 39).

4.5.4.5 Other Equipment/Operational Considerations

Various design methods are used in commercial equipment to achieve agitation and reduce the impact of concentration polarization. One manufacturer (ref. Eco-Tec file) advertises the use of convection air agitation that directs a uniform curtain of fine air bubbles across the face of the cathode and thereby bringing a constant supply of fresh solution to the cathode surface. According to the manufacturer, the improved agitation permits close anode to cathode spacing (1 in.) which reduces the IR (ohmic) resistive voltage drop across the cell, resulting in lower energy consumption. It also reduces the overall size of the electrowinning unit for a given cathode area requirement. Another manufacturer uses a fluidized bed design to achieve agitation (ref. BEWT file). With this design, mesh metal electrodes sit in a bed of inert glass beads, which is fluidized by the action of the pumped electrolyte. The scouring action of the beads against the mesh electrodes unit provides agitation to reduce concentration polarization and improves the quality of the deposit. Due to the mesh design, the deposit cannot be mechanically removed from the cathodes. Rather they are placed into specially designed anode bags and put into plating tanks, where they function as anodes. This equipment is advertised for recovery of nickel, nickel-iron, zinc, cadmium, silver and gold. Most of the electrowinning units manufactured for silver recovery for use in the photographic industry employ a rotating cylindrical cathode. Rotating the cathode provides the needed agitation at the interface between the cathode and the solution.

Several types of controls are available with electrowinning units. Inexpensive units usually have just an on/off switch as the only means of current control. Such equipment may be satisfactory if the solution variables remain relatively constant. Many units have variable current control and a meter to indicate current flow in the solution. Sensor probes are available on some units which will automatically adjust the current to the metal concentration. Microprocessor controls are also offered by many manufacturers.

Nickel, although it is one of the most frequently plated metals, has traditionally not been recovered by electrowinning. This is partially due to the fact that alternative technologies exist for nickel recovery, but is also due to the difficulty of the nickel electrowinning process. The recovery of nickel using electrowinning has become more common in recent years owing to research and development. The reason for the difficulty with nickel is that the pH of the electrolyte (typically a sulfate media) will drop as the electrowinning process proceeds due to the electrode reaction (electrolysis) that produces hydrogen ions. As this occurs, the metal deposition rate will decrease and hydrogen production will continue to increase. For this reason, it is necessary to control the pH of the electrolyte. A nickel recovery system employing ion exchange and electrowinning is shown in Exhibit 4-41. With this process, the electrolyte is continuously circulating from the cell to an adjustment tank where the critical operating parameters are controlled. This includes caustic addition for pH control. Another reference suggests the use of ammonia for adjusting the pH of Watts, Woods and sulfamate nickel baths. Note that the overall recovery system in Exhibit 4-41 includes the recovery of metal from both electroplating and electroless plating processes. A selective ion exchange column is used prior to electrowinning to separate the nickel from the chelates contained in the electroless bath and rinses.

4.5.5 Costs

4.5.5.1 Capital Costs

The capacity requirement for conventional electrowinning depends most heavily on the amount of metal to be recovered and the rate of metal deposition. Factors that influence the rate of metal deposition are (ref. 39 and 128):

- Electrode type and area
- Agitation rate (or in general, mass transfer)
- Solution chemistry
- Electrical variables
- Temperature

The cathode surface area depends on the size and number of the cathodes employed in the unit. The agitation rate, average metal concentration in the rinse solution, solution conductivity and temperature all influence the current density that can be maintained and still result in an even, homogeneous metal deposit on the cathode. The higher the current density allowed, the higher the rate of metal deposition per unit area of cathode (ref. 39). Exhibits 4-42 and 4-43 present useful data for cathode sizing exercises.

A nearly linear relationship between cost and capacity is displayed in Exhibit 4-44. Capital costs, therefore, can be estimated once capacity requirements are determined. Most vendors refer to capacity in terms of amperage; more precisely, the maximum amperage setting on a unit's rectifier. Less commonly, capacity is expressed in terms of total cathode area. The rectifier and electrodes comprise the majority of the cost for most units; other contributing components are the fluid containment tank, pumps, filters and optional metering devices.

The strategy for determining the appropriate capacity of an electrowinning unit for a specific application is straightforward in theory: match the expected plate-out rate of the unit with the application's waste metal generation rate. For dragout tank applications, such as those diagrammed in Exhibit 4-32, the rate at which metal is introduced into the tank is determined either by direct analysis, or by a method such as that proposed in Section 3.5.3.1. The plating rate of the metallic species in question is obtained from the Faraday Table presented in

Exhibit 4-41. Electrowinning System Applicable to Nickel and Electroless Nickel Plating Operations

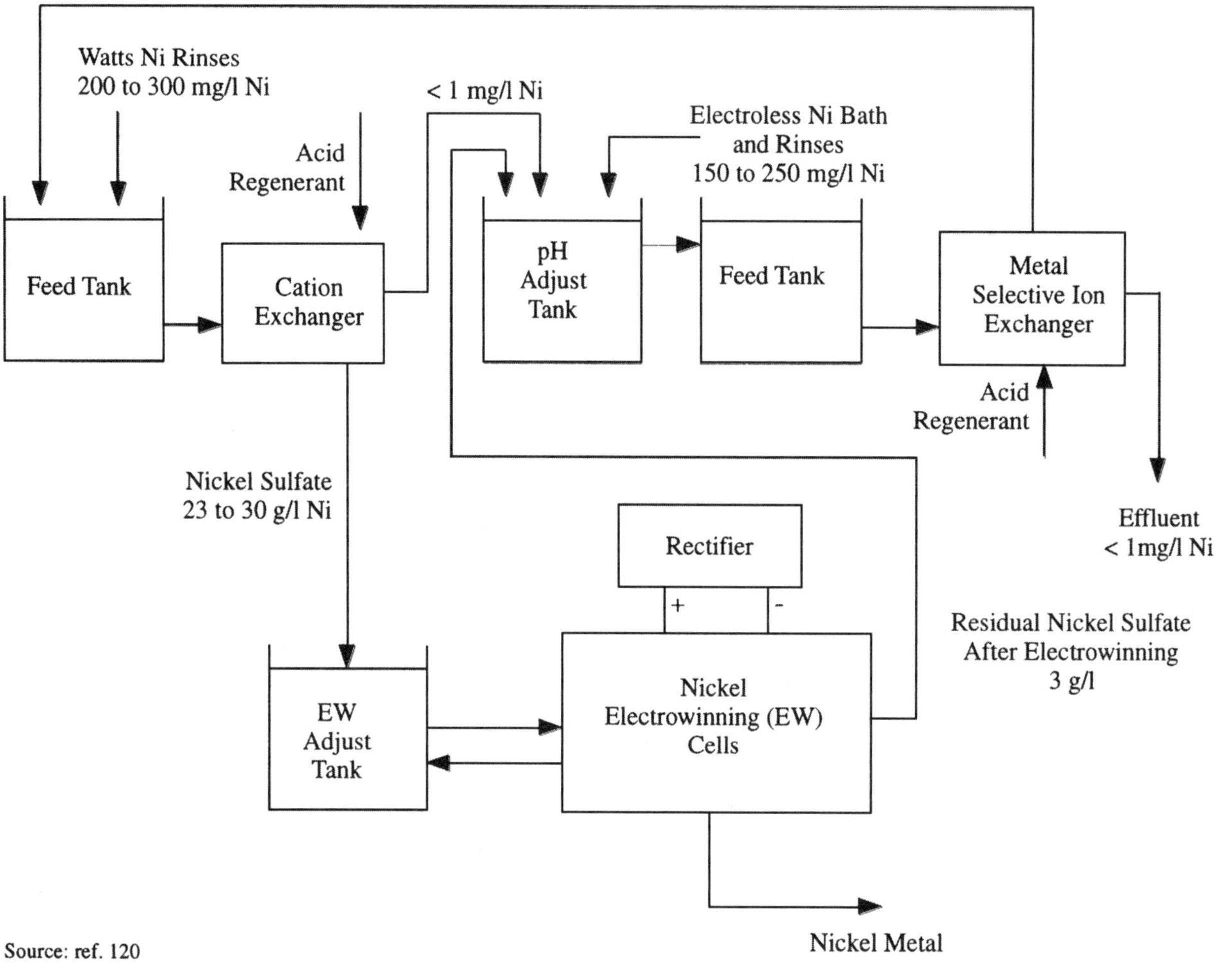

Source: ref. 120

Exhibit 4-42. Faraday Table for Common Metals

Element	Symbol	Density g/cc	Gram Atomic Weight	Valence	Gram Equivalent Weight	Deposition Rate g/amp-hr	Current Required amp-hr/g
Cadmium	Cd	8.65	112	2	56	2.10	0.476
Copper	Cu	8.96	63.5	1	63.5	2.37	0.422
Copper	Cu	8.96	63.5	2	31.8	1.19	0.840
Gold	Au	19.3	197	1	197	7.35	0.136
Gold	Au	19.3	197	3	65.7	2.45	0.408
Lead	Pb	11.3	207	2	104	3.87	0.258
Platinum	Pt	21.5	195	4	48.8	1.82	0.550
Nickel	Ni	8.90	58.7	2	29.4	1.10	0.909
Rhodium	Rh	12.4	103	3	34.3	1.28	0.781
Silver	Ag	10.5	108	1	108	4.03	0.248
Tin	Sn	7.30	119	2	59.5	2.21	0.452
Tin	Sn	7.30	119	4	29.8	1.11	0.901
Zinc	Zn	7.13	65.4	2	32.7	1.18	0.847

Conversions:
1 Faraday = 96,500 coulombs = 26.8 ampere-hours; gram equivalent weight/26.8 = grams/ampere-hour; density X 2.36 = gram/mil-square feet

Exhibit 4-42. Capacity requirement in amperes, therefore, is the introduction rate in g/hr divided by the plating rate in g/amp-hr. This quotient requires an adjustment for expected current efficiency (that portion of current available to the cell that actually is employed reducing the target metal on the cathode) before serving as a reliable guide for required capacity. Current efficiency for electrolytes with high metallic concentrations will approach theoretical levels, but it may range down below < 10% of the theoretical rate for electrolytes concentrations of metal below 100 mg/l.

Other application configurations lend themselves to similar analysis. For units employed to electrowin metal from ion-exchange regenerant, the volume and concentration of the regenerant is required for capacity sizing. These quantities will be known from the analyses required for ion exchange sizing. The time available for electrowinning is limited by the time between regenerations. Spent regenerant may be contaminated by several species of metallic ions; this will make the calculation of appropriate cell amperage less accurate. For batch dumps, concentration of metal in the spent bath and the dump period are usually available data. The following formula applies to regenerant or process batches alike.

amps = [g/batch]/[g/amp-hr]/[hours/batch or cycle]

Cell amperage is not likely to approach the maximum rectifier output. It is limited by the maximum practical cathode current density for the metallic species being electrowinned which, in turn, is a function of metallic concentration. Thus, in practice, it is the total cathode area and the concentration of metal in the electrolyte that best determines the practical amperage capacity of a unit. Exhibit 4-43 demonstrates the considerable effect of concentration (and temperature) on maximum current density for copper electrowinning; similar curves are to be expected for other metals.

Exhibit 4-43. Maximum Current Density for Copper Electrowinning

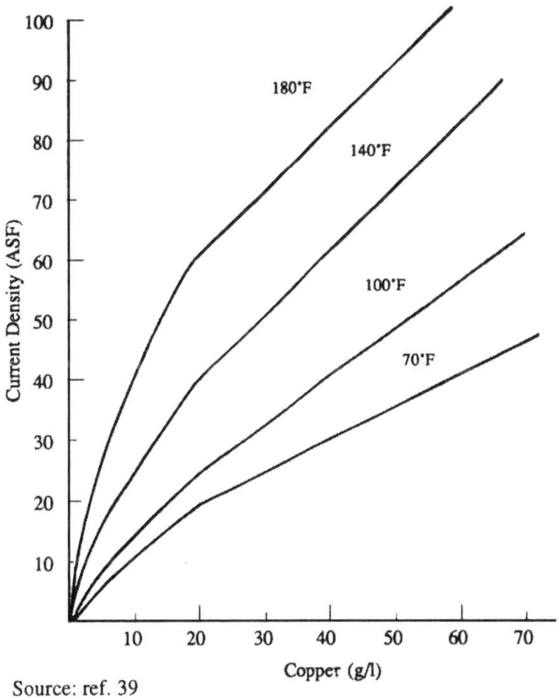

Source: ref. 39

Low concentration electrolytes present significant sizing implications. Plating proceeds at a fraction of high concentration current densities and efficiency is lost as current is wasted reducing hydrogen at the cathodes. Units lacking design features aimed at reducing the thickness

Exhibit 4-44. Capitol Costs for Electrowinning Units

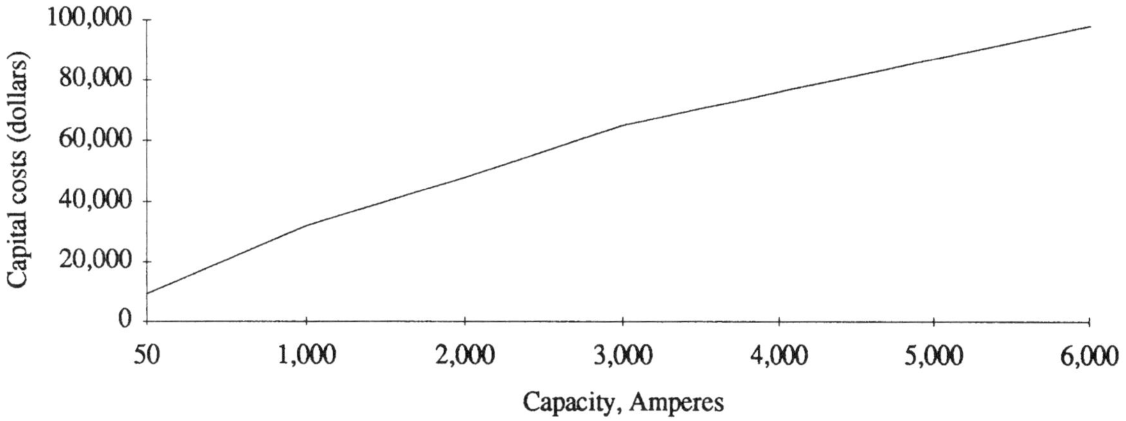

of the polarization depletion layer and not equipped with high surface area cathodes will be least efficient in these environments. At concentrations of <100 mg/l, a reasonable current efficiency estimate may be 10% or less. These factors must be offset with larger capacity units. Success at low concentrations will also depend on the metallic species being electrowinned, the presence of multi-valent cations in the electrolyte (such as tin and iron, which further lower efficiency by oxidizing to higher valence at the anode and reducing to lower valence at the cathode, thereby wasting current and yet staying in solution) and the time available for electrowinning (eventually, units so designed can reduce concentrations of certain metals to below compliance levels).

Anode and cathode construction will significantly impact the cost of an electrowinning unit. A list of cathode and anode materials is displayed in Exhibit 4-36. Materials options for a specific application are limited by the peculiarities of the electrolyte being electrowinned and by the manufacturer of the unit. Cost differences can be significant; e.g., ELTECH Systems Corporation offers it's units with either graphite or DSA® (proprietary rare earth coating) anodes. The graphite anodes were quoted in 1993 at $80, while the DSA were $335. For a Retec 25 (26 anodes) this represents a cost difference of $6,630.

4.5.5.2 Operating Costs

Typical operating costs for this technology are shown in Exhibit 4-45. Respondents employing this technology reported their annual operating costs to be only $4,100/yr on average, roughly split between the labor and non-labor categories. This technology is not labor-intensive nor expensive to run. Operating costs components are labor, electrode replacement, maintenance and energy.

Labor costs are largely installation- and application-specific. Units used in batch configurations may require considerable solution transport, pre-adjustment of the electrolyte, cleaning of salt deposits and adjustment of batch releases from the unit. On the other hand, dedicated dragout rinse units treating fluids with lower total dissolved solids may require only occasional cleaning and rare maintenance of any kind. In all, little scheduled maintenance beyond cleaning and replacement of corroded connectors is to be expected.

Energy costs will comprise only a small percentage of total operating costs for most applications. For large units, however, energy costs may more significant in relation to total operating costs as other expenses benefit from economies of scale. Predicting energy costs for a given application is complicated by the fact that several variables affecting total power consumption are difficult or impossible to know prior to actually running the unit. Conductivity, required voltage, rectifier efficiency, and current efficiency are either unknown or will vary with time or batch to batch. Once the unit is operating, energy costs will become easy to assess. In the example given in Exhibit 4-46, electrical costs were $0.4198/lb of Cu; these costs reflect metal removal to <5 mg/l and should be typical of similar applications.

Exhibit 4-45. Operating and Maintenance Costs for Electrowinning Equipment

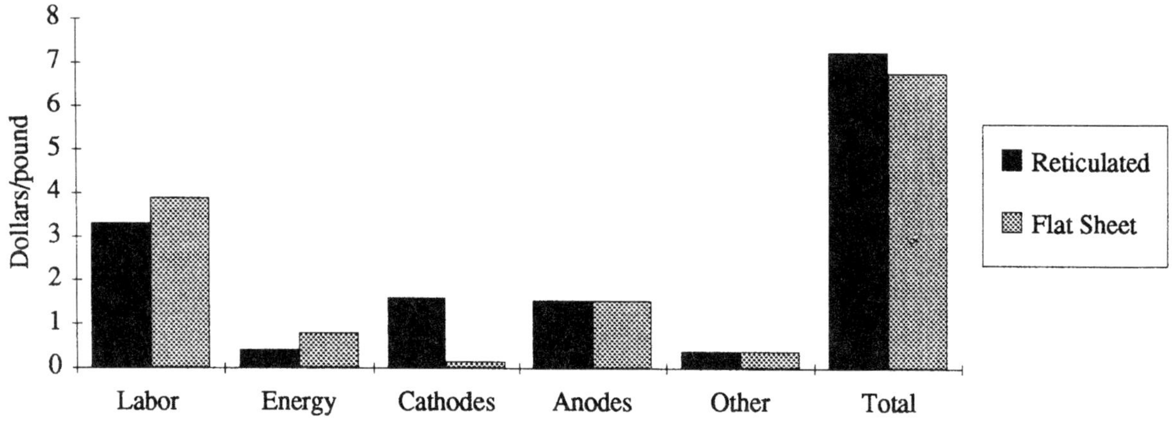

Electrode replacement costs depend on their construction and life expectancy. Stainless steel cathodes are usually peeled or scraped free of plating deposits and reused. Reticulated cathodes must be replaced after they accumulate sufficient metal to lower their effectiveness, roughly 5-10 pounds/sq. ft. of cathode area. Anodes may, as in the case of DSA® and other proprietary materials, be semi-permanent and last in excess of 5 years. Graphite anodes are well-known to "melt" or gradually exfoliate carbon particles into the electrolyte; this obviously shortens their operating life.

The example offered in Exhibit 4-46 has total operating costs at $6.81/lb of Cu. Unlike a dragout rinse application, considerable labor was involved transporting and adjusting spent process baths prior to electrowinning. Labor, at $3.64/lb, was by far the largest operating cost component.

4.5.6 Performance Experience

A partial summary of the User Survey data relative to electrowinning is presented in Exhibit 4-47. There are a number of general observations that can be made from these data and other data contained in the Users Survey database and literature:

- The average satisfaction level for the electrowinning technology is 3.1 (on a scale of 1 to 5, with 5 being most satisfactory), which is lower than the average level rating for all recovery technologies. Fifty-six percent of the shops indicated that this technology satisfied the need for which it was purchased and another 15% indicated that it partially satisfied the purchase need. The following is a breakdown of the reasons why shops purchased this technology:

To meet of help meet effluent regulations:	38
To reduce plating chemical purchases:	9
To reduce the quantity of waste shipped off-site:	20
To reduce wastewater treatment costs:	20
To improve product quantity:	1
Other (mostly to recovery valuable metals):	8

- The use of electrowinning for metal recovery generally did not impact production quality or the rate of production. The following responses were provided:

	Product Quality	Production Rate
Improved	0	1
No Change	55	50
Decreased	1	5

(Where product or production impacts occurred, the respondents did not provide any details of the impacts.)

- Most plating shops indicated that based on their experience with this technology, if given the chance, they would purchase the same type of equipment from the same vendor. The following is a breakdown of their responses:

Purchase the same technology from the same vendor:	35
Purchase the same technology from a different vendor:	4

Exhibit 4-46. An Example of Calculating Operating Costs for an Electrowinning Application

Application: Batch treatment of copper-containing spent process and dragout solution, including acid copper sulfate and potassium persulfate.

Capitol Costs:

Cost of Unit	$22,000
Installation[a]	5,100
Total	**$27,100**

Operating Costs/Year:

Energy	$450
Labor	3,900
Cathodes[b]	895
Anodes[c]	1,650
Parts	300
Chemistry[d]	750
Scrap metal value[e]	(643)
Total	**$7,302**

Savings/Year:

Waste treatment savings[f]	$2,700
Metal recovered	1,072 lb.

[a] Unit required a fume hood and exhaust system due to the generation of SO_2 gas during reduction of persulfate, and cost figure includes a pH adjust tank with mixer.
[b] Reticulated, cathode held approximately 10 lbs of Cu before replacement.
[c] Annualized, assuming 5 year life. No anodes were actually procured.
[d] Sodium hydroxide and sodium meta-bisulfite.
[e] Received average of $0.60/lb. Metal price averaged $1.00 during year.
[f] Labor-intensive batch precipitation method.

Exhibit 4-47. Summary of Users Survey Data for Electrowinning

Shop Code	Application	Vendor	Year Purchased	Equipment Costs: Equip.	Equipment Costs: Other Cap.	Equipment Costs: Total	Annual Operating Costs: Non-Labor	Annual Operating Costs: Hrs	Annual Operating Costs: Labor	Annual Operating Costs: Total	Annual Savings: Bath Chem.	Annual Savings: Treat. Chem.	Annual Savings: Disposal	Annual Savings: Other	Annual Savings: Total	Use Code	Sat. Level	Future Decision	Shop Code
008	Cadmium, Cyanide	BEWT	1989	$25,000	$8,000	$33,000	ND	500	$6,000	$6,000	ND	ND	ND	ND	ND	2	1	4	008
008	Zinc, Non-Cyanide	BEWT	1989	$25,000	$8,000	$33,000	ND	500	$6,000	$6,000	ND	ND	ND	ND	ND	2	1	3	008
009	Cadmium, Cyanide	In-House	ND	$5,000	$1,000	$6,000	ND	50	$750	$750	ND	ND	ND	ND	ND	1	3	3	009
012	Cadmium, Cyanide	HSA	1981	$80,000	$20,000	$100,000	ND	ND	ND	ND	ND	ND	ND	ND	ND	2	ND	3	012
023	Cadmium, Cyanide	Eltech	1992	$19,000	ND	$19,000	$700	20	$260	$960	ND	$5,330	$1,000	$4,290	$10,620	1	4	1	023
025	Cadmium, Cyanide	In-house	1986	$15,000	ND	$15,000	$7,000	120	$1,080	$8,080	ND	ND	ND	ND	ND	1	4	1	025
036	Copper, Cyanide	In-House	1984	$4,000	ND	$4,000	ND	ND	ND	ND	ND	ND	ND	ND	ND	2	1	3	036
039	Cadmium, Cyanide	HSA Reactors	1983	$32,500	$44,000	$76,500	$20,000	1000	$18,000	$38,000	ND	$250	$500	ND	$750	3	2	3	039
041	Copper, Sulfate	In-House	1990	$4,000	$6,000	$10,000	$2,000	100	$1,500	$3,500	$144	$250	$4,500	ND	$4,894	1	3	1	041
043	Z inc, Cyanide	In-House	ND	ND	ND	ND	ND	ND	ND	ND	ND	ND	ND	ND	ND	1	ND	4	043
053	Cadmium,Cyanide	Covofinish Co.	1989	ND	ND	ND	ND	100	$1,400	$1,400	ND	ND	ND	ND	ND	3	2	3	053
061	Zinc,Acid	not given	ND	ND	ND	ND	ND	ND	ND	ND	ND	ND	ND	ND	ND	1	ND	3	061
081	Cadmium,Cyanide	Precious Metals Pr	1991	$10,000	$2,000	$12,000	$2,900	24	$240	$3,140	ND	ND	ND	ND	ND	1	4	1	081
086	Cadmium,Cyanide	HSA	1982	$90,000	$60,000	$150,000	$7,000	250	$2,750	$9,750	ND	ND	ND	ND	ND	2	1	4	086
086	Zinc,Cyanide	HSA	1982	$90,000	$60,000	$150,000	$7,000	250	$2,750	$9,750	ND	ND	ND	ND	ND	1	ND	2	086
090	Cadmium,Cyanide	Eltech	1984	$15,000	$1,000	$16,000	$1,900	250	$6,000	$7,900	ND	ND	ND	ND	ND	2	1	3	090
091	Cadmium,Cyanide	Eltech	1986	$10,000	$16,000	$26,000	$1,050	150	$2,250	$3,300	ND	$1,000	$3,000	ND	$4,000	1	3	1	091
107	Zinc, Non-Cyanide	Eltech	1988	$6,500	$1,100	$7,600	ND	160	$12,800	$12,800	ND	$500	ND	ND	$500	3	3	1	107
114	Silver Cyanide	In-House	1983	$1,000	$100	$1,100	ND	60	$5,400	$5,400	$800	ND	ND	ND	$800	1	4	1	114
119	Gold, Non-Cyanide	Precious Metals Pr	1990	$2,600	$300	$2,900	$300	24	$288	$588	ND	ND	ND	$15,900	$15,900	1	4	1	119
123	Silver, Cyanide	In-House	1980	$1,000	$500	$1,500	$100	200	$2,400	$2,500	ND	ND	ND	ND	ND	1	4	1	123
124	Cadmium, Cyanide	HSA Reactors	1981	$64,692	ND	$64,692	ND	ND	ND	ND	ND	ND	ND	ND	ND	2	2	3	124
124	Z inc, Cyanide	HSA Reactors	1982	$37,000	$28,000	$65,000	ND	ND	ND	ND	ND	ND	ND	ND	ND	2	1	3	124
128	Passivation, Nitric	Serfilco	1988	$3,500	$1,500	$5,000	ND	ND	ND	ND	ND	ND	ND	ND	ND	2	1	3	128
139	Brass	Technic Inc.	1985	$2,500	$625	$3,125	$125	15	$240	$365	ND	ND	ND	ND	ND	1	4	1	139
139	Cadmium, Cyanide	Technic Inc.	1985	$2,500	$625	$3,125	$125	15	$240	$365	ND	ND	ND	ND	ND	1	4	1	139
139	Copper, Cyanide	Technic Inc.	1985	$2,500	$625	$3,125	$125	15	$240	$365	ND	ND	ND	ND	ND	1	4	1	139
139	Silver, Cyanide	Technic Inc.	1985	$2,500	$625	$3,125	$125	15	$240	$365	ND	ND	ND	ND	ND	1	4	1	139
146	Nickel	In-House	1992	$2,000	$1,000	$3,000	$2,000	300	$2,400	$4,400	ND	$1,000	$5,000	$1,000	$7,000	1	4	1	146
146	Zinc,Cyanide	In-House	1989	$1,000	$400	$1,400	$3,000	ND	ND	$3,000	ND	ND	ND	ND	ND	3	2	1	146
164	Nickel, Electroless Phos	Eltech	1993	$8,000	$3,000	$11,000	ND	ND	ND	ND	ND	ND	ND	ND	ND	3	ND	1	164
167	Anodize, Hardcoat	Eltech	1990	$7,500	$1,000	$8,500	$100	50	$400	$500	ND	ND	ND	ND	ND	2	2	3	167
174	Cadmium, Cyanide	In-House	1992	ND	ND	ND	ND	ND	ND	ND	ND	ND	ND	ND	ND	1	4	1	174
176	Copper, Strike (CN)	Clinton	1988	$2,000	$300	$2,300	ND	60	$600	$600	ND	ND	ND	ND	ND	1	4	1	176
176	Silver, Cyanide	Precious Metals Pr	1991	$1,500	ND	$1,500	ND	50	ND	ND	$1,500	ND	ND	ND	$1,500	3	4	1	176
179	Silver, Cyanide	RFE Industries Inc	1987	$6,000	ND	$6,000	$2,000	24	$192	$2,192	ND	ND	ND	$18,000	$18,000	1	5	1	179
184	Cadmium, Cyanide	Eltech	ND	$14,000	$31,000	$45,000	ND	250	$2,500	$2,500	ND	ND	ND	ND	ND	2	2	1	184
188	Cadmium, Cyanide	MRT	1985	$3,000	ND	$3,000	ND	30	$450	$450	ND	ND	ND	ND	ND	1	4	1	188
188	Nickel, Sulfamate	MRT	1985	$3,000	ND	$3,000	ND	20	$300	$300	ND	ND	ND	ND	ND	1	4	1	188
189	Copper, Cyanide	ND	1989	$2,000	$700	$2,700	$750	80	$1,600	$2,350	ND	ND	ND	$200	$200	1	3	1	189
196	Cadmium, Cyanide	Eltech	1989	$5,000	$2,000	$7,000	$500	50	$500	$1,000	ND	$1,500	$200	$2,100	$3,800	1	4	1	196
213	Copper, Sulfate	Napco	1988	$4,731	ND	$4,731	ND	ND	ND	ND	ND	ND	ND	ND	ND	3	3	2	213
221	Silver, Cyanide	Jaynor	1988	$2,300	ND	$2,300	$50	52	$520	570	ND	$500	$200	$200	$900	1	5	1	221
223	Brass	ND	1991	$10,000	$10,000	$20,000	$5,000	150	$4,500	9500	ND	$40,000	$20,000	$8,000	$68,000	3	3	2	223
245	Silver, Cyanide	Jaynor	1989	$1,500	$500	$2,000	$500	15	$195	695	ND	ND	ND	$3,000	$3,000	1	4	1	245
252	Copper, Cyanide	Serfilco	1986	$5,000	$200	$5,200	$1,000	40	$3,600	4600	ND	ND	ND	$200	$200	1	3	2	252
254	Cadmium, Cyanide	Eltech	1988	$14,000	$2,000	$16,000	$2,400	ND	ND	2400	ND	ND	ND	ND	ND	2	4	1	254
262	Cadmium, Cyanide	Serfilco Ltd.	1990	$2,500	$500	$3,000	$50	20	$300	350	ND	$300	$400	ND	$700	1	4	1	262
273	In-House	In-House	1950	$3,000	$5,264	$8,264	ND	ND	ND	ND	ND	ND	ND	ND	ND	1	4	1	273
276	Cadmium, Cyanide	HSA	1979	$40,000	ND	$40,000	ND	ND	ND	ND	ND	ND	ND	ND	ND	2	ND	4	276
285	Silver, Cyanide	Jaynor	1985	$2,500	ND	$2,500	ND	ND	ND	ND	ND	ND	ND	ND	ND	1	5	1	285
288	Copper, Sulfate	In-House	1990	$5,000	$2,000	$7,000	$1,000	150	$2,250	3250	$1,000	$7,200	$7,500	ND	$15,700	1	3	1	288
288	Zinc, Non-Cyanide	In-House	1982	$1,000	$200	$1,200	$600	300	$6,000	6600	ND	ND	ND	ND	ND	2	ND	1	288
290	Nickel, Sulfamate	In-House	1992	ND	ND	ND	ND	ND	ND	ND	ND	ND	ND	ND	ND	1	4	1	290
295	Gold, Non-Cyanide	Precious Metals Pr	1991	$3,000	$300	$3,300	$1,200	5	$250	1450	ND	ND	ND	ND	ND	1	4	1	295
300	Cadmium, Cyanide	HSA	1983	$96,710	$48,000	$144,710	$3,000	ND	ND	3000	ND	ND	ND	ND	ND	2	1	3	300
308	Copper, Sulfate	ND	ND	ND	ND	ND	ND	ND	ND	ND	ND	ND	ND	ND	ND	1	ND	1	308
316	Zinc, Cyanide	Clev. Fluid	1990	$35,000	$3,000	$38,000	$2,500	100	$2,000	4500	$4,000	$6,000	$19,000	$7,200	$36,200	1	4	1	316
318	Cadmium, Cyanide	HSA	1980	$20,000	$2,500	$22,500	ND	ND	ND	ND	ND	ND	ND	ND	ND	2	1	3	318
	Average		1986	$16,076	$9,119	$25,195	$2,378	139	$2,548	$4,178	$1,489	$5,319	$5,573	$5,463	$10,140		3.1		
	Average (excluding HSA)		1987	$6,717	$2,480	$9,198	$1,396	117	$2,142	$3,353	$1,489	$5,780	$6,105	$5,463	$10,690		3.4		

(1) Other capital costs include installation and ancillary equipment. (2) Use Codes: 1-currently operating; 2-not currently operating and no intension for future use; 3-not currently in use, but intend to use in future.

(3) Satisfaction levels (manufacturer and technology) 1 to 5, with 1=lowest and 5=highest.

(4) Future decision codes: 1-purchase the same technology from the same vendor; 2-purchase the same technology from a different vendor; 3-purchase a different technology; 4-do nothing.

(5) Average total equipment cost is the sum of average equipment and other capital costs. Average annual operating cost is the sum of average non-labor and labor costs. Average savings is simply the average of reported savings. This method of calculation probable understates the average annual savings of the users. (6) ND = No data

Purchase a different technology: 13
Do nothing: 4

- The major savings from the operation of electrowinning were reduced treatment chemical use, reduced sludge generation and the value of the recovered metal (especially precious metals and to a lesser extent, cadmium).

- Most of the respondents use electrowinning to recover metal from rinse water and incorporate dragout (or drag-in/dragout) tanks to concentrate the metals prior to electrowinning. The electrowinning system is either connected directly to the dragout tank or the dragout is periodically pumped to a side tank for electrowinning. Other configurations from the survey forms include: (1) an electrowinning unit recovered metal from a spray rinse (PS 184); (2) metal was recovered from a spent process solution (PS 039, PS 128, PS 164); (3) metal was recovered from a copper sulfate bath purge (i.e., used to control a buildup of metal concentration in bath) (PS 041).

- Nearly 50 percent of the respondents' applications of electrowinning were used for cadmium recovery.

- The most successful application of electrowinning appears to be precious metals recovery (based on the number of applications, the percentage of applications still in use, and the satisfaction level of the users). This includes the use of both commercial and home-made units. This is most likely due to the fact that noble metals are more easily recovered by electrowinning than common plated metals.

- All the HSA units purchased between 1979 and 1983 were purchased from a single manufacturer (HSA Reactors Ltd.). The average cost of these units was $66,360. The high capital cost was probably tolerated at the time because these units were advertised as compliance technologies rather than simple recovery methods (PS 276). As such, they were intended to fulfill a portion of a plant's conventional treatment requirements. These units received an average technology satisfaction level of 1.4. Asked what they would do if given the opportunity to repeat the technology selection process, the eight users of this technology indicated:

Purchase the same technology from the same vendor: 0
Purchase the same technology from a different vendor: 1
Purchase a different technology: 5
Do nothing: 2

Only one of these early HSA units is currently operating (PS 086) and that unit was extensively modified by the user by replacing the carbon cathodes with corrugated steel panels and by removing the heat exchanger and the cyanide destruct module. Some shops indicated that the performance of the HSA system was good, when it was operating (PS 012, PS 124, PS 276). Other shops reported complete dissatisfaction, e.g., "it has been a major expense and headache ... too much downtime" (PS 086). Another problem cited with this equipment was the competing nature of the cyanide destruction and metal removal processes. As reported by PS 039, the cyanide destruction process reduced the technology's ability to remove cadmium. PS 086 also cited this problem and reported that they abandoned this portion of the technology.

Of the various "black box" technologies utilized during the late 1970's and early 1980's, a period when plating shops were installing equipment to meet the new Federal effluent standards, the high surface area units probably had a negative impact on the utilization of advanced technology. Following this time period, the plating industry moved in the direction of conventional treatment with sludge dewatering and dehydration and off-site metals reclamation.

- Two HSA type units were purchased from a manufacturer other than HSA Reactors Ltd. in 1985 (PS 188). These are much lower cost units ($3,000) that are still in use and received higher than average satisfaction ratings.

- Eltech International Corporation, a producer of reticulated cathode units, manufactured more of the electrowinning units reported in the survey forms than any other manufacturer.

The average technology satisfaction level for these units was 2.9, slightly less than average. However, asked what they would do if given the opportunity to repeat the technology selection process, the users of this technology indicated:

Purchase the same technology from the same vendor: 8
Purchase the same technology from a different vendor: 0

Purchase a different technology: 1
Do nothing: 0

- Some performance failures of the electrowinning technology can be attributed to misapplications by the user. This is especially true with the use of home-made units and units purchased from manufacturers' representatives rather than the manufacturer. For example, PS 128 purchased an electrowinning unit from a manufacturer's representative to recover copper, nickel and chromium (chromium cannot be recovered using electrowinning because very high concentrations of chromium are required for a deposit to form) from a spent nitric acid solution (inappropriate electrolyte). The user indicated that he intended to recover the metals and reuse the nitric acid. The supplier-stated capacity of the unit, according to the user, was 8 gpm (flow-through is an inappropriate application). The result was that "it did not work" and "it fumed." Another shop (PS 146) that modified an old commercial unit indicated that they could not determine the proper electrical settings for its use. They were using the unit with a 5 gpm flow-through of zinc cyanide rinse water (flow-through is an inappropriate application because it does not permit sufficient time for the metal to be plated-out). That same shop indicated in their survey form that they are planning to use their electrowinning unit in the future for chromium recovery.

- Approximately 26% of the electrowinning units used by respondents were constructed in-house (where the manufacturer was not given, those data were not used in the percentage calculation). These units received a higher average satisfaction level than the commercial units (3.6 vs 3.1), although plating shops with home-made electrowinning units gave mixed performance reviews. The capacity and quality of the components that went into these units probably had a significant bearing on performance. Also, the lack of technical support available led to misapplications and unsolvable problems. For example, PS 025 spent $ 15,000 on their equipment in 1986. This unit was still running at the time of the survey (1993) and the shop gave the unit a satisfaction level of 4. On the other hand, PS 036 constructed a unit in 1984 using components available in-house and purchased iridium coated anodes ($4,000) from an anode supplier. The unit had a 60 to 70 percent downtime, reportedly was labor-intensive and their efforts were abandoned in four months. This unit was applied to treatment of dragout and spent bath (copper cyanide). PS 041 constructed a unit for $4,000 in 1990. Although this unit is still operating, it has a downtime of 20% and is ineffective in removing copper unless the copper sulfate concentration is 75 g/l or higher (unit used to treat bath purge, lower copper concentration and electrowinning discharge is returned to bath). PS 043 abandoned their home-made unit because of problems with conductivity and sluffing off of metal (zinc) from the cathode.

4.5.7 Operational and Maintenance Experience

The following summarizes the respondents O&M experiences and provides operating labor information.

- The average number of annual operating hours per electrowinning unit were: 140 hrs/yr. The labor categories commonly used for operating this technology are wastewater treatment plant operator and trained technician. The following is a breakdown of the responses for skill requirements (based on data from 39 respondents):

Environmental Engineer: 4
Process/Chemical Engineer: 5
Chemist: 12
Consultant: 2
Plumber/Pipe Fitter: 13
Electrician: 11
Vendor: 3
Senior Level Plater: 10
Junior Level Plater: 14
Wastewater Treatment Plant Operator: 20
Trained Technician: 17
Common Labor: 10

- The average percentage of downtime for this technology experienced by the respondents was 20 percent.

- The most frequent and significant operational and maintenance problems identified with electrowinning include: labor intensive to clean (e.g., anode cleaning, electrode contact cleaning) (PS 025, PS 053,); high level of fuming or gassing (PS 036, PS 053, PS 128); sluffing off of deposit from cathode

(PS 043); temperature build-up (PS 036); salting of the electrolyte (PS 036); anodes polarize at high current density and deteriorate or are attacked by chemicals (PS 213); anodes passivate (PS 239); and deterioration of fiber cathodes (PS 086).

- Approximately 40% of the total number of electrowinning units reported in the survey forms are no longer in use.

- Some shops reported that poor support from the manufacturer was partially the cause of their system failure. PS 008 indicated that their zinc recovery unit was removed after 4 to 5 months of operation because they could not get help with system problems ("everything went wrong ... like pulling teeth to get help").

- Two of the shops that purchased equipment from HSA Reactors Ltd. indicated that the fact the company went out of business led to the failure of their systems (PS 012, PS 276). The carbon cathodes and other equipment components used in their products were too unique to find elsewhere. Also, users cited numerous mechanical problems (e.g., pump failures-PS 086) with these units and they complained about the fragile nature, short life and high cost of electrodes (PS 039, PS 086). The labor costs for HSA systems appear to be higher than the average electrowinning system, which points to the complexity of the systems. Further, the operation of the equipment required a moderate level of expertise. As one shop (PS 276) indicated, it was "too technical for our people." Generally, it is observed that this technology was probably sound from a theoretical standpoint, but lacked good engineering design and components. A similar conclusion is presented in a report sponsored by the Canadian Branch of the AESF (ref. 351). That report concluded that such a system could be operated in a manual mode, but that more developmental work was needed before an adequate automated system could be marketed.

- Some facilities have added sodium chloride to the electrolyte to increase the efficiency of cyanide destruction (PS 036). One shop reported unsatisfactory results, because the solution temperature increased and gassing occurred and the residual cyanide level was too high for sewering (PS 036).

- Shops that reported difficulty with the electrowinning process cited two possible technical causes, including: (1) contamination in electrolyte (PS 036, PS 090) and (2) build-up of carbonates (PS 036).

- Shops using the small precious metals recovery units (both commercial and in-house) reported none or very few operational and maintenance problems.

4.5.8 Residuals Generation

The standard flat plate cathode unit recovers metals in the form of metal sheets 1/8 to 3/8 inch thick. Some shops (e.g., PS 053) described their deposit as a metal sponge rather than a metal sheet. Depending on the purity of the deposit, which in turn will depend on the purity of the solution being recovered, these residuals can be either used as anodes, sold to a scrap dealer or refiner for recycle, or disposed of in a landfill. When sold as scrap, the plater receives approximately 50 to 90% of the commodity price for the material, depending mostly on quantity and purity. Some electrowinning cathodes are specifically designed to be hung in plating tanks after they are coated with metal and serve as anodes. This eliminates the need to mechanically remove the deposit.

The reticulate cathodes cannot be used as anodes. They must be either sent off-site for recovery or disposed of. Most of the shops responding to the survey that use reticulate cathodes send them to a scrap dealer. They receive approximately the same price as for flat plate cathode material.

The high surface area electrolytic units generate a strip solution that can be recycled to a plating bath, or more frequently, is recovered using conventional flat plate electrowinning.

4.6 ELECTRODIALYSIS

4.6.1 Overview

The electrodialysis technology employs ion-permeable and selective membranes under an applied d.c. potential difference to separate ionic species from an aqueous solution. Its primary application for chemical recovery is nickel plating, where it competes with recovery rinsing (dragout tanks), evaporation technologies and ion exchange. As such, it has not seen widespread use, despite being commercialized for this application for nearly fifteen years.

When used for nickel recovery, a concentrated rinse solution is pumped into an electrodialysis unit that is comprised of a stack of alternating cation- and anion-selective membranes, each separated by a spacer through which the solutions are allowed to flow (Exhibit 4-48). Due to the electrical potential applied across the stack, cations ($Ni2^{+}$) in solution migrate toward the cathode, and anions (SO_4^{2-} and Cl^{-}) toward the anode. When a cation encounters a cation-selective membrane, it passes through. As it continues to migrate, it will subsequently encounter an anion-selective membrane through which it cannot pass and will remain in that compartment. Similarly anions migrate toward the anode, pass through anion selective membranes and their travel is halted when they encounter a cation-selective membrane. This process results in alternating compartments between consecutive membranes becoming increasingly enriched (concentrate) and increasingly depleted (diluate) in ion concentration (ref. 399). Based on data presented in the literature, the feed stream (typically from a dragout tank) usually contains 2 to 5 g/l Ni, the concentrated return stream contains 30 to 60 g/l Ni and the depleted stream contains less than 200 mg/l Ni (ref. 32). The return stream is less concentrated than the bath (typically 70 to 90 g/l Ni). However, surface evaporation from the plating bath tank is generally sufficient to volumetrically accommodate complete assimilation. The percentage of nickel recovered from the feed stream is typically 90 to 95%.

A potential advantage of electrodialysis over other concentrating and return methods of nickel recovery (e.g., evaporation and reverse osmosis) is its ability to selectively retard the recovery of certain organic materials that tends to buildup in nickel plating baths, while more freely permitting the transport of a desirable organic bath constituent (saccharin) and nickel salts. This aspect of the process could reduce the frequency of bath purification as

Exhibit 4-48. Diagram of an Electrodialysis Process for Nickel Recovery

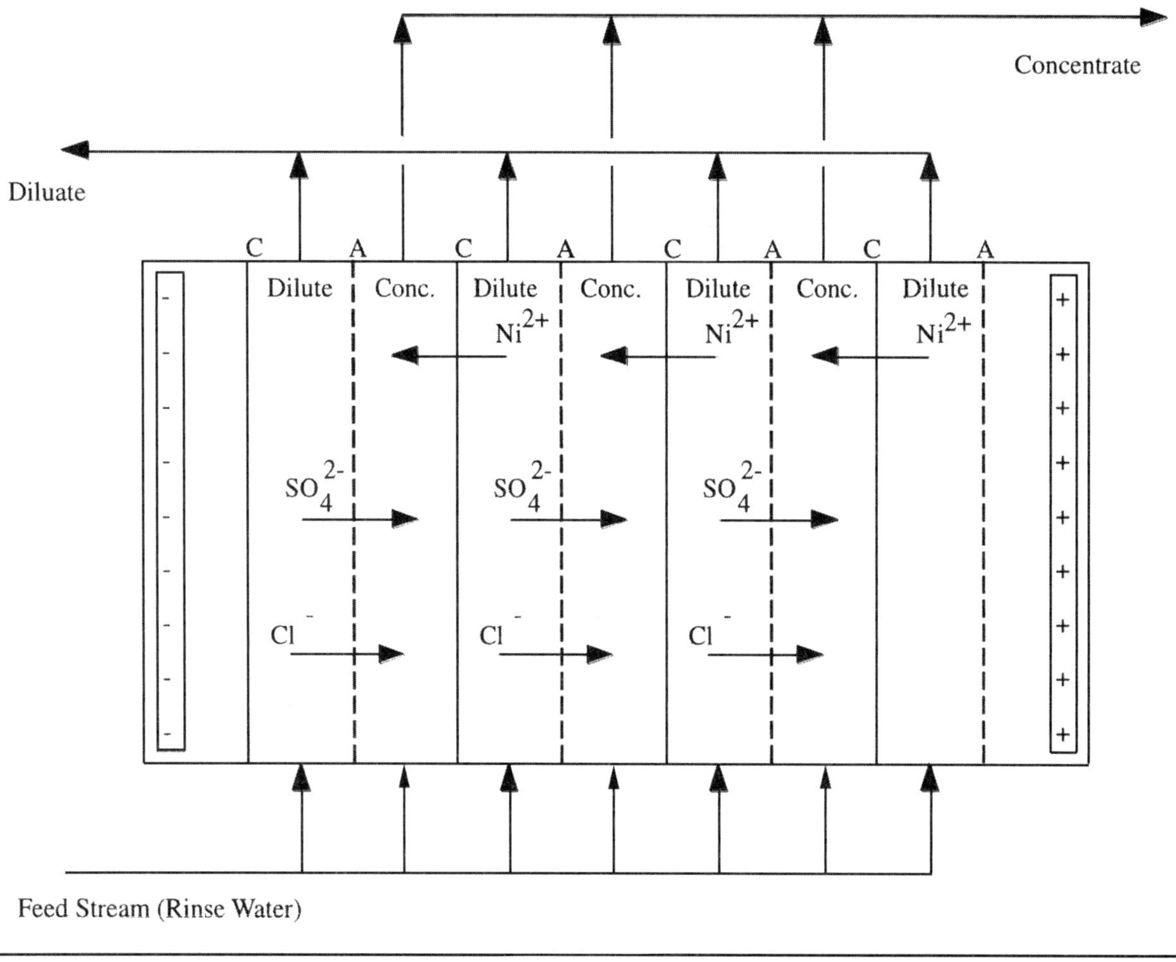

compared to other recovery schemes. Bath purification causes the loss of 10 to 15% of the bath per cycle, which significantly reduces the pollution prevention benefits of recovery (ref. 28).

4.6.2 Development and Commercialization

Although the phenomenon of electrodialysis was observed in the 1800's, it did not achieve any commercial application until durable, ion-selective membranes were developed in the 1950's (ref. 39). The two early applications of this technology, desalination and recovery of edible salt from sea water, still constitute the bulk of membrane utilization. The largest electrodialysis installations are salt recovery systems located in Japan. These systems replaced solar ponds that once occupied large tracts of land (ref. 400).

Large-scale industrial usage of electrodialysis began in the early 1970's with applications in chemical processing, biotechnology and water pollution control. In 1972, the first attempts were made to use electrodialysis to reclaim plating chemicals from rinse solutions (ref. 32). This early work was performed with unidirectional systems, where the transfer of ions is in one direction only. These efforts were plagued by membrane fouling with scale, slime and undesirable organics (ref. 32). Subsequently, the electrodialysis reversal (EDR) process was utilized, where there is a periodic reversal of direct current flow through the membrane stack and a simultaneous interchange of diluting and concentrating streams. This improvement greatly reduced the occurrence of membrane fouling. By 1984, there were more than 100 recovery units operating in the U.S. on a variety of plating solutions. The major area of application was nickel plating, with most other applications restricted to cyanide baths (e.g., gold, cadmium, silver and zinc). Additional applications at that time, mentioned in the literature, include tin and tin-lead fluoroborate and trivalent chromium (ref. 39). Since approximately 1985, there have not been any major changes to electrodialysis as applied to chemical recovery from rinse waters. However, similar technology applications for bath maintenance have significantly increased. Presently, membrane electrolysis, a related technology, is applied to the maintenance and reformulation of solutions containing chromic acid, nitric acid, sulfuric acid or sodium hydroxide (Section 5).

4.6.3 Applications and Restrictions

The application of electrodialysis for recovery of nickel plating chemicals is shown in Exhibit 4-49. This application, which includes a material balance, was taken from the literature (ref. 32), and is based on an actual case study performed by an electrodialysis equipment manufacturer. The work for this application was performed at a job shop with an automatic nickel barrel plating operation. The plating and rinsing configuration used prior to the implementation of electrodialysis is also shown. Based on a comparison of the two diagrams, the electrodialysis system recovered approximately 38 lbs of nickel per day more than the existing dragout/drag-in recovery rinsing system. The difference in plating tank evaporation rates between the original and revised systems was not discussed in the article, but presumably was due to a change in the operating temperature of the bath.

Most applications for this technology have involved nickel recovery including Watts, sulfamate and bright nickel plating solutions and nickel galvanizing. Other applications identified by vendors include: gold, copper, cadmium, platinum, silver and zinc cyanide; tin-lead, palladium, zinc phosphating and zinc galvanizing (ref. 134, Kinetic Recovery Corporation file, Graver Co. file).

4.6.4 Technology/Equipment Description

This subsection contains names of commercially available electrodialysis equipment that is manufactured and/or sold by vendor survey respondents. This is intended to provide the reader with information and data on a cross section of available equipment. Mention of trade names or commercial products is not intended to constitute endorsement for use.

Graver Water, a division of The Graver Company, markets electrodialysis units manufactured by Tokuyama Soda, a world leader in the membrane technology. Their target industries are food, photograph finishing, pulp processing, waste control and metal finishing. Their key metal finishing applications include nickel and zinc recovery from galvanizing and acid or alkali recovery.

Kinetic Recovery Corporation markets electrodialysis equipment for various applications, including: Watts nickel, copper cyanide, cadmium cyanide, and zinc phosphates. Their typical units have membrane surface areas ranging from approximately 35 ft^2 to 190 ft^2, although

Exhibit 4-49. Flow Schematic of a Nickel Plating Line Before and After Installing Electrodialysis

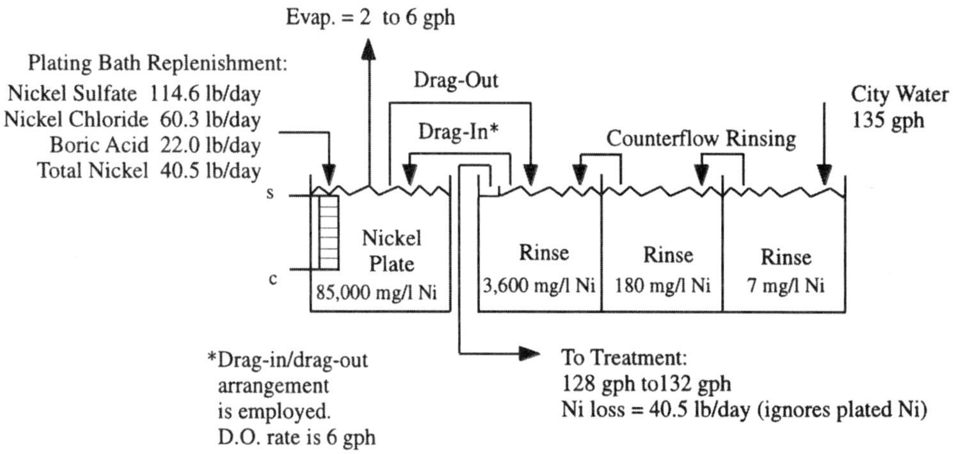

Nickel Plating Line Prior to Installing Electrodialysis

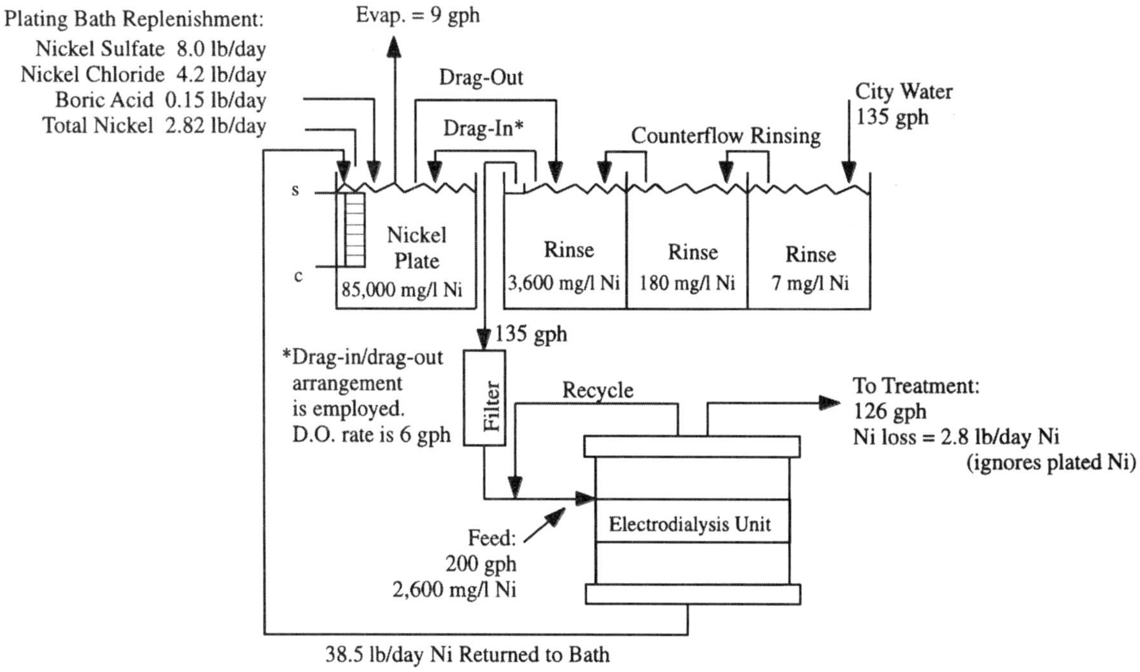

Soure: ref. 32

Nickel Plating Line After Installing Electrodialysis

all units are custom designed to meet individual process needs. Their systems are skid mounted and include: membrane stack, holding tank for concentrate, holding tank for diluate, holding tank for electrode rinse, three sets of level controls, three pumps for liquid transfer, plumbing, automated valves, flow meter, pressure gauges, cartridge filters and control panel.

4.6.5 Costs

4.6.5.1 Capital Costs

The capital cost of an electrodialysis unit is related primarily to the stack size, which can be expressed in terms of membrane area. Exhibit 4-50 shows typical membrane area requirements for Watts nickel solution and other applications in terms of the dragout rate (gph). Exhibit 4-51 shows capital costs, based on feed stream flow rate (typical feed stream concentration of 2 to 5 g/l dissolved metal.

4.6.5.2 Operating Costs

Electrodialysis operating costs include O&M labor, electricity and replacement costs for membranes. Estimated operating costs are 15 to 30 percent of investment costs.

4.6.6 Performance Experience

Only two of the user survey respondents indicated that they have used electrodialysis as a recovery technology (PS 012 and PS 125). PS 012 purchased a system from LEA in 1982 and PS 125 purchased a system from Baker Brothers in 1986. A summary of the data provided by these respondents is shown in Exhibit 4-52. Neither of these applications was successful. Failure of the technology at PS 012 was attributed to maintenance problems with the membrane and associated repair and parts replacement costs. Their system was removed from service in 1987 after 5 years of service (non-continuous due to O&M problems). They indicated that the technology did not satisfy the purpose for which they purchased it (i.e., to help meet effluent regulations and reduce operating costs). PS 125 did not provide many details on their system's performance. They simply stated: "machine returned after threatened law suit because it never worked." Presently PS 125 uses a vacuum evaporator for nickel recovery that has been relatively successful (only reported problem was a build-up of sodium chloride in the nickel bath, see Section 4.3).

4.6.7 Operational and Maintenance Problems

Only one of the two survey respondents who used this technology provided any information on O&M problems. PS 012 provided the following statement:

> *"Lack of availability of replacement membranes and long repair time by factory after membrane failure (1 year and up to 6 months down) coupled with high repair cost for each failure ($6,000) made this a non-viable technology, was abandoned circa 1987."*

Exhibit 4-50. Electrodialysis Membrane Requirements Based on Dragout Rate for Common Plating Baths

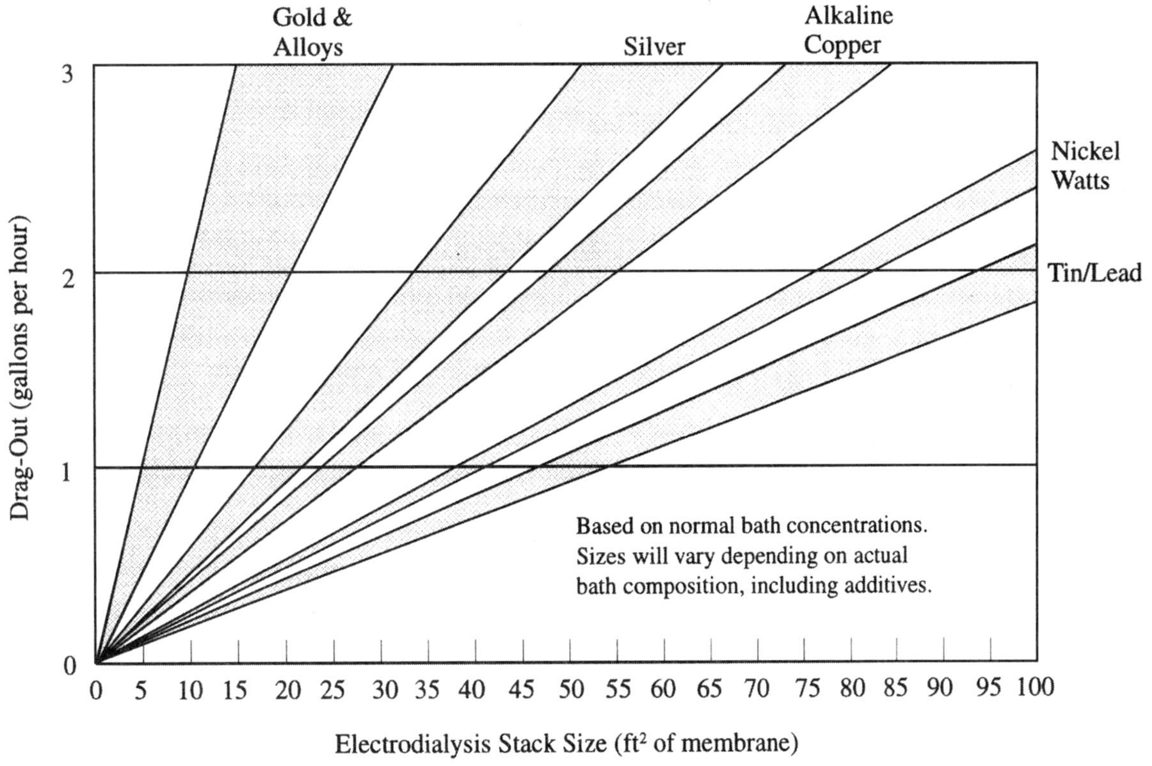

Source: Lancy International, Inc. File

Exhibit 4-51. Capital Costs for Electrodialysis Recovery Units Applied to Metal Recovery from Rinse Waters

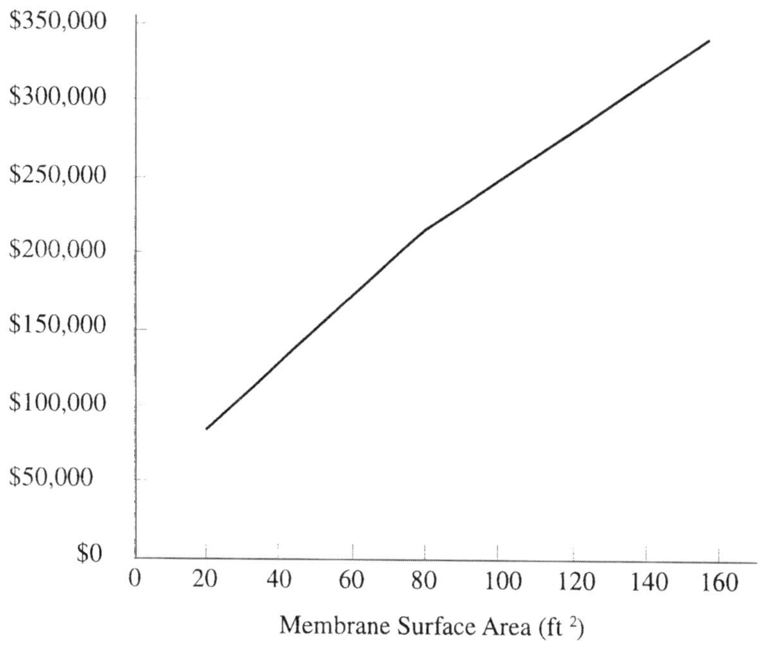

As with most membrane technologies, the life span of the membrane is a key concern to the system owner. One vendor provided estimates of membrane life span during the Vendors Survey. Kinetic Recovery Corporation estimated that their membranes (OS-FD/FS types) last three to seven years for all recovery applications. The Graver Company provided the following estimates for membranes, gaskets and electrodes for acid and alkali recovery (i.e., from concentrated baths, not rinse water).

	Acid Recovery	Alkali Recovery
Membranes	0.5 to 1.0 year	0.3 to 1.0 year
Gaskets	1 to 2 years	1 to 2 years
Electrodes	2 to 4 years	2 to 4 years

The Graver Company further indicated that the longevity of membrane life was dependent largely on the level of contaminants present.

4.6.8 Residuals Generation

When applied as a chemical recovery technology to plating rinse waters, electrodialysis produces two effluents from the unit: a concentrated stream that is recycled to the bath and a dilute stream that is either reused as rinse water or discharged to treatment.

Typically, electrodialysis recovers up to 95% of the plating chemicals in the dragout. The remaining 5% are lost when discharged from the rinsing system or directly from the recovery unit. This waste is typically treated on-site which generates a hazardous sludge (RCRA hazardous waste code F006).

Other miscellaneous wastes generated by this technology include spent pre-filtration cartridge filters and worn or damaged modules, electrodes and gaskets. No quantitative data were available for these other wastes.

4.7 REVERSE OSMOSIS

4.7.1 Overview

Reverse osmosis (RO) is a separation process that has been employed in the metal finishing industry to purify raw water (e.g., city water) before use as rinse water, recover plating chemicals from rinse water, and polish wastewater treatment effluents (usually for reuse as rinse water). Of particular interest in this section of the report is the application of RO for chemical recovery, however; end-of-pipe applications are also covered. Use of this technology as a raw water treatment technology is not covered in this report.

Exhibit 4-52. Partial Summary of Users Data for Electrodialysis Recovery Applications

Shop				Capital Costs			Operating Costs (Yr.)		Savings (Yr.)							Use	Down	Satisfaction Levels		Future
Code	Application	Manf.	Year	Equip. Cost	Other	Total	Non-labor	Labor Hrs.	Water	Process Chem.	Treat. Chem.	Sludge Disp.	Energy	Other	Total	Code	Time %	Manf. Support	Technology	Decision
012	Nickel, Watts	LEA (Lancy)	1982	$23,000	$4,000	$27,000	ND	125	√	√	√	√	$0	$0	√	2	ND	2	1	4
125	Nickel, Watts	Baker Bros.	1986	$44,000	$2,000	$46,000	ND	ND	ND	ND	ND	ND	ND	ND	ND	2	ND			3
Average for ED Recovery Applications			1984	$33,500	$3,000	$36,500		125					$0	$0				2.0	1.0	

Notes:
(1) Use Codes: 1-currently operating; 2-not currently operating and no intension for future use; 3-not currently in use, but intend to use in future.
(2) Satisfaction levels (manufacturer and technology) 1 to 5, with 1=lowest and 5=highest.
(3) Future decision codes: 1-purchase the same technology from the same vendor; 2-purchase the same technology from a different vendor; 3-purchase a different technology; 4-do nothing.
(4) "√" indicates that a savings was realized, but not quantified.
(5) Blank spaces indicate that no data were provided.
(6) ND = no data.

As a recovery technology, RO has been applied to a range of processes, including: brass, chromium, copper, nickel, tin and zinc plating solutions (ref. 269, 348), with nickel recovery being the most frequent and successful (ref. 39). Of the 318 plating shops responding to the Users Survey, only six applications of RO chemical recovery were identified. The survey also included one application for end-of-pipe treatment and several applications of raw water purification. The infrequent use of RO for chemical recovery may be due to the limited number of baths to which it has been successfully applied and the availability of competing technologies (especially for nickel plating). Further, one of the competing technologies frequently used for nickel recovery is atmospheric evaporation, which has a very low capital cost.

Reverse osmosis is often referred to as a "crossflow filtration" process. This term, which also describes most ultrafiltration and microfiltration equipment used by the metal finishing industry, distinguishes these processes from surface barrier filtration, which operates in dead-end flow. As shown in Exhibit 4-53, in dead-end filtration, all of the feed solution is forced through the membrane by an applied pressure. With crossflow filtration, the fluid to be filtered is pumped across the membrane, parallel to its surface. Because the feed and concentrate of RO flows parallel to the membrane instead of perpendicular to it (i.e., dead-end flow), the process is termed crossflow. The pressure required to drive the process is determined by the specific nature of the feed solution and the membrane pore size (ref. 380, Osmonics file).

There are several key differences between RO and ultrafiltration/microfiltration: (1) only RO has the ability to concentrate dissolved salts (e.g., plating chemicals); (2) RO cannot tolerate significant concentrations of suspended solids, whereas the other two processes can, especially microfiltration; and (3) RO operates at higher pressures and usually requires a heavy gauge stainless steel housing, whereas the other two lower pressure processes can be housed in plastic or lightweight stainless steel. It should be noted, that whereas RO is a distinctly unique filtration process, ultrafiltration and microfiltration are similar to one another and have overlapping definitions (ref. 380). Microfiltration, which has the largest pore size of the three technologies, is discussed in Section 5, as a method of alkaline cleaner maintenance. Also, both microfiltration and ultrafiltration are discussed in Section 7, as end-of-pipe polishing technologies. Ultrafiltration is also used by several survey respondents for the recovery of electrocoat (paint), an application not

Exhibit 4-53.
Diagram of Barrier vs. Cross-Flow Filtration

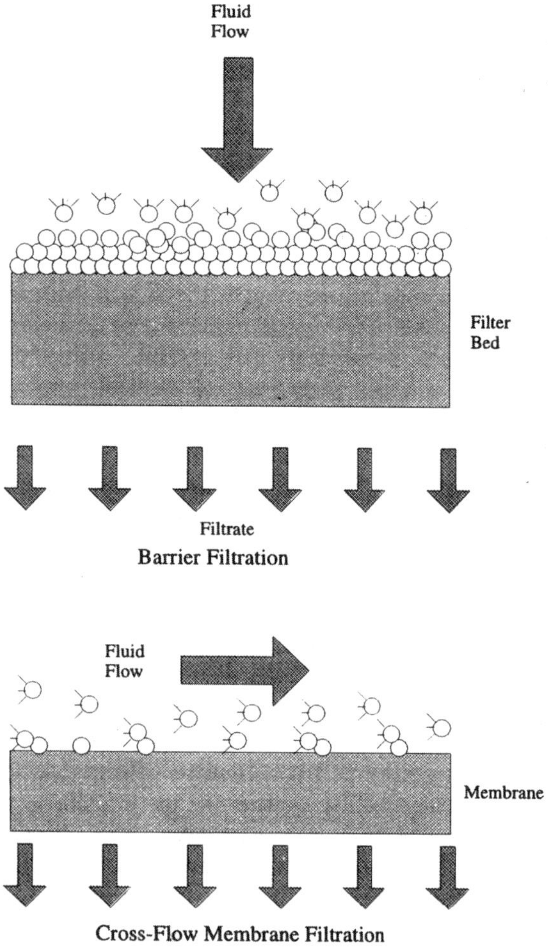

covered by this project. It is also used in the machining industry for the recovery of cutting oils.

RO theory is based on two physical processes: osmosis and ionic repulsion. Osmosis is related to diffusion, which describes the tendency of molecules in solution to move about until they are uniformly distributed. Osmosis is the tendency for diffusion to take place across a semi-permeable membrane. It occurs when a water permeable membrane separates two solutions of different concentrations of dissolved solids. Pure water will flow into the concentrated solution until an equilibrium energy state is achieved. By applying pressure to the more concentrated solution, the normal osmotic flow is reversed and pure water is forced through the semipermeable membrane into the less concentrated solution. Suspended solids are blocked by mechanical exclusion and dissolved solids are chemically repulsed by the membrane surface. Multi-charged ions are rejected at rates exceeding 99 percent and single-charged ions have rejection rates in the range of 90 to 96 percent. RO will also reject neutral solutes, although no general efficiency data are available. Besides ionic charge, rejection efficiency is also affected by the concentration gradient. As the concentration gradient increases, the rejection efficiency decreases. The flow of water through an RO membrane (flux) is determined by the pressure differential across the membrane. Higher pressure differentials generally result in higher flow rates.

The RO process is designed to operate continuously. The RO membrane is enclosed in a pressure vessel and the feed stream is pumped through the vessel under pressure, 400 to 1,000 psig, where it is separated into a clean water permeate stream and a concentrated chemical stream by selective permeation. Three important parameters describe the performance of the RO process: recovery, flux, and rejection.

Recovery is defined as the percentage of the feed that is converted to permeate and it is usually expressed as percent. Flux is the rate at which the permeate passes through the membrane per unit of membrane surface area. Rejection is the ability of the membrane to restrict the passage of dissolved salts into the permeate, and is related to particular salt species (ref. 39).

There are different types of RO membranes used (tubular, spiral wound and hollow fiber), the selection of which depends mostly on the applications and in particular the plating bath chemistry. The most common RO membranes are the hollow fiber and spiral wound configurations.

Most reverse osmosis systems are designed with a single filtration stage operating below 700 psig. With a single stage system operating in this pressure range, the practical limit for concentrating plating chemicals in rinse waters is 15 to 20 g/l. Because this concentration is below that of most plating baths, a "solution volume" problem is sometimes created with an RO recovery application, in that there is insufficient head-room in the process tank for the return of the recovered chemical solution. This problem occurs especially with ambient to low temperature baths, where the surface evaporation rate is low. This condition limits the direct reuse of the RO concentrate stream in the plating tank. An evaporator can be used to further concentrate the solution or to supplement tank surface evaporation. However, the added capital and operating costs of an evaporator often make this approach less attractive than using an alternative recovery method.

Newer reverse osmosis technology includes multiple stage systems and higher operating pressures (800 to 1,000 psig). With the multiple stage design, the concentrate stream from the first stage is passed through a second stage to further concentrate the chemicals. This permits the direct reuse of some solutions that could not be directly recovered with the less effective single pass units.

The key attributes of RO as a recovery technology are: (1) it is an ambient temperature, low energy process; (2) it generates a permeate stream that is usually of sufficient quality that it can be reused for rinse water; and (3) for some applications, it has relatively low capital and operating costs as compared to other recovery technologies. The negative aspects of this technology are: (1) RO membranes can be fouled by precipitation products and/or suspended solids; (2) membranes have a fairly limited life-span; (3) this technology does not sufficiently concentrate the chemicals for direct return in some applications; and (4) similar to most other recovery technologies, RO returns both essential plating chemicals and unwanted impurities to the bath, unless some post-treatment is performed.

4.7.2 Development and Commercialization

Reverse osmosis is a relatively mature technology, having the distinction of being the first membrane-based separation technology to be widely commercialized (ref. 380). Much of the early developmental work for this technology took place in the late 1950's and the 1960's and focused on desalination for drinking water supplies. This work was sponsored by the U.S. Department of the Interior, Office of Saline Water (ref. 380). The spiral-wound element, which is the building block of modern recovery units was developed in 1963 (ref. 380). The first large industrial application of RO occurred in 1970 when a 100,000 gpd system was placed into operation at Texas Instruments' (TI) electronic manufacturing facility in Dallas, TX. The application at TI was the purification of municipal water for use in manufacturing. Now, essentially all electronics plants in the U.S. use RO for this purpose (ref. 65). The first RO metal finishing application for chemical recovery that was identified in the literature occurred in 1974, which was applied to copper cyanide (ref. 382). Other applications during the 1970's and 1980's included: bright nickel, Watts nickel, acid copper, acid zinc, and end-of-pipe effluent polishing. In 1976 the USEPA sponsored a series of experiments to evaluate the application of RO to plating chemical recovery. These experiments evaluated the performance and life-spans of membranes that were commercially available at the time (ref. 382).

Metal finishing applications of RO are very limited in comparison to those of desalination and other industrial applications. Worldwide, there are approximately 1,500 RO desalinating plants with a total capacity of more than 750 million gallons per day. Estimated RO membrane sales were approximately $118 million in 1988. Of these sales, Osmonics, one of the largest U.S. manufacturer's of RO equipment for the plating industry, accounted for $3 million (ref. 380). Osmonic's sales figures include not only the metal finishing industry applications, but also a range of other industries, including: food, beverage, dairy, chemical processing, and textile manufacturing. There are approximately 20 companies that manufacture and/or sell RO and ultrafiltration equipment to the U.S. plating industry (ref. 421). Of these companies, five responded to the NCMS vendors survey, but only four provided data on the number of systems they have sold within the metal finishing industry. The total number of systems sold by these four companies is 15, with nine units applied to chemical recovery, five applied to reuse of wastewater and one applied to raw water treatment. These numbers significantly understate the total number of RO units in use in the metal finishing industry because two major manufacturers (Osmonics and Ionics) did not provide data.

4.7.3 Applications and Restrictions

Exhibit 4-54 shows an example of applying reverse osmosis as a chemical recovery technology. In this example, nickel salts are recovered from rinse water and returned to the bath. A three stage counterflow rinse system is used in order to concentrate and reduce the flow rate of the feed to the RO unit. The 100 gph feed rate is typical for this application. The feed is pumped (high pressure feed pump) through a cartridge filter (typically 5 micron) and through the RO system. The reject passes through a carbon filter to remove bath impurities and is returned to the plating bath. In this example, the surface evaporation of the bath is sufficient to create the necessary head-room. The permeate is returned to the counterflow rinse system.

The primary plating chemical recovery application for RO is nickel plating. This includes Watts nickel and bright nickel plating (ref. 382). A wide range of other successful applications are identified in the literature (mainly articles written by RO equipment manufacturers). Some of the more frequently discussed applications include: brass cyanide, cadmium cyanide, copper cyanide, non-cyanide alkaline zinc, and zinc cyanide (ref. 156, 157, 263). From the Users Survey, there were only two successful application of RO for chemical recovery. These involved recovery of nickel acetate seal and acid zinc dragout (PS 010, 230). Another shop successfully operates an RO unit on a cadmium cyanide process, but does not return the concentrate to the bath due to their concern over bath contamination (PS 131). This shop indicated that they operate under stringent aircraft manufacturer's specifications.

Reverse osmosis is applicable to the recycle of effluent from an end-of-pipe treatment system. End-of-pipe systems that employ hydroxide precipitation generate an effluent that is relatively free of toxic metals; however, it contains a high concentration of total dissolved solids (TDS). A typical effluent contains between 500 to 4,000 mg/l of TDS (ref. Memtek file). Although approximately 1.2% of the shops responding to the Users Survey indicated that they directly reuse this effluent (see Section 3.6), most sources indicate that it is of insufficient quality for rinsing, especially in critical rinse situations. Typically, rinse quality criteria for functional and bright plating is in the range of 100 to 700 mg/l and 5 to 40 mg/l TDS, respectively (Exhibit 3-16). Most end-of-pipe RO wastewater recycle applications consists of: preconditioning to prevent the precipitation of salts in the membrane; prefiltration; RO filtration and storage. One survey respondent used RO and other recycling technologies in place of an end-of-pipe treatment system (PS 233). Their system is described in Exhibit 4-55, where RO is incorporated into a complex, zero discharge configuration. In this application, RO is employed to upgrade the quality of the ultrafiltration permeate so that it can be reused as rinse water. Because this plant operates at zero discharge, the concentrated stream from the RO unit is hauled off-site for treatment/disposal; ordinarily, it would be treated on-site.

RO is generally not considered applicable to highly concentrated, oxidative solutions, like chromic acid, nitric acid and peroxy-sulfuric etchant. Process chemicals in these solutions can be recovered using RO; however, membrane life-span is a concern (ref. 157). The membrane life-span for these applications was demonstrated on a bench scale to be only 15 to 35% of that for nickel plating applications (ref. 157). However, one supplier advertises use of their equipment for chromate conversion coating (ref. KRC file). That company agrees that membrane life is shortened by this application, but indicated that RO is still cost effective. For this application, the primary competing technology is vacuum evaporation.

The unsuccessful chemical recovery applications from the Users Survey included: Watts nickel (PS 172), zinc cyanide (PS 008), acid zinc (PS 010), and copper cyanide (PS 089). The reasons for failure of these systems relate mostly to fouling, and are discussed in subSection 4.7.7.

4.7.4 Technology/Equipment Description

This subsection contains names and/or descriptions of commercially available reverse osmosis equipment that is manufactured and/or sold by vendor survey respondents or discussed in the literature. This is intended to provide the reader with information and data on a cross section of available equipment. Mention of trade names or commercial products is not intended to constitute endorsement for use.

Kinetic Recovery Corporation supplies an advanced reverse osmosis ("ARO") packaged system for chemical recovery from rinse waters. The system consists of: membrane modules (usually two, maximum of three); prefiltration (5 to 10 micron); conductivity controls for rinse water and concentrate quality test; miscellaneous controls (flow, pressure, level) for system automation; recycle and boost pumps for liquid transfer; pressure pumps for liquid/solids separation; internal tanks for concentrate storage; and an electrical control system, including PLC and operator's interface unit. A diagram and system specifications for the ARO system is shown in Exhibit 4-56. A diagram showing a typical layout for chromate conversion coating recovery is shown in Exhibit 4-57. The ARO typically has two membrane modules. The first membrane module is dedicated to the initial feed stream (rinse water) and the second is dedicated to increasing the concentration of the concentrate stream through successive passes. From the first module, the permeate is returned to plating rinses while the concentrate is held in the first internal storage tank. The concentrate is then passed through the second module for further concentrating. The permeate from the second module, depending on its conductivity, is returned to the rinse tank or directed to one of the internal storage tanks. The concentrate moves sequentially through the series of tanks to tank 3, where

Exhibit 4-54. Common Configuration for a Reverse Osmosis Nickel Recovery Application

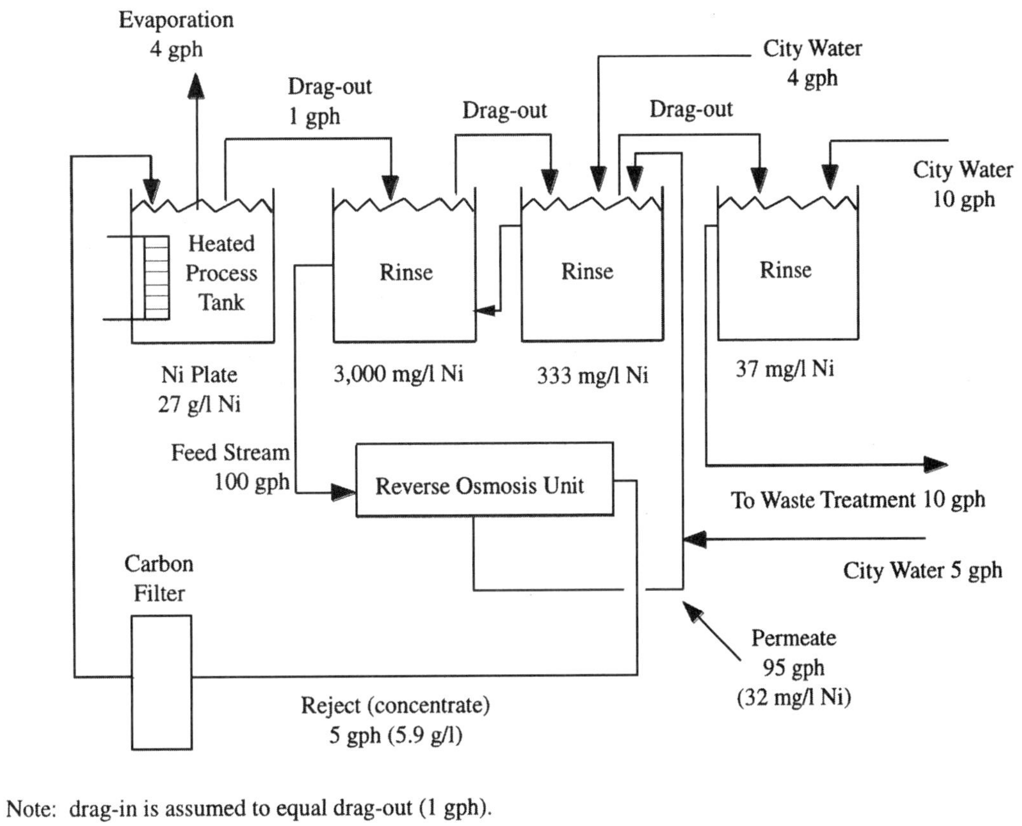

Note: Drag-in is assumed to equal drag-out (1 gph).

Source: ref. 65

Exhibit 4-55. End-of-Pipe Reverse Osmosis Configuration by PS 233

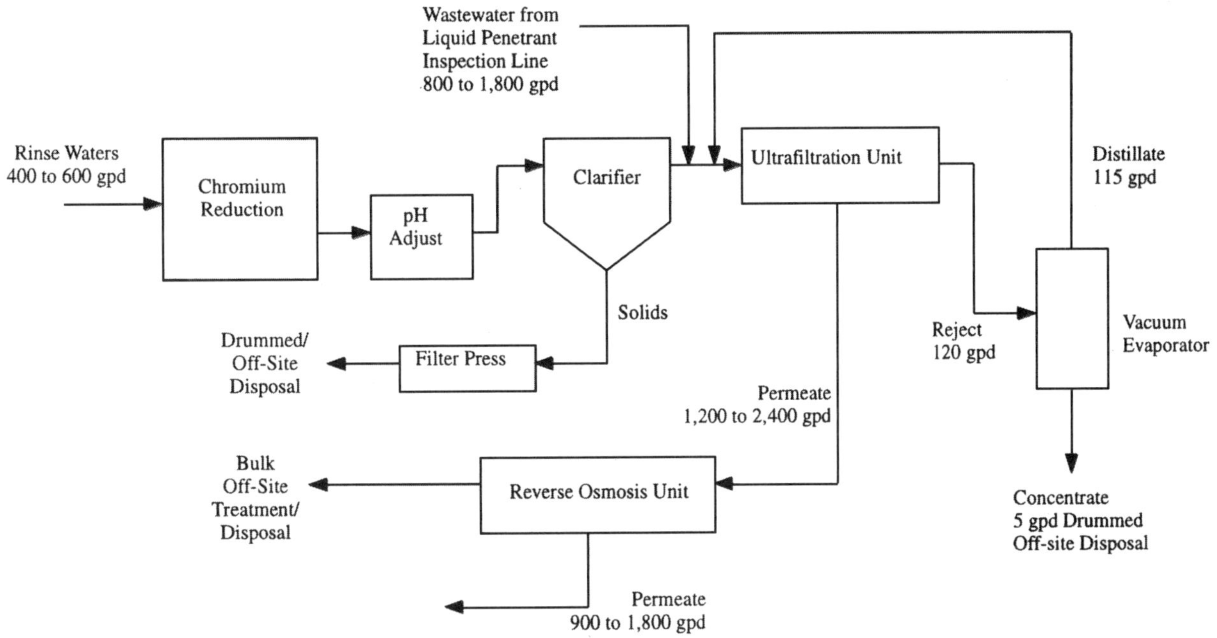

Exhibit 4-56. Operational Diagram and System Specifications for Advanced Reverse Osmosis (ARO) (Courtesy of Kinetic Recovery Corporation)

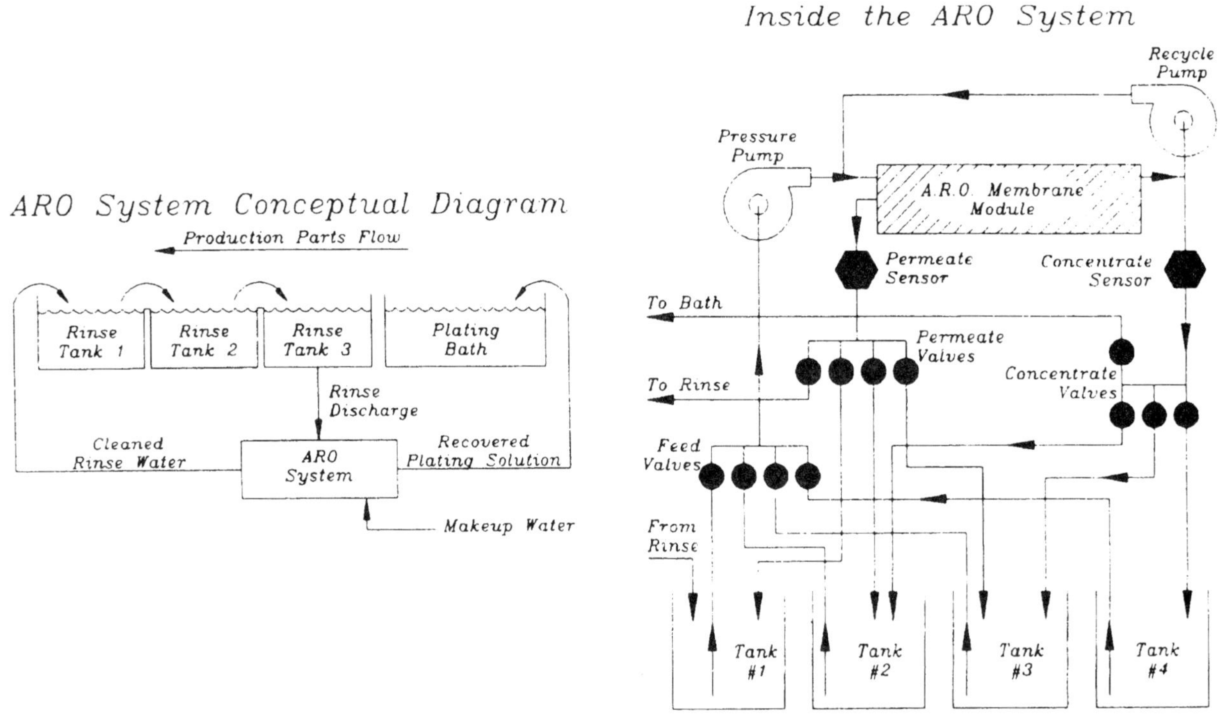

Specifications of Standard Systems

System Type	500	1500	3000	4500
Membrane Surface (ft^2)	597	1,494	2,390	3,585
Max. Operating Pressure (psi)	1,000	1,000	1,000	1,000
Avg. Operating Pressure (psi)	950	950	950	950
Process Water Feed (gpm)	5	15	30	45
Concentrate (gph)	0.75	2.3	4.5	6.75
Temperature (°F)	100	100	100	100
pH Range	0.5 to 13	0.5 to 13	0.5 to 13	0.5 to 13
Number of Modules	2	5	8	12

it is most concentrated, and then is pumped to the bath. The ARO's internal microprocessor changes operating parameters for each pass of concentrate through the second membrane. Pressures and process times are controlled in order to achieve higher concentrated solutions and improve membrane life (ref. 157, Kinetic Recovery Corporation file). The ARO operates in a pressure range of 900 to 1,000 psig, whereas most RO chemical recovery systems operate below 700 psig. According to the manufacturer, an ARO system can concentrate dilute solutions to at or near bath strength without any evaporation or additional concentration technology (however, for most applications a concentration of 40% to 70% of bath strength is targeted) (ref. 157). Results of bench scale tests on various plating and finishing solutions are shown in Exhibit 4-58.

The ARO is the only unit that was advertised for recovery of chromate conversion coating (it is also sold for traditional applications such as nickel plating). These conversion coating baths, which contain chromic acid, are usually operated at low temperatures and therefore, there is little surface evaporation to provide head room for recovered solution. According to the vendor, the multi-pass design and higher operating pressure of the ARO permit this application. Because the first membrane module of the ARO unit contains more dilute rinse waters and the second membrane module contains more concentrated solution, the life-spans of the two module sets are very different. The first module has a life-span of 1 to 2 years and the second module, 4 to 6 months.

Exhibit 4-57. Advanced Reverse Osmosis (ARO) Application to Chromate Conversion Coating Recovery (Courtesy of Kinetic Recovery Corporation)

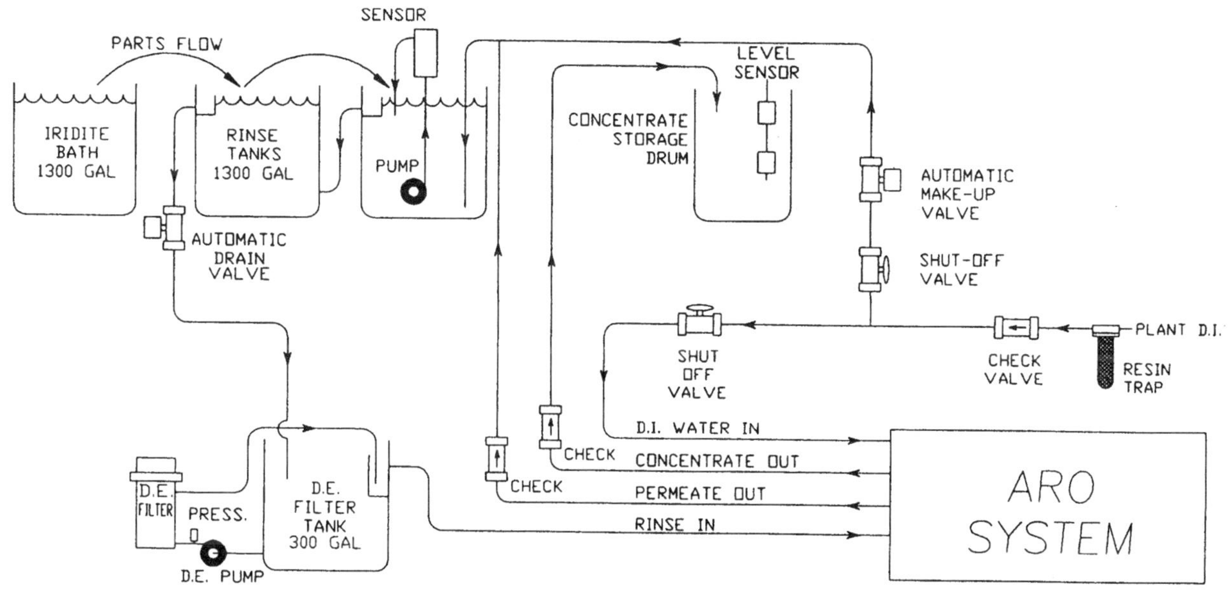

Exhibit 4-58. Reverse Osmosis Bench-Scale Test Results for Various Applications

Process Solution	Bath pH	Concentration Ratios	Passes to Reach Required Strength	Membrane Life Years
Plating Solutions:				
Copper Sulfate	0	40	3	6+
CopperPyrophosphate	8	200	2to 3	6+
Tin/Lead Fluoborate	0	10	4	6+
Tin and Tin/Lead Methane Sulfonic	3.6	10	4	6+
Electroless Copper	12	1,000*	1*	6+
Electroless Nickel	10	250	2	6+
Bright Nickel	4.3	110	2 to 3	6+
Nickel Sulfamate	4	250	2 to 3	6+
Watts Nickel	4.4	100	2 to 3	6+
Zinc Chloride	4.9	30	3	6+
Zinc Cyanide	12	25	3	6+
Copper Cyanide	13.5	30	3	6+
Cadmium Cyanide	12	25	3 to4	6+
Hexavalent Chromium	-0.14	100	2 to 3	1
Etchants:				
Peroxy-Sulfuric	0	30	4	1 to 2
Ammonium Chloride	8.0	60	3	6+
Chromic-Sulfuric Acid	-0.5	70	2 to 3	1
SulfuricAcid	0	40	3 to 4	6+
Hydrochloric Acid	0.5	20	4	6+
Cotings/Sealers/Passivators/Cleaners				
Chelated Lead Brightener	13.5	80	2	4+
Chrome Iridite	2.0	11	2 to 3	1+
Nickel acetate	5.5	150	2	6+
Nitric Acid	1.0	30	4	1
Sodium Hydroxide	13.5	20	4	3+

*Special membrane, selective separation.
Source: ref. 157

Memtek Corporation manufactures an RO wastewater recycle system (Exhibit 4-59). This system is used to upgrade wastewaters from hydroxide precipitation systems for use as rinse water. There are four stages in the system. The first stage is a pre-RO conditioning step where, as necessary, pH adjustment is performed, an anti-scalant chemical is added to prevent precipitation of salts in the membrane elements, reducing agents are added to destroy residual oxidizers, and carbon filtration (optional) removes oxidizers and organics. The second stage is prefiltration, which is achieved by a cartridge filter unit (5 micron). Optional multimedia filtration can be purchased for wastewaters with high suspended solids loadings, where cartridge filters would require frequent replacement. The third step is reverse osmosis filtration. The wastewater is pumped through the unit at 200 to 600 psig, depending on the selection of RO elements. Cellulose acetate or thin film composite elements are used for rejection of dissolved salts. The choice of a membrane element is determined by the chemical characteristics of the wastestream and by the percent recycle desired. The permeate from the RO unit is collected in a storage

Exhibit 4-59. Diagram and Equipment List for Commercial Reverse Osmosis Filtration and Related Equipment (Courtesy of Memtek Corporation)

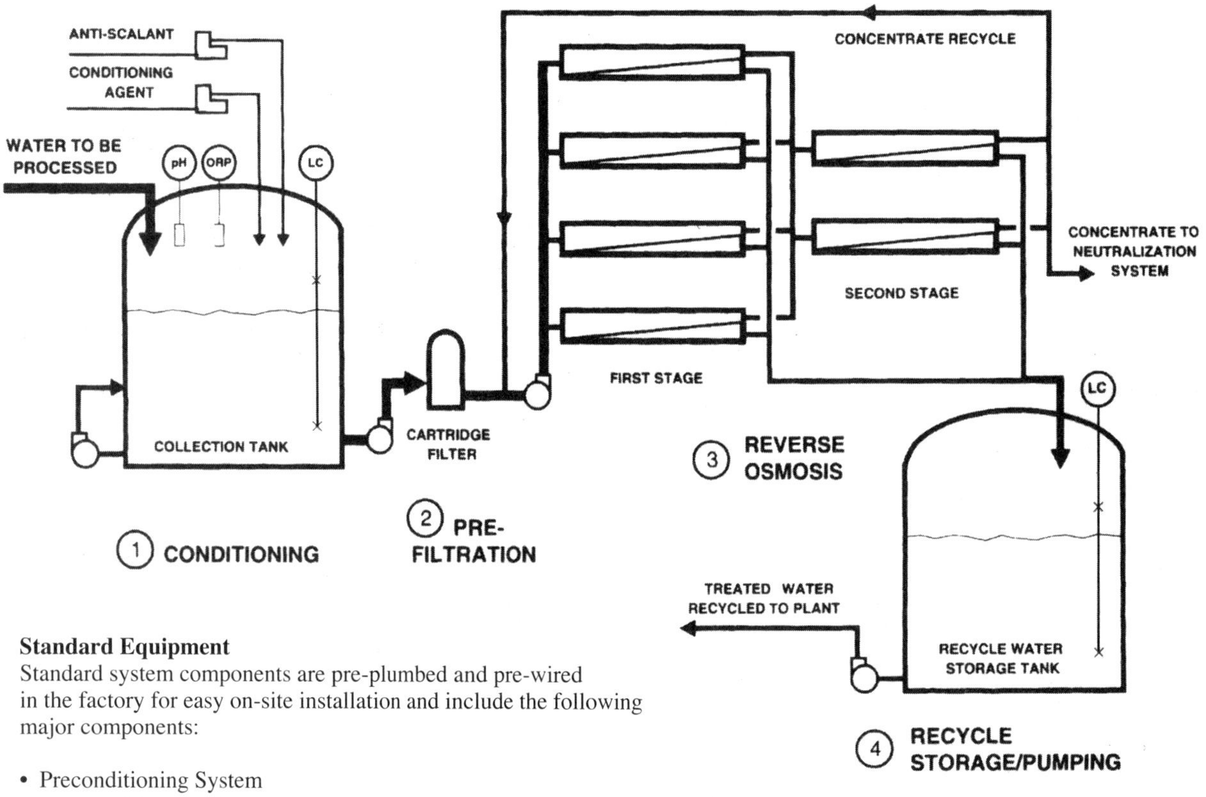

Standard Equipment

Standard system components are pre-plumbed and pre-wired in the factory for easy on-site installation and include the following major components:

- Preconditioning System
  - One HDPE tank
  - On pH monitor/controller
  - Up to three PCP-50 chemical metering injection systems (optional)
  - One all-plastic mixing pump
  - One ORP monitor/controller
  - One level control for RO pump on/off and high level alarm
  - One RO low pressure feed pump
- Reverse Osmosis Filtration
  - One epoxy coated carbon steel frame
  - One set of 5 micron filter cartridges
  - Three flow-indicator-totalizers for monitoring feed, product, and concentrate flow
  - One electrical panel for automatic control
  - One 304 stainless steel cartridge filter housing
  - One high-pressure multistage RO process pump
  - One temperature monitor
  - Temperature and pressure alarms
  - Cleaning skid (optional)
- Product Water Recycle Station
  - One product water collection tank
  - One level control for RO pump on/off & high level alarm
  - One diverter valve and piping for recycle to preconditioning tank (optional)
  - One duplex recycle water pump station rated at 60 psig
  - One recycle water pressure tank

tank equipped with water recycle pumping equipment. The RO unit is equipped with conductivity monitors to monitor product quality. Optional membrane cleaning systems are available.

4.7. 5 Costs

4.7.5.1 Capital Costs

The capital costs of reverse osmosis units are best expressed in terms of membrane surface area, where the required area for a given chemical recovery application will depend on the flux rate and the percent rejection. Flux is the volume flow of permeate per unit of membrane area, usually expressed as gal/ft^2/day or gfd. The percent rejection is defined as follows (ref. 348):

% Rejection = [(feed concentration - permeate concentration)/(feed concentration)] x 100%

Higher percent rejections will result in better quality (i.e., higher purity) permeate and a higher concentration of the plating chemicals. The permeate is typically reused for rinsing and the concentrated chemicals are typically returned to the bath.

In turn, these two factors are affected by the design of the RO equipment and the operating conditions of the system. More specifically, this includes: the type of membrane used, the applied pressure of the RO unit, the number of RO stages, the flow rate of the feed stream, the chemical concentration of the feed stream, and the required concentration of the product returned to the bath.

The feed stream flow to the RO unit is the rinse water, as shown in Exhibit 4-54. The flow rate of this stream and its chemical concentration will depend on the bath concentration, the dragout rate, the required quality of rinse water in the final rinse and the rinsing configuration employed. For example, if dragout recovery rinsing is practiced, the RO membrane surface area will be proportionally reduced by the percentage of dragout returned to the bath by the recovery rinse.

Unfortunately, there are insufficient data available to account for the various parameters that impact RO system sizing and cost. Articles and conference papers that describe this technology fail to provide all of the needed data on system components such as membrane area, flux rates, dragout rates, etc. Therefore, at this time, capital costs are simply presented in terms of the feed stream rate (Exhibit 4-60). The multi-pass data include systems with feed stream rates of 5 and 10 gpm and corresponding membrane surface areas of 597 ft^2 and 1,194 ft^2. The single pass data include systems with feed stream flows of 5 and 200 gpm. These units are applicable to end-of-pipe polishing. No corresponding membrane surface area data were provided for the single pass units.

Most RO units are sold as packaged systems that include a set of membrane modules, feed and recirculation pumps, prefilters, controls, and internal tanks.

4.7.5.2 Operating Costs

The most significant operating costs include O&M labor, energy, chemicals (cleaning) and membrane replacement. Estimates of RO operating costs for single and multi-pass systems are shown in Exhibit 4-61.

4.7.6 Performance Experience

A partial summary of the user data relative to reverse osmosis is presented in Exhibit 4-62. There are a number of observations that can be made from these data and other data contained in the database and literature:

- The average satisfaction level for chemical recovery applications is 3.2 (on a scale of 1 to 5, with 5 being most satisfactory), which is slightly less than the average rating for all recovery technologies (3.4). Only 33% of the shops using RO for chemical recovery indicated that this technology satisfied the need for which it was purchased. The following is a breakdown of the reasons why shops purchased this technology:

  To meet of help meet effluent regulations: 4
  To reduce plating chemical purchases: 4
  To reduce water usage: 2
  To reduce the quantity of waste shipped off-site: 2
  To reduce wastewater treatment costs: 3
  To improve product quantity: 0

- The only recovery applications that were fully successful for survey respondents were: nickel acetate and zinc, acid.

- The use of reverse osmosis as a recovery technology did not impact production quality or the rate of production for the survey respondents. The following responses were provided:

Exhibit 4-60(a). Capital Costs for Single Pass Reverse Osmosis Systems

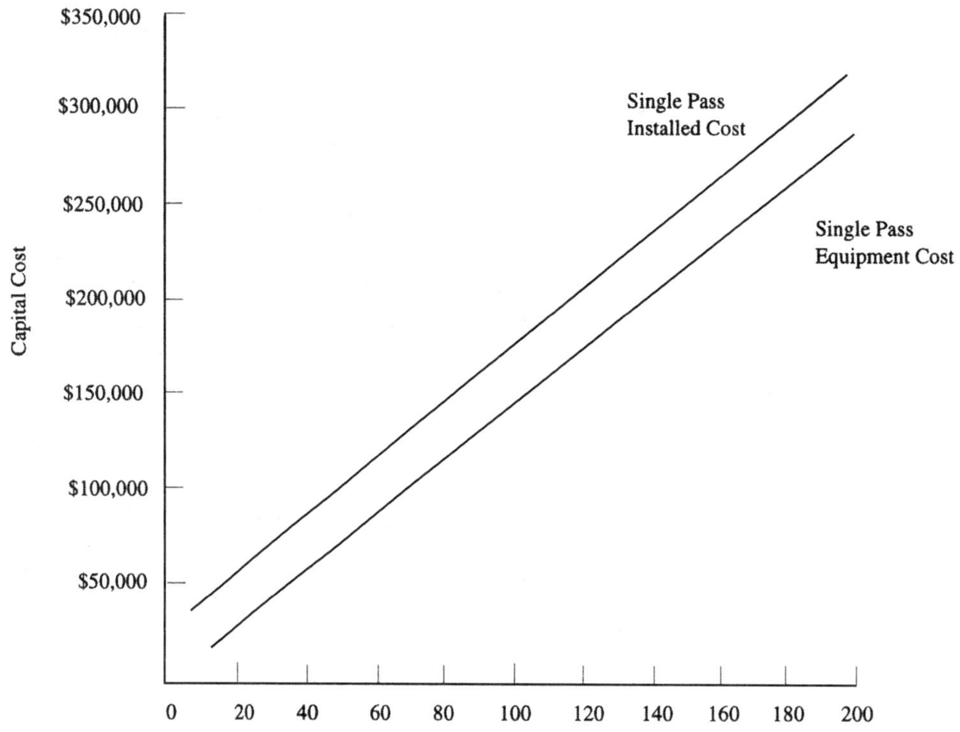

Exhibit 4-60(b). Capital Costs for Multi-Pass Reverse Osmosis Systems

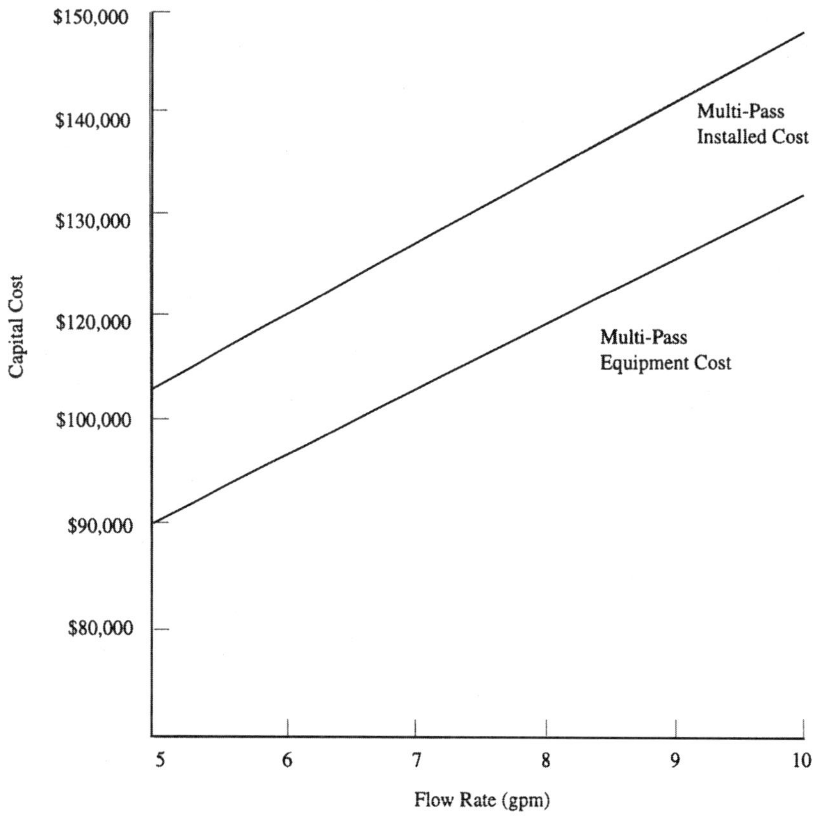

Exhibit 4-61. Operating and Maintenance Costs for Reverse Osmosis Equipment

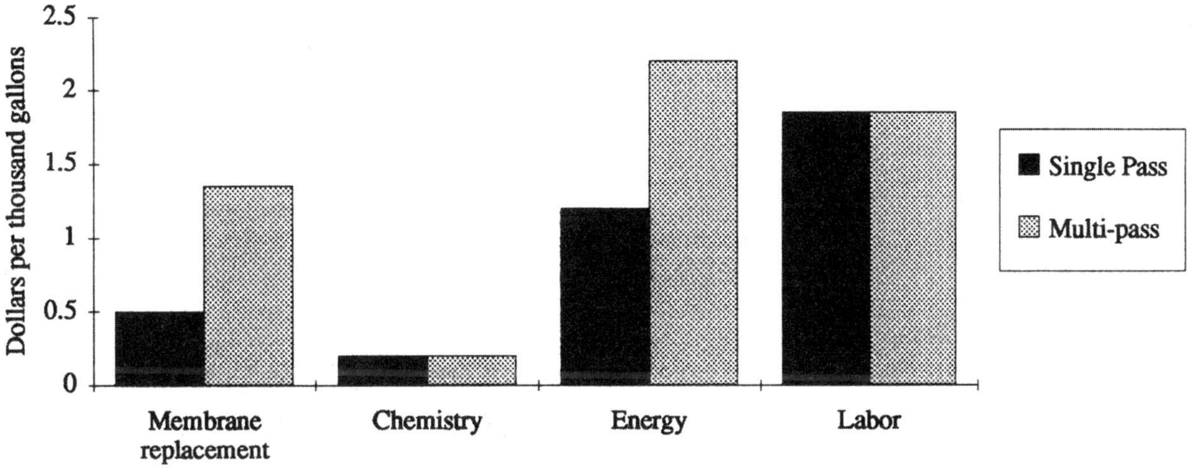

	Product Quality	Production Rate
Improved	0	0
No Change	5	5
Decreased	0	0

- The respondents indicated that, based on their experience with this technology and, if given the opportunity, they would:

Purchase the same technology from the same vendor: 2
Purchase the same technology from a different vendor: 1
Purchase a different technology: 3
Do nothing: 0

- None of the respondents indicated that their reverse osmosis system was the cause of an effluent compliance excursion.

- Although RO application to Watts nickel is a well established application, one of the survey respondents found it to be unsuccessful (PS 172). The supplier stated capacity of their unit was 120 gph and the actual capacity was 50 to 80 gph. PS 172 indicated that their unit did not concentrate the product sufficiently for return to the plating bath.

4.7.7 Operational and Maintenance Experience

The message from the Users Survey respondents with respect to operational and maintenance problems is very clear: RO membranes are highly susceptible to fouling and have relatively short life-spans. As stated by one respondent: "Membrane problems, not enough filtration with (original) unit. Had to add much, much extra filtration ahead of (the RO) unit" (PS 008). The shop using RO as an end-of pipe technology also had membrane fouling problems and frequent membrane replacement (every 3 mth.) (PS 233). They also indicated that their efforts to clean the membrane have been unsuccessful. Short membrane life was identified as a problem by another respondent (PS 172). Membrane fouling by algae was identified as a problem by one respondent (PS 172). Another identified high caustic and carbonate levels as the cause of fouling (PS 089). This shop suggested that carbonate levels must be below 45 g/1 or "some kind of caustic membrane washing must be applied."

RO membranes are susceptible to fouling by suspended solids in the feed stream or materials that precipitate during processing. Solids in the feed stream can be controlled by installing prefilters that are properly selected for the solids encountered and sized for the loading. Precipitation is inhibited by changing operational parameters such as pH. Because pH adjustment of plating chemicals is not usually possible, applications where precipitation may occur should be investigated through bench and/or pilot testing before purchasing equipment.

Two respondents reported no operational or maintenance problems and a percentage downtime of less than 1% (PS 131, PS 230).

A respondent that has eliminated use of their RO unit, indicated that the process "increased solution purification requirements with attendant solution losses, yielding questionable benefits from the technology" (PS 172). It should be noted that any recovery technology that returns both bath chemicals and contaminants to the process tank will cause this problem. However, when contaminants can be removed from the process bath (e.g., carbonates), the benefits of recovery will often outweigh any losses.

4.7.8 Residuals Generation

As discussed in Section 4.7.1, RO separates the feed stream into two components: (1) the permeate which passes through the membrane and (2) the concentrate which retains the dissolved and suspended solids. In typical chemical recovery applications the permeate is used as rinse water and the concentrate is used in the bath. In typical wastewater recycle applications, the permeate is used as rinse water and the concentrate is treated on-site by the end-of-pipe treatment system. In such cases, the quantity of residuals from RO is low, and mostly restricted to used cartridge filters and RO membranes.

The experience of the respondents to the Users Survey was generally the same as described above. One respondent indicated that quantity of used filters generated was 0.3 drums per year. That same shop sent 60 to 300 gpy of unspecified "overflow" to off-site treatment (PS 089). The wastes from all other respondents were treated on-site, except for the end-of-pipe, zero discharge application (PS 233). That shop indicated that the reject stream from their RO unit (300 to 600 gpd of concentrated waste with 2.5% total dissolved solids) is hauled off-site for treatment/disposal.

4.8 MESHPAD MIST ELIMINATORS

4.8.1 Overview

Meshpad mist eliminators are used to recover plating chemicals that become entrained in the airstream that is exhausted from the surface of a plating tank. The primary application of this technology is with chromic acid baths, particularly hard chromium plating and chromic acid anodizing. From the Users Survey, 16 respondents (or 5.0%) indicated that they have employed this technology. Of these responses, there were 16 applications to hard chromium and 2 applications to chromic acid anodizing (two shops used the technology for both applications). The 16 hard chromium applications represent 19.2% of all shops that employ the hard chromium plating process. One manufacturer that responded to the Vendors Survey indicated that they have sold approximately 475 meshpad

Exhibit 4-62. Partial Summary of Users Data for Reverse Osmosis Recovery Applications

Code	Application	Manf.	Year	Capital Costs			Operating Costs (Yr.)		Savings							Use	Down	Satisfaction Levels		Future
				Equip. Cost	Other	Total	Non-labor	Labor Hrs.	Water	Process Chem	Treat. Chem	Sludge Disp.	Energy	Other	Total	Code	Time %	Manf. Support	Technology	Decision
008	Zinc, CN	Water Technologies	1988	$11,355	$2,000	$13,355		550	$0	$0	$0	$0	$0	$0	$0	2		3	1	3
010 (1)	Zinc, acid	Unk.	1987	$65,000	$30,000	$95,000	$3,000	250	$2,500	$8,300	$5,000	$3,000	$10,000	$0	$26,300	1		4	5	1
089	Copper, CN	Water Technologies	1988	$29,650	$2,000	$31,650	$4,000	100		√	$600				$600	3	30	2	2	3
131 (2)	Cadmium, CN	Osmonics (1)	1988	$16,000	ND	$16,000	ND	ND	√							1	<1	5	5	2
172	Nickel, Watts	Osmonics	1975	$15,000	$3,000	$18,000	ND	ND	ND	ND	ND	ND	ND	ND	ND	2	10	3	1	3
230 (3)	Ni acetate seal	Ecosystems (2)	1992	ND	ND	ND	ND	ND		$624					$624	1	<1	4	4	1
Average for Recovery Applications			1985	$27,401	$9,250	$36,651	$3,500	300	$1,250	$4,150	$1,867	$1,500	$5,000	$0	$8,967		10	3.5	3.0	
233 (4)	EOP	Abcor/Koch	1984	$45,000	$10,000		$600	200	√	$0	$0	$0	$0	$0	$0	1	30	3	3	1

(1) Other capital costs include an atmospheric evaporator used to make head-room in the zinc plating tank.
(2) Equipment purchased used, original manufacture date is not known.
(3) Equipment leased for $500/mth. Nickel acetate use reduced by 3 lbs/week.
(4) End-of-pipe application.
(5) Use Codes: 1-currently operating; 2-not currently operating and no intension for future use; 3-not currently in use, but intend to use in future.
(6) Satisfaction levels (manufacturer and technology) 1 to 5, with 1=lowest and 5=highest.
(7) Future decision codes: 1-purchase the same technology from the same vendor; 2-purchase the same technology from a different vendor; 3-purchase a different technology; 4-do nothing.
(8) "√" indicates that a savings was realized, but not quantified.
(9) ND = No data

mist eliminators. Of these units, 63% are installed on hard chromium plating baths, 13% on sulfuric acid baths and the remainder on decorative chromium plating, chromic acid anodizing, caustic, cyanide, hydrochloric acid, and nitric acid baths (ref. Midwest Air Products file).

Meshpad mist eliminators are one of several technologies employed for the removal of plating chemicals from exhausted air. The other two technologies include liquid scrubbers and chevron mist eliminators. Of the three technologies, meshpad mist eliminators are considered to be the most efficient (ref. 213, 404). Other methods of control that may assist in reducing chemical emissions, but are not by themselves adequate control measures, include fume suppressants and polypropylene balls (both are used to cover the surface of the plating tank and reduce mist generation).

The EPA proposed air standards for decorative and hard chromium plating and chromic acid anodizing operations in December, 1993 (ref. 521). The standards are based on the use of scrubbers and meshpad mist eliminators.

Typically, a separate mist eliminator is used for each plating tank, although different configurations are possible. Meshpad mist eliminators are installed within the exhaust system ductwork, as near to the exhaust hood as practical, as shown in Exhibit 4-63. The mist eliminator enlarges the cross-sectional area of the duct which causes a reduced airstream velocity within this section. The reduced velocity permits the entrained droplets of plating solution to impinge on and adhere to the meshpads, thus removing them from the air stream. Having multiple pads in series increases the removal efficiency of the process. The accumulated plating chemicals are periodically washed from the pads (see Exhibit 4-64). This is usually accomplished with an integral water spray system. The liquid from pad washing drains to the bath.

The meshpad mist eliminator technology competes with wet scrubbers and chevron mist eliminators as a control device for chromic acid air emissions. However, with respect to chemical recovery, the meshpad has very little competition from scrubbers. While the recovery of plating chemicals, including chromium, from scrubber systems is possible, scrubber units generally serve multiple types of processes (e.g., cleaning and plating) and the entrapped chemicals are combined, and therefore less amenable to recovery. One recovery strategy that has been employed with scrubbers involves the use of an anion exchange column to collect hexavalent chromium from the scrubber water and to reformulate the resulting sodium chromate regenerant using membrane electrolysis. The resultant products are chromic acid, which is reused as a plating chemical, and sodium hydroxide, which is reused for the anion column regeneration (ref. 402).

4.8.2 Development and Commercialization

Meshpad mist eliminators have been manufactured for more than 30 years (ref. 403) and applied to chromium plating for more than 20 years (e.g., PS 253). However, use of this technology for plating applications only recently became widespread. Until the past few years, most hard chromium shops either had no emission control, used chevron mist eliminators or used wet scrubbers. Most decorative platers relied on fume suppressants and most chromic anodizers either had no controls or used either wet scrubbers or fume suppressants. In 1992, data were presented on the relative uses of these technologies (see Exhibit 4-65) (ref. 213).

The results from the Users Survey show that 86% of the mist eliminator units for which the purchase date was given were installed since 1988 and 71% since 1990. These data and those from the literature indicate that use ot mist eliminators is expanding at a very significant rate. The reason for this change may be due to the recognized advantages of this technology over other competing technologies plus regulatory requirements.

The high efficiency and recovery potential of meshpad mist eliminators was documented by the Navy in the mid-1980's. That work concluded that the meshpad mist eliminator proved to be more effective than a wet scrubber on chromic acid mists and that it produced a concentrated, low volume chromic acid stream with potential for recycle. Further, the Navy concluded that it was a simpler device, compared to the wet scrubber (ref. 404).

The State of California undertook regulatory development efforts in the mid-1980's to control chromium emissions from plating and anodizing (ref. 405). California enacted strict rules on chromic acid emissions in 1988 (ref. 406). The EPA generated and compiled performance data for control measures during the mid-to late 1980's (ref. 46). Included in these efforts were detailed studies of meshpad mist eliminators (ref. 407, 408). These research efforts also showed that high efficiency levels were attained by the meshpad mist eliminator.

Exhibit 4-63. Meshpad Mist Eliminator Installation Diagram (Courtesy of ChromeTech, Inc.)

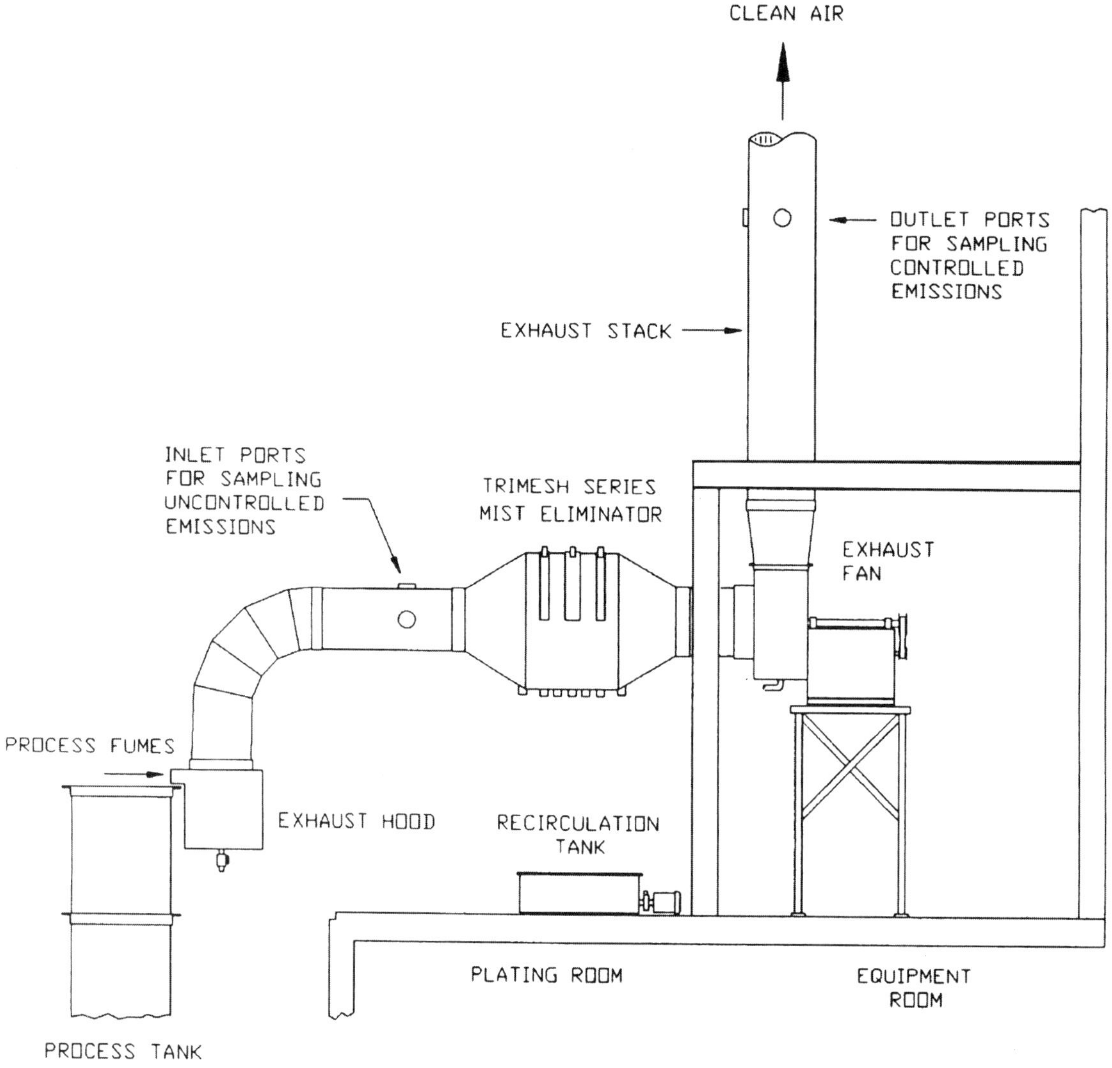

4.8.3 Applications and Restrictions

Unlike scrubbers, which are typically installed to serve multiple tanks or processes, meshpad mist eliminators are usually dedicated to individual tanks. Exhibit 4-66 shows a plan view of a hard chrome line where meshpad units are incorporated into the exhaust systems for plating and stripping.

Meshpad mist eliminators can be installed during new plating line installation or as a retrofit to an existing line. With retrofit installations, it is best to include new exhaust fans and ductwork along with the mist eliminators. This ensures proper air pressures and velocities, and allows the ventilation system to be "tuned" for maximum removal efficiency (ref. ChromeTech file).

Exhibit 4-64. Meshpad Mist Eliminator Pad and Wash-Down Arrangement (Courtesy of ChromeTech, Inc.)

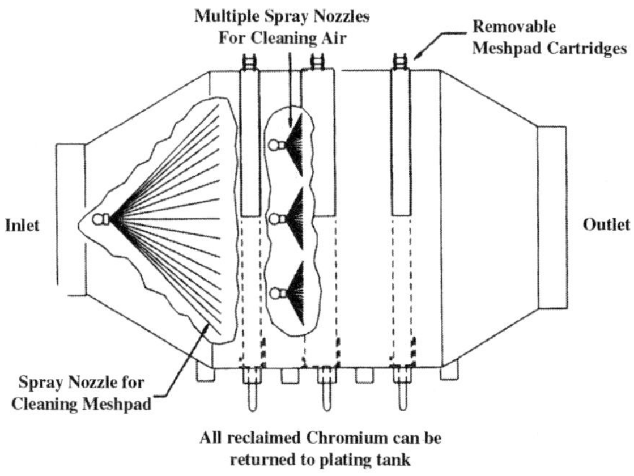

Exhibit 4-65. Frequency of Use for Various Chromium Emission Control Technologies and Methodologies in 1992

Process	Number (rel. 406)	No Control	Chevron Mist Eliminator	Wet Scrubber	Mesh Pad MistEliminator	Fume Suppressant
Hard Chrome	1,500	30%	30%	40%	<1%	—
Decorative Chrome	2,800	15%	—	5%	—	80%
Chromic Acid Anodizing	—	40%	10%	20%	—	30%

Source: ref. 213

Some existing shops may have space limitations that preclude the installation of meshpad mist eliminators or cause installation costs to be exorbitant. This is particularly a problem where the existing ductwork is crowded together or obstructed. Also, it is preferable to locate the mist eliminator at an elevation that is higher than the exhaust hood with the air flow in the horizontal direction.

4.8.4 Technology/Equipment Description

This subsection contains a description of commercially available meshpad mist eliminator equipment that is manufactured and/or sold by respondents to the Vendors Survey. This is intended to provide the reader with information and data on a cross section of available equipment. Mention of trade names or commercial products is not intended to constitute endorsement for use.

ChromeTech, Inc. developed and manufactures a Trimesh series mist eliminator (Patent Number 5,200,055) that has been applied to the recovery and control of hard chromium tank emissions. A diagram of the ChromeTech unit was previously shown in Exhibit 4-64.

The first stage is essentially a dry meshpad mist eliminator. Inlet air, heavily laden with chromium, flows through various laminations of woven polypropylene fibers which are contained in the first meshpad cartridge. Most of the chromium (typically 99%) is removed by impingement. As the condensed chrome collects, it is removed from the system through a drain network fitted to the bottom of the unit, and is automatically returned to the plating tank. The small percentage of chrome passing through the first meshpad represents the smallest droplets of the particle-size distribution curve, and is the most difficult to filter with a dry mechanism. Therefore, the second stage of the control device offers a wetted, scrubbing section. Water reservoired in a separate container beneath the device is pumped continuously during plating to the front face of the second meshpad, through atomizing nozzles provided. The pad offers a very large wetted surface area, and chromium mist still entrained in the airstream is dissolved into the recirculating liquid. At regular intervals, typically the end of each plating shift, the chrome that accumulates in this liquid can be returned to the plating tank by using it as the cleaning solvent for the front pad. A considerable amount of evaporation occurs in this stage, typically 2 gph to 5 gph. For this reason, the recirculation tank is fitted with a liquid level control system, and is automatically replenished with deionized water or clean tap water.

The third and final stage is similar to the first - i.e., a single, dry meshpad. The primary purpose of this section is to remove the scrubbing liquor that was administered to the airstream in the previous section. The efficiency of the Trimesh unit was tested by EPA (ref. 407, 408). The results of these two studies showed a consistent removal rate of 98.9 and 99.7%, respectively.

Midwest Air Products Co., Inc. manufactures a range of equipment for emission control including fume collection hoods with integral mesh pads (Terminator™), in-line mist eliminators (single and two-stage), mist eliminators utilizing either meshpads or chevron blades, and chromium scrubber units with meshpads (Enforcer III™). They also manufacture mesh pads designed as an integral mist eliminator stage within a packed bed wet scrubber. Two of their products are often combined in series to meet stringent emissions requirements. Diagrams showing their standard chromium scrubber unit (Enforcer III™) with an in-line mist eliminator and details of the mist eliminator and scrubber are presented in Exhibit 4-67. With the meshpad mist eliminator/chromium scrubber configuration, the in-line mist eliminator removes gross quantities of chromic acid and the scrubber unit further cleans the air. The mesh pad mist eliminators can be mounted in either a horizontal or vertical position. Vertical units incorporate a design that prevents washdown from draining down through the duct system. The standard meshpad is manufactured from polypropylene, however, Kynar meshpads are used for most chromic acid applications. As an alternative to the in-line mist eliminator, Midwest Air Products offers a tank exhaust hood that contains a mesh pad (Terminator™) (see Exhibit 4-67). The scrubber unit, which is itself a combined scrubber/mist eliminator, is configured with either three or four mesh pads with each stage designed using specific parameters such as particle size, velocity, removal efficiency requirement and pressure drop. The first pad removes the larger droplets or particles of chemicals and the subsequent stages remove smaller particles. Each mesh pad section is equipped with removable spray headers for periodic washdown. The Enforcer III™ recirculation system is shown in Exhibit 4-68. The first pad is washed with effluent from the scrubber unit or rinse water and subsequent pads are washed with clean water. The effluent is periodically pumped to the plating tank to make up for evaporative losses and to recover the chromic acid.

4.8.5 Costs

4.8.5.1 Capital Costs

Capital costs for chromium plating tank ventilation systems using meshpad mist eliminators are shown in Exhibit 4-69. The capital costs include the tank hood, ductwork, mist eliminator, water recirculation system and exhaust fan. The cost curves are based on two types of systems: (1) a 3-pad wet system and (2) a 2-pad dry system. Basic equipment and installation equipment costs are shown for both types of systems.

4.8.5.2 Operating Costs

Operating costs for mesh pad mist eliminators are shown in Exhibit 4-70 for 3-pad wet and 2-pad dry systems. The operating costs are based on 4,000 hr/yr of operation. Energy costs are for the exhaust fan and water recirculation pump.

4.8.6 Performance Experience

Eight respondents to the Users Survey provided some detailed data covering their experience with the meshpad mist eliminator technology. A summary of their responses are shown in Exhibit 4-71 and discussed in this section.

- Meshpad mist eliminators had the highest average satisfaction level among recovery technologies (4.1 on a scale of 1 to 5, with 5 being most satisfactory). Also, all respondents indicated that this technology satisfied the need for which it was purchased. The following is a breakdown of the reasons why shops purchased this technology (multiple responses given by some shops):

  To meet or help meet effluent regulations: 4
  To reduce plating chemical purchases: 5
  To reduce the quantity of waste shipped off-site: 2
  To reduce wastewater treatment costs: 2

Exhibit 4-66. Meshpad Mist Eliminator Floor Plan View (Courtesy of ChromeTech, Inc.)

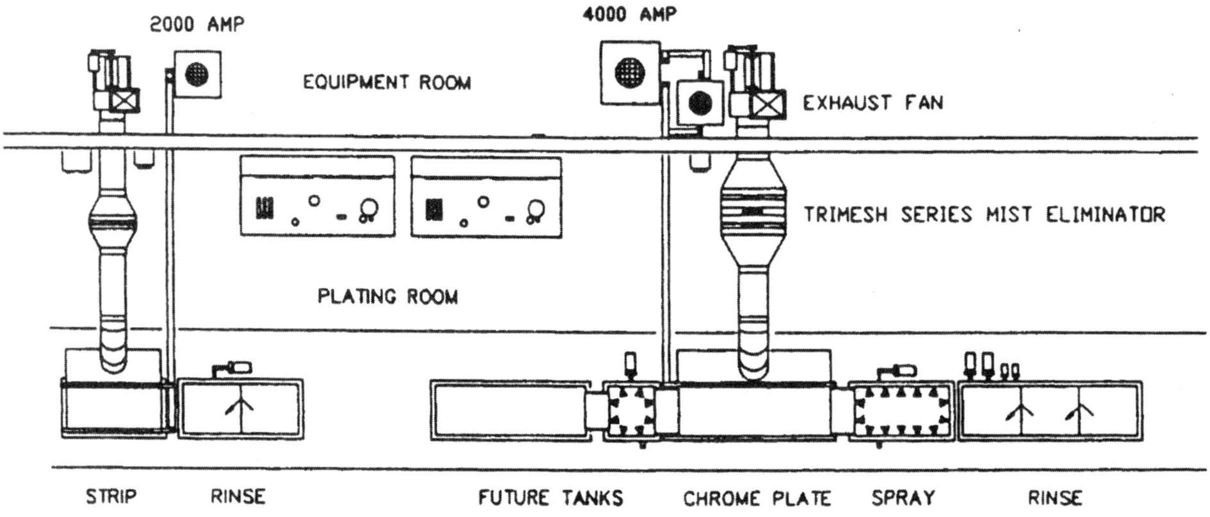

Exhibit 4-67. Meshpad Mist Eliminator and Scrubber Systems (Courtesy of Midwest Air Products Co., Inc.)

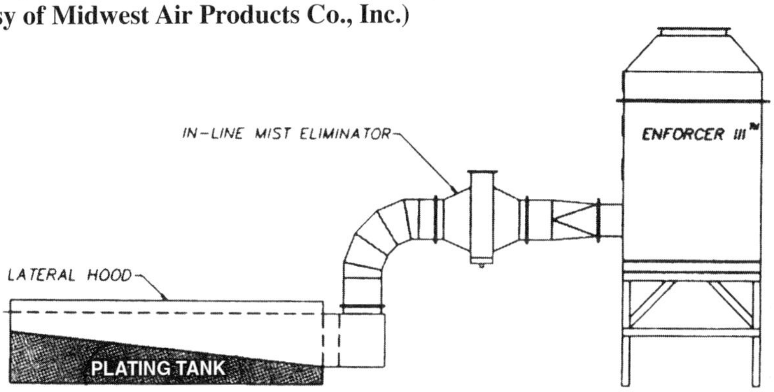

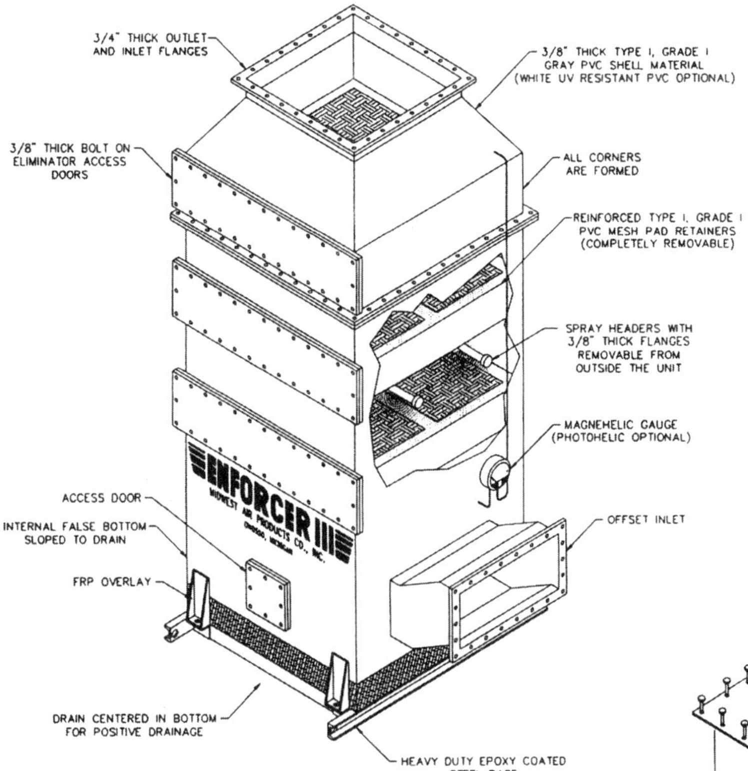

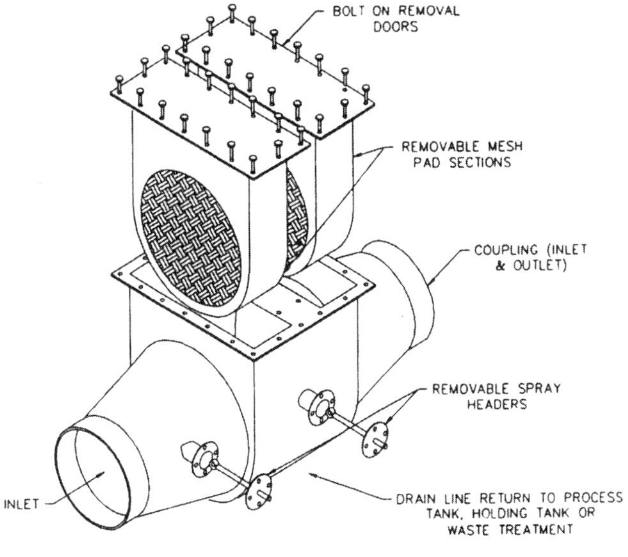

Exhibit 4-68. Enforcer III™ Recirculation System
(Courtesy of Midwest Air Products Co., Inc.)

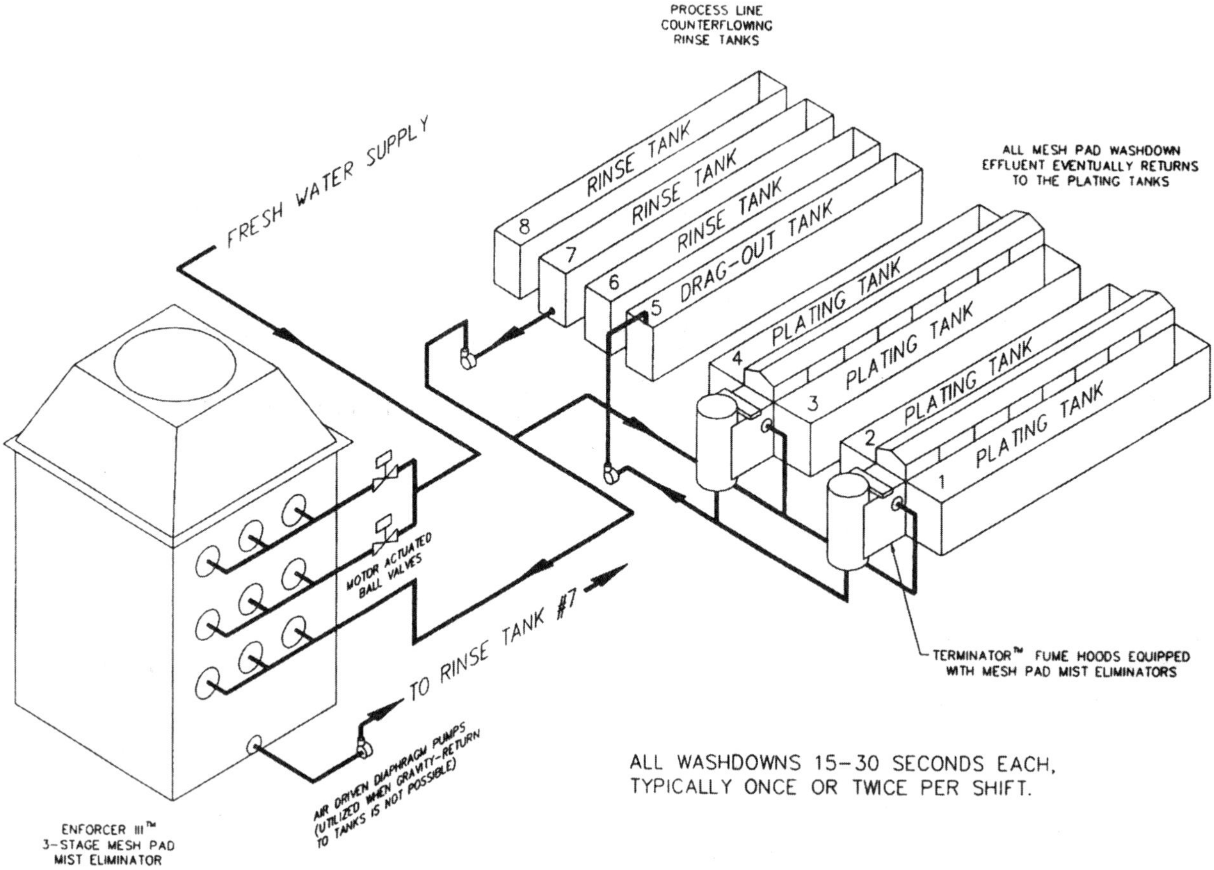

To improve product quantity: 1
Reduce emissions or meet air standards: 3

- The use of meshpad mist eliminators for chromium recovery generally did not impact production quality or the rate of production. The following responses were provided:

	Product Quality	Production Rate
Improved	1	1
No Change	5	5
Decreased	0	0

- All plating shops providing responses indicated that, based on their experience with this technology, they would purchase the same type of equipment from the same vendor. The following is a breakdown of their responses:

Purchase the same technology from the same vendor: 7
Purchase the same technology from a different vendor: 0
Purchase a different technology: 0
Do nothing: 0

- None of the shops indicated that a meshpad mist eliminator was the cause of an effluent compliance excursion.

- The major savings from the operation of a meshpad mist eliminator was the result of reductions in bath chemical use. The reported chemical savings ranged from $250 to $5,000 per year. One shop/ also reported a $1,000 per year savings due to reduced sludge disposal costs.

4.8.7 Operational and Maintenance Experience

The following summarizes the respondent's O&M experiences and provides operating labor information.

- Seven shops provided operating labor data. For these shops, the average number of annual O&M

Exhibit 4-69. Capital Costs for Meshpad Mist Eliminator (Courtesy of ChromeTech, Inc.)

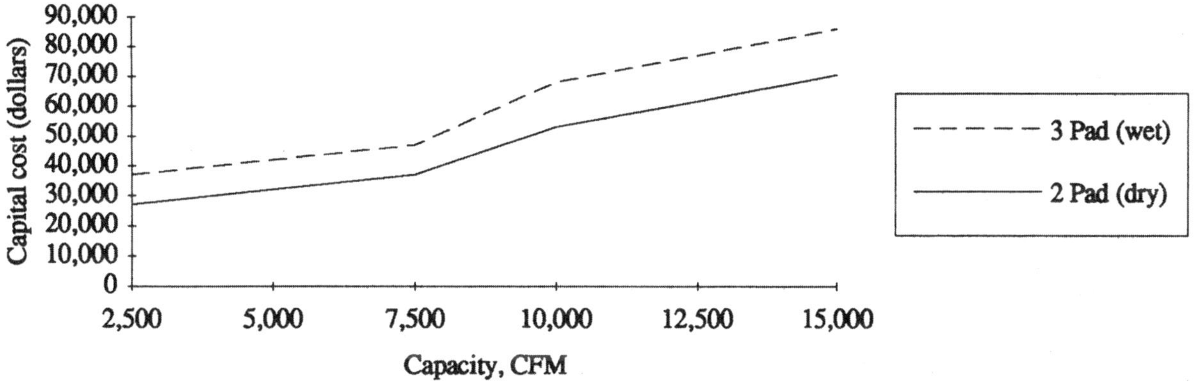

Exhibit 4-70. Operating Costs for Meshpad Mist Eliminators (Courtesy of ChromeTech, Inc.)

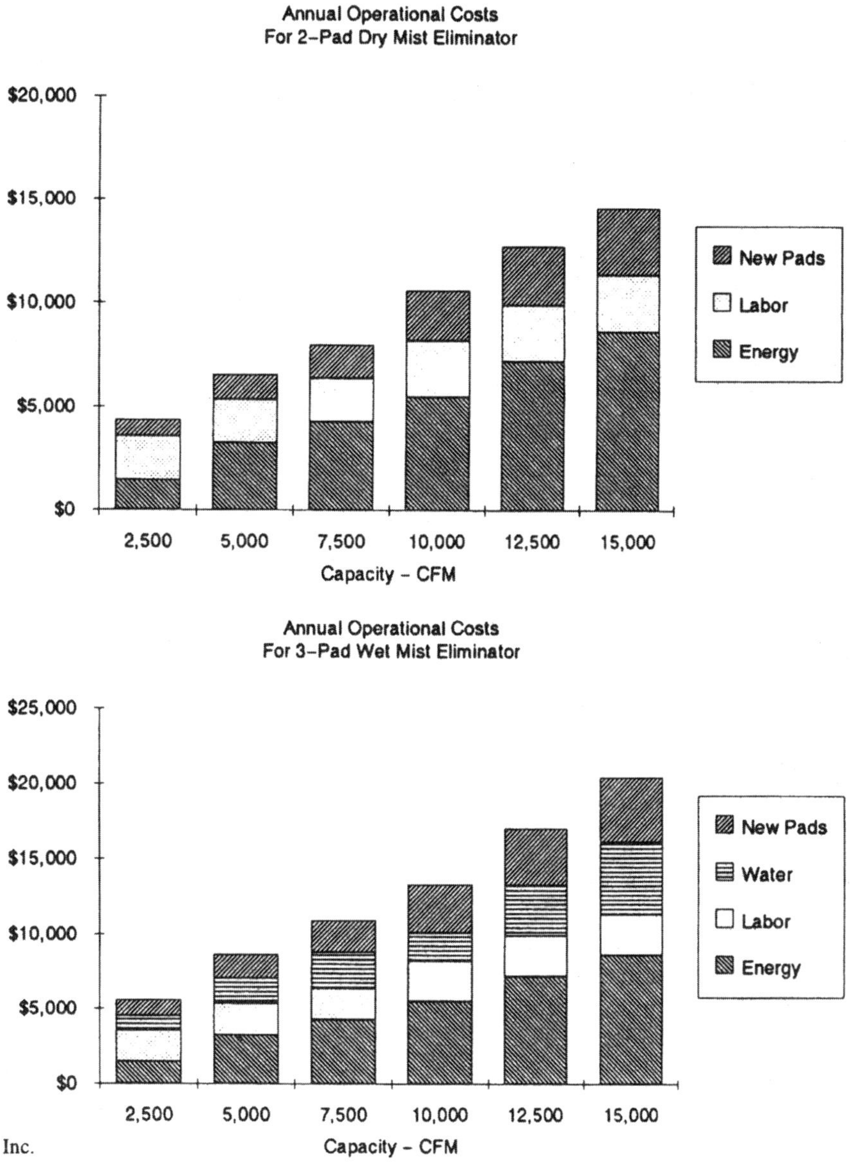

Source: ChromeTech, Inc.

Exhibit 4-71. Partial Summary of Users Data for Meshpad Mist Eliminators

				Capital Costs			Operating Costs (Yr.)		Savings				Use	Down	Satisfaction Levels (4)		Future
Code	Application	Manf.	Year	Equip. Cost	Other	Total	Non-labor	Labor Hrs.	Process Chem.	Treat. Chem.	Sludge Disp.	Total	Code (3)	Time %	Manf. Support	Technology	Decision (5)
20 (1)	hard/blk Cr	Midwest Air Products	1992	$190,000	$0	$190,000	ND	ND				ND	1		4	4	1
21	hard Cr	Midwest Air Products	1990	$50,000	$4,000	$54,000	ND	80	$1,650			$1,650	1	5	4	4	1
52	hard Cr	KCH Services	1988	$80,000	$0	$80,000	$1,000	100	$5,000			$5,000	1	2	3	4	1
67	hard Cr	Midwest Air Products	1990	$70,000	$5,000	$75,000	$150	150	$2,000			$2,000	1	0	4	4	1
105	hard Cr	Vanaire	1990	$5,000	$3,500	$8,500	$1,000	250	$250		$1,000	$1,250	1	1	4	4	1
180 (1)	hard Cr/anodize	Tri-Mer Corp.	1990	$250,000	$40,000	$290,000	$3,000	60				ND	1	1	4	4	1
253 (2)	hard Cr	Heil (Avon, OH)	1971	ND	ND	ND	$1,176	72	$546			$546	1	0	4	5	1
319	hard Cr	Midwest Air Products	1992	$23,000	$10,000	$33,000	$1,500	100			$6,000	$6,000	1	0	5	5	1
		Averages	1987	$648,286	$53,929	$702,214	$1,265	116	$1,889	$0	$1,000	$2,089		1.3	4.0	4.3	

(1) Cost data may cover multiple units.
(2) Combination meshpad/scrubber unit.
(3) Use Codes: 1-currently operating; 2-not currently operating and no intension for future use; 3-not currently in use, but intend to use in future.
(4) Satisfaction levels (manufacturer and technology) 1 to 5, with 1=lowest and 5=highest.
(5) Future decision codes: 1-purchase the same technology from the same vendor; 2-purchase the same technology from a different vendor; 3-purchase a different technology; 4-do nothing.
(6) ND = no data.

hours for meshpad mist eliminators were: 116 hrs/yr (for several of the respondents this includes O&M labor for multiple units). The skill requirement most commonly needed for operating this technology is a trained technician. The following is a breakdown of the responses for skill requirements (based on data from seven respondents, some respondents gave multiple responses):

Environmental Engineer: 0
Process/Chemical Engineer: 0
Chemist: 0
Consultant: 1
Plumber/Pipe Fitter: 1
Electrician: 1
Vendor: 0
Senior-Level Plater: 2
Junior-Level Plater: 1
Trained Technician: 5
Wastewater Treatment Plant Operator: 0
Common Labor: 2
Other: 0

- In general, meshpad mist eliminator equipment requires little maintenance and has a relatively long equipment life. Most shops indicated that the downtime for this technology was low or non-existent and none identified any major O&M problems. Long equipment life was evidenced by the fact that no shops responding to the survey indicated that they had previously used this process and subsequently discarded it. All of the identified units were in operation at the time of the survey. The oldest systems identified in the survey were purchased in 1971. All other units were purchased within the past five years.

- The following specific O&M problems were reported. PS 021 indicated that their first stage pad partially failed. PS 067 also reported a meshpad failure ("disintegration"). PS 052 reported occasional leaks in the associated ductwork, which-they patch themselves. PS 105, which reported the highest O&M labor requirement (250 hrs/yr), indicated that their unit needs daily attention, but that it is minimal.

4.8.8 Residuals Generation

When used as a recovery technology, the meshpad mist eliminator generally does not generate a residual. Chemicals captured by the meshpads are returned to the process tank during pad wash-down. None of the shops responding the survey reported a generation of residuals, other than the recovered wash-down stream.

Section 5
Chemical Solution Maintenance

5.1 INTRODUCTION

Chemical solution maintenance includes a range of pollution prevention practices and technologies that preserve or restore the operating integrity of metal finishing process solutions, thereby extending their useful lives. Some forms of solution maintenance, such as filtration, have been utilized nearly as long as metal finishing itself. However, due to rising costs for chemicals, energy and treatment/disposal and increasingly more stringent environmental requirements, solution maintenance has become a greater priority to plating shops and the methods and technologies they employ have increased in sophistication. Today, firms are willing to expend significant amounts of capital and operating funds for equipment and methods that primarily reduce the disposal frequency of their baths.

In addition to extending bath lives, solution maintenance often improves the average operating efficiency and effectiveness of a process solution and therefore has a positive impact on production rates and finish quality.

Metal finishing solutions are subjected to a variety of forces that cause them to become unusable. The key contributing factors are: (1) depletion of bath chemicals; (2) chemical break-down of process chemicals or chemical side reactions; (3) contamination from impurities in make-up water, chemicals or anodes; (4) anodic/cathodic etching of parts and inert electrodes; (5) corrosion of parts, racks, bussing, tanks, heating coils, etc.; (6) drag-in of non-compatible chemicals; (7) buildup of by-products (e.g., carbonates); (8) breakdown of maskant, fume suppressant and wetting agents; (9) errors in bath additions; and (10) airborne particles entering the tank.

Solution maintenance replaces the practices of: (1) using a chemical solution until it is degraded and replacing it with fresh solution or (2) decanting a portion of a degraded solution and replacing it with fresh solution. In both cases, the spent solution is usually either treated on-site or transported to a treatment/disposal facility. On-site treatment is not always possible because concentrated wastes may upset treatment facilities designed primarily for treating dilute rinse waters. In some cases, shops are able to reuse spent solution for either: (1) a less critical process application or (2) as a treatment reagent (e.g., spent acid cleaner used in place of sulfuric acid for pH adjustment). The former of these uses is regarded as a pollution prevention option by EPA. The latter method may reduce the overall use of chemicals at a shop, but because it involves treatment, it is not considered "pollution prevention" by EPA. EPA's definition of pollution prevention is presented and discussed in Section 3.

Two major categories of solution maintenance have been identified during the NCMS project: preventative and corrective. Within this text, preventative solution maintenance refers to the practices that avoid bath contamination or involve monitoring and adjusting of solution chemistry. Corrective solution maintenance refers to the practice of removing contaminants from the bath, whether they are dissolved or particulate, organic or inorganic. Both preventative and corrective solution maintenance involve the use of methods, techniques and technologies. Methods and techniques are typically procedural in nature or low capital items that can be implemented quickly and have an almost immediate payback. Technologies are generally equipment packages that have a moderate to high capital cost and payback periods of one year or greater. Most preventative measures are either methods or techniques. However, some technologies such as an electroless nickel bath automatic replenishment system would also fall into this category. Corrective measures include both methods/techniques such as dummy plating and technologies such as microfiltration.

Methods of preventative and corrective solution maintenance that are commonly applied by plating shops, and therefore do not require a detailed discussion, are reviewed in Section 5.2. The corrective technologies, which are generally less familiar to platers, are covered in detail in Sections 5.3 through 5.8. Exhibit 5-1 presents the results of the Users Survey that show the frequency of use for each of the common corrective bath maintenance techniques.

The information contained in Sections 5.3 through 5.8 is derived from the results of the Users Survey, Vendors Survey and literature search. The Users Survey requested platers to provide detailed technical, performance and operating cost data for bath maintenance technologies. The vendors were requested to provide technology descriptions, operating data and capital cost data. As a result of obtaining data from these two sources, plus the information from the extensive literature review, Section 5 contains a substantial quantity of information for the following corrective bath maintenance technologies: microfiltration, ion exchange, acid sorption, ion transfer, membrane electrolysis and diffusion dialysis. A separate subsection of the text is devoted to each of these technologies. Within each subsection, the following are provided: technology overview; development and commercialization; applications and restrictions (with diagrams showing different potential configurations); technology/equipment description; capital costs; operating costs; performance experience; and residuals generation. The capital cost curves contained in Section 5 are based on data collected from the technology vendors and the operating cost curves are based mainly on data from platers. Both the capital and operating cost information are expressed in 1993 dollars. A labor cost of $25 per hour (includes overhead) and an electricity cost of $0.10/kWh have been used, where applicable, in calculating operating costs.

5.2 COMMONLY APPLIED PREVENTATIVE AND CORRECTIVE BATH MAINTENANCE TECHNIQUES

5.2.1 Common Preventative Measures

Preventative methods for extending the life span of plating and other metal finishing baths were covered in Section 3.3 as part of the discussion of good operating practices. A summary of these methods, is contained in Exhibit 3-19. The following subsections address common corrective bath maintenance techniques.

5.2.2 Common Corrective Measures

5.2.2.1 Filtration

Filtration is the most commonly applied method of corrective bath maintenance. It is used to remove suspended solids from plating and other metal finishing solutions. Suspended solids in plating solutions may cause roughness and burning of deposits. Various equipment are used for filtration, with the most common being cartridge filters and precoat (diatomaceous earth) filters. Sand or multimedia filters are also employed. Cartridge filters are available with either in-tank or external configurations, with the former used mostly for small tanks and the latter for larger tanks. Most cartridges are disposable, however, washable and reusable filters have been recently commercialized. Several respondents reported use of the reusable filters (PS 180, PS 265, PS 229). Precoat filters are used mostly for large tank applications. Filter media are selected based on the chemical composition of the bath. Filtration systems are sized based on solids loading and the required flow rate (turnovers per hour). Typical flow rates for plating solution applications are 2 to 3 bath turnovers per hour. Other operating criteria for filtration processes can be found in the literature (ref. 421).

A total of 1,020 applications of bath filtration were reported by the 318 respondents of the Users Survey or an average of 3.2 per shop. Approximately two thirds of these filtration applications involved the use of the less expensive in-tank devices and one-third the external devices. The distribution of filtration applications among the different types of plating baths was approximately the same as the usage of the baths themselves (see Exhibit 5-1 which indicates for each process the total number of plating processes and the number of filtration applications identified). One notable exception was chromium plating, where filtration was not frequently applied.

Cost data for common bath maintenance techniques were not specifically requested during the Users Survey. However, four shops provided the following data for filtration.

PS 160 reported a capital cost of $14,000 (1991 costs, includes $1,000 of installation costs), annual non-labor costs of $2,000 and an annual labor requirement of 50 hours for an external sand filtration system (Techmatic) that serves three zinc chloride tanks containing 7,600 gal of solution. They reported that the sand bed requires manual stirring once per month and that the sand must be replaced once each year. They indicated that the down-

Exhibit 5-1. Summary of Users Survey Data Addressing the Use of Common Solution Maintenance Techniques

Process Name	Total No. of Processes	Carbon Treatment Batch	Carbon Treatment Continuous	Dummy Plating	Carbonate Freeze	Filtration In-tank	Filtration External	High pH	Precip.
Anodize, Boric/Sulfuric	5	0	0	0	0	1	0	0	0
Anodize, Chromic Acid	44	0	0	0	0	5	3	0	0
Anodize, Hardcoat	47	0	0	0	0	8	3	0	0
Anodize, Oxalic Acid	3	0	0	0	0	0	0	0	0
Anodize, Phosphoric Acid	4	0	0	0	0	1	0	0	0
Anodize, Sulfuric Acid	83	0	0	0	0	9	4	0	0
Bright Dip	28	0	0	0	0	3	0	0	0
Deoxidize	68	0	0	0	0	4	1	0	0
Desmut	64	0	0	0	0	3	2	0	0
Dye	69	0	0	0	0	7	1	0	0
Etch	82	0	0	0	0	3	2	0	0
Seal	54	0	0	0	0	1	3	0	0
Seal, Dichromate	37	0	0	0	0	3	1	0	0
Seal, Nickel	52	0	0	0	0	9	2	0	0
Seal, Sodium Bicarbonate	1	0	0	0	0	0	0	0	0
Abrasive Blasting	91	0	0	0	0	0	0	0	0
Acid Clean/Pickle, H2S04	131	0	0	1	0	6	0	0	0
Acid Clean/Pickle, H3P04	36	0	0	0	0	1	0	0	0
Acid Clean/Pickle, HCl	239	0	0	1	0	11	0	0	0
Acid Clean/Pickle, HF	38	0	0	0	0	0	1	0	0
Acid Clean/Pickle, HN03	51	0	0	0	0	3	0	0	0
Acid Clean/Pickle, Mixed	29	0	0	0	0	4	1	0	0
Acid Clean/Pickle,HFB03	1	0	0	0	0	0	0	0	0
Acid Salts	14	0	0	0	0	0	0	0	0
Alk., Elect.Clean	7	0	0	0	0	0	0	0	0
Alk., Elect.Clean(Non-CN)	232	0	0	0	0	18	1	0	0
Alkaline, Cyanide	26	0	0	0	0	1	0	0	0
Alkaline, Descaling	55	0	0	0	0	5	0	0	0
Alkaline, Soak(Non-Cyan)	265	0	0	0	0	21	1	0	0
Alkaline, Ultrasonic	2	0	0	0	0	0	0	0	0
Emulsion Cleaning	10	0	0	0	0	0	0	0	0
Salt Bath Descaling	3	0	0	0	0	0	0	0	0
Spray washers	28	0	0	0	0	0	0	0	0
Tumbling	49	0	0	0	0	0	0	0	0
Alkaline Resist Develop	1	0	0	0	0	0	0	0	0
Ammonium Bifluoride	2	0	0	0	0	0	0	0	0
Black oxide	55	0	0	0	0	0	0	0	0
Bright Dip of Cadmium	29	0	0	0	0	2	0	0	0
Bright Dip of Cu/Cu Alloy	61	0	0	0	0	3	0	0	0

Process Name	Total No. of Processes	Carbon Treatment Batch	Carbon Treatment Continuous	Dummy Plating	Carbonate Freeze	Filtration In-tank	Filtration External	High pH	Precip.
Bright Dip of Nickel	2	0	0	0	0	0	0	0	0
Cadmium, Vacuum	1	0	0	0	0	0	0	0	0
Chem Film	1	0	0	0	0	0	0	0	0
Chem Milling of Al	1	0	0	0	0	0	0	0	0
Chem Milling of Ti	1	0	0	0	0	0	0	0	0
Chromate on Aluminum	121	0	0	0	0	10	1	0	0
Chromate on Black Oxide	1	0	0	0	0	0	0	0	0
Chromate on Cadmium Plate	75	0	0	0	0	5	0	0	1
Chromate on Copper	1	0	0	0	0	1	0	0	0
Chromate on Magnesium	21	0	0	0	0	3	0	0	0
Chromate on Phosphate	1	0	0	0	0	0	0	0	0
Chromate on Silver	4	0	0	0	0	0	0	0	0
Chromate on Zinc	19	0	0	0	0	0	0	0	0
Chromate on Zinc Plate	114	0	0	0	0	4	0	0	0
Chromium Etch	4	0	0	0	0	0	1	0	0
Electropolish	24	0	0	0	0	0	0	0	0
Etch,Chromic Acid	1	0	0	0	0	0	1	0	0
Flame Spray	1	0	0	0	0	0	0	0	0
Galvanizing	2	0	0	0	0	0	0	0	0
Magnesium HAE	2	0	0	0	0	0	0	0	0
Mass Finishing	8	0	0	0	0	0	0	0	0
Mechanical Plating	6	0	0	0	0	0	0	0	0
Nital Etch	9	0	0	0	0	0	0	0	0
Passivation	67	0	0	0	0	2	0	0	0
Passivation, Nitric	58	0	0	0	0	1	1	0	0
Passivation, SIS	2	0	0	0	0	0	0	0	0
Phosphate, Calcium	1	0	0	0	0	0	0	0	0
Phosphate, Iron	18	0	0	0	0	2	0	0	0
Phosphate, Manganese	40	0	0	0	0	1	0	0	0
Phosphate, Titanium	1	0	0	0	0	0	0	0	0
Phosphate, Zinc	74	0	0	0	0	2	1	0	0
Tin Hot Dip	1	0	0	0	0	0	0	0	0
Tin-Lead Solder Dip	1	0	0	0	0	0	0	0	0
Titanium Anodize	1	0	0	0	0	1	0	0	0
Brass	55	5	3	4	3	12	4	1	1
Bronze	16	1	2	2	1	4	2	0	2
Cadmium, Cyanide	94	14	8	26	43	27	15	1	2
Cadmium, Non-Cyanide	8	1	1	1	1	3	1	1	1
Chrome, Black	2	0	0	0	0	0	0	0	0
Chromium, Hard	81	1	0	19	0	9	9	0	1

Exhibit 5-1 (continued). Summary of Users Survey Data Addressing the Use of Common Solution Maintenance Techniques

Process Name	Total No. of Processes	Carbon Treatment Batch	Carbon Treatment Continuous	Dummy Plating	Carbonate Freeze	Filtration In-tank	Filtration External	High pH	Precip.
Chromium,Decorative(Cr+3)	26	1	3	11	0	6	5	0	1
Chromium,Decorative(Cr+6)	91	0	0	20	0	6	7	0	1
Copper	4	1	1	1	0	1	1	0	0
Copper, Cyanide	120	21	18	18	23	41	24	0	2
Copper, Electroless	10	1	1	0	0	3	2	0	0
Copper, Fluoborate	3	0	1	1	0	1	0	0	0
Copper, Pyrophosphate	4	1	0	0	0	1	1	0	0
Copper, Strike(CN)	65	10	8	8	10	29	14	0	0
Copper, Strike(Non-CN)	13	2	2	0	1	5	3	0	0
Copper, Sulfate	61	16	10	10	0	27	17	0	0
Gold, Cyanide	64	6	7	2	1	27	5	0	0
Gold, Non-Cyanide	19	3	2	0	0	8	0	0	0
Indium	5	1	0	1	0	2	0	0	0
Iron	3	0	0	0	0	0	0	0	0
Lead	6	2	1	0	0	2	1	0	0
Lead-Tin	36	2	3	2	0	8	2	0	0
Nickel	18	4	4	6	0	4	3	0	0
Nickel, Black	2	0	0	0	0	0	0	0	0
Nickel, Brite	21	9	11	11	0	9	9	2	0
Nickel, Electroforming	3	0	0	0	0	0	0	0	0
Nickel, Electroless,Boron	4	0	0	0	0	2	0	0	0
Nickel, Electroless,Other	5	0	0	0	0	2	0	0	0
Nickel, Fluoborate	1	1	1	1	0	1	0	0	0
Nickel, Semi-Brite	3	2	2	0	0	2	1	0	0
Nickel, Sulfamate	82	38	24	48	0	36	26	11	3
Nickel, Sulfate	18	4	7	8	0	6	5	2	1
Nickel, Teflon	3	0	0	0	0	0	0	0	0
Nickel, Tin	1	0	0	0	0	0	0	0	0
Nickel, Watts	132	69	58	83	2	55	63	18	3
Nickel, Woods	32	11	9	20	0	14	6	1	0
Nickel,Electroless,Phosph	94	3	1	6	0	27	10	2	1
Palladium	2	1	0	1	0	2	0	0	0
Platinum	2	1	0	0	0	1	0	0	0
Rhodium	16	0	0	0	0	2	0	0	0
Silver, Cyanide	78	9	7	2	3	29	8	0	4
Silver, Non-Cyanide	3	2	1	1	0	2	1	0	0
Silver, Strike(CN)	2	1	1	0	0	2	0	0	0
Tin, Acid	86	5	5	11	0	22	13	0	0
Tin, Alkaline	30	1	0	5	0	6	5	0	0
Tin-Lead (PC Boards)	8	1	1	0	0	1	1	0	0

Process Name	Total No. of Processes	Carbon Treatment Batch	Carbon Treatment Continuous	Dummy Plating	Carbonate Freeze	Filtration In-tank	Filtration External	High pH	Precip.
Zinc, Acid	33	8	8	6	0	15	17	2	4
zinc, chloride	1	0	0	0	0	0	1	0	0
Zinc, Cyanide	48	5	4	11	16	8	11	1	2
Zinc, Non-Cyanide	90	10	14	23	7	29	30	0	5
Zinc/Cobalt	5	0	0	0	0	0	0	0	0
Zinc/Iron	4	0	1	0	0	1	0	0	0
Zinc/Nickel	3	0	0	0	0	0	0	0	0
Zincate	9	0	0	0	0	0	0	0	0
1,1,1-Trichloroethane	92	0	0	0	0	0	0	0	0
Acetone	4	0	0	0	0	0	0	0	0
Chlorofluorocarbons	11	0	0	0	0	0	0	0	0
Freon	1	0	0	0	0	0	0	0	0
Methyl Ethyl Keytone	15	0	0	0	0	0	0	0	0
Methylene Chloride	13	0	0	0	0	0	0	0	0
Perchloroethylene	36	0	0	0	0	0	0	0	0
Stoddard	1	0	0	0	0	0	0	0	0
Trichloroethylene	53	0	0	0	0	0	0	0	0
Xylene	1	0	0	0	0	0	0	0	0
Aluminum	1	0	0	0	0	0	0	0	0
Anodize	49	0	0	0	0	0	0	0	0
Anodize, Caustic	2	0	0	0	0	0	0	0	0
Anodize, Chrome-Phosphate	1	0	0	0	0	0	0	0	0
Brass	6	0	0	0	0	0	0	0	0
Bronze	3	0	0	0	0	0	0	0	0
Cadmium	53	0	0	0	0	0	0	0	0
Chromium	77	0	0	0	0	0	0	0	0
Copper	45	0	0	0	0	0	0	0	0
Gold	9	0	0	0	0	0	0	0	0
Magnesium HAE,Strip	1	0	0	0	0	0	0	0	0
Nickel	138	0	0	0	0	0	0	0	0
Nickel,Electroless,Strip	1	0	0	0	0	0	0	0	0
Plasma,Strip	1	0	0	0	0	0	0	0	0
Silver	31	0	0	0	0	0	0	0	0
Tin	6	0	0	0	0	0	0	0	0
Tin-Lead	32	0	1	1	0	1	0	0	0
Tungsten Carbide,Strip	1	0	0	0	0	0	0	0	0
Zinc	60	0	0	0	0	0	0	0	0
Zinc/Iron	1	0	0	0	0	0	0	0	0
Total No. of Applications	5,150	274	231	372	111	665	355	42	35

time of the unit was 2% and reported a high satisfaction level for the filter.

PS 191 reported that they have purchased multiple filter units (Serfilco) that had a total capital cost (purchased during various years dating back to 1965) of $40,940 (includes $3,000 of installation costs) to filter two nickel plating tanks totaling 7,200 gal. They employ standard filter cartridges, but rather then discard them after a single use, the spent cartridges are washed in a standard washing machining (inexpensive Sears unit). They have found that by washing the cartridges (after first soaking them in a tank), they can reuse them 3 to 4 times. They expend approximately 30 man-hrs per year washing cartridges. They first implemented this procedure in 1989 and had to replace the washer in 1992. The discharge from the soak tanks and the washer is sent to their conventional treatment system. They originally implemented this procedure in order to dispose of the filters as nonhazardous waste, but after seeing the results of the cleaning process, decided to reuse them. Used anode bags are processed in the same manner.

PS 114 reported the purchase of used filtration equipment for $500 with installation costs of $350 (1992) for filtering an electrocleaner and acid pickle (105 gal each). They indicated that annual non-labor costs are $100 and the labor requirement is only 10 hrs/yr. They estimated that by using filtration, they extend the life of the electrocleaner by 10 times and the life of the acid pickle by 5 times.

PS 253 reported cost data for a carbon treatment/filtration system (see Section 5.2.2.2).

5.2.2.2 Carbon Treatment

Carbon treatment of plating baths is a common method of removing organic contaminants. The carbon adsorbs organic impurities that are present as a result of oil introduction or the breakdown of bath constituents. It is used on both a continuous and batch basis. Various application methods are available, including carbon filtration cartridges (contain up to 8 oz of carbon and are restricted to use on small applications), carbon canisters (up to 10 lbs of carbon), precoat filters, and bulk application/agitation/filtration (ref. 421). Typical dosages are 1 to 4 pounds of carbon per 100 gallons of solution (ref. 341, 421).

A total of 505 applications of carbon treatment were reported by respondents to the Users Survey. The most frequently cited applications were nickel electroplating (mostly Watts nickel and nickel sulfamate), which accounted for 50% of all carbon treatment applications. The other most common applications were copper electroplating (18% of all applications) (mostly copper cyanide and copper sulfate), zinc plating (10% of all applications) and cadmium cyanide plating (4% of all applications). Survey respondents used continuous (46%) and batch (54%) carbon treatment methods on an almost equal basis.

Cost data were not requested in the Users Survey for common bath maintenance techniques. However, two shops provided the following data for carbon treatment. PS 253 reported a capital cost of $14,200 (installed cost for external Mefiag 6500 SS filter/carbon treatment system,1986), non-labor annual costs of $616 and 39 hours/yr of labor for maintaining a 3,600 gal Watts nickel bath. The bath is continuously pumped through a carbon filter at a rate of 55 gpm. This operation generates 125 lbs/mth of spent filter pads which are sent to off-site disposal. PS 253 indicated that they are very satisfied with the unit and that it has a downtime of 0%. PS 160 reported a capital cost of $120 (1992) for an in-tank unit (Flo-King) and annual operating costs of $240 (includes both labor and non-labor costs). They indicated that they are satisfied with the unit and that it has a downtime of only 2%.

5.2.2.3 Electrolysis (Dummying)

Dummy plating is an electrolytic treatment process in which metallic contaminants in a metal finishing solution are either plated out (low current density electrolysis) or oxidized (high current density electrolysis). Dummy plating has been documented to be in use since 1916 (ref. 339).

Low current density (LCD) dummy plating is applied to a range of plating and other metal finishing processes. The contaminant metals that are most frequently removed by dummy plating are copper, zinc, iron and lead. Dummy plating is usually performed using a corrugated steel sheet cathode with an anode to cathode spacing of approximately 4 in. The optimal current density will depend on the metal contaminants being removed. The normal range is 2 to 8 ASF. The duration of treatment is typically 2 to 5 amp-hr/gal. Agitation is essential for speedy removal of contaminants and air agitation should be used if the type of bath permits (ref. 270, 339, 340, 341).

LCD dummy plating can be performed on a batch or continuous basis. Batch treatment is usually performed in the process tank and requires down-time. Continuous treatment is usually performed in a side-tank and cathodes are typically sized to permit 0.05 amp/gal of solution (ref. 339, 341). The solution is preferably returned to the process tank through a filter (ref. 314). An example of a continuous LCD dummy plating application used by PS 118 on a closed-loop nickel plating line is shown in Section 3.4.

High current density (HCD) dummy plating typically refers to the practice of oxidizing trivalent chromium to hexavalent chromium in chromic acid baths (e.g., chromium plating and chromic acid anodizing). It is also used to gas-off chloride as chlorine.

The HCD process requires an anode to cathode ratio of between 10:1 and 30:1. Lead or lead alloy anodes are typically used in the process. A lead peroxide film is formed on the anode which functions as the oxidation agent. Current densities of between 100 to 300 ASF are used. The rate of conversion is controlled by the overall cathode and anode areas and current flow.

Respondents to the Users Survey employed dummy plating most frequently with nickel electroplating (48% of all electrolysis applications), chromium electroplating (13%), zinc electroplating (11%), copper electroplating (10%) and cadmium electroplating (7%).

5.2.2.4 Carbonate Freezing

Cyanide baths are adversely affected by carbonate buildup. Carbonates are formed by the breakdown of cyanide (especially at high temperatures), excessive anode current densities and the adsorption of carbon dioxide from the air. Excessive cabonates cause increased resistance in the bath, yielding low plating current densities, which normally accentuate the poor appearance that metallic impurities cause. An excessive carbonate concentration can affect the smoothness of deposits, plating efficiency and plating range. Both sodium and potassium baths are affected by carbonates. However, the sodium bath is affected at a much lower concentration (14 oz/gal vs 40 oz/gal) (ref. 340, 482).

Sodium cyanide baths can be treated for excessive carbonate buildup by "carbonate freezing" or crystallization. Potassium cyanide baths must be treated by precipitation, which is also applicable to sodium cyanide baths (see Section 5.2.6). Carbonate freezing takes advantage of the low solubility of carbonate salts in the sodium cyanide bath. The method involves lowering the bath temperature to approximately 26°F (-3°C) where hydrated salt/$Na_2CO_3 \bullet 10H_2O$ crystallizes out. This treatment procedure will also remove sodium sulfate and sodium ferrocyanide (ref. 339).

A total of 111 applications of carbonate freezing were reported by the respondents to the Users Survey. The vast majority of the applications were for cadmium cyanide plating, copper cyanide plating and copper cyanide strike baths.

5.2.2.5 Precipitation

Various chemical treatments of plating baths are performed to remove bath contaminants via precipitation. Precipitation is generally a batch process that is often performed in a spare tank where the solution is chemically treated and filtered and returned to its original tank. Electrolytic purification is sometimes applied following chemical treatment to remove metal contaminants less affected by precipitation (e.g., following purification of a Watts nickel bath to remove copper, zinc and cadmium). Precipitation is an alternative method to carbonate freezing for cyanide baths and is especially applicable to potassium cyanide baths. Chemicals used for this purpose include: barium cyanide, barium hydroxide, calcium hydroxide, calcium sulfate or calcium cyanide. The least expensive of these chemicals, calcium sulfate, forms a bulky precipitate that is less easily removed. Other relatively common uses of precipitation include lime addition for the removal of carbonates from silver cyanide baths, sodium sulfide treatment of cyanide baths for zinc and lead removal and nickel carbonate (1 to 2 lbs/100 gal) or nickel hydrate treatment of nickel plating baths to remove miscellaneous metal contaminants (e.g., iron, aluminum, silicon). The latter of these methods is termed "high-pH treatment." Peroxide is sometimes added during high-pH treatment to enhance precipitation and destroy organics (ref. 340). Hydrogen peroxide is also used for oxidizing soluble ferrous iron to insoluble ferric hydroxide in acid chloride zinc baths (ref. 340, 482). Two respondents reported the use of potassium permanganate (oxidizing agent) for iron removal from zinc baths (PS 076, PS 268). The precipitated iron is removed by filtration. Also, lime is sometimes used to remove phosphorus compounds from electroless nickel solutions; however, this process is not widely applied (ref. 286). One respondent reported the use of silver oxide to precipitate chloride from a decorative chromium bath (PS 162).

Respondents to the Users Survey reported 35 applications of precipitation and another 42 applications of high pH treatment. The use of high-pH treatment was restricted mostly to Watts nickel and sulfamate nickel baths. Other precipitation applications in order of decreasing frequency include: zinc plating, nickel plating, silver plating, copper plating, bronze plating, chromium plating and cadmium plating.

5.3 MICROFILTRATION

5.3.1 Overview

Microfiltration is a relatively new bath maintenance technology that is applied to aqueous and semi-aqueous degreasing and cleaning baths for the removal of oil and grease. Only one respondent to the Users Survey (or 0.3%) reported the utilization of microfiltration for this application (end-of-pipe applications of microfiltration and ultrafiltration are discussed in Section 7). Respondents to the Vendors Survey indicate that hundreds of degreasing and cleaning bath applications of microfiltration have been installed. The discrepancy between the Users Survey and the Vendors Survey with regard to the use of this technology is most likely caused by the fact that most respondents to the Users Survey are electroplating job shops. According to the vendors, most microfiltration/degreaser applications are found in captive electroplating shops and non-plating facilities such as metal-working and painting shops. Also, this technology is finding use in the printed circuit board industry for recovering semi-aqueous cleaning agents from rinse water (e.g., as part of the defluxing process) (ref. 515).

In part, electroplating job shops use solvent cleaning less frequently than captive shops because many of them receive parts for plating that have been pre-cleaned and in these cases do not require as much heavy duty cleaning or degreasing. Also, the majority of the job shop respondents that used solvents at one time eliminated their use by substituting aqueous cleaners before the time when the microfiltration technology (as applied to aqueous degreasing baths) was commercially available to them (circa 1988). Most of these shops have implemented a scheme that is technically and economically satisfactory, a condition that does not promote change. Generally, the alternative approach to maintenance is to operate pre-cleaning and cleaning baths to exhaustion and either treat them on-site or haul the spent baths to off-site treatment/disposal. Also, some shops use simple oil skimming and particulate filtration to extend bath life. Conversely, a large percentage of captive shops, painting and metal-working facilities, in response to regulations, have only recently moved to eliminate chlorinated solvent use. The pre-cleaning requirements for many of these plants are more demanding and the substitution of aqueous cleaners has therefore been slower and has coincided with the advancement of microfiltration technology. Additionally, the captive facilities generally have more capital available for the purchase of equipment.

Aqueous degreasing solutions are essentially a mixture of surfactants, alkali salts, caustic soda, phosphates, silicates and complexing agents. Aqueous degreasing and cleaning baths buildup concentrations of oil, grease and soils during use. Free oils can be removed by simple skimming and most solids can be removed by settling and/or cartridge filtration. However, emulsified oils and colloidal solids are not affected by these devices. At some point, the cleaning efficiency of the bath is impaired and the solution is discarded, despite the fact that most of the bath's constituents are still usable. In many cases, heavy duty cleaners must be replaced once per week. The microfiltration technology separates the emulsified oils from the aqueous cleaning solution, thereby extending the life of the bath. This technology is also applicable to the recovery of cleaning solution dragout from rinse waters. However, it is used much less frequently for this purpose.

In the semi-aqueous process, parts are cleaned in an organic solvent (e.g., Dupont's Axarel®, or various terpene cleaners) and subsequently rinsed in an emulsion rinse (rinse water and solvent), followed by a water rinse(s). The bulk of the cleaning solvent can be separated from the emulsion rinse by decantation and returned to the solvent cleaning tank. However, some of the solvent is carried into the water rinse. The semi-aqueous process can be operated in a near closed-loop fashion by separating the solvent from the rinse water using microfiltration, returning the solvent to the process tank and recirculating the rinse water. Although this application of microfiltration is better categorized as recovery/recycle than bath maintenance, it is discussed in this section because the equipment used for this application is similar to that used for aqueous cleaner bath maintenance.

Most commercial microfiltration systems used for this application employ ceramic filter membranes in a cross-flow filtration configuration. These membranes are a new development that permits application of microfiltration to solutions and emulsions that are both heated and corrosive. Earlier efforts using polymeric membranes were unsuccessful with this application (ref. 311). The polymeric membranes deteriorate due to the high temperatures encountered and the corrosive nature of the clean-

Exhibit 5-2. Diagram Showing Barrier Filtration vs. Cross-Flow Membrane Filtration

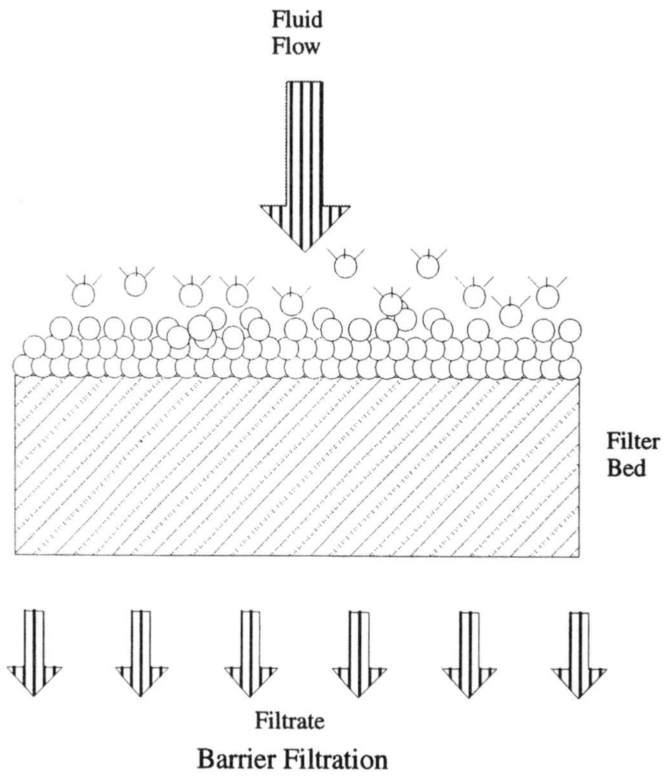

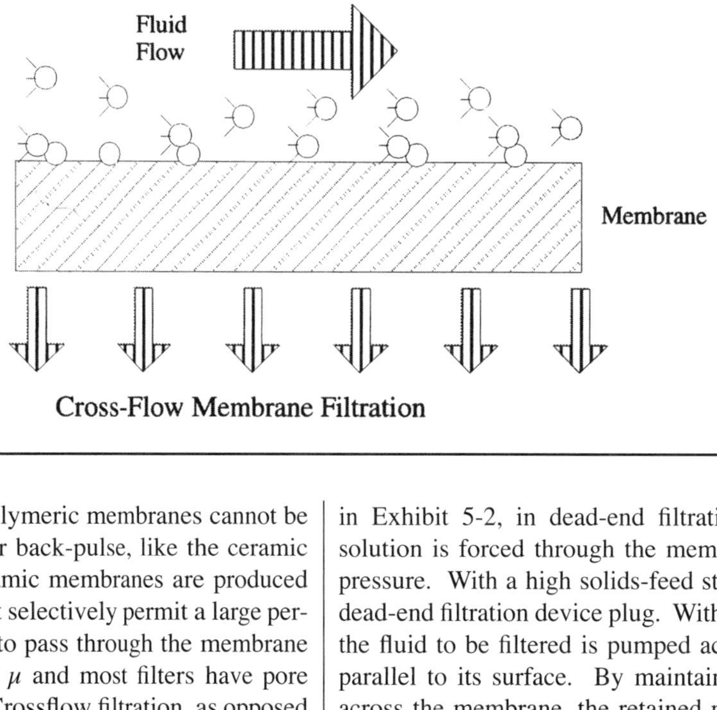

ing solutions. Also, the polymeric membranes cannot be cleaned on-line with an air back-pulse, like the ceramic membranes can. The ceramic membranes are produced in a range of pore sizes that selectively permit a large percentage of the surfactants to pass through the membrane (a typical pore size is 0.8 μ and most filters have pore sizes greater than 0.2 μ). Crossflow filtration, as opposed to barrier or "dead-end" filtration, permits the application of this technology to high solids-feed streams. As shown in Exhibit 5-2, in dead-end filtration, all of the feed solution is forced through the membrane by an applied pressure. With a high solids-feed stream, the pores of a dead-end filtration device plug. With crossflow filtration, the fluid to be filtered is pumped across the membrane, parallel to its surface. By maintaining a high velocity across the membrane, the retained material is swept off the membrane surface. This mode of operation typically requires multiple passes and consumes a greater amount of energy than with dead-end filtration. However, for high solids applications, crossflow is the only practical method (ref. 380).

5.3.2 Development and Commercialization

Although the historical roots of membrane filtration extend back to the eighteenth century, commercial and industrial applications did not emerge until the 1900's. Early investigators experimented with animal bladders serving as membrane filters. In latter work colloidal (nitrocellulose) membranes were used, and by 1906 a technique was devised to prepare nitrocellulose membranes of graded pore size. By the 1930's, mircroporous colloidal membranes were commercially available. During the next 20 years, other polymers were used, particularly cellulose acetate. These were used by the end of World War II for testing questionable drinking water supplies in Germany and other parts of Europe where water supplies had been damaged by the war. Research to develop these filters, sponsored by the U.S. Army, was later exploited by Millipore Corporation, the first and largest microfiltration membrane manufacturer. New developments in this area were limited until the U.S. Department of Interior sponsored research in reverse osmosis for desalination. These efforts also assisted the development of microfiltration, ultrafiltration and electrodialysis. A tremendous change occurred in this industry during the time period from 1960 to 1980. Advanced technologies were produced employing high performance membranes that were more chemically stable and mechanically sturdy than their predecessors. Also, membrane module design was advanced, with the development of the spiral-wound, hollow-fine-fiber, capillary and plate-and-frame modules. Application of the technology to degreasing and cleaning baths did not successfully occur until the development and commercialization of the ceramic membrane. These membranes, which were originally used in the food and beverage industry, could easily resist the temperatures and chemicals associated with the degreasing and cleaning baths. Also, they can be cleaned without shutting down the filtration process by back-pulsing the filter with air.

Although this application of microfiltration is fully commercialized, by no means is its development stagnant. Some respondents to the Vendors Survey indicated that equipment modifications are constantly being made to improve the process, expand its applicability, and to make it more cost effective.

5.3.3 Applications and Restrictions

The application of microfiltration to an aqueous degreasing operation is shown in Exhibit 5-3, which is based on a case study presented by Schwering, Golisch and Kemp (ref. 311). This application was a retrofit to an existing metal cleaning line. Prior to conversion (see top diagram), the degreasing bath was discarded after 80 hours of operation when its oil content reached 8 g/l. The bath was remade using the solution in the dragout tank as make-up. The conversion (see lower diagram) consisted of dividing the degreaser tank into two sections, changing the rinsing configuration and adding a microfiltration unit to maintain the solution in the first degreaser section. The concentration of cleaner in the first section (primary degreaser) was increased to 50 g/l and the temperature was raised to 185°F. The concentration of cleaner in the second section (secondary degreaser) was set at 20 g/l. With the installation of the microfiltration unit, a constant oil concentration of 1.0 g/l was maintained in the primary degreaser and 0.2 g/l oil in the secondary. The final rinse contained 10 mg/l oil when operated with a 3.2 gpm feed rate (67 lbs/yr oil). In addition to the rinse water discharge, 308 gpy of oil and contaminants (30% emulsion) were discarded off-site.

Microfiltration is not applicable to all aqueous degreasing and cleaning applications. Usually, shops implementing this technology must consider changing to an alternate cleaner. The most applicable type of cleaner is a simply structured non-silicate cleaner that operates in the higher temperature range (160° to 175°F or 70° to 80°C) and that can easily be replenished with additives (ref. 311). As suggested by Schwering, Golisch and Kemp, cleaners that are particularly well suited for recycling are ones that emulsify oils at the high temperature range and then have the ability to release these emulsified oils at lower temperatures and during plant shutdown (ref. 311).

Cleaning formulations with a high silicate content are generally less amenable to microfiltration recycle. These cleaners contain colloidal silicic acid, which has a tendency to plug the pores of the ceramic membrane (ref. 311). However, the only respondent to the Users Survey that employs this technology operates a bath formulated with sodium metasilicate.

Microfiltration is not applicable to aluminum cleaning solutions since the dissolved aluminum concentration will buildup over time because it will be unaffected by the filtration process.

Exhibit 5-3. Example of Microfiltration Application from the Literature

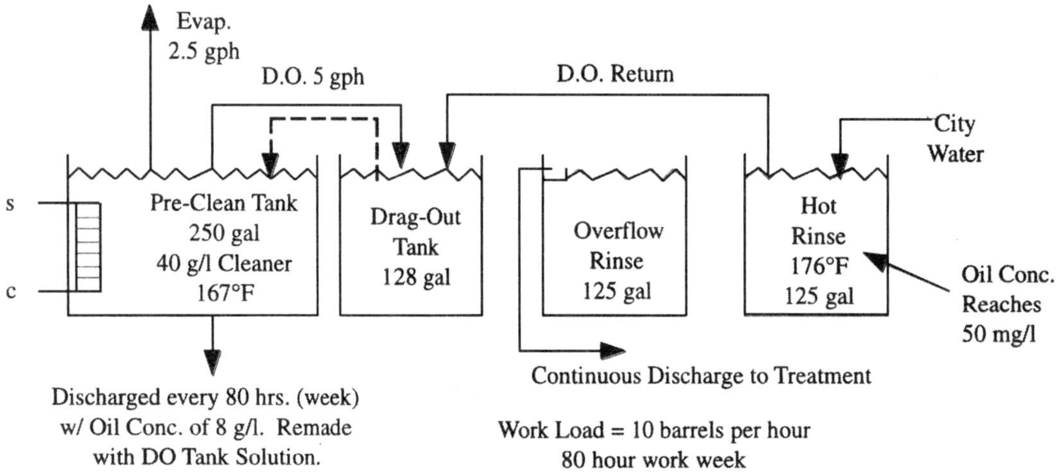

Configuration Prior to Microfiltration Installation

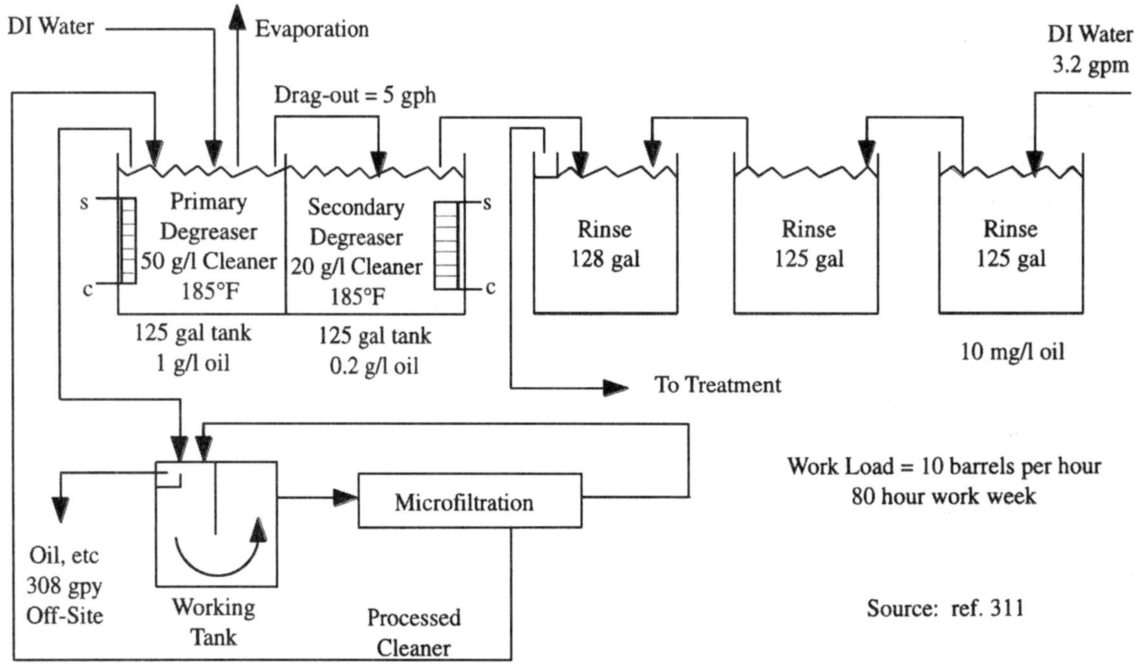

Configuration After Microfiltration Installation

An application of microfiltration to semi-aqueous cleaning is diagrammed in Exhibit 5-4. The semi-aqueous (SA) solvent tank contains 100% solvent. In this tank, soils are dislodged and/or dissolved in the solvent. Agitation (e.g., pumping) and heat can be applied to minimize the cleaning time. Following the cleaning step, the parts are transferred to the emulsion tank. Air knives are sometimes employed to minimize the carry-over of solvent. The emulsion tank contains a relatively high concentration of semi-aqueous cleaning agent. The concentration is maintained at a relatively constant level by employing a decanter/coalescer to separate the solvent and water. The solvent is returned to the solvent tank (except for periodic blowdown) and the water is returned to the

Exhibit 5-4. Example of Semi-Aqueous Cleaning Line Employing Microfiltration

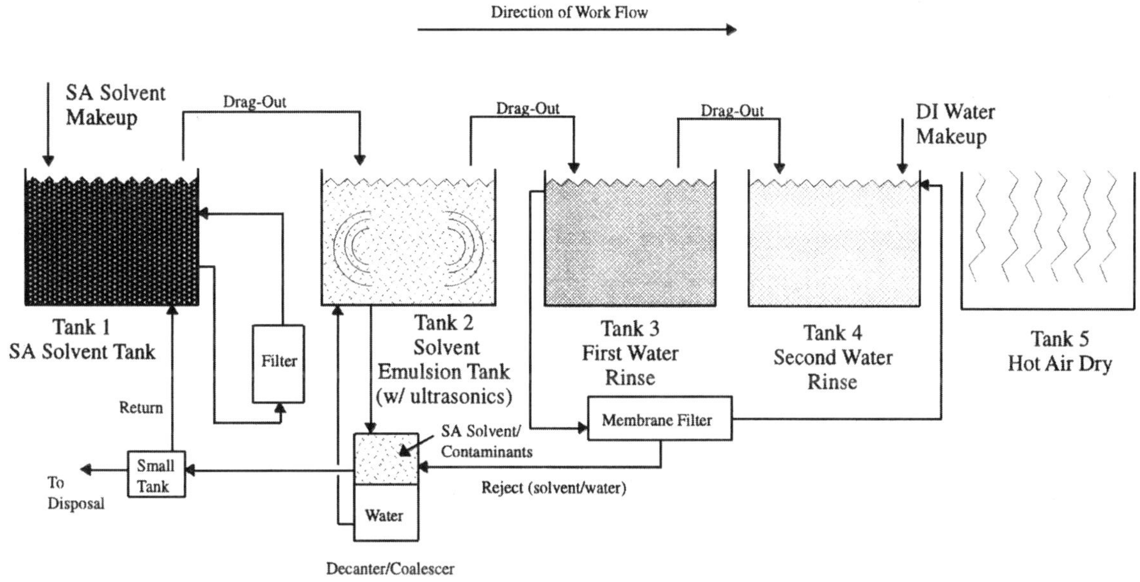

emulsion rinse. A low solvent concentration in the emulsion rinse is desired in order to minimize the carry-over of solvent to the water rinses. However, the lowest practical emulsion concentration is the minimum concentration that will separate in the decanter (e.g., the minimum concentration for DuPont's Axarel® is about 2 percent) (ref. 515). Ultrasonics and mechanical agitation (pumping) are often used in the emulsion tank to improve the efficiency of rinsing. Subsequently, parts are transferred to water rinses (usually two rinses connected in a counterflow configuration). Again, air knives are often used to minimize the carry-over of dragout. The water rinses remove the vast majority of the remaining solvent from the parts. Due mainly to the presence of the organics in the rinse water, firms sometimes separate the solvent from the rinse water using microfiltration, recover the solvent and recycle the water. Competing technologies for this application include carbon treatment and ion exchange.

5.3.4 Technology/Equipment Description

5.3.4.1 General

Equipment selected for this application should have a simple mechanical configuration and be physically sturdy and compact. The materials of construction, including gaskets and seals, must be resistant to high alkalinity, high temperature and sharp temperature fluctuations. Soils and fine metal shavings must not impede the proper functioning of the system (ref. 311).

The available systems typically have two sections within their equipment package. In the first section, free oil and settleable solids are separated from the remaining feedstream. The second section houses or is connected to the microfiltration unit. A key element of the design is the membrane cleaning function. This can be achieved by a pulsing valve fitted into the permeate line. Compressed air is periodically cycled into the permeate discharge line to generate a series of back pulses, creating a momentary (e.g., 0.5 seconds) reversal in the direction of the permeate flow (ref. 311). This prevents solids from fouling the membrane.

The performance and sizing of microfiltration systems is dependent primarily upon the specified flux. The flux is the amount of flow per unit time that will permeate a unit of area of filter space. U.S. industry usually expresses flux as gallons per square foot of filter space per day (gfd). The selection of the membrane and designation of the pressure, retenate flow rate, and concentration of oil in the influent are the most critical factors for proper system operation (ref. 477).

5.3.4.2 Commercial Equipment

This subsection contains a description of commercially available microfiltration equipment used for degreasing/cleaning bath maintenance. This is intended to provide the reader with information and data on a cross section of available equipment. Mention of trade names or commercial products is not intended to constitute endorsement for use.

Exhibit 5-5. Schematic of a Cleaner Purification System
(Courtesy of Kinetic Recovery Corporation)

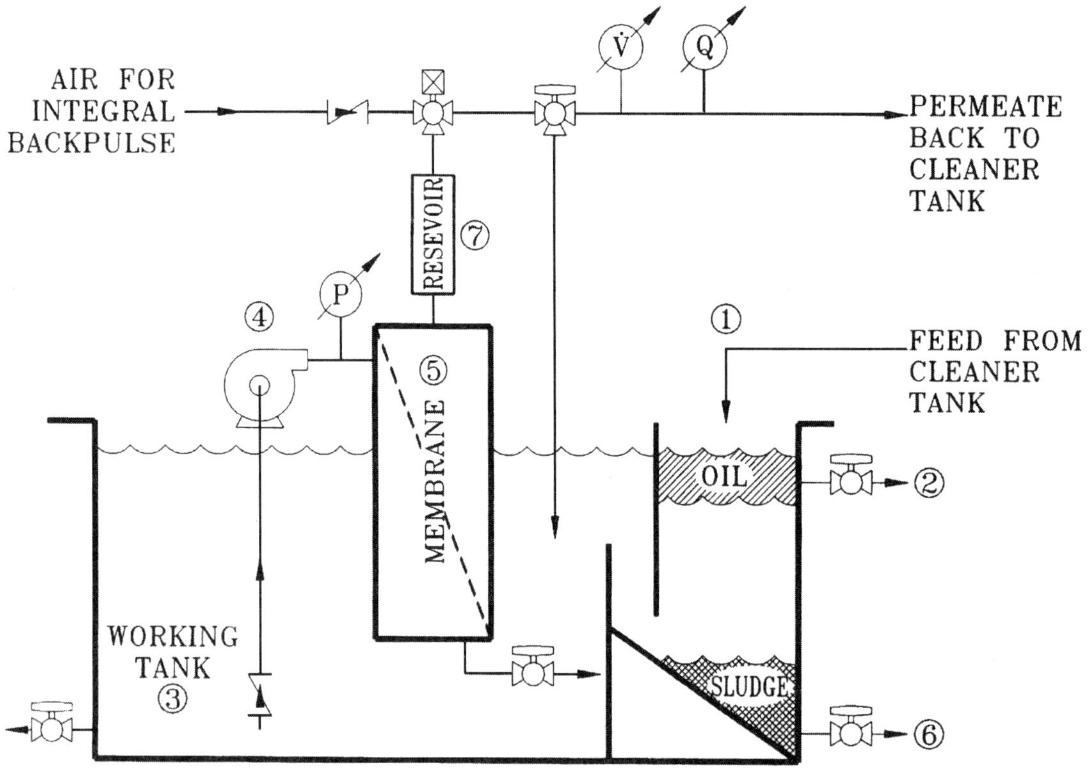

U.S. Filter markets a bath maintenance system, Membralox® 3000, which is applicable to aqueous and non-aqueous degreasing/cleaning baths. It is a packaged unit requiring a 24 in. by 40 in. floor space. It utilizes ceramic membranes with pore sizes ranging from 500 angstroms to 0.8 microns, depending upon the chemical make-up of the bath. Systems using membranes with pore sizes of 0.2 microns and larger contain an integral backpulsing system to help prevent plugging of the filter membrane. In operation, the contaminated cleaner enters a two-compartment holding tank through a bag filter which initially removes large particulate material from the feed stream. The level in the tank is maintained by a level switch, which controls the tank inlet valve and also acts as a low-level cutoff for the system pump. The oils accumulate in the initial compartment and can be drained on a periodic basis. The liquid then moves to a second tank compartment through a sub-surface passage, thereby leaving the floating oils in the first compartment. The liquid in the second compartment is pumped through the ceramic filter. The water and cleaner chemicals are forced through the ceramic membrane, while the oil and other soils are retained and recycled back to the initial tank compartment. A range of system capacities is available for feed stream flow rates of 300 gpd to 1,500 gpd. The flux rate of the membrane is approximately 50 to 100 gfd. The expected membrane life is 10 or more years.

Kinetic Recovery Corporation markets a cleaner purification system (CPS). A diagram of their equipment is shown in Exhibit 5-5. The contaminated cleaner enters the system through a tank compartment (1) that provides laminar flow conditions due to the presence of baffles. Oil accumulates in the tank and is periodically drained (2). The liquid then moves into the main tank compartment (3) (working tank) from where it is pumped by the immersed pump (4) through the ceramic membrane (5). The water with the dissolved cleaner chemicals are forced through the membrane while the oil and solids are retained and moved to the laminar flow compartment (1). The dirt and solids (sludge) which settle to the bottom of the tank can be periodically drained (6). The continuous operation of the system provides a constant stream of purified cleaner back to the cleaner bath.

The CPS utilizes microfiltration membranes (pore sizes of 0.2 μ or higher) and contains an integral back pulsing system to prevent plugging and fouling (impurity layer

Exhibit 5-6. Capitol Costs for Microfiltration Applied to Degreaser/Cleaner Bath Maintenance

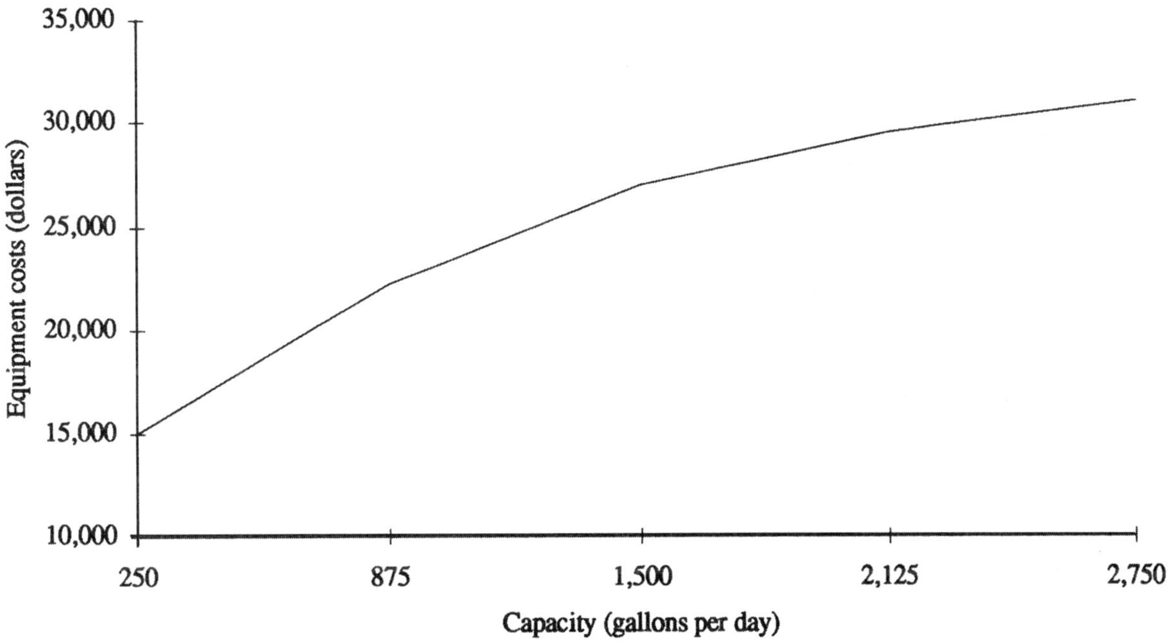

on the surface of the membrane). The integral back pulse is an air purge that forces permeate from the reservoir back through the membrane without pushing air into the membrane. This purge takes only a fraction of a second. Dirt and solid impurities end up as slurry at the bottom of the tank and are drained off.

Prosys Corporation manufactures microfiltration equipment for various applications, including: recycle of caustic cleaner, end-of-pipe treatment, polishing of treated effluent, vibratory media filtration, zyglo removal, and coolant recovery. They have been marketing equipment to the plating industry since 1987 to which they have sold over 70 microfiltration units.

5.3.5 Costs

5.3.5.1 Capital Costs

Equipment costs for microfiltration units applicable to the maintenance of cleaning/degreasing baths are presented in Exhibit 5-6. These costs are based on the use of the ceramic membrane technology with a flux rate of approximately 50 to 100 gfd. Equipment sizing is heavily dependent on the oil and soil loading of the bath. A 300 gpd microfiltration unit will typically maintain a cleaning/degreasing bath that is processing 10,000 ft^2/week of parts coated with machining oil.

Installation costs for microfiltration units are estimated to be 10% to 30% of the equipment costs.

5.3.5.2 Operating Costs

Operating costs for microfiltration have not been estimated due to the limited number of installed systems and the limited availability of operating data. Operating costs, including labor, reported by one respondent to the Users Survey are given in Section 5.3.6.

Kinetic Recovery Corporation, a vendor of microfiltration equipment, provided an example calculation of payback for a Cleaner Purification System (CPS) which is presented in Exhibit 5-7 with some modifications (to conform with standard cost items used in this book).

5.3.6 Preformance Experience and O&M Problems

Only one respondent to the Users Survey has employed microfiltration as a bath maintenance technology (PS 199). This unit was purchased from Prosys in 1992 to maintain a 600 gallon alkaline soak cleaner on a barrel line. The unit was purchased because the cleaner builds up an oil concentration of 5% (50 g/1) and requires disposal (on-site treatment) every two weeks. At the higher

Exhibit 5-7. Example of Return on Investment for a Cleaner Purification System Using Microfiltration

COSTS	
Equipment Cost:	
1 gallon per minute (960 gpd) permeate system (equipment cost plus 20% for installation)	$ 27,000
Operating Costs:	
Energy (7.5 kW/hr, 16 hr/day, 250 days/yr, $0.10/kWh)	3,000
Labor (Includes daily check and daily clean and oil drain, 0.5 hrs/day, $25/hr)	3,250
Total Operating Costs	$ 6,250
SAVINGS	
Replacement of Spent Cleaner Bath* (17 make-ups/yr, $1,008/make-up)	$ 17,136
Labor to Replace Cleaner (3 hrs/make-up, $25/hr, 17 make-ups/yr)	1,275
Neutralization Chemicals (550 gallons of 32% HCl, $0.80/gal)	440
Total Savings	$ 18,715
Return on Investment (Equipment cost/(Savings-Operating Costs)	2.1 years**

* Bath volume for this example: 2400 gallons. Typical cleaner costs are $0.42/gallon.

**Using the same assumptions and proportional operating costs and savings, the return on investment for a 3 gpm system is approximately 1 year.

Source: based on example provide by Kinetic Recovery Corporation

oil content, the bath begins to impact work quality. Their cleaner is formulated with sodium hydroxide, sodium metasilicate, sodium tripolyphosphate and mixed surfactants. This bath is used to clean approximately 10 million fasteners per year prior to zinc plating. The equipment cost of the microfiltration unit was $50,000. PS 199 estimated that non-labor operating costs were $500/yr and they projected an O&M labor requirement of 1,000 hrs/yr (plumber/pipe fitter, electrician, trained technician, and common labor).

PS 199 used the microfiltration unit for a period of only five months and then returned it to the manufacturer because of problems with membrane fouling. However, PS 199 indicated that they have considered trying a second unit from the same manufacturer that has a different back-pulse (self-cleaning) design that may eliminate the membrane fouling problem and thereby reduce labor requirements.

5.3.7 Residuals Generation

The application of microfiltration to degreaser/cleaner bath maintenance is intended to reduce the quantity of waste that is treated on-site and/or shipped off-site. The degreasing/cleaning bath is continuously maintained by removing oil and other contaminants that normally cause it to become spent. The tramp oil and other contaminants removed by skimming, settling and filtration will require off-site disposal. Depending on the volume of tramp oil generated, off-site recycling of this residual is possible.

The Users Survey did not provide any data on process residuals. In one case study, presented in the literature (see Section 5.3.3), maintenance of a 250 gallon degreasing bath with an oil loading of 100 g/hr generated approximately 300 gallons of 30 percent emulsion waste per year (based on an 80 hour work week) (ref. 311).

5.4 ION EXCHANGE

5.4.1 Overview

The use of ion exchange for bath maintenance is a relatively widespread practice; however the scope of applications is rather small. Within this text, acid sorption (also referred to as acid retardation) is presented in Section 5.5 as a separate technology from ion exchange. Although acid sorption is performed with similar equipment, as explained in Section 5.5, during the separation process ion exchange does not occur.

The results of the Users Survey show that 11 (or 3.5%) of the respondents have used ion exchange for chromic acid or trivalent chromium bath maintenance. The types of solutions treated by respondents include: hard chromium, decorative chromium (Cr^{+3}), chromic acid anodize and chromic acid copper strip solution. Hard chromium is the most frequently treated solution, making up 64% of all ion exchange bath maintenance applications identified during the Users Survey. Of the 81 shops reporting the use of hard chromium plating, 7 (or 8.6%) have employed ion exchange for bath maintenance, making it the second most popular hard chromium bath maintenance technology among survey respondents. Ion transfer (e.g., porous pot) is the most frequently employed maintenance technology (see Section 5.6).

When used to purify plating baths, ion exchange (cation only) removes impurities that buildup in baths from drag-in, corrosion of parts, racks and anodes, reduced or decomposed bath chemicals, and other sources. These contaminants reduce the performance of the bath and eventually accumulate up to a concentration where the bath must be discarded. Also, contaminated baths cause platers to increase the concentration of plating chemicals so that they are able to maintain plating efficiency. This results in higher solution viscosity which in turn increases dragout rates and overall chemical losses. Other negative aspects of operating contaminated baths include lower plating rates and higher electrical consumption (ref. 370).

A typical application for ion exchange bath purification is the removal of iron and trivalent chromium from hexavalent chromium plating solutions. Purification can be accomplished by directly treating the bath. However, in some cases, such as chromium plating, the concentrated bath has a detrimental impact on the resins, which shortens the life of the material. As an alternative method, the bath can be diluted, treated with ion exchange and reconcentrated. Reconcentration is unnecessary in instances where the surface evaporation rate of the plating bath provides sufficient headroom to return the treated solution. Also, if dragout recovery is practiced, the dragout, which is typically less concentrated than the bath, can be treated with ion exchange before it is returned to the bath. A sufficiently high dragout rate is needed for this strategy to work. Even when chromic acid is diluted, it has a detrimental effect on ion exchange resin. The short resin life (6 to 12 months) and its replacement cost are simply accepted as part of the operating costs for this process.

Ion exchange competes with ion transfer (Section 5.6) and membrane electrolysis technologies (Section 5.7) as a chromic acid bath maintenance technology for tramp metal removal. Dummy plating (high current density electrolysis, Section 5.2.4) is an alternative method for trivalent chromium oxidation, but it is ineffective for tramp metal removal. There is no clear choice between ion exchange, ion transfer and membrane electrolysis for tramp metal removal.

The porous pot type ion transfer technology is the least capital intensive technology (single tank models are less than $1,000), but it has a questionable role as a pollution prevention tool due to the high quantity of residual waste generated. In some cases, the porous pot is comparable to a "bleed and feed" method of tramp metal control. It is however, an effective method of trivalent chromium oxidation. The polyester membrane ion transfer technology may reduce residual waste quantities, however, there are insufficient data available to evaluate its performance.

Ion exchange can also generate a significant chromium waste volume. This process is unable to oxidize trivalent chromium to the desired hexavalent state, like ion transfer and membrane electrolysis. However, unlike ion transfer, it does not produce significant quantities of hexavalent chromium wastes. Also, trivalent chromium losses can be reduced by using selective resins and operating them to exhaustion (discussed in Section 5.4.3).

Membrane electrolysis, which performs both trivalent chromium oxidation and tramp metal removal, appears to be the best technology in terms the ratio of chromium residual volume generated to the volume of bath treated (ref. 370). However, some users of this technology indicate that it has significant O&M problems (see Section 5.7.7). Also, it is the most capital intensive method of the three technologies.

As such, the problem of chromic acid bath maintenance is still unresolved. One respondent to the Users Survey, who described this dilemma as "one of the biggest problems facing hard chrome platers," has purchased and operated an ion transfer unit (porous pot), an ion specific electrochemical membrane unit and ion exchange technology (PS 234). This respondent concluded that ion exchange was the best method, but indicated that it produced a high waste load (see complete comments by PS 234 in Section 5.6.6).

General background information on the ion exchange process and applications involving chemical recovery are presented in Section 4. End-of-pipe applications of ion exchange are discussed in Section 7.

5.4.2 Development & Commercialization

The development and commercialization of the ion exchange process is reviewed in Section 4.4.2. The use of this technology for chromic acid bath maintenance is one of the oldest pollution prevention applications of ion exchange for the metal finishing industry. The first reported application was at the Rock Island Arsenal in 1952. The Arsenal's unit was designed to treat 1,000 gallons of contaminated chromium plating solution or 5,000 gallons of chromic acid anodizing solution. The purification process involved passing contaminated solution through a cation resin to remove trivalent chromium and tramp metals. The solutions were diluted prior to ion exchange treatment to avoid rapid oxidation of the resin and then reconcentrated using an evaporator. The general practice has not changed significantly since the first application, however today's resins are more chemically tolerant and selective and the equipment is much more sophisticated (e.g., automatic controls) (ref. 384). The first known commercial ion exchange product designed specifically for chromic acid bath maintenance was manufactured in 1975 (ref. 385).

Two of the respondents to the Vendors Survey (Eco-Tec Inc. and Kinetico Engineered Systems, Inc.) manufacture commercial units specifically for chromic acid bath treatment and a third company produces custom units. Kinetico Engineered Systems, Inc. also manufactures ion exchange equipment for trivalent chromium bath maintenance. Eighty-two percent of the ion exchange bath maintenance systems identified in the Users Survey were manufactured by either Eco-Tec Inc. or Kinetico Engineered Systems.

5.4.3 Applications

The primary application of ion exchange, when used as a bath maintenance technology, is the removal of cations from chromic acid baths. Of the eleven respondents to the NCMS/NSMF User Survey that employed ion exchange for bath maintenance, ten of their applications were for chromic acid baths and one for a trivalent chromium bath. The chromic acid bath applications included seven hard chromium, two chromic acid anodize and one chromic acid copper strip solution.

Exhibit 5-8 shows two typical applications for this technology. In IXBM-1 the ion exchange process is applied directly to the process solution and in IXBM-2 the process is applied to the return flow from the dragout recovery tank. IXBM-1 is applicable to both hexavalent chromium and trivalent chromium baths. In order for IXBM-2 to be effective, the dragout rate would have to be sufficiently high such that the percentage of the chromic acid bath treated over a given time period is sufficient to keep bath contaminants at a tolerable level. The necessary dragout rate depends heavily on the rate of contaminant buildup. Typically, a dragout rate equal to one tank volume per year will be sufficient for most hard chrome plating processes to operate with the IXBM-2 configuration. Anodizing processes and chromium plating processes with high contamination rates may require even greater dragout rates to use this configuration. When the dragout rate is insufficient (this will be evidenced by a steady buildup in contaminant concentration), the plater can bleed the bath into the dragout tank. However, this practice may overly contaminate the rinse system and result in poor rinsing.

Several respondents to the Users Survey combine the use of ion exchange with evaporative recovery operations for chromic acid recovery. In these cases, ion exchange is applied to the recovered chromic acid solution before it is concentrated by the evaporator and returned to the plating tank (PS 018, PS 082, PS 125).

Resin selectivity is an important factor with the chromic acid bath maintenance process because it reduces trivalent chromium losses. The practice of separating ions of the same charge is referred to as selective ion exchange (SIE) (ref. 46). SIE takes advantage of the preference that ion exchange resins exhibit for some ions over others to selectively separate different ions that have the same charge and would normally be removed by the same resin. Essentially, all resins exhibit selectivity. However, some resins are more selective than others and the

Exhibit 5-8. Common Configurations for Application of Ion Exchange for Bath Maintenance

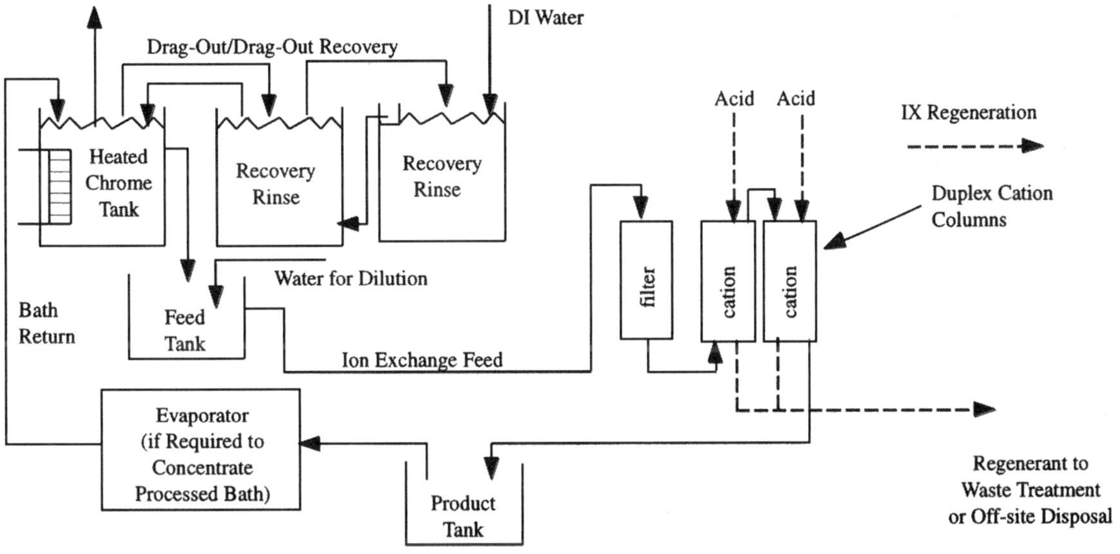

Application IXBM-1: Chromium Bath Treatment

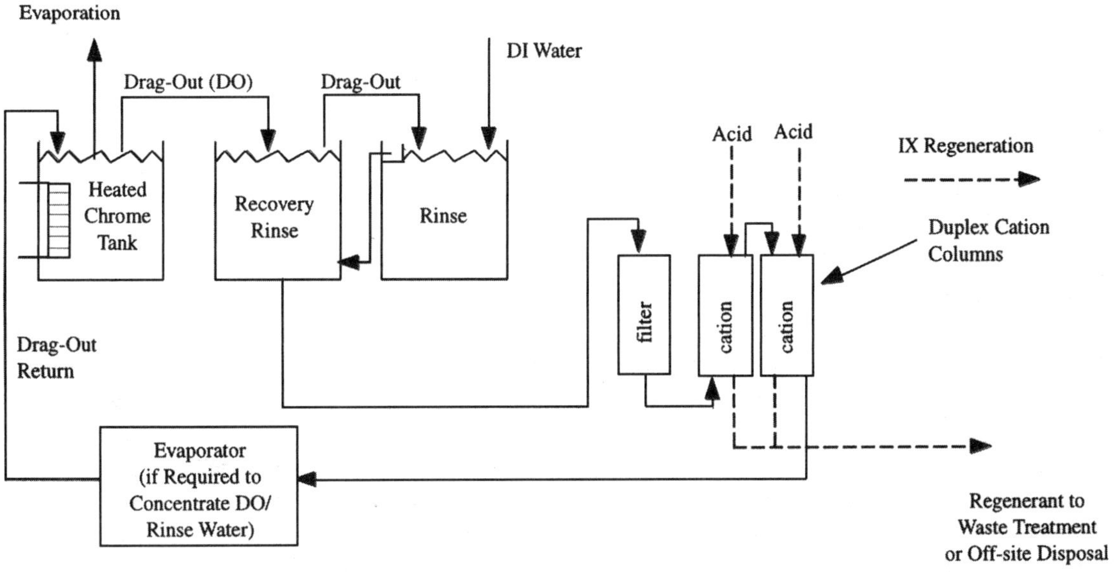

Application IXBM-2: Ion Exchange Recovery-Chromium Drag-Out Return

ordering of selectivity among metal species varies among resins. When employed for chromic acid purification, the ion exchange sites of the resin exchange hydrogen for the tramp metals and Cr^{+3}. When most of the hydrogen is displaced from the resin sites, the resin column shows its preference by displacing Cr^{+3} that has been attached to the resin with the preferred tramp metals. The Cr^{+3} goes back into solution and is returned to the bath along with the chromic acid. The process, if run to exhaustion (i.e., the point where the percentages of ions held by ion exchange sites are in proportion to the preference of the resin), will result in a high removal of the preferred tramp metals and a low removal of Cr^{+3} (ref. 384). If Cr^{+3} buildup presents a plating quality problem with a given bath, electrolysis or dummy plating (Section 5.2.2.3) can be employed.

Sulfuric acid is used to regenerate the cation exchange columns for hexavalent chromium bath applications and sulfuric acid and ammonium hydroxide (NH_4OH) are used for trivalent chromium bath applications. Ammonium hydroxide is used with the trivalent application to remove copper, which is tightly held by the resin and is not sufficiently removed by the acid. The ammonium hydroxide step is not usually performed during each regeneration cycle, but rather as needed to remove the copper (e.g., every two to four times purification is performed).

Because ion exchange is a separation technique that is dependent on the general chemistry of the process solution and electrochemical differences between the metal species (i.e., plated metal vs the tramp metal), it is not applicable to the maintenance of many plating solutions. This is due to the fact that, most frequently, the metal being plated and the contaminants are both cations (e.g., nickel baths). Although different cations can be separated by selective ion exchange, it is a more difficult and less efficient process than the separation of cations and anions. In this regard, chromic acid baths are a good candidate for ion exchange since the desired species of chromium (i.e., hexavalent chromium) is an anion and the tramp metals (including trivalent chromium) are cations. Cyanide-metal complexes in cyanide plating solutions (e.g., copper cyanide) also have the opposite charge of tramp metals. However, if ion exchange bath maintenance was attempted, the cyanide-metal complex would be destroyed by the acidic conditions in the cation resin bed causing the plated metal to be removed from solution along with the tramp metals (see Section 4.4.3). Also, for these applications, ion exchange would have to compete with low current density electrolysis (dummy plating), which is an inexpensive and effective method of controlling tramp metals for many non-chromium baths.

The literature indicates that selective ion exchange (see Section 5.4.4) has been investigated as a potential solution maintenance technology for acid etch and nickel strike baths (ref. 384). The results of the investigation showed that the SIE process was impractical for these applications because the resin could not be adequately regenerated. Additionally, like the cyanide-metal baths, these two baths are treatable by dummy plating, making the SIE application less significant.

Tinker Air Force Base (Oklahoma City, OK) is conducting research, through their contractor Science Applications Incorporated (SAIC), to evaluate the applicability of ion exchange to the maintenance of cadmium stripper solution (ammonium nitrate). Early project results are very promising.

5.4.4 Technology/Equipment Description

This subsection contains a description of commercially available ion exchange equipment that is manufactured and/or sold by vendor survey respondents. This is intended to provide the reader with information and data on a cross section of available equipment. Mention of trade names or commercial products is not intended to constitute endorsement for use.

Eco-Tec Inc. manufactures a Decationzation Unit (DCU) that consists of a patented Recoflo™ ion exchange bed (see description in Section 4.4.1.3), filter, piping, valves and control panel. The DCU is an integrated ion exchange system as defined in Section 4.4.1.2. Depending on the model selected, the unit treats 50 to 200 gal (200 to 750 l) of chromic acid plating bath per run. The process is initiated by transferring the solution to be treated into a feed tank. The DCU is then activated and the process proceeds automatically. The unit automatically dilutes the acid to between 100 and 150 g/l (13 to 20 oz/gal) to reduce oxidation of the resin before the contaminated chromic acid passes through the bed. The DCU continuously operates a three step cycle where contaminated chromic acid is fed through the unit and the bed is rinsed and regenerated (sulfuric acid) in a counterflow direction. The treated solution is transferred to another holding tank and is returned to the bath. The surface evaporation of the bath makes up for the small quantity of water added during the process. According to Eco-Tec, the process removes 80 to 90% of the bath contaminants and generates about 3 gal/hr of residual (regenerant). The expected resin life is 6 to 12 months (ref. 385). Eco-Tec also markets a batch treatment system for treatment of the regenerant from this process.

Kinetico Engineered Systems, Inc. (Kinetico) manufactures a modular ion exchange system (see definition in Section 4.4.1.3) for hexavalent and trivalent chromium bath maintenance. The modular design is referred to as "point of use" by Kinetico. The ion exchange service unit consists of a bath purification ion exchange station and an ion exchange tank. The station consists of a bath recirculation pump, cartridge filter unit, timer mechanism and controls mounted on a skid assembly. The ion exchange tank is a separate unit that connects to the service station. The service units are located at the point of use (near the plating tank) for bath treatment and the tanks are transported to a central regeneration station after use. The central regeneration station is similar in appearance to the ion exchange station, but performs the function of regeneration. The central regeneration station can be rigged to treat both hexavalent and trivalent baths. After regeneration, the tanks are returned to the point of use. Specifications for a range of Kinetico models are shown in Exhibit 5-9.

5.4.5 Costs

5.4.5.1 Capital Costs

A diagram showing the equipment costs for the application of ion exchange to chromic acid baths is presented in Exhibit 5-10. Installed costs for these systems are approximately 10 to 30 percent higher than basic equipment costs.

5.4.5.2 Operating Costs

Operating cost data provided by respondents to the Users Survey are presented in Exhibit 5-11. The annual costs and savings for the respondents are shown graphically in Exhibit 5-12, which also expresses capital costs in terms of 1993 dollars. This graphic indicates that a relatively rapid capital payback is possible with this technology (see bar graph for PS 049 and PS 191).

5.4.6 Performance Experience

Eleven respondents to the Users Survey provided some detailed data on their experience using ion exchange for chromium bath maintenance. All of these applications involved hexavalent baths, except for one, which was applied to a trivalent decorative chromium bath. A summary of the Users Survey data for these applications is presented in Exhibit 5-11.

The following information and data summarize the performance experience of the eleven survey respondents.

- In general, shops using ion exchange for bath maintenance gave the technology a high rating. One exception to the high satisfaction level was PS 244. This shop only operated the process for eight months during 1983 and 1984. Their low level of satisfaction was due to a high residuals generation rate (see Section 5.4.8). Also, they indicated that the supplier stated capacity of the unit was 208 gallons per 12 hours and that the actual capacity was approximately fifty percent of this rate. Another shop that gave this technology a low rating was PS 273. This shop indicated that they discontinued use of the technology shortly after startup for administrative and technical reasons (discussed later in this section). The one shop using ion exchange for trivalent chromium bath maintenance had a satisfaction level of 5. The average satisfaction level for ion exchange applied to bath maintenance is 3.9 (on a scale of 1 to 5, with 5 being most satisfactory), which is higher than the average level rating for ion exchange chemical recovery (3.2). Seven of the shops (or 64%) indicated that this technology satisfied the need for which it was purchased. The following is a breakdown of the reasons why shops purchased this technology (multiple responses were permitted):

To meet of help meet effluent regulations: 4
To reduce plating chemical purchases: 8
To reduce the quantity of waste shipped off-site: 8
To reduce wastewater treatment costs: 4
To improve product quantity: 9

- The use of ion exchange for bath maintenance generally improved the production quality and to a lesser extent improved the rate of production. The following responses were provided:

	Product Quality	Production Rate
Improved	8	4
No Change	2	6
Decreased	0	0

- Most plating shops indicated, that based on their experience with this technology, they would purchase the same type of equipment from the same vendor. The following is a breakdown of their responses:

Purchase the same technology from the same vendor: .. 7
Purchase the same technology from a different vendor: .. 1
Purchase a different technology: 3
Do nothing: 0

Exhibit 5-9a).
Commercial Ion Exchange Bath Maintenance System Specifications for Chromic Acid Bath Maintenance Systems (Courtesy of Kinetico Engineered Systems)

Model Number	CAIX-40 Exchange Station	CAPU Regeneration Station
System Dimensions	19" x 24"	21" x 24"
Media Tank Dimensions	14" x 47"	none
Resin Volume	3.0 cubic feet	none
Maximum Flow Rate	4 gpm	2 gpm
Power	230/460 volts 4/2 amps	100 volts 4 amps
Regenerant Volume/Cycle (30% sulfuric acid)	none	6.5 gallons
Regenerant Time	none	120 minutes
Regenerant Volume/Cycle	none	200 gallons

Exhibit 5-9(b).
Commercial Ion Exchange System Specifications for Trivalent Chromium Bath Maintenance Systems (Courtesy of Kinetico Engineered Systems)

Model Number	TCIX-10	TCIX-20	TCIX-30	TCIX-40	TCPU-1
System Dimensions	19" x 24"	19" x 24"	19" x 24"	19" x 24"	21" x 24"
Media Tank Dimensions	8" x 10"	10" x 40"	12" x 42"	12" x 42"	none
Resin Volume	0.5 cubic feet	1.0 cubic feet	1.5 cubic feet	2.0 cubic feet	none
Maximum Flow Rate	1 gpm	2 gpm	3 gpm	4 gpm	2 gpm
Regenerant Volume/Cycle (30% H_2SO_4)	3.8 gallons	3.8 gallons	4.5 gallons	4.5 gallons	none
Regenerant Volume/Cycle (25% NH_4OH)	5.9 gallons	5.9 gallons	5.9 gallons	7.0 gallons	none
Regeneration Time	70 minutes	70 minutes	90 minutes	90 minutes	none

- One shop indicated that they are able to recycle spent chromic acid bath as another product and have therefore eliminated the need for the ion exchange system (PS 244).

- The major cost savings from the operation of ion exchange for bath maintenance were due to reductions in bath chemical use and disposal costs.

- One respondent used ion exchange to treat a chromic acid copper strip solution (PS 273). Their efforts have been unsuccessful thus far (project stopped due to change in personnel and other reasons). This facility uses, on an annual basis, 25,900 lbs/yr of chromic acid to strip approximately 3,200 lbs of copper in a 1,158 gallon strip tank. The process is operated at 160°F with a chromic acid concentration of 120 g/l. This shop purchased a water softening ion exchange unit and was told by the manufacturer's representative that they could process the bath at full strength at its operating temperature. The result was that the "copper removal efficiency dropped quickly after startup." The respondent felt that the hot, concentrated chromic acid had oxidized the resin.

- At different times, one of the respondents used ion exchange, the porous pot technology and membrane electrolysis for chromic acid bath maintenance. This shop indicated that ion exchange provided the best results in terms of impurity removal (PS 234). The complete statement provided by this shop, which provides a great deal of insight to this problem, is presented in Section 5.6.6.

Exhibit 5-10. Capitol Costs for Ion Exchange Bath Maintenance Equipment

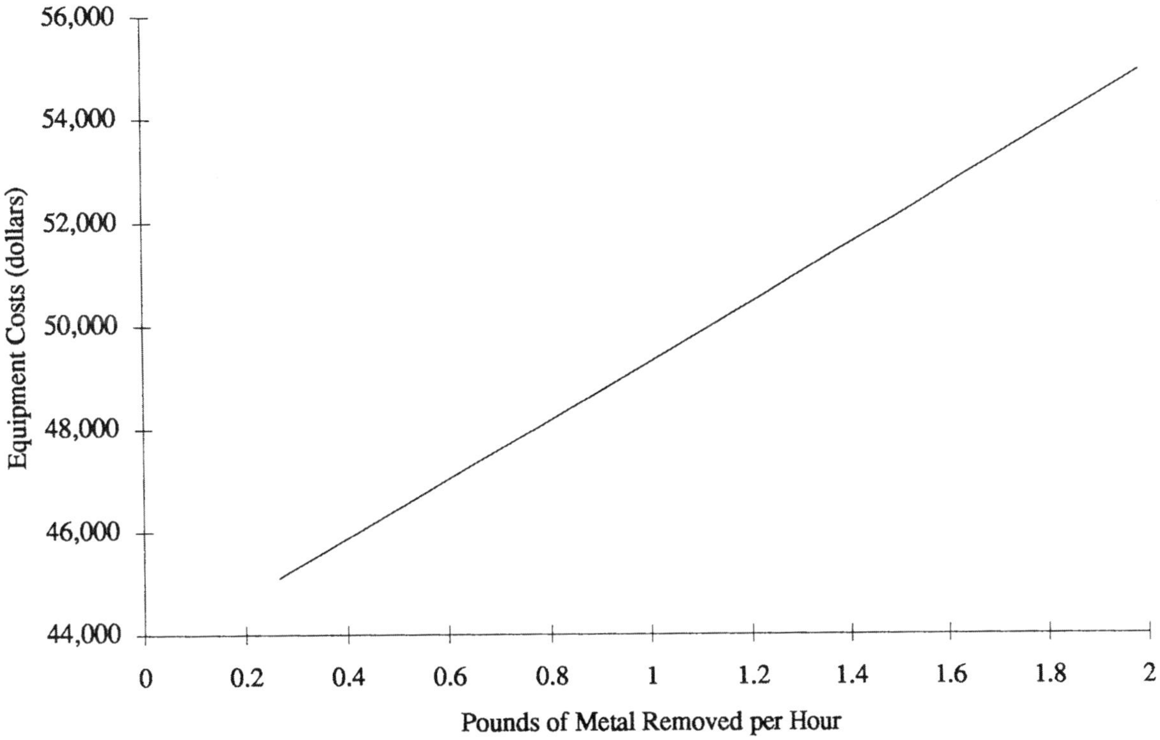

5.4.7 Operational and Maintenance Experience

The following summarizes the respondent's O&M experiences and provides operating labor information.

- Six shops provided operating labor data. For these shops, the average number of annual operating hours per ion exchange system were 220 hrs/yr. The labor category most frequently used for O&M was a trained technician or a wastewater treatment plant operator. The following is a breakdown of the responses for skill requirements:

  Environmental Engineer: 0
  Process/Chemical Engineer: 0
  Chemist: 2
  Consultant: 0
  Plumber/Pipe Fitter: 0
  Electrician: 0
  Vendor: 0
  Senior Level Plater: 2
  Junior Level Plater: 0
  Wastewater Treatment Plant Operator: 4
  Trained Technician: 6
  Common Labor: 0
  Other: 0

- In general, the commercial equipment used for chromium bath maintenance requires little maintenance and has a relatively long equipment life. Most shops indicated that the downtime for this technology was low and none identified any major O&M problems except for PS 273.

- Long equipment life was evidenced by the fact that only one shop (PS 273) in the survey indicated that they used this process and subsequently discarded it due to deterioration (note that one shop purchased their equipment in 1983 and later sold it and that PS 244 eliminated the need for this equipment). The oldest system identified in the survey that is still operating was purchased in 1975 (PS 131). Shops estimated the life of this equipment to be between 5 and 20 years and the average estimate was 11.5 years.

Exhibit 5-11. Summary of Users Survey Data for Ion Exchange Bath Maintenance Applications

Code	Application	Manf.	Year	Capital Costs			Operating Costs (Yr.)		Savings						Use Code (3)	Down Time %	Satisfaction Levels (4)		Future Decision (5)
				Equip. Cost	Other Cap. Costs	Total Cap. Costs	Non-labor	Labor Hrs	Process Chem.	Treatment Chem.	Sludge Disp.	Labor	Other	Total			Manf. Support	Tech.	
049	Anodize, Chromic	Kinetico	1990	$18,000	$1,000	$19,000	$1,500	60	$2,000	$1,000	$10,800	$1,000		$14,800	1	10	1	5	2
067	Chromium, Hard	Eco-Tec	1993	$49,000	$3,000	$52,000	ND	ND	ND	ND	ND	ND	ND	ND	1		4	5	1
077	Chromium, Hard	Eco-Tec	1991	$60,000	$40,000	$100,000	ND	ND	√	√	√	√	$0	√	1	10	4	4	1
131	Chromium, Hard	Eco-Tec	1975	$26,000	$3,500	$29,500	ND	ND	ND	ND	ND	ND	ND	ND	1	1		4	1
150	Chromium, Hard	Eco-Tec	1992	$65,000	$16,000	$81,000	ND	145	ND	ND	ND	ND	ND	ND	1	1	4	4	1
165	Chromium, Hard	Eco-Tec	1987	$31,000	$12,000	$43,000	$5,513	132	ND	ND	ND	ND	ND	ND	1	1	4	5	1
191	Chromium, Cr+3	Kinetico	1991	$34,121	$1,500	$35,621	$1,000	150	$16,000	$1,000	$2,500	$8,000	$500	$28,000	1	1	5	5	1
198	Chromium, Hard	Aqua Line Tech.	1990	$10,000	$800	$10,800	$1,000	50	$2,000	$0	√	$0	$0	$2,000	1	1	4	4	1
234	Chromium, Hard	Eco-Tec	1983	$17,000	$0	$17,000	ND	ND	ND	ND	ND	ND	ND	ND	2				3
244	Anodize, Chromic	Eco-Tec	1982	$14,647	$1,500	$16,147	ND	780	$0	$0	$0	$0	$0	$0	2		3	1	3
273	CrO3 Copper Strip	Ill. Water Treat	1987	$58,000	$0	$58,000	ND	ND	ND	ND	ND	ND	ND	ND	2		3	2	3
	Average-All		*1987*	*$34,797*	*$7,209*	*$42,006*	*$2,253*	*220*	*$5,000*	*$500*	*$4,433*	*$2,250*	*$125*			*4*	*3.6*	*3.9*	

(1) Other capital costs include installation and ancillary equipment.
(2) Use Codes: 1-currently operating; 2-not currently operating and no intension for future use; 3-not currently in use, but intend to use in future.
(3) Satisfaction levels (manufacturer and technology) 1 to 5, with 1=lowest and 5=highest.
(4) Future decision codes: 1-purchase the same technology from the same vendor; 2-purchase the same technology from a different vendor; 3-purchase a different technology; 4-do nothing.
(5) "√" indicates that a savings was realized, but not quantified.
(6) ND = no data.

- The percentage of down time for this technology was low. Nearly one-half of the respondents indicated that downtime was 1% or less.

- Several shops pointed out that there are two key technical restrictions for this technology: chromic acid concentration and temperature (PS 049, PS 150, PS 165, PS 273). Two shops indicated that the solution temperature must be 90°F or less before ion exchange treatment is performed (PS 165). Two shops pointed out that the chromic acid concentration must be below 142.5 g/l (19.0 oz/gal) CrO_3 for treatment (PS 165). One respondent indicated that their operating conditions (160°F and 16 oz/gal CrO_3) may have contributed to resin damage (PS 273).

- One shop indicated that elements and filters must be changed two times per year (PS 165).

- The shop operating an ion exchange unit for trivalent chromium bath maintenance reported that they experience mechanical breakdowns about four times per year. They also indicated that the downtime of their unit was less than one percent.

- PS 131, the respondent that has used this process the longest (since 1975), indicated that the process dilutes the hard chrome bath and raises the sulfate level, causing them to adjust the solution chemistry following treatment.

5.4.8 Residuals

The quantity and nature of the residuals generated from ion exchange treatment of chromium baths appears to be the most significant concern of platers using this technology (PS 198, PS 234, PS 244). The process generates a concentrated acidic waste during regeneration that contains mostly trivalent chromium and tramp cations such as iron and aluminum and dilute streams from backwashing and cleansing of the bed. Only two shops provided quantity data for this residual. PS 045 indicated that they generate 700 gallons per month from the maintenance of a 650 gallon chromic acid anodizing tank with a production rate of 363,670 amp-hrs/year. PS 244 indicated that they generated 7.7 gallons of mixed water and acid for each gallon of chromic acid anodizing solution treated.

Nearly all of the shops that reported residuals data indicated that regenerant is treated in-house and that the

Exhibit 5-12. Selected Users Survey Cost Data for Ion Exchange Bath Maintenance Applications Currently in Use

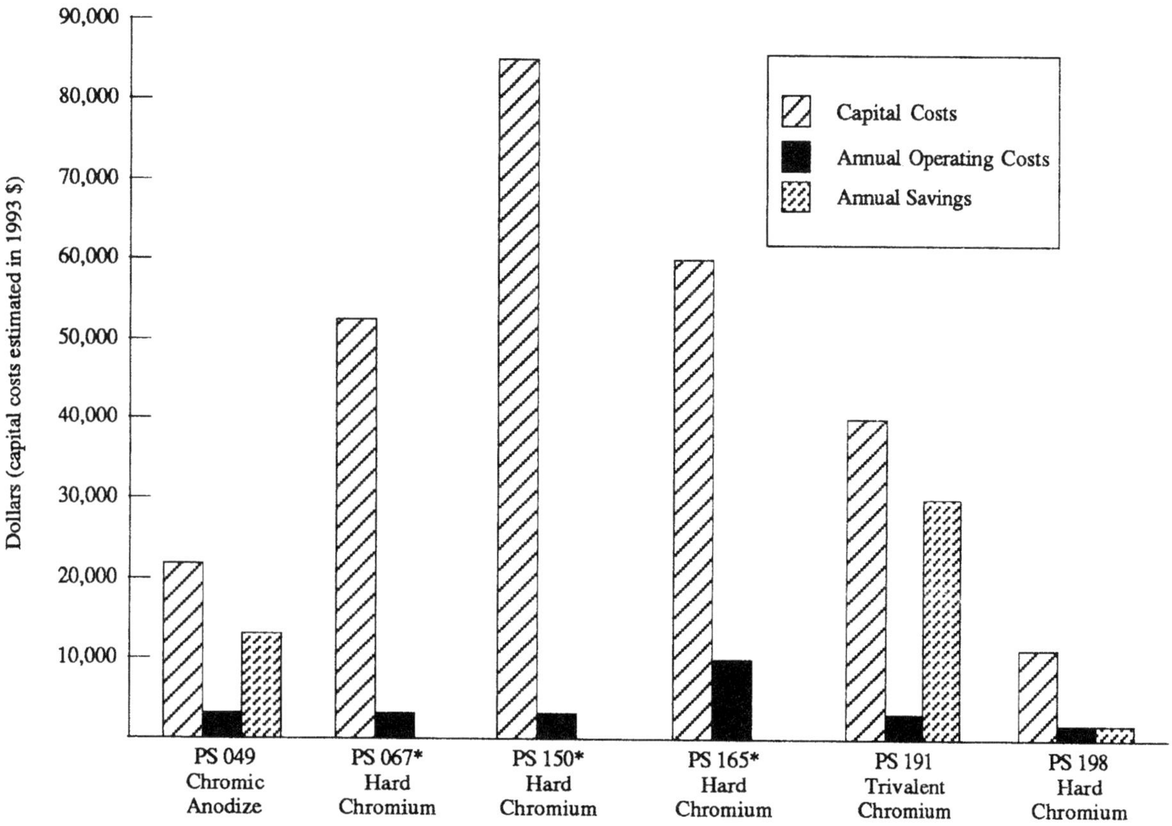

*No annual savings data provided by user.

resultant sludge is sent off-site for disposal or recovery. One shop indicated that the regenerant is sent off-site for deepwell injection (PS 198).

It appears that a batch treatment system is the best method for processing the regenerant on-site. Shops that attempted to treat the regenerant in their continuous treatment systems experienced overloading problems. For example, PS 244 purchased the technology in 1983 and discontinued its use eight months later (currently recycle chromic acid as "another product"). This shop, which has a total wastewater flow of approximately 26,000 gpd, was unable to assimilate the regenerant into their general wastewater flow. PS 244 provided the following input: "Adds a large load to the pollution control unit. To reduce labor it needs it's own automatic pH adjust and chrome reduction unit for it's effluent [sic]."

The need for a separate treatment system is due to the non-continuous nature of the waste stream and its high concentration of acid and metals. The required batch treatment process consists of chromium reduction with a sulfur compound (to reduce any hexavalent chromium present), pH adjustment (8.0 to 9.5), flocculation (polymer addition), settling/clarification, and solids dewatering. Eco-Tec recommends the use of magnesium hydroxide for pH adjustment. Alternatively, sodium hydroxide or lime could be used.

PS 049 generates 700 gal/mth of regenerant plus 2,300 gal/mth of other chromium wastewater. These wastes are batch treated in a 750 gal treatment tank. The wet sludge generation rate from this batch treatment process is high, 1,400 gal of sludge from the treatment of only 3,000 gal of wastewater. The high sludge production rate is due to the concentrated nature of the regenerant waste. The wet sludge is dewatered using a filter press (4,800 lbs/mth of filter cake) and the resultant filter cake is dehydrated with a sludge dryer (1,200 lbs/mth of dry sludge). The 1,200 lbs/mth of dry sludge is sent off-site for disposal. PS 049 is able to directly sewer their supernate from the batch treatment tank and the filtrate from the filter press that dewaters the sludge. Most shops will find it necessary to return the filtrate to the batch treatment process because it will contain residual concentrations of pollutants. Since much of the water from the original feed stream is contained in the filtrate, the quantity of this solution should be closely evaluated and considered when sizing the batch treatment tank.

Another residual generated by this process is spent cartridge filters. Cartridge filtration is used to prevent suspended solids from entering and fouling the ion exchange bed. No quantity data were provided by the respondents for the spent cartridge filter waste.

5.5 ACID SORPTION

5.5.1 Overview

Acid sorption is a purification technology applicable to dilute to moderately concentrated acid solutions such as anodizing and pickling baths. The term sorption, which includes both adsorbtion and absorption, is a general expression for a process in which a component moves from one phase to another, where it is accumulated, particularly for cases in which the second phase is a solid (ref. 435). Acid sorption is not a widely used technology by the metal finishing industry, although it has been commercially available in North America for approximately 15 years.

Acid sorption is one of several processes where resins are used to absorb chemicals present in surrounding solutions and the chemicals are subsequently desorbed with water. These reversible sorption processes include ion exclusion (cation resin), ion retardation (special resin), and acid retardation (anion resin). Of particular interest in metal finishing is acid retardation (ref. 340). This is a separation process where an acid is separated from its salts by using a column containing a strongly basic anion exchange resin of a specific porosity and particle size. This separation occurs because at high concentration the acid crosses the Donnan potential barrier (Donnan invasion) and is taken up by the resin, whereas the salts are excluded from it. The acid is thus "retarded" and the salts pass through the resin. This is not an ion exchange process, because the acid is desorbed from the resin with plain water.

The acid sorption or retardation process is employed to remove dissolved metal contaminants from acid baths. It is most often applied to the purification of sulfuric acid anodizing baths and sulfuric acid and hydrochloric acid pickling baths. When these solutions are contaminated with dissolved metal, the free acid concentration decreases and the anodizing or pickling efficiency drops. Additions of fresh acid are possible up to a point, but eventually, the bath must be either purified or dumped.

Diagrams of the acid sorption process are presented in Exhibit 5-13. These diagrams illustrate the equipment of a particular commercial acid sorption manufacturer. During the sorption step, the acid and metal salt mixture is fed up through the resin bed. Acid is sorbed into the resin while the remaining dissolved metal salts are rejected as mildly acidic solution leaving from the top of the bed. Depending on the metal salt, this solution may be waste-treated or diverted to an electrowinning cell for recovery of the metal. During the desorption step, water flows down through the resin bed. Acid is desorbed from the resin and displaced from the bottom of the bed. City water is typically adequate for this step. The resin is stable under normal operating conditions for many years without the need for regular replacement or any special treatment.

Acid sorption does not recover all of the acid in a treated bath. Rather, it recovers only a percentage (typically 80% to 90%) of the "unused" or free acid (i.e., that acid which is not chemically bonded to the dissolved metal). Typically, 40% to 70% of the total acid is free acid. Therefore, if a shop's current method of operation involves dumping and treating spent acid baths and replacing the bath with fresh solution, then acid sorption can be expected to reduce their total acid usage by approximately 30% to 65% (ref. 363, 364).

In addition to reducing acid usage there are several benefits from using acid sorption. These include: (1) reduces neutralization treatment reagent usage (e.g., caustic or lime); (2) reduces interruptions in production (i.e., when used on a continuous basis as opposed to batch purification); and (3) reduces process control variability caused by fluctuations in bath composition (i.e., when used on a continuous basis).

Exhibit 5-13. Diagram of APU™ Operating Cycle

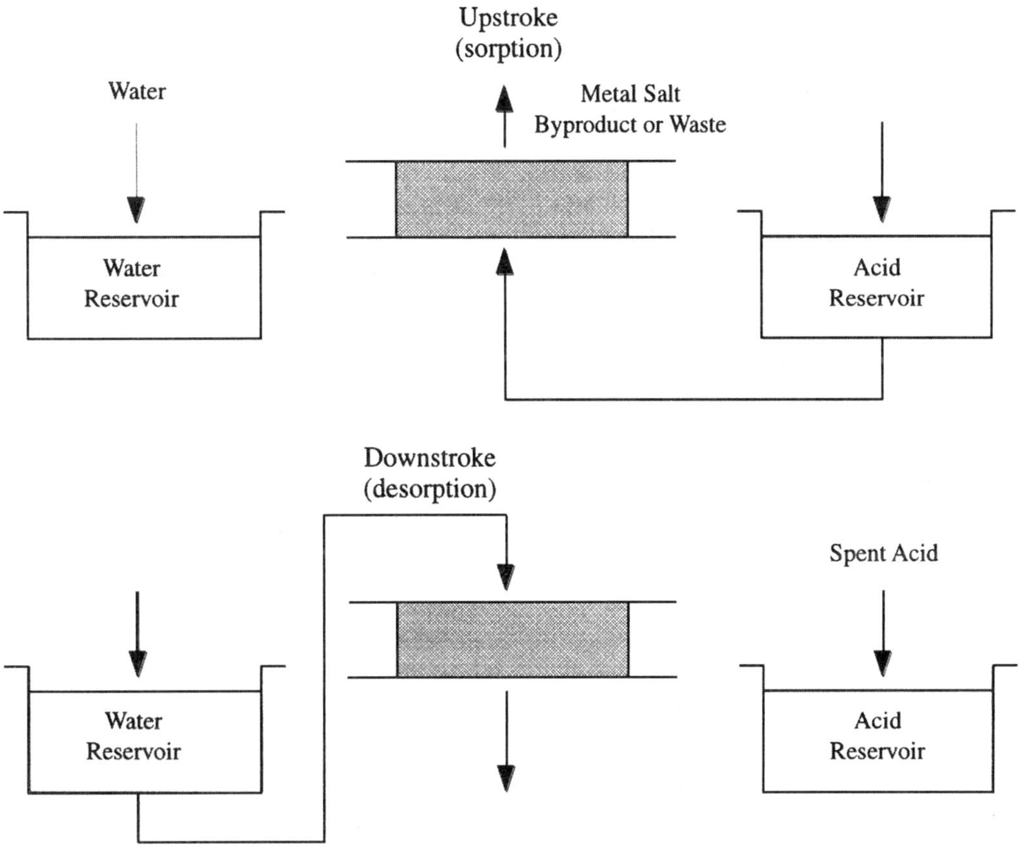

Source: ref. 364

5.5.2 Development & Commercialization

This technology is not widely used by the plating industry and it is not well documented in most common electroplating references. Therefore, tracing its origin has been difficult. The Metals Handbook, Vol. 5 (1982) indicated that acid sorption "has recently seen extensive application in the metal finishing industry for acid recovery" (ref. 340). Examples of applications given in that reference include the purification of sulfuric, hydrochloric and nitric acids. The only manufacturer of acid sorption equipment identified during the Vendors Survey, Eco-Tec, commercialized their equipment (APU™) in 1977 (ref. 363). As of 1993, Eco-Tec has sold approximately 200 units. The APU™ was originally manufactured for the recovery of sulfuric acid anodizing solutions. However, as early as 1978 the technology was also applied to the purification of nitric acid rack stripping (nickel) baths (ref. 363).

In addition to the metal finishing industry, acid sorption is applicable to other industries that use strong acids, including the primary metals manufacturing and metals forming industries.

5.5.3 Applications and Restrictions

Exhibit 5-14 shows the basic configuration for acid purification using acid sorption. The process is typically performed on a continuous basis, although the same configuration could be used for batch treatment. The process is usually not performed on a batch basis because it is relatively time consuming to perform and would require that the process bath be inactive for days. For example, to purify a 2,600 gallon sulfuric acid anodizing bath from 15 g/l to 8 g/l Al would take approximately 200 hours using a resin bed with a volume of 0.42 ft^3 (12 liters) (ref. Eco-Tec file).

Exhibit 5-14. Typical Layout for Acid Sorption Process

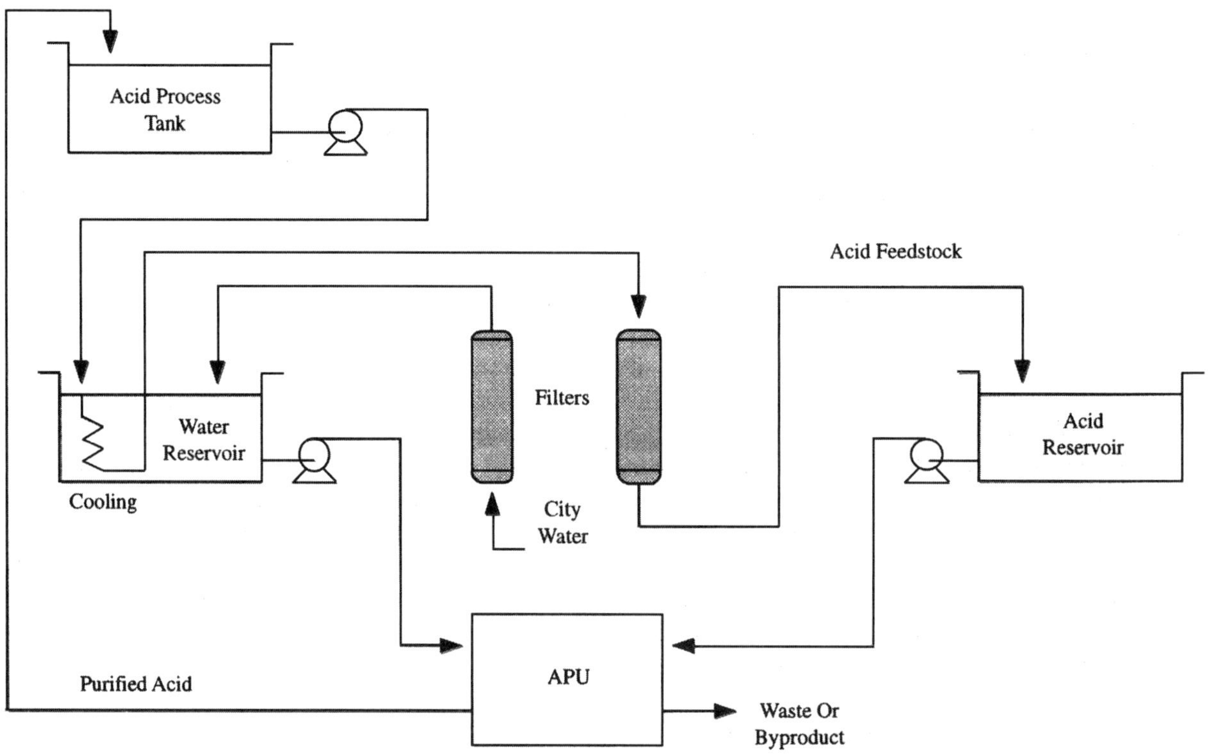

Source: ref. 364

The acid sorption technology is applicable to a range of acids and applications, including the following (ref. Eco-Tec file):

- Aluminium anodizing, using sulfuric acid
- Steel and galvanized steel pickling, using sulfuric or hydrochloric acid
- Copper and brass pickling, etching and brightening, using sulfuric or nitric acid
- Hydrometallurgical ore leach circuits
- Stainless steel and titanium pickling using nitric and hydrofluoric acids
- Electroplating rack or rework stripping, using nitric acid
- Aluminum bright dipping or electropolishing, using phosphoric and nitric acids or phosphoric and sulfuric acids
- Stainless steel electropolishing, using phosphoric and/or sulfuric acids
- Spent regenerants from acid cycle cation exchangers
- Continuously removing acid from electrowinning solutions to prevent acid interference.

Five respondents to the Users Survey indicated that they use or have used acid sorption. Their responses included the following applications: two applications for the purification of sulfuric acid anodizing solutions; one application to a hard anodize solution and two applications to hydrochloric acid pickle baths.

The five shops that use acid sorption represent 1.6% of the total number of shops responding to the Users Survey. It is interesting to note that approximately 200 acid sorption units have been sold by the only known North American manufacturer of this technology. Assuming that each of their units were sold to a different plating shop in North America, then approximately 1.5% of all North American shops have purchased this technology (based on combined EPA and Environment Canada estimates of 13,640 shops).

There are several important restrictions for acid sorption. First, the process cannot be applied to highly concentrated acids. For example, the concentration limit for nitric acid is 35% (by weight). Second, acids containing chromates should not be purified with this process. Chromates, as anions, will consume resin sites and prevent acid recovery. Third, hydrochloric acid solutions containing zinc and lead should not be purified using acid sorption. The zinc and lead form a strong chloride complex that preferably attach to the resin and prevent the sorption of acid. In these cases, it is possible to remove the zinc and lead from the acid using a process where the contaminated acid solution is passed through the resin bed two times. During the first pass, the zinc and lead chloride complex attach to the resin and are subsequently eluted with water. Then the acid would be passed through a second time to remove the acid and again be eluted with water. The key problem with this procedure is that the quantity of water needed to elude the zinc and lead chloride complex is approximately ten times greater than the quantity needed for a normal acid removal step, and therefore a large volume of metal bearing wastewater is generated.

5.5.4 Technology/Equipment Description

This subsection contains a description of commercially available acid sorption equipment that is manufactured and sold by a respondent to the Vendors Survey. This is intended to provide the reader with information and data on available equipment. Mention of trade names or commercial products is not intended to constitute endorsement for use.

The information contained in this section describes the Eco-Tec APU™, which was the only commercial acid sorption process identified during the Vendors Survey.

A standard APU™ consists of a cartridge filter, a vessel containing a resin bed, plastic piping, valving and a control system, all mounted on an epoxy coated steel frame. There are four standard APU™ sizes. Small APU™ models use hydro-pneumatic tanks to meter the flow of acid and water through the resin bed and larger units use a microprocessor-controlled flow sensor. The smallest unit has the dimensions 42 in. (107 cm) x 63 in. (160 cm) x 80 in. (204 cm) (high) and the largest unit has the dimensions 80 in. (204 cm) x 142 in. (361 cm) x 80 in. (204 cm) (high). Options include additional pre-filters, pumps, and high-temperature-resistant construction (ref. Eco-Tec file).

The operational steps of an APU™ were previously shown in Exhibit 5-13. There are two basic steps of the cycle, the upstroke and the downstroke. During the upstroke, metal-contaminated acid is pumped into the bottom of the APU™ resin bed. Acid is adsorbed by the resin bed, and the deacidfied metal salt solution, referred to as the "by-product," is collected from the top of the bed. During the downstroke, water is pumped into the top of the bed, desorbing the purified acid from the resin so that a purified acid product is collected from the bottom of the bed. The entire upstroke/downstroke cycle takes only about five minutes to complete. The equipment is fully automated to allow the cycles to repeat continuously (ref. 364). Eco-Tec has published "typical results for various applications." These are shown in Exhibit 5-15.

In addition to the standard APU™ components, some acid sorption installations may require cooling and additional filtration equipment. Cooling is needed when oxidizers such as nitric acid or hydrogen peroxide are present. These types of applications require cooling to approximately 90°F (32°C). Sulfuric and hydrochloric acid can be processed at temperatures up to 158°F (70°C). Eco-Tec indicated that city water is typically used for cooling and, to reduce water consumption, it can be reused for regeneration of the acid bed. Filtration is needed for most applications to minimize the pressure drop across the resin bed. The manufacturer recommends using a multimedia filter followed by a polishing cartridge filter for most applications.

5.5.5 Costs

5.5.5.1 Capital Costs

The required size of an APU™ depends mainly on: (1) the volume of the acid bath; (2) the type and concentration of acid; (3) the dissolved metal buildup rate; and (4) the permissible operating concentration of dissolved metal. With regard to the fourth sizing criteria, most pickling applications are sized to operate at 40 g/l to 50 g/l of dissolved metal, most anodizing applications are sized to operate at 10 g/l of aluminum, and most nickel or copper stripping (nitric) applications are sized to operate at 70 g/l. Exhibit 5-16 shows the capital costs (APU™) for the maintenance of sulfuric acid anodizing baths (based on a 10 g/l aluminum feed concentration. The equipment costs include the APU™, prefiltration, feed and product tanks, and a cooling system. Installation costs are estimated to range from 10% to 35% of equipment costs. Exhibit 5-16 covers the most common range of applications. A larger APU™ unit is available that would provide 70 kg/hr of aluminum removal.

Exhibit 5-15. Typical Results for Application of an APU™

By-Product	Bath Component, g/l	Bath, g/l	Purified Product, g/l
Sulfuric Acid	190	182	13
Aluminum	10	5.5	6
Sulluric Acid	127	116	10
Iron	36	11	21
Nitric Acid	514	581	10
Nickel and Copper	99	48	71
Sulfuric Acid	128	113	18
Hydrogen Peroxide	41	35	7
Copper	13	5.9	9.2
Nitric Acid	150	139	4.5
Hydrofluoric Acid	36	29	7.2
Iron	29	8.7	20
Nickel	7	2.1	4.9
Copper	7.3	2.2	5.1
Hydrochloric Acid	146	146	10
Iron	34	25	15
Sulfuric Acid	62	55	5.9
Sodium	7.8	0.8	5.6

Source: Eco-Tec file

Exhibit 5-17 presents equipment costs for other applications.

5.5.5.2 Operating Costs

The primary operating cost elements for this technology include: O&M labor, parts replacement, resin replacement and electricity. Water use costs are negligible for most applications.

Estimated operating costs for the sulfuric acid anodizing application are $16.00/kg ($7.27/lb) of aluminum removed. Operating cost data and annual savings data provided by respondents to the Users Survey are presented graphically and compared to capital costs (estimated in 1993 dollars) in Exhibit 5-18.

5.5.6 Performance Experience

Only three shops (PS 020, PS 037, and PS 167) completed evaluations of this technology (a total of five users were identified). A partial summary of the users data relative to acid sorption is presented in Exhibit 5-19. There are a number of observations that can be made from these data and the literature:

- All of the shops providing detailed data use the technology for sulfuric acid purification (two-sulfuric acid anodize, one-hard anodize). These three shops gave the vendor and technology high grades in terms of technical support and overall satisfaction. A11 three companies indicated that the technology satisfied the need for which it was purchased. The following is a breakdown of the reasons why the three shops purchased the technology:

  To meet or help meet effluent regulations: 2
  To reduce plating chemical purchases: 2
  To reduce the quantity of waste shipped off-site: 2
  To reduce wastewater treatment costs: 1
  To improve product quantity: 3
  To meet customer specification: 1

- The respondents indicated that use of acid sorption for bath purification improved product quality and in one case, improved the production rate. The following responses were provided:

Exhibit 5-16. Capitol Costs for Sulfuric Acid Anodizing Bath Maintenance Using Acid Sorbtion

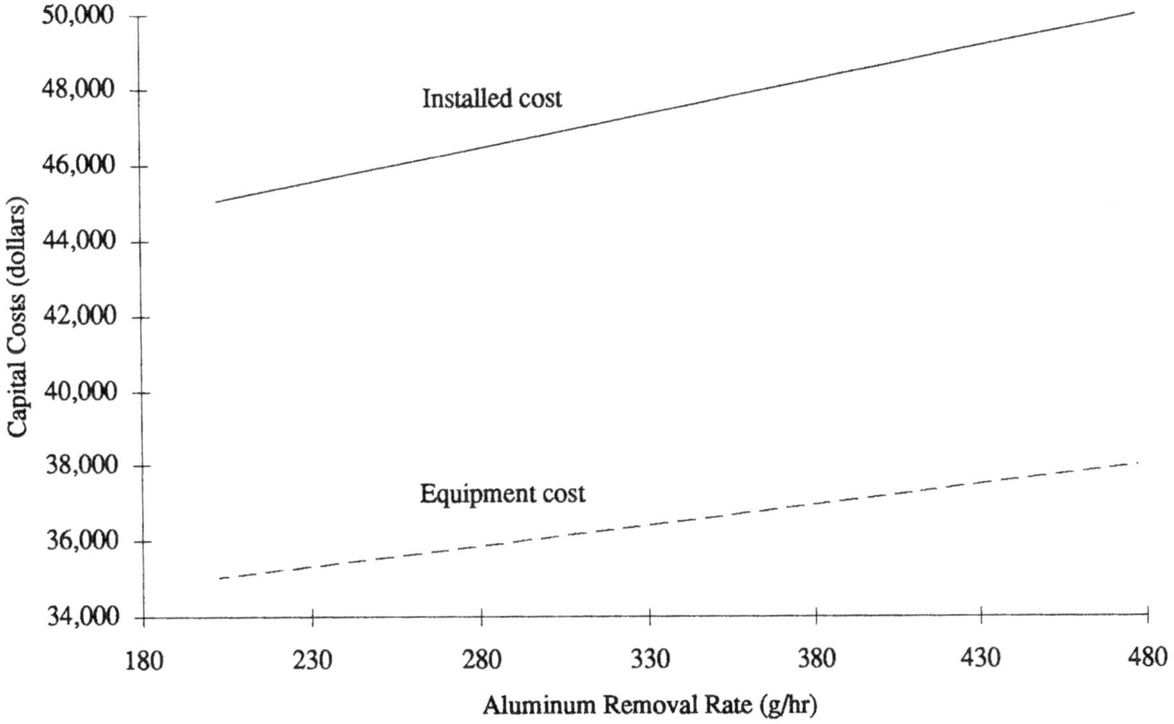

Exhibit 5-17. Equipment Costs for Acid Sorption Systems for Non-Sulfuric Acid Anodizing Applications

Type of Acid	Metal Concentration of Feed Solution	Removal Rate of Acid Sorption Unit	Equipment Cost
Hydrochloric Acid Pickle	50 g/l	15 kg/hr	$160,000
Hydrochloric Acid Pickle	50 g/l	98 kg/hr	$404,000
Sulfuric Acid Pickle	50 g/l	16 kg/hr	$152,000
Sulfuric Acid Pickle	50 g/l	102 kg/hr	$401,000
Nitric/HF Acid Pickle	40 g/l	12 kg/hr	$155,000
Nitric/HF Acid Pickle	40 g/l	75 kg/hr	$417,000

	Product Quality	Production Rate
Improved	3	1
No Change	0	0
Decreased	0	0

- Although all three shops indicated that they were satisfied with the technical support provided by the manufacturer, one shop indicated that they would use a different vendor in the future.

- The literature was also reviewed for performance experience. The only articles that were located were written by employees of the manufacturer (Eco-Tec). These include several case studies covering nitric/hydrofluoric acid and sulfuric acid pickling (ref. 364 and 365), sulfuric acid anodizing (ref. 364 and 366) and nitric acid nickel strip (ref. 363).

5.5.7 Operational and Maintenance Experience

Three shops provided operating and maintenance (O&M) information regarding acid sorption. The following summarizes the respondents O&M experience:

Exhibit 5-18. Selected Users Survey Cost Data for Acid Sorbtion Applications Currently in Use

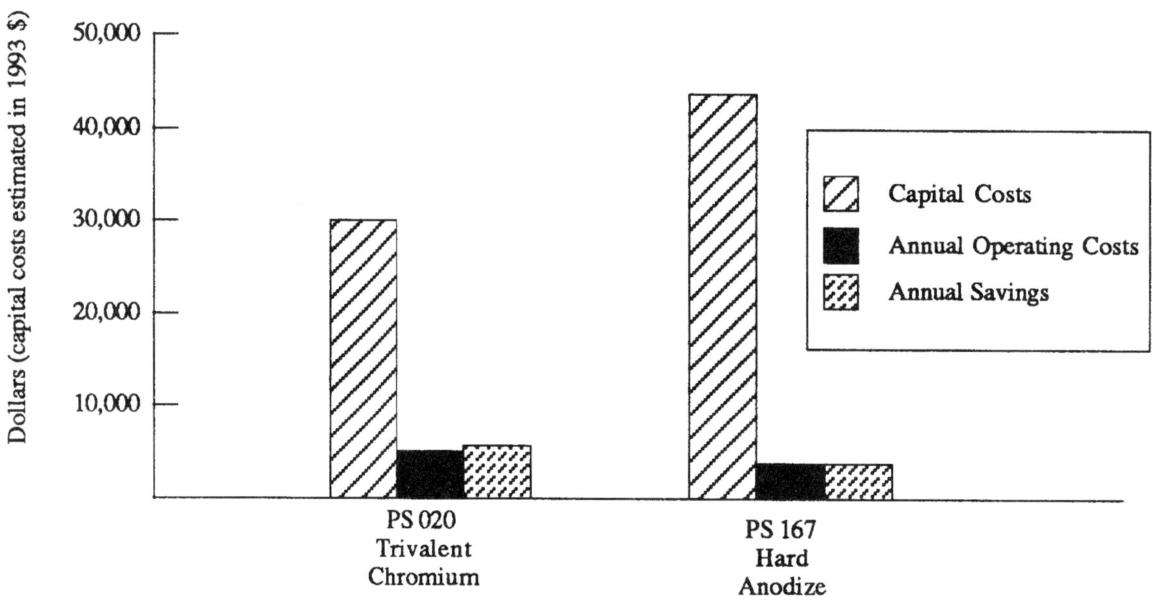

*No annual savings data provided by user.

- The average percentage of time this technology was unavailable for use due to O&M problems (i.e., downtime) was 4%. Two shops provided comments on their O&M experience. PS 037 indicated that O&M problems were minimal and cited relay replacement as the only specific example. PS 167 stated: "Equipment is well designed, with very low maintenance and high reliability [sic]."

- All three respondents indicated that the only labor commonly needed to operate this technology was a trained technician. The average number of man-hours needed to operate this technology is 103 hrs/yr.

- All three of the respondents are still operating their acid sorption systems. One system was purchased in 1978 (PS 037), one in 1984 (PS 167) and the other system was purchased in 1990 (PS 020). It should be noted that the earlier system was purchased only one year after the technology was commercialized in North America. Oddly, although this unit has been operational for 15 years, this same respondent indicated that, if given the chance to repeat the technology selection process, they would purchase the same technology from a different vendor.

- Eco-Tec provided the following data on the expected life spans of resins for common applications of acid sorption (species of dissolved metal is shown in parenthesis):

  - Sulfuric acid (aluminum): 10 to 15 years
  - Nitric/HF acids (stainless steel): 2 to 5 years
  - Sulfuric acid (iron): 3 to 7 years
  - Hydrochloric acid (iron): 5 to 10 years
  - Nitric acid (nickel/copper): 2 to 5 years

5.5.8 Residuals Generation

The acid sorption process only recovers a portion of the free or unused acid and it does not recover any of the combined acid (salt). As a result, approximately 35 to 70 percent of the total acid used is incorporated into a by-product. The by-product or waste stream from the process will require treatment. Depending on the metal involved, treatment will range from simple neutralization

Exhibit 5-19. Summary of Users Survey Data for Acid Sorption

Shop				Capital Costs			Operating Costs (Yr.)		Savings (Tr.)						Satisfaction Levels			Shop
Code	Application	Manf./Model	Year	Equip. Costs	Other Cap. Costs	Total	Non-labor	Labor Hrs.	Process Chem.	Treat. Chem.	Sludge Disp.	Total	Use Code	Down Time %	Manf. Support	Tech-nology	Future Decision	Code
020	Anodize, sulfuric acid	Eco-Tec (AP06-12)	1990	$26,000	$1,500	$27,500	$1,400	125	√	$1,000	$775	$1,775	1	5	4	4	1	020
037	Anodize, sulfuric acid	Eco-Tec	1978	$50,000	$5,000	$55,000			√	√			1	5	4	5	2	037
167	Anodize, hard	Eco-Tec (4365-C)	1984	$20,000	$12,500	$32,500	$750	80	$1,000	$1,000		$2,000	1	1.5	4	4	1	167
Average for ED Recovery Applications			1984	$32,000	$6,333	$38,333	$1,075	103	$1,000	$1,000	$775	$1,888		4	4.0	4.3		

(1) Use Codes: 1-currently operating; 2-not currently operating and no intension for future use; 3-not currently in use, but intend to use in future.

(2) Satisfaction levels (manufacturer and technology) 1 to 5, with 1=lowest and 5=highest.

(3) Future decision codes: 1-purchase the same technology from the same vendor; 2-purchase the same technology from a different vendor; 3-purchase a different technology; 4-do nothing.

(4) "√" indicates that a savings was realized, but not quantified.

(5) Blank spaces indicate that no data were provided.

(pH adjustment with caustic) to metals removal (e.g., precipitation). Eco-Tec also suggests that electrowinning can be used; however, no specific applications were identified. Sludges resulting from waste treatment will be a listed hazardous waste (F006 or F019). PS 167 indicated that treatment of their by-product results in the generation of 2 drums of sludge per year.

Eco-Tec indicates that work is currently under way to recover the by-product streams from the acid sorption system. A demonstration site has been set up for the recovery of nitric and hydrofluoric acid from the metal salt by-product, and work is proceeding on the recovery of a usable by-product from an APU™ system used with an anodizing process (ref. 364).

5.6 ION TRANSFER

5.6.1 Overview

The term ion transfer is used in this text to refer to a small group of technologies that are applied primarily to the maintenance of chromic acid baths, most notably hard and decorative chrome plating, chromic acid anodizing and chromic acid etch baths. Used for this purpose, ion transfer is a competing technology with ion exchange (cation) and membrane electrolysis. Ion transfer is also a chromium recovery technology, although it is used less frequently for this purpose. Competing technologies for chromium recovery include evaporation and ion exchange.

The basic ion transfer technology involves the use of a membrane, typically a porous ceramic pot or a polyfluorocarbon membrane (e.g., Teflon®). The unit consists of an electrolytic cell with an anode and cathode (or sets of each) that are separated by the membrane. When energized, trivalent chromium present at the anode is oxidized to hexavalent chromium, and cations (e.g., dissolved iron) present in the anolyte migrate through the membrane into the cathode compartment. The catholyte is periodically discarded and the cathode cleaned of any deposits.

As a chromium recovery technology, ion transfer equipment has been applied by survey respondents to chromium strip solutions and rinse waters. The chromium strip application uses the same equipment (porous pot) as is used for bath maintenance, but it is operated with reverse polarity. A specially designed ion transfer system for chromium recovery from rinse water was introduced in the early 1980's. Although that particular equipment

(ChromeNapper) is no longer manufactured, a similar device is presently available.

5.6.2 Development and Commercialization

Chromic acid bath contamination has been a concern to platers since its development in the 1920's (ref. 367) and it continues to be a problem today. One of the surveyed shops offered this statement:

One of the biggest problems facing hard chromium platers, such as ourselves, is the purification of the plating baths. Contaminants in the plating tanks reduce the efficiency and quality of the chromium deposit, and once the contaminant level reaches a certain level, the solutions must be disposed of [sic]. (ref. PS 234)

The two key problems that plague chromic acid bath users are: (1) reduction of hexavalent chromium (Cr^{+6}) to trivalent chromium (Cr^{+3}) causing an unwanted buildup of Cr^{+3} and (2) buildup of dissolved metals (ref. 368, 369 and 370). The first of these two problems can be solved using the simple process of auxiliary electrolysis or high current density dummying (Section 5.2). However, dummy plating is a relatively slow and inefficient method of oxidizing Cr^{+3} (ref. 370). Since production plating cannot be performed during dummy plating, the tank being treated would be unusable during this time period. Therefore, for shops using this method, dummying is usually performed during weekends.

The shortcomings of dummy plating led to investigative work by Seegmoller and Lamb in 1948 (ref. 368) involving use of the porous pot. Prior to that time, and as early as 1840, the porous pot was employed in the plating industry as a "bladder" that separated zinc anodes from gold cyanide plating solution (ref. 371). The porous pot is also mentioned in U.S. patent literature in 1918 (ref. 371). The technology, however, was not widely applied until the late 1970's and 1980's with the commercialization of the Cosmos Mineral Corporation unit (latter sold by Pfaudler Co. and subsequently by E.P. Technology) and the Hard Chrome Plating Consultants' Porous Pot (PPS1 and PPS2). Before this time period, it is believed that homemade versions of the porous pot were in use. Hard Chrome Plating Consultants indicated during the Vendors Survey that they have sold approximately 1,500 porous pot units to approximately 1,100 different plating shops.

Also, in the early 1980's, a polyfluorocarbon membrane version of the porous pot was developed (CatNapper). The polyfluorocarbon membrane acted like the ceramic pot, separating the anode and cathode reactions and products while permitting cations to migrate from the anolyte to the catholyte. The polyfluorocarbon membrane of the CatNapper is not anion or cation selective; the selectivity (i.e., migration of cations only into the catholyte) is a result of the electrical driving force. This fact distinguishes the technology from the more recently developed membrane electrolysis units that employ ion exchange membranes (Section 5.7). Another very significant difference between these technologies is that chromic acid is used as the catholyte with ion transfer technologies, whereas ion specific electrochemical membrane technology units typically use a non-chromic acid solution. This is a disadvantage of the ion transfer technologies if the catholyte, which accumulates the contaminants, is discarded. In such cases, ion transfer technologies would create a significant quantity of chrome-bearing residual. However, the use of a chromic acid catholyte would be viewed as an advantage in the event of a membrane leak. With a non-chromic acid catholyte, a membrane leak would result in significant bath contamination. The original CatNapper is no longer marketed.

It should be noted that ion exchange is a competitive technology for chromic acid bath maintenance. Ion exchange was used for this purpose as early as 1952 (ref. 368) and is still commercially available (ref. Eco-Tec file, Kinetico file). Use of ion exchange for chromic acid purification is discussed in Section 5.4.

Also, some hard chrome plating shops perform the preliminary reverse etching step (the surface of the part that is to be chrome plated is etched by reversing the normal polarity of the process tank which makes the part an anode and the lead alloy anodes become cathodes) in a dedicated etch tank. This practice reduces the introduction of iron (from the base metal) into the plating bath and delays the need to treat the bath (ref. 482).

The use of ion transfer as a chromium recovery technology probably began in the 1940's to 1950's, when the porous pot was being used for bath maintenance. However, there is no record of its use until the 1980's with the commercialization of the ChromeNapper by Innova (same manufacturer that made the CatNapper).

5.6.3 Applications and Restrictions

Exhibit 5-20 shows the two basic configurations used for application of ion transfer technologies for bath maintenance. Configuration IT-1 shows the use of the less expensive in-tank porous pot. This unit does not have its own rectifier, but is powered by the plating tank rectifier (alternatively, PS 118 connects to their electrocleaner rectifier). Therefore, its use is restricted to the time period when the plating tank rectifier is energized. The unit must be removed from the tank during idle periods, otherwise the contaminates will migrate from the catholyte back into the bath. One of the survey respondents indicated that they remove the unit from the tank whenever it is idle for more than 8 hours (PS 118).

Configuration IT-2 shows an external unit that is powered by its own rectifier. Either the multi-pot porous pot unit or the external polyfluorocarbon membrane unit would be used in this manner. Plating solution is pumped into the tank containing the cells, circulated and returned by gravity to the plating bath.

Although the ion transfer technology has its roots in non-chromium applications, and one reference indicated that other applications are possible (ref. 371), no non-chromium applications were found during the course of the NCMS project. The specific applications that were identified were all for chromic acid solutions except for one application to caustic chromium strip. The chromic acid applications include hard chromium plating (single and dual catalyst), decorative chromium plating (single and dual catalyst), chromic acid anodizing and chromic acid etch.

Two important restrictions should be noted for the ion transfer technology. First, this technology should be considered as a bath maintenance method and not a means of quickly rejuvenating a spent bath. Chromic acid baths that are laden with dissolved metal contaminants will take months to correct with ion transfer and a significant volume of chromium waste can be generated in the process. The correct application of this technology is as a continuous maintenance method that is first applied before the bath is overly contaminated. Second, the ion transfer technology is not practical as a bath maintenance method where the desired tramp metal contamination level (excluding consideration of Cr^{+3}) is less than 4 g/l. To reach a lower point would require frequent changes of the catholyte solution, resulting in a very high waste volume.

Exhibit 5-21 shows the basic configuration (IT-3) used for application of ion transfer for chromium recovery from rinse water. The IT-3 configuration takes advantage of the significant amount of surface evaporation that normally occurs in chromium plating baths operating between 115°F and 140°F. By using countercurrent recycle of rinse water through multiple-rinse tanks and a chromium recovery unit on the final rinse, the chromium plating process becomes a closed-loop operation. Chromic acid is removed from the rinse water and concentrated into the ion transfer system modules where it can reach a concentration of approximately 150 g/l. The recovered chromic acid is periodically drained from the ion transfer system modules and transferred to the plating bath (ref. 39).

Exhibit 5-22 presents vendor supplied data that shows the performance of an ion transfer unit (APS-3, see specifications in Section 5.6.4) applied to a decorative chromic acid plating bath (ref. i^3 file). During the data collection period, the unit was operated at 110 amps and 6.5 volts. The data indicate there is an average removal rate of 385 grams of cations (copper, nickel and zinc) per day (0.85 lbs/day) when the bath (6,018 l) is treated from a concentration of 12.2 g/l to 10.6 g/l.

The PPS1 or the PPS2 can also be used as a chromium recovery device when placed into a caustic strip solution and reverse polarity is applied. Chromic acid will accumulate within the ceramic pot and is periodically removed and transferred to the plating bath. One of the survey respondents used the equipment in this way and reported that it was the best application for this equipment (i.e., they also tried using the PPS1 as a bath maintenance technique) (ref. PS 006).

5.6.4 Technology/Equipment Description

This subsection contains a description of commercially available ion transfer equipment that is manufactured and/or sold by respondents to the Vendors Survey. This is intended to provide the reader with information and data on a cross section of available equipment. Mention of trade names or commercial products is not intended to constitute endorsement for use.

The PPS1 and PPS2 (Hard Chrome Plating Consultants) are the simplest commercial versions of the ion transfer technology. Both units are single cell systems, the difference being their size and capacity. The PPS1 is the larger

Exhibit 5-20. Two Common Ion Transfer Configurations for Chromium Bath Purification

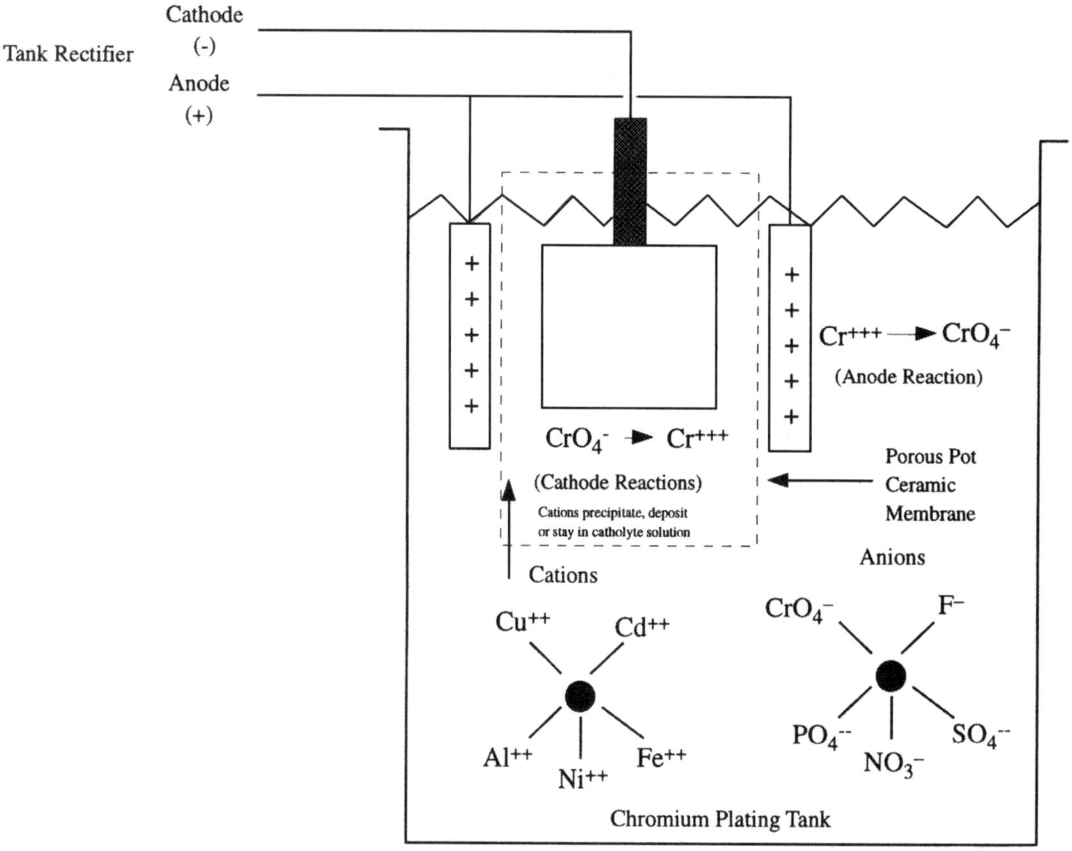

Configuration IT-1: In-Tank Porous Pot Arrangement

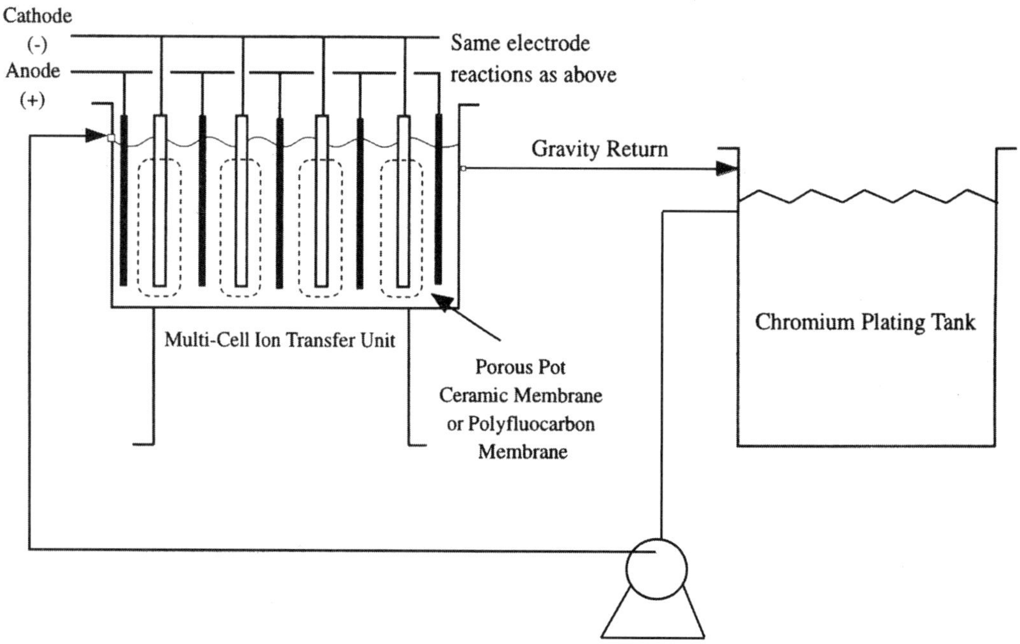

Configuration IT-2: External Tank Porous Pot or Polyester Membrane Arrangement

Exhibit 5-21. Application of Ion Transfer for Chromium Recovery

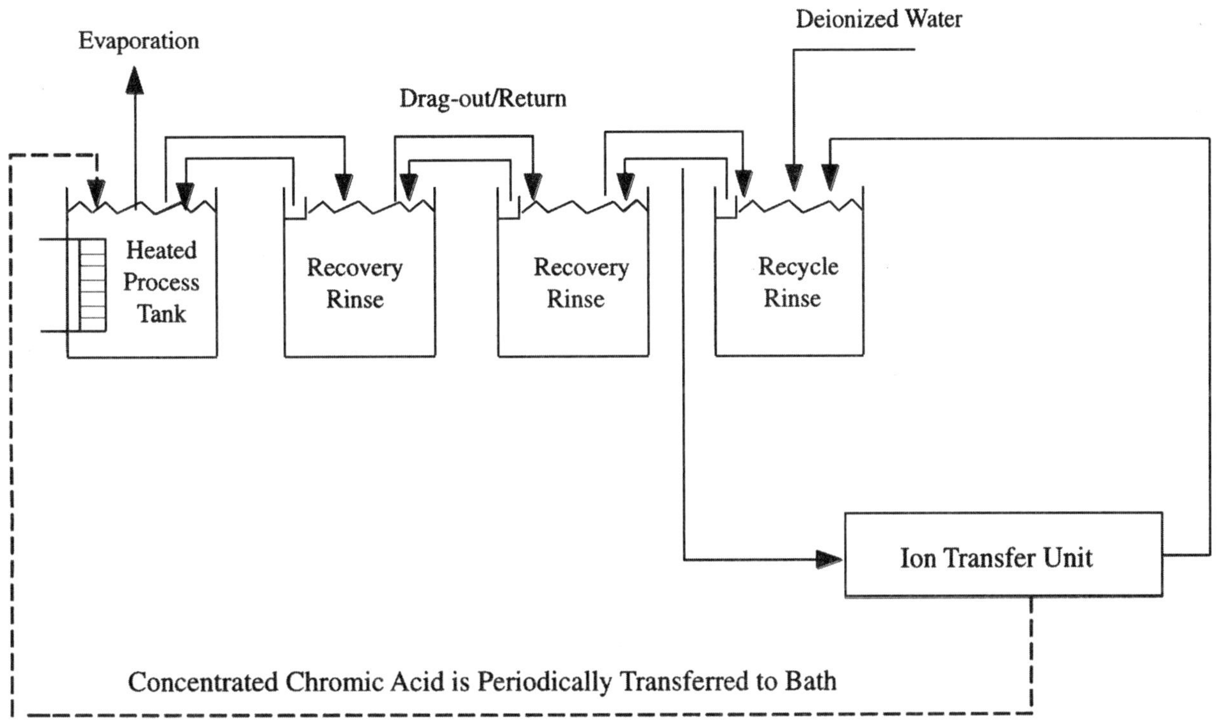

Configuration IT-3: Chromium Recovery from Rinse Water

Source: ref. 39

of the two units and is recommended by the manufacturer for tanks containing up to 3,000 gal. of chromium plating solution. The PPS2 is recommended by the manufacturer for shops with less than 700 gal. of plating solution. Each unit consists of a porous pot, anode, cathode and connecting bars. The pots have a 40% porosity and a pore size of 1 micron or less. The cathode is made from lead alloy mesh (ref. 371). These units are designed for use with the 2-bus bar, reversible rack plating tank configuration. However, they can be rigged for other bus bar configurations. The units are located directly in the plating tank during use, where they consume approximately six to twelve inches of tank length and have a maximum working depth of 25 inches. No pumps, piping or power supplies are usually needed to operate the unit. The porous pot is powered by the tank rectifier. The manufacturer also sells a PPS-4 unit that is applicable to cadmium, nickel copper and other solutions.

Before using the PPS1 or PPS2, the porous pot is filled with plating solution or fresh chromic acid solution, without catalyst (approximately 2 gallon for the PPS1). The unit is then lowered into the plating bath. By design, the lip of the pot is above the solution level of the plating tank to avoid having direct exchange of solution. When the unit is energized (smaller unit draws about 70 amps and the larger unit 210 amps) (ref. 371), Cr^{+3} in the anolyte (plating tank) is oxidized to Cr^{+6} and cations in the anolyte are electrically driven through the pores of the pot into the catholyte. Some cations are deposited on the cathode, most remain in solution. It is unclear whether the cations are electrolytically deposited or precipitated. One source suggested the latter method of deposition occurs because hydrogen evolution is occurring at the cathode which will result in a rise in pH (ref. 371). The chrome present in the catholyte is reduced to Cr^{+3}. As these changes occur, the effectiveness of the unit diminishes and eventually the catholyte is replaced. The frequency of catholyte replacement is usually in the range of 8 hours to 3 days (ref. 370).

The E.P. Technology unit (patent formally owned by Cosmos Minerals Corporation and subsequently by Pfaudler Co. and later by E.P. Technology) consists of a tank holding four to eight pots and an equal number of sets of anodes and cathodes. Basically, this is the same

Exhibit 5-22(a). APS-3 Copper Removal

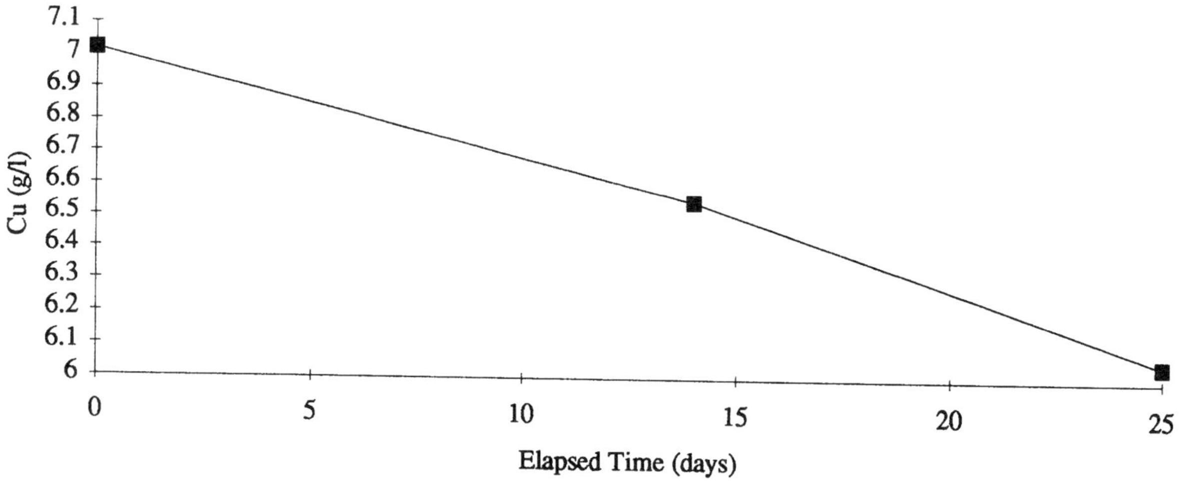

Exhibit 5-22(b). APS-3 Nickel Removal

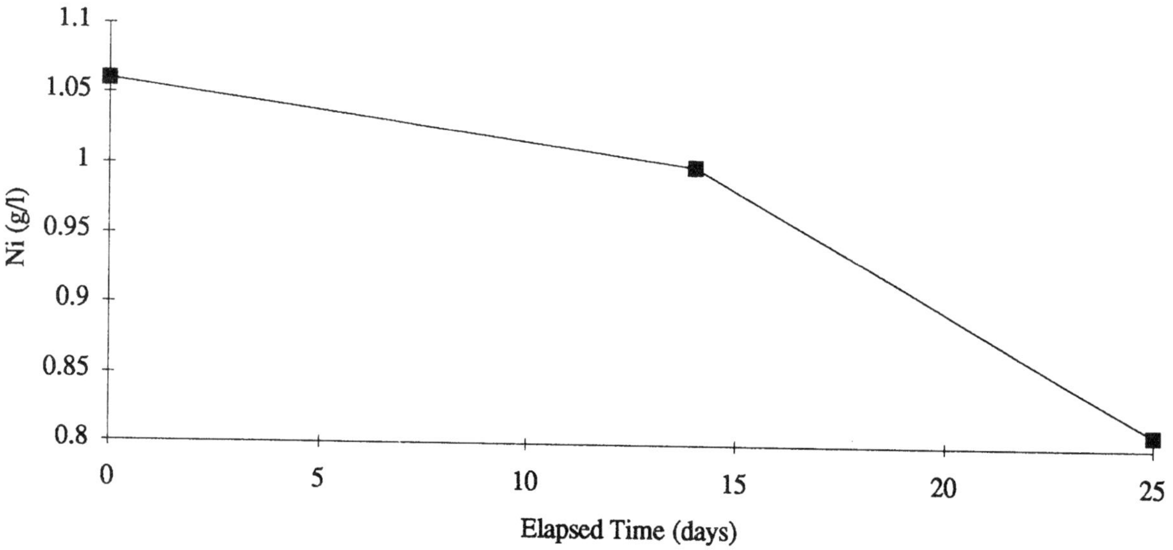

technology as the Hard Chrome Plating Consultants PPS1 or PPS2, but on a much larger scale. The pots hold the catholyte solution and cathode. Plating solution is pumped to the tank on a continuous basis; it circulates around the pots and is returned by gravity flow to the plating tank. The cells are powered by a rectifier (1,000 to 2,000 amps depending on the number of pots) dedicated to the purification unit. The unit requires lateral exhaust ventilation similar to a chrome plating tank, since chromic acid mist is generated by the release of oxygen at the anode and hydrogen at the cathode. It is not known if the EP Technology unit is still marketed. No recent advertising or installations were located during this project.

Artec manufactures four models of the polyfluorocarbon membrane ion transfer technology for chromium bath maintenance (APS series) and seven models for chromium recovery from rinse water (ARS series) (marketed by i^3). Three of the APS models are progressively larger units of similar design (Exhibit 5-23) and the fourth is a Unicell model. The three larger Artec units operate in a manner very similar to the EP Technology porous pot unit. Solution is pumped from the plating tank to the unit, circulated past the cells and returns by gravity to the plating tank. Within the unit, the chromic acid solution circulates around modules (3 to 12), which extract cationic impurities. The solution collected in the membrane compartment which contains iron, trivalent chromium,

Exhibit 5-22(c). APS-3 Zinc Removal

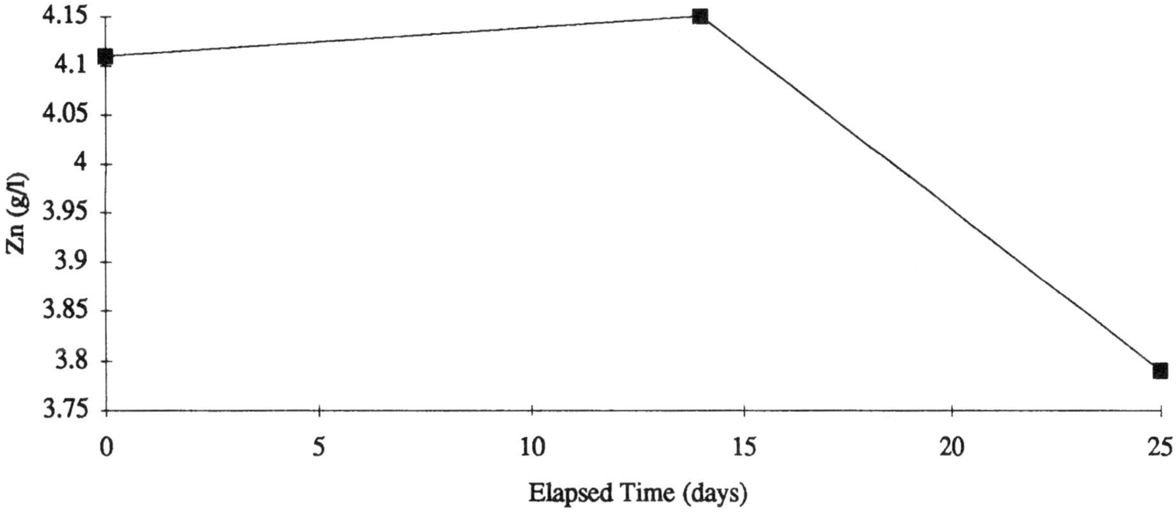

Exhibit 5-22(d). APS-3 Total Cation Removal

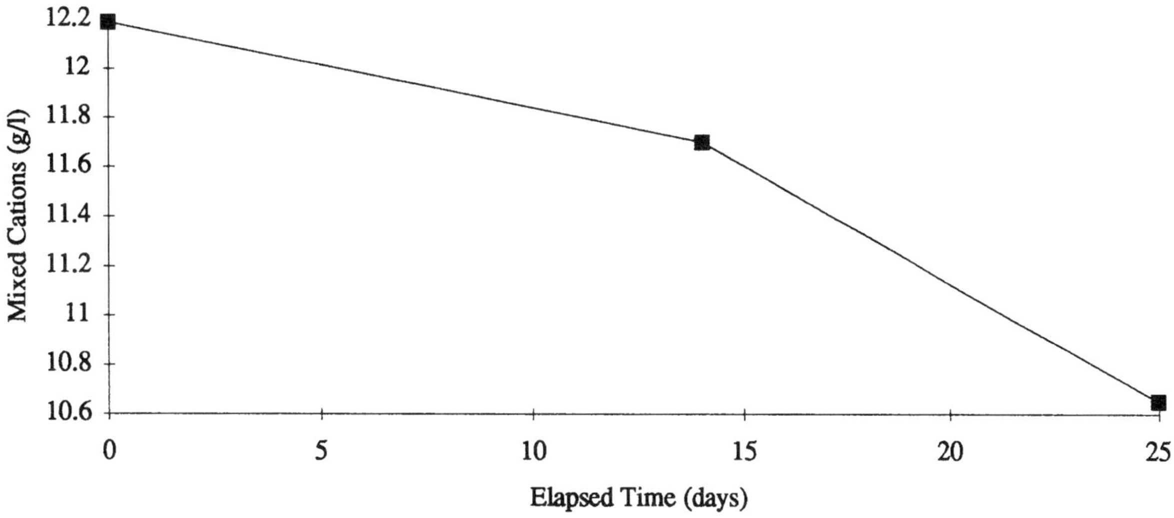

and other cation impurities is discharged to waste treatment. System specifications for the Artec units are given in Exhibit 5-24. The Artec Unicell is a single module unit that fits into a plating tank, and therefore functions like the single porous pot cell units.

5.6.5 Costs

5.6.5.1 Capital Costs

The PPS1 and PPS2 are sold for $800 and $460, respectively. The capacity of these units is not documented, so no attempt has been made in this report to show capital costs over a range of removal capacities. The manufacturer suggests that the PPS2 is applicable to tank volumes of 700 gallons or less and the PPS1 should be used for higher volumes, up to approximately 3,000 gal. (ref. Hard Chrome Plating Consultants, Inc. file)

The capital costs for the Artec polyfluorocarbon membrane technology ranges from approximately $10,000 to $30,000 for standard units with larger custom units available.

Exhibit 5-23. APS-3-SA Ion Transfer System
(Courtesy of i^3)

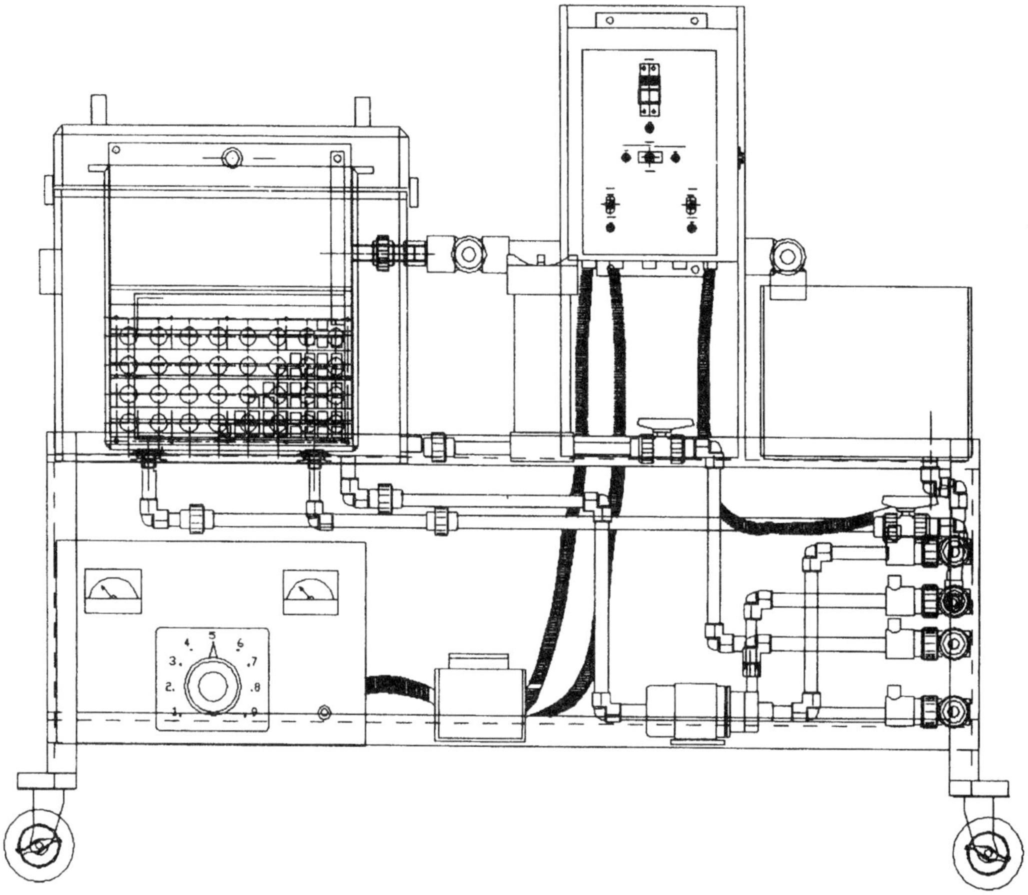

Exhibit 5-24. System Specification for APS Ion Transfer Systems (Courtesy of i^3)

Model Number[a]	APS-3/ARS-3	APS-6/ARS-6	APS-12/ARS-12	UNICELL
Dimensions	40"L x 24"W x 50"H	60"L x 24"W x 36"H	80"L x 24"W x 36"H	18"L x 2"W x 50"H
Tank Capacity	30 gallons	45 gallons	60 gallons	none
Rectifier	150 amp (100 A, 7.5 V max operating level)	250 amp (200 A, 7.5 V max operating level)	500 amp (400 A, 7.5 V max operating level)	not included (30 A, 7.5 V max operating level)
Cathodes	3 (stainless steel)	6 (stainless steel)	12 (stainless steel)	1 (stainless steel)
Anodes	4 (lead)	7 (lead)	13 (lead)	1 (lead)
Membranes	6	12	24	1
Average Cation Removal Capacity[b]	395 grams/24 hours (Fe)	800 grams/24 hours (Fe)	1600 grams/24 hours (Fe)	100 grams/24 hours (Fe)
Maximum Cation Removal Capacity[b]	2000 grams/24 hours (Fe)	4000 grams/24 hours (Fe)	8000 grams/24 hours (Fe)	400 grams/24 hours (Fe)

[a] APS/ARS series available with HCL option, recommended for hydrochloric based solutions and SA option for semi-automatic start-up and shutdown sequences for non-continuous operations.
[b] removal rates dependent on bath concentrations

Exhibit 5-25. Selected Users Survey Cost Data for Ion Transfer Bath Maintenance Applications Currently in Use

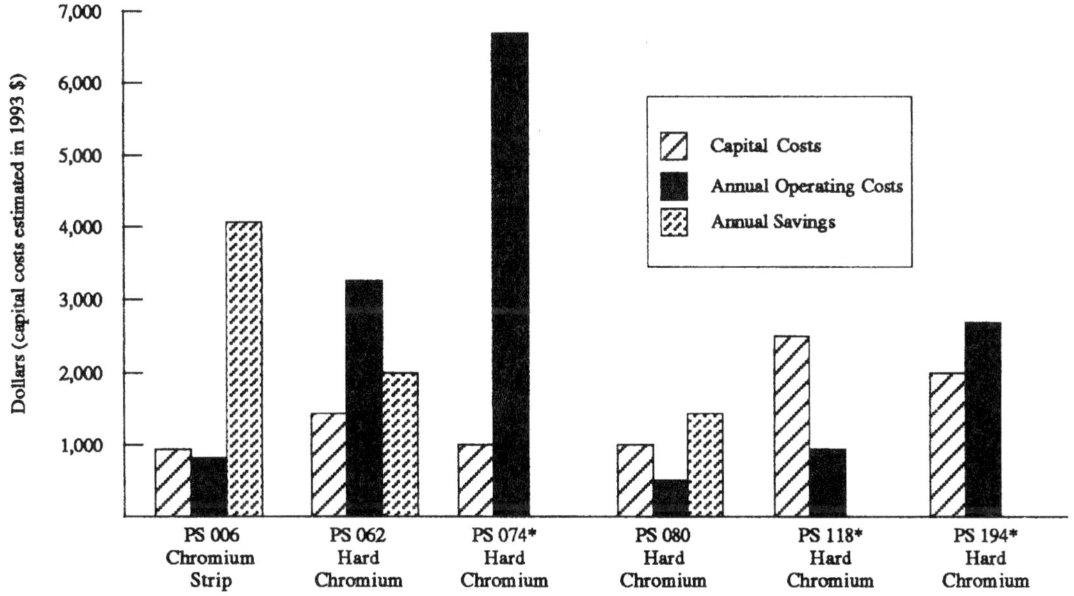

*No annual savings data provided by user.

5.6.5.2 Operating Costs

Operating costs for the Artec polyfluorocarbon membrane technology are estimated by the manufacturer to range from $5 to $27 per pound of iron removed, depending on the level of bath contamination.

Respondents to the Users Survey reported that O&M labor requirements for the porous pot (PPS1 and PPS2) range from 26 hr/yr to 260 hr/yr and that non-labor costs range from $100/yr to $500 (excluding anolyte disposal costs).

Selected Users Survey cost data for the ion transfer technology is presented in Exhibit 5-25.

5.6.6 Performance Experience

A partial summary of the Users Survey data relative to ion transfer technology is presented in Exhibit 5-26. There are a number of observations that can be made from these data, other data contained in the database and from the literature.

- In general, shops using ion transfer for chromic acid bath maintenance gave the technology a high rating. The average satisfaction level for ion transfer applied to bath maintenance is 3.8 (on a scale of 1 to 5, with 5 being most satisfactory). Seventy percent of the shops indicated that this technology satisfied the need for which it was purchased. The following is a breakdown of the reasons why shops purchased this technology:

To meet of help meet effluent regulations:	2
To reduce plating chemical purchases:	8
To reduce the quantity of waste shipped off-site:	2
To reduce wastewater treatment costs:	1
To improve product quantity:	5

- The use of ion transfer for bath maintenance generally improved the production quality but did not impact the rate of production. The following responses were provided:

	Product Quality	Production Rate
Improved	6	2
No Change	2	4
Decreased	0	1

- Most plating shops indicated that, based on their experience with this technology, they would purchase the same type of equipment from the same vendor. The following is a breakdown of their responses:

Purchase the same technology from the same vendor:	8
Purchase the same technology from a different vendor:	1
Purchase a different technology:	2
Do nothing:	0

- The major savings from the operation of ion transfer for bath maintenance was the result of reductions in bath chemical use. The reported savings were in the range of $500 to $2,000 per year.

- The ion transfer process is relatively slow with respect to removal of dissolved metals. For example, a PPS2 was tested on a 50 gal solution of chromic acid contaminated with 13.05 g/1 of copper, iron and nickel (i.e., combined concentration). The contaminant level was decreased to 4.08 g/1 in 192 hours (8 days). The maximum tank volume applicable to the PPS2 is 700 gal (ref. 371). For this size tank the time needed for purification would be 112 days. If the tank were used for production while purification was conducted, the time for purification could be much longer and possibly unachievable (e.g., if the introduction rate of contaminants exceeded the removal rate).

One survey respondent has used three different types of chromic acid bath maintenance techniques since 1980 (ref. PS 234). This shop did not provide the data requested in the survey form, but attached the following useful insight:

Over the years, we have tried three different technologies in order to solve this problem. We moved into new offices 2 years ago, and I have been unable to locate my files relating to the specifics of each system. Therefore, the following information is based on my recollection.

1. Short-bed ion exchange. Purchased from EcoTec (Canada) in 1983 - 1985 at a cost of $ 17,000 plus. Of the three systems, this technology provided the best results as far as removing impurities. The major problem, which resulted in our selling the equipment, was the overloading of our waste treatment facilities with hex chrome - producing large quantities of F006 sludge.

2. Porous Pot. Purchased from Cosmos (California) somewhere around 1980 - 1981. Purchase price unknown. This experiment was a disaster. The cleaning of the system was an incredibly messy operation which took 4 to 5 hours to complete. This system also produced abundant amounts of hex cr. waste, and the porous pots were susceptible to breakage. This system gave only marginal results for purification of the bath, although it did a good job of removing trivalent chromium.

3. Membrane. Unable to recall name of company. They were out of the Atlanta, GA area. The cost of the unit was around $8,000. This system was similar to the porous pot technology with the substitution of a membrane for the porous pots. Like its predecessor, however, it was dreadful to clean out the collected material. It also generated tremendous amounts of hex cr. and seemed limited to the removal trivalent chromium only. The other impurities such as cu. and iron were largely unaffected.

All of these systems were sold, and we are in the market for suitable (proven) technology [sic].

[The third technology discussed by the respondent was actually a membrane electrolysis technology (Section 5.7). The particular system they purchased is no longer being sold. As indicated in Section 5.7, some membrane electrolysis technology applications have been successful.]

- There was very little performance experience documented in the survey forms with respect to the polyfluorocarbon membrane units. Of the two shops providing data, one was satisfied with its performance, but discontinued its use due to maintenance problems (PS 134) and the other respondent was dissatisfied with the performance of the unit (PS 194).

5.6.7 Operational and Maintenance Experience

The following summarizes the respondent's O&M experiences and provides operating labor information.

- The average number of labor hours for ion transfer usage was: 148 hrs/yr. The skill category most commonly needed for operating this technology is a plater or common labor. The following is a breakdown of the responses for skill requirements:

Environmental Engineer: 0
Process/Chemical Engineer: 0
Chemist: 0
Consultant: 0
Plumber/Pipe Fitter: 0
Electrician: 0
Vendor: 0
Senior Level Plater: 4
Junior Level Plater: 3
Wastewater Treatment Plant Operator: 1
Trained Technician: 2
Common Labor: 3
Other: 0

- In general, the commercial ion transfer equipment used for chromium bath maintenance requires little maintenance, has a relatively small downtime and a relatively long equipment life.

- The porous pot technology is free of complex mechanical operational and maintenance problems due to its simple design and lack of moving parts. For the in-tank units, the following O&M items were identified by survey respondents:

  - Liquid and sludge must be removed from the pots periodically (range of frequency was 1 day to 1 week) (ref. PS 006, PS 062, PS 194, PS 074, PS 080, PS 1 18).

  - Unit must be dismantled and cleaned every week (PS 021).

  - The ceramic pots are fragile and/or must be periodically replaced due to breakage (PS 006, PS 062).

  - The unit must be removed from the tank during idle plating times and/or it is difficult to move from tank to tank (PS 062, PS 118).

  - Requires frequent observation to assure proper operation (PS 006).

  - Use of technology causes catalysts to become unbalanced (PS 006).

  - Unit consumes plating tank space (PS 062).

- Only one of the respondents to the Users Survey used a multiple-porous pot unit. This respondent reported pot breakage to be an O&M problem. Similar experience is found in the literature (ref. 372).

Exhibit 5-26. Summary of the Users Survey Data for Ion Transfer

Shop Code	Solution/ Application	Manufacturer	Year	Capital Costs: Equip. Costs	Capital Costs: Other Cap. Costs	Capital Costs: Total	Operating Costs (Yr.): Non-labor	Operating Costs (Yr.): Labor Hrs.	Savings: Treat. Chem.	Savings: Process Chem.	Savings: Other	Savings: Total	Use Code	Down Time %	Satisfaction Levels: Manf. Support	Satisfaction Levels: Tech-nology	Future Decision
006	Cr, Caustic strip (Cr recovery)	Hard Chrome Plating Conts.	1990	$750	$0	$750	$0	30	$2,000	$2,000		$4,000	1		3	4	1
006	Cr, hard (F catalyst, bath maintenance)	Hard Chrome Plating Conts.	1990	$750	$0	$750		100		√			3		3	3	1
021	Cr, hard (bath maintenance)	Cosmos Minerals Corp.	1983	$16,000	$8,000	$24,000		312		$1,500		$1,500	1	5	1	4	1
062	Cr, hard (bath maintenance)	Hard Chrome Plating Conts.	1990	$785	$500	$1,285	$510	102		$2,000		$2,000	1	10	3	3	1
074	Cr, hard (bath maintenance)	Hard Chrome Plating Conts.	1989	$800	$0	$800		260		√			1		2	4	1
080	Cr, hard (bath maintenance)	Hard Chrome Plating Conts.	1990	$700	$0	$700	$500			$500	$1,000	$1,500	1	0	3	3	1
118 (1)	Cr, hard (bath maintenance)	Hard Chrome Plating Conts.	1989	$1,300	$750	$2,050	$200	26					1	10	2	4	2
131	Cr, hard (bath maintenance)	In-House		$1,000	$0	$1,000				√			1	1.5		5	1
134	Cr, decorative (Cr recovery)	ChromeNapper Jr (Innova)	1981	$8,000	$500	$8,500		200		$500		$500	2	15	3	4	3
194	Cr, hard (Cr recovery)	ChromeNapper (Inova)	1985	$12,000	$2,500	$14,500	$200	200		$1,000	$2,000	$3,000	2	5	2	1	3
194	Cr, hard (bath maintenance)	Hard Chrome Plating Conts.	1986	$1,500	$500	$2,000	$100	100					1	1	4	4	1
		Average	1987	$3,962	$1,159	$5,121	$252	148		$1,250	$1,500			6	2.6	3.5	

(1) Two units.
(2) Use Codes: 1-currently operating; 2-not currently operating and no intension for future use; 3-not currently in use, but intend to use in future.
(3) Satisfaction levels (manufacturer and technology) 1 to 5, with 1=lowest and 5=highest.
(4) Future decision codes: 1-purchase the same technology from the same vendor; 2-purchase the same technology from a different vendor; 3-purchase a different technology; 4-do nothing.
(5) "√" indicates that a savings was realized, but not quantified.
(6) ND = no data

- Users of the polyfluorocarbon membrane units for bath maintenance and chromium recovery reported that equipment replacement was frequently necessary and expensive (PS 194, PS 134).

5.6.6 Residuals Generation

Due to a lack of data, the true pollution prevention value of the ion transfer technology cannot be fully ascertained. Some evidence indicates that the porous pot technology is inefficient in the sense that a significant quantity of chromium waste is often generated relative to the volume of bath being maintained. This observation is especially evident for the large-scale porous pot units (i.e., Cosmos Minerals Corporation, Pfaudler, EP Technology). For example, a report prepared for the Air Force described an application where 16,830 gal. of hard chromium plating bath was maintained with two large porous pot systems (8 pots each). Annually, these systems generated 10,000 gallon of chromic acid waste. Also, equipment replacement (electrodes, pots) costs were $60,000 annually (ref. 372).

Similarly, one survey respondent indicated that use of a PPS1 generated 4,800 gpy of waste (combination of catholyte and wash-down water that is sent off-site for recycle) while maintaining a 3,000 gal. bath (PS 074).

For these applications, it appears that a "bleed and feed" strategy would be more efficient than use of a porous pot. However, some other survey respondents reported less waste generation and in some cases very small volumes (e.g., PS 194 and PS 118).

No data were located with respect to the waste generation rates of the polyfluorocarbon membrane technology.

5.7 MEMBRANE ELECTROLYSIS

5.7.1 Overview

This group of bath maintenance technologies was used by 5 (or 1.6%) of the respondents to the Users Survey. These technologies, applied primarily to the maintenance of chromic acid solutions such as decorative and hard chrome plating baths, chromic acid etch solutions and chromic acid anodizing baths, compete with ion exchange (Section 5.4) and ion transfer (Section 5.6) technologies. All of the applications identified in the Users Survey involve these solutions. Membrane electrolysis is also applicable to non-chromium solutions, including those formulated with sulfuric acid and sodium hydroxide, but are less frequently used for them.

Membrane electrolysis is related to the ion transfer technology discussed in Section 5.6, however, commercial membrane electrolysis units employ ion specific membranes rather than the ceramic or polyfluorocarbon materials used in ion transfer.

With membrane electrolysis, an electrical current is passed through electrolytes that are separated by an ion specific membrane. Two reactions typically occur as the result of using membrane electrolysis: (1) ions of a given species are electrically driven across a selective membrane and (2) chemical changes occur (e.g., oxidation/reduction such as the electro-oxidation of Cr^{+3} to Cr^{+6}) at the electrodes.

The term membrane electrolysis is found in German literature (membran-elektrolyse) (ref. 520). Some U.S. manufacturers of this technology prefer to categorize their equipment as an electrodialysis technology. This latter term however, is more often reserved for an electrochemical device with alternating anion- and cation-exchange membranes (referred to as stacks) that alter the concentration and/or composition of an electrolyte simply as the result of electromigration (ref. 401, 423).

The primary function of membrane electrolysis, when applied as a bath maintenance technology, is to lower or maintain at an acceptable level the concentration of metallic impurities in plating, anodizing, etching, stripping and other metal finishing solutions. This is accomplished through the use of an ion exchange membrane(s) and an electrical potential applied across the membrane(s). The membranes employed with these technologies are ion-permeable and selective, permitting ions of a given electrical charge to pass through. Cation membranes allow only cations, such as copper, nickel, and aluminum to pass from one electrolyte to another, while anion membranes allow only anions, such as sulfates, chromates, chlorides, or cyanide to pass through. Bath maintenance units can be configured with only cation or anion membranes or both. The oxidation of Cr^{+3} to Cr^{+6} occurs at the anode, in the same manner as high current density dummy plating described in Section 5.2.2.3. The anode material and/or coating must be properly selected in order for this reaction to occur.

The most common applications for this technology within the plating industry are the purification of chromium plating (especially hard chromium) and chromic acid anodizing baths. These baths become contaminated with various metallic impurities that are introduced mostly as a result of: drag-in; anodic etching; corrosion of electrodes, bus bars, racks, fixtures and parts; and reduction. The metal impurities combine with the acid in the bath and form metal salts. Eventually, the concentration of metallic impurities exceeds operational limits and the baths must be discarded.

Hard chromium plating solutions are most often fouled by iron, aluminum, copper and trivalent chromium. Iron is contributed during the anodic etch of steel parts, which is usually performed in the plating bath prior to plating. Etching prepares the surface of the parts and improves adhesion. Iron is also contributed by the corrosion of steel parts that have been dropped in the tank and not retrieved. Aluminum is contributed to hard chromium baths mostly from aluminum fixtures that hold the parts during plating. Some facilities chromium plate aluminum parts, which also contributes aluminum to the bath. Copper is added mostly from corrosion of copper bus bars and fixtures. Trivalent chromium is generated in chromium plating baths when hexavalent chromium (i.e., the desired chromium species) is reduced. This occurs partly due to the introduction of organic chemicals to the bath, but is mostly due to an imbalance in the plating process. During plating, hexavalent chromium is reduced to chromium metal on the surface of the part (cathode). However, some hexavalent chromium is incompletely reduced to trivalent chromium. Simultaneously, trivalent chromium is oxidized at the anode to hexavalent chromium. An imbalance is created when the surface area of the plated part approaches or exceeds the surface area of the anode. This condition is most prevalent with inside diameter (ID) plating since the anode, which is inserted into the part, must be physically smaller than the part.

The deleterious effect of metallic impurities on the chromium plating process is widely acknowledged, but the tolerable limits are often debated. Metallic impurities are known to limit the permissible cathode current density, resulting in slower plating to obtain comparable deposit quality (ref. 369). Various defects in chromium plated deposits, including burning, roughness, pitting, and reduced adhesion, brightness and hardness are also attributed to metallic impurities. The actual effects of metallic impurities depends on many factors including the overall chemistry of the bath, plating procedures used and the type of parts being plated (ref. 370).

Various control limits for chromium plating baths are in use, with a 4 g/l combined concentration of iron, aluminum and copper, and other tramp metals as a common standard. Experts are in less agreement over trivalent chromium control limits and many report advantages from the presence of some Cr^{+3}. However, a 4 g/l control limit for Cr^{+3} is typical of that used in industry (ref. 370).

The effects of metallic impurities on chromic acid anodizing baths and acceptable limits are not as well documented as hard chromium plating. The key metallic impurities are trivalent chromium and aluminum. Trivalent chromium is generated by the same methods discussed for hard chromium plating (i.e., introduction of organics and excessive cathode area in relation to anode area). One widely recognized source recommends controlling Cr^{3} below 3 g/l to prevent darkening of the anodic film. A 4 g/l limit is used by the Air Force (ref. 384).

Aluminum is added to the bath by the electrolytic etching of the parts during the process (in anodizing, the part is the anode) and causes reduced efficiency. Chromic acid in the anodizing bath is neutralized by aluminum dissolving in the solution. Knowing the neutralizing effect of aluminum, the baths are typically controlled by pH rather than direct aluminum concentration. This practice is implemented by the Air Force, where the upper control limit is pH 0.85 (ref. 384). In practice, a new chromic acid anodizing bath is formulated by the Air Force with 48 g/l of chromic acid. The control limits for chromic acid are 40 to 105 g/l. As chromic acid is neutralized by dissolving aluminum, chromic acid additions are made to keep the pH below 0.85. These additions increase the chromic acid concentration of the bath. In theory, once the bath reaches a chromic acid concentration of 105 g/l, the bath is discarded. However, more frequently, the pH of the bath becomes difficult to control at approximately 75 g/l of chromic acid and it is discarded (ref. 384).

5.7.2 Development & Commercialization

Membrane electrolysis is the most recent technology applied to the maintenance or purification of plating baths. Although the membrane electro-transport phenomenon was observed in the 1800's, it did not achieve any commercial application until durable, ion-selective membranes were developed in the 1950's. The first applications were for desalination of brackish water using electrodialysis. In the mid 1970's, the first attempts in the U.S. were made to use electrodialysis to reclaim plating chemicals from rinse solutions. The earliest commercial success was the recovery of nickel. The implementation of membrane electrolysis for bath purification proceeded less quickly due to operational problems with membrane fouling and electrode maintenance. Early work on bath maintenance was performed by the U.S. Bureau of Mines for rejuvenation of spent chromic acid pickling solution in the late 1970's. The Bureau built an industrial scale unit using Du Pont's Nafion® ion exchange membrane and a sulfuric acid catholyte and tested it at several plating shops in the mid-west in 1979 and 1980. A commercial unit (COPS), based on the Bureau of Mines concept and design was developed and applied to a sodium dichromate/sulfuric acid pickle for brass forgings and a chromic/sulfuric acid copper strip solution (following carburization) (ref. 389). The use of membrane electrolysis for plating bath maintenance did not accelerate until the mid-1980s when a second generation of technology, also using the Nafion® membrane, was introduced for the purification of hard chromium plating baths (ref. 370). That system (Ionsep) uses a patented approach employing a caustic catholyte solution, rather than sulfuric acid, to precipitate the tramp metals as they cross the membrane. Also, during the early 1980's, commercial efforts were also directed at an ion transfer technology using a non-selective polyfluorocarbon membrane (CatNapper) (see Section 5.6). A French company manufactured a membrane electrolysis unit that was marketed in the U.S. in the mid- to late-1980's (marketed by Pentol Incorporated USA). That system was previously applied to chromic acid anodizing baths by aerospace firms in France (ref. 390). Other plating bath and pickling solution applications of membrane electrolysis including non-chromium applications, have since been commercially proven. Increased commercial activity in membrane electrolysis was spurred by cost increases for chemicals and waste disposal and environmental laws such as the requirements in the Resource Conservation and Recovery Act (RCRA) for waste minimization. These considerations have urged plating shops to search for methods of reducing the disposal frequency of process solutions.

5.7.3 Applications and Restrictions

Potential applications of this technology for plating shops are shown in Exhibit 5-27. Application ME-1 is the typical bath purification scheme and the most widely used application for this technology. Application ME-2 shows the use of this process for several functions related to bath maintenance and chromic acid recovery.

The bath maintenance process in ME-1 is typically operated on a continuous basis. Process solution is pumped to a tank holding the electrolytic cell (alternatively the cell can be placed into the process tank) or through a stacked membrane system and returned to the bath. With each pass of the process solution, a portion of the cations are transported through the membrane and into the catholyte solution. This application describes the use of membrane electrolysis systems that operate with either an acidic or caustic catholyte (see catholyte discussion in Section 5.7.4).

Application ME-2 is a theoretical configuration showing: (1) conventional bath maintenance; (2) reformulation of cation regenerant; and (3) reformulation of a combined anion regenerant/caustic strip solution. The first two applications use the two cell membrane electrolysis configuration (one anode located in an anolyte and one cathode located in a catholyte) and the third uses a more complicated three cell arrangement (anode, reactor and cathode compartments). With this particular application of a three cell arrangement, sodium dichromate solution is reformulated into sodium hydroxide and chromic acid. Other applications requiring three compartments include phosphoric acid and conversion coating bath purification (ref. 384).

The functions of the two and three compartment cells are described in Exhibit 5-28. With the two cell arrangement, the chromic acid bath is the anolyte and the tramp metals present in that solution are transported into the catholyte, where they precipitate as metal hydroxides. With the three compartment cell, the anolyte is the sodium dichromate solution. The cations in the anolyte are transported into the center compartment (reactor compartment) where the multi-valent cations are precipitated as metal hydroxides. The mono-valent cations are further transported to the cathode compartment where the caustic is reformed. This process, which includes the precipitation of metals, is patented by Ionsep.

For the plating industry, membrane electrolysis is most applicable to the maintenance of chromic acid solutions, including hard chromium and decorative chromium plating, chromic acid etching, chromic acid anodizing and chromic acid stripping (e.g., copper). These are the only applications that were identified during the NCMS/NAMF Users Survey. Other potential applications include sulfuric and nitric acid and sodium hydroxide based solutions (e.g., pickle, etch, strip, rust removal), chromate conversion coating, and sodium dichromate deoxidizer (ref. 384).

Exhibit 5-27. Configurations for Application of Membrane Electrolysis for Bath Maintenance

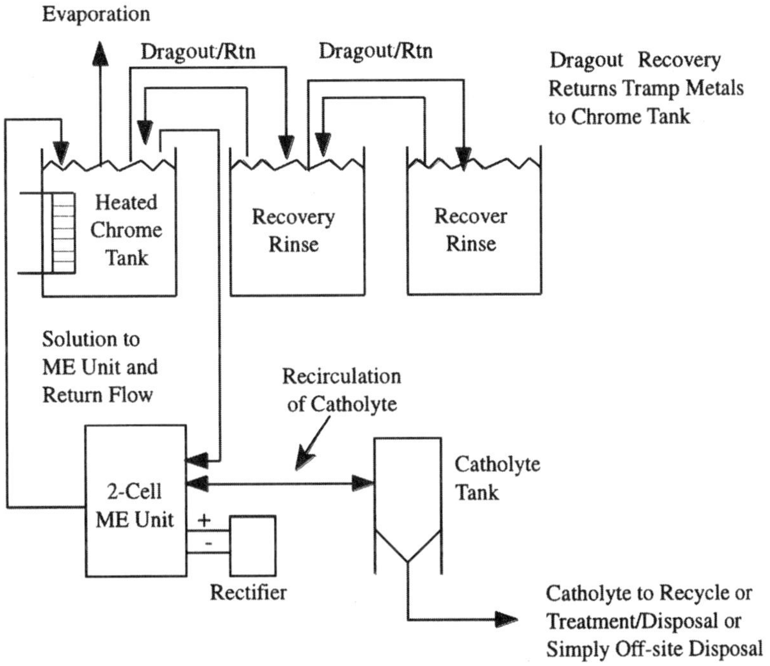

Application ME-1: Chromium Bath Purification

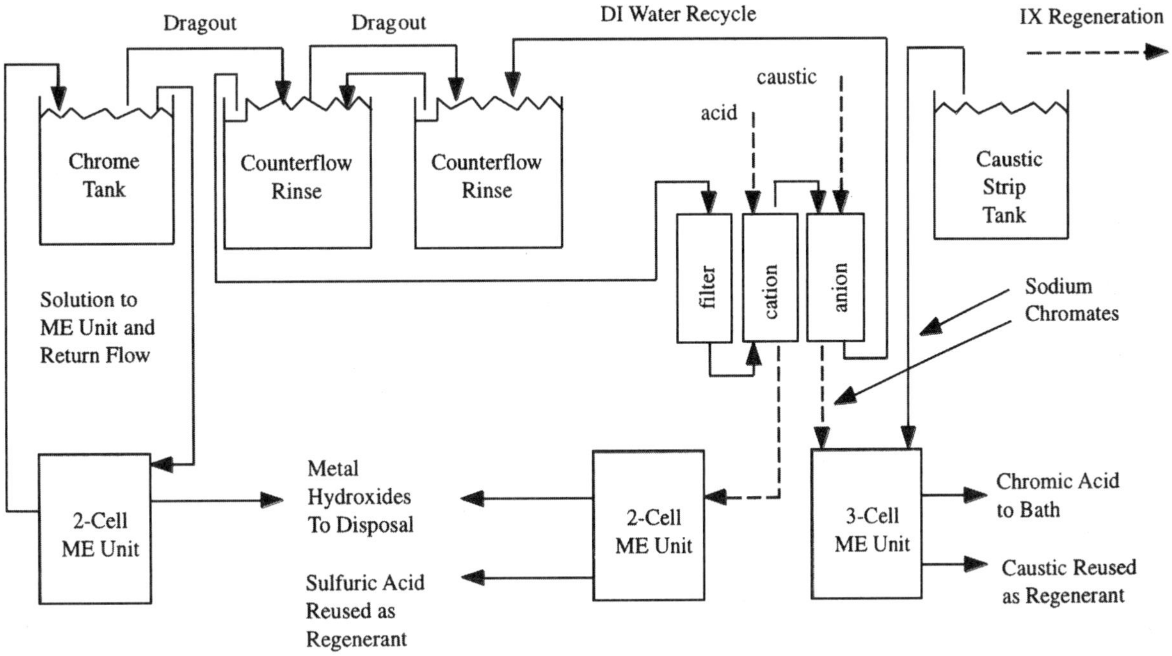

Note: rectifiers and catholyte tanks are not shown in order to simplify the ME-2 drawing.

Application ME-2: Bath Maintenance and Recovey from Rinsing and Stripping

Exhibit 5-28. Comparison of Two and Three Compartment Membrane Electrolysis Cells

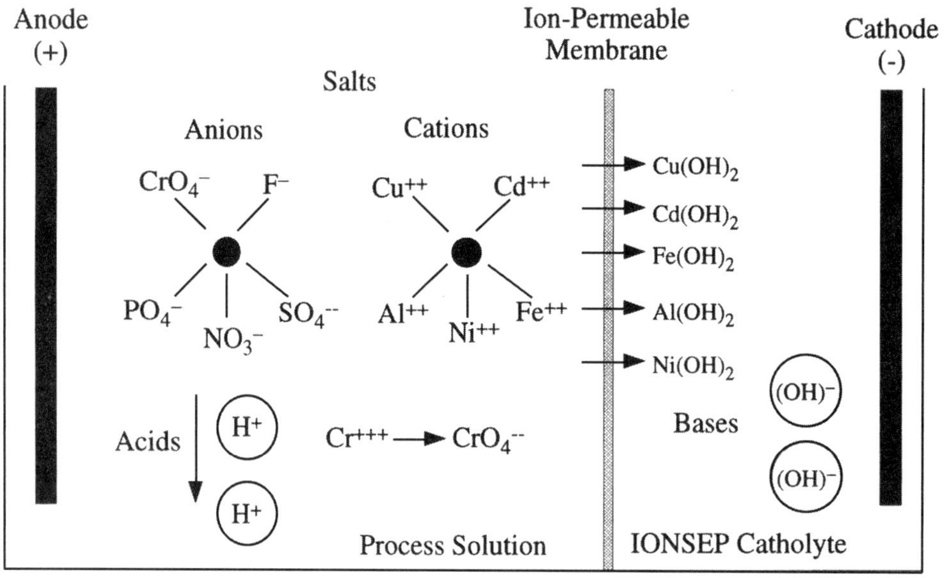

Two Compartment Cell

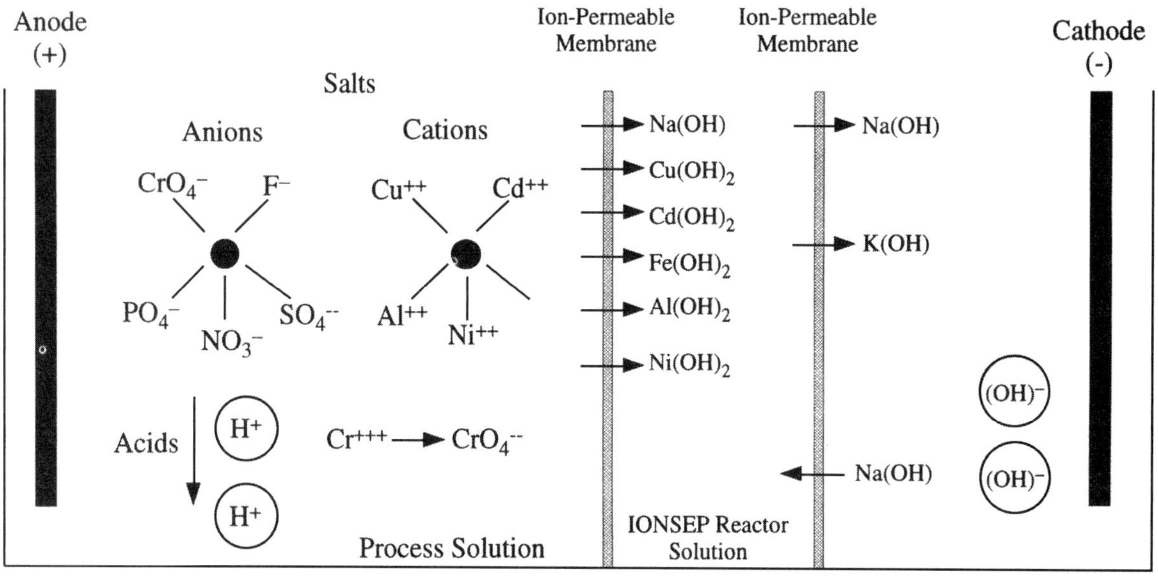

Three Compartment Cell

5.7.4 Technology/Equipment Description

5.7.4.1 General

Membrane electrolysis is the newest of the chromium bath purification technologies. In the most basic sense, these units consist of: (1) a tank containing an anode and cathode compartment separated by a selective membrane(s), and (2) a power source. The membranes are ion specific in a manner similar to ion exchange resin. They allow the passage of only positive or negative ions, depending on the type of membrane. This design element is used in this text to distinguish these units from ion transfer technologies that employ non-ion permeable membranes such as ceramic pots or a polyfluorocarbon material (Section 5.6). Cation specific membranes are used for chromic acid purification, the most common application for this technology. When current is applied to the cell, cations present in the anolyte flow through the cation permeable membrane. The anions in the catholyte are restricted by the membrane and remain in that compartment. Trivalent chromium present in the anolyte (plating bath) is reoxidized to the hexavalent state. Membrane electrolysis units can be configured with only cation or anion membranes or both.

It is important to note that Cr^{+3} is a cation, while Cr^{+6} combines with oxygen in solution to form CrO_4 and acts like an anion. Therefore, in a membrane electrolysis system, the Cr^{+3} that is not oxidized to Cr^{+6} will behave like iron and aluminum ions and be transported through the cation membrane while the Cr^{+6} remains in the plating solution. The loss of Cr^{+3} in this manner is undesirable since Cr^{+3} can be restored to a useful state by simple oxidation at the anode. Further, the catholyte solution, into which the cations are transported, becomes more quickly spent when Cr^{+3} passes through the membrane. The cathode solution is subsequently discarded and its chromium content will increase disposal costs. Therefore, units with a rapid Cr^{+3} oxidation rate will generate less process residual and the residual will have a lower chromium content.

The available commercial membrane electrolysis units are discussed in Section 5.7.3.1. Four different membrane electrolysis manufacturers were identified during the literature search, Users Survey and Vendors Survey. A comparison of these technologies shows that there are several basic variations among the units. The following discussion highlights these differences.

The choice of catholyte solution is one major difference between the various commercially available units. Most of the units use mineral acid (typically sulfuric acid) as the catholyte, which was used by the Bureau of Mines in their early work. With an acidic catholyte solution, cations flow through the permeable membrane and into the acidic catholyte. The electroplatable cations deposit on the cathode and the non-electroplatable cations stay in that solution as salts. Periodically, the cathode is cleaned and the catholyte solution is discarded and replaced (370).

One of the available technologies uses a patented caustic catholyte (combination of alkali reagents). The caustic catholyte causes the precipitation of cations as they pass through the membrane. This overcomes any problem of plating-out the cations on the cathode. This catholyte converts multi-valent metal cations entering the catholyte solution into insoluble hydroxides. The hydroxyl ions needed to react with the metal cations are formed at the cathode of the cell. Precipitation of the cations prevents a loss of conductivity and eliminates the buildup of a deposit on the cathode; consequently, the operational period is extended to two or more weeks and there is no need to clean the cathode.

The overall appearance of the commercially available units varies significantly. Some of the units are configured similarly to the Bureau of Mines' prototype units. The chromic acid bath is pumped to a tank on either a batch or continuous basis. The tank contains multiple cathode cells, around which the anolyte (chromic acid bath) is circulated. Another commercial unit has a conventional electrodialysis design with "stacks" of membranes that are kept together in a clamping unit which looks similar to a filter press. Anode and cathode electrodes are located at opposite ends of the stack. The plating solution is pumped through the stack. The pumps and other support equipment are located on a skid. A third configuration, places the cathode in a "membrane sock," which is lowered into the plating tank. The catholyte is circulated from a side tank into the cathode compartment (ref. 384).

The capacities and capabilities of membrane electrolysis units should be measured in terms of the tramp metal removal rate (g/hr), which will vary over a range of tramp metal concentrations in the anolyte. Unfortunately, there is a limited amount of such data available. Exhibit 5-29 shows the results of a comparative test of a caustic catholyte membrane electrolysis system and a single cell porous pot unit.

5.7.4.2 Commercially Available Equipment

This subsection contains a description of commercially available ion specific electrochemical membrane technology equipment. This is intended to provide the reader with information and data on a cross section of available equipment. Mention of trade names or commercial products is not intended to constitute endorsement for use.

Ionsep Corporation has manufactured electrochemical processing equipment since 1985 and has sold more than 150 units to metal finishing firms. Approximately two-thirds of their units were sold for chromic acid applications. The basic Ionsep unit consists of a electrolytic cell, a catholyte tank (located near and plumbed into the electrolytic cell), and a power source (typically a 500 amp rectifier). The cell is inserted in a protective 6 to 9 in. diameter PVC pipe approximately 36 in. long that is kept in the plating bath during operation (alternatively, a side tank is used, especially with larger applications). The pipe is perforated to permit plating solution to enter and flow around the cell membrane. Within the PVC pipe is a cylindrical shaped anode, which is exposed to the plating bath, and a selective membrane "sock" that creates an inner cathode compartment. A cathode and catholyte solution (an alkaline solution) are located in the sock. This configuration is termed a two compartment cell where the anode compartment is the plating bath and the cathode compartment is the space inside the sock.

In operation, a direct current is passed to the cell electrodes and the catholyte is continuously circulated from the catholyte tank (typically a 55-gal. plastic drum with a small circulation pump located on top) into the sock formed by the membrane. The cations (e.g., iron) are electrically driven from the plating bath through the selective membrane and into the cathode compartment. The metal cations precipitate as hydroxides. Most of the trivalent chromium present in the bath is converted to hexavalent chromium at the cell's anode. Some trivalent chromium, which is a cation, passes from the bath, through the membrane, and into the catholyte solution. The tramp metal removal rate for a small Ionsep unit (130 amp unit) is shown in Exhibit 5-29.

The two compartment Ionsep cell is applicable to hard chromium and chromic acid anodizing baths and other baths with similar chemistry. Baths with more complex chemistry often require cells with three or more compartments. For example, a caustic chromium stripper can be reformulated into sodium hydroxide and chromic acid with a three compartment cell. The resultant sodium hydroxide is reused as stripper and the chromic acid is reused in the plating bath.

Large Ionsep systems are configured differently from that described above. With the larger systems, the electrolytic cell is not located in the plating tank, but rather, inside of a skid mounted PVC piping system. With this configuration, the bath is pumped to the cell and circulates back to the plating tank. Many of the large systems also have a catholyte recycle capability, while in the small systems the catholyte solution is discarded weekly. The catho-

Exhibit 5-29. Performance of Membrane Electrolysis Unit for Chromic Acid Bath Maintenance

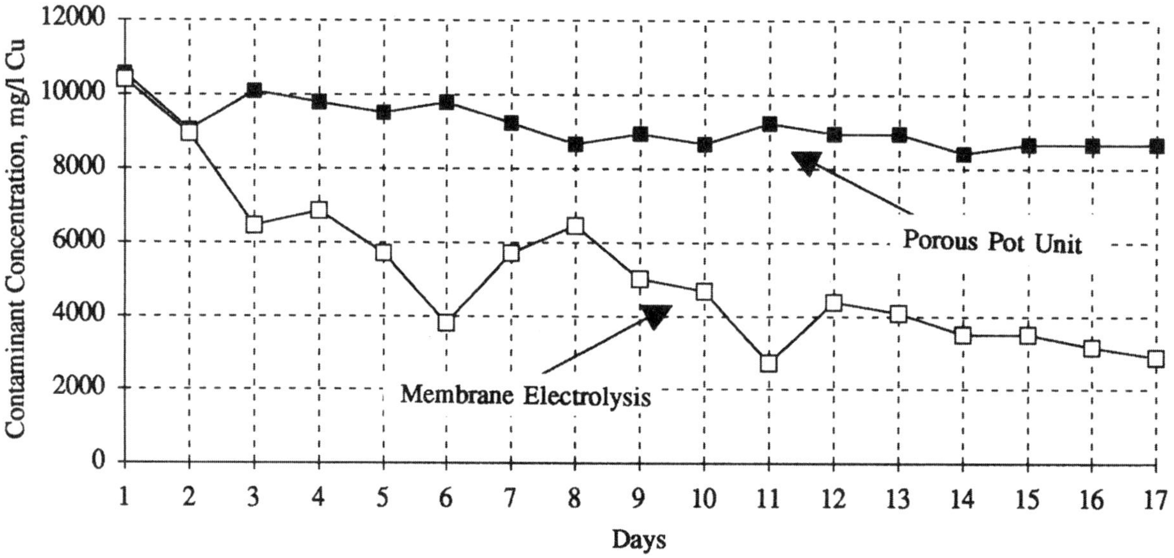

Test Conditions: (1) single porous pot unit (70 amp); caustic catholyte membrane electrolysis unit (130 amp), (3) anolyte- 240 g/l CrO_3 and 2.4 g/l SO_4; (4) anolyte volume 40 to 45 gal; and (5) 100 to 105°F anolyte operating temperature.

Source: ref. 370.

lyte recycle system causes the separation of precipitated metals from the catholyte which permits the reuse of the catholyte.

Ionsep units and systems are available with capacities from 150 amps to 15,000 amps. In addition to the chromic acid bath maintenance (chromium plating, chromic acid anodizing and chomate conversion coating), their equipment has been applied to nitric acid strip, nitric-HF pickle, sulfuric acid pickle, hydrochloric acid, and caustic chem mill etch maintenance. They also manufacture a closed-loop rinse water recovery system that employs ion exchange and electrochemical regeneration to recover chemicals and enable the reuse of water.

GOEMA, a German-based company, manufactures a family of membrane technologies for water, wastewater, and bath maintenance. These technologies include membrane electrolysis, diffusion dialysis, ultrafiltration, and microfiltration. Their membrane electrolysis units are designed with a stacked membrane configuration. A "membrane stack" consists of a number of thin membrane compartments created by alternating layers of ion-specific membranes and spacers. The spacers are present to permit solutions (e.g., plating solution and catholyte) to flow between the membrane layers. Either anion- or cation-specific membranes or both are used to create a stack. For hard chromium plating and chromic acid anodizing bath purification, only cation membranes are used. The membranes and flow spacers are held together by a cell frame and clamping unit, giving the membrane stack the appearance of a plate and frame filter press.

In operation, plating solution and catholyte solution are pumped into the stacked cell. Separate inlets and outlets and flow channels are present for each solution. Anode (lead alloy) and cathode (stainless steel) electrodes are located at the ends of the stack in their associated electrolyte. When an electrical potential is applied across the stack, Cr^{+3} present in the anolyte (plating solution) is oxidized to Cr^{+6} and dissolved tramp metals migrate toward the cathode and pass through the cation membrane into the catholyte. Sulfuric acid (20 percent) is typically used as a catholyte for these applications.

GOEMA units can be purchased in various capacities. The key sizing variable is the number of compartments that make up the membrane stack. GOEMA's basic unit (KRME-100) has a rated metals removal capacity of 100 grams per hour. The membranes are approximately 4 ft^2 each with a total membrane area of 20 ft^2. The DC power source is a 1,000-amp rectifier. The flow rate of anolyte through the stack is approximately 6 gph (144 gpd). A single solution pass through the stack provides a high cation removal rate. Approximately one to two tank volumes of treatment will adequately purify a moderately contaminated bath. Assuming this rate of bath purification, a 500-gallon bath could be treated in approximately 1 week (two tank volumes x 500 gallons/144 gpd = 6.9 days) (Based on the manufacturer's rating for these units.) (ref. 384).

Graver Water Division of the Graver Company (Graver) is the distributor of Tokuyama Soda membrane electrolysis equipment. These units have been utilized for the following applications: nickel and zinc recovery from galvanizing, caustic recovery from aluminum etch, acid recovery from metal pickling, and aluminum recovery from nitric acid solutions.

Kinetic Recovery Corporation (KRC) manufactures the Membrane Electrolysis System 125 (Exhibit 5-30). This unit is designed for the oxidation of Cr^{+3} to Cr^{+6} and the removal of tramp metals from chromic acid. The basic unit is skid mounted and consists of: electrolysis container with a drip pan; anode compartment including anode (Ti/Pt expanded metal), inflow-assembly and electrical connections; stainless steel cathode; ion exchange membrane; anolyte and catholyte feed tanks; anolyte filter; catholyte pH control system; rectifier (15V, 150A); and control panel. The dimensions of the skid are 48" x 21" x 26" (H).

5.7.5 Costs

5.7.5.1 Capital Costs

Equipment costs for two different manufacturers of membrane electrolysis equipment are shown in Exhibit 5-31, based on their estimated removal capacity for iron from a bath containing 10 g/l of iron. Installed costs are approximately 5% to 20% higher than equipment costs.

5.7.5.2 Operating Costs

Operating costs for the membrane electrolysis technology are presented in Exhibit 5-32. These are based on the operation of units designed for the removal of 400 g/day of iron from chromic acid solutions containing 4 to 10 g/l of iron.

5.7.6 Performance Experience

Only five membrane electrolysis applications were identified during the Users Survey. These included two

Exhibit 5-30. Diagram of Membrane Electrolysis System 125 (Courtesy of Kinetic Recovery Corporation)

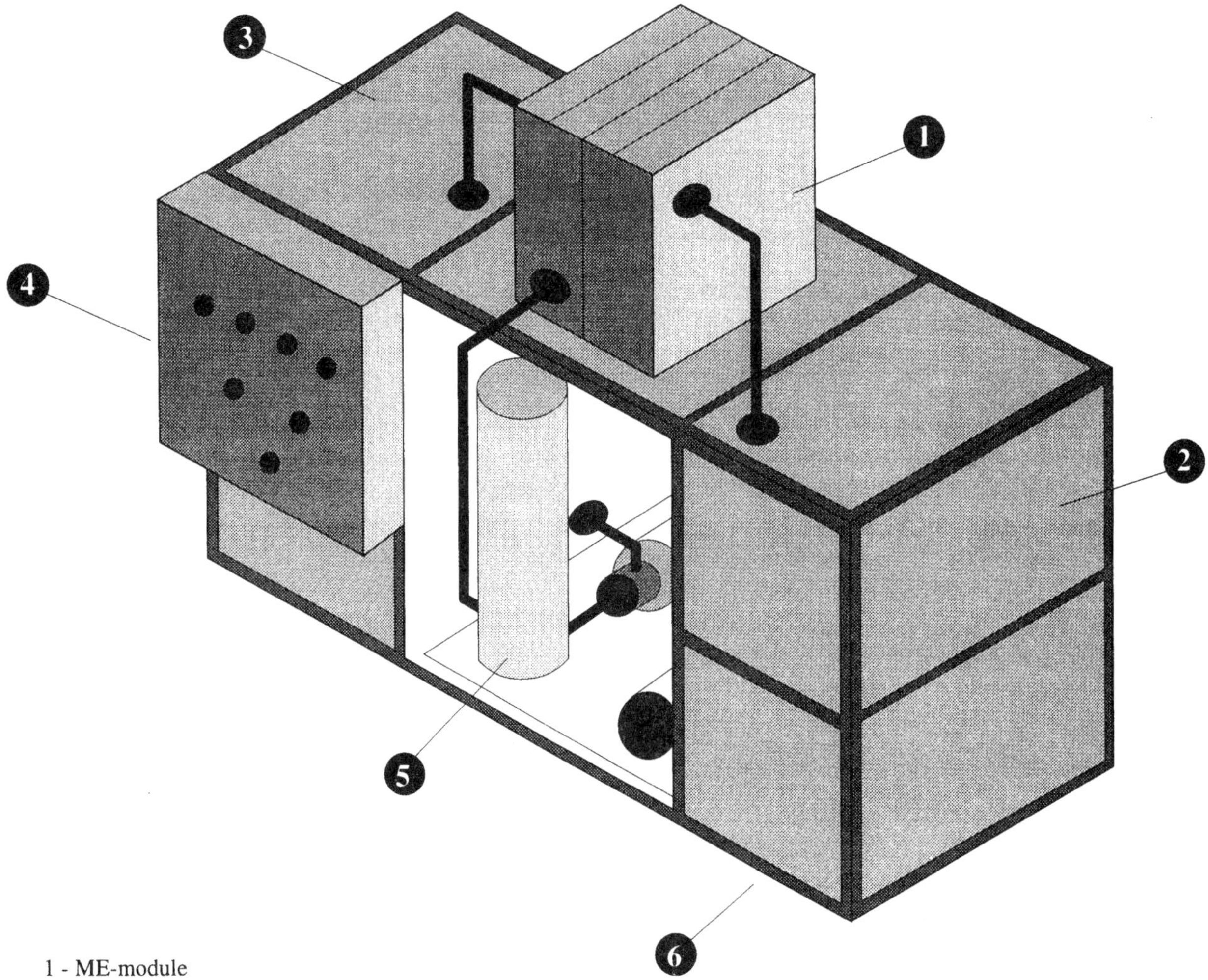

1 - ME-module
2 - process solution tank
3 - maintenance solution tank
4 - electrical control panel
5 - pump-filter assembly
6 - skid

applications of in-house units built by PS 275, two applications of Ionsep units and one application of a Pentol CAPS system (manufactured in France). Detailed data were provided for all systems except for the Pentol unit. The user of that system, however, provided a short informative description of their experience. A summary of the Users Survey data are provided in Exhibit 5-33.

The following information and data summarize the performance experience of the survey respondents.

- Shops using membrane electrolysis for chromic acid bath maintenance gave the technology mixed ratings, even for the same manufacturer. PS 006 and PS 052, both of which purchased Ionsep units, had general satisfaction levels of 1 and 5, respectively (1= lowest and 5 = highest). The low satisfaction level from PS 006 may have been influenced or otherwise related to the manufacturer's technical support, which they also rated 1. That shop also provided the following: "Technical support info

Exhibit 5-31. Capital Costs for Chrome Plating Bath Maintenance Using Membrane Electrolysis

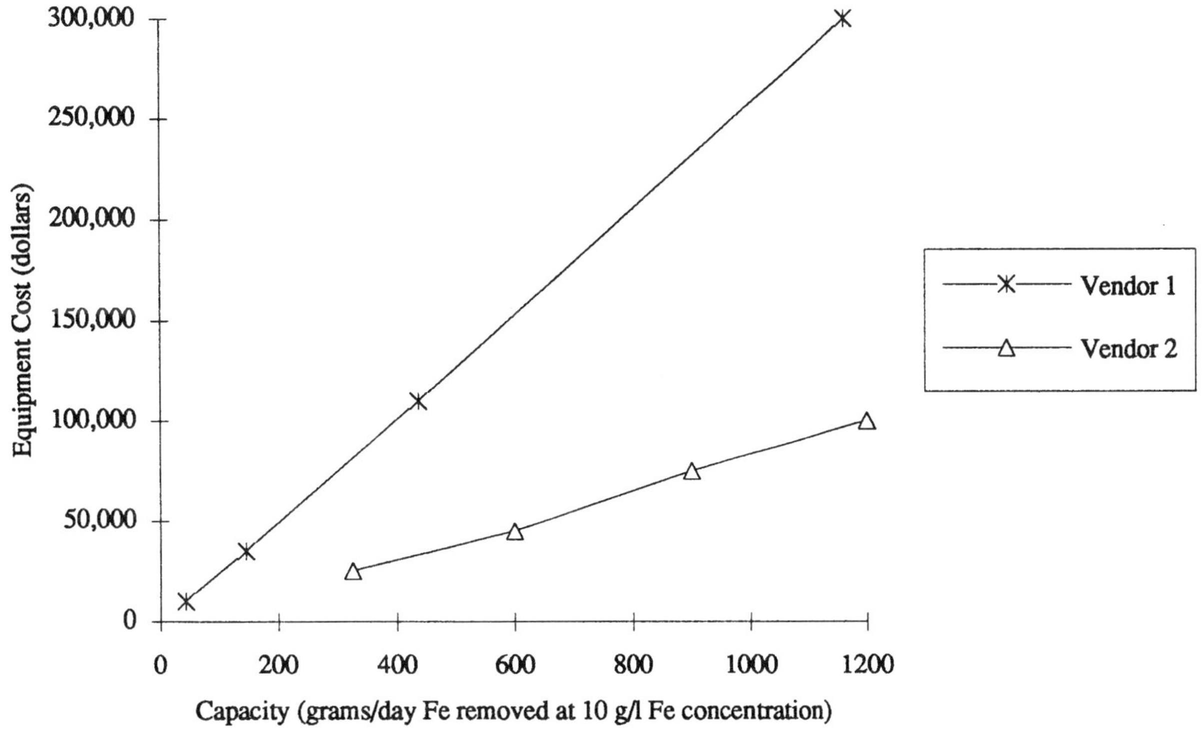

Exhibit 5-32. Operating Costs* for Chromic Acid Bath Maintenance Using Membrane Electrolysis

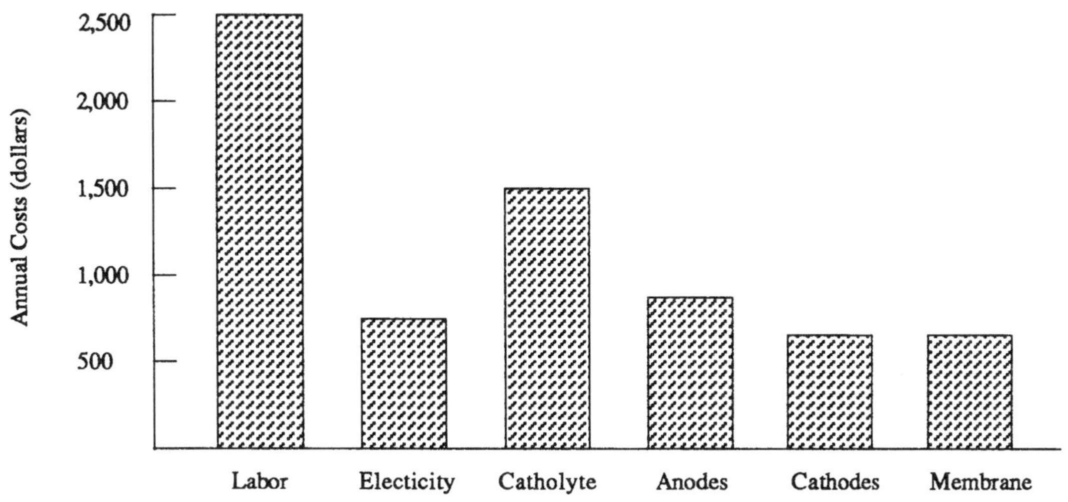

*Estimated operating costs for a membrane electrolysis unit removing 400 g/day of cations from a chromatic acid bath containing 4 to 6 g/l of cations (excluding trivalent Cr). Actual costs for anodes, cathodes, and membrane replacement can vary up to 500%, depending on the manufacturer selected. Catholyte costs include chemical remake costs and disposal costs.

very difficult to get from vendor. Process not compatible with our bath, which we had to learn on our own." PS 052 gave a manufacturer's support rating of 4. The owner of the Pentol unit was displeased with its performance (PS 234). They wrote: Seemed limited to the removal (of) trivalent chromium... other impurities such as Cu and iron were largely unaffected."

- The following is a breakdown of the reasons why shops purchased this technology:

To meet or help meet effluent regulations: 0
To reduce plating chemical purchases: 1
To reduce the quantity of waste shipped offsite: 2
To reduce wastewater treatment costs: 0
To improve product quantity: 3
Other: 0

- The use of membrane electrolysis had a variable impact on production quality and the rate of production. The following responses were provided:

	Product Quality	Production Rate
Improved (PS 275)	1	1
No Change(PS 052)	1	1
Decreased (PS 006)	1	1

- Although PS 052 indicated in the survey form that product quality and the production rate were not impacted, they also stated that purifying their plating solutions "reduces our reject rates which eliminates the costly need to strip the part and replate."

- None of the shops provided any estimates of savings from the operation of membrane electrolysis units.

- Membrane electrolysis may be more applicable to simple chromic acid bath chemistry (e.g., generic chromic acid/sulfuric acid hard chromium bath) more so than baths with complex chemistry (e.g., dual catalyst plating baths). This is evidenced by the difficulties of PS 006, who operated their unit on a proprietary fluoride bath (M&T CR 110), whereas PS 052, successfully operates their unit on a generic hard chromium bath.

- Two of the survey respondents have purchased and used alternative technologies for chromic acid bath maintenance. PS 006 previously used the porous pot (an ion transfer technology) and PS 234 previously used the porous pot and ion exchange. PS 006 indicated that the porous pot was "moderately successful", whereas they were unsatisfied with the membrane electrolysis unit. They indicated that both technologies caused bath chemistry problems with their proprietary plating solution. PS 234 has not been satisfied with any of the chromic acid bath maintenance technologies that they have tried and subsequently sold each of them. However, they indicated that ion exchange provided the best results of the three technologies. A complete discussion of their comments is presented in Section 5.6.8.

5.7.7 Operational and Maintenance Experience

The following summarizes the respondent's O&M experiences and provides operating labor information.

- Two shops provided operating labor data. For these shops, the average number of annual operating hours per membrane electrolysis unit were: 200 hrs/yr (assumes 250 hrs/yr for each of the two units operated by PS 275). The following is a breakdown of the responses for skill requirements (based on data from PS 006, PS 052 and PS 275):

Environmental Engineer:...0
Process/Chemical Engineer:...0
Chemist: 1
Consultant: 0
Plumber/Pipe Fitter: 0
Electrician: 0
Vendor: 0
Senior Level Plater: 1
Junior Level Plater: 0
Trained Technician: 2
Common Labor: 0
Other: 0

- Reported O&M problems for this technology include: (1) occasional clogging of pipes (PS 052); short life span of membrane and anodes (PS 006); and (3) very difficult to clean out collected residuals from unit.

- Ionsep reported the following expected life spans of system components: membrane > 12 mth, 12 to 36 mth, and cathode > 60 mth.

Exhibi 5-33. Summary of Users Survey Data for Membrane Electrolysis Bath Maintenance Applications

Shop Code	Application	Manf.	Year	Capital Costs: Equip. Costs	Capital Costs: Other Cap. Costs (2)	Capital Costs: Total	Operating Costs (Yr.): Non-labor	Operating Costs (Yr.): Labor Hrs.	Savings: Process Chem.	Savings: Reduced Rejects	Savings: Sludge Disp.	Use Code (3)	Down Time %	Satisfaction Levels (4): Manf. Support	Satisfaction Levels (4): Tech-nology	Future Decision (5)
006	hard Cr	IONSEP	1989	$15,000	$0	$15,000			ND	ND	ND	2		1	1	4
052 (1)	hard Cr	IONSEP	1988	$50,000	$0		$2,300	100	ND	√	ND	1	2	4	5	1
275	Cr etch	In-House	1984	$32,500	$3,750	$36,250	$35,600	250	ND	ND	ND	1				1
275	Dec. Cr	In-House	1992	$32,500	$3,750	$36,250	$35,600	250	ND	ND	ND	1				1
234	hard Cr	Pentol (CAPS)	ND	ND	ND	ND	ND	ND	ND	ND	ND	2				3
		Average	1988	$32,500	$1,875		$24,500	200					2	2.5	3	

(1) Capital cost includes three units.
(2) Other capital costs include installation and ancillary equipment.
(3) Use Codes: 1-currently operating; 2-not currently operating and no intension for future use; 3-not currently in use, but intend to use in future.
(4) Satisfaction levels (manufacturer and technology) 1 to 5, with 1=lowest and 5=highest.
(5) Future decision codes: 1-purchase the same technology from the same vendor; 2-purchase the same technology from a different vendor; 3-purchase a different technology; 4-do nothing.
(6) "√" indicates that a savings was realized, but not quantified.
(7) ND = no data

5.7.8 Residuals Generation

The primary residuals generated by the use of membrane electrolysis are spent catholyte and associated sludge. During the process, cations are electrically transported from the anolyte (process bath) through the ion specific membrane into the catholyte. The nature of the residual generated during the process depends mostly on the selection of the catholyte, either acid or caustic. With an acidic electrolyte (typically sulfuric acid), the electroplatable cations deposit on the cathode and the non-electroplatable cations stay in that solution as salts. Periodically, the cathode is cleaned of the sludge-type deposit and the catholyte solution is discarded and replaced (ref. 370). The smallest GOEMA unit has a catholyte volume of 10 gal (20% sulfuric acid). It is typically replaced every 4 months (30 gal/yr) (ref. 384). With a caustic catholyte (a mixture of alkalis used in a patented approach by Ionsep), the high pH converts multi-valent metal cations that enter the catholyte solution into insoluble hydroxides. The hydroxyl ions needed to react with the metal cations that enter are formed at the cathode of the cell. The precipitated metals buildup in the catholyte. With smaller systems, it is usually preferable to replace the catholyte (approximately every one to four weeks). With larger systems, the precipitated metals are separated using a filter press and the catholyte is returned to service (this practice greatly reduces the volume of waste generated by the plating/purification process). Alternatively, a conical bottom catholyte tank can be employed and the solids (with some catholyte) can be drained from the cone of the tank and disposed of as a slurry. In some cases, the metal hydroxides are reusable for other purposes (e.g., ferrous hydroxide, which can be used for making pigments, is produced during applications with steel pickling solutions). The Ionsep catholyte tank typically has a volume of 55 gal (plastic drum), although large installations may have tanks with volumes up to 300 gal.

Two shops operating the Ionsep system provided data on residuals production. PS 006 operated the process on a 2,800 gal hard chromium bath (3 plating tanks). They generated 55 gal/wk (2,860 gal/yr) of spent catholyte. PS 052 operated the process with 12,700 gal of hard chromium solution (7 plating tanks) and generated 55 gal/month (660 gal/yr) of spent catholyte.

If a plating shop using membrane electrolysis has a conventional wastewater treatment system in place, the spent catholyte could be bled to this system or batch treated and then discharged to the general treatment system. Treatment of the catholyte will generate a sludge that under most conditions will be considered a RCRA F006 sludge. The sludge volume will depend mainly on the metal content of the catholyte and the dryness (percent solids) of the resultant sludge.

Spent catholyte may meet the characteristics of a hazardous waste due to its pH and or toxic metals content (particularly chromium). Treatment of the catholyte may be interpreted as a hazardous waste treatment process (RCRA controlled) rather than wastewater treatment (Clean Water Act controlled), especially for shops that have achieved zero discharge. Before developing a treatment and disposal scheme, the plater must investigate applicable RCRA and Clean Water Act rules, including state and local regulations, relative to the treatment process.

5.8 DIFFUSION DIALYSIS

5.8.1 Overview

Diffusion dialysis is an ion exchange membrane technology that competes with acid sorption (Section 5.5) as a purification/recovery method for acids that have become contaminated with metals (e.g., pickling, anodizing, stripping, etching, and passivation baths).

The diffusion dialysis process separates acid from its metal contaminants via an acid concentration gradient between two solution compartments (contaminated acid and deionized water) that are divided by an anion exchange membrane. Acid is diffused across the membrane into the DI water whereas metals are blocked due to their charge and the selectivity of the membrane. A key difference between diffusion dialysis and other membrane technologies such as electrodialysis or reverse osmosis is that diffusion dialysis does not employ an electrical potential or pressure across the membrane. Rather, the transport of acid is caused by the difference in acid concentration on either side of the membrane. As such, the energy requirements for this technology are low.

Exhibit 5-34 describes the diffusion dialysis process. The process uses ion exchange membranes which are assembled in a membrane stack. The membrane separates two liquids: (1) acid contaminated with metal and (2) deionized water. The physical laws of diffusion and electroneutrality cause material in high concentration to move to an area of low concentration without an imbalance of electrical charge. Because of the presence of the anion membrane, the metals in the concentrated solution are unable to pass from the concentrate to the DI water. However, anions in the concentrate (e.g., chlorides, sulfates, nitrates, phosphates) are permitted passage. Also, hydrogen ions, although positively charged, diffuse along with the disassociated acid (anions). The passage of hydrogen, which is key to the success of this process, is due to the small size of the hydrogen molecules and their mobility. The passage of the positively charged hydrogen ions satisfies the law of electroneutrality, preventing an imbalance of ionic charge on either side of the membrane (ref. 192, 336).

Diffusion dialysis, like other membrane technologies, is not 100 percent efficient; not all of the acid will be recovered and some leakage of metal will occur. In the laboratory, the process has yielded acid recovery efficiencies as high as 99% with 98% metal removal. In the manufacturing environment, the practical limits are 80% to 95% acid recovery with 60% to 90% of the metal contaminants removed. Also, the recovered acid may be of insufficient concentration to permit direct reuse. In such cases, vacuum evaporation may be needed to increase its concentration (ref. 338), although the economics of a concentration step are questionable. One source indicates, based on 1.5 years of experience with diffusion dialysis, that it is more efficient and economical than acid sorption for certain applications (e.g., recovery of mixed acid pickling baths) (ref. 336).

A related predecessor to this technology, Donnan dialysis, was a popular research topic for the metal finishing industry in the late 1970's (ref. 39). Donnan dialysis is more often associated with metal recovery applications (e.g., nickel and copper) using cation membranes than with acid bath maintenance. However, anion membrane applications of Donnan dialysis, including metal recovery from cyanide solutions are also discussed in the literature (ref. 39, 380). The recovery of nickel from rinse water was one of the primary targets of this technology. This was accomplished by separating nickel bearing rinse water from a low pH solution using a cation membrane. Nickel ions from the rinse water would replace hydrogen ions in the cation membrane matrix and pass from the nickel rinse water to the low pH solution while the hydrogen ions passed into the rinse water. Simultaneously, hydrogen ions must be replaced in the low pH solution from an outside source to maintain electroneutrality and the driving force for ion transfer. To maintain electroneutrality, two hydrogen ions ($2H^+$) must move from the low pH solution to the rinse water when one nickel ion (Ni^{++}) is transferred. Anions present in either compartment are restricted from migration due to the use of the cation membrane (ref. 39). No commercial metal finishing applications of Donnan dialysis are known to exist.

Exhbiit 5-34. Diagram of Diffustion Dialysis Membrane Process for Acid Bath Maintenance

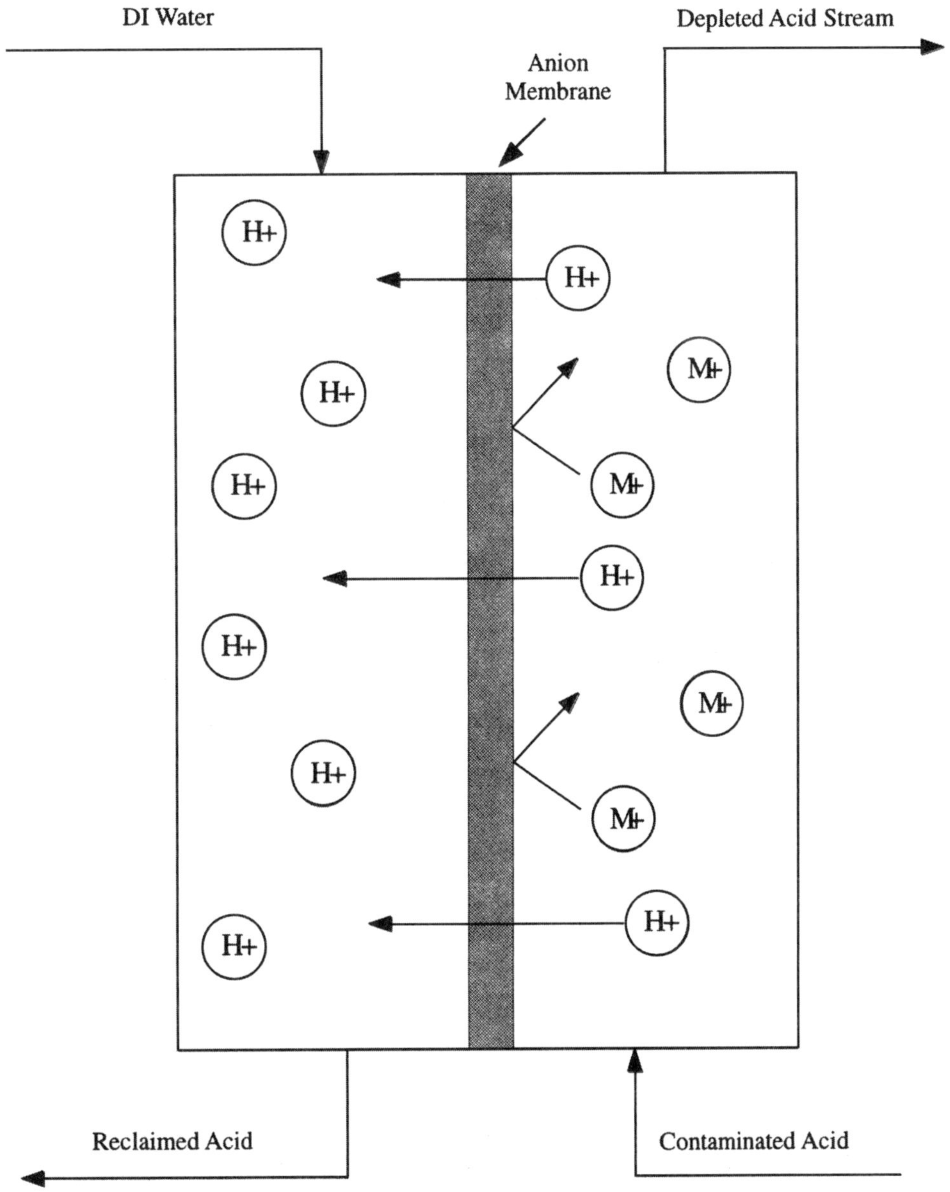

Source: ref. 192

5.8.2 Development and Commercialization

The development of diffusion dialysis as a practical recovery technology took place in the early 1980's in Japan. In 1984 a device for the recovery of nitric and hydrofluoric acids from a mixed acid waste was installed at the Kashima works of Nippon Stainless Steel Co. (ref. Graver file.) The first European commercial application of diffusion dialysis identified in the literature took place in Sweden in 1989. That system, which is applied to the regeneration of a stainless steel pickling acid (HF/HNO_3), was still in operation in 1993. Only one U.S. application of the technology was identified during the project (caustic recovery from an aluminum etch bath at Caspian, Inc., San Diego, 1992). That system processes the etch solution from three milling tanks with a total solution volume of 72,000 gal. Three U.S. vendors of this technology were identified during the Vendors Survey (Kinetic Recovery Corporation, Baker Brothers and Graver Water Division of the Graver Company).

5.8.3 Applications and Restrictions

The diffusion dialysis process has been applied for the purification of hydrochloric, nitric, hydrofluoric, phosphoric, and sulfuric acids, including mixed acids; and ferric chloride etchant (ref. 192, 252, 336)

The most common applications of this technology include:

- Recovery of mixed acids of stainless steel pickling baths (ref. 336).
- Recovery of H_2SO_4/HC_l or H_2SO_4/HNO_3 of pickling plants for non-ferrous metals such as zinc and aluminum (ref. 336).
- Recovery of H_2SO_4 and HCl of steel and non-ferrous metal pickling baths (ref. 336).
- HC_l and HNO_3 acid rack strips (ref. 192).
- H_2SO_4 anodizing solutions (ref. 192).
- Regeneration of battery acids (ref. 252).

Like acid sorption, diffusion dialysis is limited in applicability by the strength of acid that it can produce. If highly concentrated acids are processed, the resultant product will not be of sufficient strength for direct reuse and the addition of concentrated acid is not practical because it increases the total volume of the solution beyond the capacity of its tank. A vacuum evaporator could be employed to concentrate the acid to bath strength; however, this would most likely make the overall recovery process uneconomical. Examples of successful applications in the literature are shown in Exhibit 5-35, where the original acid concentrations are in the range of 3.4 to 4.7 N.

The diffusion dialysis membrane material is relatively resistant to chemicals commonly used in the plating shop. However, contact with solvents could cause swelling of the membrane and strong oxidizing agents such as chromic acid can deteriorate the membrane material (ref. 252). The process is tolerant of feed solution temperatures up to 50°C (ref. 252).

A process diagram for diffusion dialysis is shown in Exhibit 5-36. Acid is pumped through one side of an ion exchange membrane stack and water flows through the other side of the membrane in the opposite direction. Prefiltering of the acid is needed to prevent fouling of the membrane. Most of the acid diffuses through the membrane and into the clean water producing a clean acid stream. The metal and a percentage of the acid remains in the original acid stream, which is directed to treatment. The acid product is monitored for strength and adjusted with concentrated acid as necessary.

Exhibit 5-35. Reported Recovery Results from Diffusion Dialysis Applications

Application/Parameter	Acid (mg/l)	Aluminum(mg/l)	Chrome (mg/l)	Copper (mg/l)	Iron (mg/l)
Hydrochloric Rack Strip					
Initial Concentration	4.7	NA	900	NA	1,700
Recovered Concentration	4.3	NA	200	NA	400
Spent Concentration	0.4	NA	700	NA	1,300
Removal Efficiency	91.5%	NA	77.8%	NA	76.4%
Sulfuric Acid Anodizing					
Initial Concentration	3.4	5,200	NA	35	150
Recovered Concentration	3.0	30	NA	1	5
Spent Concentration	0.4	5,150	NA	34	140
Removal Efficiency	88.2%	99.0%	NA	97.1%	93.3%

Source: ref 192

Exhibit 5-36. Example of Diffusion Dialysis Application to an Anodizing Bath

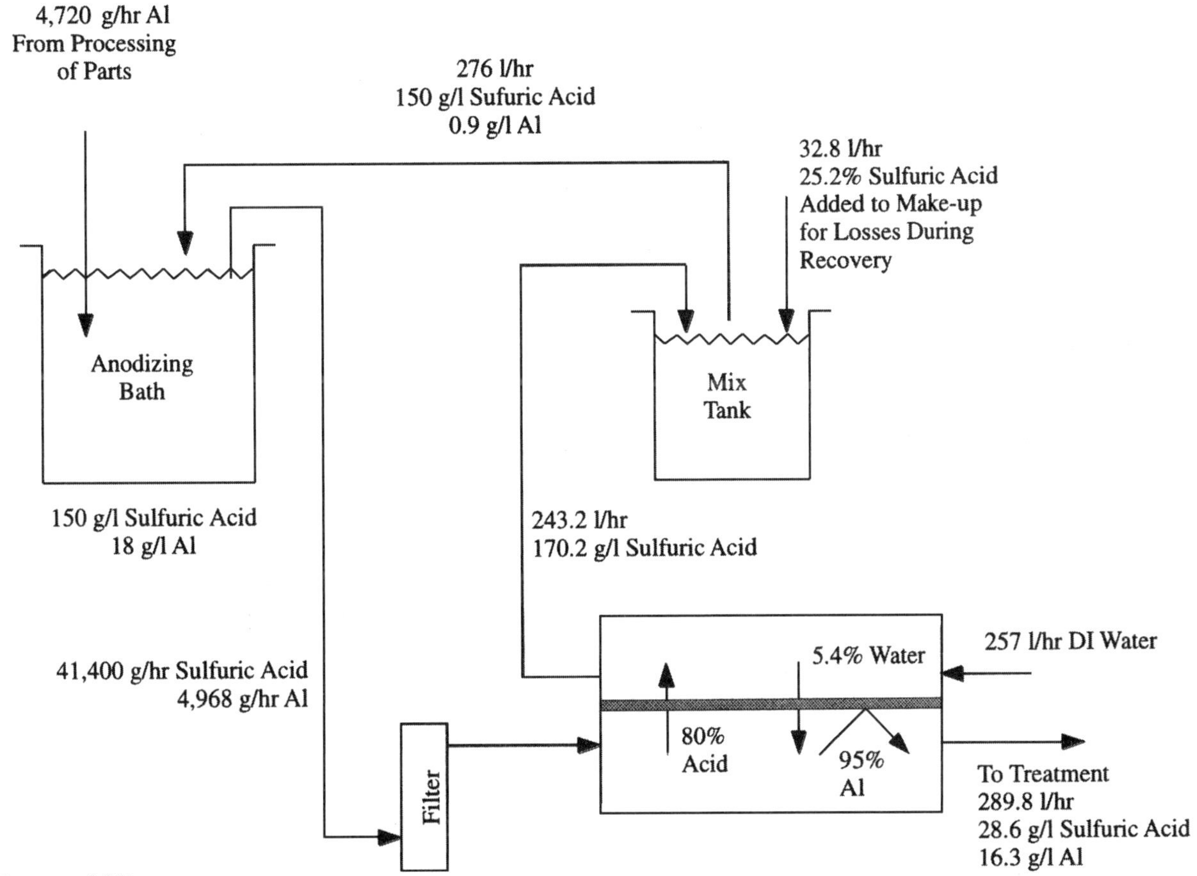

Source: ref. 252

5.8.4 Technology/Equipment Description

5.8.4.1 General

Diffusion dialysis equipment resembles that of other membrane technologies such as electrodialysis. These units are designed with a stacked membrane configuration. A "membrane stack" consists of a number of thin membrane compartments created by alternating layers of ion-specific membranes and spacers. The spacers are present to permit solutions (i.e., contaminated acid and DI water) to flow between the membrane layers. The membranes and flow spacers are held together by a cell frame and clamping unit, giving the membrane stack the appearance of a plate and frame filter press (ref. 252, Kinetic Recovery Corporation file).

The commercial diffusion dialysis units usually employ co-polymer membranes of polystyrene and divinylbenzene. These units permit the diffusion of highly disassociated acids through the polymer structure while cations (metals) remain behind because of their positive electrical charge. The mass transfer through the membrane was described by Reinhard (ref. 252) as:

$$j_n = K \times A \times C_n$$

where, j_n is the overall transport rate of solute n,
C_n is the average solute concentration difference between the fluids,
A is the membrane area, and
K is the overall rate constant (dialysis coefficient).

The proper average solute (acid) concentration difference is the logarithmic mean of the inlet and outlet differences. The dialysis coefficient is a function of: (1) chemical characteristics of the acid; (2) concentration of the acid; (3) chemical characteristics of the salt; and (4) temperature of liquid (ref. 252).

The temperature of the feed appears to be a significant variable. One vendor indicates that a temperature drop of 3.6 °F (2.0° C) will reduce the recycling rate by approximately 1.5%. The membranes can typically withstand temperatures up to 122°F (50°C). They also indicate that heating of the feed stream may be required for winter operation (ref. Kinetic Recovery Corporation file).

The diffusion dialysis technology is commercially available in a wide range of capacities. Existing systems were identified with inlet acid flows ranging from 100 gpd to 2,000 gpd (ref. 192, 336).

5.8.4.1 Commercial Equipment

This subsection contains a description of commercially available diffusion dialysis equipment that is manufactured and/or sold by Vendor Survey respondents. This is intended to provide the reader with information and data on a cross section of available equipment. Mention of trade names or commercial products is not intended to constitute endorsement for use.

Kinetic Recovery Corporation markets a series of diffusion dialysis units that have been used in Germany and Sweden for more than two years on the following applications: HNO_3/HF (stainless steel); H_2SO_4/HCl (zinc or aluminum); H_2SO_4/HNO_3 (zinc or aluminum); H_2SO_4 (iron or zinc); and HCl (iron or zinc). They have also tested a unit at United Technologies on a HNO_3/HF (titanium). Each of their systems are custom made and include: tanks for acid and DI water; level controls; membrane stack, internal plumbing and connections to tanks; automatic valves; feed pump; cartridge filter; and control panel. The units are skid mounted and factory assembled. The normal range of capacities is from 30 to 415 gpd of acid feed.

5.8.5 Costs

5.8.5.1 Capital Costs

Diffusion dialysis units can be operated on either a batch or continuous basis. When operated on a batch basis the units are sized to process the contaminated acid bath within a period permissible downtime period. More frequently, the units are operated on a continuous basis and are sized based on the contaminant introduction rate. In

Exhibit 5-37. Capital Costs for Diffustion Dialysis Systems

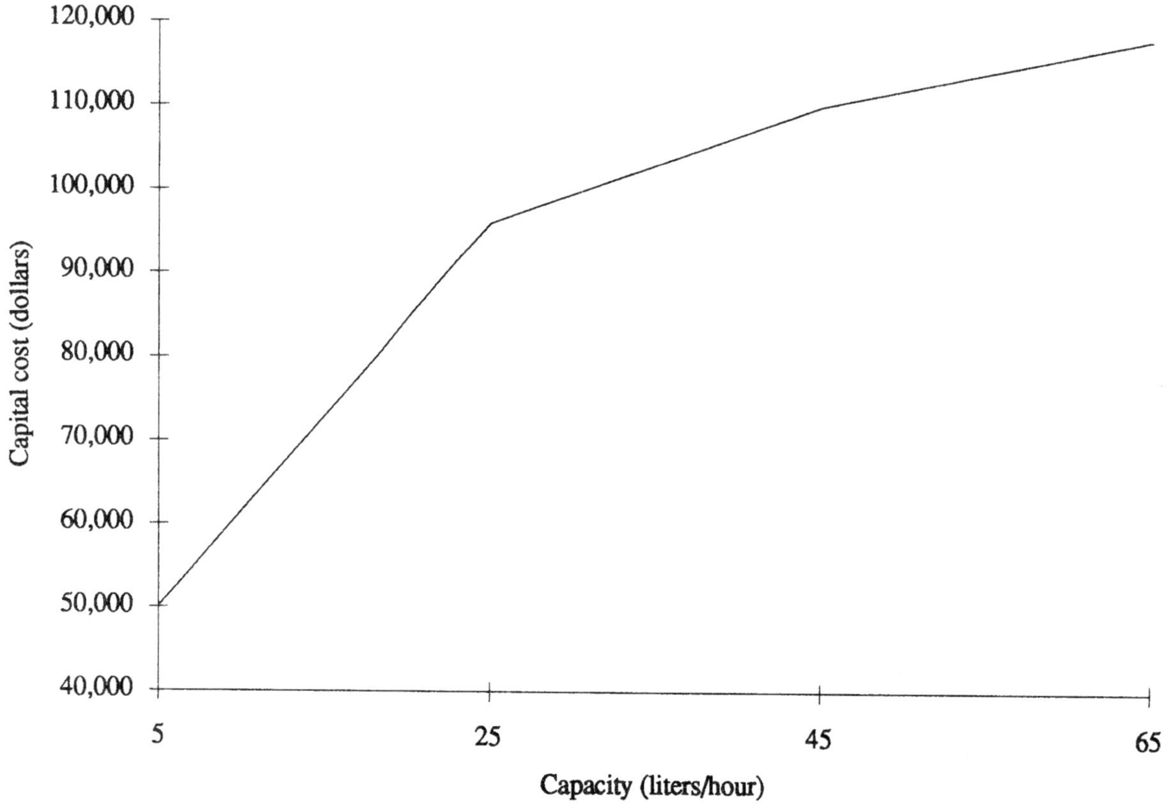

this case, the minimum sized system is one that is capable of processing the entire bath volume in the same time interval that dumping had been previously scheduled (ref. 192). Owing to the newness of the technology, it may be prudent for the user to validate design expectations through pilot testing (ref. 192).

Capital costs for diffusion dialysis equipment is shown in Exhibit 5-37.

5.8.5.2 Operating Costs

The primary operating cost items for the diffusion dialysis technology include: O&M labor, energy, DI water, and membrane replacement. The operating costs were estimated by a vendor to be $0.78 per gallon of acid feed (ref. Kinetic Recovery Corporation file).

5.8.6 Performance Experience and Operational and Maintenance Problems

As indicated previously, none of the respondents to the Users Survey indicated that they have employed this technology and therefore no firsthand performance experience is available. Various literature sources have reported on the efficiency of the diffusion dialysis process (ref. 192, 252, 336). These articles and conference papers, written by equipment manufacturers and suppliers, indicate that 80 to 95 percent of the initial acid can be recovered with 60 to 95 percent of the contaminant metal removed (i.e., a residual metal content of 5 to 40 percent of original concentration remains in the recovered acid).

Vendors of the technology suggest that the extremely low pressure feed of diffusion dialysis puts minimal stress on the stack components and therefore, increases the longevity of the units. However, as with any membrane technology, there is concern over membrane fouling and deterioration. Membrane fouling is most frequently caused by the presence of solids in the acid feed solution. It is reported (ref. 252, 336) that prefiltering will effectively eliminate the potential fouling problem. With prefiltering, membrane cleaning is required approximately two times per year (ref. 252).

Organic substances (e.g., oil and grease) can also cause membrane fouling. This problem can be overcome using a pretreatment of the raw acid solution with activated carbon.

Membrane life is reported to be three to seven years (ref. 252, 336). Membrane life will be shortened when the technology is used with hydrogen peroxide, chromic acid, and nitric acid with concentrations over 20 percent, as well as temperatures over 45°C and organic substances such as oil and grease, solvents or cleaners (ref. 336).

5.8.7 Residuals Generation

Use of the diffusion dialysis technology will generate a depleted acid stream that is equal in volume to the original contaminated acid bath and contains approximately 5 to 20% of the original acid content and 5 to 40% of the original metal content.

Section 6
Substitute Technologies

6.1 INTRODUCTION

As discussed in Section 3, pollution prevention includes a broad range of approaches and methods. Section 6 focuses on two aspects of pollution prevention: product changes and input material changes. The pollution prevention efforts and experiences of the NCMS/NAMF survey respondents with regard to these two topics have been categorized into six groups:

- Product changes
- Chlorinated solvent use reduction/elimination
- Cyanide use reduction/elimination
- Cadmium use reduction/elimination
- Chromium use reduction/elimination
- Other changes

Product changes made by survey respondents generally involved modifications to parts that resulted in a reduction of dragout (e.g., adding drain holes to parts). A relatively low percentage (8%) of the shops surveyed have utilized this pollution prevention tool. Its limited use is most likely due to the fact that most of the shops surveyed were job shops, which usually have little or no control over the design of the parts they process.

However, the results of the Users Survey did show that respondents have made significant strides in reducing or eliminating the use of chlorinated solvents, cadmium, cyanide and chromium. Sometimes referred to as the four Cs, these materials have been identified by EPA as key targets for control within the metal finishing industry (ref. 415, 416). Although job shops do not have much control over part design, they usually have control over process chemicals. Some exceptions exist, such as instances where aerospace or military specifications require specific finishes, solutions and processing procedures; however, in spite of any limitations, approximately 60 percent of the shops attempted material input changes that potentially reduce or eliminate the use of one or more of the four Cs or another pollutant problem. Based on the comments received from respondents, these changes were made in an effort to reduce the impacts of their processes on the environment and worker health, to help meet environmental regulations and to reduce operating costs.

Although most of the material input changes attempted by survey respondents have been successful, there have been some failures and in many cases, even with successful changes, there have been adverse production impacts. An attempt has been made in this section to summarize the status of change in these areas and to convey the attitudes and concerns of the respondents.

One additional factor to keep in mind when reviewing this section, is that not all of the changes discussed in this section may be improvements. There currently exists a general feeling among regulators, the military and some private industry that all uses of cyanide, cadmium and chromium should be targets for substitution or elimination to meet health and safety and environmental goals. In some cases, the substitutes selected provide a lower quality finish and create new health and safety or environmental problems. In such cases, the question of whether it makes sense to improve the existing process through engineering or through making a substitution must be investigated.

As an example, the hard chromium process has been the target of reduction and/or elimination by the military and private industry (ref. 304, 418, 436). Potential substitutes for this process include: electroless nickel plating (ref. 418, 74), brush plating (ref. 69, 418), nickel alloys and metal spray coatings (ref. 418). In all cases, there is no single substitute that will meet all the requirements of hard chromium plating. Therefore, multiple substitutes must be used, and even then there will remain some applications for which no substitute is available.

Substitutes such as brush plating and nickel alloy plating may reduce the use of chromium, but they usually increase the net quantity of waste generated. These types of solutions are not amenable to recovery or bath maintenance and their by-products must be treated rather than recovered or reused. Conversely, the basic hard chromium bath (chromic acid/sulfate) can be operated in a closed-loop or zero discharge. Chromium can be recovered from rinse water (e.g., recovery rinsing, ion exchange, evaporation), air emissions (mesh pad mist eliminator), and strip solutions (membrane electrolysis) and the baths can be operated indefinitely without discharges (e.g., membrane electrolysis) (Sections 3, 4 and 5 provide information on methods and technologies for closed-loop operation of hard chromium plating). Additionally, parts plated with chromium generally last significantly longer than those plated with substitute coatings (ref. 304). Over time, this reduces the total number of parts that needs to be processed and further reduces the quantity of waste generated.

EPA recognizes the importance of considering all inputs and outputs in manufacturing and recommends a process termed life-cycle assessment to objectively identify and evaluate opportunities to reduce the environmental impacts associated with a specific product, process or activity. Life-cycle assessments take a holistic approach by analyzing the pollutant quantities and environmental impacts associated with: extraction and processing of raw materials, manufacturing, transportation, distribution, use/reuse/maintenance, recycling and final disposal (ref. 302). The life-cycle process should be kept in mind when reading this section, especially with regard to alternatives to cyanide, cadmium and chromium.

6.2 PRODUCT CHANGES

EPA defines product changes for pollution prevention purposes as: changes in the composition or use of the intermediate or end products performed by the manufacturer with the purpose of reducing waste from manufacture, use, or ultimate disposal of the products (ref. 302).

The responsibility of product changes falls mostly on the manufacturer because they have control over the design and specification of the product. Unfortunately, when products and product components are processed by subcontractors, such as electroplating job shops, the effects of poor designs are felt by others. As discussed in Section 2, the majority of plating shops surveyed during the NCMS project were job shops. Not surprisingly, most of these shops either did not respond to the question of product changes or they indicated that this section of the survey was not applicable to them. This attitude is an important environmental disconnect between designers and platers which has been identified in several documents relating to environmental impacts of the electronics industry (ref. 518, 519). However, a number of job shops indicated that they do have some input on the design of the parts that they plate. In an effort to convey the actual attitudes of the respondents, actual responses to the product changes question are printed in this section in their entirety. To differentiate between the responses provided by captive and job shops, the captive shop responses are printed in bold type (includes all shops that declared themselves 1% or more captive).

The responses are divided into two groups, negative and positive responses.

6.2.1 Negative Responses to Product Changes

- Our customers are rather disinterested in pollution prevention that they can employ (PS 039).
- N/A [not applicable]—We are a job shop (PS 055).
- N/A—We are a job shop and have no control over what the customer wants us to plate (PS 066).
- No control over parts—We are a job shop (PS 098).
- Because this company is a job shop we have no control over the design of the parts to be plated (PS 116).
- We are a job shop. The customer designs the part, therefore we have no control over product design. Some customers are cognizant of problems concerning plating, but the majority just tell us to plate it or they will take it elsewhere (PS 128).

- N/A-All product is customer supplied (PS 174).
- We don't manufacture parts (job shop) (PS 175).
- N/A—Contract only shop (PS 199).
- We are a job shop only. We rarely have input into a customer's designs to improve plating and reduce pollution (PS 204).
- N/A—We are a job shop anodizer (PS 230).
- Products supplied by customers, no control (PS 233).

6.2.2 Positive Responses to Product Changes

- We have gotten customers to change their plating specification from brass or nickel-brass to "simulated brass." This is zinc plating with a blue chromate dyed to a brass-tone. As a result, we have converted 50,000 lbs to 75,000 lbs [of parts] per month from cyanide-brass to acid zinc (PS 004).
- If drain holes are not present, we will not plate the parts. Usually the customer will add adequate holes with no problem (PS 010).
- We have reviewed the design of parts in an effort to reduce dragout and contacted our customers with recommended changes (e.g., use of easily cleaned material and change to non-petroleum machine lubricants) (PS 041).
- On proprietary furniture, we have optimized orientation and drain holes (PS 058).
- We insist on drain holes (PS 082).
- Redesign of racking and adding drain holes to reduce dragout (PS 089).
- Had customer redesign some weldments to avoid pockets that trap solution and added drain holes to other parts (PS 099).
- At one time we copper plated bars used in the textile industry. Later, we would buy this material already plated. Now, this product is manufactured from stainless steel and copper plating is not needed (PS 102).
- We request and usually get permission to drill additional drain holes in tubular products (PS 109)
- Whenever necessary, we ask our customers to include drain holes or change the design to improve drainage (PS 124).
- We work with customers to control the drainage of parts (PS 125).
- As a job shop we do not have much control on design changes, but there have been times when we were successful in making suggestions to our customers where a change was possible (PS 156).
- We provide constant customer education for part modification and design. In some cases we refuse jobs involving problem designs. Our success is difficult to determine since this is our business philosophy, but we feel we are generally successful (PS 172).
- We communicate with customers on part design to include drain holes (PS 176).
- We refuse work that is highly oiled or soiled and return them to the customer for cleaning. We have the customer provide adequate drain holes, and we recommend corrosion resistant parts when feasible (PS 188).
- We have implemented drain holes, material substitution and racking modifications (PS 192).
- Auto companies have changed part finish specifications to use processes perceived to reduce pollution problems (PS 214).
- Provide input on part design, when possible (PS 228).
- Some customers have implemented changes to allow for adding additional drain holes and additional "dimples" to separate barrel plated work (PS 245).
- We advise our customers on proper drainage in the design of their parts (PS 251).
- Redesigning of parts to add or re-position drain holes has been successful (PS 275).
- Successfully replaced hot rolled steel with cold rolled steel to reduce pickling (PS 288).

- We "no-quote" work that could generate excessive dragout. Tooling has been re-configured for more evenly distributed plating. Plastisol coating of tooling is increased to avoid stripping/plating of our racks (PS 298).

6.3 CHLORINATED SOLVENTS USE REDUCTION/ELIMINATION

Chlorinated solvents have been used extensively in the metal working and metal finishing industry for degreasing and cleaning. The most commonly used degreasing solvents include: 1,1,1-trichloroethane (or TCA), trichloroethylene (or TCE), perchloroethylene (or PERC), chlorofluorocarbons, (or solvent 113) and methylene chloride. These solvents are used in a variety of methods, but most frequently in vapor degreasers, immersion or spray operations, and hand wiping. Solvent use has declined during recent years and especially since 1990, due to regulations restricting the use of ozone-depleting substances, voluntary actions by industry under EPA's 33/50 Program (Industrial Toxics Project) (ref. 203), rising costs for the purchase of solvents and the disposal of solvent wastes, and concerns over the health effects of solvents. However, as late as 1989, an estimated 73,000 U.S. firms (includes all types of industry) still used vapor degreasing with TCA (ref. 79).

Solvent use among the respondents to the Users Survey is summarized in Exhibit 6-1. Column 1 indicates the shop code number (PS 001 to PS 318). Column 2 shows the type of solvent used. Column 3 indicates the percentage of the shop's work that is job shop-type. Where 100% is shown, that business is entirely a job shop. Where 0% is shown, that business is entirely a captive shop. Where a different percentage is shown, that shop does both job and captive-type work. Column 4 indicates the year that the business commenced plating operations. Column 5 indicates the year that solvent use was eliminated. Where no year is shown in column 5, the solvent process was still in use at the time of the survey. Columns 6 and 7 indicate the quantity of solvent (in gallons or pounds) that is currently used or was used in cases where the applications were eliminated. Column 8 indicates the type of application (vapor degreasing, cold dip, spray or hand wipe). The last column indicates the most sophisticated method that is used (or was used in the case of eliminated applications) for controlling solvent emissions. The following abbreviations are used (shown in increasing levels of sophistication):

- NONE — no controls
- FREEBOARD — increased freeboard of degreaser
- ROLLTOP — automatic rolltop for degreaser
- LIFTSPEED — controls to limit vertical speed of degreaser hoist
- REFRIGER — refrigeration zone to supplement conventional cooling coils

Exhibit 6-2 shows that among respondents to the Users Survey, the number of solvent users has changed since 1980. In Exhibit 6-2, the shops are divided into three groups: (1) those in existence in 1980; (2) those established from 1981 to 1985 (inclusive); and (3) those established from 1986 to 1993 (inclusive). For the older shops, the number of solvent users remained approximately the same from 1980 to 1985 and then declined substantially from 1986 to 1993. In 1980, 53% of the shops used solvent and by 1993 only 39% used solvent. Therefore, 26% of the solvent users in 1980 have eliminated its use. For shops established from 1981 to 1985, the frequency of solvent use was below that of the older shops in 1985 and then from 1986 to 1993, the percentage declined similarly to the declining use rate of the older shops. Thirty-one percent of the shops established from 1981 to 1985 that originally used solvent have eliminated its use. The most recently established shops (1986 to 1993) presently have approximately the same percentage of solvent use as the shops established in 1981 to 1985.

Exhibit 6-2 indicates that shop age had only a small bearing on the frequency of solvent use. This is somewhat surprising since older shops are often thought of as being less likely to change their operational procedures and processes.

Exhibit 6-3 shows that the quantity of solvent use per application has also declined during the time period 1980 to 1993 and that this decline has been restricted to the time period of 1986 to 1993. A solvent application is defined as the use of a given type solvent by a given shop. Therefore, shops can have more than one solvent application. Combined with the fact that there are fewer applications, overall solvent use has declined significantly, probably by 50 percent or more from 1986 to 1993. All of the solvent use data are contained in the database and users are encouraged to utilize the database and draw their own conclusions.

Exhibit 6-1. Solvent Use Data from the Users Survey

Shop Code	Type of Solvent	% Job Shop	Year Shop Estab.	Year Solvent Eliminated	Solvent Use Rate lbs/yr	Process Type	Control Method
015	TCA	100	1984		132,471	NONE	NONE
019	TCA	100	1988		2,400	WIPE	NONE
020	TCA	100	1961		32,400	VAPOR	FREEBOARD
024	TCA	100	1984		46,440	VAPOR	LIFTSPEED
025	TCA	0	1930		660	WIPE	NONE
027	TCA	100	1965		18,800	VAPOR	NONE
033	TCA	100	1975			VAPOR	ROLLTOP
034	TCA	100	1967		7,920	VAPOR	ROLLTOP
037	TCA	100	1972	1992	2,400	WIPE	NONE
038	TCA	100	1949		6,000	VAPOR	REFRIGER
041	TCA	100	1947	1982		VAPOR	NONE
047	TCA	100	1937		660	COLD	NONE
049	TCA	100	1980		29,040	VAPOR	LIFTSPEED
053	TCA	100	1913	1992		VAPOR	NONE
064	TCA	100	1960	1988		VAPOR	ROLLTOP
074	TCA	100	1965		23,760	WIPE	NONE
080	TCA	100	1940		1,980	VAPOR	LIFTSPEED
091	TCA	100	1960		14,400	VAPOR	NONE
092	TCA	100	1976			VAPOR	REFRIGER
105	TCA	100	1973	1992	18,000	VAPOR	LIFTSPEED
108	TCA	0	1975	1992	9,600	VAPOR	LIFTSPEED
109	TCA	100	1962		30,000	VAPOR	LIFTSPEED
111	TCA	100	1958		14,400	VAPOR	ROLLTOP
114	TCA	100	1983		900	COLD	NONE
116	TCA	100	1952	1991	2,000	VAPOR	NONE
120	TCA	100	1987			VAPOR	NONE
122	TCA	100			3,000	VAPOR	REFRIGER
123	TCA	100	1958		15,833	VAPOR	REFRIGER
127	TCA	0	1972		1,200	VAPOR	NONE
128	TCA	100	1980	1990	84,000	COLD	NONE
130	TCA	60	1978			VAPOR	NONE
131	TCA	100	1953		3,744	VAPOR	ROLLTOP
138	TCA	100	1947	1990	660	SPRAY	NONE
142	TCA	100	1961		660	VAPOR	LIFTSPEED
146	TCA	100	1981	1992		VAPOR	NONE
149	TCA	100	1966	1993	3,000	VAPOR	NONE
163	TCA	100	1984	1992	5,832	COLD	NONE
165	TCA	100	1977		13,860	VAPOR	REFRIGER
166	TCA	100	1952	1992	36,000	VAPOR	NONE
167	TCA	100	1964	1992	30,000	VAPOR	NONE
170	TCA	100	1955		16,000	VAPOR	REFRIGER
171	TCA	100	1980	1991		VAPOR	NONE
174	TCA	100	1980	1992	4,800	VAPOR	ROLLTOP
178	TCA	100	1965	1983	10,000	VAPOR	NONE
180	TCA	100	1989		16,000	VAPOR	LIFTSPEED
184	TCA	100	1979		27,000	VAPOR	ROLLTO
187	TCA	100	1978	1991	6,000	VAPOR	REFRIGER
189	TCA	100	1983		3,000	VAPOR	NONE
190	TCA	100		1992	31,680	VAPOR	NONE
191	TCA	100	1968		15,588	VAPOR	ROLLTOP
194	TCA	100	1940	1986	1,200	VAPOR	NONE
196	TCA	100	1980		72,000	VAPOR	NONE
197	TCA	0	1951		42,000	VAPOR	REFRIGER
203	TCA	100	1952		600	VAPOR	LIFTSPEED
206	TCA	100	1958		24,000	VAPOR	REFRIGER
207	TCA	100	1937		42,000	COLD	NONE
208	TCA	100	1988		7,920	VAPOR	NONE
210	TCA	100	1960	1990		NONE	NONE
213	TCA	0	1965	1988	500	VAPOR	NONE
218	TCA	100	1975		7,200	VAPOR	ROLLTOP
219	TCA	100	1941		7,100	VAPOR	NONE
220	TCA	100	1978		12,600	VAPOR	NONE
221	TCA	0	1987	1991	2,400	VAPOR	NONE
223	TCA	100	1950	1990		VAPOR	NONE
228	TCA	100	1940	1993		VAPOR	REFRIGER
232	TCA	100	1920	1992	19,000	VAPOR	NONE
233	TCA	100	1973		18,750	VAPOR	ROLLTOP
238	TCA	100	1911		17,160	VAPOR	LIFTSPEED
242	TCA	100	1959		1,200	VAPOR	ROLLTOP

Shop Code	Type of Solvent	% Job Shop	Year Shop Estab.	Year Solvent Eliminated	Solvent Use Rate lbs/yr	Process Type	Control Method
245	TCA	100	1968	1991	7,920	VAPOR	NONE
247	TCA	55	1952			VAPOR	NONE
248	TCA	100	1966	1990	3,960	VAPOR	NONE
249	TCA	100	1983		1,200	COLD	NONE
258	TCA	100				NONE	NONE
259	TCA	100	1946		600	VAPOR	NONE
261	TCA	0		1985		VAPOR	NONE
262	TCA	100	1960	1991	12,000	VAPOR	NONE
264	TCA	100	1972	1992	660	VAPOR	REFRIGER
266	TCA	80	1975		7,920	WIPE	NONE
271	TCA	100	1959		4,200	COLD	NONE
273	TCA	0	1942		133,488	VAPOR	NONE
276	TCA	100	1946	1989	2,000	VAPOR	REFRIGER
278	TCA	0	1967	1993	36,000	VAPOR	NONE
281	TCA	0	1930	1992	36,000	COLD	NONE
283	TCA	0	1987	1993	12,000	VAPOR	NONE
285	TCA	0	1981	1989	3,960	VAPOR	REFRIGER
287	TCA	0	1905	1988		VAPOR	NONE
295	TCA	0	1949	1991	3,600	VAPOR	NONE
296	TCA	0		1993	30,000	VAPOR	REFRIGER
316	TCA	100	1930			VAPOR	ROLLTOP
317	TCA	100	1944	1988		VAPOR	FREEBOARD
318	TCA	100	1960	1984	6,000	VAPOR	REFRIGER
037	Acetone	100	1972	1992	2,680	WIPE	NONE
144	Acetone	100	1963		402	COLD	NONE
166	Acetone	100	1952		26,800	COLD	NONE
230	Acetone	100	1973			COLD	NONE
020	CFC	100	1961		7,920	VAPOR	REFRIGER
064	CFC	100	1960	1990		COLD	NONE
085	CFC	95	1957		4,140	COLD	NONE
187	CFC	100	1978	1991	6,000	VAPOR	REFRIGER
223	CFC	100	1950			COLD	NONE
257	CFC	0	1986	1990		SPRAY	NONE
261	CFC	0		1992		VAPOR	NONE
283	CFC	0	1987		8,970	VAPOR	FREEBOARD
292	CFC	0	1960	1977	52,800	VAPOR	NONE
295	CFC	0	1949	1991	360,000	COLD	NONE
295	CFC	0	1949	1991	600,000	NONE	NONE
178	CFC	100	1965		6,210	VAPOR	REFRIGER
007	MEK	100	1940			WIPE	NONE
019	MEK	100	1988		670	WIPE	NONE
049	MEK	100	1980		67	WIPE	NONE
067	MEK	100			1,005	WIPE	NONE
070	MEK	100	1950			WIPE	NONE
077	MEK	0	1959		40,500	WIPE	NONE
149	MEK	100	1966		3,660	COLD	NONE
152	MEK	100	1984			WIPE	NONE
173	MEK	100			1,000	WIPE	NONE
176	MEK	100	1977		670	WIPE	NONE
184	MEK	100	1979		7000	VAPOR	ROLLTOP
242	MEK	100	1959		101	WIPE	NONE
244	MEK	100	1971		302	WIPE	NONE
248	MEK	100	1966		1,608	WIPE	NONE
318	MEK	100	1960		1,206	WIPE	NONE
007	METH	100	1940			WIPE	NONE
038	METH	100	1949		1,320	COLD	NONE
157	METH	0	1978		60	WIPE	NONE
169	METH	0	1990	1987		VAPOR	NONE
173	METH	100			3,960	COLD	NONE
176	METH	100	1977	1992	480	WIPE	NONE
227	METH	0	1973	1992	660	COLD	NONE
232	METH	100	1920	1990		VAPOR	NONE
264	METH	100	1972		660	VAPOR	REFRIGER
272	METH	100	1959			COLD	NONE
287	METH	0	1905	1985		NONE	NONE

Solvent Key:
TCA = 1,1,1-Trichloroethane METH = Methylene Chloride CFC = Chlorofluorocarbons MEK = Methyl Ethyl Keytone TCE = Trichloroethylene PERC = Perchloroethylene

Exhibit 6-1 (continued). Solvent Use Data from the Users Survey

Shop Code	Type of Solvent	% Job Shop	Year Shop Estab.	Year Solvent Eliminated	Solvent Use Rate lbs/yr	Process Type	Control Method
294	METH	100	1947		180	WIPE	NONE
308	METH	100	1985		1,800	NONE	NONE
007	PERC	100	1940		7,450	COLD	NONE
026	PERC	100	1968		15,984	VAPOR	LIFTSPEED
027	PERC	100	1965		13,200	VAPOR	NONE
036	PERC	100	1978		600	VAPOR	REFRIGER
053	PERC	100	1913	1975		NONE	NONE
064	PERC	100	1960	1990		VAPOR	ROLLTOP
069	PERC	100	1964		13,200	VAPOR	NONE
077	PERC	0	1959		85,415	VAPOR	ROLLTOP
078	PERC	100	1977		6,960	VAPOR	ROLLTOP
091	PERC	100	1960		9,600	VAPOR	NONE
101	PERC	100	1978	1985	600	VAPOR	NONE
105	PERC	100	1973		9,600	VAPOR	LIFTSPEED
111	PERC	100	1958	1990		NONE	NONE
121	PERC	100	1978		480	VAPOR	FREEBOARD
123	PERC	100	1958	1982		VAPOR	NONE
149	PERC	100	1966		65	COLD	NONE
153	PERC	10		1992	3000	VAPOR	REFRIGER
155	PERC	100	1976			VAPOR	REFRIGER
158	PERC	100	1968		18,000	VAPOR	FREEBOAR
164	PERC	0	1972		3000	VAPOR	REFRIGER
173	PERC	100			9,800	VAPOR	NONE
194	PERC	100	1940		600	WI PE	NONE
201	PERC	100	1962		4,668	VAPOR	NONE
209	PERC	o	1966			VAPOR	LIFTSPEED
210	PERC	100	1960		48,000	VAPOR	REFRIGER
216	PERC	100	1929		24,000	VAPOR	ROLLTOP
223	PERC	100	1950	1990		VAPOR	NONE
224	PERC	100	1986		600	COLD	NONE
230	PERC	100	1973	1986		VAPOR	NONE
233	PERC	100	1973	1991	7,500	VAPOR	NONE
240	PERC	100	1982	1993	660	VAPOR	LIFTSPEED
245	PERC	100	1968		13,200	VAPOR	NONE
246	PERC	100	1935		2,400	VAPOR	NONE
261	PERC	0		1992		VAPOR	NONE
262	PERC	100	1960		6,000	VAPOR	LIFTSPEED
287	PERC	0	1905			VAPOR	NONE
281	Stoddard	0	1930		120000	COLD	NONE
004	TCE	100	1956	1988	4,534	COLD	NONE
007	TCE	100	1940	1989	44,245	COLD	NONE
008	TCE	100	1945		42,000	COLD	NONE
009	TCE	100	1973		30,000	VAPOR	NONE
012	TCE	100	1940		1,500	WI PE	REFRIGER
016	TCE	100	1969		10,523	VAPOR	NONE
022	TCE	100	1964		39000	VAPOR	REFRIGER

Shop Code	Type of Solvent	% Job Shop	Year Shop Estab.	Year Solvent Eliminated	Solvent Use Rate lbs/yr	Process Type	Control Method
024	TCE	100	1984		109,056	VAPOR	LIFTSPEED
039	TCE	100	1963		144	WI PE	NONE
051	TCE	100	1966	1972		VAPOR	NONE
063	TCE	0	1973		6,600	VAPOR	NONE
064	TCE	100	1960	1990		VAPOR	ROLLTOP
066	TCE	0	1980	1992	10,000	VAPOR	NONE
071	TCE	0	1954		4,800	WI PE	NONE
085	TCE	95	1957		5,400	VAPOR	REFRIGER
089	TCE	100	1960		16,200	VAPOR	LIFTSPEED
093	TCE	100	1927		21,000	VAPOR	LIFTSPEED
095	TCE	100	1975	1983		VAPOR	NONE
098	TCE	100	1973			VAPOR	FREEBOARD
105	TCE	100	1973		1,860	VAPOR	LIFTSPEED
111	TCE	100	1958	1985		VAPOR	NONE
124	TCE	100	1919		70,600	VAPOR	LIFTSPEED
150	TCE	100	1969	1990		VAPOR	NONE
156	TCE	100	1972		31,680	VAPOR	ROLLTOP
161	TCE	100	1966		600	COLD	NONE
162	TCE	100	1956		7,920	VAPOR	LIFTSPEED
172	TCE	100	1945		7,920	VAPOR	REFRIGER
178	TCE	100	1965		7,710	VAPOR	REFRIGER
179	TCE	100	1983		7,920	COLD	NONE
187	TCE	100	1978			VAPOR	REFRIGER
192	TCE	100	1983		240	VAPOR	LIFTSPEED
210	TCE	100	1960		9,600	COLD	NONE
217	TCE	O	1951			VAPOR	NONE
223	TCE	100	1950	1990		VAPOR	NONE
237	TCE	100	1981		30000	VAPOR	REFRIGER
244	TCE	100	1971		13872	VAPOR	FREEBOARD
253	TCE	100	1967	1988	9600	VAPOR	NONE
258	TCE	100				VAPOR	LIFTSPEED
266	TCE	80	1975			WIPE	NONE
272	TCE	100	1959			VAPOR	REFRIGER
278	TCE	O	1967		12000	VAPOR	LIFTSPEED
281	TCE	0	1930	1992	60000	VAPOR	NONE
283	TCE	0	1987			VAPOR	FREEBOARD
286	TCE	0	1967	1990	60000	VAPOR	NONE
292	TCE	0	1960	1977	7920	COLD	NONE
298	TCE	100	1950	1988	18000	SPRAY	NONE
301	TCE	100	1968		2640	WIPE	NONE
304	TCE	100	1971		480	VAPOR	NONE
307	TCE	100	1972	1992	6948	COLD	NONE
308	TCE	100	1985	1990	1800	NONE	NONE
310	TCE	100	1985	1992	972	VAPOR	NONE
316	TCE	100	1930	1992		VAPOR	ROLLTOP
317	TCE	100	1944			VAPOR	FREEBOARD
268	Xylene	100	1969	1988	523	WIPE	NONE

Solvent Key:
TCA = 1,1,1-Trichloroethane METH = Methylene Chloride CFC = Chlorofluorocarbons MEK = Methyl Ethyl Keytone TCE = Trichloroethylene PERC = Perchloroethylene

Solvent use reduction has been achieved by survey respondents by improving the designs of vapor degreasers and the operating practices used with these units and by substituting aqueous and semi-aqueous cleaners for solvent degreasing. The most frequent vapor degreaser design and operating changes made by survey respondents focus on reducing solvent emissions, and include: increasing freeboard, adding a rolltop, controlling hoist speed and adding a refrigeration zone (see Exhibit 6-1). Of the 90 vapor degreasing applications in use in 1993, at least 75 percent were equipped with one or more of these design and operational changes. Various substitutions for solvent have been made by the survey respondents and these are discussed later in this section.

Although the numerical survey data provide insight to the changes regarding solvent use in the metal finishing industry, the comments submitted by respondents give an even clearer picture. The remainder of this subsection is devoted to their comments.

Exhibit 6-2. Distribution of Solvent Usage From 1980 to 1993 by Shop Age

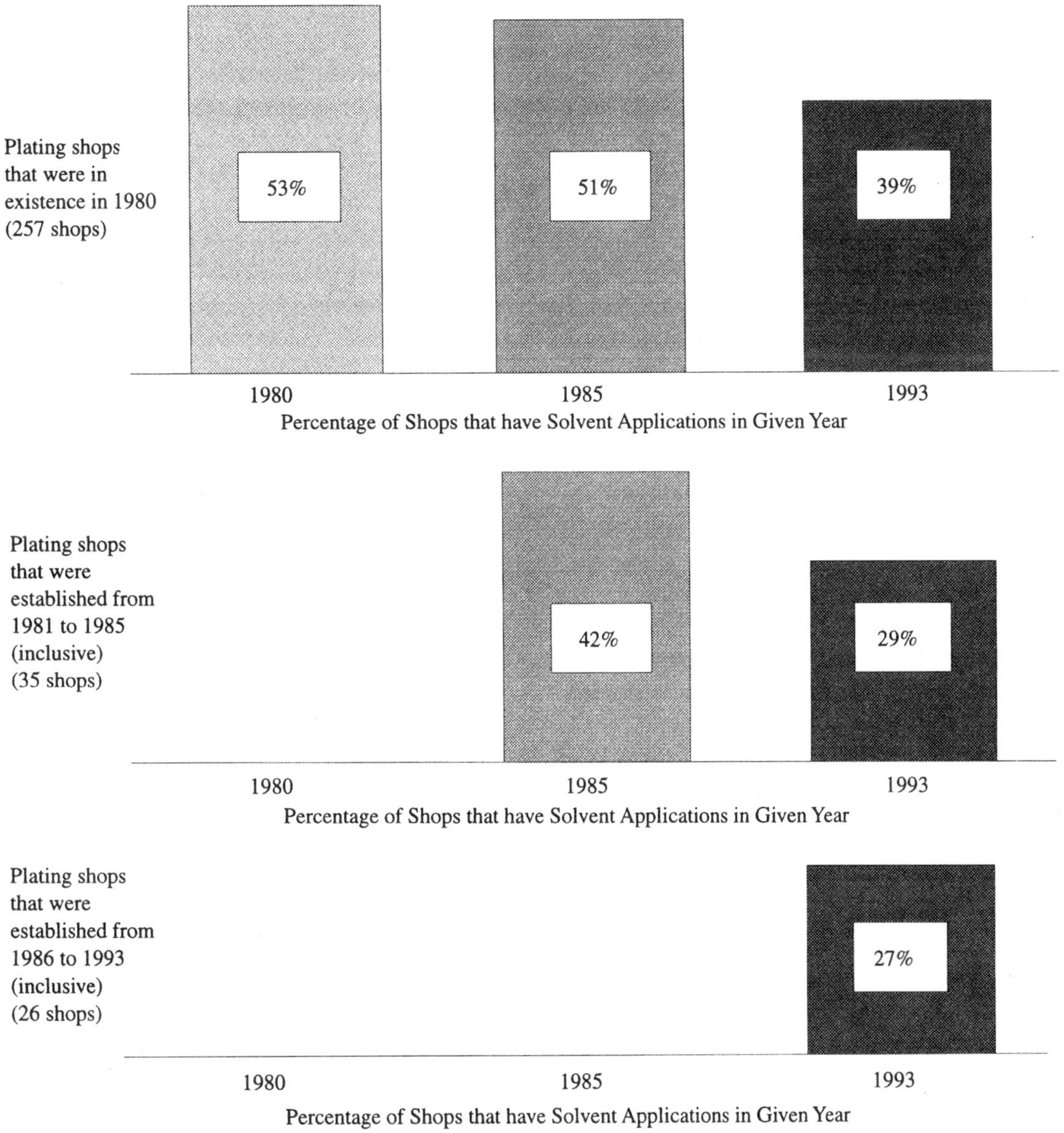

Sixty-eight percent of the respondents that are presently using solvent identified that its elimination is a pressing environmental problem. These shops are at various stages in the process of elimination. Some shops indicated that they are just beginning to evaluate solvent substitution alternatives (PS 156, PS 165, PS 184) and other shops indicated that they have made progress and are continuing to work toward elimination (PS 124, PS 166, PS 176, PS 178, PS 181, PS 191). Only 11.7 percent of the solvent users gave an indication that they plan to continue the use of solvent. These shops gave a range of reasons why solvent use is necessary. These reasons include:

- Solvent use is necessary to meet aerospace requirements (PS 121).
- Unsuccessful at attempts to clean zinc die castings without solvent (PS 089).
- Solvent is needed to remove polishing or buffing compounds (PS 098, PS 170, PS 171, PS 191, PS 205).
- No single answer for replacing solvent degreasing (PS 049).

- Some non-solvent products work well, but there is always a problem with long drying times for parts.
- Have replaced TCA vapor degreasing with alkaline cleaning, but can't replace acetone used to strip maskant (PS 166).
- Substitutes are too expensive (PS 217).
- Citrus cleaners did not work (PS 248).
- Tried six different safety solvents (citric based). None were satisfactory due to residue or unacceptable evaporation time (PS 271).

A number of problems with solvent substitution were cited by survey respondents. These include the following comments from both successful and unsuccessful substitution efforts:

- Alkaline cleaners increase the amount of waste treatment performed and/or the labor for these efforts (PS 114).
- More difficult to remove masking wax without solvent (PS 108).
- Citric acid based emulsion cleaner leaves a residue that inhibits good plating and causes poor adhesion (PS 109, PS 271).
- "Orange-Peel Cleaners" caused skin rashes on employees and excessive foaming in the waste treatment system (PS 021).
- Caused oil fouling of ion exchange resins (PS 130).
- Increased labor for degreasing and/or bath maintenance (PS 130, PS 262).
- Created a new wastestream (e.g., oily waste, spent cleaner) in some cases that is more difficult to deal with than solvent waste (PS 153, PS 171).
- Solvent substitutes simply do not perform as well as solvents (PS 163).
- Substitutes are too expensive (PS 217).
- Citrus smell of substitute cleaners is objectionable.
- Many substitutes contain chelating agents or surfac-

Exhibit 6-3. Average Solvent Usage Rate Per Application from 1980 to 1993

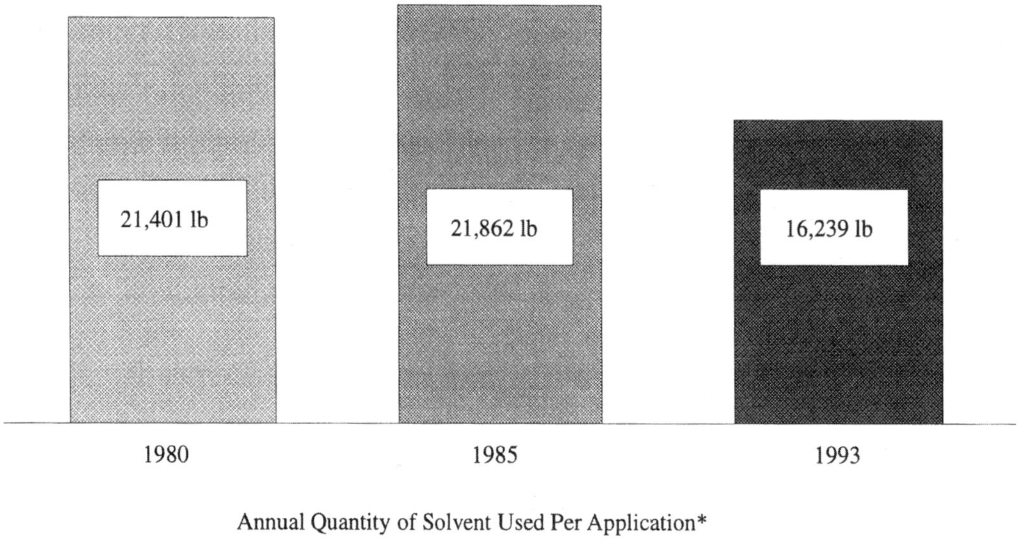

*An application is defined as the use of a given type of solvent by a given shop. Some shops had multiple applications. The total number of applications by year, for all 318 respondents were: 1980—197 applications; 1985—202 applications; and 1993—143 applications. The quantity of solvent used by a shop was assumed to be constant over time. Therefore, if an application was in existence in 1993, the annual quantity used was assumed to be the same amount in 1980 and 1985 (unless the shop was not in existence, whereas it was not used in calculating the average). As such, the usage trend shown above is believed to be accurate; however, the actual usage values should not be taken from context to estimate total usage throughout this industry.

tants that interfere with metal hydroxide precipitation (PS 296).

- PS 021 indicated that they had a "difficult transition" that took approximately 24 months and "a lot of patience." Their efforts included trials with terpene cleaners, "terpene-like cleaners and non-ionic surfactant packages," and finally they settled on a mildly alkaline aqueous cleaner with a non-ionic surfactant. They also use barium carbonate for post cleaning but eventually switched to pumice. These changes combined with "enough (cleaning) time" and "elbow grease" have resulted in "few cleaning problems."

Only two shops specifically indicated that there were no production related problems associated with the substitution for solvent degreasers (PS 153, PS 190).

Some respondents provided information with regard to successful substitute degreasers that they use in place of chlorinated solvents. The following responses were given:

- Aqueous cleaner with glycol ether (PS 064).
- Aqueous cleaner (PS 124, PS 166, PS 187, PS 292, PS 268).
- Citri-Solv product and Safety Clean (PS 105).
- Mildly alkaline, biodegradable aqueous cleaner with a non-ionic surfactant (PS 021).
- Global 6 (PS 169).
- Aqueous, citric, terpene cleaners (PS 174).
- Emulsion cleaner (PS 181).
- Alkaline power washer (PS 079, PS 197).
- Alkaline non-etch ultrasonic clean, Enthone NS-35 (PS 221).
- Increased concentration of existing cleaner (PS 253, PS 295), e.g., PS 295 eliminated solvent use (80,030 gal/yr CFC 113 and TCA) by increasing cleaner concentration from 10 oz/gal to 12 oz/gal.
- Converted degreasers to soak cleaners and use 7-9 oz/gal Midstate 4310, 140 - 180 °F, alkalinity ratio 0.40 minimum, soil load 10% maximum (PS 273).
- Water soluble solvents (PS 276).

A number of miscellaneous methods were employed to reduce solvent use; these include:

- Use safety solvent cans with hand wipe operation (PS 069, PS 176).
- Use low vapor pressure solvent blend in place of MEK for hand wiping.
- Increase level of training and reviewed good operating practices in safety meetings (PS 191).
- Increased use of hand wiping and decreased use of vapor degreaser (PS 191).
- Added inspection of degreaser to regular monthly procedures (PS 191).
- Working with customer on the cleanliness of parts (PS 248).
- Increased agitation in cleaning bath.
- Subcontract the 5% of workload that requires solvent degreasing.
- Install solvent recovery stills (PS 124, PS 149).

6.4 CYANIDE USE REDUCTION/ ELIMINATION

Cyanide, in the form of either sodium or potassium cyanide, has been a key component of plating solutions for many years. Some of the metals commonly plated from cyanide baths include: cadmium, zinc, copper, brass and precious metals. When used, cyanide forms a moderately stable complex with metal ions that permits the deposition of the metal under the influence of a suitable electrical potential.

Although cyanide-containing baths are eminently satisfactory as far as plating results go, pollution control poses a problem in terms of both compliance and economics (ref. 350). As a result, there has been a significant effort to find and implement cyanide-free plating processes since approximately 1975.

Survey respondents indicated concerns over the use and control of cyanide and expressed a desire to find alternative non-cyanide plating solutions. For example, 14 percent indicated that they currently have compliance difficulties with cyanide. Eleven percent indicated that improved cyanide plating alternatives or controls was a major environmental technology need for their company. Twenty-six percent of the respondents indicated that there exists a technology transfer insufficiency with respect to non-cyanide finishing. This was the highest percentage of response for any topic listed in the survey form (see summary of results in Section 2).

The earliest and most complete cyanide substitution that has taken place in the plating industry is the conversion from zinc cyanide plating to zinc chloride and zinc alkaline (non-cyanide) plating. Although some respondents indicated that they did not make this substitution until the 1990's, most respondents converted during the period of the late 1970's to mid-1980's. A total of 28 respondents (or approximately 17% of the zinc platers) successfully converted to zinc chloride, 18 respondents (or approximately 11% of the zinc platers) converted to zinc alkaline (non-cyanide) and 8 respondents (or approximately 5% of the zinc platers) converted to an unspecified zinc non-cyanide plating process (see Exhibit 6-4). Another 9 (or approximately 6% of the zinc platers) converted from a high cyanide to a low cyanide concentration bath. Perhaps even more impressive than the number of successful conversions is that only 2 respondents of all those that attempted conversions were unable to successfully make a substitution.

One significant drawback with regard to zinc cyanide plating substitution is that some shops found it necessary to install both zinc chloride and zinc alkaline (non-cyanide) baths to replace the single cyanide bath (PS 053, PS 059, PS 087, PS 204, PS 251, PS 288, PS 300). On a positive note, in addition to cyanide use reduction, some platers enjoyed production benefits from the substitution, including better and brighter plating.

In addition to zinc plating, a significant number of respondents have reduced or eliminated cyanide usage with copper plating and nickel stripping, although the success rates (i.e., ratios of successes to attempts) are not as high as for zinc plating. Other plating and finishing cyanide solutions for which substitutes were identified by respondents included cadmium, silver and brass plating and chromate conversion coatings. The number of successes was either limited or non-existent for these solutions.

In addition to the plating shops that indicated they have substituted a non-cyanide process for a cyanide process, there were 12 shops that simply eliminated approximately 22 cyanide plating processes in order to meet environmental regulations.

6.5 CADMIUM USE REDUCTION/ ELIMINATION

Cadmium, which is most often plated from a cyanide solution, was identified by the respondents of the Users Survey to be of significant concern. At the time of the survey, 30 percent of the respondents performed cadmium plating using a cyanide bath and 2 percent used a non-cyanide bath. Ten percent of all respondents indicated that they have compliance difficulty with cadmium. Seventeen percent indicated that they felt there was a technology transfer insufficiency with respect to cadmium and 8 percent indicated that there was a need for better cadmium alternatives or control technologies.

Based on the results of the Users Survey, it is obvious that cadmium-cyanide plating is much more difficult to substitute for than cyanide-based plating in general (see Exhibit 6-5). Only 8 of the respondents (or approximately 8% of the cadmium platers) have successfully made this substitution, with another 8 (or approximately 8% of the cadmium platers) indicating partial success. Another 3 respondents (or approximately 3% of the cadmium platers) have successfully implemented a low-cyanide, low-cadmium substitute. No single cadmium substitute stood out as the most successful, with tin and tin alloys, zinc (non-cyanide), cobalt-zinc and zinc-iron being used. The primary problems with cadmium substitutes were: (1) customer acceptance; (2) the quality of the finish; and (3) their higher cost.

In addition to the plating shops that indicated they have substituted non-cadmium or low-cadmium processes for a conventional cadmium process, there were ten shops that indicated they have eliminated cadmium plating in order to meet environmental regulations.

6.6 CHROMIUM USE REDUCTION/ ELIMINATION

Chromium is heavily used in electroplating and aluminum finishing. The most common chromium bearing solutions include: decorative and hard chromium, aluminum conversion coating, bright dipping of copper and copper alloys, chromic acid anodizing, deox/desmut and chro-

Exhibit 6-4. Chemical Substitutions to Reduce or Eliminate Cyanide Usage

Original Process	New Process	Successes	Partial Successes	Failure	Other Information
Copper, cyanide	Copper, non-cyanide, (type not specified)	PS 212 PS 309	PS 028	PS 005 PS 089 PS 101	• Tried Udylite process. Bath is more expensive to operate and plating too brittle. Also, bath precipitates during long idle periods. • Blistering of base metal (PS 089). • Problem with adhesion and plating efficiency (PS 101).
	Copper, non-cyanide, alkaline	PS 039 PS 190 PS 267	PS 036 PS 045		• Coppers and steels are easily plated in both rack and barrel operations. Limited success with aluminum alloys (PS 036). • Works for rack plating only not barrel (PS 045).
	Copper, acid	PS 062 PS 182 PS 260	PS 129		• Still requires a Cu - CN strike (PS 062). • Cannot plate some zinc die castings (PS 129).
	Zinc caustic solution (Bondal)	PS 076			
	Copper, cyanide, low conc.	PS 081 PS 109			
	Zinc, non-cyanide	PS 102			
	Copper, neutral pH, non-cyanide		PS 229		• Won't plate effectively on certain hard metals or polishded metal. Requires close monitoring (PS 229).
Copper, cyanide strike	Copper, non-cyanide, alkaline strike	PS 130 PS 224	PS 036 PS 176	PS 119 PS 213	• Does not work well on lead or steel parts, only brass (PS 119). • M&T - Harshaw product (PS 130). • Some adhesion problems (PS 130). • Blistering (PS 213). • Enthone Cupral (PS 224).
	Copper, sulfate	PS 154			
Copper, cyanide, electroless	Copper, non-cyanide, electroless	PS 257			
Copper, strip, cyanide	Copper, strip, non - CN alkaline	PS 283			
Nickel, CN strip	Nickel, non - CN strip	PS 024 PS 086 PS 161 PS 190	PS 093 PS 163		• More expensive and does not work as fast or as well (PS 093). • Sodium nitro benzonate does not work as well as CN strip (PS 163).
Nickel, electroless strip, cyanide	Nickel, electroless strip, non-cyanide	PS 074			
Cadmium, cyanide	Cadmium, non-cyanide	PS 025 PS 309	PS 229 PS 316	PS 300	• Requires close monitoring, agitation and cooling. Created pH control problems with wastewater (PS 229). • Ongoing project (PS 316) • Neutra - Cad (PS 300).
	Cadmium, acid	PS 053 PS 190	PS 245		• Would not throw into extreme low current density areas, some rusting in masked areas (PS 245).
	Cadmium, low cyanide	PS 090 PS 081 PS 268	PS 229		• Quality and polarization problems (PS 229). • From 3.0 oz/gal Cd, 16.0 oz/gal CN to 2.4 oz/gal Cd, 10.0 oz/gal CN (PS 268).
	Cobalt-zinc		PS 124		• Customer acceptance is sometimes tough (PS 124).
Silver, cyanide	Silver, non-cyanide		PS 068	PS 101	• Cannot be applied on nickel (PS 068). • Problem with adhesion and deposit color (PS 101).

Exhibit 6-4 (continued). Chemical Substitutions to Reduce or Eliminate Cyanide Usage

Original Process	New Process	Successes	Partial Successes	Failure	Other Information
Brass, cyanide	Brass, cyanide, low conc.	PS 081			
Zinc, cyanide	Zinc, chloride	PS 008 PS 014 PS 053 PS 059 PS 061 PS 087 PS 088 PS 098 PS 099 PS 117 PS 118 PS139 PS 141 PS 162 PS 181 PS 186 PS 201 PS 204 PS 225 PS 242 PS 247 PS 248 PS 251 PS 272 PS 277 PS 288 PS 289 PS 300	PS 150	PS 245	• Reduced CN usage by several hundred lbs/yr (PS 014). • Better and brighter plating with substitute (PS 061, PS 088, PS 139). • Requires better precleaning and tighter bath control (PS 087, PS 204). • Higher cost (PS 098). • Barrel plating (PS 141). • Problems with proper chromating (PS 150). • More recent bath better than 10 years ago (PS 272).
	Zinc, non-cyanide, alkaline	PS 033 PS 034 PS 043 PS 053 PS 059 PS 072 PS 087 PS 095 PS 106 PS 181 PS 190 PS 204 PS 207 PS 245 PS 251 PS 267 PS 288 PS 300 PS 353	PS 050 PS 058	PS 086	• Tight control needed (PS 034). • Higher reject rate after change (PS 050). • Embrittlement during delayed post forming (PS 058). • Iron/Zn codeposited causing blistering (PS 086). • Requires better precleaning and tighter bath control (PS 087, PS 204).
	Zinc, non-cyanide (not specified)	PS 014 PS 041 PS 042 PS 110 PS 142 PS 145 PS 203 PS 229 PS 302 PS 309 PS 316			• Zinc, non-cyanide is a better technology (PS 041). • Successful at 0.6 to 0.8 oz/gal Zn. Reduces load on ion exchange system (PS 229).
	Zinc, low cyanide	PS 031 PS 051 PS 081 PS 100 PS 124 PS 133 PS 143 PS 175 PS 226 PS 268			• Minimum conc. requirements: CN - 4.2 oz/gal; Zn - 2.0 oz/gal (PS 124). • Includes both rack and barrel (PS 133). • From 2.3 oz/gal Zn, 6 oz/gal CN to 1.0 oz/gal Zn, 1.2 oz/gal CN (PS 268).
Chromate conversion, with ferroocyanide	Conv. w/out ferrocyanide			PS 142	• Did not work as well (PS 142).

mium stripping (see Section 2 for data on the use of each of these processes). Due to its high toxicity and the cost for treatment and disposal of chromium bearing wastes, it has been a target for reduced usage or elimination.

The results of the Users Survey indicate that 18 percent of the respondents have compliance difficulty with chromium. Twenty percent of the respondents indicated that there exists a technology transfer insufficiency with regard to non-chromium metal finishing.

Efforts to reduce chromium usage in the aerospace industry and for military applications have focused on non-chromium aluminum finishing and hard chromium plating substitution (ref. 75, 77, 211, 418). However, among survey respondents, which are mostly job shops, efforts were directed more at substituting trivalent chromium decorative plating and zinc conversion coating solutions for the traditional hexavalent baths. Both of these substitutes offer cost savings in terms of lower wastewater treatment chemical usage since the hexavalent chromium reduction step is eliminated. Also, trivalent chromium plating solutions are formulated with as little as 6 g/l of chromium, significantly less than their hexavalent counterparts (ref. 89). The lower chromium concentration in the baths results in less chromium in the

Exhibit 6-5. Chemical Substitutions to Reduce or Eliminate Cadmium Usage

Original Process	New Process	Successes	Partial Successes	Failure	Other Information
Cadmium, cyanide	Tin or tin alloys	PS 041 PS 292			• Currently used on trial basis (PS 025). • Offers solderable finish (PS 041).
	Zinc, non-cyanide	PS 042 PS 181 PS 292	PS 139 PS 318		• Couldn't substitute for some parts (PS 139). • Currently implementing (PS 318)
	Cadmium, cyanide, low conc.	PS 055 PS 081 PS 268			• From 3.0 oz/gal Cd, 16.0 oz/gal CN to 2.4 oz/gal Cd, 10 oz/gal CN (PS 268).
	Cobalt-zinc	PS 215	PS 124 PS 214 PS 229		• Requires close control (PS 229). • Customer acceptance is sometimes difficult (PS 124). • Higher cost sacrafice of deposit properties (PS 214).
	Nickel-zinc		PS 214	PS 173	• Can't implement because military customers require cadmium (PS 173). • Under investigation (PS 188). • Higher cost, sacrifice of deposit properties (PS 214).
	Zinc-iron	PS 215 PS 235	PS 214 PS 229		• Requires close control (PS 229). • Higher cost, sacrifice of deposit properties (PS 214). • First try was unsuccessful. Changed chemical supplier. R&D required to make it work (PS 235).
	Phos/organic		PS 214		• Requires close control (PS 229). • Higher cost, sacrifice of deposit properties (PS 214). • First try was unsuccessful. Changed chemical supplier. R&D required to make it work (PS 235).

dragout. The substitution of trivalent plating solutions was generally successful for the survey respondents (15 shops), although some exceptions were identified (see Exhibit 6-6). Sixty percent of these respondents indicated that the process substitution was successful, 20 percent indicated it was partially successful and 20 percent indicated that their substitution efforts failed. The most frequent problem with the process was its acceptability to the shop's customers; a problem that was primarily related to the color of the deposit.

Switching from a hexavalent chromium zinc conversion coating (both clear and blue) to a trivalent coating was successful in most cases. Seventy-six percent of the shops attempting this substitution were successful. Another 8 percent were partially successful and only 15 percent resulted in failure. The limited success of one respondent and the failures of others were related to the characteristics of the coating.

In addition to the plating shops that indicated they have substituted a non-chromium, trivalent chromium or low-chromium process for a conventional chromium process, there were four shops that simply eliminated chromium plating and one shop that eliminated chromating of magnesium in order to meet environmental regulations.

6.7 OTHER CHEMICAL SUBSTITUTIONS

Substitutions made for environmental reasons, not involving the previously discussed four Cs chemicals, are shown in Exhibit 6-7. These other substitutions were implemented for two main purposes: (1) to reduce the mass of metal or other undesirable constituents in the wastewater and (2) to improve the performance of their treatment operations by reducing the use of chelating compounds. As indicated in Exhibit 6-7 nearly all of these substitutions were successful.

Exhibit 6-6. Chemical Substitutions to Reduce or Eliminate Chromium Usage

Original Process	New Process	Successes	Partial Successes	Failure	Other Information
Chromium, decorative (Cr^{+6})	Chromium, decorative (Cr^{+3})	PS 028 PS 034 PS 150 PS 169 PS 191 PS 262 PS 261 PS 237 PS 103 PS 301 PS 305 PS 308 PS 311	PS 124 PS 125 PS 251	PS 058 PS 092 PS 089	• Currently used on a trial basis (PS 025). • Unacceptable to customer (PS 058, PS 092). • Never implemented, color not acceptable to customers (PS 089). • Experienced rusting at weld interfaces (PS 058). • Not completely pleased with deposit color (PS 124). • Color is excellent, reduced number of rejects (PS 305) • Limited use on small hand line (PS 150). • Also switched to ion exchange Cr recovery at the same time (PS 191). • One customer still requires Cr^{+6} bath, so it has been retained (PS 251). • Difficult to control (PS 034, PS 262, PS 301). • Better throwing (PS 262). • Tricrolyte (PS 103). • Reduced reject rate (PS 103). • Reduced Cr in air emissions (PS 301) • Reduced risk to plating personnel (PS 305)
	Chromium, decorative (Cr^{+6}), low conc.	PS 145 PS 172 PS 124			• From 40 - 45 oz/gal to 32 oz/gal CrO_3 (PS 124, PS 172).
Chromate, hexavalent (on zinc)	Chromate, trivalent (on zinc)	PS 013 PS 039 PS 135 PS 160 PS 168 PS 175 PS 219 PS 237 PS 276 PS 298	PS 124	PS 218 PS 086	• Higher quality and lower cost (PS 013). • Cr coating was too soft and there is insufficient shop space to store parts for required 24 hrs (PS 086). • Longer life (PS 013, PS 237). • Clear chromate (PS 160, PS 168, PS 276). • Blue chromate (PS 175, PS 237, PS 276). • Does not meet one QC test, cyclic temp./humidity test, as well as Cr^{+6} (PS 124). • Didn't pass customer's lead acetate test (PS 218).
Chromium, hard	Chromium, hard, lower CrO_3 conc.	PS 021			• From 35 oz/gal to 25 oz/gal CrO_3. • Lower cost to maintain and easier to scrub exhaust (PS 021). • From 40 oz/gal to 33 oz/gal CrO_3 (PS 253).
	Vacuum plating				• Success/failure not indicated (PS 157).
Desmut, chromium	Desmut, non-chromium	PS 029 PS 017			• Haviland AD-84.
Boiler treatment, chromium	Boiler treatment, non-chromium	PS 029			
Deox, chromium	Deox, non-chromium	PS 231 PS 086			• De-ox NC (PS 231).
Desmut/deox, chromium	Desmut/deox, HF/Sulfuric	PS 278			
Chromate	Black anodize	PS 261			• Org. process of mass finish/Irridite/Solvt. spray paint replaced by black anodize. Eliminated VOC from painting and chromic acid from Al processing (PS 261).

Exhibit 6-7. Other Chemical Substitutions to Made for Environmental Purposes

Original Process	New Process	Successes	Partial Successes	Failure	Other Information
Desmut, 50% nitric	Desmut, 2% nitric, ferric nitrate	PS 014			• Reduced nitric usage by 50 percent (> 10,000 lbs/yr)(PS 014). • Reduced sulfuric usage by 67 percent on this line. Required employee retraining (PS014). • Changed cleaners to reduce concentration of BOD in discharge (PS 015).
Desmut, 50% sulfuric, 50% nitric	Desmut, 15% sulfuric, 45% nitric, fluoride	PS 014			
Cleaner	Not specified	PS 015			
Cleaner, phosphated and/or chelated	Cleaner, non-phosphated (Hurri-Kleen)	PS 017 PS 064 PS 250	PS 139		• Hurri-Kleen (PS 017). • Treatment costs are lower than hauling costs of old bath (PS 064). • More rejects, but improved effluent (PS 139).
Akaline cleaner, Oakite 166	Alkaline cleaner, Novanae A1 Det	PS 049			• Eliminated zinc in wastewater (PS 049).
Zinc, chloride	Zinc, non-cyanide, alkaline	PS 066			• Better throwing power, lower zinc concentration (PS 066).
Bright dip (brass), nitric	Bright dip, no-nitric	PS 086			
Seal, nickel acetate	Seal, low conc. nickel acetate	PS 073			
Etch, caustic	Etch, no dump etch	PS 073			
Bright dip, nitric acid and sulfuric	Bright dip, hydrogen peroxide and sulfuric	PS 153			• Works extremely well on parts that have portions iron plated. It brightens the iron and removes less copper (PS 153). • Less fuming and lower copper in wastestream (PS 153).
Tin-lead, 60-40	Tin-lead, 90-10	PS 171			• Too early to tell if it will be successful (PS 171).
Copper, electroless, copper sulfate	Copper, electroless, copper chloride	PS 275			• Change reduced bail out (PS 275).
Nickel, brite/semi-brite	Nickel, brite/semi-brite lower conc.	PS 275			
Cleaner, Al etch	Etch, non-etch	PS 299			• Reduced conc. of Al in wastewater (PS 299).

Section 7
Wastewater Treatment

7.1 INTRODUCTION

Section 7 covers end-of-pipe (EOP) wastewater treatment. Most of the respondents to the Users Survey employ conventional EOP wastewater treatment and sludge handling processes. These processes, which are so named because of their wide usage throughout the plating industry, are defined and discussed in Section 7.2. Capital and operating costs for conventional treatment are addressed in Section 7.3. End-of-pipe treatment data from the Users Survey are presented in Section 7.4. Some shops employ non-standard methods of end-of-pipe treatment. For metals removal/concentration, the non-standard methods that are used by survey respondents are ion exchange, evaporation and microfiltration. These three technologies are discussed in Section 7.5.

7.2 CONVENTIONAL TREATMENT TECHNOLOGIES

7.2.1 Overview

Conventional treatment is a series of unit processes used extensively by industry that have provided reliable treatment for many electroplating operations. Exhibit 7-1 is a schematic of a conventional treatment system for electroplating wastes containing chromium and cyanides in addition to other heavy metals, acids, and alkalis.

The configuration of conventional treatment is relatively standard. It consists generally of the following unit processes:

- Chromium reduction of segregated chromium waste streams to reduce the chromium from its hexavalent form to the trivalent state, which can be subsequently precipitated as chromium hydroxide by additions of alkali.

- Cyanide oxidation of segregated cyanide-bearing waste streams to oxidize the toxic cyanides to harmless carbon and nitrogen compounds.

- Metals removal of the combined metal-bearing wastewaters using hydroxide precipitation techniques.

- Sludge dewatering using gravity thickening followed by a mechanical dewatering device to increase the solids content of the sludge and therefore reduce its volume. Within the past five years, thermal dehydration using sludge dryers has also become a conventional unit operation. This equipment further increases the solids content of the sludge.

The Federal electroplating and metal finishing pretreatment wastewater standards (Exhibits 7-2 and 7-3) were developed by EPA by identifying commonly used treatment practices and determining their effectiveness by collecting effluent data from well operated systems. Conventional treatment was selected by EPA as the standard system. Therefore, for most plating shops, use of conventional treatment will provide sufficient pollutant removal to meet discharge standards. There are two major exceptions to this rule. First, many plating shops are regulated by local discharge standards that are more stringent than the Federal standards and conventional treatment may be insufficient to meet these limitations. Second, the treatment systems selected by EPA for establishing the Federal standards were those systems that

Exhibit 7-1. Conventional End-of-Pipe Treatment System

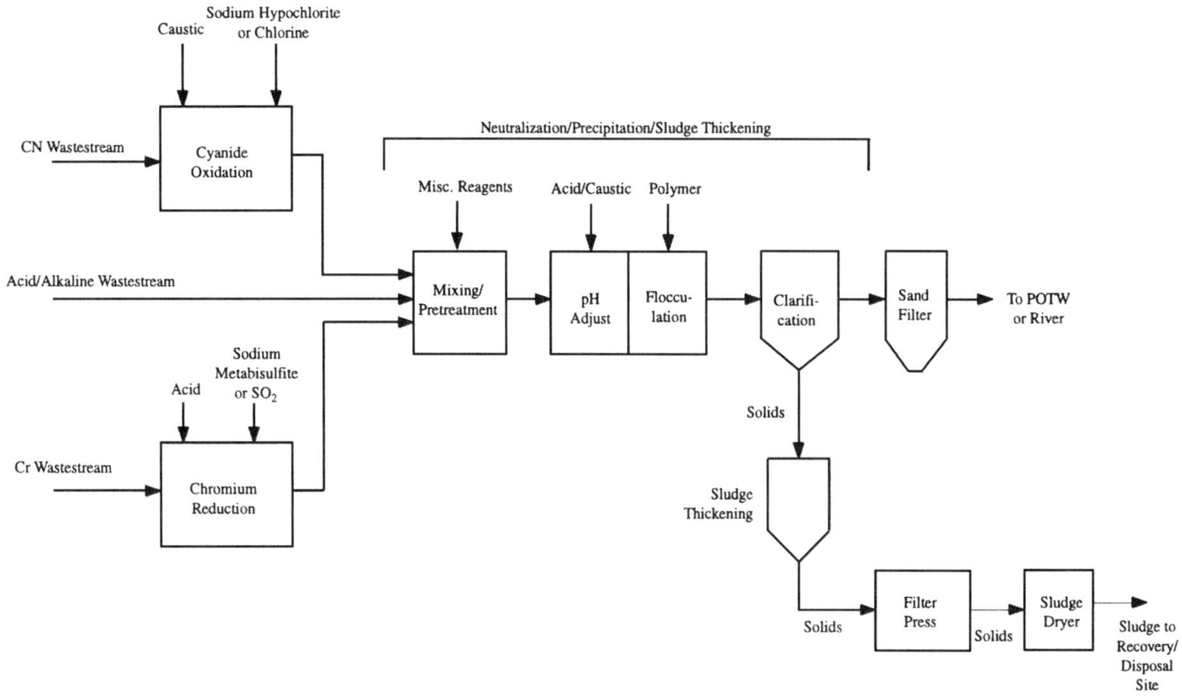

EPA determined to be "properly operating facilities." For example, EPA omitted facilities that: (1) did not have well operated treatment processes; (2) had complexing agents (e.g., non-segregated wastes from electroless plating); and (3) had dilution from non-plating wastewaters. As a result, some plating facilities may not meet the properly operated facility criteria used by EPA and may have difficulty meeting Federal standards using conventional treatment.

In cases where conventional treatment is insufficient to meet discharge limitations for a given facility, there are three basic choices for attaining compliance: (1) correct or upgrade the existing processes; (2) make internal changes (e.g., improve rinsing, add recovery, segregation of waste streams) to "normalize" the wastewater, (3) use conventional treatment plus additional treatment (i.e., polishing), and (4) use alternative treatment processes. Information on treatment system operation is reviewed in this section and details can be found in the literature (e.g., ref. 38). Methods for internal changes are discussed in Sections 3, 4, 5 and 6. Additional and alternative treatment processes are discussed in this section.

7.2.2 Chromium Reduction

Various metal finishing processes contribute chromium to the wastewater. Among these are chromium plating, chromating, bright dipping, chromic acid anodizing and chromium stripping. Although chromium is present in both the trivalent (Cr^{+3}) and hexavalent (Cr^{+6}) state in process solutions and wastes, the dominant species is Cr^{+6}. Unlike most heavy metals which are precipitated readily as insoluble hydroxides by pH adjustment, Cr^{+6} must first be reduced to the trivalent state because it forms the chromate complex which behaves as an anion and cannot form an insoluble hydroxide (ref. 38).

Conventional chromium reduction is achieved by reaction of Cr^{+6} with a reducing agent. The most commonly used reducing agents are sulfur dioxide gas and sodium metabisulfite (dry granular power). Alternative reducing agents include: sodium hydrosulfite, ferrous sulfate and iron or steel scrap (ref. 38). The chemical equations applicable to reduction with sulfur dioxide and sodium metabisulfite (actually sodium bisulfite, which is formed when sodium metabisulfite is added to water) are (ref. 38, 413):

Sulfur Dioxide Gas:

$$3SO_2 + 2H_2CrO_4^- \rightarrow Cr_2(SO_4)_3 + 2HOH$$

Sodium Bisulfite:

$$3NaHSO_3 + 2H_2CrO_4 + 3H_2SO_4 \rightarrow Cr_2(SO_4)_3 + 5HOH + 3NaHSO_4$$

Exhibit 7-2. Pretreatment Standards for the Electroplating Category (40 CFR 413)

Facilties Discharging <10,000 gpd

Pollutant	Daily Maximum, mg/l	Maximum 4 Day Average, mg/l
Cadmium	1.2	0.7
Lead	0.6	0.4
Cyanide (A)	5.0	2.7
Total Toxic Organics	4.57	—

Facilities Discharging > 10,000 gpd

Pollutant	Daily Maximum, mg/l	Maximum 4 Day Average, mg/l
Cadmium	1.2	0.7
Chromium (T)	7.0	4.0
Copper	4.5	2.7
Lead	0.6	0.4
Nickel	0.6	2.6
Zinc	4.1	2.6
Silver	1.2	0.7
Total Metals	0.6	0.4
Cyanide (T)	5.0	2.7
Total Toxic Organics (TTO)	4.57	—

Exhibit 7-3. Pretreatment Standards for the Electroplating Category (40 CFR 433)

Pretreatment Standards for Existing Sources (PSES)

Pollutant	Daily Maximum, mg/l	Maximum 4 Day Average, mg/l
Cadmium	0.69	0.26
Chromium (T)	2.77	1.71
Copper	3.38	2.07
Lead	0.69	0.43
Nickel	3.98	2.38
Zinc	2.61	1.48
Silver	0.43	0.24
Cyanide (T)	1.20	0.65
Cyanide (A)*	0.86	0.32
Total Toxic Organics (TTO)	2.13	—

Pretreatment Standards for New Sources (PSNS)

Pollutant	Daily Maximum, mg/l	Maximum 4 Day Average, mg/l
Cadmium	0.11	0.07
Chromium (T)	2.77	1.71
Copper	3.38	2.07
Lead	0.69	0.43
Nickel	3.98	2.38
Zinc	2.6]	1.48
Silver	0.43	0.24
Cyanide (T)	1.20	0.65
Cyanide (A)*	0.86	0.32
Total Toxic Organics (TTO)	2.13	—

Cyanide (A) = Cyanide amenable to chlorination
Cyanide (T) = Cyanide, Total
* Alternative to total cyanide that can be applied by Control Authority under given conditions.

A typical chromium reduction process is shown in Exhibit 7-4. The basic equipment includes a reaction tank, mixer, chemical feed system, pH meter/controller, an oxidation reduction potential (ORP) meter/controller, transfer pumps and level controls.

The type of chemical feed system selected will depend on the reduction reagent selected. Sulfur dioxide gas usage is restricted mostly to large treatment plants and sodium bisulfite is used at most small and medium-sized systems, but it is also used at large systems. Sulfur dioxide gas is the reagent of choice for larger systems due mainly to its lower chemical cost (at high feed rates it is nearly 50% less expensive, see Section 7.3.2). However, its use has drawbacks that cause it to be less desirable for smaller systems; these are: (1) requires a more expensive chemical feed system (sulfonator) and (2) presents a health hazard. The latter of these factors is the result of the toxic nature of sulfur dioxide gas. To overcome the health hazard, many large facilities install a vacuum system to prevent losses of sulfur dioxide to the working area, which further increases the capital cost of using this reagent. The results of the Users Survey indicate that approximately 6% of all respondents use sulfur dioxide and approximately 60% of all respondents use sodium bisulfite (approximately 34% do not have a chromium treatment process or use an alternative chemical). Shops using sulfur dioxide had average and median industrial wastewater flow rates of 153,814 gpd and 55,000 gpd, respectively. For all shops surveyed, the average and median flows were 34,600 gpd and 14,000 gpd.

The efficiency of the reduction reaction is highly dependent on pH (Exhibit 7-5), and most conventional processes are operated between pH 2.0 and 3.0. Since the incoming wastewater is usually above this desired range, sulfuric acid is added to reduce the pH. This is a drawback of the conventional process since the subsequent metal removal step is performed at an elevated pH (typically 7.0 to 9.5), and therefore the introduction of acid increases the subsequent need for alkali reagent.

The conventional chromium reduction processes are capable of producing an effluent with less than 0.1 mg/l Cr^{+6} (ref. 39).

The theoretical quantity of sodium bisulfite or sulfur dioxide required to reduce one pound of hexavalent chromium is approximately 3 lbs and 2 lbs, respectively. One source estimates that 0.2 lbs of sulfuric acid are required to lower the pH to the required range of 2.0 to 3.0 (ref. 392). The results of the Users Survey indicate that actual chemical requirements for chromium reduction and pH adjustment are much greater than the theoretical or esti-

Exhibit 7-4. Chromium Reduction Treatment Unit

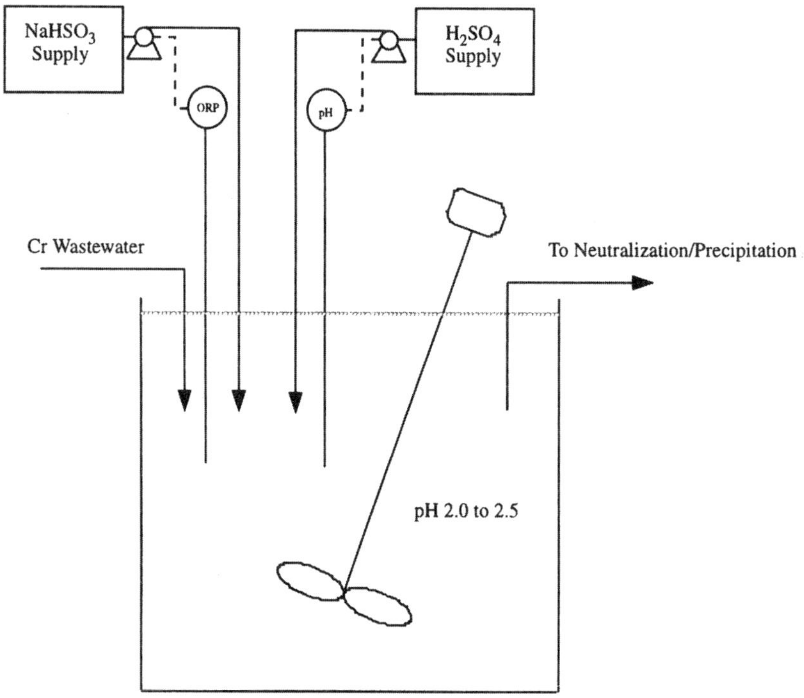

Exhibit 7-5. Effect of pH on Chromium Reduction

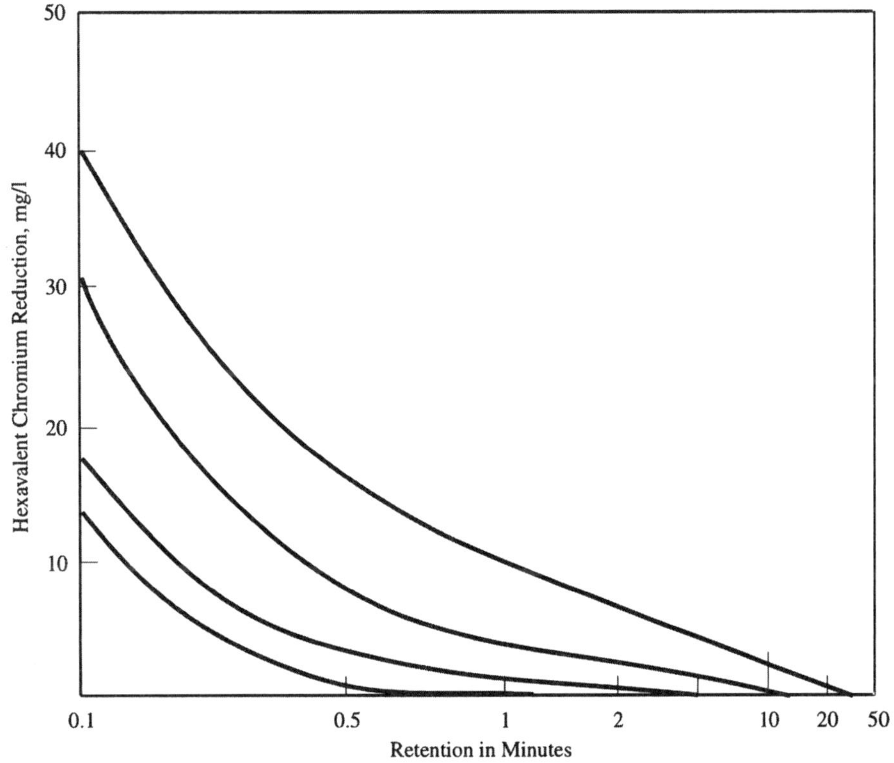

Source: ref. 39

mated quantities presented by most references. In fact actual dosages are often five to ten times or more than theoretical dosages. Treatment chemical usage data from survey respondents is presented and discussed in Section 7.3.2.

The following is a discussion of two alternative methods to the conventional sulfur compound chromium reduction process that are in use by the metal finishing industry: the sacrificial iron anode and sulfate reduction methods.

The sacrificial iron anode technology makes use of an electrochemical reaction in which an electrical current is applied to consumable iron electrodes. The electrochemical cell consists of a number of 1/8 in. cold rolled carbon steel plates separated by 1/8 in. gaps. A DC power supply is connected across the two end plates. Wastewater flows through the gaps in contact with the electrodes allowing the current to flow from electrode to electrode. The current flow causes the anode to dissolve slowly, thereby generating ferrous ions into the wastewater stream. The ferrous iron chemically reacts with the hexavalent chromium and reduces it to the trivalent state. During this reduction process, the iron is converted to trivalent iron hydroxide which results in a co-precipitation effect, where the iron hydroxide adsorbs heavy metal cations onto its surface (ref. 412). The process has the advantage of being able to reduce chromium at neutral pH. A drawback is that the process produces ferric hydroxides, which increases the quantity of sludge (ref. 39). Five respondents to the Users Survey (1.6%) use this technology (PS 083, PS 181, PS 240, PS 244 and PS 300). These systems, which were purchased between 1983 and 1991, were given satisfaction ratings of 3 and 4 (on a scale of 1 to 5, with 5 being the highest level of satisfaction) (three respondents did not provide a satisfaction rating). One of the technology users indicated that they purchase approximately $2,000 of replacement electrodes per year for their sacrificial iron anode unit (PS 083). This shop has a chromium bearing wastewater flow of 8,000 to 10,000 gpd. Another shop purchases approximately 2,900 lbs/yr of iron ($1.35/lb). This shop has a total plating flow of 25,810 gpd.

Ferrous sulfate reduction has been used to reduce chromate in an acid environment for a number of years. The primary advantage for many facilities was an inexpensive, abundant supply of ferrous sulfate, which is a waste product from steel pickling. The disadvantage of the process is the considerable increase in sludge generation owing to the precipitation of ferric hydroxide in the neutralization phase of the treatment. Pilot evaluations during the early 1980's of alkaline ferrous sulfate reduction of chromate were relatively successful. The process was proven capable of rapid reduction at pH levels between 8 to 10. The advantage of the alkaline process is the savings in sulfuric acid and caustic that are used with the conventional reduction process to lower and subsequently raise the pH. Also, the reduction process can be accomplished in the same reaction vessel as neutralization, which reduces equipment requirements (ref. 39). However, the process still has unfavorable sludge generation characteristics and a lack of control in chemical metering (ref. 348). The use of ferrous sulfate as a hexavalent chromium reducing compound was reported by only two respondents to the Users Survey (PS 020 and PS 135). At least one of those shops performs reduction in an alkaline environment (PS 135). Also, ferrous sulfate is used by approximately 3.1% of all respondents for co-precipitation. Sodium hydrosulfite is also used by one respondent as a chromium reducing agent in an alkaline environment (PS 173 and PS 253). PS 253 treats a 28,300 gpd chromium wastestream containing an average of 68.3 mg/l Cr. This shop uses 12,400 lbs/yr of sodium hydrosulfite in their chromium reduction process.

7.2.3 Cyanide Oxidation

Nearly all electroplating shops that generate dilute cyanide bearing wastewaters employ alkaline chlorination treatment. This process, which has been in commercial use for over 35 years, is suitable for destroying free dissolved hydrogen cyanide and for oxidizing all simple and some complex inorganic cyanides in aqueous media (ref. 348). If properly designed, maintained and operated (good pH and oxidation reduction potential control), the process will oxidize cyanides which are amenable to chlorination (i.e., the cyanide that can be oxidized by the alkaline chlorination process), to less than 1.0 mg/l cyanide (ref. 39). Extensive sampling by EPA showed that the average effluent concentration from a well operated cyanide destruction system contains 0.18 mg/l of total CN and 0.06 mg/l of amenable CN.

The cyanide in very stable complexes, such as ferrocyanides or ferricyanides, is basically unaffected by chlorination (ref. 243). Cyanide that is complexed with copper, nickel and precious metals is amenable to chlorination, but reacts more slowly than free cyanide and therefore requires excess chlorine for efficient cyanide destruction. Concentrated cyanide wastes, such as spent plating or stripping solutions, should not be reacted with hypochlorite because the reaction can be violent, with the emission of chlorine gas. These wastes can be batch treated by electrolytic oxidation and thermal destruction (ref. 242).

A schematic of an alkaline chlorination process is shown in Exhibit 7-6. Most often the process is operated in two stages, with separate tanks for each stage. A common exception is batch treatment systems, where one tank is typically employed (ref. 38). Destruction of dilute solutions of cyanide by chlorination can be accomplished by direct addition of sodium hypochlorite (NaOCl), or by addition of chlorine gas plus sodium hydroxide (NaOH) to the wastewater. With direct chlorine gas addition, sodium hydroxide reacts with the chlorine to form sodium hypochlorite. Selection between the two methods is based primarily on economics and safety. The chemical costs for chlorine gas treatment are less than

Exhibit 7-6. Schematic of a Two-Stage Treatment Process for Cyanide Destruction

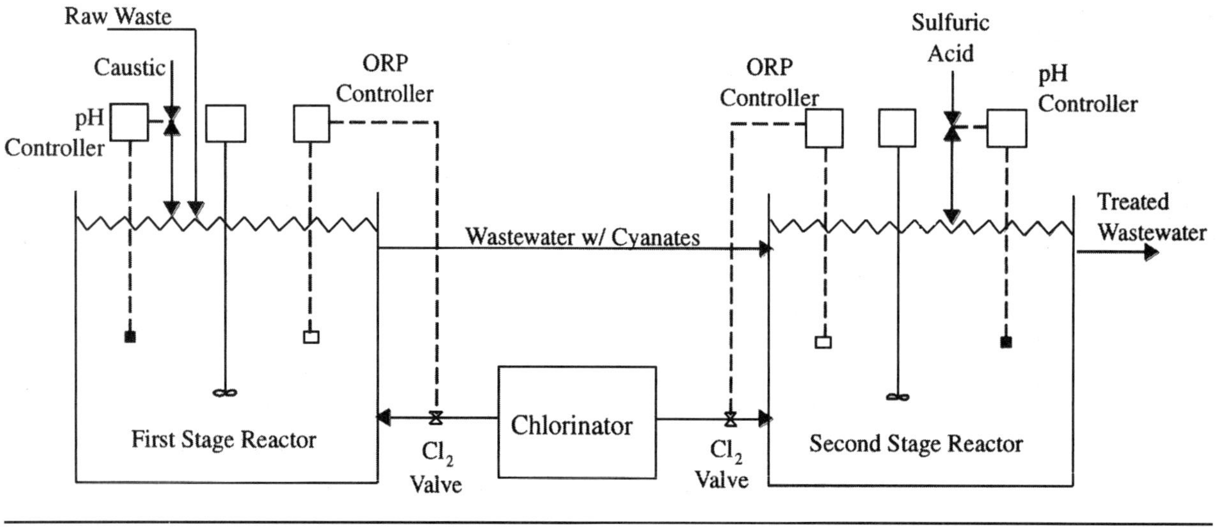

half those of direct hypochlorite addition (see Section 7.3.2), but handling is perceived to be more dangerous and equipment costs are higher (ref. 39). Relative to the safety issue, one expert source argues that modern gas handling equipment is highly reliable and designed to be relatively fail safe (ref. 38). The results of the Users Survey indicate that 95% of the respondents with alkaline chlorination processes use sodium hypochlorite and 5% use chlorine gas plus sodium hydroxide. The shops that use chlorine gas tend to be the same shops that select sulfur dioxide for chromium reduction (e.g., PS 025 and PS 093). Generally, these shops have substantial cyanide wastestream flow rates.

In the first stage of treatment, hypochlorite oxidizes cyanide to cyanate. This reaction is accomplished most completely and rapidly under alkaline conditions at pH 10 or higher (preferred range is 11.0 to 11.5) (ref. 38). An oxidation period of 10 to 15 minutes is usually adequate (ref. 38); however retention times up to 60 minutes are routinely used (ref. 39, 348). An ORP set-point of approximately +325 millivolts is adequate for most operations. A higher set-point may be needed, depending on the composition of the wastewater. A set-point of +400 millivolts is considered a maximum point (ref. 38). Potassium iodide-starch test paper, which indicates residual chlorine, is sometimes used to determine if the reaction is complete (ref. 243). To avoid producing solid cyanide precipitates, which may resist chlorination, the wastewater should be continuously and vigorously mixed during treatment (ref. 38, 39). The resulting cyanate is further oxidized to carbon dioxide and nitrogen in the second stage. In this stage, the pH is lowered to approximately 8.5 and additional hypochlorite is added. An ORP reading of +600 to +800 generally signals a complete reaction. The retention time of the second stage is typically 30 to 60 minutes; however, times of 120 minutes are sometimes specified (ref. Delta Pollution Control File).

Although a two-tank system is preferred, complete cyanide oxidation to carbon dioxide and nitrogen can be accomplished in a single-stage unit, provided close pH control is maintained (ref. 39).

When sodium hypochlorite is used, the reaction in the first stage is:

$$NaCN + NaOCl \text{ - -> } NaCNO + NaCl$$

and in the second stage,

$$2NaCNO + 3NaOCl + H_2O \text{ - -> } 3NaCl + N_2 + 2NaHCO_3$$

Sodium hypochlorite consumption is usually estimated to be 25 to 100 percent greater than the stoichiometric requirement (approximately 7 lbs of Cl_2 or 7.5 lbs of NaOCl per lb of CN); where the excess is consumed by oxidation of organics and raising the valences of metals in the wastewater (ref. 39). The results of the Users Survey indicate that most shops are using substantially higher dosages (up to five hundred percent or more). Higher dosages by respondents may be the result of the formation of metal complexes (e.g., due to inadvertently combining cyanide and nickel or iron bearing wastewaters, use of unlined steel tanks, use of steel anode baskets and not retrieving fallen parts from tank bottoms) and/or poor pH control during treatment. The latter reason is especially apparent for several shops that operate single stage cyanide destruction processes and do not add acid to lower the pH. As a result, the second part of the oxidation reaction is slow and operators probably add higher dosages of NaOCl to compensate for the speed of the reaction. Shops operating under these conditions include: PS 058, PS 135, PS 204 and PS 210. It should be noted that acid additions must be made under the correct conditions or a severe safety problem could arise. Dilute acid is sometimes used in place of concentrated acid to reduce the danger of operating this process (ref. 243). Also, some equipment vendors provide a caustic feed capability that is controlled by an independent set-point, usually at a pH of 7.5 (ref. 38). The use of acid in the alkaline chlorination process is discussed by Roy (ref. 38).

Alkaline chlorination systems have generally proven reliable if well maintained. Use of a well-designed ORP control system is highly recommended. Most problems with the system focus on failures of this element. Exhibit 7-7 shows the response of various electrodes to the cyanide-to-cyanate reaction end point. The graph shows that the gold-plated electrode, although more expensive, gives much better reagent addition control (ref. 39).

There are several alternatives to the alkaline chlorination process. These are discussed in the following paragraphs.

Ozone oxidation of cyanide has been practiced as a substitute technology for alkaline chlorination. It is effective in destroying cyanide to the levels required by EPA. The advantage of using ozone lies in reduced operating costs. Ozone is generated on-site (typically by the silent electrical discharge method) and is less expensive than chlorine or sodium hypochlorite. The equipment cost is significantly higher, however, owing to the expense of an ozone generator. Another disadvantage of the process is that it

Exhibit 7-7. Response Curves of Available Electrodes for Cyanide to Cyanate Oxidation

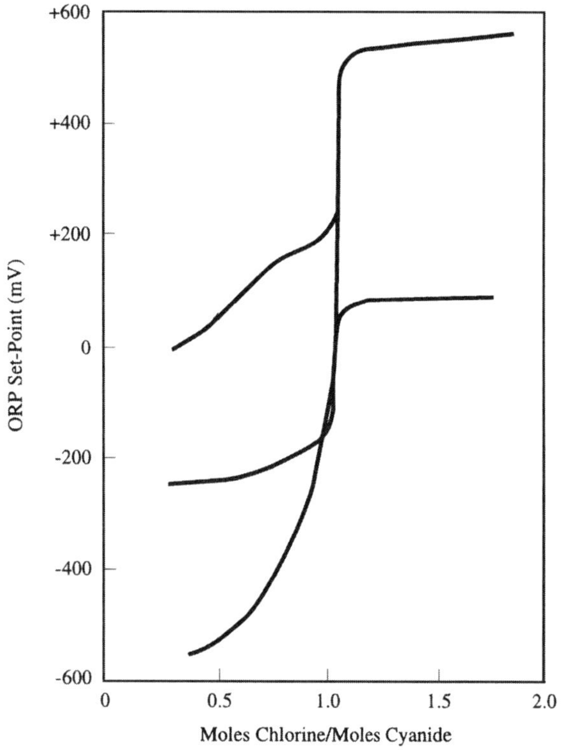

Source: ref. 38

does not oxidize cyanide past the cyanate stage, unless excess ozone is used. Ozone oxidation requires 1.8 to 2.0 lbs of ozone per pound of CN to reach the cyanate stage and 4.6 to 5.0 lbs to reach complete oxidation. An advantage of ozone oxidation is the absence of chlorine that can combine with organics present in the wastewater to produce toxic compounds (ref. 39). Another advantage is the ability of ozone to destroy zinc, copper and nickel cyanide complexes (ref. 243). The ozonation process has also been combined with UV radiation for the treatment of halogenated organics (ref. 348, 386). None of the respondents to the Users Survey employ ozone oxidation of cyanides.

Other alternatives to alkaline chlorination include: alternative chemistries; electrochemical oxidation; thermal oxidation; and precipitation.

Alternative chemistries include hydrogen peroxide (PS 020 and PS 273) (ref. 386) and calcium hypochlorite (PS 174).

Electrochemical oxidation is sometimes used to destroy concentrated cyanide wastes (e.g., >50,000 mg/l CN). A description of the required equipment, including a small commercial unit, is presented in AESF literature (PS 243).

Thermal treatment of cyanide wastes is performed by two of the respondents to the Users Survey (PS 142 and PS 245). Their equipment is provided on a rental basis by Cyanide Destruct Systems (Canada). The equipment can be purchased for $40,000 or rented for $600 per week. PS 142 also indicated that they required a facility modification/ancillary equipment purchase costing of $700. The equipment is intended for concentrated cyanide wastes (at both shops dilute CN wastes are treated by alkaline chlorination). The rental unit is essentially a heated pressure chamber (35 gal, 450 to 470°F, 600 psi) with an agitator. PS 142 used the unit to treat cyanide wastes containing 100,000 mg/l CN to approximately 25 mg/l. Disposal costs for these wastes were previously $800 to $1,000 per 55-gal drum. In operation, CN wastes are transferred into the vessel and treated on a batch basis for 10 hours and then discharged to conventional treatment. Ammonia gas generated by the unit is vented to the atmosphere. PS 142 reported operating costs, excluding labor, of $4,000 per year and a labor requirement of 750 man-hours. They also indicated that they experienced temperature and pressure control problems due to a seal leak (N_2 and oil seal) at the point of agitator entry into the vessel. PS 142 indicated that the unit had a downtime of 50% and that they have discontinued its use.

Chemical precipitation of cyanide can be achieved using ferrous sulfate. This process precipitates the cyanide as a ferrocyanide, which can be removed in a subsequent sedimentation process (ref. 386). Results of EPA sampling of such a process showed that an average influent concentration of 2.7 mg/l CN was treated to 0.023 mg/l CN.

7.2.4 Metals Removal

Hydroxide precipitation is the standard method of removing heavy metals from wastewater. This is achieved by adjusting the pH of the wastewater with an alkaline reagent to reduce the solubility of the dissolved metals and settling and removing the resultant metal hydroxide precipitates. Seventy-five percent of the respondents to the Users Survey employ this process. The metals removal process is often divided into four steps: (1) pretreatment; (2) precipitation; (3) flocculation; and (4) settling. Each of these steps will be discussed in this subsection, with reference to the methods and chemicals employed by survey respondents.

7.2.4.1 Pretreatment

Prior to metals precipitation, many shops use "pretreatment" processes to affect substances that require special treatment or interfere with subsequent treatments. Common metal complexing compounds which will cause interference are listed in Exhibit 7-8. Some shops combine this step with the precipitation and/or flocculation steps. Exhibit 7-9 (PS 064, PS 116, PS 287) shows several configurations for pretreatment or co-treatment using various chemicals. Approximately 45% of the survey respondents use some form of precipitation aid (not including polymer, which is employed by nearly every shop to aid in flocculation). Chemicals used by survey respondents prior to or during metals precipitation include ferrous sulfate, sodium hydrosulfite, aluminum sulfate, soda ash and sodium dithiocarbamate (DTC). Ferrous sulfate is sometimes added to function as a reducing agent, but more often to provide co-precipitation for the removal of metals from chelated wastewaters. With the latter application, complexed metals are co-precipitated with ferric hydroxides. While being an effective metals removal method, this process generates approximately four times as much sludge as sodium hydroxide alone (ref. 144). Calcium chloride and aluminum sulfate are typically used during a pretreatment step (they are often used together[1] because the combined results are better than for either one used alone) for the removal of fluoride, phosphate, silicates and/or emulsified oil (ref. 38).

7.2.4.2 Precipitation

Precipitation is the process by which dissolved metals are made insoluble, usually as metal hydroxides. With metal finishing wastewaters this is brought about by the addition

Exhibit 7-8. Common Complexing Agents Found in Electroplating Wastewaters

Ammonia
Ammonium Chloride
Ammonium Hydroxide
Ammonium Bifluoride
Acetylacetone
Citric Acid
Chromotropic Acid (DNS)
Cyanide
DTPA
Dipyndyl
Disulfopyrocatechol (PDS)
Dimethylglyoxime
Disalicylaldehyde 1,2-propylenediimine
Dimercaptopropanol (BAL)
Dithizone
Diethyl Dithiophosphoric Acid
Ethylenediaminetetraacetic Acid (EDTA)
Ethylenebis (hydroxyphenylglycine) (EHPG)
Ethylenediamine
Ethylenediaminetetra (methylenephosphoric Acid) (EDTPO)
Glycenc Acid
Glycolic Acid
Gluconic Acid
Hydroxyethylethylenediaminetriacetic Acid (HEDTA)
Hydroxyethylidenediphosphonic Acid (EDEDP)
HEDDA
Lactic Acid
Malic Acid
Monosodium Phosphate
Nitrilotriacetic Acid (NTA)
N-Dihydroxyethylglycine
Nitrilotrimethylenephosphonic Acid (NTPO, ATMP)
N-Hydroxyethylethylenediamine
O-phenanthroline
Oxine, 8-Hydroxyquinoline (Q)
Oxinesulphonic Acid
Phthalocyanine
Potassium Ethyl Xanthate
Phosphoric Acid
Polyethyleneimine (PEI)
Polymethacryloylacetone
Poly (p-vinylbenzyliminodiacetic Acid)
Rochelle Salts
Sodium Gluconate
Sodium Pyrophosphate
Succinic Acid
Sodium Tripolyphosphate
Sulphosalicylic Acid (SSA)
Salicylaldehyde
Salicylaldoxime
Sodium Hydroxyacetate
Sodium Citrate
Sodium Fluonde
Sodium Malate
Sodium Amino Acetate
Tartaric Acid
Trisodium Phosphate (TSP)
Trifluoroacetylacetone
Triethylenetetramine
Triaminotriethylamine
Triethanolamine (TEA)
Toluene Dithiol
Thioglycolic Acid
Thiourea

Source: ref. 386

Exhibit 7-9. Wastewater Treatment Configurations Used by Several Respondents to the Users Survey

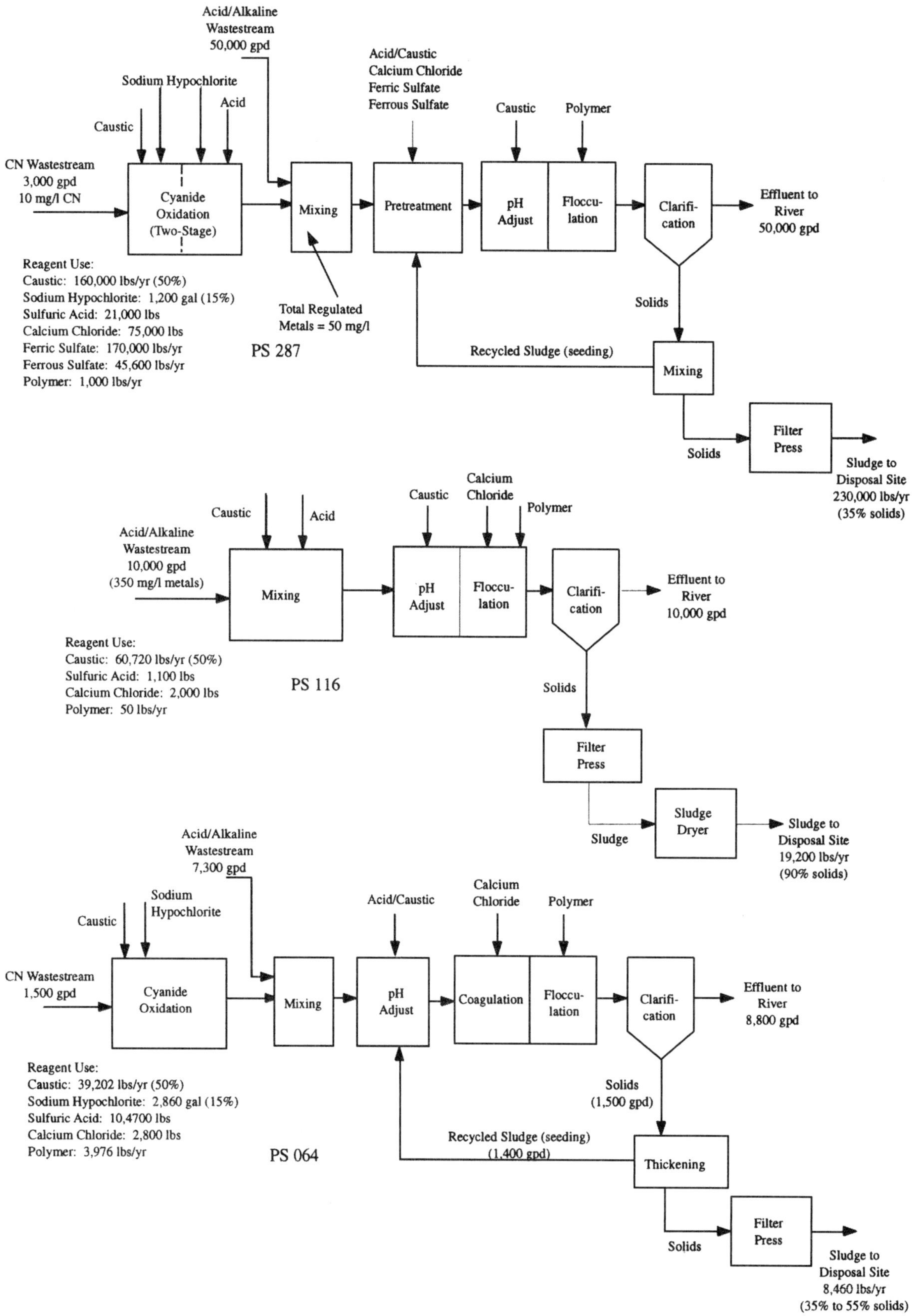

of alkali treatment reagents. Two alkalis are most commonly used for hydroxide precipitation: sodium hydroxide (i.e., NaOH, or, as it is commonly referred to, caustic soda or simply caustic) and lime (i.e., calcium hydroxide or hydrated lime, $Ca(OH)_2$). Of the survey respondents that have metals precipitation processes, 84% use caustic only, 5% use lime only and 9% use a combination of caustic and lime. Several other treatment reagents are used as either the sole alkali ingredient or as an adjunct to caustic or lime. The other primary alkalis are magnesium hydroxide (used by 4.3% of all respondents that perform metals precipitation, but not as the sole source of alkali) and calcium chloride (used by 11.9% of respondents that perform metals precipitation, but only 2% use it as their sole source of alkali). Other alkalis used by metal finishers are sodium carbonate and sodium bicarbonate (ref. 38). When magnesium hydroxide and calcium chloride are used in conjunction with either caustic or lime, this is sometimes done as a pretreatment step prior to metals precipitation, as discussed previously.

Each alkali has advantages and disadvantages. For example, of the two most common alkalis, hydrated lime has the advantage over caustic soda of lower cost per unit of neutralizing capacity. Also, the metal hydroxide precipitants produced with the use of lime have much faster settling rates because of co-precipitation of calcium solids. Further, the settled sludge from lime treatment is higher in solids content and much more amenable to dewatering. On the other hand, lime takes longer to react in the neutralizer than caustic soda, has a more complicated feed system and, most significantly, it generates a considerably higher mass of sludge solids (ref. 39). Sulfide precipitation, which precipitates metals as sulfides instead of hydroxides, has been found capable of achieving low levels of metal solubility in highly chelated wastestreams. However, sulfide precipitation is not used by any of the respondents to the Users Survey. This process has been proven as an alternative to hydroxide precipitation or as a method for further reducing the dissolved metal concentration in the effluent from a hydroxide precipitation system. Two processes are used

Exhibit 7-10. Solubility of Metal Hydroxides as a Function of pH

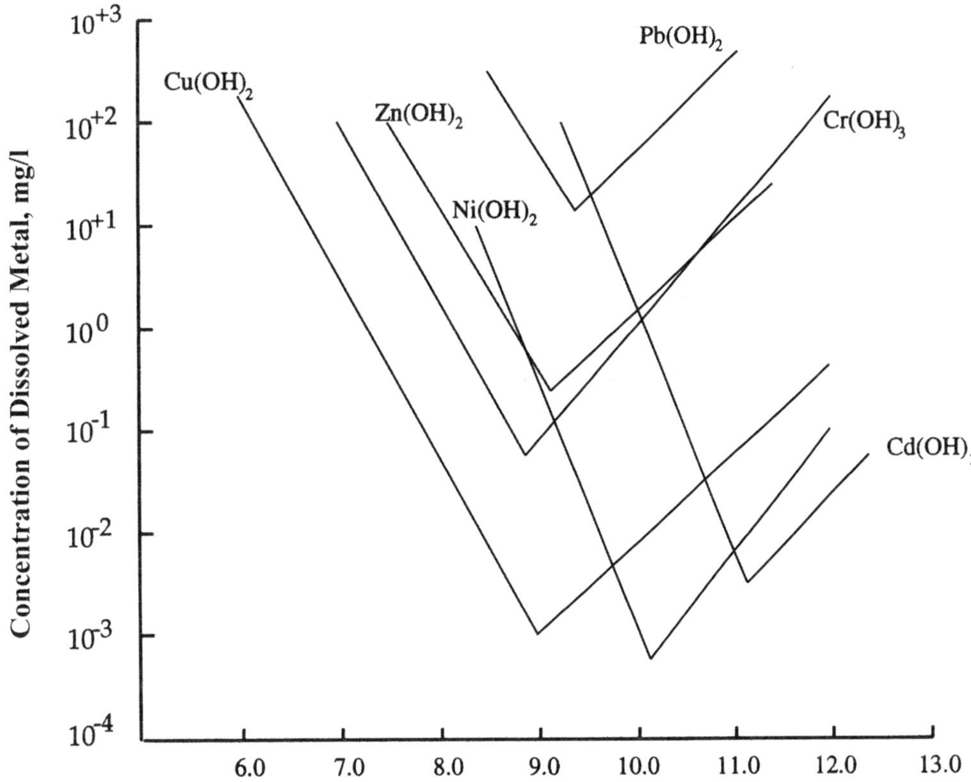

Exhibit 7-11. Process Schematic of a Single Stage Neutralizer

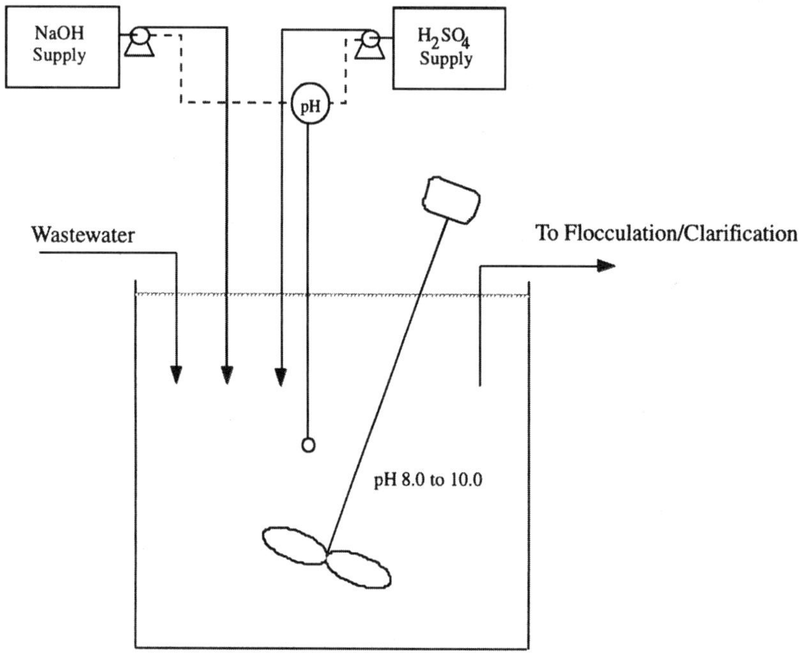

for sulfide precipitation: the soluble sulfide process uses sodium sulfide as the treatment reagent, and the insoluble sulfide process uses ferrous sulfide. Both processes generate metal sulfide sludge and the sludge may be less amenable to off-site recovery than hydroxide sludge. An additional drawback of the insoluble sulfide precipitation process is that it generates a significantly larger volume of sludge compared with conventional hydroxide treatment. The large sludge volume is caused by the liberation of ferrous ions from the treatment reagent. These ions are converted to ferrous hydroxide and add to the sludge volume (ref. 39). Additional information and data on treatment reagents can be found in the literature (e.g., ref. 38, 39, 348, and 409).

The metals precipitation process begins with pH adjustment, as previously shown in Exhibit 7-1. The pH is usually adjusted to between 8.5 and 10.0, with 9.2 being the most frequent target. The optimal pH for precipitating metals from a wastestream will depend mostly on the species of regulated metals present. Each metal hydroxide has a characteristic solubility that is dependent on pH. Exhibit 7-10 shows the solubility curves for the hydroxides of commonly regulated metals (metal sulfides have different curves with minimum solubilities at much lower concentrations than hydroxides, ref. 348). The lowest point of each curve corresponds to the pH at which that metal species will be removed to its minimum solubility point. For example, copper will be removed to its minimum concentration if the pH of the precipitation step is held at 9.0. Since the low points of the curves occur at different pHs, it is necessary to make a compromise when selecting a target pH (pH control set-point). Often, the pH control set-point selection process is based on the need to reach a low concentration for a particular metal species (e.g., cadmium) that is regulated to a lower concentration than other species.

Exhibit 7-11 shows the process schematic of a single-stage neutralizer with provisions for adding both acid and base. Since most combined waste streams are acidic when entering the metals removal step, many facilities do not provide for the addition of acid. However, it is useful in instances when caustic has been inadvertently added in excess and therefore is present in most system designs. With a single-stage system, proportional control for reagent addition is required to maintain a reasonably constant effluent pH. The elevation of the reagent storage and gravity feeding through a proportioning valve is one mode of control. Use of variable speed pumps will achieve the same level of control. If incoming wastes are subjected to wide swings in reagent demand, a two-stage neutralization system should be used. The first stage would control at a pH of approximately 6 to 7 (ref. 39).

Required residence volume in the neutralizer depends on the reagent used. Assuming good mixing, a minimum of 15 minutes residence volume is required for sodium hydroxide; with lime as the alkali, a minimum of 30 minutes is required (ref. 39).

The level of residual dissolved metals after pH adjustment depends on the pH control set-point, the mixture of metals in the wastewater, and whether any compounds are present that interfere with metal hydroxide precipitation.

7.2.4.3 Flocculation

Nearly every metals removal system uses chemical aids to foster particle growth before the wastewater enters the clarifier. By adding coagulating/flocculating agents in slow-mix reactors, the solids in a wastewater can be agglomerated into sturdy, fast-settling particles easily separated in the clarifier. Coagulants/flocculants in broad commercial use include inorganic chemicals such as alum and ferrous sulfate (previously discussed as pretreatments and co-precipitants) and a highly diverse range of organic polyelectrolytes with varying characteristics suitable for different wastewaters. Organic polyelectrolytes are superior to inorganic compounds because both the charge density and valence have been synthetically introduced to the large polymer molecule. The length of the polymer also allows the particles to "knit" together. Normally the appropriate polyelectrolyte is selected by testing a range of different polymers and observing settling behavior after mixing.

7.2.4.4 Clarification

Removal of solids by gravity settling (clarification) is the most common method of separating insoluble particles from a waste stream before discharge. Clarification is a relatively simple process that relies on a density difference between the particles and water and the presence of gravity. However, it is often the unit operation of the waste treatment process most subject to upsets. With effluent limits placing strict control on the level of suspended solids in the wastewater, many modifications to the original circular clarifier have resulted from research and development activities. The two most successful approaches include the sludge blanket clarifier and the plate settler. In the sludge blanket unit, the clarifer inlet first passes through a sludge blanket of agglomerated particles. The mixing tends to promote particle growth and reduce the concentration of slow-settling particle fines. The plate settler (Exhibit 7-12) relies on a series of inclined plates between which the wastewater flows in an upward direction. In essence, the particles must only settle a few inches before impinging on the plate surface. The particles then slide down the plate surface to the base of the separator. In a chamber of equal size, plate settlers can provide considerably greater effective settling volume than a conventional clarifier.

Performance of clarifiers varies significantly depending on the type of waste, the design of upstream components, and the design of the clarifier itself. According to EPA, a "properly operating system" has a total suspended solids (TSS) concentration of 50 mg/l or less and the mean concentration of such systems is 16.8 mg/l TSS (ref. 386). A well-operated clarifier will have, at best, 5 to 10 mg/l of suspended solids in the overflow (ref. 477). Frequently, turbidity in the overflow contributes to the metals content of the discharge and makes strict pollutant guidelines difficult to meet. Consequently, many advanced waste treatment systems employ a polishing filter that uses a sand bed or mixed media filter to remove suspended solids not effectively removed by clarification (ref. 39). Another option is the use of membrane filtration, which either supplements or replaces the clarification step (see Section 7.4.4).

Granular media polishing filters are used for the removal of suspended solids in the 5 to 50 mg/l range where an effluent of less than 1 Jackson Turbidity Unit[2] (JTU) is desired (ref. 362). These units typically contain graded sand or multimedia such as garnet, ground anthracite and silica sand. The principal solids removal mechanism of these units is straining. When a single medium such as sand is employed, it will classify in the filtration tank according to size, with the smallest grains at the top. When water flows downward through the sand, which was common with older designs, the solids would form a mat on the surface and filtration only occurred in the top few inches. The sand bed was then cleaned by an upward washing of the bed with water or with water and air (ref. 362). A patented alternative design permits upward feed flow and a continuous backwash of the sand bed as described in Exhibit 7-13.

Multimedia filters employ two or more filter media with different grain size and densities. The media are selected such that the smaller particles are the most dense (e.g., garnet with a specific gravity of 4.5 and particle size of 0.2 to 0.4 mm), the medium-sized particles have an average density (e.g., sand grains with a specific gravity of 2.65 and a grain size of 0.5 mm) and the largest particles are the least dense (e.g., anthracite grains with a specific gravity of 1.6 and a grain size of 1.0 mm). When mixed and permitted to settle, the multimedia bed will grade

Exhibit 7-12. Diagram and Description of a Plate Settler (Courtesy of Parkson Corporation)

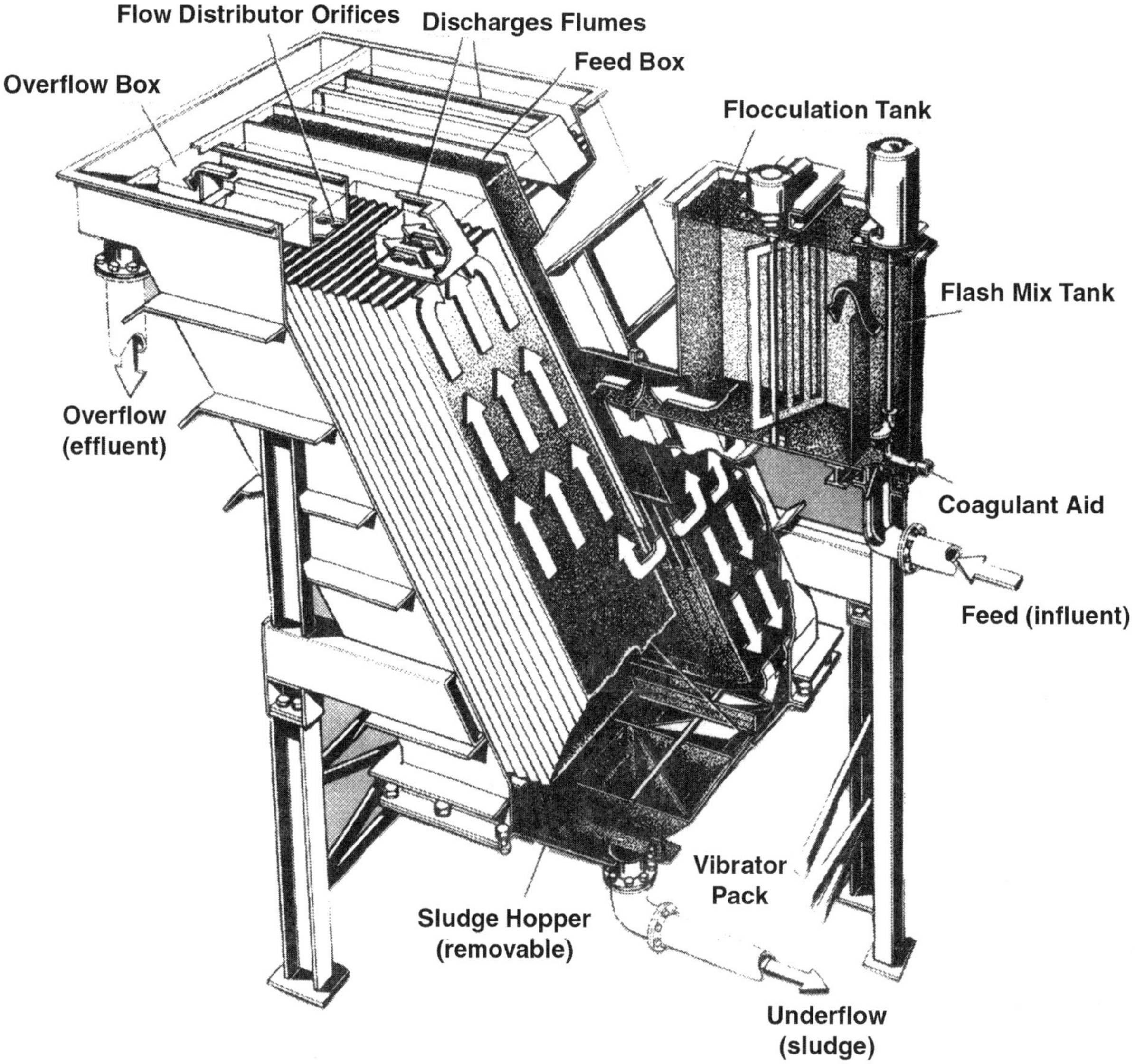

The feed stream is introduced into the Lamella® Gravity Settler by means of a feed duct to the Lamella® Gravity Settler feed box which is a bottomless channel between plate sections. The flow is then directed downward toward the individual, side-entry plate slots. The feed is distributed across the width of the plates and then flows upward under laminar flow conditions. The solids settle on the plates while the supernatant exits from the plates through orifice holes. These holes are placed immediately above each plate and are sized to induce a calculated pressure drop to ensure that the feed is hydraulically distributed equally among the plates. The solids, meanwhile, slide down the plates into the sludge hopper. Further thickening of the sludge is achieved in the hopper due to compression in the quiescent zone made possible by feeding the plates from the side rather than from the bottom. In some applications, a low amplitude vibrator pack can enhance thickening performance.

In applications where coagulants and/or flocculating aids are used, flash mix and/or flocculation tanks can be supplied; they are mounted adjacent to the feed duct to avoid the possibility of floc breakup.

Total effective settling area is based on projected area of each plate. The plates are proportioned to clarification and thickening area by adjusting the plate feed point. This ratio of clarification-to-thickening area is determined from laboratory settling tests on representative feed samples or, in cases where samples are not available, from past experience with similar applications. Plate spacing is 2".

Lamella® Glavity Settler and Dynasand® Filter are registered trademarks of Parkson Corporation

Exhibit 7-13. Diagram and Description of a Polishing Filter (Courtesy of Parkson Corporation)

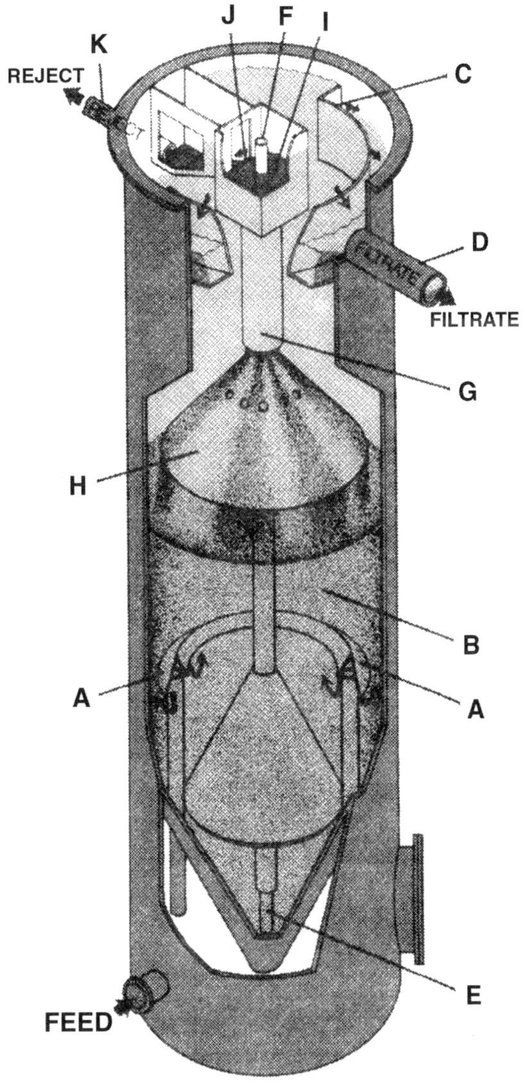

The DynaSand® Filter is a continuous back-wash, upflow, deep bed granular media filter. The filter media is continuously cleaned by recycling the sand internally through an air lift pipe and sand washer. The regenerated sand is redistributed on top of the sand bed allowing for a continuous uninterrupted flow of filtrate and reject water.

Feed is introduced into the bottom of the filter, then flows upward through a series of riser tubes and is evenly distributed into the sand bed through the open bottom of an inlet distribution hood (A). The influent flows upward through the downward moving sand bed (B) with the solids being removed. The clean filtrate exits from the sand bed, overflows a weir (C) and is discharged from the filter (D). Simultaneously the sand bed, along with the accumulated solids, is drawn downward into the suction of an air-lift pipe which is positioned in the center of the filter. A small volume of compressed air is introduced into the bottom of the airlift (E). The sand, dirt, and water are transported upward through the pipe at a rate of about 200 gpm/ft2. The impurities are scoured loose from the sand during this violently turbulent upward flow. Upon reaching the top of the airlift (F), the dirty slurry spills over into the central reject compartment (1). The sand is returned to the sand bed through the gravity washer/ separator (G) which allows the fast settling sand to penetrate, but not the dirty liquid. The washer/separator is placed concentrically around the upper part of the airlift and consists of several stages to prevent short circuiting. By setting the filtrate weir (C) above the reject weir (J) a steady stream of clean filtrate flows upward, countercurrent to the sand, through this washer section and acts as a liquid barrier that carries away the dirt and reject water (K). Since the sand has a higher settling velocity than the dirt particles, it is not carried out of the filter. The sand is redistributed by the means of a sand distribution cone (H). The sand bed is continuously cleaned while both a continuous filtrate and a continuous reject stream are produced.

Lamella® Gravity Settler and Dynasand® Filter are registered trademarks of Parkson Corporation.

itself according to the density of the material. Therefore, the smallest particles will locate at the bottom and the largest particles at the top. When the feed stream flows from top to bottom, the courser suspended solids will be removed in the upper layers of the filter and the smaller suspended solids near the bottom. Backwashing is performed in the opposite direction.

The results of the Users Survey indicate that 58 respondents (or 19.3%) employ polishing filters.

Some respondents to the Users Survey use alternative methods of removing metal hydroxides rather than using standard clarification. PS 036 and PS 168 pump the entire volume of treated wastewater containing precipitated metals through a filter press (use of the filter press is normally reserved for settled sludge). Both of these shops have wastewater flow rates below 3,000 gpd, which is the reason they are able to use this method of separation. As indicated by Roy (ref. 38), when direct filtration is used, the gelatinous character and the shear bulk of solids results in premature blinding or blockage of the filter, causing minimal hydraulic throughput. One respondent to the survey uses dissolved air flotation (DAF) rather than settling (PS 240). A DAF unit generates fine air bubbles that attach to the hydroxide particles and cause them to rise to the surface of a tank, where they are removed by skimming. It is a treatment technology commonly used by industries with oily waste-streams (e.g., petroleum refining and industrial laundries). Several shops use membrane filtration to separate precipitated metals. This method of treatment is a viable alternative to conventional clarification for plating shops, but currently is more common to the printed circuit board industry. Microfiltration is discussed in Section 7.5.3.

7.2.4.5 Metals Removal Process Limitations

As a summary for the conventional metals removal subsection, the following limitations of the hydroxide precipitation/clarification process are presented:

- The process cannot precipitate metals to low levels of solubility in the presence of chelating compounds unless additional treatment reagents are employed. The use of additional treatment reagents may significantly add to the volume of sludge generated by the process.

- The precipitation/clarification process cannot remove suspended solids below approximately 5 to 10 mg/l TSS and TSS can range up to 50 mg/l for a well operated system unless polishing filtration is employed. Because the suspended solids remaining in the effluent contain metals, this

Exhibit 7-14. Diagrams of Plate and Frame Filter Press Components (Courtesy of JWI, Inc.)

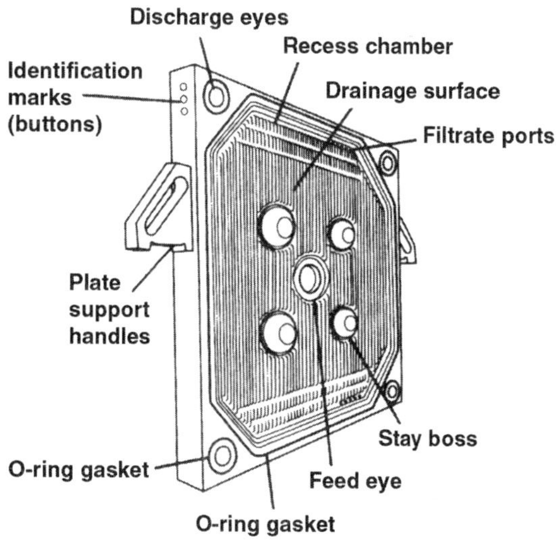

Typical Molded Recessed Chamber Plate

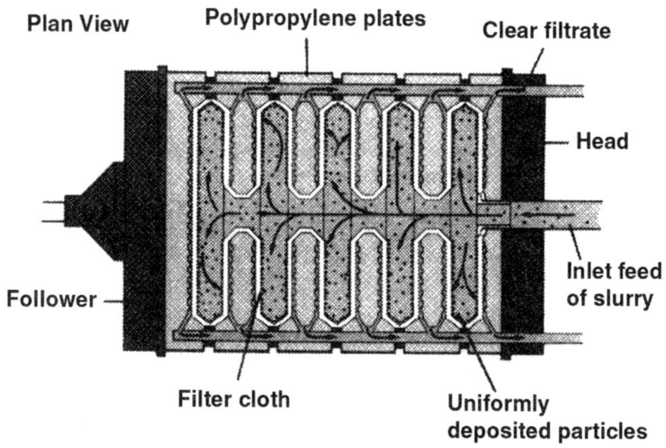

Cross Section of Recessed Plate Filter Press Chamber Area During Fill Cycle factor limits the attainable effluent concentration for metals.

- Certain metal hydroxides are amphoteric; that is, solubility is at a minimum at a specified pH and increases if pH is higher or lower. The solubility minimum for different metals occurs at different pH levels (ref. 39). Therefore, a compromise is necessary with mixed metal wastes that limits attainable effluent concentrations.

- The metal hydroxide sludge resulting from treatment of electroplating wastewater has been classified as a hazardous waste (F006) and must be managed, handled, transported, and recovered or disposed of under the provisions of RCRA.

7.2.5 Sludge Dewatering

Dewatering is primarily a physical operation that separates the liquid and solid portions of the dilute sludge generated during the precipitation/clarification process. The resulting material is a stiff fluid. Dewatering is usually performed in a series of steps utilizing two or more pieces of equipment, with each subsequent step reducing the percentage of liquid in the sludge as previously shown in Exhibit 7-1.

In the treatment of metal-bearing wastes, the dewatering process begins once the dilute sludge is removed from the clarifier. The sludge from the clarifier of most precipitation processes will have a solids concentration of 0.5 to 3 percent (ref. 39). Further sludge concentration is first accomplished by the use of thickening equipment which will increase the solids content to between 2 and 5 percent. Sludges are thickened primarily to decrease the capital and operating costs of subsequent sludge processing steps by substantially reducing the volume. There are three general methods of thickening:

- Gravity Thickening

- Flotation Thickening

- Centrifugal Thickening

Gravity thickening is by far the most commonly used method in the metal finishing industry. It can be achieved in a separate tank or within the clarifier, if it is so designed (ref. 331). Thickening within the clarifier is achieved at the lowest part of the clarifier within a sludge storage zone or hopper. Within the hopper, the sludge is slowly mixed with a motorized rack to enhance the release of water. Often, a secondary tank is used to supplement or replace the thickening zone of the clarifier, especially when extended thickening times are required. These units are typically designed to store the accumulated solids for at least 24 hours. Chemicals are sometimes added to aid the thickening process (e.g., iron and aluminum salts, polymers).

The main design variables to be considered in selecting a thickening process are:

- Solids concentration and volumetric flow rate of the feed system

- Chemical demand and cost if chemicals are employed

- Suspended and dissolved solids concentrations and volumetric flow rate of the clarified stream

- Solids concentration and volumetric flow rate of the thickened sludge.

Other variables that impact the selection of a thickening process are: subsequent processing steps; operation and maintenance (O/M) cost; and the reliability required to meet successful operational requirements. Detailed selection, design and operating parameters for thickening devices can be found in the literature (e.g., ref. 331).

After thickening, further dewatering of the sludge is achieved through the use of mechanical dewatering and thermal dehydration devices. The most widely used mechanical dewatering device is the filter press. Other technologies employed in the metal finishing industry include vacuum filters (used by only two respondents), centrifuges and belt presses (latter two not presently used by any of the respondents). Mechanical dewatering will produce a sludge with an approximately 10 to 60 percent solids content. Sludge dryers (Section 7.2.6) are used to further remove moisture from the sludge and are capable of producing a material with a 90 percent solids content. Increasing the solids content of the sludge greatly reduces its volume, which in turn reduces transportation costs and often disposal/recovery costs. The following discussion focuses on the filter press.

The filter press has a number of advantages over other filtration equipment such as vacuum filters and centrifuges. Filter presses can operate well at variable or low feed solids conditions. They can also produce a relatively dry cake because of the high pressure differential they can exert on the sludge. A typical filter press operating at 100 psig will produce a sludge with a solids content of 25 to 60 percent solids, depending on the chemicals used for precipitation. As a comparison, the basket centrifuge produces a sludge with a solids content of 10 to 25 percent and the vacuum filter produces a sludge with 15 to 40 percent solids (ref. 395).

The disadvantages of the filter press include its batch operating cycle, the labor associated with removing the cakes from the press, and the downtime associated with finding and replacing worn or damaged filter cloths.

The original filter press design consisted of alternating plates and frames and these types of units were referred to as the plate-and-frame filter press. The modern technology is the recessed plate filter press (see Exhibit 7-14). The plates (usually constructed of polypropylene) are recessed on each side to form cavities and they are covered with a filter cloth. The two types of presses work in basically the same manner. At the start of a cycle, a hydraulic pump clamps the plates tightly together and a feed pump forces a dilute sludge slurry into the cavities of the plates. The liquid (filtrate) escapes through the filter cloth and grooves molded into the plates and is transported by the pressure of the feed pump to a discharge port. The solids are retained by the cloth and remain in the cavities. This process continues until the cavities are packed with sludge solids. The hydraulic pressure is then released and the plates are separated. The sludge solids or cake is loosened from the cavities and falls into a hopper or drum.

Under normal to ideal conditions, a filter press subjected to a 100 psi pumping pressure will produce a cake with a dryness of approximately 25 to 40 percent solids for caustic soda precipitated metal hydroxides and 35 to 60 percent solids for lime precipitated hydroxides (ref. 410). The level of dryness attained depends heavily on the length of the drying cycle. Altmayer suggests that most shops only achieve 20 to 25% dry solids and that to reach higher dryness levels requires a sludge dehydration unit (see Section 7.2.6) (ref. 482).

Modern recessed plate filter presses can be specified with the following design enhancements (ref. JWI file):

- Lightweight polypropylene plates that exhibit good chemical resistance and provide a long service life.

- Gasketed plates that reduce leakage during the filtration cycle. These replace non-gasketed types where the filter cloth extends beyond the plate to form the seal between plates.

- An air blow-down manifold that is employed at the end of the filtration cycle to drain remaining liquid in the system, thereby improving sludge dryness and aiding in the release of the cake.

- Microprocessor control which permits unattended operation throughout the filtration cycle. Capable of automatically adjusting the feed pressure and deactivating the pump whenever hydraulic pressure falls below preset limits.

- Manual, semi-automatic or automatic plate shifters that are used to separate the plates prior to releasing the sludge cake.

Filter presses are available in a very wide range of capacities. Respondents to the Vendors Survey sell equipment with a capacity range of 0.6 ft^3 to 200 ft^3. A typical operating cycle is from 4 to 8 hr, depending on the dewatering characteristics of the sludge. Units are usually sized based on one or two cycles per day.

7.2.6 Sludge Dehydration

The solids content of the sludge dewatered on a filter press is usually in the range of 25 to 60 percent. This represents an approximately 20 to 1 volume reduction from the original sludge volume discharged from the clarifier/thickener. Additional dewatering can be accomplished with sludge dehydration equipment that can produce a waste material with a solids content of approximately 90%. This represents an approximately 4 to 1 waste volume reduction above that achieved by the filter press. Sludge dehydration equipment is available in both batch and continuous operating modes. Batch units are typically designed with a heated sludge chamber with rotating blades that consistently break up and blend the sludge cake, thereby exposing it to the heat source. Continuous units have a feed hopper and employ an auger or conveyor system that moves a thin layer of sludge through a drying region and discharges it into a hopper. Some continuous units can be located such that the sludge cake discharge from a filter press drops into the feed hopper of the dehydration unit, making the overall dewatering process more automated. System capacities range from less than 1 ft^3/hr to more than 20 ft^3/hr of feed. Various heat sources are used for sludge drying, including: electric, electric infrared, steam and gas. Energy requirements are mostly dependent on the water content of the feed stock and the efficiency of a given unit. Sludge dehydration equipment will require an air exhaust system.

During the past five years, sludge dehydration equipment has been one of the most frequently purchased of all pollution prevention and control devices.

7.2.7 Sludge Generation and Disposal Data from the Users Survey

A summary of sludge generation data from the Users Survey is presented in Exhibit 7-15. Column 1 indicates the plating shop code. Column 2 indicates the type of sludge that is generated (e.g., hydroxide, sulfide). In all cases, shops either generated a hydroxide treatment sludge or no sludge (zero discharge shops and shops that do not plate regulated metals). Examples of the latter case are sulfuric acid anodizing shops that simply adjust the pH of the wastewater and discharge it (e.g., PS 001 and PS 037). The average sludge generation rate for all respondents that reported a sludge quantity is 158,272 lb/yr. Column 4 shows the percent solids of the sludge. The average percent solids for all shops was 54%. Columns 5 and 6 show the unit cost and total annual cost of sludge disposal or recovery for each shop (includes transportation, stabilization, disposal/processing and disposal tax). Columns 7 and 8 show the disposal/recovery site name and the distance between the shop and the site. The average and median distances to disposal/recovery sites are 457 and 300 miles, respectively. On the average, recovery sites are further away from shops than disposal sites. For disposal sites alone, the average and median distance to the sites are 330 and 200 miles, respectively. Additional details on recovery sites are presented in Section 8.

The average unit cost for sludge disposal was $0.53/lb and the median cost was $0.25/lb. Unit disposal costs for shops varied, depending mostly on the quantity of sludge generated. The following range of average unit costs were given for various annual sludge generation rates:

lbs/Yr of Sludge	Unit Cost $/lb
<10,000	$1.34
10,001 to 50,000	$0.51
50,001 to 100,000	$0.43
100,001 to 250,000	$0.25
250,001 to 500,000	$0.22
500,001 to 1,000,000	$0.05

Data showing the chemical constituents of sludges generated by some survey respondents are presented in Section 8 (see Exhibit 8-5).

The following methodology can be used to estimate the sludge generation rate for a particular plating shop. It is based on a model presented in reference 392 that was revised, based on the chemical constituent data collected during the Users Survey (see Exhibit 8-5).

Sludge Generation:

(1) Calculate lbs/yr of dry sludge solids:

First, for each metal in the wastewater, multiply the lbs/yr of metal precipitated (this value can be assumed to be equal to the mass of metal in the influent) by its "solids generation factor," and sum.

Metal	Solids Generation Factor
Cr	1.98
Ni	1.58
Cu	1.53
Cd	1.30
Fe	1.61
Zn	1.52
Al	2.89
Other*	1.77 (avg.)

**see Exhibit 8-5 in Section 8 for identification of "other" metals contained in plating sludges.*

(1a) If the sludge will be dewatered above 50% solids, multiple by:

[(0.029 x % Solids*) - 0.41]

(1b) The "solids generation factors" are based on the use of caustic as an alkali reagent. If lime will be used instead of caustic, multiply by 1.13.
(2) Lbs/yr of wet sludge = 100 (lbs dry solids/% Solids*)

**express as a whole number, e.g., 50% = 50*

Exhibit 7-15. Sludge Generation and Disposal Data from the Users Survey

Shop Code	Sludge Type	Sludge Quantity, lbs/yr	Solids %	Disposal Cost, $/lb	Disposal Cost, $/yr	Disposal/Recovery Site	Distance to Site, mi.
001	NONE						
002	HYDROX	42,000	99	0.25	$10,500		
003	HYDROX	550,000	50	0.19	$101,750		
004	HYDROX	240,000	85	0.09	$21,000		
005	HYDROX	240,000	35	0.08	$18,600		
006	HYDROX	40,000	10	0.06	$2,400		
007	HYDROX	24,000	98	0.19	$4,500	Stabilex	
008	HYDROX	10,000	98	0.10	$1,000		
009	HYDROX	200,000	25	0.13	$25,900	Peoria Disposal	<500
010	HYDROX	26,000	40	0.45	$11,700		
011	HYDROX	66,000	60	0.21	$14,025		
012	HYDROX	40,000	90	0.30	$11,920	Michigan Disp.	501 to 1,000
013	HYDROX	244,000	35	0.23	$55,930	Stabilex	
014	HYDROX	40,000	45	0.26	$10,500	Mich. Disp., Inc.	
015	HYDROX	370,000	20	0.18	$65,000		
016	HYDROX	200,000	5	0.04	$8,000	World Resources	501 to 1,000
017	HYDROX	19,250	40	0.55	$10,560	Michigan Disp.	<500
018	HYDROX	14,000	75	0.27	$3,834	World Resources	<500
019	HYDROX	5,000	30	1.10	$5,500	Kettleman Hills	<500
020	HYDROX	36,000	80	0.39	$14,175	World Resources	<500
021	HYDROX	1,107	20	6.14	$6,800	Envirite Corp.	<500
022	HYDROX	520,000	15	0.05	$26,534	Envirite Corp.	
023	HYDROX	32,000	78	0.42	$13,460	World Resources	501 to 1,000
024	HYDROX	164,000		0.03	$4,822	Envirite	<500
025	HYDROX	214,000	78	0.21	$45,380		
026	HYDROX	360,000	61	0.02	$6,276	CID Landfill #2	<500
027	HYDROX	50,000		0.80	$40,000	World Resources	<500
028	ND						
029	HYDROX	200,000	99	0.04	$8,850	Clean Harbors	
030	HYDROX	16,000	70	0.26	$4,190	Chem. Waste Mangt.	<500
031	NONE						
032	HYDROX	120,000	40			Stabilex	
033	HYDROX	40,000	40	0.15	$6,000	Chemwaste	<500
034	HYDROX	15,000	40	0.35	$5,250	Eticam	<500
035	ND						
036	HYDROX	74,000	40	0.11	$8,437	Encycle	1,001 to 1,500
037	NONE						
038	ND						<500
039	ND						<500
040	HYDROX	25,000	60	0.50	$12,500	Encycle	<500
041	HYDROX	100,000	30	0.38	$38,000	Envirite	<500
042	HYDROX	5,000	43	0.20	$1,000		501 to 1,000
043	HYDROX	68,000	80	0.18	$12,000	Horsehead	<500
044	HYDROX	12,000		1.00	$12,000		
045	HYDROX	560,000	70	0.15	$84,000	Peoria Disposal	<500
046	NONE						<500
047	HYDROX	500		0.70	$350	Cyno-Chem	
048	HYDROX	460,000	26	0.11	$50,000	Envirite Corp.	<500
049	HYDROX	14,400		1.01	$14,500	Arlington	
050	HYDROX	106,000	34	0.17	$18,500	Michigan Disp.	
051	HYDROX	462,000	40	0.13	$58,000	Envirite	<500
052	NONE						
053	HYDROX	17,100		0.47	$8,000	Stablex	<500

Shop Code	Sludge Type	Sludge Quantity, lbs/yr	Solids %	Disposal Cost, $/lb	Disposal Cost, $/yr	Disposal/Recovery Site	Distance to Site, mi.
054	HYDROX	120,000	35	0.21	$24,800	Michigan Disp.	<500
055	HYDROX	3,360	40	1.25	$4,200	World Resources	
056	HYDROX	6,000	30	0.50	$3,000	Michigan Disp.	501 to 1,000
057	ND		60				
058	HYDROX	180,000	20	0.21	$36,900	Word Resources	<500
059	HYDROX	12,000	75	0.49	$5,854	Hwy.36 Land Dev	<500
060	HYDROX	380	99	1.97	$750	Landfill	<500
061	HYDROX	16,000	40	0.38	$6,100	Eticam	<500
062	HYDROX	9,500	47	0.60	$5,725	World Resources	501 to 1,000
063	HYDROX	3,600		3.67	$13,200	Enviroserve Co.	1,001 to 1,500
064	HYDROX	10,000	45	0.35	$3,500		
065	HYDROX	606,666	25	0.11	$63,917	Envirite	<500
066	HYDROX	12,000	80	0.83	$10,000	Clean Harbors	<500
067	HYDROX	28,000	90	0.14	$3,800	Clinton Landfil	<500
068	HYDROX	15,000	55	0.96	$14,400	Eticam	501 to 1,000
069	HYDROX	12,000	40	0.50	$6,000	Chem Tech	<500
070	HYDROX	5,000	99	1.88	$9,400	Eticam	501 to 1,000
071	NONE						
072	ND						
073	NONE						
074	HYDROX	26,400	40	0.50	$13,200	Encycle	<500
075	HYDROX	6,000		0.32	$1,920	Inmetco	<500
076	HYDROX	90,000		0.18	$15,868	Encycle	1,001 to 1,500
077	HYDROX	240,000	30	0.17	$40,800	World Resources	<500
078	HYDROX	15,300	38	0.40	$6,165	Encycle	1,001 to 1,500
079	HYDROX	2,000	40	1.25	$2,500	Waste Research	<500
080	NONE						
081	NONE						
082	HYDROX	800,000	30	0.11	$87,000	Peoria Disposal	
083	HYDROX	14,000	60	0.48	$6,675	Stablex	<500
084	HYDROX	1,000,000	55	0.01	$6,000	New Georgia	<500
085	HYDROX	66,040	20	0.13	$8,585	W.R.C.	
086	HYDROX	180,000	90	0.18	$31,500		
087	HYDROX	2,780,000		0.02	$42,728	B.F.I.	<500
088	HYDROX	26,000	99	0.41	$10,725	Confidential	<500
089	HYDROX	260,000		0.14	$36,800	U.S.Filt.Recov.	<500
090	HYDROX	53,456	16	0.17	$9,125	Envirite	<500
091	HYDROX	80,000	45	0.26	$20,800	World Resources	<500
092	HYDROX	70,000	99	0.48	$33,600	W.R.C.	<500
093	HYDROX	112,000	52	0.13	$14,000	Peoria Disposal	<500
094	HYDROX	16,000	60	0.38	$6,000	Michigan Dispos	<500
095	HYDROX	164,000	27	0.16	$25,719	U.S. Ecology	<500
096	HYDROX	240,000	28	0.02	$5,892	B.F.I.	<500
097	HYDROX	40,000	30	0.14	$5,400	Peoria Disposal	<500
098	HYDROX	ND	50		$40,000		
099	HYDROX	102,000	38	0.12	$12,416	Peoria Disposal	<500
100	HYDROX	40,000	70	0.49	$19,600	Encycle	<500
101	HYDROX	323,800	30	0.19	$114,000	Encycle	501 to 1,000
102	HYDROX	600,000	80	0.21	$19,100	Encycle	
103	HYDROX	90,500					
104	HYDROX	8,800	80	0.55	$4,800	Wayne Disposal	<500
105	HYDROX	40,000	62	0.25	$10,000	Encycle	
106	HYDROX	240,000	98	0.06	$13,200	Heritage	<500

Exhibit 7-15 (continued). Sludge Generation and Disposal Data from the Users Survey

Shop Code	Sludge Type	Sludge Quantity, lbs/yr	Solids %	Disposal Cost, $/lb	Disposal Cost, $/yr	Disposal/Recovery Site	Distance to Site, mi.
107	HYDROX	26,000		0.60	$15,600	Eticam	501 to 1,000
108	HYDROX	24,000	40	0.38	$9,000	Envirite	>1,500
109	HYDROX	38,000	90	0.36	$13,680	Alpha Omega	<500
110	HYDROX	54,000	73	0.26	$14,000	Michigan Disp.	<500
111	HYDROX	15,000	40	0.80	$12,000	World Resources	501 to 1,000
112	HYDROX	200,000	40	0.09	$18,000	Envirite	<500
113	HYDROX				$15,000	Last Chance	<500
114	HYDROX	2,100	40	1.81	$3,800	Ashland Chem.Co	501 to 1,000
115	HYDROX	400,000	60	0.17	$67,000	Envirite	<500
116	HYDROX	19,200	99	0.63	$12,144	Heritage	501 to 1,000
117	HYDROX	2,000	60			C.W.M.	1,001 to 1,500
118	HYDROX	228,000	65	0.19	$44,200	Encycle	<500
119	HYDROX	6,000	43	0.30	$1,800	World Resources	1,001 to 1,500
120	HYDROX	7,000		0.36	$2,500		
121	HYDROX	45,000	40	0.20	$8,800	Artek	<500
122	HYDROX	4,560,000	90	0.02	$78,000	B.F.I.	<500
123	HYDROX	380,000	40	5.00	$19,570	Cyprus Miami Mi	<500
124	HYDROX	523,000	52	28.00	$146,440	World Resources	>1,500
125	HYDROX	36,500	20	0.21	$7,665	Inmetco	501 to 1,000
126	HYDROX	10,000	40	1.48	$14,800	Encycle	<500
127	HYDROX	500		1.10	$550	Haz.waste spec.	501 to 1,000
128	HYDROX	520,000	69	0.02	$12,242	Winthrop-Habor	<500
129	NONE						
130	HYDROX	76,727	40	0.36	$27,775	Encycle	501 to 1,000
131	HYDROX	56,000	75	0.19	$10,694		<500
132	HYDROX	960,000	25	0.13	$120,000	Stablex	<500
133	HYDROX	40,000	32	0.34	$13,700	Michigan Disp.	501 to 1,000
134	ND					Amele	
135	HYDROX	18,000	98	0.56	$10,000	Eticam	<500
136	NONE						
137	ND					County Landfill	<500
138	HYDROX	18,000		0.69	$12,420	Encycle	>1,500
139	HYDROX	117,000	22	0.26	$31,000	World Resources	501 to 1,000
140	HYDROX	50,000		0.04	$2,000	Inmetco	501 to 1,000
141	HYDROX	80,000	25	0.02	$1,600	Landfill	<500
142	HYDROX	40,000	72	0.30	$12,000	Encycle	501 to 1,000
143	HYDROX	120,000		0.33	$40,000	Encycle	
144	ND						
145	NONE			0.26	$20,800	Kettleman Hills	<500
146	HYDROX	80,000	60				
147	NONE						
148	ND						
149	NONE						
150	HYDROX	425,000	63	0.24	$100,000		>1,500
151	NONE						
152	ND						
153	HYDROX	27,000	40	1.30	$35,000	Encycle	501 to 1,000
154	NONE						
155	ND						
156	HYDROX	29,000	30	0.26	$7,600	W.R.C.	501 to 1,000
157	ND						
158	HYDROX	80,000		Encycle			
159	NONE						
160	HYDROX	210,000	40	0.14	$29,400	Horsehead Recyc	<500
161	HYDROX	550	75	0.56	$309	Environ.Enterpr	<500
162	HYDROX	100,000	70	0.15	$15,000	World Resources	501 to 1,000
163	HYDROX	3,000	75	5.81	$17,425	U.S.Filt.Recov.	501 to 1,000
164	HYDROX	44,000		0.37	$16,390	World Resources	501 to 1,000
165	HYDROX	42,000	53	0.25	$10,500	Encycle	1,001 to 1,500
166	HYDROX	26,000	90	0.19	$5,000	Chem.Waste Met.	<500
167	NONE						
168	HYDROX	120,000					
169	HYDROX	8,000	23	0.88	$7,000		
170	HYDROX	2,200	80	0.46	$1,020	Sovil	501 to 1,000
171	HYDROX	8,000	80	1.25	$10,000	Federal Envir	1,001 to 1,500
172	HYDROX	74,000	70	0.22	$16,411	Encycle	1,001 to 1,500
173	HYDROX	40,000	35	0.33	$13,000	W.R.C.	<500
174	HYDROX	40,000	80	0.23	$9,000	Envirosafe	1,001 to 1,500
175	HYDROX	25,000	30	0.40	$10,000	World Resources	501 to 1,000
176	HYDROX	600,000	20	0.02	$11,200	Lancaster Cty.	<500
177	HYDROX	16,000	99	0.50	$8,000	World Resources	501 to 1,000
178	HYDROX	12,000	30	0.43	$5,100	Clean Harbors	<500
179	HYDROX	2,000				Behr Metals	<500
180	HYDROX	50,000	80			Chem Waste Mgt .	<500
181	HYDROX	74,000	35	0.31	$22,900	Stanlex	<500
182	ND					World Resources	
183	HYDROX	30,000	50	0.39	$11,600	World Resources	501 to 1,000
184	HYDROX	62,400	30	0.23	$14,300	Grand View	501 to 1,000
185	NONE						
186	HYDROX	2,500	60	0.93	$2,335	Michigan	501 to 1,000
187	HYDROX	16,000	30	0.50	$8,000		501 to 1,000
188	NONE						
189	ND						
190	HYDROX	32,000	70	0.21	$6,800	Michigan Disp.	501 to 1,000
191	HYDROX	43,380	99	0.25	$10,974	World Resources	501 to 1,000
192	NONE						
193	NONE						
194	NONE					Immetco	501 to 1,000
195	NONE						
196	HYDROX	160,000	40	0.10	$15,600	World Resources	<500
197	HYDROX	556,000	99	0.09	$47,370	World Resources	<500
198	NONE						
199	HYDROX	1,040,000	33	0.02	$20,800	Zion	<500
200	NONE						
201	HYDROX	400,000		0.15	$60,000		
202	HYDROX	6,000		5.00	$30,000	Encycle	501 to 1,000
203	ND					Chem.Waste Mgt.	
204	HYDROX	1,152,000	24	0.10	$116,905	Envirite	
205	HYDROX	800	98	3.50	$2,800	Ashland Chemic.	501 to 1,000
206	HYDROX	1,000		11.00	$11,000	World Resources	<500
207	HYDROX	16,000	80	0.21	$3,400	W.R.C.	501 to 1,000
208	HYDROX	200,000	2	0.06	$11,700	Encycle	<500
209	HYDROX	150,000	40	0.27	$41,000	Stablex	1,001 to 1,500
210	HYDROX	86,000	70	0.19	$16,378		
211	NONE						
212	HYDROX	2,115	99	1.02	$2,155	W.R.C.	<500
213	HYDROX	85,000	65	0.16	$13,600	World Resources	<500
214	HYDROX	1,700,000	75	0.10	$175,000	Envirite	<500

Exhibit 7-15 (continued). Sludge Generation and Disposal Data from the Users Survey

Shop Code	Sludge Type	Sludge Quantity, lbs/yr	Solids %	Disposal Cost, $/lb	Disposal Cost, $/yr	Disposal/Recovery Site	Distance to Site, mi.
215	HYDROX	520,000	40	0.03	$13,559	Michigan Disp.	<500
216	HYDROX	70,000		0.05	$3,200	Kettleman Hills	<500
217	ND						
218	HYDROX	280,000	25	0.18	$50,000	Encycle	>1,500
219	HYDROX	154,464	20	0.17	$26,000	Encycle	>1,500
220	HYDROX	48,398	40	0.25	$12,000	Encycle	>1,500
221	HYDROX	5,000	50	1.00	$5,000	Chem . Conservat .	501 to 1,000
222	HYDROX	30,000	40	0.12	$3,500	Peoria Disposal	<500
223	HYDROX	44,000	50	0.78	$34,200	Stablex	<500
224	ND						
225	HYDROX	7,500	60	0.53	$4,000	Envirite	
226	HYDROX	120,000	70	0.23	$27,000	W.R.C.	501 to 1,000
227	HYDROX	300	99	0.50	$150	Haz . Control	<500
228	HYDROX	330,000	25	0.12	$40,000	Envirite	<500
229	NONE					City Environmen	<500
230	ND						
231	HYDROX	25,000	40	0.43	$10,800	Peoria Disposal	<500
233	HYDROX	27,000	35	4.15	$112,000	T . S . D.	
234	HYDROX	40,000	20	0.25	$9,800	G.S.X.	<500
235	HYDROX	80,000	50	O . 60	$48,000	World Resources	<500
237	NONE					M.R.S.	
238	HYDROX	1,500,000	65	0.10	$151,000	Cyprus Miami Mi	<500
239	HYDROX	40,000	95	0.38	$15,000	Remtec	<500
240	HYDROX	20,000	40	0.28	$5,500	N.Shelby Landfi	<500
241	ND						
242	HYDROX	8,000	60	0.50	$4,000	World Resources	501 to 1,000
243	HYDROX	5,000		0.50	$2,500	World Resources	501 to 1,000
244	HYDROX	32,000	80	0.44	$13,920	T . S . D.	<500
245	HYDROX	17,000	90	0.32	$5,440	Encycle	501 to 1,000
246	HYDROX	440,000	40	0.09	$40,000	Peoria Disposal	<500
247	HYDROX	370,000	60	0.11	$41,000	Envirite	<500
248	HYDROX	62,000	35	1.00	$62,000	Encycle	501 to 1,000
249	NONE						
250	HYDROX	32,000	85				
251	HYDROX	120,000		1.80	$216,000	Etichem	501 to 1,000
252	NONE						
253	HYDROX	126,000	50	0.20	$25,060	World Resources	<500
254	HYDROX	408,000	40	0.15	$60,316	Envirite	<500
255	HYDROX	45,000	90	0.30	$13,500	World Resources	<500
256	HYDROX	324,000	70	0.66	$213,000	World Resources	501 to 1,000
257	HYDROX	5,500	30	0.46	$2,520	Encycle	1,001 to 1,500
258	HYDROX	240,000	33	0.09	$22,000	Envirite	<500
259	HYDROX	4,000	10	0.50	$2,000	Norris Indust.	
260	HYDROX	30,000	64	0.10	$2,867	Heritage	<500
261	HYDROX	12,000	40			Michigan Disp.	501 to 1,000
262	HYDROX	2,000	50	0.80	$1,600	Camercon	501 to 1,000
263	HYDROX	440,000	75	0.04	$19,700	Envirite	<500
264	HYDROX	143,400	54	0.13	$18,600	World Resources	<500
265	HYDROX	120,000	15	0.79	$95,000	Kettleman Hills	
266	HYDROX	380,000	83	0.20	$74,195	Encycle	<500
267	HYDROX	4,300	22	0.29	$1,253	Worls Resources	501 to 1,000
268	HYDROX	160,000	50	0.23	$37,000	Horsehead	<500

Shop Code	Sludge Type	Sludge Quantity, lbs/yr	Solids %	Disposal Cost, $/lb	Disposal Cost, $/yr	Disposal/Recovery Site	Distance to Site, mi.
269	ND						
270	NONE						
271	NONE						
272	HYDROX	88,000	60	0.23	$20,000	N.W.Envi.oserv.	<500
273	HYDROX	13,000,000	9	0.05	$630,000	Heritage	<500
274	HYDROX	3,000	35	0.25	$750	C.W.M.Chem.Svc.	<500
275	HYDROX	240,000	85	0.10	$24,000	Tomar Services	>1,500
276	HYDROX	40,000	95			Michigan Disp.	501 to 1,000
277	HYDROX	600,000	20	0.12	$70,000	Envirite	<500
278	HYDROX	200,000	30	0.01	$2,000	Heritage	501 to 1,000
279	HYDROX	100,000	50	0.14	$13,700	Peoria Disposal	<500
280	HYDROX	312,000	35	0.12	$36,600	Heritage	<500
281	HYDROX	81,760	90	0.19	$15,562	Envirite	<500
282	NONE						
283	HYDROX	9,000	45	0.30	$2,677	World Resources	<500
284	HYDROX	1,500	80	2.00	$3,000	Chemical Consv.	<500
285	HYDROX	15,600	70				<500
286	HYDROX	76,000	95	0.28	$21,000	World Resources	<500
287	HYDROX	70,000	35	0.14	$9,625	Envirite	<500
288	HYDROX	125,000	30	0.28	$35,000	Envirite	<500
289	HYDROX	30,000	99			Emelle	<500
290	HYDROX	25,000	40	0.76	$19,000	Emelle	<500
291	HYDROX	87,000					
292	HYDROX	90,000	75	0.24	$22,000	Envirosafe	<500
293	HYDROX	200,000	70	0.15	$30,000	Encycle	>1,500
294	NONE						
295	HYDROX	760,000	50	0.03	$26,000		
296	HYDROX	800,000	15	0.10	$80,000	Envirite	<500
297	NONE						
298	HYDROX	240,000	25	0.15	$35,000	Michigan Disp.	<500
299	NONE						
300	HYDROX	1,350,000	32	0.06	$86,625	Envirite	<500
301	HYDROX	800	85	1.59	$1,275	Eltex Chemical	<500
302	HYDROX	160,000	100			Wayne Disposal	<500
303	HYDROX	530,000	70	0.25	$130,000		
304	ND						
305	HYDROX	120,000					
306	NONE						
307	HYDROX	24,000	60	0.13	$3,000		
308	NONE						
309	HYDROX	44,000	65	0.34	$15,000	Stablex	<500
310	NONE						
311	HYDROX	252,000	8	0.10	$25,000		<500
312	HYDROX	120,000	30			Stablex	501 to 1,000
313	ND						
314	HYDROX	240,000	85	ND	$800		<500
315	ND						
316	ND						
317	NONE						
318	ND						

ND = no data

Exhibit 7-16. Capital Costs for Conventional End-of-Pipe Unit Operations

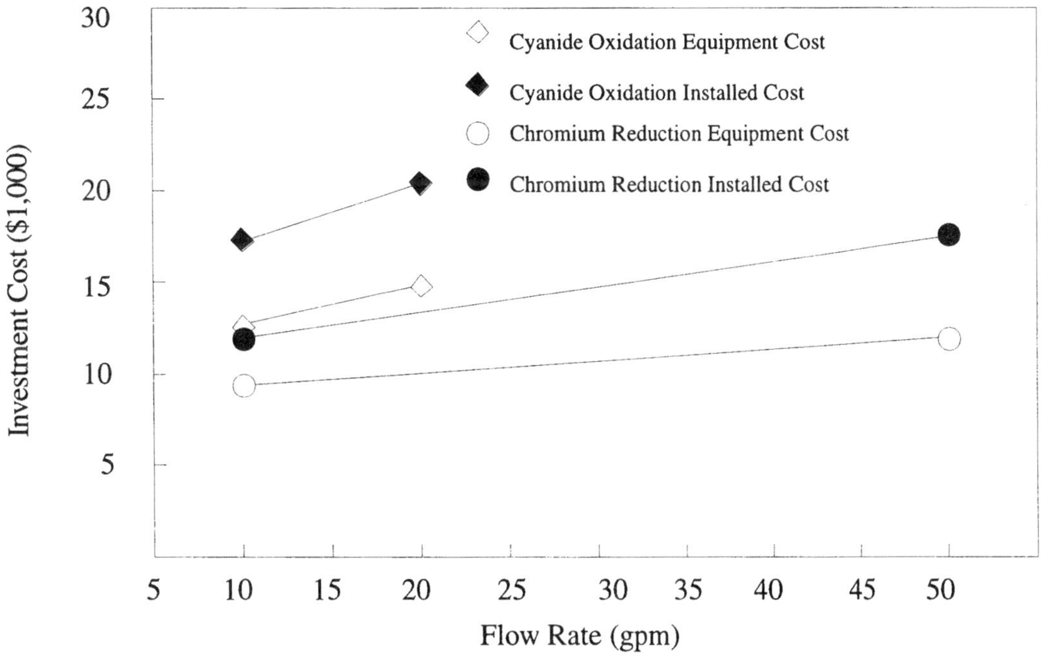

Investment Costs for Cyanide Oxidation and Chromium Reduction

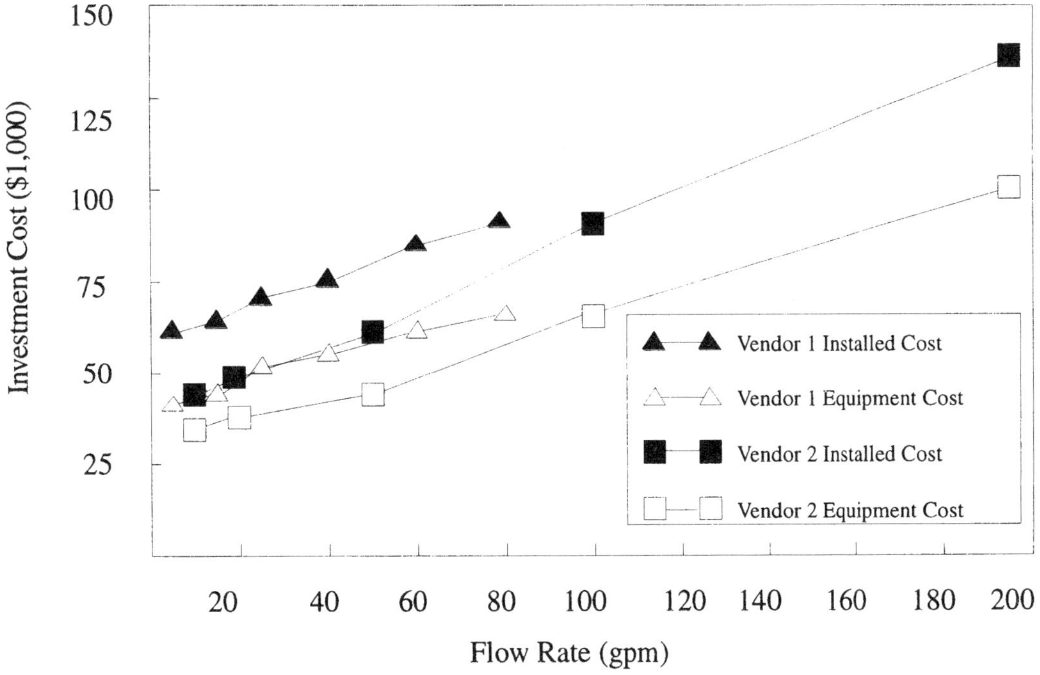

Investment Costs for Precipitation/Clarification Systems

Exhibit 7-17. Capital Costs for Multimedia Polishing Filters

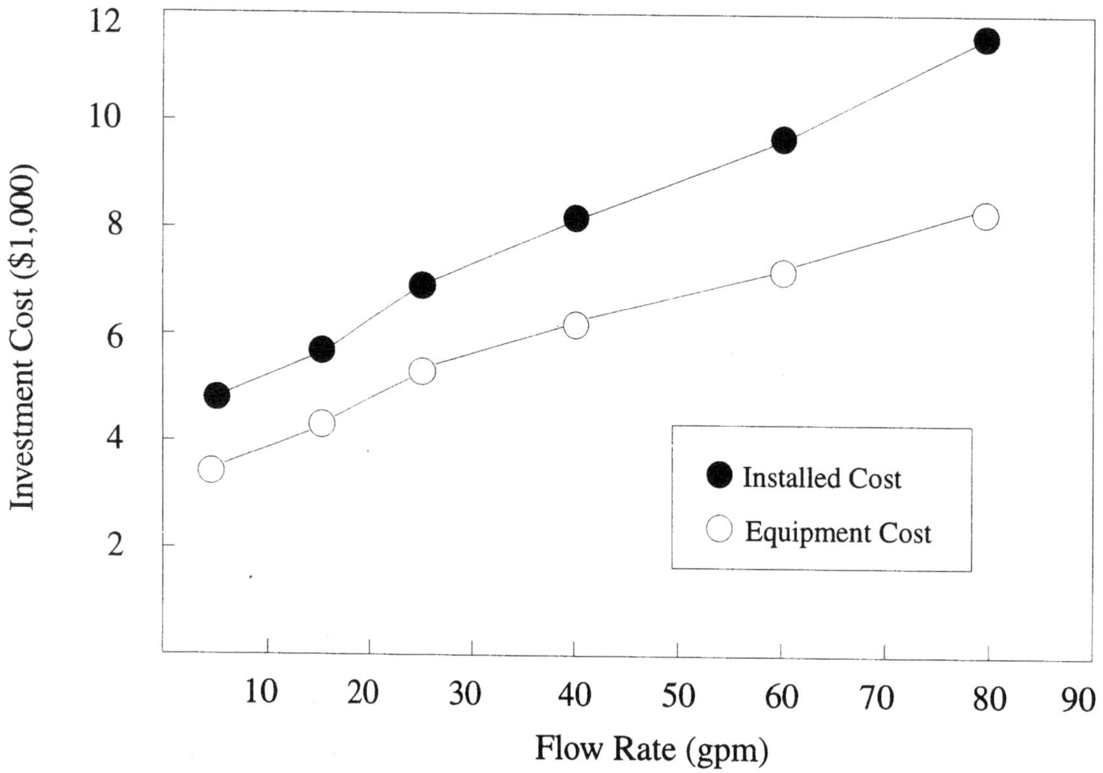

Exhibit 7-18. Filter Press Volume or Daily Sludge Dryer Capacity (cubic feet)

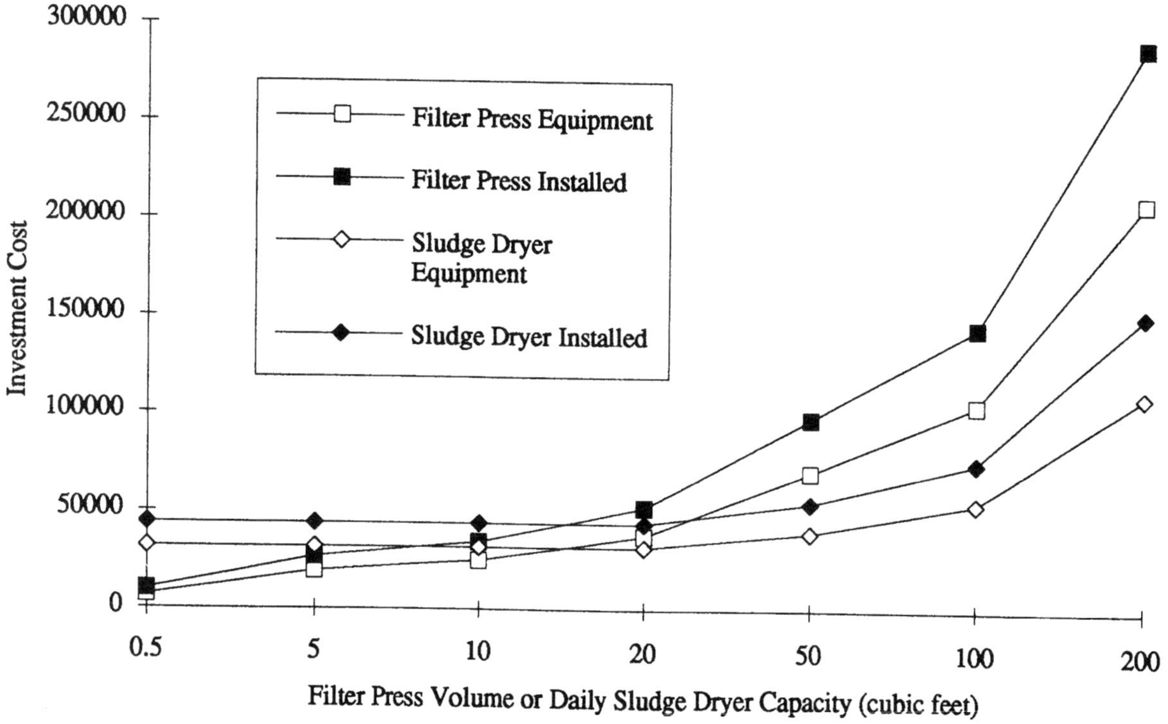

Exhibit 7-19. Operating Costs for Conventional Wastewater Treatment

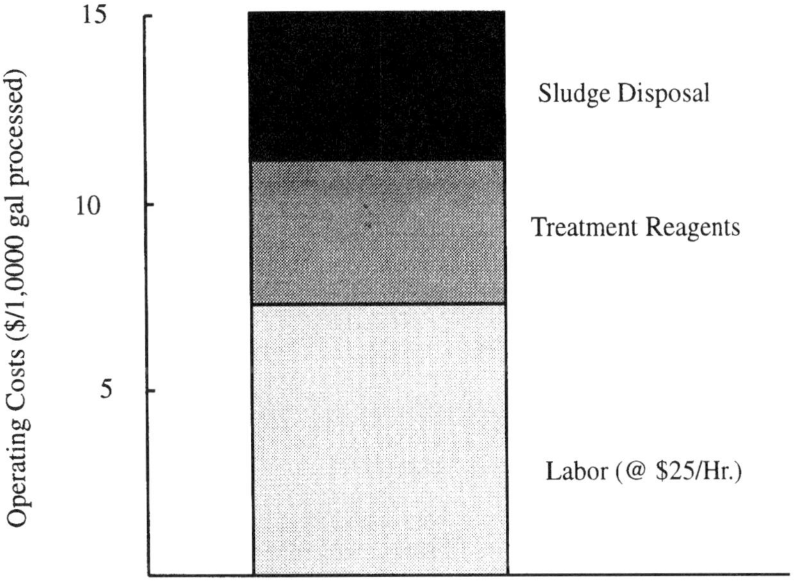

Based on a typical plant responding to the Users Survey with a flow rate of 34,600 gpd and a sludge generation rate of 160,000 lbs/yr (disposal cost of $0.25/lb). Actual operating costs will vary widely depending on flow rates and concentrations of pollutants. Methods for calculating more accurate estimates are presented in Section 7.3.

7.3 COSTS OF CONVENTIONAL WASTEWATER TREATMENT

7.3.1 Capital Costs

Capital costs for conventional end-of-pipe treatment equipment are presented in Exhibits 7-16, 7-17 and 7-18. For each unit operation, both basic equipment costs and installed equipment costs are presented. Installation costs (38% over basic equipment costs) include: engineering (10%), shipping (5%), piping (8%) and electrical/instrumentation (15%). All equipment costs are for packaged, skid mounted systems. The unit operations are based on the following retention times:

- cyanide oxidation first stage: 60 minutes
- cyanide oxidation second stage: 120 minutes
- chromium reduction: 30 minutes
- neutralization: 30 minutes
- precipitation: 30 minutes
- sludge thickening: 24 hours
- polishing filter: 10 minutes

7.3.2 Operating Costs

The major operating cost items for conventional treatment (labor, treatment reagents and sludge disposal) are shown in Exhibit 7-19. This chart is based on the results from the NCMS/NAMF Users Survey, which are discussed in Section 7.4. An individual shop can use the results of the survey and other data presented in this text to accurately estimate their chemical treatment reagent costs, based on the individual wastestream (i.e., Cr, CN, acid/alkaline) flow rates and characteristics.

Treatment reagent prices have risen dramatically over the past 20 years due to inflation and increased manufacturing costs (Exhibit 7-20). During this time period, most reagent costs have increased by a factor of two to five times. The actual reagent unit costs paid by respondents is significantly impacted by the annual quantity of reagent that is purchased. This is due to the packaging costs for small quantities vs the lack of packaging for bulk quantities. Exhibit 7-21 presents average treatment reagent prices paid by respondents, based on usage rates.

The quantity of reagents used by respondents was somewhat unexpected. As discussed in Section 7.4, on the average, respondents use much greater amounts

Exhibit 7-20. Cost Data[a] for Common Wastewater Treatment Reagents

Wastewater Treatment Reagent	1972 Cost ($/lb)	1978 Cost ($/lb)	1986 Cost ($/lb)	1994 Cost ($/lb)
Chromium Reduction				
Sodium Bisulfite (anhyd. bags)	0.066	0.130	0.303	0.300
Sulfur Dioxide	0.038	0.085	0.108	0.115
Cyanide Oxidation				
Chlorine (Cl_2)	0.038	0.075	0.099	0.120
Pretreatment				
Calcium Chloride (80% flake, 100 lbs bags)	ND	ND	0.098	0.113
Ferrous Sulfate (monohydrate)	ND	ND	0.088	0.108
Sodium Hydrosulfite (anhyd. bags)	ND	ND	0.455	0.665
Alumium Sulfate (100 lbs bags)	ND	ND	0.101	0.119
pH Adjustment				
Sodium Hydroxide (50% liquid)	0.036	0.080	0.093	0.158
Calcium Hydroxide ($Ca(OH)_2$)	0.010	0.017	0.024	0.027
Sulfuric Acid (66 Be)	0.017	0.023	0.029	0.034

[a] Cost data from the Chemical Marketing Reporter (Schnell Publishing Company, NY). Prices are for bulk quantities, average F.O.B. east and west coasts. ND = no data.

of reagents than needed to satisfy chemical equations governing the treatment process reactions (i.e., stoichiometric requirements). This effect is due to several factors (based on conversations with survey respondents): (1) wastewaters contain buffering, chelating and other types of substances that interfere with the treatment processes and cause higher than expected reagent requirements; (2) wastestreams can be highly variable both in terms of flow and pollutant concentration, which causes dramatic swings in chemical feed and results in some wasted reagent; (3) shops must meet effluent regulations 100% of the time, rather than be in compliance "on the average," which causes them to add excess reagents; (4) local effluent requirements for cyanide or a particular metal species are often below the Federal limitations and can only be met by overdosing the wastewater with reagents; (5) treatment system designs may not be ideal; (6) instruments (e.g., ORP and pH) and chemical feed systems may not function perfectly all the time; and (7) operator ignorance or error.

The following are typical ranges of treatment reagent usage rates for survey respondents for each of the steps in the conventional treatment process[3]. Examples of calculations using these factors are presented in Exhibit 7-22.

Chromium Reduction
- H_2SO_4: 1 to 10 lbs/1,000 gal, and 2 lbs/lb Cr^{+6}
- $NaHSO_3$: 3 to 15 lbs/lb Cr^{+6}, or
- SO_2: 2 to 10 lbs/lb Cr^{+6}
- NaOH (50%): 3 to 6 lbs/1,000 gal

Cyanide Oxidation
- NaOCl (15%): 8 to 40 gal/lb CN

Precipitation
- NaOH (50%): 2 to 6 lbs/1,000 gal, and
  - 5 lbs/lb Cr, and
  - 4 lbs/lb other metals

Flocculation
- Polymer: 0.1 to 0.3 lb/1,000 gal

7.4 END-OF-PIPE TREATMENT SYSTEM DATA FROM THE USERS SURVEY

Exhibit 7-23 presents a summary of the end-of-pipe (EOP) treatment equipment purchased by the survey respondents. Most respondents purchased their EOP equipment in stages. Exhibit 7-23 shows the initial and final purchase dates (years) of equipment and the total capital cost of all equipment purchased. Also shown in Exhibit 7-23 are the types of treatment operations and processes that are present. Appendix B presents further details on these equipment purchases including the manufacturer of each major equipment item purchased (includes information on the initial system and up to three system additions), its cost and the shop's satisfaction level (shops were asked to rate on a scale of 1 to 5 their level of satisfaction with the equipment with 1 being the lowest satisfaction level and 5 being the highest).

Exhibit 7-21. Cost Data for Common Wastewater Treatment Reagents Used by Respondents to the Users Survey

Wastewater Treatment Reagent	Annual Usage Rate Per Plating Shop	Average Unit Cost ($/lbs or gal)
Chromium Reduction Reagents		
Sodium Bisulfite	≥50,000 lbs/yr	0.305
Sodium Bisulfite	≥10,000 lbs/yr, <50,000 lbs/yr	0.349
Sodium Bisulfite	≤10,000 lbs/yr	0.371
Sulfur Dioxide	≥10,000 lbs/yr	0.220
Sulfur Dioxide	<10,000 lbs/yr	0.543
Cyanide Oxidation Reagents		
Sodium Hypochlorite	≥50,000 gal/yr	0.509
Sodium Hypochlorite	≥10,000 gal/yr, <50,000 gal/yr	0.725
Sodium Hypochlorite	≤10,000 gal/yr	0.934
Chlorine (Cl_2)	≥5,000 lbs/yr	0.230
Chlorine (Cl_2)	<5,000 lbs/yr	0.540
Pretreatment Reagents		
Calcium Chloride	≥50,000 lbs/yr	0.154
Calcium Chloride	<50,000 lbs/yr	0.205
Ferrous Sulfate	≥10,000 lbs/yr	0.164
Ferrous Sulfate	<10,000 lbs/yr	0.200
Sodium Hydrosulfite	≥10,000 lbs/yr	0.805
Sodium Hydrosulfite	<10,000 lbs/yr	0.936
Alumium Sulfate	≥10,000 lbs/yr	0.135
Alumium Sulfate	<10,000 lbs/yr	0.690
DTC	All	0.870
pH Adjustment Reagents		
Sodium Hydroxide	≥50,000 lbs/yr	0.155
Sodium Hydroxide	≥10,000 lbs/yr, <50,000 lbs/yr	0.181
Sodium Hydroxide	≤10,000 lbs/yr	0.243
Calcium Hydroxide (Lime)	≥50,000 lbs/yr	0.070
Calcium Hydroxide (Lime)	≥10,000 lbs/yr, <50,000 lbs/yr	0.090
Calcium Hydroxide (Lime)	<10,000 lbs/yr	0.095
Sulfuric Acid	≥50,000 lbs/yr	0.078
Sulfuric Acid	≥10,000 lbs/yr, <50,000 lbs/yr	0.081
Sulfuric Acid	≤10,000 lbs/yr	0.130
Polymer (organic)	All	3.134

All unit costs on a dry weight basis except for sodium hypochlorite ($/gal of 15% solution) and sodium hydroxide ($/lb of 50% solution).

For plating shops reporting treatment system costs, the average cost was $252,141 (purchased between 1967 and 1993). The following indicates the percentage of shops that reported the purchase of each type of unit operation:

- CN Cyanide Destruction 44.7%
- Cr Chromium Reduction 58.3%
- pH pH Adjustment 90.0%
- HP Hydroxide Precipitation 85.7%
- SP Sulfide Precipitation 0.0%
- PF Polishing Filter 19.3%
- EV Evaporation 3.7%
- FP Filter Press 75.0%
- SD SludgeDryer 29.3%
- Oth Other (e.g., ion exchange) 2.7%

Exhibit 7-24 presents a summary of operating data submitted by the respondnents. Columns 1 and 2 show the shop code and the average electroplating wastewater flow rate. Columns 3, 4 and 5 provide treatment system labor information. Shown are the number of annual hours spent for system operation and maintenance, the hourly rate paid to the operators (in some cases the shops provided loaded rates that included overhead) and the total dollars spent annually for labor. Columns 6 and 7 show the annual costs for treatment chemicals and sludge disposal, and column 8 shows the sum of the reported annual operating costs (labor, treatment chemicals and sludge disposal). The final column indicates if a shop reported a compliance excursion ("Yes") or not ("No") or did not answer ("NA") that question on their survey form.

Exhibit 7-22. Example Calculations of Estimated Treatment Reagent Use

Example 1

Calculate the expected range of treatment reagent requirements for PS 135 using reagent usage rates from Section 7.3.2:

Conditions:

Operating days/yr:	260 days/yr
Cr Wastestream:	4,000 gpd, 50 mg/l Cr^{+6}
CN Wastestream:	3,000 gpd, 25 mg/l CN
Metals Precipitation:	7,000 gpd, 29 mg/l Cr^{+3}, 66 mg/1 other metals

(There is not a separate acid/alkaline wastestream. The metals precipitation wastestream is made up of the pretreated Cr and CN wastestreams.)

Chromium Reduction:	**Low Range**	**High Range**
H_2SO_4 lbs/yr:	1,040	10,400
H_2SO_4 1bs/yr:	868	868
$NaHSO_3$, lbs/yr:	1,302	6,510
NaOH (50%):	3,120	6,240

Cyanide Oxidation:	**Low Range**	**High Range**
NaOCl (15%):	1,296	6,480

Precipitation:	**Low Range**	**High Range**
NaOH (50%):	3,640	10,920
lbs/yr	2,170	2,170
	4,004	4,004

Flocculation:	**Low Range**	**High Range**
Polymer lbs/yr:	364	546

Reagent Use Summary:

Reagent	**Estimated**	**Actual**
H_2SO_4, lbs/yr:	1,908 to 11,268	5,000
$NaHSO_3$, lbs/yr:	1,302 to 6,510	not used
Ferrous Sulfate:	—	10,000 lbs/yr
NaOCl (15%):	1,296 to 6,480	5,000 gal/yr
NaOH (50%):	12,934 to 23,334	15,000 lbs/yr
Polymer, lbs/yr:	182 to 546	750

Example 2

Calculate the expected range of treatment reagent requirements for PS 142 using reagent usage rates from Section 7.3.2:

Conditions:

Operating days/yr:	260 days/yr
Cr Wastestream:	1,000 gpd, 482 mg/l Cr^{+6}
CN Wastestream:	1,000 gpd, 11.5 mg/l CN
Metals Precipitation:	12,000 gpd, 40.2 mg/l Cr^{+3}, 346 mg/l other metals

(The metals precipitation wastestream is made up of the acid/alkaline wastestream and the pretreated Cr and CN waste streams.)

Chromium Reduction:	**Low Range**	**High Range**
H_2SO_4 lbs/yr:	260	2,600
H_2SO_4 lbs/yr:	2,090	2,090
$NaHSO_3$, lbs/yr:	3,135	15,675
NaOH (50%):	780	1,560

Cyanide Oxidation:	**Low Range**	**High Range**
NaOCl (15%):	800	4,000

Precipitation:	**Low Range**	**High Range**
NaOH (50%):	6,240	18,720
lbs/yr	5,225	5,225
	40,168	40,168

Flocculation:	**Low Range**	**High Range**
Polymer lbs/yr:	312	936

Reagent Use Summary:

Reagent	**Estimated**	**Actual**
H_2SO_4, lbs/yr:	2,250 to 4,690	8,525
$NaHSO_3$, lbs/yr:	3,135 to 15,675	6,500
NaOCl (15%):	800 to 4,000	2,035 gal/yr
NaOH (50%):	52,413 to 65,673	62,625 lbs/yr
Polymer, lbs/yr:	312 to 936	100

Exhibit 7-25 presents the chemical use data for survey respondents (not all shops were able to submit these data). Available wastestream characterization data for these shops are presented in Exhibit 7-26 (even fewer of the shops had characterization data). These data were used in developing the reagent use formulas in Section 7.3.2. As previously discussed, the data in Exhibits 7-25 and 7-26 indicate that actual reagent use is significantly higher than theoretical use. This is especially true for sulfuric acid. Higher than expected sulfuric acid usage rates are most common during the pH adjustment step in the chromium reduction process. This is most likely due to the buffering capacity of the wastestreams, which is not accounted for in commonly accepted models.

7.5 ALTERNATIVE TECHNOLOGIES FOR METALS REMOVAL

7.5.1 Overview

For many plating shops, the conventional hydroxide precipitation/clarification process will be the most economical and reliable end-of-pipe method of meeting metals discharge limitations. However, for various reasons, shops may want to employ an alternative metals removal process. Some of the more common reasons include:

- Some local jurisdictions have adopted discharge standards more stringent than the Federal regula-

Exhibit 7-23. End-of-Pipe Treatment Equipment Data of Survey Respondents

Shop Code	Average Plating Discharge, gpd	BOP Treatment System Installation Dates: Initial	Last	Total Capital Cost, $	Treatment Operations and Processes: CN	Cr	pH	HP	PF	EV	FP	SD
001	9,000						X	X				
002	80,000	1984	1992	$840,000	X		X	X	X		X	X
003	225,000	1981	1989	$1,300,000	X		X	X	X		X	X
004	160,000	1982	1992	$540,000	X		X	X			X	X
005	100,000	1970	1986	$162,000	X		X	X			X	
006	0	1972	1988	$305,000			X			X		
007	20,000	1991	1991	$263,772	X		X	X	X		X	X
008	58,000	1974	1984	$125,000			X	X			X	
009	190,000	1973	1987	$180,000	X		X	X	X		X	
010	6,000	1985	1985	$125,000			X	X			X	
011	4,500	1975	1991	$74,000	X		X	X			X	
012	31,000	1983	1989	$1,050,000	X		X	X			X	X
013	160,000	1969	1969		X		X	X	X		X	X
014	22,000	1985	1989	$450,000	X		X	X	X		X	
015	30,000	1992	1992	$235,000			X	X	X		X	
016	50,000	1969	1990	$250,000	X		X	X				
017	1,200		1990	$5,500	X		X	X			X	
018	8,500	1988	1988	$125,000			X	X			X	X
019	7,200	1988	1988	$20,000			X	X			X	
020	35,000	1991	1991	$450,000	X		X	X			X	X
021	0	1988	1991	$70,300			X	X			X	
022	70,000	1978	1992	$405,000			X	X				
023	42,000		1986	$120,000			X	X			X	X
024	20,000	1984	1984	$125,000	X		X	X			X	
025	55,000	1969	1990	$1,700,000	X		X	X			X	
026	5,000	1975	1990	$36,457			X	X			X	
027	29,000				X		X	X			X	
028	8,000											
029	10,000	1991	1991	$15,000			X	X				X
030	500		1983		X		X	X			X	
031	7,800	1986	1988	$12,600	X		X	X			X	
032	13,000		1989				X	X	X			
033	22,000		1988	$201,000	X		X	X			X	
034	70,000	1980	1987	$480,000	X		X	X	X		X	
035	0											
036	2,704	1978	1986	$255,600	X		X	X			X	
037	5,000											
038	1,400	1992	1992	$30,000								
039	75,000	1975	1992	$365,000	X		X	X	X		X	
040	20,500	1972	1988	$130,000			X	X			X	X
041	18,000	1970	1992	$249,500			X	X	X		X	
042	5,000	1990	1990	$200,000			X	X			X	
043	50,000	1980	1992	$130,000			X	X			X	X
044	3,000						X	X				
045	90,000	1984	1992	$394,000	X		X	X			X	
046	6,500	1992	1992	$25,500			X	X				
047	1,400	1988	1988	$65,000			X	X				
048	30,000	1984	1987	$180,000			X	X			X	
049	5,500						X	X			X	X
050	20,000	1984	1991	$168,000	X		X	X			X	X
051	30,371	1989	1992	$435,000	X	X	X	X	X		X	
052	0											
053	15,000	1985	1985	$200,000	X	X	X	X			X	X
054	44,872	1983	1990	$113,000			X	X			X	

Shop Code	Average Plating Discharge, gpd	BOP Treatment System Installation Dates: Initial	Last	Total Capital Cost, $	Treatment Operations and Processes: CN	Cr	pH	HP	PF	EV	FP	SD
054	44,872	1983	1990	$113,000			X	X			X	
055	5,000	1989	1989	$6,200	X	X	X	X			X	
056	1,000	1983	1987	$29,000		X	X	X	X		X	
057	8,000	1985	1985	$25,000			X	X			X	
058	78,000	1986	1986	$450,000	X	X	X	X			X	
059	10,000	1989	1989	$140,000		X	X	X	X		X	
060	10,000		1991	$10,000			X	X				X
061	9,000	1990	1990	$15,000			X	X			X	X
062	1,500	1988	1988	$18,500			X	X			X	
063	67,700		1989	$350,000	X		X	X	X			
064	10,759	1980	1987	$130,000	X		X	X			X	
065	70,000	1987	1990	$480,000			X	X			X	
066	50,000	1989	1992	$74,500	X	X	X	X	X		X	X
067	2,200	1989	1989	$100,000		X	X	X			X	X
068	4,300	1986	1990	$103,000	X		X	X	X		X	
069	26,880	1992	1992	$143,500		X	X	X			X	
070	32,000	1983	1992	$267,500	X	X	X	X			X	X
071	0	1975	1975							X		
072	115,000					X	X	X			X	X
073	6,000	1992	1992	$20,000			X	X				
074	10,600	1989	1989	$19,000			X	X			X	
075	0	1980	1980	$30,000								
076	50,000	1986	1990	$288,125		X	X	X			X	X
077	55,000	1983	1992	$424,500	X	X	X	X			X	
078	0	1988	1988	$20,000		X	X	X			X	
079	18,045	1987	1987	$90,000		X	X	X			X	X
080	0	1987	1987	$10,000						X		
081	6,000	1976	1992	$13,000	X		X	X				
082	100,000					X	X	X			X	
083	8,000	1987	1987	$140,000		X	X	X			X	
084	25,000	1980	1992	$534,000		X	X	X	X		X	
085	28,407	1990	1991	$241,741	X		X	X			X	
086	70,000	1986	1987	$100,000	X	X	X	X			X	X
087	300,000	1982	1990	$660,000		X	X	X			X	
088	103,000	1981	1985	$298,491	X	X	X	X			X	X
089	12,500	1984	1989	$43,810		X	X	X	X			
090	4,334	1986	1986	$38,000	X		X	X			X	
091	5,500				X	X	X	X	X		X	X
092	12,000	1982	1992	$65,000	X	X	X	X			X	X
093	38,500	1985	1992	$299,000	X	X	X	X			X	X
094	11,500	1991	1991	$125,000		X	X	X			X	X
095	50,000	1990	1990	$75,000		X	X	X			X	
096	41,000	1991	1991	$500,000			X	X			X	
097	9,416	1985	1992	$80,500	X	X	X	X			X	
098	76,000	1980	1980	$100,000	X	X	X	X	X		X	X
099	496	1980	1980	$9,500	X	X	X	X	X		X	
100	0	1990	1991	$59,000			X			X	X	
101	30,000	1978	1991	$335,000	X		X	X	X		X	
102	50,000	1985	1985	$650,000	X	X	X	X			X	X
103	26,000						X	X			X	X
104	17,600	1986	1991	$242,420	X	X	X	X	X		X	
105	22,000	1988	1992	$305,000	X	X	X	X	X		X	X
106	30,000	1985	1985	$350,000			X	X			X	X
107	3,000	1988	1988	$10,000		X	X	X			X	

Exhibit 7-23 (continued). End-of-Pipe Treatment Equipment Data of Survey Respondents

Shop Code	Average Plating Discharge, gpd	BOP Treatment System Installation Dates		Total Capital Cost, $	Treatment Operations and Processes							
		Initial	Last		CN	Cr	pH	HP	PF	EV	FP	SD
108	ND	1987	1987	$60,680	X	X	X	X			X	
109	38,000	1981	1992	$175,000	X	X	X	X			X	X
110	8,600	1987	1991	$78,000		X	X	X			X	X
111	6,200	1985	1989	$86,500	X	X	X	X			X	
112	9,500	1978	1978		X	X	X	X			X	
113	15,000	1986	1990	$25,000			X	X				
114	1,300	1983	1989	$35,000	X		X	X	X		X	
115	7,500	1980	1992	$70,000			X	X			X	
116	10,000	1980	1989	$55,000			X	X	X		X	X
117	750	1987	1987	$50,000			X	X			X	
118	42,630	1983	1990	$489,000		X	X	X			X	X
119	1,500	1989	1989	$90,000		X	X	X			X	
120	5,000	1987	1987	$400,000	X	X	X	X			X	X
121	6,000		1988	$5,000		X	X	X			X	
122	175,000	1989	1992	$1,488,000			X	X			X	X
123	26,500	1985	1991	$715,000	X	X	X	X			X	
124	354,956				X	X	X	X			X	X
125	230,000	1982	1982		X		X	X			X	
126	12,000	1984	1984	$65,000	X	X	X	X			X	
127	4,000											
128	4,600	1980	1980	$85,000	X		X	X			X	
129	27,000	1974	1974		X	X	X	X			X	X
130	9,938	1990	1990	$110,000		X	X	X			X	
131	2,000				X	X	X	X			X	X
132	90,000	1982	1982		X	X	X	X			X	
133	0	1981	1988	$265,000	X		X			X	X	
134	2,000	1981	1983	$17,500			X	X				
135	5,000	1986	1988	$50,000	X	X	X	X	X		X	X
136	19,800						X		X			
137	100						X	X			X	
138	500											
139	52,000	1987	1987	$235,000	X	X	X	X			X	
140	10,000	1988	1988	$350,000		X	X	X	X		X	
141	30,000	1984	1992	$145,000		X	X	X			X	
142	11,500	1984	1988	$170,000	X	X	X	X			X	X
143	40,000	1980	1991	$238,000	X	X	X	X	X		X	X
144	5,500											
145	4,500	1984	1984	$40,000	X		X	X				
146	7,200	1981	1991	$80,000	X	X	X	X	X		X	
147	600											
148	1,100											
149	35,000						X	X	X			
150	100,000	1970	1989	$1,805,000	X	X	X	X			X	X
151	10,000											
152	32,500	1989	1992	$2,395			X	X				
153	2,200	1984	1984	$40,000		X	X	X			X	
154	8,000	1982	1987	$6,000			X	X				
155	10,000					X	X	X			X	X
156	16,800	1984	1984	$300,000	X	X	X	X			X	
157	3,000	1990	1990	$150,000		X	X	X			X	
158	48,000											
159	7,000	1987	1987	$500,000					X			
160	60,000	1976	1992	$465,000		X	X	X			X	X
161	11,000	1972	1972	$1,800			X	X				
162	44,000	1982	1990	$310,000		X	X	X			X	X
163	9,047	1987	1987	$30,000		X	X	X			X	
164	2,000	1985	1985	$27,000			X	X			X	
165	2,029	1985	1991	$98,495		X	X	X			X	X
166	35,000	1991	1991	$33,200			X			X		
167	60,000	1978	1978	$15,000			X	X				
168	ND	1988	1988	$80,000	X	X	X	X			X	
169	1,500	1989	1989	$70,000		X	X	X			X	
170	12,500	1978	1990	$162,000			X	X				
171	12,000	1980	1980	$80,000			X	X				
172	70,000	1985	1989	$528,000	X	X	X	X			X	X
173	42,400	1983	1991	$185,000	X	X	X	X	X		X	
174	26,000	1989	1989	$350,000	X	X	X	X			X	
175	8,000	1990	1992	$225,000	X	X	X	X			X	
176	99,484	1977	1992	$794,932		X	X	X	X		X	
177	9,163	1985	1985	$50,000		X	X	X			X	X
178	12,000	1984	1990		X	X	X	X			X	
179	7,930	1988	1988	$50,000	X		X	X			X	
180	35,000				X	X	X	X			X	X
181	38,110	1981	1987	$230,000		X	X	X			X	X
182	20,000	1985	1985				X	X			X	
183	9,000	1988	1988	$5,000	X	X	X	X			X	
184	11,000	1980	1988	$350,000	X	X	X	X			X	
185	700						X	X				
186	7,500	1984	1986	$13,000	X		X	X				
187	3,000	1985	1985	$50,000	X		X	X			X	
188	5,280	1987	1992	$21,300		X	X	X				
189	8,000	1985	1989	$5,000			X	X				
190	18,000	1990	1992	$95,000		X	X	X			X	X
191	6,000	1979	1983	$74,096			X	X			X	
192	40,000	1988	1988	$81,810			X		X			
193	0									X		
194	1,000											
195	0									X		
196	90,800	1980	1990	$345,000	X	X	X	X			X	X
197	ND				X	X	X	X			X	
198	0									X		
199	20,000	1981	1981	$75,000		X	X	X			X	
200	4,800	1984	1984	$1,000			X					
201	19,448	1983	1992	$173,000	X		X	X			X	X
202	0						X			X	X	
203	22,368				X	X	X	X			X	
204	208,720	1984	1993	$615,000	X	X	X	X			X	
205	5,000	1983	1983	$75,000	X	X	X	X				
206	9,000	1986	1986	$95,000	X	X	X	X				
207	18,000	1980	1992	$675,000	X	X	X	X			X	X
208	0											
209	11,500	1978	1993	$365,000	X	X	X	X			X	X
210	20,000	1989	1989	$548,666	X	X	X	X			X	
211	2,000	1991	1991	$3,000			X	X				
212	5,400	1986	1986	$36,790		X	X	X			X	X
213	50,000	1980	1988	$777,000	X	X	X	X	X		X	X

Exhibit 7-23 (continued). End-of-Pipe Treatment Equipment Data of Survey Respondents

Shop Code	Average Plating Discharge, gpd	BOP Treatment System Installation Dates: Initial	BOP Treatment System Installation Dates: Last	Total Capital Cost, $	CN	Cr	pH	HP	PF	EV	FP	SD
215	49,000	1982	1985	$400,800		X	X	X			X	X
216	3,000	1985	1985	$50,000	X	X	X	X	X		X	
217	130,000					X	X	X	X		X	
218	51,820	1986	1986	$163,000	X	X	X	X			X	
219	87,900	1980	1980	$425,000	X	X	X	X			X	
220	17,420	1980	1980	$170,000		X	X	X			X	
221	12,000	1986	1986	$250,000	X		X	X			X	
222	6,000	1979	1990	$23,500			X	X			X	
223	45,000	1986	1986	$105,000	X		X	X	X		X	
224	6,000	1986	1986	$5,500			X	X	X			
225	40,000	1988	1988	$150,000								
226	25,000	1976	1976	$150,000	X	X	X	X	X		X	
227	4,000	1987	1987	$50,000								
228	40,000	1985	1989	$133,500	X	X	X	X			X	
229	45,000	1982	1988	$385,000			X		X			
230	15,000	1993	1993				X	X			X	
231	64,800	1991	1991	$111,000		X	X	X			X	
233	0	1979	1987	$160,000		X	X				X	
234	1,174	1985	1985	$150,000		X	X	X			X	
235	10,650	1983	1992	$345,000	X	X	X	X			X	
237	76,000	1988	1988	$100,000			X		X			
238	420,000	1979	1990	$2,200,000	X	X	X	X	X		X	X
239	18,000	1980	1990	$170,000	X		X	X			X	X
240	1,000	1982	1990	$148,000			X	X	X		X	
241	7,500						X	X				
242	800	1987	1987	$10,000			X	X			X	
243	47,600	1984	1991	$167,000	X	X	X	X	X		X	X
244	25,810				X	X	X	X			X	X
245	24,400	1984	1989	$366,800	X	X	X	X	X		X	X
246	ND	1990	1990	$600,000	X	X	X	X			X	
247	109,000	1983	1983	$100,000		X	X	X			X	
248	75,000	1989	1989	$190,000		X	X	X			X	
249	0	1991	1991	$60,000			X					
250	11,000	1987	1991	$200,000		X	X	X			X	X
251	40,000	1982	1982	$200,000			X	X			X	
252	14,000					X	X	X				
253	31,582	1967	1988	$246,000	X	X	X	X			X	X
254	54,800	1977	1978	$338,000		X	X	X	X		X	X
255	3,000	1983	1986	$85,000	X	X	X	X			X	X
256	140,000	1987	1987	$200,000	X		X	X			X	
257	2,000	1990	1990	$1,000,000	X		X				X	
258	170,000	1985	1988	$630,000	X	X	X	X	X		X	X
259	750	1984	1984	$14,000	X		X	X				
260	6,589		1989	$31,035		X	X	X	X		X	X
261	20,000	1980	1980				X	X			X	
262	30,000	1993	1993	$25,000			X	X			X	
263	40,000	1987	1992	$475,000		X	X	X	X		X	X
264	77,900	1985	1991	$339,500	X	X	X	X	X		X	X
265	20,000		1993	$160,000	X		X	X			X	
266	56,607	1992	1992	$105,000		X	X	X				
267	9,000	1981	1981	$85,000	X		X	X			X	
268	17,000	1985	1989	$545,000	X	X	X	X			X	X
269	2,700	1991	1991	$10,000	X		X	X				
270	0											
271	0											
272	35,000	1976	1985	$111,000	X	X	X	X			X	X
273	813,000	1953	1993	$900,000	X	X	X	X			X	
274	47,000	1976	1989	$104,900	X	X	X	X			X	
275	150,000	1985	1990	$689,000		X	X	X			X	X
276	42,000			$540,000	X	X	X	X			X	X
277	60,000	1974	1981	$300,000		X	X	X			X	
278	40,000	1987	1987	$100,000		X	X	X			X	
279	50,000	1985	1991	$291,000		X	X	X			X	X
280	142,000	1990	1990	$1,000,000		X	X	X			X	X
281	73,000	1987	1990	$320,000		X	X	X				X
282	0	1989	1989	$400,000	X					X		
283	16,844	1987	1987		X		X	X			X	
284	600	1990	1990	$60,000		X	X	X	X		X	
285	8,000	1981	1989	$16,000	X	X	X	X	X		X	
286	8,000	1982	1982	$400,000			X	X			X	
287	50,000	1981	1981		X		X	X			X	
288	15,000	1972	1990	$165,000		X	X	X			X	
289	48,000	1982	1982	$300,000		X	X	X			X	X
290	700	1987	1987	$120,000		X	X	X			X	
291	350											
292	60,000	1985	1985	$730,000	X	X	X	X			X	X
293	70,000				X	X	X	X			X	X
294	0											
295	17,000				X		X	X			X	
296	1,700,000	1973	1973	$700,000			X	X			X	
297	0				X	X	X	X			X	X
298	70,000	1975	1989	$300,000		X	X	X			X	
299	1,400				X		X	X	X		X	
300	90,000	1983	1991	$1,495,000	X		X	X	X		X	
301	50	1988	1991	$29,425		X	X	X			X	
302	86,400	1989	1991	$105,000			X	X			X	X
303	37,000	1980	1980	$150,000		X	X	X			X	X
304	3,700											
305	11,000	1991	1992	$330,000		X	X	X	X			
306	50,000											
307	58,606	1991	1991	$243,000	X		X	X	X		X	X
308	17,500				X	X	X		X		X	X
309	1,450	83	83	$45,000			X	X	X		X	X
310	0	1989	1989	$500								
311	115,000	1980	1991	$47,000			X	X	X			
312	1,200	1982	1982	$24,000		X	X	X	X		X	
313	35,000											
314	40,000	1978	1992	$435,000		X	X	X	X		X	X
315	75,000	1979	1989	$117,000	X			X				
316	360,000				X	X	X	X	X			
317	0									X		
318	13,000	1960	1960	$20,000			X	X				

ND = no data

Exhibit 7-24. End-of-Pipe Treatment Operating Costs of Survey Respondents

Shop Code	Average Plating Discharge, gpd	Operating Costs: Labor Hr/Yr	Labor Rate	Labor Cost, $	Treatment Chemicals	Sludge Disposal	Total Op Cost	Compliance Excursion
001	9,000							NA
002	80,000				$81,268	$10,500	$91,768	YES
003	225,000					$101,750	$101,750	NA
004	160,000				$33,292	$21,000	$54,292	NO
005	100,000				$68,480	$18,600	$87,080	NO
006	0	260	$20.00	$5,200	$3,600	$2,400	$11,200	NO
007	20,000				$17,500	$4,500	$22,000	NO
008	58,000				$22,690	$1,000	$23,690	YES
009	190,000				$26,070	$25,900	$51,970	YES
010	6,000				$14,454	$11,700	$26,154	NO
011	4,500				$15,400	$14,025	$29,425	NO
012	31,000				$31,411	$11,920	$43,331	NO
013	160,000				$110,708	$55,930	$166,638	NO
014	22,000				$25,331	$10,500	$35,831	NO
015	30,000				$103,728	$65,000	$168,728	NO
016	50,000				$6,000	$8,000	$14,000	NO
017	1,200	40	$10.00	$400	$14,413	$10,560	$25,373	NO
018	8,500	2,400	$15.00	$36,000	$13,114	$3,834	$52,948	NO
019	7,200	4,800	$1.80	$8,640		$5,500	$14,140	NO
020	35,000	2,470	$26.00	$64,220	$9,689	$14,175	$88,084	NO
021	0				$2,820	$6,800	$9,620	NO
022	70,000	2,000	$15.00	$30,000	$71,537	$51,174	$152,711	NO
023	42,000	120	$21.50	$2,580	$17,872	$13,460	$33,912	NO
024	20,000	1,200	$15.00	$18,000	$9,744	$4,822	$32,566	NO
025	55,000				$71,654	$45,380	$117,034	NO
026	5,000	1,040	$15.00	$15,600	$4,223	$6,276	$26,099	NO
027	29,000					$40,000	$40,000	YES
028	8,000							NA
029	10,000	1,800	$10.00	$18,000	$5,900	$8,850	$32,750	NO
030	500	1,400	$6.00	$8,400	$2,764	$4,190	$15,354	NO
031	7,800	400	$10.00	$4,000	$6,292		$10,292	NO
032	13,000							NO
033	22,000				$9,068	$6,000	$15,068	NO
034	70,000	1,000	$20.00	$20,000	$3,537	$5,250	$28,787	YES
035	900							NA
036	2,704	4,160	$10.00	$41,600	$103,795	$8,437	$153,832	NO
037	5,000							NA
038	1,400							NA
039	75,000							NO
040	20,500	2,080	$9.00	$18,720	$3,900	$12,500	$35,120	YES
041	18,000	4,000	$12.00	$48,000		$38,000	$86,000	YES
042	5,000	250	$10.00	$2,500	$15,000	$1,000	$18,500	YES
043	50,000	3,000	$10.00	$30,000	$26,535	$12,000	$68,535	YES
044	3,000	200	$6.00	$1,200	$200	$12,000	$13,400	NO
045	90,000	10,400				$84,000	$84,000	YES
046	6,500	520	$15.00	$7,800			$7,800	YES
047	1,400	100	$25.00	$2,500	$3,000	$350	$5,850	NO
048	30,000	1,500	$11.00	$16,500	$46,500	$50,000	$113,000	NO
049	5,500					$14,500	$14,500	NA
050	20,000	4,000	$14.00	$56,000	$24,111	$18,500	$98,611	YES
051	30,371	10,286	$12.50	$128,575	$120,107	$58,000	$306,682	NO
052	0							NA
053	15,000					$8,000	$8,000	NA

Shop Code	Average Plating Discharge, gpd	Operating Costs: Labor Hr/Yr	Labor Rate	Labor Cost, $	Treatment Chemicals	Sludge Disposal	Total Op Cost	Compliance Excursion
054	44,872				$3,000	$24,800	$27,800	YES
055	5,000	500	$10.00	$5,000		$4,200	$9,200	NO
056	1,000	100	$6.00	$600	$8,824	$3,000	$12,424	NO
057	8,000	1,000	$10.00	$10,000	$1,780		$11,780	NO
058	78,000	3,500	$15.00	$52,500	$63,200	$36,900	$152,600	YES
059	10,000	1,700	$10.00	$17,000	$20,770	$5,854	$43,624	NA
060	10,000					$750	$750	NO
061	9,000	520	$8.00	$4,160	$18,800	$6,100	$29,060	YES
062	1,500	200	$10.00	$2,000		$5,725	$7,725	NO
063	67,700	1,500	$12.00	$18,000	$24,681	$20,900	$63,581	NO
064	10,759	200	$50.00	$10,000	$13,761	$3,500	$27,261	NO
065	70,000				$12,430	$63,917	$76,347	NO
066	50,000	3,000	$12.00	$36,000	$27,500	$10,000	$73,500	NO
067	2,200	1,560	$12.00	$18,720	$7,500	$3,800	$30,020	NO
068	4,300	1,200	$7.50	$9,000	$1,962	$27,000	$37,962	NO
069	26,880	1,248	$20.00	$24,960	$18,447	$6,000	$49,407	NO
070	32,000	750	$30.00	$22,500	$17,000	$9,400	$48,900	NO
071	0	24	$20.00	$480			$480	NO
072	115,000							NA
073	6,000	100	$8.00	$800	$2,680		$3,480	NO
074	10,600	1,040	$8.00	$8,320		$13,200	$21,520	NO
075	0					$1,920	$1,920	NA
076	50,000	2,580	$9.35	$24,123		$15,868	$39,991	NO
077	55,000				$92,630	$40,800	$133,430	YES
078	0	832	$15.00	$12,480	$865	$6,165	$19,510	NO
079	18,045	100	$20.00	$2,000		$2,500	$4,500	NO
080	0	50	$10.00	$500	$850		$1,350	NO
081	6,000	1,500	$10.00	$15,000			$15,000	NO
082	100,000	7,500	$15.00	$112,500	$106,000	$87,000	$305,500	NO
083	8,000		$30.00		$2,350	$6,675	$9,025	NO
084	25,000	4,160			$65,340	$6,000	$71,340	YES
085	28,407				$15,694.	$8,585	$24,279	NO
086	70,000	2,000	$10.00	$20,000	$23,566	$31,500	$75,066	NO
087	300,000	7,340	$15.07	$110,614	$165,298	$42,728	$318,640	YES
088	103,000	2,000	$10.00	$20,000	$109,600	$10,725	$140,325	YES
089	12,500	500	$15.00	$7,500	$1,340	$49,476	$58,316	NA
090	4,334	260	$8.60	$2,236	$5,730	$9,125	$17,091	NO
091	5,500				$7,117	$20,800	$27,917	NO
092	12,000	2,600	$12.00	$31,200	$24,900	$33,600	$89,700	YES
093	38,500	4,000	$12.00	$48,000	$19,692	$14,000	$81,692	YES
094	11,500	220	$9.50	$2,090	$6,530	$6,000	$14,620	YES
095	50,000	6,000	$8.00	$48,000	$85,985	$25,719	$159,704	NO
096	41,000	520	$12.00	$6,240	$17,733	$5,892	$29,865	NO
097	9,416	780	$13.50	$10,530	$3,945	$5,400	$19,875	YES
098	76,000				$76,750	$40,000	$116,750	NO
099	496				$15,294	$12,416	$27,710	YES
100	0	24	$11.50	$276	$55	$19,600	$19,931	NO
101	30,000	780	$35.00	$27,300	$43,960		$71,260	NO
102	50,000	500	$10.00	$5,000	$305,658	$114,000	$424,658	YES
103	26,000	5,000	$12.00	$60,000	$17,320	$19,280	$96,600	NO
104	17,600	1,200	$14.00	$16,800	$21,220	$4,800	$42,820	NO
105	22,000	3,800	$19.75	$75,050	$88,959	$19,600	$183,609	YES
106	30,000	2,600	$9.00	$23,400	$4,679	$13,200	$41,279	YES

Exhibit 7-24 (continued). End-of-Pipe Treatment Operating Costs of Survey Respondents

		Operating Costs						
Shop Code	Average Plating Discharge, gpd	Labor Hr/Yr	Labor Rate	Labor Cost, $	Treatment Chemicals	Sludge Disposal	Total Op Cost	Compliance Excursion
107	3,000	528	$8.00	$4,224	$2,090	$15,600	$21,914	NO
108	ND		$25.00		$7,861	$9,000	$16,861	NO
109	38,000	2,080	$10.00	$20,800	$68,000	$13,680	$102,480	YES
110	8,600	7,512	$8.66	$65,054	$29,523	$14,000	$108,577	NO
111	6,200	700	$15.00	$10,500	$3,676	$12,000	$26,176	YES
112	9,500	2,000	$8.00	$16,000	$9,700	$18,000	$43,700	NO
113	15,000	780	$10.00	$7,800	$16,250	$15,000	$39,050	NO
114	1,300	400	$20.00	$8,000		$3,800	$11,800	YES
115	7,500	2,500	$10.00	$25,000	$26,272	$67,000	$118,272	NO
116	10,000	750	$16.00	$12,000	$8,559	$29,144	$49,703	NO
117	750	416	$21.25	$8,840	$1,928		$10,768	NO
118	42,630				$52,188	$44,200	$96,388	YES
119	1,500	780	$12.00	$9,360	$239	$1,800	$11,399	YES
120	5,000	2,100	$20.00	$42,000		$2,500	$44,500	NO
121	6,000	520	$10.00	$5,200	$2,440	$8,800	$16,440	NO
122	175,000	8,000	$8.00	$64,000	$60,800	$78,000	$202,800	N0
123	26,500	5,000	$25.00	$125,000	$257,373	$19,570	$401,943	N0
124	354,956				$156,566	$146,440	$303,006	NO
125	230,000	2,080	$13.00	$27,040	$10,447	$7,665	$45,152	NO
126	12,000	780	$15.00	$11,700	$11,137	$14,800	$37,637	NO
127	4,000					$550	$550	NA
128	4,600	2,000	$15.00	$30,000	$32,977	$12,242	$75,219	YES
129	27,000							YES
130	9,938	4,040	$5.75	$23,230	$35,880	$34,111	$93,221	YES
131	2,000				$7,472	$10,694	$18,166	NO
132	90,000	4,888	$10.00	$48,880	$64,800	$120,000	$233,680	NO
133	0	1,000	$18.00	$18,000	$34,317	$26,300	$78,617	NO
134	2,000							NO
135	5,000	2,300	$20.00	$46,000	$13,000	$10,000	$69,000	NO
136	19,800				$152		$152	YES
137	100,000							NO
138	500					$12,420	$12,420	NA
139	52,000	2,600	$18.00	$46,800	$36,000	$31,000	$113,800	NO
140	10,000	1,000	$15.00	$15,000	$50,000	$2,000	$67,000	NA
141	30,000	1,040	$8.50	$8,840	$13,629	$1,600	$24,069	YES
142	11,500	500	$15.00	$7,500	$15,657	$12,000	$35,157	NO
143	40,000				$94,360	$40,000	$134,360	NO
144	5,500							NA
145	4,500	100	$10.00	$1,000	$920		$1,920	NO
146	7,200	6,000	$6.00	$36,000	$11,700	$20,800	$68,500	NO
147	600							NA
148	1,100							NA
149	35,000							NA
150	100,000				$150,000	$100,000	$250,000	YES
151	10,000							NA
152	32,500	126	$9.00	$1,134	$1,648		$2,782	NA
153	2,200	2,000	$20.00	$40,000	$4,953	$39,000	$83,953	YES
154	8,000	1,200	$20.00	$24,000			$24,000	N0
155	10,000							NA
156	16,800	1,000	$10.00	$10,000	$12,342	$7,600	$29,942	NO
157	3,000							YES
158	48,000							NA
159	7,000							N0

		Operating Costs						
Shop Code	Average Plating Discharge, gpd	Labor Hr/Yr	Labor Rate	Labor Cost, $	Treatment Chemicals	Sludge Disposal	Total Op Cost	Compliance Excursion
160	60,000	5,000	$14.50	$72,500		$29,400	$101,900	NO
161	11,000	10	$20.00	$200		$309	$509	N0
162	44,000	140	$10.00	$1,400	$34,000	$15,000	$50,400	NO
163	9,047	2,080	$20.00	$41,600	$5,541	$34,850	$81,991	N0
164	2,000	950	$10.50	$9,975	$6,144	$16,390	$32,509	NO
165	2,029	662	$28.00	$18,536	$10,194	$10,500	$39,230	N0
166	35,000	2,000	$8.00	$16,000		$5,000	$21,000	N0
167	60,000	50	$8.00	$400	$2,900		$3,300	NO
168	ND	2,000	$18.00	$36,000			$36,000	NO
169	1,500				$621	$7,000	$7,621	NO
170	12,500	260	$20.00	$5,200	$660	$2,400	$8,260	NO
171	12,000	400	$20.00	$8,000	$4,650	$10,000	$22,650	YES
172	70,000	2,750	$11.75	$32,313	$23,619	$16,411	$72,343	NO
173	42,400	1,040	$15.00	$15,600	$19,541	$13,000	$48,141	YES
174	26,000	2,300	$12.00	$27,600	$14,800	$9,000	$51,400	NO
175	8,000					$10,000	$10,000	NO
176	99,484	2,600	$11.00	$28,600	$28,344	$11,200	$68,144	YES
177	9,163					$8,000	$8,000	N0
178	12,000	2,300	$15.00	$34,500	$6,969	$5,100	$46,569	N0
179	7,930	100	$8.00	$800	$2,240		$3,040	YES
180	35,000	4,200	$25.00	$105,000	$8,736		$113,736	NO
181	38,110	4,680	$12.50	$58,500	$7,882	$22,900	$89,282	NO
182	20,000							N0
183	9,000	260	$8.00	$2,080	$6,070	$11,600	$19,750	N0
184	11,000	4,600	$11.00	$50,600	$15,095	$14,300	$79,995	NO
185	700							NA
186	7,500	200	$15.00	$3,000	$2,700	$2,335	$8,035	YES
187	3,000					$8,000	$8,000	NO
188	5,280				$300		$300	N0
189	8,000	100	$15.00	$1,500			$1,500	N0
190	18,000	500	$15.00	$7,500	$1,911	$6,800	$16,211	YES
191	6,000				$8,415	$10,974	$19,389	NA
192	40,000	2,000	$15.00	$30,000			$30,000	NO
193	0							NA
194	0	160	$6.00	$960			$960	NO
195	0							NA
196	90,800	4,500	$10.00	$45,000	$105,795	$15,600	$166,395	NO
197	ND	8,760				$47,370	$47,370	NO
198	0							NA
199	20,000	3,000	$10.00	$30,000	$19,776	$20,800	$70,576	YES
200	4,800				$135		$135	NO
201	19,448	1,500	$12.00	$18,000	$6,604	$60,000	$84,604	N0
202	0					$30,000	$30,000	NA
203	22,368							NO
204	208,720	14,500	$15.00	$217,500	$228,892	$116,905	$563,297	NO
205	5,000					$2,800	$2,800	N0
206	9,000	500	$20.00	$10,000	$3,779	$11,000	$24,779	YES
207	18,000	5,000	$15.00	$75,000	$23,055	$3,400	$101,455	YES
208	0					$21,830	$21,830	NA
209	11,500	6,292	$12.00	$75,504	$15,927	$41,000	$132,431	YES
210	20,000	6,656	$35.00	$232,960	$37,080	$16,378	$286,418	NO
211	2,000							NO
212	5,400	225	$40.00	$9,000	$2,377	$2,155	$13,532	YES

Exhibit 7-24. End-of-Pipe Treatment Operating Costs of Survey Respondents

Shop Code	Average Plating Discharge, gpd	Operating Costs: Labor Hr/Yr	Labor Rate	Labor Cost, $	Treatment Chemicals	Sludge Disposal	Total Op Cost	Compliance Excursion
213	50,000	1,000	$25.00	$25,000	$26,113	$13,600	$64,713	NO
214	72,000	5,400	$12.00	$64,800	$86,400	$175,000	$326,200	NO
215	49,000				$125,252	$26,425	$151,677	NO
216	3,000	1,500	$12.00	$18,000	$12,645	$3,200	$33,845	NO
217	130,000							NA
218	51,820	8,000	$10.00	$80,000	$37,000	$50,000	$167,000	NO
219	87,900	8,000	$10.00	$80,000	$38,000	$26,000	$144,000	NO
220	17,420	800	$10.00	$8,000	$22,000	$12,000	$42,000	NO
221	12,000	4,500	$10.00	$45,000	$9,840	$5,000	$59,840	NO
222	6,000	520	$7.00	$3,640		$3,500	$7,140	NA
223	45,000	5,200	$15.00	$78,000	$96,500	$66,550	$241,050	NO
224	6,000	260	$10.00	$2,600			$2,600	YES
225	40,000	6,500	$13.00	$84,500		$4,000	$88,500	YES
226	25,000	6,240	$10.00	$62,400		$27,000	$89,400	YES
227	4,000	100	$15.00	$1,500		$150	$1,650	NO
228	40,000	1,000	$12.00	$12,000	$55,500	$40,000	$107,500	YES
229	45,000	4,000	$46.00	$184,000	$24,000		$208,000	NO
230	15,000							NO
231	64,800	900	$15.00	$13,500		$10,800	$24,300	NO
233	0	1,560	$12.00	$18,720		$161,500	$180,220	NO
234	1,174	2,162	$14.00	$30,268	$4,690	$9,800	$44,758	NO
235	10,650	6,240	$9.00	$56,160	$88,267	$48,000	$192,427	NO
237	76,000	700	$15.00	$10,500			$10,500	NO
238	420,000	1,320	$50.00	$66,000	$89,351	$160,000	$315,351	NO
239	18,000	8,000	$10.00	$80,000	$35,000	$15,000	$130,000	NO
240	1,000	1,200	$30.00	$36,000	$21,980	$5,500	$63,480	NO
241	7,500							YES
242	800	780	$7.50	$5,850	$1,290	$4,000	$11,140	NA
243	47,600				$31,329	$2,500	$33,829	NO
244	25,810				$15,078	$13,920	$28,998	NO
245	24,400	1,500	$13.50	$20,250	$30,732	$5,440	$56,422	NO
246	ND	2,100	$40.00	$84,000	$50,180	$40,000	$174,180	NO
247	109,000	3,000	$8.50	$25,500	$22,342	$41,000	$88,842	YES
248	75,000	2,100	$8.00	$16,800	$25,683	$62,000	$104,483	YES
249	0	500	$25.00	$12,500	$460		$12,960	NO
250	11,000	4,470	$15.00	$67,050	$11,659		$78,709	YES
251	40,000					$216,000	$216,000	NA
252	14,000	600	$10.00	$6,000	$880		$6,880	NO
253	31,582	7,064	$15.00	$105,960	$48,326	$25,060	$179,346	NO
254	54,800	800	$10.00	$8,000	$38,350	$60,316	$106,666	NO
255	3,000	2,000	$10.00	$20,000	$3,076	$13,500	$36,576	NO
256	140,000	2,080	$13.00	$27,040	$396,661	$213,000	$636,701	NO
257	2,000	2,080	$49.05	$102,024	$11,696	$2,520	$116,240	NO
258	170,000	6,240	$15.00	$93,600	$22,445	$22,000	$138,045	YES
259	750	500	$12.00	$6,000	$674	$2,000	$8,674	NA
260	6,589				$12,611	$3,332	$15,943	NA
261	20,000	2,000	$20.00	$40,000			$40,000	NO
262	30,000	50	$15.00	$750	$1,500	$1,600	$3,850	NO
263	40,000	3,000	$9.00	$27,000	$27,000	$19,700	$73,700	NO
264	77,900	3,000	$14.30	$42,900	$86,670	$18,600	$148,170	NO
265	20,000	8,000	$15.00	$120,000	$46,000	$95,000	$261,000	YES

ND = no data
NA = no answer given by respondent.

Shop Code	Average Plating Discharge, gpd	Operating Costs: Labor Hr/Yr	Labor Rate	Labor Cost, $	Treatment Chemicals	Sludge Disposal	Total Op Cost	Compliance Excursion
266	56,607	2,000	$12.00	$24,000	$151,955	$74,195	$250,150	NA
267	9,000	780	$7.00	$5,460	$11,005	$1,253	$17,718	NO
268	17,000				$56,379	$37,000	$93,379	NA
269	2,700	50	$25.00	$1,250	1,250NO			
270	0							NA
271	0							NA
272	35,000	3,000	$15.00	$45,000	$39,308	$20,000	$104,308	NO
273	813,000				$184,842	$630,000	$814,842	YES
274	47,000	800	$20.00	$16,000	$8,101	$750	$24,851	NA
275	150,000	4,887	$14.85	$72,572	$464,000	$24,000	$560,572	NO
276	42,000				$63,950		$63,950	YES
277	60,000				$102,000	$70,000	$172,000	NO
278	40,000	1,000	$10.00	$10,000	$12,500	$5,000	$27,500	NO
279	50,000	9,360	$10.00	$93,600	$66,931	$13,700	$174,231	NO
280	142,000	6,250	$10.50	$65,625	$95,307	$36,600	$197,532	NO
281	73,000	2,200	$25.00	$55,000	$58,950	$15,562	$129,512	YES
282	0							NO
283	16,844				$11,812	$2,677	$14,489	YES
284	600	270	$15.00	$4,050	$4,646	$3,000	$11,696	NO
285	8,000	416	$16.00	$6,656	$6,854		$13,510	NO
286	8,000	1,500	$47.00	$70,500	$11,185	$21,000	$102,685	NO
287	50,000				$51,666	$9,625	$61,291	N0
288	15,000	5,376	$15.00	$80,640		$35,000	$115,640	YES
289	48,000	3,000	$15.00	$45,000	$13,920		$58,920	NO
290	700	1,560	$12.00	$18,720	$2,656	$19,000	$40,376	NO
291	350							NA
292	60,000				$38,758	$22,000	$60,758	NO
293	70,000	6,000	$7.50	$45,000		$30,000	$75,000	YES
294	0							NA
295	17,000				$105,400	$26,000	$131,400	NO
296	1,700,000					$80,000	$80,000	NO
297	0							NA
298	70,000	6,300	$13.00	$81,900	$42,020	$35,000	$158,920	YES
299	1,400							NA
300	90,000	4,600	$16.50	$75,900	$249,980	$86,625	$412,505	NO
301	50	545	$13.00	$7,085	$1,020	$1,275	$9,380	YES
302	86,400				$30,000		$30,000	YES
303	37,000				$20,000	$130,000	$150,000	YES
304	3,700							NA
305	11,000	3,000	$12.00	$36,000			$36,000	YES
306	50,000							NA
307	58,606	2,340	$17.00	$39,780	$33,100	$3,000	$75,880	YES
308	17,500							NA
309	1,450	2,000				$15,000	$15,000	NO
310	0							NA
311	115,000					$32,500	$32,500	YES
312	1,200	2,000	$15.00	$30,000	$18,912		$48,912	NO
313	35,000							NA
314	40,000	2,500	$20.00	$50,000		$800	$50,800	NO
315	75,000							YES
316	360,000							NO
317	0							NA
318	13,000	60	$20.00	$12,000	748		$1,948	NO

Exhibit 7-25. Annual Treatment Reagent Use Data

	Flow Rates, gpd					Common Chromium Reduction Chemicals		Common Cyanide Oxidation Chemicals		Common Precipitation Pretreatments					Common pH Adjustment Chemicals			Flocculant	
PS Code	Chromium Wastewater Flow	Cyanide Wastewater Flow	Acid/Alkaline Wastewater Flow	Other Wastewater Flows	Total Wastewater Flow	Sodium Bisulfite lbs	Sulfur Dioxide lbs	Sodium Hypochlorite (15% solution) gal	Chlorine Gas lbs	Calcium Chloride lbs	Ferrous Sulfate lbs	Sodium Hydrosulfite lbs	Aluminum Sulfate lbs	DTC lbs	Sodium Hydroxide (50% solution) lbs	Lime lbs	Sulfuric Acid lbs	Polymer lbs	Additional Reagents Used
002	20,000	20,000	40,000	0	80,000	28,710		19,600							200,466		34,443		
004	12,800	6,400	140,800	0	160,000			3,636							192,242		836,586		
007	7,000	7,000	6,000	0	20,000	20,400		5,400							18,000	10,200			
008	0	0	58,000	0	58,000										178,080		3,062	825	
009	0	4,500	185,500	0	190,000			21,000								81,000	45,924		
010					6,000	11,400								4,240	536,987			1,650	
011	500	2,000	6,500	0	9,000	5,200				1,200		1,200							√
012	7,000	8,000	15,000	0	30,000	12,000		15,196							38,576	12,000	661,921		
013					160,000	51,200		55,723				11,250			232,236				√
014	2,500	1,500	16,500	0	20,500	9,300		3,000		11,140	6,250				68,000		225,029		
015	0	0	25,000	50,000	75,000					360,000	142,000		28,000		201,612		2,520	4,500	
017					1,200	4,880		48							91,000				
018	4,250	0	4,250	0	8,500	5,000				5,000					56,500		23,000	563	
020					35,000						6000*				50,402	4,000	5,280		√
022	70,000	0	0	0	70,000		904,000								228,000			2,460	
023	10,000	0	10,000	0	20,000	4,200					10,650				72,250	700	17,810		
024	5,760	7,200	10,800	0	23,760	20,000		1,925							15,401		5,052	4,950	
025	10,780	21,570	22,650	0	55,000		31,200		26,000						168,000		40,413		
030	100	500	500	0	1,100	2,400		800							15,274				
033	8,000	14,000	0	0	22,000	770		1,500							11,840		3,062	4,057	
036					2,704			550		4,594					218,560		996,050	3,156	
040	6,765	0	13,735	0	20,500	2,500									7,150		10,000	415	
043	1,000	0	49,000	0	50,000	4,000									15,000	5,000		1,500	
045					90,000			6,000					52,000		649,128			4,700	
048	30,000	0	0	0	30,000	30,000									180,000		45,000	2,300	√
050	8,000	3,000	9,000	0	20,000	31,200		2,530				3,195			80,500		35,406		
051	18,000	22,000	0	0	40,000	28,600		45,760			3,328				182,010		122,200	39,000	
057					8,000											8,000		200	
058	6,000	1,500	62,500	0	70,000	35,000		30,000							175,000		140,000		
059					10,000	1,600									84,500		54,000		√
061	0	0	8,000	0	8,000						600				7,000			188	
064	0	1,500	7,300	0	8,800			2,860		2,800					39,202		10,470	3,976	
065					70,000										91,000			200	
067	2,200	0	0	20	2,220	4,200				4,200					12,218		3,674	3,600	√
068					43,000			100							3,818		1,684	10	√
069	26,880	0	0	0	26,880										36,402		43,781		

Exhibit 7-25 (continued). Annual Treatment Reagent Use Data

	Flow Rates, gpd					Common Chromium Reduction Chemicals		Common Cyanide Oxidation Chemicals		Common Precipitation Pretreatments					Common pH Adjustment Chemicals			Flocculant	
PS Code	Chromium Wastewater Flow	Cyanide Wastewater Flow	Acid/Alkaline Wastewater Flow	Other Wastewater Flows	Total Wastewater Flow	Sodium Bisulfite lbs	Sulfur Dioxide lbs	Sodium Hypochlorite (15% solution) gal	Chlorine Gas lbs	Calcium Chloride lbs	Ferrous Sulfate lbs	Sodium Hydrosulfite lbs	Aluminum Sulfate lbs	DTC lbs	Sodium Hydroxide (50% solution) lbs	Lime lbs	Sulfuric Acid lbs	Polymer lbs	Additional Reagents Used
077	3,840	6,720	44,440	0	55,000	17,000									600,000		19,000		√
082					100,000	100,000									250,000		90,000		
086					70,000	41									95,460			150	
087	50,000	0	250,000	0	300,000	61,360									1,216,542			6,490	√
088	40,000	60,000			100,000	50,000		40,000							150,000		95,675	3,750	√
089	0	560	3,780	0	4,340										17,264				√
091	3,685	1,815	0	0	5,500	6,000		800							16,800		1,800	10,800	
093					38,500		4,200		14,000						49,639		21,431	900	
094	3,000	0	9,000	0	12,000										16,500		4,592		√
095	3,000	0	17,000	30,000	50,000	2,500									456,960		214,200	100	
096	0	0	14,000	0	14,000										58,804		50,517	3,623	
097	0	9,500	0	0	9,500			1,900							8,400				
101	0	13,640	16,360	0	30,000			26,000							165,464		58,171		
102					650,000	235,000		170,439			20,500				930,894	9,200	428,626	1,012	√
103	0	0	26,000	0	26,000										153,600		20,240	540	√
104					17,600			1,040				10,250			23,848		2,000	1,470	
105	0	0	22,000	0	22,000										364,020	51,480	22,962		√
110	2,880	0	5,760	0	8,640	16,200				26,200					66,504		21,584	700	
111					6,200										2,800			113	√
116	0	0	10,000	0	10,000					2,000					60,720		1,100	50	
118	17,052	0	25,578	0	42,630	20,800									159,400		73,100	1,550	√
122	0	0	175,000	0	175,000										78,000	628,000		10,600	√
123	4,800	4,800	15,400	0	25,000	√		29,475							96,250		79,602		√
124					354,956	69,900		57,200							1,010,604		28,800		
128	0	0	4,100	400	4,500			6,466		66,200							231,000		
130	9,938	0	0	0	9,938	26,200				29,600					105,290		49,911	14,728	
131	2,000	60	0	0	2,060		9,000	385							4,602		9,273		
132	25,000	25,000	40,000	0	90,000	18,000		36,000							600,000		91,848	450	
135	4,000	3,000	0	0	7,000			5,000			10000*				15,000		5,000	750	
139	4,000	8,000	40,000	0	52,000	7,260		6,837		31,190					40,625		58,880	290	√
141	6,500	0	24,000	0	30,500	5,000					5,000				90,000		10,200	225	
142	1,000	1,000	10,000	0	12,000	6,500		2,035		28,680					62,625		8,525	100	
143	6,000	4,000	30,000	0	40,000	128,430		38,000							30,000		4,041	400	
145	0	1,500	3,000	0	4,500			500							1,272		765		
153	2,200	0	0	0	2,200	3,600									24,300		750	50	
156	6,400	4,000	6,400	0	16,800	6,000		3,560							29,700		14,630		

Exhibit 7-25 (continued). Annual Treatment Reagent Use Data

	Flow Rates, gpd					Common Chromium Reduction Chemicals		Common Cyanide Oxidation Chemicals		Common Precipitation Pretreatments					Common pH Adjustment Chemicals			Flocculant	
PS Code	Chromium Wastewater Flow	Cyanide Wastewater Flow	Acid/Alkaline Wastewater Flow	Other Wastewater Flows	Total Wastewater Flow	Sodium Bisulfite lbs	Sulfur Dioxide lbs	Sodium Hypochlorite (15% solution) gal	Chlorine Gas lbs	Calcium Chloride lbs	Ferrous Sulfate lbs	Sodium Hydrosulfite lbs	Aluminum Sulfate lbs	DTC lbs	Sodium Hydroxide (50% solution) lbs	Lime lbs	Sulfuric Acid lbs	Polymer lbs	Additional Reagents Used
165	3,765	0	0	0	3,765	8,580									37,400				√
172					70,000	18,800		1,300							41,688		8,280		
173					42,400			1,155				9,000			70,000			750	
174	10,000	5,000	15,000	0	30,000	300		400				100			5,092		1,148		√
176	15,000	0	85,000	0	100,000		13,500									192,400	51,450	75	
179	0	4,400	3,600	0	8,000			1,430							2,800			150	
180	25,000	500	0	0	25,500		16,000	2,450							√			30	
190	0	0	18,000	0	18,000										18,202	2,500			√
191	6,000	0	0	0	6,000	3,400									49,192				
199	2,000	0	18,000	0	20,000	18,300									140,008				√
204					208,720	113,500		8,055		597,632					1,202,020		498,599	75,000	√
206	7,200	1,800	0	0	9,000	3,000		660		1,100					7,636		4,210	30	
207					18,000	10,000		2,500							127,280		35,880		
210					20,000	30,000		1,350		18,000					120,000		40,000		
213	7,800	4,000	38,000	0	49,800	10,000		10,800							127,280		20,207	3,000	
215	0	0	49,000	0	49,000											135,800		2,700	
218					51,820				2,000					√	99,406				
228	15,000	600	26,900	0	42,500	57,000		7,000							180,000		40,000	1,350	
243	34,000	13,600	0	0	47,600	3,750		15,290							59,700		2,694	4,163	√
244					25,810			9,733							80,352		15,522	190	
245	24,400	0	0	0	24,400	1,200									51,102			12,975	
247	9,000	0	95,000	5,000	109,000	276									62,000	38,000			
248	32,140	0	42,900	0	75,040	3,600									168,010		42,097		√
250	2	0	11,000	0	11,002	330							3,306		33,600		27,000	75	√
253	28,282	3,300	0	0	31,582			9,619				12414*			180,636		24,568	660	
254	3,200	0	54,000	0	57,200	10,000									170,000		54,000	12,000	
256					140,000			42,000							1,069,152		321,469		
258					170,000	58,000		52,000							47,250	12,000	3,000	1,600	
264	40,000	20,000	17,900	0	77,900	23,600		48,441							175,000	41,200	75,200	2,420	
272	0	30,000	0	0	30,000	10,000		12,652		12,820					62,510		109,906		√
273					813,000	150,400	112,227				646,950					511,730		500	
274					47,000	1,300									24,502		29,468	780	√
276					42,000										10,433				
277	20,000	0	30,000	20,000	70,000					278,400			160,000		458,208		30,616		
278	10,000	0	30,000	0	40,000	6,500									99,278		59,701	19,500	
280					142,000	57,783				77,500				3,740	450,000		36,617		√

Exhibit 7-25 (continued). Annual Treatment Reagent Use Data

	Flow Rates					Common Chromium Reduction Chemicals		Common Cyanide Oxidation Chemicals		Common Precipitation Pretreatments					Common pH Adjustment Chemicals			Flocculant	
PS Code	Chromium Wastewater Flow	Cyanide Wastewater Flow	Acid/Alkaline Wastewater Flow	Other Wastewater Flows	Total Wastewater Flow	Sodium Bisulfite	Sulfur Dioxide	Sodium Hypochlorite (15% solution)	Chlorine Gas	Calcium Chloride	Ferrous Sulfate	Sodium Hydrosulfite	Aluminum Sulfate	DTC	Sodium Hydroxide (50% solution)	Lime	Sulfuric Acid	Polymer	Additional Reagents Used
						lbs	lbs	gal	lbs	lbs	lbs	lbs	lbs	lbs	lbs	lbs	lbs	lbs	
281	40,000	0	30,000	0	70,000	12,000									208,640		122,082	9,285	
284	600	0	0	0	600	800									8,910		4,592	1,875	
285	960	2,560	4,480	0	8,000	14,400		2,640							17,280		10,400		
286	0	0	8,000	0	8,000										20,000		13,000		√
287	0	3,000	50,000	0	53,000			120		75,000	45,600	600	6,500		75,000		21,000	500	√
289	15,000	0	35,000	0	50,000	21,000							20,000		6,000		3,750		√
292	7,000	9,000	42,000	0	58,000	13,000		16,500		33,000					133,644			2,588	
296	0	0	120,000	0	120,000						13,000					35,000			
307	0	4,100	54,506	0	58,606			√					√		60,000		36,739	638	
312	6,500	0	0	0	6,500	24,000									48,000		48,000		
314	7,000	0	31,000	2,000	40,000	15,000									570,456		80,641	1,065	
318	0	0	13,800	0	13,800										6,085				√

*Ferrous sulfate is used for hexavalent chromium reduction.
√ indicates reagent used, but quantity unknown

Numerous additional reagents were used by survey respondents.
The following are the most significant in terms of quantity.
PS 011--5,200 lbs/yr of calcium hypochlorite
PS 013--206,436 lbs/yr of HCl
PS 020--650 gal/yr of hydrogen peroxide
PS 077--5,000 lbs/yr of calcium hypochlorite
PS 086--28,800 lbs/yr of magnesium hydroxide
PS 087--135,936 lbs/yr of magnesium hydroxide
PS 089--18,672 lbs/yr of ferric chloride
PS 102--4,512 lbs/yr of potassium permagnate and 3,200 lbs/yr of sodium sulfide
PS 105--1,040 gal/yr of hydrogen peroxide
PS 111--1,500 lbs/yr of calcium hypochlorite and 1,200 lbs/yr of soda ash
PS 118--131,000 lbs/yr of calcium hypochlorite
PS 123--18,000 gal/yr of magnesium carbonate
PS 199--5,701 lbs/yr of HCl
PS 244--2,900 lbs/yr of iron anodes
PS 248--52000 lbs/yr of soda ash
PS 273--180,587 gal/yr of hydrogen peroxide
PS 280--24,236 gal/yr of HCl
PS 286--25,000 lbs/yr of aluminum chloride
PS 287--170,000 lbs/yr of ferric sulfate
PS 289--16,000 lbs/yr of magnesium hrdroxide

Exhibit 7-26. Available Wastestream Characterization Data from the Users Survey

Shop/Stream	Flow gpd	Conc. mg/l	Mass lbs/day	Mass lbs/yr	Shop/Stream	Flow gpd	Conc. mg/l	Mass lbs/day	Mass lbs/yr
PS 058 Cr	6,000	48	2.4	624	PS 024 A/A	23,760	600	119	30,913
PS 058 CN	1,500	30	0.4	98	PS 018 Cr	4,250	107.7	3.8	993
PS 058 A/A	70,000	135	108	28,073	PS 018 CN	ND	ND	ND	ND
PS 204 Cr	36,000	40	12.0	3,122	PS 018 A/A	8,500	189.2	13.4	3,487
PS 204 CN	2,880	20	0.5	125	PS 043 Cr	1,000	159.3	1.3	345
PS 204 A/A	38,880	150	66.6	17,325	PS 043 CN	ND	ND	ND	ND
PS 289 Cr	15,000	3.5	0.4	114	PS 043 A/A	50,000	135	56.3	14,648
PS 289 CN	0	0	0.0	0	PS 111 Cr	0	0	0.0	0
PS 289 A/A	50,000	3.0	1.7	446	PS 111 CN	ND	ND	ND	ND
PS 012 Cr	7,000	32	1.9	486	PS 111 A/A	6,200	349	18.0	4,687
PS 012 CN	8,000	78	5.2	1,353	PS 142 Cr	1,000	482	4.0	1,045
PS 012 A/A	30,000	146	50.0	13,012	PS 142 CN	1,000	ND	ND	ND
PS 009 Cr	0	0	0.0	0	PS 142 A/A	12,000	426	42.6	11,087
PS 009 CN	4,500	400	15.0	3,903	PS 243 Cr	34,000	2.0	0.6	147
PS 009 A/A	190,000	1,100	2,388	620,878	PS 243 CN	13,600	ND	ND	ND
PS 022 Cr	70,000	35	20.4	5,313	PS 243 A/A	47,600	16.6	6.6	1,713
PS 022 CN	0	0	0.0	0	PS 207 Cr	0	0	0.0	0
PS 022 A/A	70,000	45	36.0	9,358	PS 207 CN	ND	ND	ND	ND
PS 048 Cr	50,000	8.5	3.5	922	PS 207 A/A	18,000	110	16.5	4,290
PS 048 CN	0	0	0.0	0	PS 264 Cr	40,000	70.2	23.4	6,089
PS 048 A/A	50,000	78.3	44.7	11,630	PS 264 CN	20,000	ND	ND	ND
PS 118 Cr	17,052	33.1	4.7	1,224	PS 264 A/A	77,900	252	163	42,500
PS 118 CN	0	0	0.0	0	PS 253 Cr	28,282	68.3	16.1	4,189
PS 118 A/A	42,630	520	253	65,853	PS 253 CN	3,300	ND	ND	ND
PS 176 Cr	15,000	90	11.3	2,927	PS 253 A/A	31,582	470	124	32,173
PS 176 CN	0	0	0.0	0	PS 087 Cr	50,000	127	53.0	13,769
PS 176 A/A	105,000	475	570	148,164	PS 087 CN	ND	ND	ND	ND
PS 199 Cr	2,000	700	11.7	3,036	PS 087 A/A	300,000	1,535	3,841	998,548
PS 199 CN	0	0	0.0	0	PS 116 Cr	0	0	0.0	0
PS 199 A/A	20,000	400	91.4	23,766	PS 116 CN	0	0	0.0	0
PS 287 Cr	0	0	0.0	0	PS 116 A/A	10,000	90	10.3	2,674
PS 287 CN	3,000	10	0.3	65	PS 210 Cr	45,000	50	18.8	4,879
PS 287 A/A	53,000	50	309	80,210	PS 210 CN	2,000	2.0	0.0	8.7
PS 023 Cr	10,000	23.6	2.0	512	PS 210 A/A	47,000	100	53.7	13,962
PS 023 CN	0	ND	ND	ND	PS 274 Cr	8,000	4.5	0.3	78
PS 023 A/A	20,000	89	14.8	3,860	PS 274 CN	0	0	0.0	0
PS 095 Cr	3,000	20	0.5	130	PS 274 A/A	29,000	3.5	1.2	302
PS 095 CN	0	0	0.0	0	PS 061 Cr	0	0	0.0	0
PS 095 A/A	20,000	60	13.7	3,565	PS 061 CN	0	0	0.0	0
PS 135 Cr	4,000	50	1.7	434	PS 061 A/A	8,000	225	20.6	5,347
PS 135 CN	3,000	25	0.6	163	PS 153 Cr	2,220	6.9	0.1	33
PS 135 A/A	7,000	95	7.6	1,976	PS 153 CN	0	0	0.0	0
PS 280 Cr	200	100	0.2	43	PS 153 A/A	2,720	90	2.8	727
PS 280 CN	0	0	0.0	0	PS 181 Cr	38,111	20	6.4	1,653
PS 280 A/A	143,800	300	493	128,156	PS 181 CN	0	0	0.0	0
PS 253 Cr	28,282	4.5	1.1	276	PS 181 A/A	38,111	70	30.5	7,925
PS 253 CN	3,300	15	0.4	107	PS 296 Cr	0	0	0.0	0
PS 253 A/A	31,582	4.6	1.7	432	PS 296 CN	0	0	0.0	0
PS 245 Cr	18,000	9.2	1.4	359	PS 296 A/A	120,000	50	68.6	17,824
PS 245 CN	0	0	0.0	0	PS 122 Cr	0	0	0.0	0
PS 245 A/A	18,000	100	15.0	3,903	PS 122 CN	0	0	0.0	0
PS 273 Cr	123	122,000	125	32,539	PS 122 A/A	175,000	50	100	25,994
PS 273 CN	73	9,400	5.7	1,488	PS 296 Cr	0	0	0.0	0
PS 273 A/A	813,000	65	439	114,237	PS 296 CN	0	0	0.0	0
PS 024 Cr	5,760	150	7	1,873	PS 296 A/A	120,000	50	69	17,824
PS 024 CN	7,200	40	2.4	624					

ND = no data

Notes:

(1) A/A flow is the combined flow into the metals precipitation treatment step. Therefore, it may include the chromium and cyanide streams as well as non-pretreated streams.

(2) Concentration and mass data indicate the quantity of chromium, cyanide or total regulated metals, depending on the wastestream. Where ferrous reagents are used, the iron content of the reagents was added to the metal content of the A/A stream. Total metals values were calculated from regulated metals values, plus a multiplication factor of 1.37 to account for non-regulated metals (based on respondents sludge data in Section 8).

(3) A/A flow for PS 273 includes 2 mg/l Cr. Multiplication factor was not used due to completeness of data provided.

tions. Because the Federal regulations are based on the technical capabilities of conventional treatment, these processes are often unable to achieve the lower concentration limits.

- Metal finishing wastewaters often contain compounds that interact with dissolved metals and interfere with their precipitation as metal hydroxides. Such compounds as ammonia, phosphates, tartrates, and ethylene-diaminetetraacetic acid (EDTA) are commonly used in plating and printed circuit board operations and consequently find their way into the wastewater. These compounds, called chelates, combine with the dissolved metal ion to form a complexed ion that is relatively soluble in neutral or slightly alkaline solutions. In many cases, waste streams containing high levels of chelates (e.g., printed circuit board shops) cannot be treated with conventional precipitation to the level required by Federal regulations.

- In some cases, metal discharge requirements are not being met, even though the level of dissolved metals in the effluent is low. In cases of this kind, the solids separation component of the process may be allowing too much suspended matter, including precipitated metals, to pass into the discharge. This condition can result from overloaded or poorly designed clarifiers, ineffective conditioning (coagulation or flocculation) of the clarifier feed, or poor pH control.

- Some shops may find that the capital and/or operating costs of conventional processes are too high. Alternative technologies may reduce capital costs under certain conditions (e.g., if a centralized waste treatment plant is located nearby) and/or specific components of operating costs (e.g., shops located long distances from sludge disposal sites may focus on reducing sludge volumes).

The alternative technologies discussed in this section are the ones that have been installed by the respondents to the Users Survey; they include: ion exchange, microfiltration and evaporation. Each of these alternatives offers solutions to one or more of the technical problems encountered with conventional treatment or, in some cases, offers cost savings through a reduction in equipment requirements or operating expenses such as chemical reagent purchases. In all cases, however, alternative technologies offer a trade-off. Their advantages are gained at the expense of other benefits. Such trade-offs are often site-specific and must be evaluated case by case.

In addition to the three alternative technologies for metals removal discussed in this report, there exists many other alternatives that have been reported in the literature and/or utilized in plating shops. Examples of these include sulfide precipitation, sodium borohydride precipitation, dissolved air flotation, freeze crystallization, and insoluble starch xanthate treatment. Of these technologies, sulfide precipitation is the most widely applied in the metal finishing industry. Although it is not used by any respondents to the Users Survey, it is used in the plating industry, as documented in the literature (e.g., ref. 39, 393, 348). Also, the sulfide precipitation chemistry is sometimes used with end-of-pipe membrane filtration systems. This application is discussed in Section 7.5.3.

7.5.2 Ion Exchange

7.5.2.1 Overview

This section deals with ion exchange as an end-of-pipe technology. Ion exchange is a versatile pollution prevention and control tool for the metal finisher. It is utilized for raw water treatment, metals recovery, water recycle, bath maintenance and end-of-pipe treatment. Use of ion exchange for metals recovery/water recycle and process solution maintenance are discussed in Sections 4 and 5, respectively.

Approximately 6% of the respondents to the Users Survey use ion exchange as an end-of-pipe technology. Several different configurations are being employed by survey respondents and these are discussed in Section 7.5.2.3.

A general description of the ion exchange process is presented in Section 4.

7.5.2.2 Development and Commercialization

The development and commercialization of the ion exchange technology is discussed in Section 4. No direct references were found in the literature to indicate when ion exchange was first applied as an end-of-pipe technology. However, it can be assumed from the fully developed nature of the technology and the need to meet local discharge standards in certain locations of the U.S. that the technology was applied for this purpose in the

1950's. The earliest application from the Users Survey was 1972 (PS 161). The EPA's regulatory document for the metal finishing industry indicated that 63 of the plants in their database used ion exchange (ref. 386). Their data collection efforts were performed in the late 1970's and early 1980's. The EPA document is unclear as to the total number of plants included in the database and therefore the percentage of shops that used ion exchange at that time cannot be calculated. However, as a comparison, EPA indicated that 154 of the plants in their database used hydroxide precipitation, a very common end-of-pipe technology. This would indicate that ion exchange had a very prominent role in pollution control at the time of the EPA study. (Note that a portion of the ion exchange users in the EPA report may have used the technology for raw water treatment, metals recovery or bath maintenance. EPA does not specify which applications were used by the shops in their database.)

Ion exchange is a very important end-of-pipe technology for shops utilizing the services of a centralized waste treatment (CWT) facility. Ion exchange resin beds can be shipped to a CWT facility more economically and safely than liquid wastes. CWT systems that utilized ion exchange were present in Europe and Asia in the 1970's (ref. 387). Small U.S. operations were present in the 1970's; however, the only major U.S. CWT plant (Minnesota Recovery Systems) was not operating until 1988 (ref. 419).

The Users Survey indicates that the role of ion exchange as an end-of-pipe treatment technology has continued through the 1980's and into the 1990's. Several shops have installed end-of-pipe equipment after 1990. Also, several of the respondents are using CWT services for regeneration of the columns, including four shops using the new CWT plant in Minnesota.

7.5.2.3 Applications and Restrictions

The shops that are discussed in this section use ion exchange for treatment of a mixed metal waste stream and/or they use it as the sole end-of-pipe method of meeting effluent limitations for metals.

In some cases, it is difficult to differentiate between recovery and end-of-pipe treatment applications of ion exchange and therefore several shops are discussed in both Section 4 and this section. These are shops that treat a single metal wastestream using ion exchange, but do not recover the metals on-site. Rather than recovering the metals on-site, these shops use a centralized waste treatment system for regeneration or they regenerate on-site and send the regenerant to an off-site metals recovery facility. In either case, their ion exchange system is the sole means of metals removal from their wastewater (i.e., no conventional treatment is present). Shops that fall into this category include: PS 061, PS 161, PS 136, PS 192, and PS 196.

Some shops in the Users Survey that employ ion exchange without any chemical recovery were placed solely into the recovery category (i.e., Section 4.4). This is a group of shops that does not perform chemical recovery; however, they treat a single metal rinse water using ion exchange, and therefore there exists a high potential for recovery. Also, each shop in this group has another technology present for end-of-pipe treatment (e.g., hydroxide precipitation). Because ion exchange is not used as their primary treatment system, this group of shops is only included in Section 4. As a result of this grouping, readers interested in ion exchange treatment should review both sections to benefit fully from the results of the Users Survey.

Ion exchange is employed as an end-of-pipe technology using three common schemes, as depicted in Exhibit 7-27. Variations of these three configurations were also observed. The first configuration employs ion exchange to remove the metals from large volumes of miscellaneous wastewaters and concentrate them into a smaller volume for subsequent treatment by conventional means (see application IXEOP-1). This was the most frequently used method of end-of-pipe ion exchange treatment by respondents to the Users Survey (PS 021, PS 036, PS 089, PS 121, PS 170, PS 187, PS 200, PS 221, PS 223, PS 209). This strategy reduces the capacity requirement of the conventional system and therefore may result in capital savings. With this configuration, all dilute rinses are sent to integrated ion exchange units (see definition in Section 4). The number of units needed will depend on the flow rate and the mixture of pollutant parameters present. As with conventional treatment, segregation of cyanide, chromium bearing and miscellaneous metal bearing rinse waters is needed. Pretreatment of the rinses may include one or more of the following processes: pH adjustment, cyanide destruction, filtration and carbon treatment. The pH adjustment step, which was performed by most shops using this configuration, is needed to ensure that the pH is within the operating range of the resin. Cyanide destruction, which was used by only

Exhibit 7-27. Common Configurations for Application of Ion Exchange to End-of-Pipe Treatment

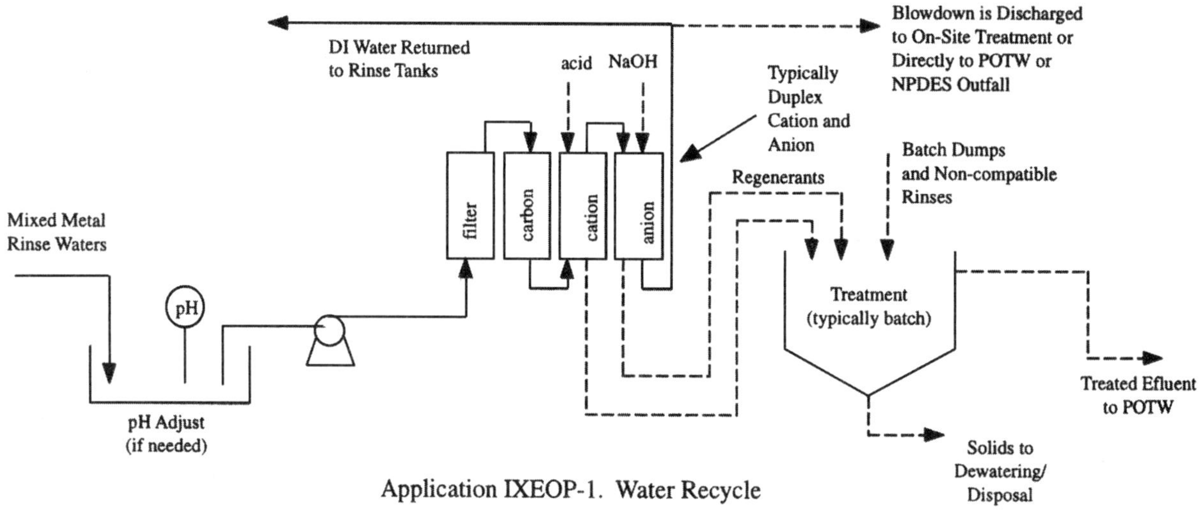

Application IXEOP-1. Water Recycle

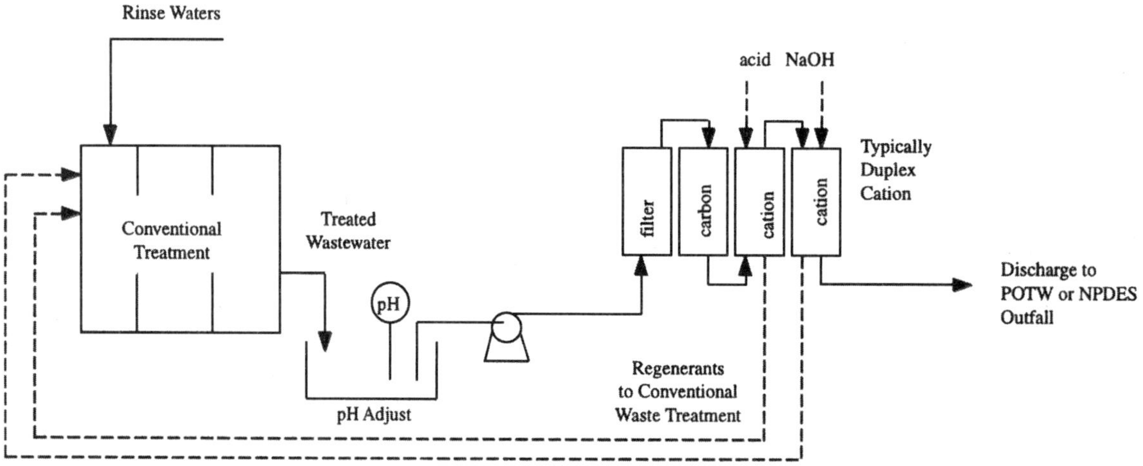

Application IXEOP-2. Effluent Polishing

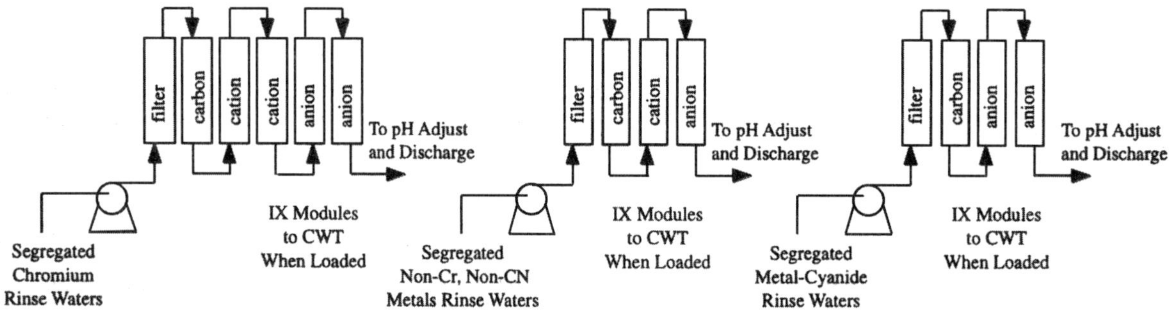

Application IXEOP-3. Centralized Waste Treatment User

one of the survey respondents (PS 223), will reduce the number of steps of the ion exchange process when cyanide complexes are present (see Section 4, Section 4.4.3 for a discussion of ion exchange treatment of cyanide-metal wastewaters). Filtration and carbon treatment, which are the most common pretreatment technologies, are needed to remove suspended solids and organics that may foul the resin beds. If water recycle is desired, the ion exchange system must consist of both anion and cation columns. Cation only systems can be used if water recycle is not desired and there are no chromates or cyanide present (these two pollutants are removed by anion resins). Concentrated wastes generated in the plating shop (e.g., batch dumps), should be discharged directly to the batch treatment system, rather than into the ion exchange feed stream. This practice prevents overloading of the ion exchange beds. Also, it should be noted that ion exchange is simply not a good technology for the treatment of concentrated wastes. If used for this purpose, the regenerant may be less concentrated than the original waste.

From the Users Survey, 11 respondents (or 3.5% of all respondents) utilized application IXEOP-1 for treatment of mixed metals waste streams. Of these, 7 shops recycle the water treated by ion exchange to rinse tanks (PS 021, PS 036, PS 170, PS 185, PS 187, PS 200, PS 209) and 4 shops do not recycle water, but rather discharge it directly to a sewer (sometimes after pH adjustment) or further process it with additional metals removal treatment (e.g., hydroxide precipitation) and then discharge it (PS 089, PS 121, PS 221, PS 223).

Shops that employed application IXEOP-1 with water recycle generally had a lower than average discharge rate. The average for all shops in the survey (318 shops) was 34,600 gpd and the average for the 11 shops using application IXEOP-1 was 10,064 gpd. One the shops using ion exchange for water recycle attained zero-discharge of effluent (PS 021). This shop batch treats regenerant in a 1,500 gallon tank (chromium reduction and metals precipitation), transfers the liquid waste to an atmospheric evaporator for concentration and dewaters the concentrate and sludge with a filter press. Approximately 1,100 lbs of sludge (20% solids) is sent off-site per year.

In the second configuration (application IXEOP-2), ion exchange is used as a polishing technology to remove residual pollutants following conventional treatment. Such applications are most common to facilities that directly discharge wastewaters to rivers or streams and are required to meet stringent metals limitations (ref. 39). Respondents to the Vendors Survey indicated that they have installed 21 polishing systems. However, only one Users Survey respondent used ion exchange in this manner (PS 068). The regulations for this particular plating and anodizing shop are below the 40 CFR 433 limitations for most parameters. PS 068 does not regenerate their spent resins on-site. Rather, they send them off-site for incineration and metals recovery, which is an unusual practice for this ion exchange application, especially considering the moderately high cost of resin. Their spent resins contain nickel, copper, silver and zinc. Presumably, the quantity of silver present in their resins justifies this practice.

IXEOP-2 is not a good configuration to use for recycling water. Water recycling requires the use of both anion and cation columns (i.e., deionization). Because the wastewater treatment process adds such an abundance of sulfates and chlorides to the wastewater, it is economically impractical (except under unusual conditions) to recycle it using ion exchange. The loading on the anion column and the resultant regeneration frequency would simply be too great. If water recycle is the goal, it is better to segregate wastewaters and recover the water as shown in application IXEOP-1 (ref. NAPCO file).

Ion exchange polishing applications generally use metal selective cation resins that have a high preference for heavy metal cations over the more common alkali or alkaline earth cations such as sodium, calcium, magnesium and potassium that are added by the conventional treatment process.

In the third configuration (application IXEOP-3), ion exchange is used as the primary end-of-pipe treatment for segregated metal and cyanide bearing rinse waters and the resin columns are sent to a CWT facility for regeneration. Because the wastestreams were segregated before ion exchange, the regenerants contain metals that can be recovered by relatively easy means (e.g., electrowinning). Application IXEOP-3 is based on sketches submitted by PS 192 and PS 237. Other shops using CWT had similar configurations (PS 136, PS 249), with only one significant variation, where resins were regenerated on-site and the regenerant was sent to an off-site metals recovery firm.

7.5.2.4 Technology/Equipment Description

This subsection contains a description of commercially available ion exchange equipment that is manufactured and/or sold by vendor survey respondents. This is intended to provide the reader with information and data on a cross section of available equipment. Mention of trade names or commercial products is not intended to constitute endorsement for use.

Descriptions of ion exchange system designs and commercially available equipment are presented in Section 4.4.4. Systems for applications IXEOP-1 and -2 will typically be integrated units and modular units are used for IXEOP-3. One additional commercial system is described in this section that is advertised specifically as end-of-pipe treatment. It would be employed as shown in Exhibit 7-27, application IXEOP-1.

Memtek Corporation manufactures the Rinse Water Maintenance System (RMS™), which is advertised as an end-of-pipe treatment/water recycle system. The unit has duplex cation and anion beds, employed in a parallel/standby mode of operation (see Section 4, Section 4.4.1.3 for definitions). It is advertised to be applicable to chromium, cyanide and acid/alkaline rinse waters. Optional equipment includes a batch treatment system for regenerants as well as an atmospheric evaporator or electrowinning unit for further processing the regenerants.

7.5.2.5 Costs

7.5.2.5.1 Capital Costs

Capital costs for end-of-pipe ion exchange equipment will be similar to those presented in Section 4, Section 4.4.5.1 for integrated and modular systems.

7.5.2.5.2 Operating Costs

Operating costs for end-of-pipe ion exchange equipment will be similar to those presented in Section 4.4.5.2 for integrated and modular systems. End-of-pipe applications may experience higher resin costs (approximately 25%) due to a shorter life span and higher fines loss (ref. Kinetic Recovery Corporation file).

Several respondents to the Users Survey provided operating cost data for using a CWT service (PS 161, PS 192, PS 237). The following data were provided by these respondents.

PS 161 leased their ion exchange columns that contain 3.2 ft^3 of mixed (anion and cation) resin. During 1992 PS 161 sent 18 nickel columns and 8 chromium columns to Dayton Water Systems. The cost of the service was $336 per column ($105/ft^3) for regeneration plus a $70 per month rental fee for the equipment (therefore, the total annual cost for 1992 is $6,888 for nickel and $3,528 for chromium). According to the respondent, Dayton Water Systems sent the regenerant to Inmetco for recovery of the metal (see Section 8 for information on off-site recycling).

PS 192 and PS 237 use the Metropolitan Recovery Systems CWT facility in Minnesota. Their operating costs are shown in Exhibit 7-28.

7.5.2.6 Performance Experience

Different sections and questions were contained in the Users Survey form for recovery (Section 6 of the survey form) and end-of-pipe systems (Section 7 of the survey form). Although respondents that use ion exchange for end-of-pipe were not asked to complete Section 6 for these applications, all of them did. Therefore, there are more data available for ion exchange end-of-pipe appli-

Exhibit 7-28. Centralized Waste Treatment Charges for Two Respondents to the Users Survey

PS 237:

Waste Stream	ft^3 resin/yr	$/$ft^3$	$/yr
Cr/Nitric Acid	1,250	$ 46	$ 57,000
Nickel	275	33	9,000
Mixed Cyanide	1,900	79	] 50,000
Pickles	1,100	29	32,000

PS 192:

Waste Stream	ft^3 resin/yr	$/$ft^3$	$/yr
Cr/Nitric Acld	600	$ 51	$ 30,500
Electroless Ni	66.5	90	6,000
Zinc (non-CN)	66.5	83	5,500
Zinc Cyanide	126	103	13,000

Exhibit 7-29. Summary of Users Survey Data for Ion Exchange EOP Treatment

Shop				Capital Costs			Operating Costs (Yr.)		Savings						Use	Down	Satisfaction Levels		Future
Code	Application	Manf.	Year	Equip.	Other	Total	Non-labor	Labor Hrs.	Water	Treat. Chem.	Sludge Disp.	Energy	labor	Other	Code	Time %	Manf. Support	Technology	Decision
187	IXEOP-1	Compliance Recycling	1986	$20,000	$20,000	$40,000	$30,000	2,000	$1,400						1	5	2	3	2/3
170	IXEOP-1	L&T Technologies	1990	$20,000	$24,000	$44,000	$1,500	20	$8,000						3	15	3	2	3
121	IXEOP-1	Culligan Water Systems	1977	$5,000	$1,000	$6,000	$500	50	$12,500	$2,000	$10,000				1	5	3	4	2
209	IXEOP-1	Amerdec	1987	$150,000	$6,000	$156,000	$1,800	300	$24,000	$3,000	$5,000	$500	$2,000		1	5	3	3	3
021	IXEOP-1	Palm Commodities	1992	$40,000	$5,000	$45,000	ND	416		√					1	5	3	3	2
036	IXEOP-1	DiSep International	1987	$113,728	$15,000	$128,728	$23,177	2,080	√	√					1	5	5	5	1
200	IXEOP-1	Compliance Recycling	1989	$50,000	$10,000	$60,000	ND	ND							1	5	3	4	1
221	IXEOP-1	Structural Fibers	1986	$10,000	$2,000	$12,000	$3,000	500							1			4	2
223	IXEOP-1	Ambi Corporation	1986	$90,000	$15,000	$105,000	$200,000	2,000		√			√	√	1	1	4	4	4
185	IXEOP-1	Unk	1991	$20,000	$10,000	$30,000		150	$10,000	$500	($3,000)	($5,000)		($5,000)	1	3	2	3	3
		Average IXEOP-1	1987	$51,873	$10,800	$62,673	$37,140	835	$11,180	$1,833	$4,000	($2,250)	$2,000	($5,000)		5.4	3.1	3.5	
068	IXEOP-2	In-House	1990	$40,000	$12,000	$52,000	$10,000	600			$2,000			$1,000	1	5	ND	5	1
061	IXEOP-3	Bio Recovery	1989	$123,000	$10,000	$133,000	$20,000	1,040		doubled	$0	up 20%	up 50%		2	75	3	1	3
089	IXEOP-3	MRS	1988	$8,627	$1,883	$10,510	$82,000	80		√	√				1	1	3	4	2
136	IXEOP-3	MRS	ND	leased	ND	ND	ND	ND							1	15	1	1	3
161	IXEOP-3	Dayton Water System	1992	$39,810	$3,000	$42,810	$12,500	350			$200				1	5	4	5	1
192	IXEOP-3	MRS	1988	$100,000	$28,000	$128,000	$55,000	525		$0	$0	$0	$0	$0	1	1	4	4	1
196	IXEOP-3	Dayton Water System		$52,630	$3,000	$55,630	$3,000	150		$1,000	$1,000	$0	$0	$0	1	2	4	3	3
237	IXEOP-3	MRS	1988	$60,000	$46,455	$106,455	$216,000	600		$0	$0	$0	$0	$0	1	1	4	4	1
249	IXEOP-3	Douglas International	1990	$64,011	$25,000	$89,011	$2,000	500		$0	$0				1	1	4	4	1
		Average IXEOP-3	1989	$64,011	$16,763	$80,774	$55,786	464		$250	$200					12.6	3.4	3.3	

(1) Application codes are defined in Exhibit 6-38.

(2) MRS is the Metropolitan Recovery Systems, a CWT facility in Minnesota.

(3) Other capital costs include installation and ancillary equipment.

(4) Use Codes: 1-currently operating; 2-not currently operating and no intension for future use; 3-not currently in use, but intend to use in future.

(5) Satisfaction levels (manufacturer and technology) 1 to 5, with 1=lowest and 5=highest.

(6) Future decision codes: 1-purchase the same technology from the same vendor; 2-purchase the same technology from a different vendor; 3-purchase a different technology; 4-do nothing.

(7) "√" indicates that a savings was realized, but not quantified.

(8) ND = no data

cations than for than for most other end-of-pipe applications. A summary of the data for ion exchange applied to endof-pipe treatment is presented in Exhibit 7-29.

The following information and data summarize the perfonnance experience of the survey respondents.

- The performance of ion exchange, as an end-of-pipe treatment technology, was rated approximately the same as for recovery applications. The average satisfaction level for ion exchange applied to end-of-pipe treatment is 3.5 for IXEOP- 1, 5.0 for IXEOP-2 (only one respondent) and 3.4 for IXEOP-3, whereas the average level rating for ion exchange used for chemical recovery is 3.5. Fourteen of the shops (or 74 %) indicated that this technology satisfied the need for which it was purchased (includes all applications), 4 (or 21%) indicated that it partially satisfied the need for which it was purchased and 1 (or 5%) indicated that it did not satisfy the need for which it was purchased (these data are not shown on Exhibit 7-29, but can be found in the database). The following is a breakdown of the reasons why shops purchased this technology (most shops had multiple reasons):

To meet or help meet effluent regulations: 15
To reduce the quantity of waste shipped off-site: 9
To reduce wastewater treatment costs: 11
To improve product quantity: 6
To reduce or eliminate effluent discharge: 3
To improve rinse water quality: 1

- The use of ion exchange for water recycle generally did not impact production quality or the rate of production. The following responses were provided (only for shops that recycle water):

	Production Rate	Product Quality
Improved	1	1
NoChange	10	8
Decreased	1	2

- Product quality was decreased for PS 121 and the production rate was decreased for PS 121 and PS 185. No explanations were given for these impacts.

- Based on their experience with this technology, 37% indicated that they would purchase the same technology from the same vendor. The following is a breakdown of all responses:

Purchase the same technology from
the same vendor: 7 (37%)
Purchase the same technology from
a different vendor: 4.5 (24%)
Purchase a different technology: 6.5 (34%)
Do nothing: .... 1(5%)

- Two shops indicated that an ion exchange unit applied to end-of-pipe treatment was the cause of an effluent compliance excursion (PS 061, PS 136). Three shops did not respond to the question (PS 185, PS 187, PS 223).

- The major savings for this technology was water use reduction. PS 209 reported the greatest water reduction. In their survey form, PS 209 indicated that they achieved a flow reduction of 20,000 gpd since 1988 (current average flow is 11,500 gpd).

- The average plating shop effluent discharge for all shops that use ion exchange for end-of-pipe treatment and water recycling (i.e., application IXEOP- 1) is 10,064 gpd. The average for all shops responding to the survey (318 shops) is 34,600.

7.5.2.7 Operational and Maintenance Problems

Reported O&M problems for end-of-pipe ion exchange applications relate mostly to resin fouling and complaints about the frequency of regenerating the resin beds. Of the 18 shops providing data, 16 (or 89%) were still operating their ion exchange equipment for end-of-pipe treatment. The average age of the operating systems was 5 years for IXEOP-1, 3 years for IXEOP-2 (only one respondent), and 4 years for IXEOP-3. The average percentage of downtime experienced by the respondents was 5.4% for IXEOP-1, 5.0% for IXEOP-2 and 14.1% for IXEOP-3 (however, just 4.0% excluding PS 061).

The following summarizes the respondents' O&M experiences and provides operating labor information.

- The average number of annual operating hours per ion exchange system were: 835 hrs/yr for IXEOP-1, 600 hrs/yr for IXEOP-2, and 464 hrs/yr for IXEOP-3. The skill requirement commonly needed for operating this technology is a wastewater treatment plant operator and/or a trained technician. The following is a breakdown of the responses for skill requirements (multiple responses given by most respondents):

Environmental Engineer:	0
Process/Chemical Engineer:	3
Chemist:	5
Consultant:	3
Plumber/Pipe Fitter:	4
Electrician:	3
Vendor:	1
Senior-Level Plater:	2
Junior-Level Plater:	4
Trained Technician:	8
Wastewater Treatment Plant Operator:	8
Common Labor:	3
Other:	0

- Six shops (or 33%) indicated that resin fouling was an O&M problem (includes all applications). The causes of fouling were not always given, however, two respondents gave the following reasons: bacterial growth during warm weather (PS 170) and oil in the feed stream (PS 209). One shop indicated that they must frequently wash the resin (PS 223) and another estimated that fouling must be dealt with approximately four times per year. Two shops indicated that prefiltration was very important.

- Several shops complained that the resin beds load too quickly and must be regenerated (PS 068, PS 187, PS 170). One shop attributed a past problem with resin bed loading to an inefficient water softener that caused their bed to load with hardness causing ions.

- Only one shop indicated that resin life was an O&M problem (PS 223). One shop indicated that resin life was 4 to 5 years (PS 036).

7.5.2.8 Residuals

The primary residuals from ion exchange end-of-pipe treatment applications are the regenerants (eluates) and backwash waters. Other residuals include cartridge filters and spent carbon. The regenerants are concentrated solutions and the backwash is dilute. Both solutions are either caustic or acidic, depending on the resin type and application. Regenerant or backwash from cation regenerations will typically contain acid, nickel, trivalent chromium, zinc, copper and/or other positively charged metals depending on the feed stream. The anion regenerant or backwash will typically contain caustic, cyanide, and hexavalent chromium. Regenerants are typically treated on-site or hauled to a metals recovery/disposal site. Backwash is typically treated on-site.

The volume of regenerant produced will depend on the regeneration requirement (e.g., lbs of acid per ft^3 of resin) and the concentration of acid used (typically 1 to 5%). The regeneration requirement will depend on the resin type, application (metal or complex being recovered) and the configuration (co-current vs counterflow). Typical volumes of regenerant are 20 to 50 gal/ft^3 of resin per regeneration. Resin loading capacities are discussed in Section 4. The volume of regenerant waste is sometimes reduced by reusing the last portion of the regenerant, which will be less contaminated with metal and contains free acid or caustic. Backwash volumes depend mostly on the equipment design and the application. Typically, backwashing generates 25 to 75 gal/ft^3. The backwash is partly reused by some equipment vendors as make-up water for regenerant, in an effort to reduce the total waste volume generated. Because backwash contains only dilute concentrations of pollutants, it is typically not a major concern and is treated on-site and discharged. However, for shops working toward zero-discharge, the backwash volume could present a significant problem. Both backwash and regenerant can be processed by evaporation to reduce the volume requiring disposal. However, this increases the capital and operating costs of the system (see Sections 4.2 and 4.3). Also, if evaporation units are not directly connected to wastewater treatment systems, they may require a RCRA permit to operate them (see Section 3.7).

Waste treatment processes for these wastes generate sludge that is an EPA listed hazardous waste (F006). Unless chemical recovery is practiced, ion exchange (application IXEOP-1) with on-site batch treatment of regenerants will produce approximately the same quantity of F006 sludge as that produced from conventional treatment without ion exchange. Since application IXEOP-2 is an extra polishing step following conventional treatment, combined with the conventional system, it will result in slightly greater sludge generation than with conventional treatment alone.

Some plating shops send their ion exchange columns to an off-site CWT facility for the regeneration process. This practice can eliminate the on-site generation of

F006 sludge. However, shops considering CWT should investigate the regulations and liabilities associated with the transport of the ion exchange columns and off-site treatment/recovery/disposal of the resultant materials.

7.5.3 Membrane Filtration

7.5.3.1 Overview

Membrane filtration, which includes microfiltration and ultrafiltration (see definitions in Section 5) is applied as an alternative technology to conventional precipitation/clarification or as a polishing technology. When used as an alternative technology to conventional precipitation/clarification the membrane filter separates treated wastewater from precipitated metals and concentrates the precipitated metals to approximately 2 to 5% solids. When used as a polishing technology, it typically follows a conventional clarifier and it removes residual solids and organics from the wastewater. The polished effluent can be discharged and will usually meet strict limitations, or it can be reused as rinse water. The membrane filtration technology is used by four (or 1.3%) of the respondents to the Users Survey.

Membrane filtration equipment is typically more capital intensive than its conventional treatment equipment counterparts. However, membrane filtration has found use in several specific areas where the conventional technology lacks sufficient performance ability. These applications include: (1) where a shop is required to meet effluent limitations significantly below the Federal standards; (2) at printed circuit board shops where metal complexing agents are heavily used; and (3) for meeting organic parameter discharge limitations such as biochemical oxygen demand (BOD). The first two of these applications were identified from the Users Survey and are believed to be the most frequently used purposes for this technology by the metal finishing industry. The third application was identified from the literature which referenced a microfiltration installation at an industrial laundry facility (ref. 414). Presumably, similar applications are present within the metal finishing industry.

Of the four respondents using this technology (PS 007, PS 105, PS 233, and PS 257), three respondents used microfiltration as a replacement for conventional precipitation/clarification and one used it as part of a complex zero-discharge configuration involving conventional precipitation, followed by ultrafiltration and reverse osmosis, with vacuum evaporation applied to the concentrate discharged from the ultrafiltration unit (PS 233). Of the remaining three shops, two had discharge limitations below the Federal limits and one had CFR 433 standards, but indicated that they had difficulty meeting the nickel limitation (PS 105). One of the two shops with low limitations was a direct discharger (PS 007). Three of the shops were job shops performing common metals plating and one was a captive printed circuit board manufacturer (PS 257).

An end-of-pipe treatment system employing microfiltration is similar to a conventional hydroxide precipitation system with the exception that the clarifier/thickener is replaced by the microfiltration unit. Also, since microfiltration systems do not depend on the ability of the precipitated metals to settle, a polymer addition/flocculation step is not present. In fact, polymers cannot be used with membrane filtration because they will foul the membrane.

Membrane filtration is capable of producing an effluent with a much lower suspended solids concentration than conventional precipitation/clarification. This is an important consideration for shops trying to achieve low metals limitations. Data from the literature (Exhibit 7-30) indicate that the contribution of TSS to metal discharges can

Exhibit 7-30. Contributions of Total Suspended Solids to Metal Discharges

Parameter	Residual Soluble Metal After Ca(OH): Precipitation mg/l	Metal Contribution from 5.0 mg/l TSS	Metal Contribution from 1.0 mg/l TSS	Metal Contribution from 0.2 mg/l TSS
Nickel	0.11	0.188	0.037	0.0075
Copper	0.02	0.32	0.064	0.0128
Zinc	0.05	0.24	0.049	0.0098
Calcium	14	—	—	—
Chromium	0.001	0.038	0.018	0.003
Iron	0.05	0.28	0.056	0.011

Source. ref 477

Exhibit 7-31. End-of-Pipe Treatment System Employing Microfiltration

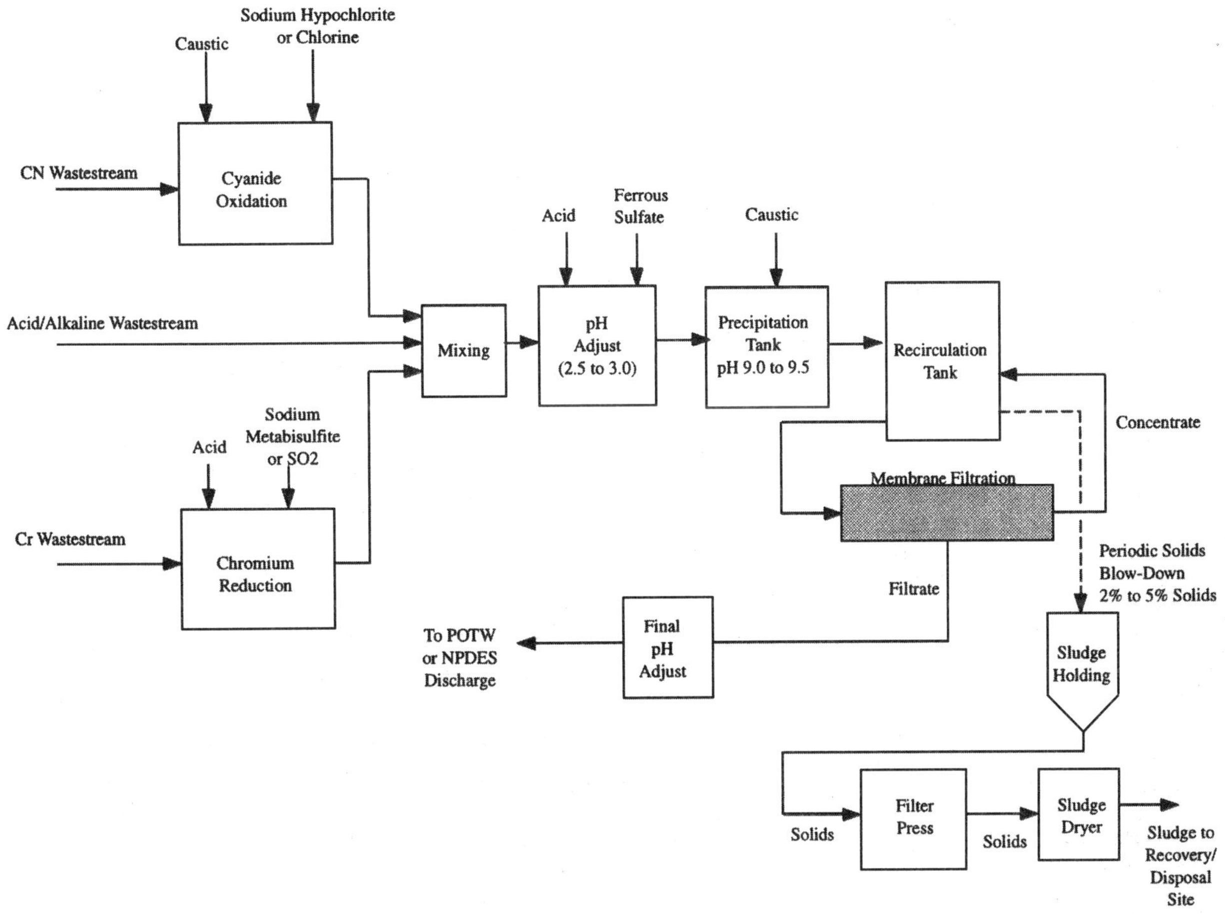

be significant and will exceed the soluble contribution at a TSS level as low as 1 mg/l. This indicates that, for plants needing to meet low limitations, the TSS metals fraction is a bigger and better target than the soluble fraction for treated wastewaters. When complexed metals are present and conventional hydroxide precipitation is employed, TSS will make-up a smaller fraction of the total metals concentration than that shown in these data. In such cases, or even in the absence of complexing chemicals, the effect of the membrane technology is enhanced by using additional treatment reagents. Reagents that are commonly used to enhance metals removal with membrane systems include: calcium chloride, sodium sulfide, ferrous sulfate and DTC.

7.5.3.2 Development and Commercialization

The development and commercialization of membrane filtration is reviewed in Section 5.3. The literature indicates that in 1988, more than 100 full-scale industrial end-of-pipe systems had been installed by Memtek, the leading manufacturing company of this technology. These systems ranged in capacity from 10 to 400 gpm and were installed in facilities in the following industries: electroplating, printed circuit board manufacturing, battery manufacturing and photographic processing. Projecting the results from the Users Survey to the entire electroplating population, the expected number of systems would be approximately 170 (based on 4 of 318 respondents with systems and a total U.S. plating population of 13,500 plants).

7.5.3.3 Applications and Restrictions

The general applications for this technology were discussed in Section 7.5.3.1. Exhibit 7-31 shows a generic EOP microfiltration system that could substitute for conventional treatment. For this application, the process chemistry will vary depending on the type and concentration of the constitutes in the feed stream and the target effluent levels. As discussed in Section 7.5.3.1, addi-

Exhibit 7-32.
EOP Membrane System Performance Data

Contaminants	Feed (mg/l)	Effluent (mg/l)
Aluminum	10 - 1,000	0.5
Arsenic	1 - 43	0.05
Cadmium	25 - 115	0.05
Chromium	3 - 275	0.1
Copper	1 - 1,525	0.1
Cyanide	5 - 300	0.1
Fluoride	18 - 5,000	1.0
Gallium	4 - 20	0.5
Germanium	20 - 110	0.5
Gold	1 - 12	0.15
Iron	2 - 1,500	0.02
Lead	2 - 25	0.05
Manganese	1 - 10	0.02
Mercury	3 - 30	0.02
Nickel	4 - 300	0.02
Radium*	1 - 10	0.6
Rhodium	20 - 500	0.1
Silver	10 - 200	0.1
Tin	20 - 75	0.1
Uranium	1 - 15	0.001
Zinc	2 - 400	0.1
BOD	50 - 5,000	**
COD	20 - 3,500	**
Suspended Solids	—	non-detectable

*Concentration given in picocuries/liter.
**95 percent removal.

Source ref 348

tional reagents may be used to enhance the precipitation of metals. Membrane system performance data from the literature, for this type of application, are presented in Exhibit 7-32.

The presence of oil and grease in the feed stream can cause fouling of the membrane. If present, additional pretreatment may be necessary. Most systems are installed with integral chemical cleaning systems that handle routine cleaning requirements. These systems are typically utilized once per week for a period of two hours (ref. 348).

7.5.3.4 Technology/Equipment Description

This subsection contains a description of commercially available microfiltration equipment that is manufactured and/or sold by vendor survey respondents. This is intended to provide the reader with information and data on a cross section of available equipment. Mention of trade names or commercial products is not intended to constitute endorsement for use.

Memtek Corporation manufactures complete end-of-pipe wastewater treatment systems that employ microfiltration for solids separation. These are modular systems to which various reactor skids (e.g., chromium reduction and cyanide oxidation) can be added, depending on the characteristics of the wastewater and the treatment requirements. For metals removal, precipitation is performed in a reactor/recirculation tank and the mixture of particles and water are pumped through the membrane filtration tubes at 40 to 50 psi and back to the reactor/recirculation tank. The filtrate passes through the pores of the filter to a final neutralization tank. Concentrated solids (2% to 5%) are periodically drawn from the reactor/recirculation tank to a sludge holding tank and subsequently dewatered.

The Memtek filtration membranes are contained in one-inch diameter porous plastic tubes. The membrane has uniform 0.1 micron pores that are funnel-shaped to minimize clogging. Filtration rates range up to 200 GFD. The membrane is chemically resistant to both acid and bleach cleaning solutions (ref. Memtek file). Every system includes a cleaning loop, consisting of a pump, tanks and the necessary piping and valving to permit in-place cleaning of the modules.

7.5.3.5 Costs

7.5.3.5.1 Capital Costs

Capital costs for end-of-pipe membrane filtration systems are presented in Exhibit 7-33. These systems include a two-tank reactor skid for pretreatment, filtration reactor skid and final pH adjustment skid. Installation costs (38% over basic equipment costs) include: engineering (10%), shipping (5%), piping (8%) and electrical/instrumentation (15%).

7.5.3.5.2 Operating Costs

Major operating costs for end-of-pipe systems that employ membrane filtration for solids separation are presented in Exhibit 7-33. The labor and chemical costs are based on the results from the Users Survey. Vendor data indicate that lower chemical costs may be experienced. Data presented by Memtek show chemical costs of $2.26 to $4.95 per 1,000 gal for two EOP applications (ref. Memtek file).

Exhibit 7-33. Capital and Operating Costs (1994) for EOP Microfiltration Systems

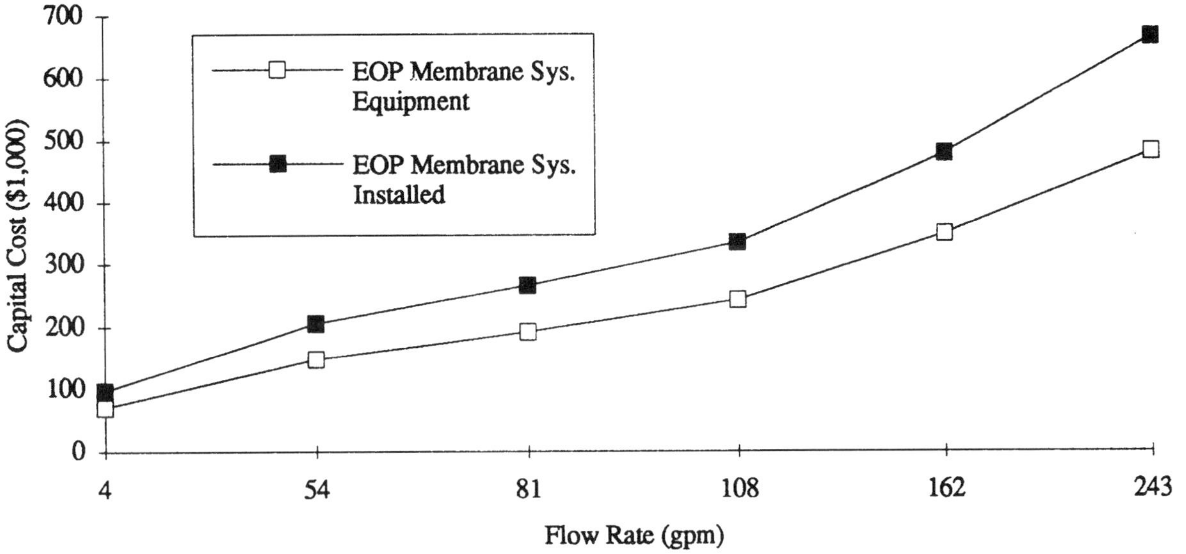

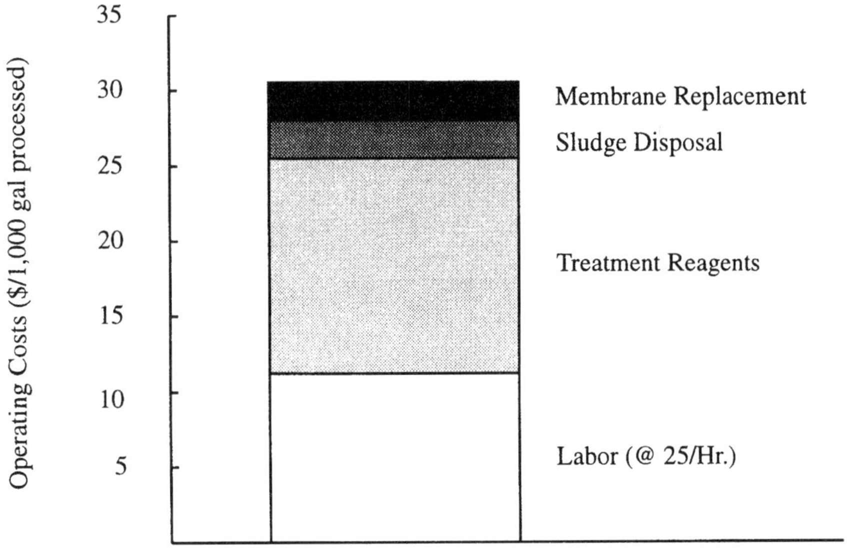

7.5.3.6 Performance Experience

Available information and data from the Users Survey is presented in Exhibit 7-34. Available diagrams of the respondents' treatment systems are presented in Exhibits 7-35, 7-36 and 7-37.

Generally, plating shops using this technology were satisfied with its performance. The average satisfaction level of the four respondents is 3.75. Two of the shops indicated that failure of their treatment system resulted in a compliance excursion.

7.5.3.7 Operational and Maintenance Problems

Only one of the four respondents to the Users Survey that employ microfiltration for an EOP application reported any operational and maintenance problems. PS 105 reported that a rupture of their membrane occurred that resulted in a compliance excursion.

7.5.3.8 Residuals Generation

Membrane filtration, when used in an end-of-pipe treatment scheme, separates precipitated metals from the

Exhibit 7-34. Partial Summary of the Users Survey Data for EOP Microfiltration Systems

	Manf. of			Membrane	Other	Operating Labor hrs/yr		Treatment System Chemical Reagents					Satisfaction	Compliance	Future
Code	Membrane Unit	Year	Flow Rate	Unit Cost	Cap. Costs	Membrane Sys.	Total Sys.	Chemical	Usage Rate	Units	Unit Cost	Chemical Cost	Level (5)	Excursion (6)	Decision (7)
7 (1)	Ionics	1991	42.5 gpm	$41,000	$222,772		1,000	NaOH	18,000	lbs.	0.31	$5,580	4	2	1
								NaOCl	5,400	lbs.	0.80	$4,320			
								NaS2O5	19,200	lbs.	0.29	$5,568			
								H2SO4	10,800	lbs.	0.07	$756			
								Ca(OH)2	10,200	lbs.	0.09	$918			
								NaS2O4	1,200	lbs.	0.30	$360			
											total	$17,502			
105 (2)	Memtek	1988	22,000 gpd		$305,000		3,800	DTC (Ni reducer)	900	gal.	7.27	$6,543	3	1	
								NaOH	28,600	gal.	1.97	$56,342			
								H2SO4	1,500	gal.	1.01	$1,515			
								CaCl2 50%	8,580	gal.	1.05	$9,009			
								Peroxide 35%	1,040	gal.	14.95	$15,548			
											total	$88,957			
233 (3)	Abcor/Koch	1980	2.5 gpm	$42,000	$162,000	150	1,560	none used				$0	4	2	1
257 (4)	Memtek	1990	2,000 gpd		$1,000,000		2,080	NaOH	660	gal.	1.92	$1,267	4	1	
								NaOCl	20	drums	52.00	$1,040			
								CaCl2	500	lbs.	0.18	$90			
								H2SO4	330	gal.	1.00	$330			
								DTC	36	drums	249.00	$8,964			
											total	$11,691			

(1) Unit was purchased used in 1991. Unit was originally purchased in 1988 at a cost of $250,000. Other capital costs include pretreatment equipment and sludge dewatering/dehydration equipment. The supplier stated capacity of the unit is 40 gpm, the actual capacity is 40 to 45 gpm. The average and maximum daily flows at this shop are 20,000 gpd and 26,000 gpd, respectively.

(2) The average and maximum daily flow rates for this shop are 22,000 gpd and 30,000 gpd respectively. The cap. cost includes $250,000 for the original system and $55,000 for an in-house upgrade (1991).

(3) Ultrafiltration unit. Supplier-stated capacity of technology is 3 gpm and the actual capacity is 2.5 gpm. The equipment cost of the UF unit was $35,000. The installation and ancillary equipment costs for the UF unit were $2,00 and $7,000, respectively. The remaining cap. costs include RO ($45,000) (1984) and vacuum evaporator ($40,000) (1987).

(4) Flow rate for the system was not given. The average and maximum flow rate from plating operations is 2,000 and 3,500 respectively.

(5) Satisfaction level codes for the treatment equipment: 1 to 5, with 1=lowest and 5=highest.

(6) Compliance excursion codes: 1-treatment system failure has resulted in a compliance excursion; 2-treatment system failure has not resulted in a compliance excursion; 3-no answer given.

(7) Future decision codes: 1-purchase the same technology from the same vendor; 2-purchase the same technology from a different vendor; 3-purchase a different technology; 4-do nothing; blank-no answer.

Exhibit 7-35. End-of-Pipe Treatment System Employing Microfiltration Used by PS 007

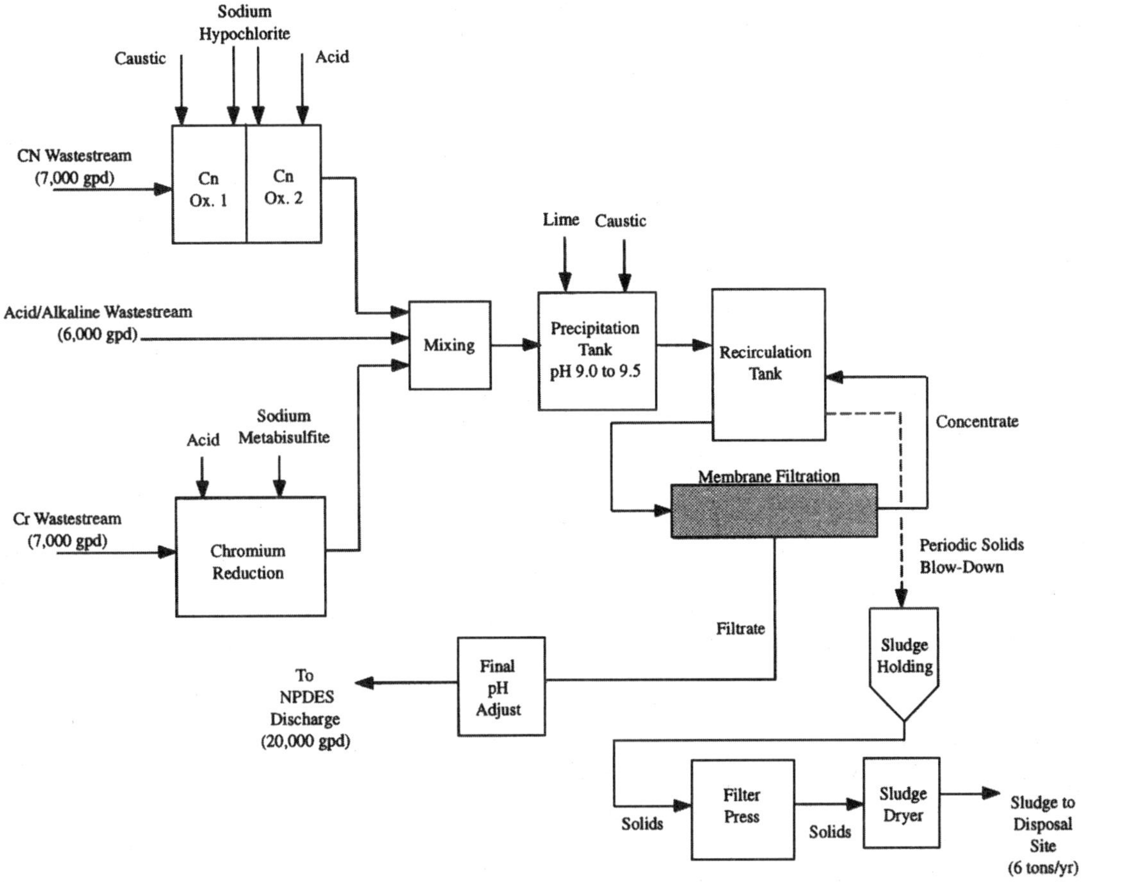

Exhibit 7-36. End-of-Pipe Treatment System Employing Microfiltration Used by PS 105

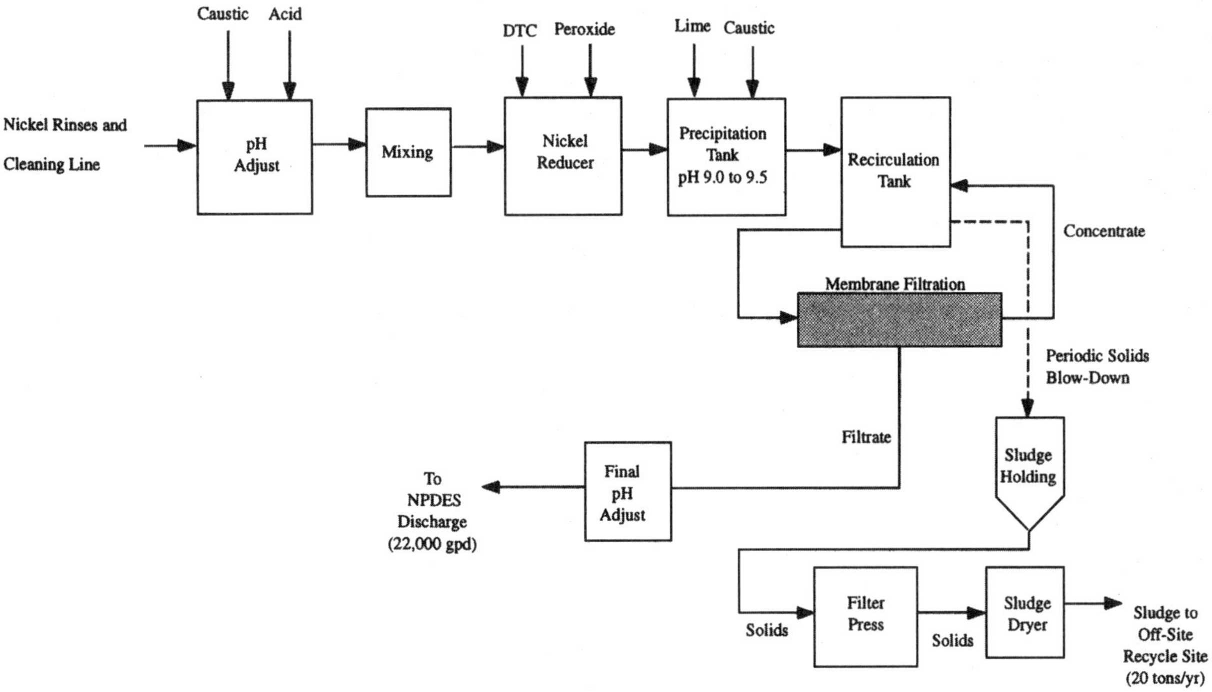

Exhibit 7-37. End-of-Ppe Treatment System Employing Ultrafiltration Used by PS 233 (zero discharge)

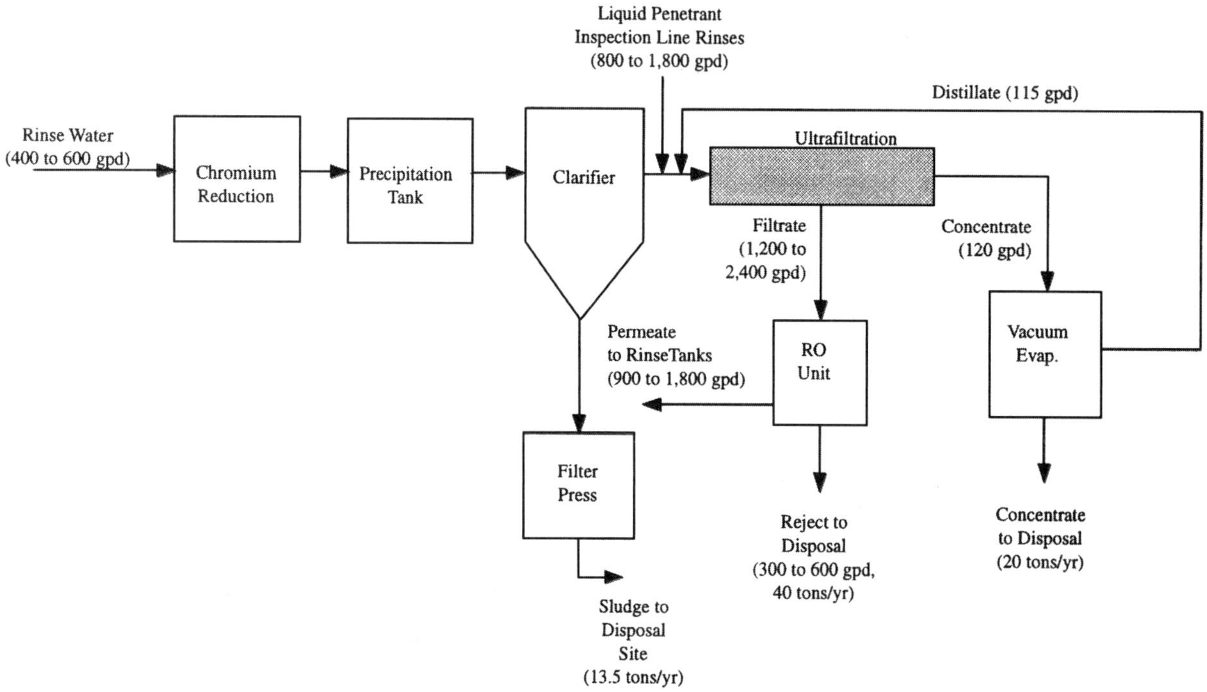

treated wastewater. The solids are subsequently dewatered and transported off-site as a hazardous waste (F006 or F019) for recovery or disposal. The mass (as dry solids) and characteristics of the sludge and its potential for recovery will depend on the characteristics of the original wastestream and the type of chemical reagents used in treatment. Membrane filtration, used as a substitute for clarification/thickening can be operated with conventional hydroxide precipitation reagents (e.g., caustic soda or lime). More often, due to the fact that membrane filtration is usually installed to meet stringent effluent limitations that cannot be attained with conventional clarification/thickening, additional treatment reagents are used to enhance the precipitation of metals, including the fraction that is normally soluble with the use of lime or caustic soda alone.

As shown in Exhibits 7-35 and 7-36, the two shops employing microfiltration as a substitute for clarification/ thickening used additional reagents to supplement precipitation. These additional reagents included dithocarbamates (DTC) and lime. Ferrous sulfate and sodium sulfide, although not used by any of the respondents, are other reagents that are commonly used to enhance precipitation. The use of additional reagents, especially ferrous sulfate, will add to the mass of sludge generated by the treatment process. The use of sulfur based compounds will result in the generation of sulfide sludges and may reduce the potential of using off-site recycle as a disposal option.

7.5.4 Evaporation

7.5.4.1 Overview

Evaporation is a commonly used technology for the recovery of plating chemicals from rinse waters, as discussed in Sections 4.2 and 4.3. In contrast, this section discusses the use of evaporation as an end-of-pipe technology. These applications are differentiated from those in Section 4 by the fact that no plating chemicals are recovered with the end-of-pipe applications. Rather, the evaporation technology is employed solely to reduce or eliminate the discharge of wastewater. It should be noted that several survey respondents used evaporation as a recovery technology and achieved zero-discharge (PS 071, PS 080, PS 193, PS 195). In three of these cases (all except PS 193), the evaporation needed to recover the rinse water/drag-out and attain zero-discharge was provided by heated process tanks (i.e., recovery rinsing). The fourth shop, PS 193, used atmospheric evaporators to provide the needed headroom in their process tanks (nickel and chrome plating) for attaining zero-discharge (their experience is discussed in Section 4.2).

A total of 8 (or 2.5%) of the respondents employ evaporation as an end-of-pipe technology. Of these eight respondents, seven have attained zero-discharge. The one plant with a discharge (PS 036) has an average daily discharge flow of 2,704 gpd. That shop indicated that they installed the evaporator as a treatment process for difficult to treat wastes (copper strip, mixed metal wastes and spent electroless nickel solution). Different types of equipment are used by the respondents, with vacuum evaporators being the most predominant. Five of the respondents use vacuum evaporators, two use atmospheric evaporators and one uses a boiler.

Most of the shops that attained zero-discharge did so recently. Four of the seven zero-discharge shops reached this goal in 1992. One shop reached zero-discharge in 1985, one in 1988 and the remaining shop did not indicate a date. All but one of these shops was discharging less than 1,000 gpd the year prior to reaching zero-discharge. The other shop appears to have jumped from 8,000 gpd to zero-discharge (PS 100). Several of the zero-discharge shops operated a limited number of plating processes. PS 282 and PS 006 only perform chromium plating and PS 021 performs chromium and electroless nickel plating. Only two of the shops gave any reasons for going to zero-discharge. One shop indicated that there was a $12,000 a year savings in self-monitoring costs. Another shop provided the following explanation (PS 202):

"To avoid ever increasing costs and goon squad regulation by the City of _____ Department of Sanitation-We severed our connection to the POTW. We recycle all rinse water that doesn't become contaminated. We use evaporation and a filter press to reduce waste to 55 gal DOT barrels and ship off-site for recycle. Crude-but effective for our operation."

7.5.4.2 Development and Commercialization

The development and commercialization of atmospheric and vacuum evaporators are discussed in Sections 4.2.2 and 4.3.2, respectively.

7.5.4.3 Applications and Restrictions

Four basic configurations of end-of-pipe evaporator use were employed by the eight survey respondents. These configurations are described in Exhibit 7-38. Application EVEOP-1 employs an atmospheric evaporator and there is no reuse of evaporated water. The other three configurations reuse the evaporated water for rinsing or, in the case of the boiler system (application EVEOP-4), the steam condensate is reused.

Evaporators not directly connected to a wastewater system may require a RCRA permit to operate them. Installation and operation of an atmospheric evaporator may require an air permit. Plating shops contemplating this type of treatment process should investigate the regulatory requirements for their specific application.

7.5.4.4 Technology/Equipment Description

This subsection contains references to commercially available evaporation equipment that is manufactured and/or sold by vendor survey respondents or identified during the literature search. This is intended to provide the reader with information and data on a cross section of available equipment. Mention of trade names or commercial products is not intended to constitute endorsement for use.

The evaporation equipment used for end-of-pipe treatment is usually the same equipment used for chemical recovery. Atmospheric and vacuum evaporation units are discussed in Sections 4.2.4 and 4.3.4. Some atmospheric and vacuum evaporation equipment is specifically designed for concentrating wastes. Applicable examples of atmospheric evaporation equipment include the Technotreat and SAMCO products discussed in Section 4.2.4. A type of vacuum evaporation equipment that is used for end-of-pipe treatment applications, but not usually applied to recovery, is the flash evaporator (e.g., LICON Flashvap). This equipment is discussed in Section 4.3.4.

7.5.4.5 Costs

Capital and operating costs for atmospheric and vacuum evaporators are presented in Sections 4.2.5 and 4.3.5. Capital costs can be expected to be approximately the same for recovery and end-of-pipe applications. In some circumstances, pretreatment of the feed stream may be warranted to prevent corrosion of the evaporator, necessitating additional tankage and chemical control equipment (e.g., pH control). Operating costs will be somewhat higher for end-of-pipe applications because solutions will typically be concentrated to a higher solids concentration in an effort to reduce the volume of concentrate produced and therefore lower off-site hauling costs. This will result in increased scaling within the evaporation unit and result

Exhibit 7-38. EOP Evaporator System Configurations Used by Survey Respondents

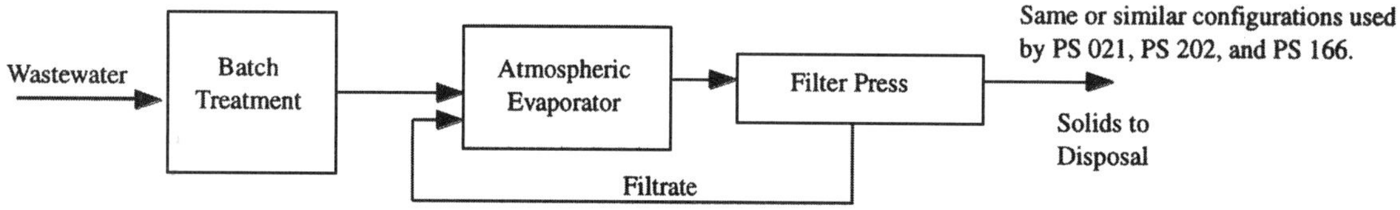

Application EVEOP-1, Zero-Discharge Using an Atmospheric Evaporator

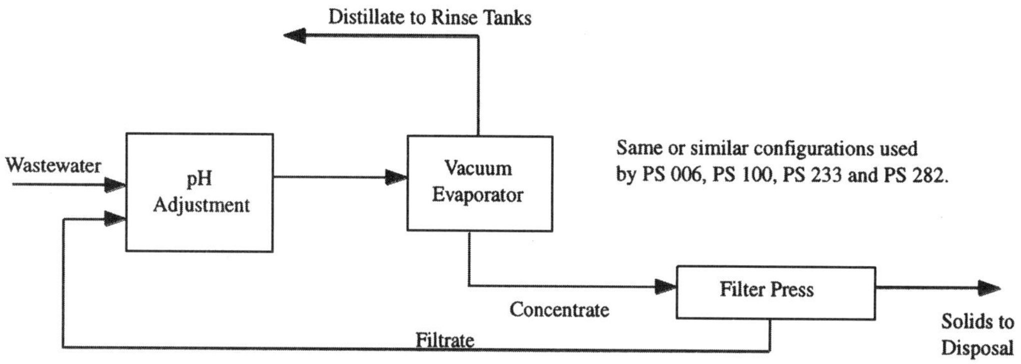

Application EVEOP-2, Zero-Discharge Using a Vacuum Evaporator

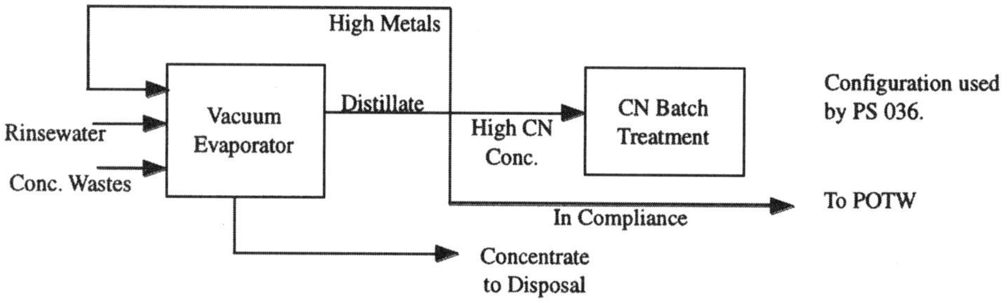

Application EVEOP-3, Low-Discharge Using a Vacuum Evaporator

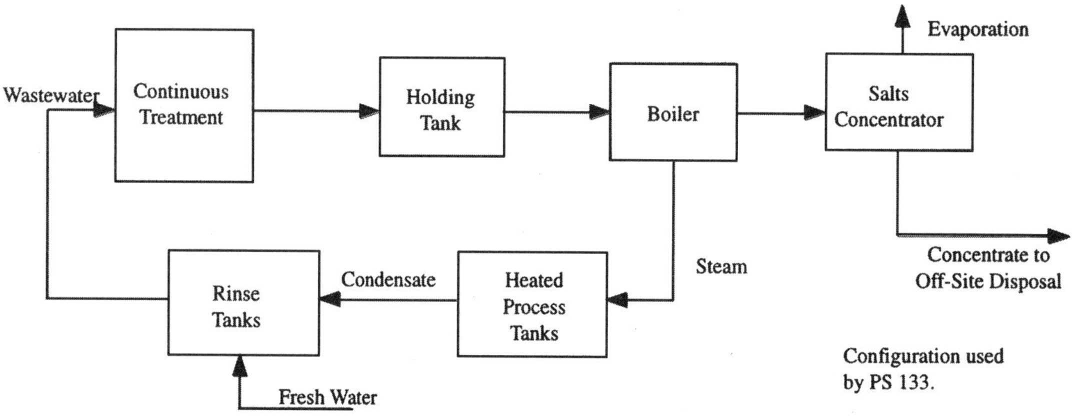

Application EVEOP-4, Zero-Discharge Using a Boiler System

in higher maintenance costs. These higher costs are specified in the operating costs for vacuum evaporators (Section 4.3.5.2).

7.5.4.6 Performance Experience

A summary of the Users Survey data is provided in Exhibit 7-39. The following information and data summarize the performance experience of the survey respondents.

- Shops using evaporation for end-of-pipe treatment gave mixed satisfaction levels for manufacturers' support and the technology itself. The low satisfaction level from PS 282 is mostly due to their inability to use their evaporator for chromium recovery. This problem may have been caused by an inappropriate selection of materials of construction. They stated: "The Water Technology system was purported to be a totally recyclable system, generating chrome acid concentrate back to the plating tank and distilled water. The chromic acid was contaminated and poisoned the plating baths. The fluoride catalyst attacks the titanium heat exchanger and the water sometimes gets contaminated with chrome. The system is a total nightmare."

- The following is a breakdown of the reasons why shops purchased this technology (includes data from six respondents):

To meet or help meet effluent regulations:	3
To reduce plating chemical purchases:	1
To reduce the quantity of waste shipped off-site:	3
To reduce wastewater treatment costs:	4
To improve product quantity:	1
Other:	0

- The use of evaporation had a little impact on production quality and the rate of production. The following responses were provided (includes data from six respondents):

	Production Rate	Product Quality
Improved	1	0
No Change	4	4
Decreased	1	1

- None of the shops indicated that evaporation applied as an end-of-pipe treatment method was the cause of an effluent compliance excursion (includes data from six respondents).

- Respondents reported operating cost savings from reduced water use, reduced treatment chemical use, reduced sludge disposal and from the elimination of discharge monitoring.

- PS 006 reported: "Unit will not process our effluent as well as Co [manufacturer/vendor] lead us to believe when purchasing unit. Technology was grossly oversold." The supplier stated capacity of the unit was 800 gal/24 hr and the actual capacity was 250 to 300 gal/24 hr.

- PS 233 reported that the supplier stated capacity of their unit was 450 gpd and that the actual capacity was 125 gpd.

7.5.4.7 Operational and Maintenance Problems

The following summarizes the respondent's O&M experiences and provides operating labor information.

- Five shops provided operating labor data. For these shops, the average number of annual operating hours per evaporator were: 468 hrs/yr (645 hr/yr excluding PS 100). The following is a breakdown of the responses for skill requirements (includes data from six respondents):

Environmental Engineer:	0
Process/Chemical Engineer:	0
Chemist:	0
Consultant:	0
Plumber/Pipe Fitter:	1
Electrician:	2
Vendor:	0
Senior-Level Plater:	3
Junior-Level Plater:	1
Trained Technician:	4
Wastewater Treatment Plant Operator:	2
Common Labor:	0
Other:	0

- PS 133 reported that use of their boiler for end-of-pipe treatment has resulted in the need to replace it.

Exhibit 7-39. Partial Summary of the Users Survey Data for EOP Evaporation Technology

Shop					Equipment	Other	Operating Costs (Yr.)		Savings				Use	Down	Satisfaction Levels		Future
Code	Application	Manf.	Model	Year	Cost	Cap. Costs	Non-labor	Labor Hrs.	Water	Treat. Chem.	Sludge Disp.	Other	Code	Time %	Manf. Support	Technology	Decision
PS 006	EVEOP-2	Calfran International	STU 1000	1988	$55,000	$2,500	$3,600	260	√	√			1		2	2	2
PS 021	EVEOP-1	not given		1991	$3,000								1			3	
PS 036 (1)	EVEOP-3	Calfran International	PTU 100	1986	$34,490	$10,000		520		√			2	20	4	4	1
PS 100	EVEOP-2	Water Vap Systems	VB 1	1990	$50,000	$41,000	$12,000	48	$8,200	$65,000	$32,000		1	<1	5	5	1
PS 133 (3)	EVEOP-4	ZERPOL Inc.		1981	$250,000		$45,000	520	$3,500			$12,000	1	5	3	5	2
PS 202	EVEOP-1	not given															
PS 233	EVEOP-2	Calfran International	PTU 450	1987	$40,000	$11,000	$900	925			$6,500		1	45	2	2	2
PS 282 (4)	EVEOP-2	Water Technology/Licon		1989			$12,000	1,000	√				1	50	1	1	3

(1) Unit purchased to concentrate difficult to treat wastes prior to off-site disposal. Have since substituted chemical batch treatment.

(2) Purchased two VB 1 units @ $25,000 each. Data shown is for both units.

(3) Other savings are for elimination of discharge monitoring due to zero discharge.

(4) Unit purchased from Water Technology includes both Water Technology and Licon Inc. equipment. Originally purchased to recover chromium plating solution, but now serves as waste concentrator.

(5) Use Codes: 1-currently operating; 2-not currently operating and no intension for future use; 3-not currently in use, but intend to use in future.

(6) Satisfaction levels (manufacturer and technology) 1 to 5, with 1=lowest and 5=highest.

(7) Future decision codes: 1-purchase the same technology from the same vendor; 2-purchase the same technology from a different vendor; 3-purchase a different technology; 4-do nothing.

(8) "√" indicates that a savings was realized, but not quantified.

Exhibit 7-40. Residuals Data for EOP Evaporation Technology

Shop Code	Description of Waste	Disposal Method
PS 006	8,060 gal/yr concentrate liquid	off-site treatment/disposal
PS 036	copper and nickel sludges	off-site recycle w/F006 (Encycle)
PS 100	filter cake	off-site recycle (Encycle)
PS 133	40,000 lbs/yr F006 sludge	off-site treatment/disposal
PS 133	10,000 gal/yr F007 salts	off-site treatment/disposal
PS 233	120 gpd concentrate	off-site treatment/disposal
PS 282	40,000 lbs/yr conc. acid	off-site treatment/disposal

- PS 036 indicated that ammonia, contained in their spent copper strip and electroless nickel solutions, is found in their distillate and that a second pass through the evaporator is necessary for complete removal.
- PS 282 reported: "Maintenance nightmare. No support from vendor-screwed up plating baths for at least one year."
- PS 233 reported: "Material failure: i.e., valves, piping, seals, vacuum pump, product pumps, refrigeration system, overflow protection."

7.5.4.8 Residuals Generation

Evaporators generate a concentrated waste product that can be directly shipped off-site for recovery or disposal or can be further processed (e.g., filter press) to increase its solids content. The concentration of solids from the discharge of the evaporator will depend on the type of unit selected. Common atmospheric evaporators used for chemical recovery are unable to concentrate waste much beyond typical plating bath strength. Some vacuum evaporators, such as the climbing film type, can achieve concentrations of 500,000 mg/l or more. Residuals data provided by respondents using evaporation as an end-of-pipe technology are shown in Exhibit 7-40.

Setion 7 Footnotes:

[1] The two chemicals can be used in the same tank, but must be fed from independent systems to prevent the formation of insoluble calcium sulfate (gypsum).

[2] The Jackson Turbidity Unit is a common measurement of turbidity (fine suspended particles). The test is performed by passing a light beam through a water sample and measuring the interference caused by the suspended particles.

[3] The treatment reagent usage rates are typical ranges, based on the usage rates of survey respondents (approximately 80% of the usage rates reported by survey respondents fall into these ranges). Not all shops will experience usage rates that correspond to these ranges. However, they are a useful comparison to determine if the usage rate for a particular shop is "normal."

Section 8
Off-Site Metals Recycling

8.1 INTRODUCTION

Section 8 covers off-site recycling of non-precious, metal-bearing wastes. Off-site recycling is becoming a popular alternative to land disposal for metal finishing wastes such as wastewater treatment sludge. Approximately 31% of the plating shops that responded to the Users Survey currently send this type of sludge to off-site recycling firms. The sludges are processed thermally or chemically and are converted to a usable form such as feedstock for smelting operations.

Exhibit 8-1 presents a summary of the Users Survey data covering off-site recycling of wastewater treatment plant sludges (RCRA waste code F006 or F019). Column 1 shows the shop code (PS 001 to PS 318). Columns 2 and 3 show the name of the recycle company and the distance from the plating shop to the recycling company (ranges of distances are used to conceal the identity of the respondents). Column 4 indicates the annual quantity of sludge that is shipped for recycling. Columns 5 and 6 show the unit cost and total cost for recycling, including transportation. The average and median values for waste quantity, unit cost and total cost are shown at the bottom of Exhibit 8-1.

Exhibit 8-2 presents a summary of the Users Survey data covering off-site recycling of spent metal finishing solutions and miscellaneous wastes. Some respondents to the survey listed wastes within their completed questionnaire that are disposed of rather than recovered. An effort was made to eliminate these data from the database; however, in some cases, the fate of the waste could not be determined. For example, many West Coast platers send their spent baths to Norris Industries. This company was contacted in an effort to determine if these wastes are treated/disposed or recycled. Norris Industries did not respond to the request for information and the fate of the wastes could therefore not be determined.

8.2 AVAILABILITY OF METALS RECYCLING SERVICES

8.2.1 General

Off-site recycling options for metal finishers were previously limited to spent solvents, precious metal wastes, and high purity common metal wastes. Since 1985, there has been a steady increase in the use of off-site recycling, primarily due to the availability of recycling services for wastewater treatment sludges (e.g., RCRA F006 and F019 sludge), rising costs for land disposal, and increased generator concern over the liability associated with land disposal.

Land disposal costs are increasing due to the loss of capacity (i.e., closure of existing sites), the impact of "land ban" restrictions (RCRA amendments that require treatment of certain wastes prior to disposal) and increased disposal taxes. As of 1990, 87% of the RCRA-permitted landfill facilities that were in operation in 1980 were closed or in the process of closing (ref. 23). Over a similar time period disposal fees increased by 160% and between 1986 and 1990 the Superfund Waste Tax (levied on hazardous waste disposed of in landfills) has increased by 27% (ref. 23). The liability of land disposal has become a costly element for many waste generators. From 1984 to 1989, the estimated contingent liability (based on clean-up of Superfund sites) grew from \$11 to \$38 per ton (ref. 23).

Exhibit 8-1. Summary of Users Survey Data for Off-Site Recycling of Wastewater Treatment Plant Sludges (F006)

I.D. Code	Recycler Name	Distance miles	Quantity lb/yr	Cost $/lb	Total Cost $/yr
002	Encycle	501 to 1,000	42,000	0.25	$10,500
003	Imetco	≤500	550,000	0.27	$148,500
008		501 to 1,000	10,000	0.10	
010	Eticam	≤500	28,000	0.28	$7,840
016		501 to 1,000	200,000	0.10	$20,000
018	WRC	≤500	15,000	0.27	$4,050
020	WRC	≤500	36,000	0.40	$14,400
023	WRC	1,001 to 1,500	23,000	0.32	$7,360
027	WRC	≤500	50,000	0.80	$40,000
034	Eticam	≤500	15,000	0.35	$5,250
036	Encycle	1,001 to 1,500	74,000	0.11	$8,140
039			68,000	0.12	$8,160
040	Encycle	≤500	20,000	0.63	$12,600
043	Horsehead	≤500	68,000	0.18	$12,240
044	Eticam	≤500	12,000	1.00	$12,000
055	WRC	≤500	3,360	0.65	$2,184
058	WRC	≤500	190,000	0.17	$32,300
061	Eticam	≤500	16,000	0.38	$6,080
062	WRC	≤500	9,500	0.55	$5,225
064	WRC	≤500	8,640	0.40	$3,456
068	Eticam	≤500	11,000	0.80	$8,800
074	Encycle	≤500	26,400	0.50	$13,200
075	Inmetco	≤500	6,000	0.11	$654
076	Encycle	1,001 to 1,500	90,000	0.21	$18,900
077	WRC	≤500	288,000	0.17	$48,960
078	Encycle	1,001 to 1,500	15,300	0.40	
085	WRC	≤500	66,040	0.13	$8,585
091	WRC	≤500	60,000	0.30	$18,000
092	WRC	≤500	70,000	0.48	$33,600
098	WRC	≤500	150,000	0.10	$15,000
100	Encycle	≤500	40,000	0.50	$20,000
102	Encycle	501 to 1,000	600,000	0.19	$114,000
103	Encycle	1,001 to 1,500	90,500	0.18	$16,290
105	Encycle	1,001 to 1,500	40,000	0.25	$10,000
105	Inmetco	≤500	12,000	0.80	$9,600
107	Eticam	≤500	26,000	0.62	$16,120
109	Alpha-Omega	≤500	38,000	0.36	$13,680
111	WRC	≤500	15,000	0.80	$12,000
118	Encycle	≤500	226,000	0.20	$45,200
119	WRC	1,001 to 1,500	6,000	0.30	$1,800
124	WRC	1,501 to 2,000	523,000	0.29	$151,670
125	Inmetco	501 to 1,000	37,500	0.21	$7,875
126	Encycle	≤500	20,350	0.73	$14,856
130	Encycle	501 to 1,000	76,727	0.36	$27,622
135	Eticam	≤500	18,000	0.56	$10,080
138	Encycle	1,501 to 2,000	18,000	0.69	$12,420
139	WRC	501 to 1,000	117,000	0.26	$30,420
140	Inmetco	501 to 1,000	50,000	0.04	$2,000
142	Encycle	501 to 1,000	40,000	0.30	$12,000
143	Encycle		120,000	0.33	$39,600
150		2,001 to 2,500	425,000	0.24	$102,000
153	Encycle	501 to 1,000	27,000	1.28	$34,560

Exhibit 8-1 (continued). Summary of Users Survey Data for Off-Site Recycling of Wastewater Treatment Plant Sludges (F006)

I.D. Code	Recycler Name	Distance miles	Quantity lb/yr	Cost $/lb	Total Cost $/yr
156	WRC	501 to 1,000	29,100	0.26	$7,566
158	Encycle	1,001 to 1,500	80,000		
160	Horsehead	≤500	210,000	0.14	$29,400
162	WRC	501 to 1,000	100,000	0.15	
164	WRC	501 to 1,000	44,000	0.37	$16,280
165	Encycle	1,001 to 1,500	42,000	0.25	$10,500
172	Encycle	1,001 to 1,500	74,520	0.22	$16,394
173	WRC	≤500	40,000	0.33	
175	WRC	≤500	24,218	0.40	$9,687
177	WRC	≤500	16,000	0.50	$8,000
182	WRC	501 to 1,000			
183	WRC	501 to 1,000	30,000	0.39	$11,700
187	WRC	≤500	16,000	0.50	$8,000
191	WRC	≤500	43,380	0.25	$10,845
194	Inmetco	501 to 1,000	200	2.00	$400
196	WRC	≤500	160,000	0.10	$16,000
197	WRC	≤500	556,000	0.09	
206	WRC	≤500	1,000		
207	WRC	≤500	16,000	0.21	$3,360
212	WRC	≤500	2,115	1.02	$2,157
213	WRC	≤500	85,000	0.16	$13,600
218	Encycle	2,001 to 2,500	279,493	0.18	$50,309
219	Encycle	2,001 to 2,500	154,464	0.17	$26,259
220	Encycle	2,001 to 2,500	48,398	0.25	$12,100
226	WRC	501 to 1,000	120,000	0.23	$27,600
235	WRC	≤500	80,000	0.60	$48,000
238	WRC	≤500			
242	WRC	≤500	8,000	1.00	$8,000
243	WRC	501 to 1,000	5,000	0.50	$2,500
245	Encycle	501 to 1,000	17,000	0.32	$5,440
248	Encycle	501 to 1,000	62,000	1.00	$62,000
251	Eticam	501 to 1,000	120,000	1.80	$216,000
253	WRC	≤500	125,140	0.20	$25,028
255	WRC	≤500	45,000	0.30	$13,500
256	WRC	501 to 1,000	324,000	0.66	
257	Encycle	1,001 to 1,500	5,500	0.46	$2,530
264	WRC	≤500	143,400	0.13	$18,642
266	Encycle	≤500	380,000	0.20	$76,000
267	WRC	1,501 to 2,000	4,300	0.29	$1,247
268	Horsehead	≤500	94,860	0.11	$10,435
268	Encycle	1,001 to 1,500	64,364		
283	WRC	≤500	9,000	0.30	$2,700
286	WRC	≤500	76,000	0.28	
293	Encycle	1,501 to 2,000	200,000	0.15	
303	WRC	≤500	533,320	0.25	$133,330
307	WRC	501 to 1,000	24,000	0.13	$3,120
Averages		697	96,668	0.40	$25,158
Weighted Average				0.26	
Median Value		500	42,000	0.30	$12,240

Exhibit 8-2. Summary of Users Survey Data for Off-Site Recycling of Spent Baths and Miscellaneous Wastes

I.D. Code	Waste Description	Recycler Name	Distance miles	Quantity lb/yr	Cost $/lb	Total Cost $/yr
001	D007	Inmetco	≤500	64,000	0.08	$5,120
003	Ni strip/sludge	Inmetco	≤500	22,000	0.27	$5,940
007	Silver-cyanide			3,465		
020	F009 (Ag)	Van Guard	≤500	15,000		
020	D002 (Ni, Sn)	CP Chemicals	≤500	130,000		
025	Zn and Cd strip	Encycle	1,001 to 1,500	36,850	0.32	$11,792
038	Gold plate soln.	Technic	≤500	10,000		
064	F009 (Au)	RFE Industries	501 to 1,000	3,860		
064	D003 (Au)	RFE Industries	501 to 1,000	3,860		
068	Metal strip soln.	Norris Ind.	≤500	1,200	0.80	$960
069	Cr solution	Norris Ind.	≤500	9,000	0.39	$3,510
069	Nitric/HF	Norris Ind.	≤500	900		
075	Cr rinse water	Inmetco	≤500	6,000	0.11	$660
095	Bright dip soln.	Holtra Chem Wt.		320,000		
111	Cleaner/acid	WRC	≤500	3,500	0.50	$1,750
111	Cd and Ni wastes		≤500	12,000	0.55	$6,600
119	Acid	Norris Ind.	≤500	2,000	0.75	$1,500
119	Ni strip soln.	Norris Ind.	≤500	2,000	0.75	$1,500
121	Acid	Norris Ind.	≤500	50,000	0.24	$12,000
123	Ni strip soln.	Norris Ind.	≤500	40,460	0.30	$12,138
130	Cr and Cu soln.	Encycle	501 to 1,000	7,024	0.90	$6,322
145	Acids	Norris Ind.	≤500	6,000	1.00	$6,000
148	Sludge	Norris Ind.	≤500	10,000	0.23	$2,300
149	Anode strip soln.	Laidlaw	≤500	6,050	0.74	$4,477
154	Acid Cu soln.	Solvent Service	≤500	495	0.65	$322
154	Acid Sn soln.	Solvent Service	≤500	1,030	0.71	$731
154	NH4SO4	Solvent Service	≤500	970	0.56	$543
166	Sulfuric acid	Norris Ind.	≤500	100,000	0.16	$16,000
166	Cr soln.	Norris Ind.	≤500	50,000	0.15	$7,500
173	Sn/Pb strip soln.	Enthone	≤500	1,000		
176	Chrome soln. from pits	Encycle	1,001 to 1,500	4,000	2.50	$10,000
176	Hard Cr tank sludge	Encycle	1,001 to 1,500	5,000	2.50	$12,500
176	Ni waste after carbon treat	Encycle	1,001 to 1,500	2,500	2.50	$6,250
185	Sulfuric acid	Norris Ind.	≤500	1,500	1.33	$1,995
185	Cr waste	Norris Ind.	≤500	900	1.39	$1,251
197	Ni waste	WRC	≤500			
198	Cyanide waste	Cyano Corp.	≤500	1,500,000	0.30	$450,000
202	Cr waste	Encycle	1,001 to 1,500	6,000	5.00	$30,000
206	Acid copper waste	Norris Ind.	≤500	3,334	2.40	$8,002
230	Acid	Noris Ind.	≤500	20,000	0.20	$4,000
235	Electroless Ni soln.	Dynocol	≤500	384,000	0.04	$15,360
249	Electroless Ni soln.	Chem-Tech Sys	≤500	9,000	0.50	$4,500
249	Metal strip soln.	Chem-Tech Sys	≤500	1,000	0.75	$750
249	IX regenerant	Chem-Tech Sys	≤500	2,500	0.65	$1,625
252	Metal strip soln.	Encycle	1,001 to 1,500			
261	Ni soln.	Inmetco	≤500			
266	Cr tank bottoms	Encycle	≤500	8,500	0.45	$3,825
270	Ferric chloride		≤500	960,000	0.10	$96,000
270	Zinc chloride		≤500	480,000	0.35	$168,000
287	Gold-cyanide soln.	RFE Industries	≤500	1,600		
287	Pd waste	Vanguard	≤500	30,000		
305	Spent Cr soln.	Inmetco	≤500	4,000		
310	Spent Cr soln.	Noris Ind.	≤500	3,400	0.60	$2,040
318	Cd waste	Encycle	≤500	2,000	1.75	$3,500
318	Ni hydroxide	Encycle	≤500	1,200	1.40	$1,680
318	Cr waste	Encycle	≤500	2,860	1.60	$4,576
Averages			291	82,112	0.89	$22,769
Median Value			150	6,000	0.60	$2,100

8.2.2 Locations of Off-Site Recycling Facilities

During the NCMS project, off-site recycling firms that accept non-precious-metal-bearing wastes from electroplating shops were identified from the Users Survey and other sources. Each of the recycling companies identified during these efforts was contacted and surveyed using a questionnaire to collect information and data on their services. From this process, ten companies were identified and confirmed as metal recycling firms. Eight of the confirmed recycling companies responded to the questionnaire survey.

Exhibit 8-3 shows the locations of all ten metals recycling companies. As shown, several of these companies have multiple plants in different regions of the country. Some of these companies accept wastes in very specific forms or within a narrow analytical range, while other accept a broad range of wastes. Also, as will be discussed, several of these companies do not accept F006 sludge.

8.2.3 U.S. Off-Site Metals Recycling Capacity

Wastewater treatment plant sludge (RCRA waste code F006) is the waste that is most commonly sent to off-site metals recycling plants by electroplating shops. The survey results from the off-site metals recycling companies indicate that there are 10 to 15 plants in the U.S. that are capable of processing F006 sludge and that a typical plant has a processing capacity of 25,000 to 150,000 tons of waste per year. Assuming there are 15 plants with an average processing capacity of 75,000 tons/yr, the U.S. capacity is approximately 1.1 million tons/yr. This estimation overstates the available capacity for the plating industry because most of these plants also accept metal-bearing wastes from other industrial sectors.

Data presented in Section 7 indicate that the average electroplating plant generates approximately 79 tons/yr of F006 sludge.[1] Using EPA's estimate of 13,470 electroplating/metal finishing plants, the total U.S. F006 generation rate is estimated to be 1.1 million tons/yr. Comparing this estimation to the present U.S. capacity of off-site metals recycling plants indicates that a rather close match exists between F006 generation rates and processing capacity. However, this comparison does not consider the waste contributions from other industrial sectors.

8.3 OFF-SITE METALS RECOVERY PROCESSES

This section presents a short description of each of the metals recycling companies for which information is available. This is intended to provide the reader with information and data on a cross section of available services. Mention of commercial names is not intended to constitute endorsement for use.

8.3.1 Horsehead Resource Development Company

Horsehead Resource Development Company (HRD) operates six metals recycling operations located in Illinois, Pennsylvania, Tennessee, Oklahoma, and Texas. HRD accepts metal-bearing sludges, filter cakes, bag house dusts and soils and processes them using two kiln technologies: (1) Waelzing (the volatilizing, or fuming of metals in a reducing atmosphere, applied as a first step to all feeds) and (2) calcining (the second refining step that separates the zinc, lead and cadmium in the crude zinc oxide and increases the overall metals values).

Only the Chicago, IL and Rockwood, TN facilities accept electroplating wastes, where they currently process wastes from approximately 100 metal finishing companies. The main business of HRD is processing electric arc furnace dust (RCRA waste K061) from the steelmaking industry and other feed sources (e.g., F006 and F019) with recoverable zinc, lead, cadmium and iron. The technology employed at Chicago and Rockwood is the Waelz kiln technology. The Rockwood facility has an annual waste processing capacity of 90,000 tons and the Chicago plant will bring another kiln on-line early in 1994, boosting it's capacity to 180,000 tons/yr.

The Waelz kiln technology was developed in 1910. At the heart of the process is a rotary kiln which is typically up to 180 ft in length and up to 12 ft in internal diameter. Kiln feed consists of blended metal-bearing sludges and oxidic dusts mixed with an appropriate amount of a suitable reducing agent (coal, petroleum coke or metallurgical coke fines). During passage through the kiln, the mixed charge is heated under the resulting reducing conditions to a temperature of 1,300°C, sufficient to volatilize the reduced metal species from the charge. The volatilized metal values are subsequently re-oxidized in the gas stream above the charge, and this product is finally collected in bag houses as an enriched fume product. The process derives its name from the German word, "walzen," meaning to trundle or roll. These words

Exhibit 8-3. Locations of Off-site Recycle Companies that Accept Electroplating Waste

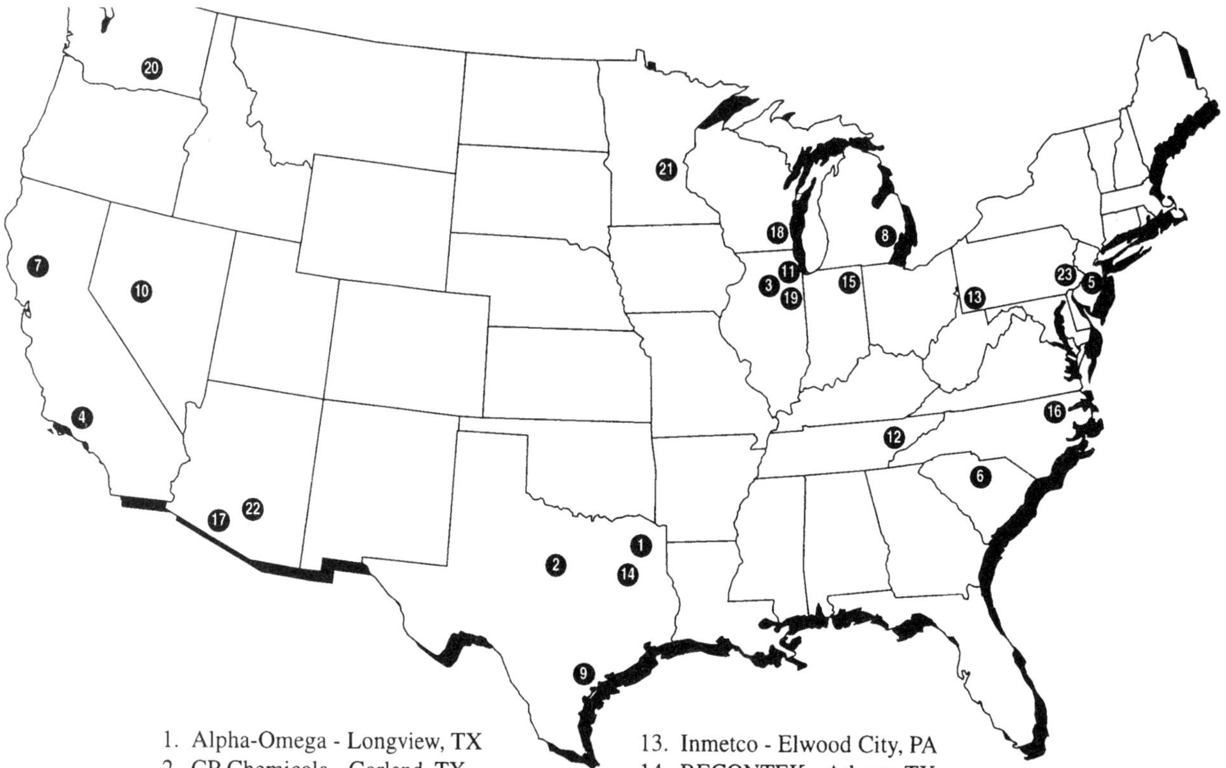

1. Alpha-Omega - Longview, TX
2. CP Chemicals - Garland, TX
3. CP Chemicals - Joliet, IL
4. CP Chemicals - Santa Fe Springs, CA
5. CP Chemicals - Sewaren, NJ
6. CP Chemicals - Sunter, SC
7. CP Chemicals - Union City, CA
8. Cyano Corporation - Detroit, MI
9. Encycle - Corpus Christi, TX
10. Eticam - Fernley, NV
11. Horsehead - Chicago, IL
12. Horsehead - Rockwood, TN
13. Inmetco - Elwood City, PA
14. RECONTEK - Athens, TX
15. RECONTEK - Butler, IN
16. RECONTEK - Enfield, NC*
17. RECONTEK - Gila Bend, AZ
18. RECONTEK - Jefferson, WI*
19. RECONTEK - Newman, IL
20. RECONTEK - Richmond, WA
21. U.S. Filler - Roseville, MN
22. World Resources - Phoenix, AZ
23. World Resources - Pottstown, PA

*Planned or under construction

accurately describe the movement of the charge through the rotating kiln (ref. Horsehead file).

Two products are generated from the Waelzing process: crude zinc oxide and iron rich material (IRM). The crude zinc oxide produced at the Chicago and Rockwood facilities (i.e., locations of electroplating waste processing) is further processed at other HRD facilities since these plants do not perform the calcining process. The IRM (45-55% Fe, <1% Zn), which typically amounts to 50-60% of the feed rate, reportedly passes the TCLP because the iron fraction encapsulates any other metals present. IRM is valued for it's strength, density and high iron content and is marketed for a variety of construction materials (e.g., cement manufacture and asphalt production).

The calcining process generates two products: zinc calcine (55-66% Zn, 4-9% Fe) and lead/cadmium concentrate (35-50% Pb, 15-25% Cl, 5-10% Zn, 1-2% Cd). The zinc calcine product is sold to smelters for zinc metal production. The lead/cadmium product is further refined in an acid leaching process to generate lead cake (50-55% Pb), copper cake (35-50% Cu), cadmium sponge (50-60% Cd), and zinc oxide which are sold for feed materials for primary metal manufacturing.

8.3.2 Inmetco

Inmetco, located in Ellwood City, PA, operates a pyrometallurgical process that is capable of annually processing 56,000 tons of feed material. The market focus for this company is on wastes containing chromium, nickel and iron. They currently process wastes from approximately 150 metal finishing companies as well as other segments of industry.

The Inmetco process utilizes equipment commonly found in the primary metals producing industry. The process consists of three main steps. In the first step, the wastes are blended with coke or coal and water in a disc pelletizer. In the second step, the pellets are processed in a rotary hearth furnace where a portion of the carbon in the pellets reacts with oxygen in the waste to produce reduced metal. A portion of the zinc, lead and halogens contained in the flue dust are exhausted into the off-gas treatment system. The hot, metallic, sintered pellets are transferred to an electric arc smelting furnace where the third operation is performed. In this stage, the pellet is melted and chromium oxides are reduced by the residual carbon in the pellet. Lime, silica, alumina and magnesia separate to form a liquid slag that assists in cleansing the metal bath. Metal and slag are tapped periodically from the furnace. The metal is cast from a refactory lined ladle into pigs which are sold to steel mills. Approximately 42% of the original feed material is contained in the pigs. The slag is treated to obtain a sized material that is used locally as fill or ballast. The process water is treated and yields a filter cake that is recycled in a separate processing plant to recover the zinc values (ref. Inmetco file).

Inmetco's pyrometallurgical process has been modified to recover metals from plating wastes. Inmetco accepts nickel and chromium containing wastewater treatment sludge, nickel strip solutions, waste chromic acid and concentrated rinse waters. In order to produce the quality product known as stainless steel remelt alloy pigs, the concentration of elements such as phosphorus and copper in the wastes must be relatively low. During processing, liquid plating wastes are combined with dry, metal bearing wastes prior to entry into the pelletizing disk. Wastewater treatment sludges (e.g., F006) by-pass the pelletizing disk and are mixed with carbon fines. The various wastes are then delivered to the rotary hearth furnace.

8.3.3 RECONTEK

RECONTEK operates a pilot or demonstration recovery plant in Newman, IL, where they are fine tuning their processes in preparation for the construction of six other plants around the country. They have received permits for planned facilities in Gila Bend, AZ and Butler, IN and they are in various stages of permitting in: Athens, TX; Enfield, NC; Richland, WA; and Jefferson, WI.

RECONTEK accepts zinc, copper and precious metal bearing wastes from the electronics and metal finishing industries and recovers these metals using a hydrometallurgical process. Their operations are divided into two lines: copper and zinc. The copper line includes acid distillation for recovery of hydrochloric acid and copper sulfate crystals; alkaline digestion to make copper oxide, sodium chloride and calcium sulfate; and an electrolytic cellhouse which produces cathode copper from waste copper sulfate. The zinc line is not yet operational but is expected to be demonstrated in the near future. This process consists of alkaline digestion, cementation and electrolysis that produces copper and lead concentrates for the smelter, zinc powder, sodium chloride and calcium sulfate.

In addition to the end uses already mentioned, RECONTEK products are used in the production of animal feed, micro nutrient mixes, soaps, wallboard, chlor-alkali products, and for the regeneration of ion exchange resins.

The current processing rate on the copper demonstration line is 300 tons per month of sludges and 500 tons per month of copper sulfate. The zinc demonstration will be initiated with a feed of 250 tons per month and they anticipate increasing this rate up to 2,000 tons per month.

8.3.4 CP Chemicals

CP Chemicals, a major U.S. producer of inorganic metallic salts in the U.S., accepts metal bearing wastes from over 1,000 clients, through their Environmental Recovery Services Division (ERS), at six U.S. plant locations. All six plants are RCRA permitted (four sites have final Part B permits). CP has been recycling metallic wastes since 1950. Details of their metals recovery methods were not provided during the survey; however, CP Chemicals indicated that their processes include hydrometallurgical steps. CP Chemicals mostly accepts and processes segregated metal bearing wastes of the following types: spent plating baths, etchants, pickling solutions, and strippers containing brass, cobalt, copper, nickel, tin, solder or zinc. The primary EPA waste codes accepted are: D002, D004, D006, D007, D008 and F006. Wastes are accepted in different forms including liquid, solid or slurry. The market focus of CP Chemicals includes both printed circuit board manufacturers and common metals plating shops. They have a processing capability of 120,000 tons per year and currently accept 100,000 tons per year. CP Chemicals indicated that their prices for processing wastes ranges from $75 to $210 per drum, or more depending on the waste type and analyses. The output of CP Chemicals recycling processes are technical grade metallic salts and etchants which are sold to industry on a value-added basis and meet the same purity requirements as non-waste derived products. The end use of their products includes: wood treatment chemicals; algaecides and fungicides; catalysts; and ceramics (ref. CP Chemicals file).

8.3.5 World Resources Company

World Resources Company (WRC) operates metals recovery processes at Pottstown, PA and Phoenix, AZ. They have been in the hazardous waste recycling business since 1980. WRC's processing methods are proprietary, as is their capacity. WRC indicated that their recovery methods include hydrometallurgical and pyrometallurgical processes. Their end products were not specified in their survey form. WRC indicted that they have accepted metal bearing wastes from 800 metal waste-generating companies during 1992.

For plating shops responding to the Users Survey, WRC is the most frequently used off-site metals recycling company for F006 sludge. Fifty percent of the respondents using off-site metals recycling for this waste send it to WRC.

8.3.6 Encycle/Texas, Inc.

Encycle/Texas, Inc. operates a metals recycling facility in Corpus Christi, TX. Their client list for 1992 includes approximately 150 electroplating shops. Encycle accepts liquid and solid wastes containing copper, lead, zinc, nickel and to a lesser extent, other metals. They describe their processes as chemical and hydrometallurgical. Their products are used by primary smelters and others (ref. Encycle file).

For plating shops responding to the Users Survey, Encycle/Texas, Inc. is the second most frequently used off-site metals recycling company for F006 sludge. Thirty percent of the respondents using off-site metals recycling for this waste send it to Encycle/Texas, Inc.

8.3.7 Alpha Omega Recycling, Inc.

Alpha Omega Recycling Inc. operates a metals recycling plant located in Longview, TX. Their client list in 1992 included approximately 100 electroplating companies. They primarily accept wastewater treatment sludges containing chromium, chromium/nickel, and copper. Their recovery process consists of acid leaching and selective precipitation and it has a capacity of approximately 5,500 tons/yr. The end use of their recovered metals includes ferrochrome, material for stainless steel manufacturing and blister copper. Alpha Omega Recycling indicated that their average price for waste processing is $350 per ton (excludes transportation costs) (ref. Alpha Omega Recycling file).

8.3.8 Cyano Corporation of Michigan

Cyano Corporation of Michigan operates recovery processes for cyanide bearing wastes (e.g., F007, F008 and F009) in Detroit, MI. They presently service approximately 50 electroplating shops. Their processes include electrochemical recovery of metal and oxidation of cyanide and they have a capacity of 2,200 tons/yr. The metals that are recovered electrochemically are sold as scrap. Cyano Corporation of Michigan indicated that their average price for waste processing is $2 to $10 per gal (ref. Cyano Corporation of Michigan).

8.3.9 Eticam

Based on the response to the Users Survey, it is believed that Eticam operates a metals recycling operation in Fernley, Nevada. However, Eticam did not respond to the metals recycling survey and no additional information is available.

8.3.10 Metro Recovery Systems

A centralized waste treatment (CWT) facility, located in Roseville, MN, is owned by Metro Recovery Systems (MRS), a partnership between Metropolitan Recovery Corporation and the U.S. Filter and Recovery Company (ref. 419). Although this facility did not respond to the metals recycling survey, some information was gathered from the literature and the Users Survey.

The MRS differs from most other off-site metals recycling companies in that it offers an alternative to in-plant conventional treatment of metal bearing wastewaters. The other firms only provide final treatment/recovery for concentrated wastes, particularly sludges. To utilize the MRS system, a plating shop installs sets of modular ion exchange units to capture the metals and cyanides from their wastewaters. Separate sets of ion exchange columns would be used for each type of metal plated so that the metal can be captured in a segregated manner. The facility also accepts bulk loads of concentrated wastes, which are not amenable to ion exchange recovery. Section 7.4.2 describes the ion exchange systems used for this purpose and the costs incurred by two respondents that subscribe to the MRS service. The CWT plant recovers metals from the ion exchange regenerant using electrowinning. Residual wastewaters are treated on-site and the sludge is sent off-site for recovery/disposal (ref. 419).

8.4 WASTE CHARACTERISTICS REQUIREMENTS

A broad range of waste types and characteristics are accepted by the recycling companies. However, each company has certain constraints and a set of preferred waste characteristics that are based on: (1) RCRA permit conditions; (2) material handling capabilities; (3) material processing capabilities; and (4) end product requirements.

Exhibit 8-4 presents data provided by several recycling firms that show the preferred waste characteristics and the actual analyses of certain waste streams that are accepted. These values represent guidelines only. The actual acceptance of a waste depends on the recycling firm's total analysis, waste quantities available and other information that the plater must provide in order to receive a quotation for the recycling service.

The characteristics of a waste will have a profound impact on the cost of using a recycling service and the costs for a given waste will vary from recycler to recycler, depending mostly on the recycling processes employed and the nature of the end products generated. The cost implications of waste characteristics are discussed further in Section 8.5. Before a waste is accepted for recycling, the recycling company will perform an analysis that is paid for by the plater. The analysis (at a cost of approximately $500) is usually repeated once per year.

Exhibit 8-5 presents analytical data provided by plating shops showing the analyses of their sludges (F006) that are sent to off-site metals recycling firms. Exhibit 8-6 presents the results of over 500 selected analyses found in the literature. These sludges were considered acceptable for recovery at Encycle and provide further guidance on the characteristics of wastes that are amenable to off-site recycling.

8.5 OFF-SITE RECYCLING COSTS

8.5.1 General

The Users Survey requested platers to provide cost data on recycling. Also, the recycling firms were requested to provide pricing formulas and typical charges. Data provided by the users were previously shown in Exhibits 8-1 and 8-2. The recycling companies were generally reluctant to provide any data relative to pricing. However, some companies provided general guidelines, some additional data were found in the literature and supplemental information was provided by platers. That information is summarized in this section and in Section 8.3.

8.5.2 Basis of Off-Site Metals Recycling Costs

The price of waste recycling is generally non-standard because it is negotiated between the plating shop and recycling company. As such, there exists an undefinable cost element. However, there are certain factors that are known to contribute to pricing. These factors can be examined for a given situation, and in some cases, plating shops can change their processing operations to make their waste products more valuable to recycling firms, thereby reducing recycling costs. These factors are:

- **Metal Constituents of the Waste**. Although some recycling processes and end uses are indifferent to the number of metals present in a waste stream, most metal recycling companies would prefer a mono-metal solution or sludge (ref. 23). Plating shops could segregate their sludges to increase the value of the material. However, this would require separate treatment of wastes which is not practical for most shops. This strategy was not used by any of the respondents to the Users Survey for F006 sludge (one shop, PS 246, operates separate filter presses for dewatering chromium and non-chromium sludges, but the dewatered sludge is sent to land disposal rather than off-site recycling). The preferred metal content of a sludge is generally a minimum of 5% to 10%.

- **Sludge Type**. Hydroxide sludges are generally preferred to sulfide sludges because the later can make reclamation more difficult. Sodium hydroxide is usually preferred to lime as the alkali reagent because it adds the least amount of inert material to the waste. Sludges generated with anionic polymers used for flocculation are usually preferred to alum and ferric compounds, also due to a lower

Exhibit 8-4. Preferred Characteristics[1] and Actual Analyses for Wastes Accepted by Off-Site Metals Recycling Firms

Parameter	Recontek Zinc Sludges	Recontek Copper Sludges	Recontek Copper Sulfate Solution/ Crystals	Inmetco General Solid Waste[2]	Inmetco General Liquid Waste	Inmetco Chromic Acid Waste[3]	Inmetco Nickel Solution[3]	Inmetco Nickel and Chromium Sludge[3]	Horsehead General Wastes[2]
Zn	≥ 1%	≤ 15%	NS	≤ 20.0%	NS	NA	NA	0.02-0.5%	2-100%
Cu	≥ 5%	≥ 10%	≥ 1%/5%	≤ 2.0%	< 1,800 mg/l	0.02 to 0.12%	4-174 mg/l	0.11-0.44%	unlimited
Pb	unlimited	≤ 1.5%	≤ 2%	≤ 10.0%	NS	NA	NA	0.04-0.19%	<20%
Sn	≤ 2%	unlimited	NS	≤ 0.03%	< 400 mg/l	NA	NA	NA	NS
Al	≤ 1%	≤ 2%	NS	≤ 15.0%	NS	NA	NA	0.21-0.68%[5]	<50%
Fe	≤ 3%	unlimited	≤ 5%	unlimited	< 50%	0.5-1.4%	0-0.3%	20-35%	0 - 100%
Ni	≤ 1%	≤ 1.5%	NS	> 1.3%	> 100 mg/l	32-215 mg/l	2.3-5.8%	2.3-4.7%	<10%
SO_4	≤ 20%	≤ 20%	NS	NS	NS	NA	NA	NA	NS
Cd	≤ 0.5%	≤ 0.2%	≤ 100 mg/l	≤ 20.0%	NS	NA	NA	0.03-0.07%	<10%
Cr	≤ 5%	≤ 3%	≤ 1%	> 5.0%	> 5 mg/l	9.8-16.1 %	32-131 mg/l	4.2-11.2%	<5%
Mn	≤ 1%	unlimited	NS	NS	NS	NA	NA	0.16-1.0%	<25%
Co	NS	NS	NS	≤ 2.0%	< 1,800 mg/l	5-65 MG/L	19-460 mg/l	0.08-0.37%	NS
Mo	NS	NS	NS	≤ 10.0%	NS	NA	NA	0.12-0.35%	NS
As	NS	NS	≤ 5%	NS	< 5.0 mg/l	NA	NA	NA	NS
Ba	NS	NS	NS	NS	< 100 mg/l	NA	NA	NA	NS
Ag	NS	NS	≤ 1%	NS	< 1.0 mg/l	NA	NA	NA	NS
Hg	NS	NS	≤ 0.2%	NS	NS	NA	NA	NA	NS
Se	NS	NS	≤ 1%	NS	NS	NA	NA	NA	NS
CN	NS	NS	NS	NS	none	NA	NA	NA	NS
PO_4	≤ 62%	≤ 2%	NS	NS	NS	NA	NA	NA	NS
Cl	≤ 6%	≤ 6%	≤ 4%	NS	NS	NA	NA	NA	NS
P	NS	NS	NS	< 0.05%	< 860 mg/l	14-50 mg/l	6-53 mg/l	0.02-0.04%	NS
Calcium-Oxide	NS	NS	NS	unlimited	NS	NA	NA	0.8-24.6%	<100%
Magnesium-Oxide	NS	NS	NS	unlimited	NS	NA	NA	0.05-2.02%	NS
Na, Cl, K	NS	NS	NS	≤ 20.0%	NS	NA	NA	NA	NS
Silica	NS	NS	NS	≤ 15.0%	NS	NA	NA	1.0-3.4%[4]	<40%[4]
S	NS	NS	NS	≤ 5.0%	NS	NA	NA	NA	NS
F	NS	NS	≤ 0.2%	NS	NS	NA	NA	NA	NS
H_2O	≤ 70%	≤ 70%	NS	NS	NS	NA	NA	NA	0-100%
Oil and Grease	NS	NS	NS	NS	NS	NA	NA	NA	<5%

[1] These values represent desired levels only. Acceptance of a waste and its processing depends on the recycler's total analysis, on information supplied by the generator and on handling costs.
[2] Percentages shown are on a dry weight basis.
[3] Actual range of chemical analyses for Inmetco clients (provided by Inmetco).
[4] SiO_2
[5] Al_2SO_3
NS = No specification given for this parameter.
NA = Not analyzed.

Exhibit 8-5. Analytical Data for F006 Sludges Provided by Respondents to the Users Survey

Parameters	PS 018	PS 043	PS 061	PS 062	PS 106*	PS 111	PS 142*	PS 207	PS 212	PS 243	PS 253	PS 264	PS 268
% Solids	98.0%		32.0%	47.0%	58.3%	31.8%	71.8%	52.0%	38.0%	88.0%	50.0%	54.0%	47.3%
Aluminum	1,057		841	6,032	2,600	3,865	3,800	10,862	35,447	243	1,957	6,272	5,900
Antimony	278		63	264		113		164	406	572	47.0	1,030	
Arsenic	≤.26		5.0	2.4	<1,000	54.0	<200	197.0	256.0	104.0	11.2	76.0	<1,000
Barium	151		49	394	<1,000	60	<200	69.0	611	45.0	185	444	<1,000
Beryllium	1.6		<0.5	<1.1		<1.6		13.0	8.0	<0.11	2.4	0.4	
Bismuth	52.0		42	26.0		<13		<5.77	<7.89	20.0	15.0	5.6	
Cadmium	176	39.0	1008.0	26.0	<0.01	651	10,700	175	<0.53	6,276	367	0.37	8,600
Calcium					23,000		9.0						4,500
Chromium (T)	70,907	5,080	8,667	61,526	13,900	7,613	36,400	6,408	19,479	29,704	33,220	42,487	35,900
Chloride*	2,820		56,900	16,400		83,600		6,500	5,100	14,400	7,670	4,600	
Copper	36,358		2,400	14,043	1,500	8,805	1,160	17,446	6,955	88,989	770	83,000	<1,000
Cyanide (T)			15	0.0		3.1	45.0	61.3		20.3	1,358	1,211	
Iron	54,661	68,900	102,789	34,123	86,400	128,377	69,500	100,673	27,316	51,261	35,260	19,722	32,000
Fluoride*	0		200	200		57,500		1,800	3,000	2,000			
Lead	3,739	57.0	411.0	2,298	<1,000	321	210	10,050	1,403	2,344	440	382	<1,000
Manganese					1,000		800						1,000
Magnesium					11,000		<200						78,700
Mercury	0.15		1.60	<0.004		<0.006		<0.38	<1.8	<5.0	1.08	0.37	
Nickel	71,816	1,260	2,500	48,065	<1,000	159,748	5,800	34,712	75,552	58,932	9,696	118,556	<1,000
Selenium	<.30		0.2	6.6	<1,000	<0.006	<200	<9.61	<13.0	<5.68	<0.4	9.3	<1,000
Silver	2.0		332.0	0.0		0.0	<0.02	250.0		<0.3	716.0	130.0	
Sodium					7,100		3,000						12,100
Tin	203		750	3,500	<1,000	2,000	1,260	39,827	<13.0	8,751	38.0	8,370	1,000
Zinc	9,642	140,000	215,055	2,585	362,500	792	256,400	47,269	11,440	95,511	172,640	15,900	356,600

All values in mg/l except for percent solids.
* Dry weight basis.
Disposal costs and quantity data for these sludges are given in Exhibit 7-1.

Exhibit 8-6. Average Analysis of 513 Selected F006 Sludges[1] Accepted for Recycling

Constituent	Copper Waste	Lead Waste	Zinc Waste	Nickel Waste	Chromium Waste	Tin Waste
No. of Samples	171	33	99	148	57	5
Copper	16.4	3.7	2.9	2.8	1.1	3.9
Nickel	7.5	0.1	0.4	16.1	1.7	4.8
Lead	0.6	36.0	3.7	0.1	4.8	1.1
Zinc	3.6	6.4	28.8	1.8	1.8	1.3
Chromium	1.4	0.5	1.6	3.3	24.4	0.3
Cyanide	383	30	583	204	131	0.1
Silver	228	744	335	250	214	43
Aluminum	1.2	1.8	1.1	1.1	2.6	0.5
Cadmium	0.5	0.4	0.4	0.8	0.3	0.0
Iron	4.7	1.1	4.8	3.8	3.5	5.4
Manganese	0.2	0.1	0.2	0.1	0.1	0.0
Tin	0.7	1.1	0.4	0.3	0.2	19.1
Magnesium	1.0	0.3	1.0	1.1	1.3	0.6
Sodium	1.7	1.1	1.1	2.8	2.6	4.8
Calcium	9.3	0.7	2.4	3.6	3.5	2.6
% Water	43.3	16.5	31.7	41.4	34.7	50.1

[1] Samples sorted by predominant metal.
All results given in % wt (wet) except for silver and cyanide, which are given in mg/l.

Source: ref. 420

inert material content. The use of dithiocarbamate (DTC) is usually acceptable.

- **Tramp metals and Other Constituents**. No recycling processes are designed to handle all possible constituents. The presence of some constituents (e.g., cyanide, arsenic, solvents) may preclude acceptance of the waste or significantly increase processing costs. The waste should be free of miscellaneous debris such as rubber gloves, cartridge filters, mop heads, etc. which add to the complexity of the treatment and handling process (ref. 420).

- **Moisture Content of the Waste**. The preferred moisture content of a waste will vary from recycler to recycler and will depend on their material storage, handling and processing facilities and equipment.

- **Waste Volume**. High volume waste streams are preferred by recycling companies. Small volume waste streams may be penalized. Also, unit transportation costs (see following) are impacted by volume.

- **Transportation**. Hazardous wastes must be transported by RCRA permitted companies. The unit transportation costs ($/ton) depend on the quantity hauled per load and the distance. Some companies offer "milk runs" that collect small volumes of waste from different generators that may reduce a shop's transportation costs.

- **Chemical Consistency**. Waste recycling firms prefer waste streams that are chemically consistent from shipment to shipment.

Relative to metal content and the presence of tramp metals and other unwanted constituents, the following example will help to demonstrate the cost impact of these factors.

- The contract of one Southeast plater (PS 043) for recycling their zinc-bearing wastewater treatment sludge (F006) is charged a processing fee of $201 per ton (as received) with the following penalty fee structure:

  - Zinc content below 8.5%: $5.00 per one percent per ton
  - Lead and cadmium above 2.0%: $15.00 per one percent per ton
  - Oil and grease above 0.0%: $5.00 per one percent per ton

8.5.3 Comparison of Off-Site Metals Recycling and Land Disposal Costs

This project has collected data on the actual costs incurred by platers for both off-site recycling and the land disposal of wastewater treatment sludge. From Exhibit 8-1, the average unit cost for off-site recycling of F006 waste is \$0.40/lb (weighted average is \$0.26/lb) and the median cost is \$0.30/lb (includes transportation and processing fees). The lower weighted average and median costs are due mostly to the high unit cost charged for small volume generators. From Section 7, the average unit cost for the disposal of F006 waste is \$0.53/lb (includes transportation, stabilization, disposal and disposal tax) and the median cost is \$0.25/lb. Therefore, the difference in cost between off-site recycling and land disposal appears to be small. This cost comparison does not consider the possibility that, due to their chemical compositions, some wastes that are presently landfilled may not qualify as cadidates for off-site recycling or that the generator of such wastes may be charged higher than average processing cost rates.

In addition to the obvious cost elements associated with recycling and disposal options (e.g., transportation, processing, taxes) a comparison of total costs for these options should include a factor for liability. The liability factor adds consideration of potential future cleanup costs that would be incurred if a disposal or recovery site became a Superfund site. Because recovery sites reclaim most of the waste material they accept, presumably the liability factor for this option would be considerably lower than the land disposal option. The liability for land disposal was estimated by one source to be \$38 per ton (\$0.02/lb) in 1989, up from \$11 per ton in 1986. These values were arrived at by dividing the average cost of cleanup at a Superfund site by the average quantity at a Superfund site (ref. 23).

Appendix A
NCMS/NAMF
Data for End-of-Pipe Equipment

The data presented in this appendix are contained in the NCMS/NAMF database and are discussed in Section 7.4 of the text. These data are linked to the data presented in Exhibit 7-23, which provide a summary of treatment equipment purchased by the respondents to the Users Survey. The data in Appendix A show additional details of EOP equipment purchases for the initial system and up to three subsequent equipment purchases. In most cases, the initial purchase is a complete EOP treatment system and the subsequent purchases are upgrades (e.g., improved controls) or added equipment (e.g., filter press or sludge dryer). Column 1 indicates the plating shop code (1 to 318). Columns 2 and 3 indicate the year the purchase was made and cost (actual price paid in the year given in column 2). Columns 4 and 5 indicate the manufacturer/supplier (in some cases the designer) of the equipment or system and the satisfaction level (SL) of the shop. Satisfaction levels are based on a scale of 1 to 5, with 1 being the least satisfactory and 5 being the most.

Initial System

Shop Code	Description	Year Purchased	Equipment Cost, $	Manufacturer	Satisfaction Level
001	Initial System			Homemade	
002	Initial System	1984	$700,000	TSR Engineering (Peter Veit)	4
003	Initial System	1981	$1,000,000	Water & Wastewater Equip. Co.	4
004	Initial System	1982	$500,000	Mabbett & Capaccio (design)	4
005	Initial System	1970	$85,000	Scientific Control Labs	4
006	Initial System	1972	$250,000		1
007	Initial System	1991	$263,772	A&T Engineering/Ionics for MF	3
008	Initial System	1974	$30,000		4
009	Initial System	1973	$80,000		4
010	Initial System	1985	$125,000	Homemade	5
011	Initial System	1975	$30,000		3
012	Initial System	1983	$1,000,000	Camp Dresser & McKee, Inc.	4
013	Initial System	1969		Homemade	4
014	Initial System	1985	$250,000	Manchester Corp.	4
015	Initial System	1992	$235,000	Aqualogic	3
016	Initial System	1969	$50,000	Homemade	4
017	Initial System			Homemade	
018	Initial System	1988	$125,000	Parkson Lamella/Neutzch Press	4
019	Initial System	1988	$20,000	Homemade	5
020	Initial System	1991	$450,000	Anoplate/Malcolm Pirnie	5
021	Initial System	1988	$25,000	Palm Commodities/IX D.I. Syst.	4
022	Initial System	1978	$80,000	Homemade	3
023	Initial System			Water Management, Inc.	2
024	Initial System	1984	$125,000	Water Management, Inc.	4
025	Initial System	1969	$50,000		4
026	Initial System	1975	$18,957	Arrington	1
027					
028					
029	Initial System	1991	$15,000	Aquatreat	4
030	Initial System				4
031	Initial System	1986	$9,000	Plating Products	2
032	Initial System				2
033	Initial System			Parkson	2
034	Initial System	1980	$450,000	Homemade	4
035	Initial System				
036	Initial System	1978	$75,900		3
037					
038	Initial System	1992	$30,000	Aqualogic	
039	Initial System	1975			3
040	Initial System	1972	$50,000	Baker Bros.	1
041	Initial System	1970	$100,000	Lancy	1
042	Initial System	1990	$200,000	Northcoast Environmental, Inc.	1
043	Initial System	1980	$75,000	Parkson & J.W.I.	3
044	Initial System			Homemade	
045	Initial System	1984	$52,000		3
046	Initial System	1992	$25,500	Homemade	5
047	Initial System	1988	$65,000	Koucour	3
048	Initial System	1984	$150,000	D.M.P.	3
049	Initial System			Delta Pollution Control	
050	Initial System	1984	$120,000	Haviland	1
051	Initial System	1989	$410,000	D.M.P.	4
052	Initial System				
053	Initial System	1985	$200,000	Haviland	4
054	Initial System	1983	$86,000	Water Management	3
055	Initial System	1989	$6,200	Homemade	5
056	Initial System	1983	$20,000	Homemade	4
057	Initial System	1985	$25,000	Production Concepts	3
058	Initial System	1986	$450,000	Manchester Corp.	2
059	Initial System	1989	$140,000	Ensotek, Inc.	2
060	Initial System		$10,000	Homemade	5
061	Initial System	1990	$15,000	Homemade	5
062	Initial System	1988	$13,500	Simpson	1
063	Initial System		$150,000		2
064	Initial System	1980	$30,000	Enthone	4
065	Initial System	1987	$450,000	Treatment Technology/I.M.P.	4
066	Initial System	1989	$50,000	Lancy	3
067	Initial System	1989	$100,000	D.M.P.	4
068	Initial System	1986	$36,000	W.T.S./Simons	3
069	Initial System	1992	$143,500	Atlantes Chemical	4
070	Initial System	1983	$150,000		1
071	Initial System	1975		Homemade	5
072					
073	Initial System	1992	$20,000	Treatment Technologies	5
074	Initial System	1989	$19,000	Water-Vap Systems	3
075	Initial System	1980	$30,000	Homemade	
076	Initial System	1986	$237,000	D.M.P.	4
077	Initial System	1983	$400,000	Homemade	3
078	Initial System	1988	$20,000	Homemade	4
079	Initial System	1987	$50,000	U.F.S.	4
080	Initial System	1987	$10,000		3
081	Initial System	1976	$5,000	Homemade	4
082	Initial System			Sperry Parkson	4
083	Initial System	1987	$140,000	Andco	4
084	Initial System	1980	$30,000	Homemade	1
085	Initial System	1990	$210,755		4
086	Initial System	1986	$70,000		2
087	Initial System	1982	$600,000	D.M.P.	2
088	Initial System	1981	$270,173	Advanced Chemical Systems	3
089	Initial System	1984			2
090	Initial System	1986	$38,000	Wastewater Management	5

Initial System (continued)

Shop Code	Description	Year Purchased	Equipment Cost, $	Manufacturer	Satisfaction Level
091	Initial System			Great Lakes	5
092	Initial System	1982	$40,000	Flowtronics	2
093	Initial System	1985	$275,000	Advance Mechanical	3
094	Initial System	1991	$125,000	Eagle Tech	2
095	Initial System	1990	$75,000		5
096	Initial System	1991	$500,000	Triad	4
097	Initial System	1985	$80,000	Scientific Control	5
098	Initial System	1980	$100,000	Homemade	3
099	Initial System	1980	$9,500	Homemade	4
100	Initial System	1990	$22,000	Water Vap Systems	5
101	Initial System	1978	$250,000	Homemade	3
102	Initial System	1985	$650,000	W.M.I. Inc.	2
103					
104	Initial System	1986	$177,010	Homemade	2
105	Initial System	1988	$250,000	Memtek	3
106	Initial System	1985	$350,000	Scienific Control	4
107	Initial System	1988	$10,000	Homemad	4
108	Initial System	1987	$60,680	Great Lakes	4
109	Initial System	1981	$100,000	Advance Inc.	3
110	Initial System	1987	$65,000		4
111	Initial System	1985	$75,000		4
112	Initial System	1978			1
113	Initial System	1986	$20,000	Homemade	3
114	Initial System	1983	$27,000	NCA Systems	3
115	Initial System	1980	$60,000	D.M.P.	2
116	Initial System	1980	$40,000	Haviland	5
117	Initial System	1987	$50,000	Corotec, Inc.	5
118	Initial System	1983	$464,000	D.M.P.	3
119	Initial System	1989	$90,000	Lancy	3
120	Initial System	1987	$400,000		3
121					
122	Initial System	1989	$1,300,000	Advanced Chemical	4
123	Initial System	1985	$700,000	Homemade	5
124	Initial System			Met Pro	4
125	Initial System	1982		Homemade	4
126	Initial System	1984	$65,000	MKT Interface	3
127					
128	Initial System	1980	$85,000	Homemade	1
129	Initial System	1974		Haviland	3
130	Initial System	1990	$110,000	Haviland	4
131	Initial System			Homemade	5
132	Initial System	1982		D.M.P.	3
133	Initial System	1981	$250,000	Zerpol	4
134	Initial System	1981	$12,500	Homemade	4
135	Initial System	1986	$40,000	Memtek	5
136	Initial System			U.S.F.R.S./Ion Exchange	1
137	Initial System			J.W.I.	3
138	Initial System				
139	Initial System	1987	$235,000	Enthone	3
140	Initial System	1988	$350,000	Parkson	5
141	Initial System	1984	$130,000		3
142	Initial System	1984	$150,000	J-Mark	4
143	Initial System	1980	$150,000		4
144	Initial System				
145	Initial System	1984	$40,000	Ecology Protection	4
146	Initial System	1981	$30,000	Homemade	1
147					
148					
149					
150	Initial System	1970	$1,200,000		3
151					
152	Initial System	1989	$1,200	Signet	1
153	Initial System	1984	$40,000	Eimco	4
154	Initial System	1982	$3,000	Homemade	4
155					
156	Initial System	1984	$300,000	Turner Technology	5
157	Initial System	1990	$150,000	Southern Cal Equipment Co.	1
158					
159	Initial System	1987	$500,000	Dornier	4
160	Initial System	1976	$250,000		2
161	Initial System	1972	$1,800	Homemade	4
162	Initial System	1982	$250,000		2
163	Initial System	1987	$30,000	Bartix Engineering Systems	5
164	Initial System	1985	$27,000	Alar Engineering Corp.	4
165	Initial System	1985	$87,500	D.M.P.	5
166	Initial System	1991	$33,200		4
167	Initial System	1978	$15,000	Homemade	3
168	Initial System	1988	$80,000		4
169	Initial System	1989	$70,000	Advanced	3
170	Initial System	1978	$100,000	Waste Saver	3
171	Initial System	1980	$80,000	Homemade	2
172	Initial System	1985	$500,000	Lancy	5
173	Initial System	1983	$160,000	A.C.S. Environmental	4
174	Initial System	1989	$350,000	Ampac	4
175	Initial System	1990	$200,000	Homemade	4
176	Initial System	1977	$751,124	Industrial Filter & Pump	4
177	Initial System	1985	$50,000	Homemade	
178	Initial System	1984		Manchester Corp.	4
179	Initial System	1988	$50,000	Atlantes Chem Systems	4
180					

Initial System (continued)

Shop Code	Description	Year Purchased	Equipment Cost, $	Manufacturer	Satisfaction Level
181	Initial System	1981	$50,000	Homemade	1
182	Initial System	1985		Haviland	3
183	Initial System	1988	$5,000	John Husband Co.	5
184	Initial System	1980	$60,000	Homemade	2
185					
186	Initial System	1984	$5,000	Homemade	
187	Initial System	1985	$50,000		4
188	Initial System	1987	$1,300	Serfilco	3
189	Initial System	1985	$3,000		3
190	Initial System	1990	$80,000	Cleveland Fluid Systems	2
191	Initial System	1979	$57,056		
192	Initial System	1988	$81,810	Metropolitan Recovery Systems	4
193					
194					
195					
196	Initial System	1980	$250,000	Great Lakes	3
197					
198					
199	Initial System	1981	$75,000	Homemade	4
200	Initial System	1984	$1,000	Homemade	
201	Initial System	1983	$125,000	Homemade	3
202					
203					5
204	Initial System	1984	$500,000	D.M.P.	3
205	Initial System	1983	$75,000	J. Mark Systems	5
206	Initial System	1986	$95,000	E.P.S.	4
207	Initial System	1980	$500,000	Manchester	2
208					
209	Initial System	1978	$350,000		4
210	Initial System	1989	$548,666	Haviland	5
211	Initial System	1991	$3,000		4
212	Initial System	1986	$36,790	Compliance Recovery Systems	4
213	Initial System	1980	$120,000	I.P.S.I.	1
214	Initial System	1984	$700,000	Andco	1
215	Initial System	1982	$321,000	Barth Engineering	4
216	Initial System	1985	$50,000	Homemade	4
217					
218	Initial System	1986	$163,000	Delta Pollution	4
219	Initial System	1980	$425,000	Delta Pollution	4
220	Initial System	1980	$170,000	Delta Pollution	3
221	Initial System	1986	$250,000		4
222	Initial System	1979	$15,000	Homemade	3
223	Initial System	1986	$105,000	Homemade	3
224	Initial System	1986	$5,500	Delta Chemicals	5
225	Initial System	1988	$150,000		3

Shop Code	Description	Year Purchased	Equipment Cost, $	Manufacturer	Satisfaction Level
226	Initial System	1976	$150,000	Homemade	
227	Initial System	1987	$50,000		5
228	Initial System	1985	$125,000	Water Management, Inc.	4
229	Initial System	1982	$50,000	Douglas America	3
230	Initial System	1993		K.R.C.	4
231	Initial System	1991	$111,000	Homemade	5
233	Initial System	1979	$40,000		4
234	Initial System	1985	$150,000	Equipment Associates	5
235	Initial System	1983	$300,000		4
237	Initial System	1988	$100,000	Lancy International	4
238	Initial System	1979	$2,000,000		4
239	Initial System	1980	$150,000		5
240	Initial System	1982	$140,000	E.T.C.	4
241	Initial System			Endquist	4
242	Initial System	1987	$10,000	Milhorn Chemical	3
243	Initial System	1984	$167,000	Advanced Chemical	3
244	Initial System			Delta Pollution Control, Inc.	4
245	Initial System	1984	$321,800	D.M.P.	3
246	Initial System	1990	$600,000		3
247	Initial System	1983	$100,000	Lancy	3
248	Initial System	1989	$190,000	Enthone	4
249	Initial System	1991	$60,000	Douglas International	3
250	Initial System	1987	$85,000	Advanced Chemical Systems	4
251	Initial System	1982	$200,000		1
252					
253	Initial System	1967	$198,000		5
254	Initial System	1977	$136,000		3
255	Initial System	1983	$60,000	Robins/Craig	1
256	Initial System	1987	$200,000		
257	Initial System	1990	$1,000,000	Memtek	4
258	Initial System	1985	$600,000	D.M.P.	4
259	Initial System	1984	$14,000		
260					
261	Initial System	1980		Parkson	4
262	Initial System	1993	$25,000	Homemade	4
263	Initial System	1987	$250,000	P.P.I	3
264	Initial System	1985	$295,000		2
265	Initial System			Lancy	3
266	Initial System	1992	$105,000	Enchem, Inc.	4
267	Initial System	1981	$85,000	Delta/Strange Engineering	5
268	Initial System	1985	$500,000	D.M.P.	3
269	Initial System	1991	$10,000	Homemade	4
270					
271					
272	Initial System	1976	$60,000		4

Initial System (continued)

Shop Code	Description	Year Purchased	Equipment Cost, $	Manufacturer	Satisfaction Level
273	Initial System	1953	$800,000		
274	Initial System	1976	$90,000	Pfaudler	5
275	Initial System	1985	$544,000	Industrial Filter	4
276	Initial System		$500,000	D.M.P.	3
277	Initial System	1974		Homemade	1
278	Initial System	1987	$100,000	D.M.P.	2
279	Initial System	1985	$250,000	Homemade	3
280	Initial System	1990	$1,000,000		5
281	Initial System	1987	$260,000	Water Management	1
282	Initial System	1989	$400,000	Licon	1
283	Initial System	1987			4
284	Initial System	1990	$60,000	D.M.P.	4
285	Initial System	1981		A.C.S.	3
286	Initial System	1982	$400,000	Homemade	5
287	Initial System	1981		Aqualogic	4
288	Initial System	1972	$100,000	Lancy	1
289	Initial System	1982	$300,000		3
290	Initial System	1987	$120,000	Mid South Testing	3
291					
292	Initial System	1985	$718,000	D.M.P.	2
293					
294					
295	Initial System				4
296	Initial System	1973	$700,000		4
297	Initial System				
298	Initial System	1975	$200,000	Homemade	3
299	Initial System				4
300	Initial System	1983	$1,400,000	Andco	3
301	Initial System	1988	$24,000	P. I. Environmental	2
302	Initial System	1989	$90,000		3
303	Initial System	1980	$150,000	World Resources Company	
304					
305	Initial System	1991	$300,000	ERM	1
306					
307	Initial System	1991	$225,000	DMP Corp.	2
308					
309	Initial System	83	$45,000		4
310	Initial System	1989	$500		
311	Initial System	1980			4
312	Initial System	1982	$20,000	Built in-house	3
313					
314	Initial System	1978	$350,000	n/a	1
315	Initial System	1979	$100,000		4
316	Initial System				3
317					
318	Initial System	1960	$20,000		4
Avg-Initial Sys			**$202,392**		**3.39**

First Addition

Shop Code	Description	Year Purchased	Equipment Cost, $	Manufacturer	Satisfaction Level
001					
002	Sludge Dryer	1990	$70,000	OSI	
003	Filter Press +	1989	$300,000		
004	Sludge Dryer	1992	$40,000	Conatminant Recovery Sys	4
005	Tank changes	1981	$22,000		
006	Vac. Distill.	1988	$55,000	Calfran	1
007					
008	New Tankage	1980	$10,000		
009	Upgrades	1984	$40,000		
010					
011	Misc. Add'n	1977	$17,000		
012	Sludge Dryer	1988	$35,000	Fenton Co.	
013					
014	Microfiltration	1989	$200,000	Ionics Corp.	4
015					
016	Tanks & Ctrl	1990	$200,000	Homemade	4
017	Filter Press	1987	$5,000		5
018					
019					
020					
021	IX D.I. Incr.	1991	$40,000	Palm Commodities	3
022	Tanks/pumps	1980	$300,000	Homemade	4
023	Clar./Filter Press	1986	$120,000		
024					
025	Clarifiers	1986	$200,000		
026	Filter Press	1977	$2,500	Shriver	3
027					
028					
029					
030	Upgrade	1983		Netzsch	5
031	Instruments	1988	$1,800	Great Lake	
032	Ultrafilt.	1989		Memtek Corp.	4
033	Clarifier	1982	$36,000		4
034	Ultrafiltration	1987	$30,000	Memtek Corp.	5
035					
036	Filter Press	1984	$31,500		4
037					
038					
039	Electrowin	1983	$90,000	HSA	
040	Upgrades	1977	$50,000	Homemade	4
041	Filter Press	1980	$25,000	Hoesch	4
042	Upgrade			Homemade	5
043	Upgrades	1991	$25,000	JWI	3
044					
045	2nd System	1987	$42,000		3

Shop Code	Description	Year Purchased	Equipment Cost, $	Manufacturer	Satisfaction Level
046					
047					
048	Upgrades	1987	$30,000	DMP	4
049					
050	Sludge Dryer	1988	$28,000	JWI	2
051	Sand Filter	1991	$15,000	Delta Chemicals	5
052					
053					
054	#2 Clarifier	1990	$27,000		
055					
056	Filter Press	1984	$7,000	JWI	5
057					
058					
059					
060	Sludge Dryer	1991			
061					
062	Evaporator		$5,000		3
063	IX Ni Canist		$200,000	Enviroserve Co	3
064	Metal Precip		$100,000	Haviland	4
065	#2 Filter Pres	1990	$30,000	IMP/Micronics	5
066	Polish Filter	1991	$12,000	Conrec	3
067					
068	Upgrades	1989	$12,000	Homemade	4
069					
070	CN Destruct	1984	$30,000		3
071					
072					
073					
074					
075					
076	H20 Rcy. Sys.	1987	$2,500	DMP	2
077	2nd Contain.	1989	$20,000	Homemade	5
078					
079	Sludge Dryer	1987	$20,000	Michigan	5
080					
081	Upgrades	1980	$3,000	Homemade	4
082					
083					
084	Lamella Clar	1985	$150,000	Homemade	2
085	Upgrades	1991	$30,986		5
086	Sludge Dryer	1987	$30,000		3
087	Mg(OH)2 Equip.	1987	$40,000		5
088	Sludge Dryer	1985	$28,318		4
089	Evaporator	1988	$4,300	JWI	4
090					

First Addition (continued)

Shop Code	Description	Year Purchased	Equipment Cost, $	Manufacturer	Satisfaction Level
091	Electrowin			Eltech	3
092	Evaporator	1992	$25,000		2
093	Sludge Dryer	1991	$15,000		5
094					
095					
096					
097	Cr Red. tank	1992	$500	Homemade	5
098					
099					
100	#2 Evap &Pres	1991	$25,000	Water Vap Systems	5
101	Polish Filt. 1985	1985	$35,000	Homemade	4
102					
103					
104	Air System	1987	$29,550	Homemade	3
105	Upgrades	1992	$55,000		
106					
107					
108					
109	Upgrades	1992	$75,000	Homemade	5
110	Sludge Dryer	1991	$13,000		3
111	Sludge Dryer	1988	$6,500	Homemade	2
112					
113	Upgrades	1990	$5,000	Homemade	4
114	Upgrades	1989	$8,000	Homemade	3
115	Upgrades	1992	$10000	Homemade	5
116	Filter Press	1986	$10,000		5
117					
118	Sludge Dryer	1990	$25,000	Homemade	3
119					
120					
121	Filter Press	1988	$5,000	Netzsch, Inc.	3
122	Upgrades	1992	$188,000	Environmental Eng. Services	5
123	Water Reclaim	1991	$15,000	Homemade	4
124	Sludge Dryer			J.W.I.	4
125					
126					
127					
128					
129					
130					
131					
132					
133	Filter Press	1988	$15,000	J.W.I.	5
134	Filter Press	1983	$5,000		5
135	CN Destruct	1988	$10,000	Ionex	5

Shop Code	Description	Year Purchased	Equipment Cost, $	Manufacturer	Satisfaction Level
136					
137					
138					
139					
140					
141	Filter Press	1992	$15,000		4
142	Sludge Dryer	1988	$20,000	Day Mixing Company	5
143	Filter Press	1987	$36,000		5
144					
145					
146	Upgrade	1987	$30,000	Homemade	3
147					
148					
149					
150	Upgrades	1986	$546,000		
151					
152	pH controls	1991	$495	Cole Palmer	3
153					
154	Upgrades	1987	$3,000	Homemade	4
155					
156					
157					
158					
159					
160	Controls	1990	$5,000		4
161					
162	Filter Press	1984	$10,000		4
163					
164					
165	Sludge Dryer	1989	$9,500	Durion	5
166					
167					
168					
169					
170	Upgrades	1984	$12,000	Waste Water Systems Eng.	4
171					
172	Sludge Dryer	1989	$28,000		
173	Polish Filt.	1991	$25,000		4
174					
175	Upgrades	1992	$25,000	Homemade	4
176	Cement Mixer	1985	$5,745		
177					
178	Chrome Red.	1990		Manchester Corp.	4
179					
180	Filter Press			Industrial Pump & Filter	3

First Addition (continued)

Shop Code	Description	Year Purchased	Equipment Cost, $	Manufacturer	Satisfaction Level
181	Upgrades	1985	$170,000	Andco	
182					
183					
184	Upgrades	1988	$290,000		4
185					
186	pH Control	1986	$8,000		
187					
188	Upgrades	1992	$20,000	Serfilco	4
189	Upgrades	1989	$2,000		3
190	Sludge Dryer	1991	$5,000	Aries Chem	2
191	Filter Press	1983	$17,040		
192					
193					
194					
195					
196	Upgrades	1988	$50,000	Honeywell	4
197					
198					
199					
200					
201	Sludge Dryer	1987	$40,000	J.W.I.	4
202					
203					
204	2nd Filter Press	1987	$40,000	D.M.P.	4
205					
206					
207	Upgrade 1	1991	$100,000	L & T Technology	2
208					
209	Sludge Dryer	1993	$15,000		3
210					
211					
212					
213	Upgrades	1985	$250,000	Napco	5
214	CN Oxidation	1986	$25,000	Homemade	5
215	Upgrades	1985	$79,800	Barth Engineering	5
216					
217					
218					
219					
220					
221					
222	Upgrades	1985	$4,000	Homemade	3
223					
224					
225					

Shop Code	Description	Year Purchased	Equipment Cost, $	Manufacturer	Satisfaction Level
226					
227					
228	Filter Press	1989	$8,500		
229	Upgrades	1984	$105,000	H.S.A.	1
230					
231					
233	Ultrafiltrat	1980	$35,000	Abcor/Koch	4
234					
235	Upgrades	1984	$10,000	Andco	
237					
238	Filter Press	1980	$30,000		4
239	Sludge Dryer	1990	$20,000		5
240	pH Controls	1990	$8,000		4
241					
242					
243	Sand Filter	1991		Techmatic	3
244					
245	Filter Press	1985	$5,000	DMP	4
246					
247					
248					
249					
250	pH Controls	1989	$30,000	Duvall & Associates	4
251					
252					
253	Filter Press	1987	$18,000	J.W.I.	5
254	Ultrafiltration	1978	$202,000	Baker Bros	1
255	Filter Press	1986	$10,000		3
256					
257					
258	Sludge Dryer	1988	$30,000		1
259					
260	Sludge Dryer	1986	$30,535		
261					
262					
263	Upgrades	1992	$225,000	J.W.I./Parkson	4
264	Sludge Dryer	1990	$36,000	J.W.I.	5
265	CN Oxidation	1993	$160,000		
266					
267					
268	Sludge Dryer	1989	$45,000	J.W.I.	3
Z69					
270					
271					
272	Filter Press	1978	$7,000		

First Addition (continued)

Shop Code	Description	Year Purchased	Equipment Cost, $	Manufacturer	Satisfaction Level
273	Vacuum Filter	1993	$100,000	Alar Engineering	5
274	Filter Press	1989	$14,900		5
275	Sludge Dryer	1987	$70,000	J.W.I.	5
276	Sludge Dryer		$40,000	J.W.I.	
277	Upgrades	1981	$300,000	Aqualogic	5
278					
279	Filter Press	1988	$8,000	J.W.I.	3
280					
281	Sludge Dryer	1990	$60,000	Fenton	4
282					
283					
284					
285	Filtration	1989	$16,000	Serfilco	4
286					
287					
288	Upgrades	1982	$25,000	Lancy	3
289					
290					
291					
292	Upgrades		$12,000	DMP	1
293					
294					
295					
296					
297					
298	Upgrades	1989	$100,000		4
299					
300	Sand Filter	1990	$45,000	Parkson	4
301	Pre-filter	1989	$500	Gull Industries	4
302	Sludge dryer	1991	$15,000		5
303					
304					
305	IX polisher	1992	$30,000	Hazardous Material Mangt.	5
306					
307	Sludge dryer	1991	$18,000		
308					
309					
310					
311	Clarifier				4
312	Polish filter	1982	$4,000	Built in-house	
313					
314	Instruments	1991	$10,000	L&N	4
315	new pumps	1989	$12,000		4
316					4
317					
318	pH controller			Sunbelt Systems	3
Avg-First Addition			**$51,960**		**3.81**

Second Addition

Shop Code	Description	Year Purchased	Equipment Cost, $	Manufacturer	Satisfaction Level
002	Polish Filt.	1992	$70,000	Self-made (Peter Veit)	
004					
005	Filter press	1986	$55,000		
006					
007					
008	New System	1984	$85,000	Chemical Recovery Systems	
009	Press/Polish	1987	$60,000	J.W.I.	
010					
011	HSA CN Dest.	1983	$12,000	HSA	1
012	Dump Tankage	1989	$15,000		
013					
014					
015					
016					
017	pH Control	1990	$500	Homemade	5
018					
019					
020					
021	Atmos. Evap.	1991	$3,000	Water Vap Systems	3
022	Centrifuges	1985	$15,000		4
023					
024					
025	Upgrade WWT	1988	$450,000		
026	Mov. Bed Filt	1979	$3,000	Serfilco	4
027					
028					
029					
030					
031	Pumps	1988	$1,800	C.M.I.	4
032					
033	Lamella/Press	1984	$145,000		4
034					
035					
036	IX D.I. H20	1985	$113,700	DiSep International	5
037					
038					
039	Instr. & Cont.	1985	$75,000		
040	Press/Dryer	1988	$30,000		
041	Tanks & Ctrls	1991	$118,000	Prochem Tech	5
042					
043	Upgrades	1992	$30,000	J.W.I.	4
044					
045	New Lg. Syst.	1992	$300,000		5
046					
047					

Shop Code	Description	Year Purchased	Equipment Cost, $	Manufacturer	Satisfaction Level
048					
049					
050	Filter Press	1989	$15,000	Lanco	4
051	Instr. Upgrade	1992	$10,000	Delta Chemicals	5
052					
053					
054					
055					
056	Bag Polish Filter	1987	$2,000	Setcho	2
057					
058					
059					
060					
061					
062					
063	IX CN Canist	1989		Enviroserve Co.	5
064					
065					
066	Sludge Dryer	1992	$12,500	Conrec	3
067					
068	IX Polish	1990	$55,000	Homemade	4
069					
070	Sludge Dryer	1985	$22,500		5
071					
072					
073					
074					
075					
076	Filter Press	1988	$9,875	J.W.I.	5
077	H20 Sampler	1992	$4,500	ISCO	4
078					
079	Filter Press	1987	$20,000	J.W.I.	5
080					
081	Equip. Replace	1980	$4,000	Homemade	5
082					
083					
084	pH Controls	1990	$54,000	Homemade	3
085					
086					
087	#2 Filter Pres	1989	$10,000		4
088					
089	R.O.on CN	1989	$29,000	Water Technologies	2
090					
091	Filter Press			Sperry	4
092					

Second Addition (continued)

Shop Code	Description	Year Purchased	Equipment Cost, $	Manufacturer	Satisfaction Level
093	CN Treat. Eq.	1992	$9,000		4
094					
095					
096					
097					
098					
099					
100	#3 Evap.	1991	$12,000	Water Vap Systems	5
101	#2 Filt.Pres	1991	$50,000	Homemade	5
102					
103					
104	Clar./Polish	1991	$35,860	Homemade	4
105					
106					
107					
108					
109					
110					
111	Filter Press	1989	$5,000	John Husband	4
112					
113					
114					
115					
116	Carbon Filter	1989	$5,000	Summit Filter	5
117					
118					
119					
120					
121					
122					
123					
124	Sand Filter			Parkson	4
125					
126					
127					
128					
129					
130					
131					
132					
133					
134					
135					
136					
137					

Shop Code	Description	Year Purchased	Equipment Cost, $	Manufacturer	Satisfaction Level
138					
139					
140					
141					
142					
143	Sludge Dryer	1991	$52,000		5
144					
145					
146	Upgrade	1991	$20,000	Homemade	4
147					
148					
149					
150	Sludge Dryer	1989	$59,000		3
151					
152	pH cont. pump	1992	$700	Cole Palmer	4
153					
154					
155					
156					
157					
158					
159					
160	Controls	1992	$10,000		5
161					
162	Sludge Dryer	1990	$50,000		4
163					
164					
165	Pump & Filter	1991	$1,495	Conrec	4
166					
167					
168					
169					
170	Ion Exchange	1990	$50,000	L & T Technologies	2
171					
172					
173					
174					
175					
176	Polish Filter	1988	$10,281	Industrial Filter & Pump	1
177					
178					
179					
180	Sludge Dryer			Fenton	4
181	Sludge Dryer	1987	$10,000		
182					

Second Addition (continued)

Shop Code	Description	Year Purchased	Equipment Cost, $	Manufacturer	Satisfaction Level
183					
184					
185					
186					
187					
188					
189					
190	Upgrades	1992	$10,000	Homemade	3
191					
192					
193					
194					
195					
196	Sludge Dryer	1990	$45,000	Cyclops	4
197					
198					
199					
200					
201	Dryer Vent	1990	$5,000	Microair	3
202					
203					
204	Upgrades	1993	$75,000		
205					
206					
207	Upgrade 2	1992	$75,000	Mabbett - Cappacio	2
208					
209					
210					
211					
212					
213	Upgrades	1988	$407,000	Napco	5
214	pH Control	1988	$5,000	Homemade	5
215					
216					
217					
218					
219					
220					
221					
222	Upgrades	1990	$4,500	Homemade	3
223					
224					
225					
226					
227					

Shop Code	Description	Year Purchased	Equipment Cost, $	Manufacturer	Satisfaction Level
228					
229	Ion Exchange	1986	$30,000	Baker Bros.	1
230					
231					
233	Reverse Osmosis	1984	$45,000	Abcor/Koch	3
234					
235	Upgrades	1985	$15,000	Andco	
237					
238	Sludge Dryer	1988	$80,000		2
239					
240					
241					
242					
243					
244					
245	Sludge Dryer	1989	$40,000	Fenton Company	4
246					
247					
248					
249					
250	Sludge Thick	1990	$70,000	Haviland	3
251					
252					
253	Sludge Dryer	1988	$30,000	J.W.I.	5
254					
255	Upgrades	1986	$15,000	Homemade	4
256					
257					
258					
259					
260	Sand Filter	1989	$500		
261					
262					
263					
264	Polish Filter	1991	$8,500	Orival	3
265					
266					
267					
268					
269					
270					
271					
272	Upgrades	1Y85	$44,000		
273					
274					

Second Addition (continued)

Shop Code	Description	Year Purchased	Equipment Cost, $	Manufacturer	Satisfaction Level
275	Sludge Dryer	1990	$75,000	Fenton	1
276					
277					
278					
279	Sludge Dryer	1989	$30,000	Fenton	3
280					
281					
282					
283					
284					
285					
286					
287					
288	Upgrades	1987	$15,000	Prochem Tech	4
289					
290					
291					
292					
293					
294					
295					
296					

Shop Code	Description	Year Purchased	Equipment Cost, $	Manufacturer	Satisfaction Level
297					
298					
299					
300	Upgrades	1991	$50,000	Homemade	5
301	Building	1990	$4,500	In House	4
302					
303					
304					
305					
306					
307					
308					
309					
310					
311	Holding Tank	1991	$47,000	TSR Engineering	4
312					
313					
314	Sludge dryer	1992	$5,000		5
315					4
316					
317					
318					
Avg-Second Addition			**$46,889**		**3.80**

Third Addition

Shop Code	Description	Year Purchased	Equipment Cost, $	Manufacturer	Satisfaction Level
010					
011	Sand Filter	1991	$15,000		5
012					
013					
014					
015					
016					
017					
018					
019					
020					
021	Filter Press	1991	$2,300		4
022	Level Control	1992	$10,000		
023					
024					
025	IX D.I. H20	1990	$1,000,000	Kinetico	
026	#2 Filt. Pres	1990	$12,000	Shriver	5
027					
028					
029					
030					
031					
032					
033	Hold Tanks	1988	$20,000		
034					
035					
036	Evaporator	1986	$34,500	Cal Fran International	3
037					
038					
039	Improvements	1992	$200,000		
040					
041	Sand Filter	1992	$6,500	Parkson	2
042					
043					
044					
045					
046					
047					
048					
049					
050	Settling Tank	1991	$5,000	Pavco	5
051					
052					
053					
054					

Shop Code	Description	Year Purchased	Equipment Cost, $	Manufacturer	Satisfaction Level
055					
056					
057					
058					
059					
060					
061					
062					
063					
064					
065					
066					
067					
068					
069					
070	Holding Tanks	1992	$65,000		5
071					
072					
073					
074					
075					
076	Sludge Dryer	1990	$38,750	J.W.I.	5
077					
078					
079					
080					
081	Tank Add'n.	1992	$1,000	Bomemade	4
082					
083					
084	Filter Press	1992	$300,000	Homemade	4
085					
086					
087	#2 Clarifier	1990	$10,000	Homemade	4
088					
089	I.X. on Ni	1988	$10,510	Lancy	3
090					
091	Sludge Dryer			J.W.I.	5
092					
093					
094					
095					
096					
097					
098					
099					

Third Addition (continued)

Shop Code	Description	Year Purchased	Equipment Cost, $	Manufacturer	Satisfaction Level
100					
101					
102					
103					
104					
105					
106					
107					
108					
109					
110					
111					
112					
113					
114					
115					
116					
117					
118					
119					
120					
121					
122					
123					
124					
125					
126					
127					
128					
129					
130					
131					
132					
133					
134					
135					
136					
137					
138					
139					
140					
141					
142					
143					
144					
145					
146					
147					
148					
149					
150					
151					
152					
153					
154					
155					
156					
157					
158					
159					
160	Upgrades	1992	$200,000		
161					
162					
163					
164					
165					
166					
167					
168					
169					
170					
171					
172					
173					
174					
175					
176	Controls	1992	$27,782	Simone Engineering	
177					
178					
179					
180					
181					
182					
183					
184					
185					
186					
187					
188					
189					

Third Addition (continued)

Shop Code	Description	Year Purchased	Equipment Cost, $	Manufacturer	Satisfaction Level
190					
191					
192					
193					
194					
195					
196					
197					
198					
199					
200					
201	Dryer Mods	1992	$3,000	S&R Industrial	4
202					
203					
204					
205					
206					
207					
208					
209					
210					
211					
212					
213					
214					
215					
216					
217					
218					
219					
220					
221					
222					
223					
224					
225					
226					
227					
228					
229	Ion Exchange	1988	$200,000		
230					
231					
233	Evaporator	1987	$40,000	Calfran	3
234					
235	Sludge Dryer	1992	$20,000	Adnco	

Shop Code	Description	Year Purchased	Equipment Cost, $	Manufacturer	Satisfaction Level
237					
238	Sand Filter	1990	$90,000		4
239					
240					
241					
242					
243					
244					
245					
246					
247					
248					
249					
250	Sludge Dryer	1991	$15,000	Aries Chemical	5
251					
252					
253					
254					
255					
256					
257					
258					
259					
260					
261					
262					
263					
264					
265					
266					
267					
268					
269					
270					
271					
272					
273					
274					
275					
276					
277					
278					
279	Upgrades	1991	$3,000	Homemade	3
280					
281					

Third Addition (continued)

Shop Code	Description	Year Purchased	Equipment Cost, $	Manufacturer	Satisfaction Level
282					
283					
284					
285					
286					
287					
288	Upgrades	1990	$25,000	Prochem Tech	4
289					
290					
291					
292					
293					
294					
295					
296					
297					
298					
299					
300					
301	Pumps	1991	$425	Wildon	5
302					
303					
304					
305					
306					
307					
308					
309					
310					
311					
312					
313					
314					
315					
316					
317					
318					
Avg-Third Addition			**$87,214**		**4.10**

Appendix B
List of Abbreviations

AR	acid retardation (acid sorption)
BOD	bio-chemical oxygen demand
cfm	cubic feet per minute
CWT	commercial waste treatment
d	day
DC	direct current
DTC	dimethyldithiocarbamates
ED	electrodialysis
EOP	end-of-pipe
eq/l	equivalents per liter
EV	evaporation
EW	electrowinning
g	grams
g/hr	grams per hour
g/l	grams per liter
gal	gallons
gfd	gallons per square foot of filter area per day
gph	gallons per hour
gpm	gallons per minute
gpd	gallons per day
hp	horse power
ft	feet
ft^2	square feet
ft^3	cubic feet
HCD	high current density
IF	ion flotation
IT	ion transfer
IX	ion exchange
hr	hour
in.	inches
kg	kilogram
kW	kilowatt
KWh	kilowatt-hour
l	liter
lb	pound
lbs	pounds
lb/ft	pounds/foot
LCD	low current density
m	meter
m^2	square meter
ME	meshpad mist eliminator
MF	microfiltration
mg	milligram
mg/l	milligrams per liter
min	minute
mth	month
μmho	micromhos
O&M	operating and maintenance
oz	ounce
PPA	Pollution Prevention Act
ppm	parts per million
psi	pound per square inch
RCRA	Resource Consevation and Recovery Act
RFIE	reciprocating flow ion exchange
RO	reverse osmosis
SIE	selective ion exchange
TDS	total dissolved solids
TSS	total suspended solids
TTO	total toxic organics
UF	ultrafiltration
wk	week
yr	year

References

1 Kushner, Joseph B., "Water and Waste Control for the Plating Shop," *Gardner Publications, Inc.*, 1976.

2 Mohler, J.B. , "Survey Your Rinse Tanks," *Metal Finishing*, 1977, July.

3 Mohler, J.B., "Dual Purpose Rinsing," *Plating and Surface Finishing*, 1979, September.

4 Roy, Clarance and Shapiro, Mike, "Automated Direct Dragout Recovery: A Novel Approach to Reclamation of Metals from Electroplating Rinsewaters," *Plating and Surface Finishing*, 1979, March.

5 Mohler, J.B., "Equilibrium and Nonequilibrium Rinsing," *Plating and Surface Finishing*, 1979, October.

6 "The Rinse Master," *Poly Products Corp.*, 1990.

7 Yates, William, "Rinsing With Less Water," *Metal Finishing*, 1989, April.

8 Favilla, John R., "No-Rinse Treatment for Aluminum," *Products Finishing*, 1990, November.

9 Yates, William, "Ten Minutes to Better Rinsing," *Products Finishing*, 1991, December.

10 Mohler, J.B., "Calculating Water Flow for Rinsing," *Plating and Surface Finishing*, 1980, November.

11 Svoboda, Clarence C., "Operation of Rinse Water (Conductivity) Controls," *McCllellan AFB*, 1990.

12 Vertes, George, "Dragout Recovery with a Double-Use Rinse," *Plating and Surface Finishing*, 1983, April.

13 Stein, Bob, "Recuperative Rinsing - A Mathematical Approach," *Metal Finishing*, 1988, January.

14 Stein, Berl, "Developing the Recuperative Rinsing Concept," *Metal Finishing*, 1989, July.

15 Mooney, Ted, "Cutting Water Usage in Plating Shops," *Plating and Surface Finishing*.

16 Ryder, George A., "Rinse Management Techniques," *Metal Finishing*, 1986, November.

17 CENTEC Corporation, "CENTEC Rinse Control," 1982.

18 None, "Plato's Process Planner," *Integrated Technologies, Inc.*, 1992.

19 Company, Myron L., "Dual Range Controlstick II," *Carlsbad, CA*, 1987.

20 Gallerani, Peter, "Waste Minimization and Pollution Prevention Metal Finishing. A Self Audit Manual," *Connecticut Hazardous Waste Management Service*, 1990, September.

21 CH2MHILL, "Engineering Evaluation of Plating Shop Renovation," *McClellan AFB Contract No. F046999000035*, July, 1991.

22 Craig, Jim, "Pollution Prevention Using Mandatory Pollution Prevention Data," *Pollution Engineering*, 1991, December.

23 Brooks, Clyde S., "Metal Recovery from Industrial Waste," *Lewis Publishers, Chelsea, MI*, 1990.

24 Hindin, David, et. al., "Pollution Prevention Options in Metal Fabricated Products Industries," *U.S. EPA Contract No. 68-WO-0027*, 1992, January.

25 U.S. EPA, "Guide to Pollution Prevention. The Fabricated Metal Products Industry (EPA/625/7-90/006)," *Office of Research and Development, Washington, D.C.*, 1990, July.

26 U.S. EPA, "Waste Minimization Opportunity Assessment Manual, EPA/625/7-88/003," *Hazardous Waste Engineering Laboratory, Cincinnati, OH*, 1988, July.

27 Pekin, Tarik, "Pretreatment and Recovery at a Plating Shop," *AESF, 72nd Annual Conference*, 1985, July.

28 Steward, Fred, et. al., "Electrodialysis for the Recovery of Nickel Salts," *AESF, 72nd Annual Technical Conference*, 1985, July.

29 Gingras, Robert K., "Environmental Design Considerations for Metal Finishing, Electroplating, and Printed Circuit Board Manufacturing Facilities," *AESF, 72nd Annual Technical Conference*, 1985, July.

30 Spearot, Rebecca M., Peck, John V., "Environmental and Safety Consequences - The Hidden Value in New Metal Finishing Processes," *AESF, 72nd Annual Technical Conference*, 1985, July.

31 Werbicki, Joseph J., "Waste Treatment or Recovery? An Introduction to Resource Recovery," *AESF, 73nd Annual Technical Conference*, 1986, June.

32 Tram, Tam V. and Clemans, Paul B., "Recovery of Nickel Salts by Electrodialysis Reversal Process," *AESF, 73nd Annual Technical Conference*, 1986, June.

33 Singelyn, Dan and Singlyn, Paul, "Low Cost Recovery," *AESF, 73nd Annual Technical Conference*, 1986, June.

34 Lindsay, M.J. and Stennick, R.S., "Pollution Control Using Sodium Borohydride," *AESF, 73nd Annual Technical Conference*, 1986, June.

35 Ellis, Norbert H., "Buoyant Media Filtration of Metal Finishing Waste," *AESF, 73nd Annual Technical Conference*, 1986, June.

36 Dahnke, D.R., et.al., "Selective Recovery of Metal Values from Electroplating Sludges by the Phosphate Process," *AESF, 73nd Annual Technical Conference*, 1986, June.

37 Vervaert, Alfred E., et. al., "Preliminary Assessment of Chromium Emissions from Chromium Electroplating Facilities," *AESF, 73nd Annual Technical Conference*, 1986, June.

38 Roy, Clarance H., "The Operation and Maintenance of Surface Finishing Wastewater Treatment Systems," *AESF, Orlando, FL*, 1988.

39 Cushnie, George C., "Navy Electroplating Pollution Control Technology Assessment Manual," CENTEC Corporation, Prepared for the Air Force Engineering and Services Laboratory (RDVA) and the Naval Civil Engineering Laboratory (NCEL) under Contract 086-35-81-C-0285, 1983.

40 Ragosta, Joseph M., "Using Your Ion Exchange Wastewater Treatment Equipment," *Metal Finishing*, 1991, November.

41 Kirman, Lyle and Kovach, Jerome, "Recycling Anodizing Rinsewater Using Ion Exchange," *Plating and Surface Finishing*, 1991, April.

42 Wood, David A., "The Use of Simple Material Balances to Solve Problems in a Circuit Board Manufacturer's Wastewater Treatment System," *Plating and Surface Finishing*, 1991, April.

43 Dow Chemical, "Idea Exchange: Countercurrent Ion Exchange, Vol. 4, No. 1," *Midland, MI*, 1990.

44 Dow Chemical, "Idea Exchange: Countercurrent Regeneration for Dowex Cation Resin Operated in the Hydrogen Cycle, Vol. 7, No. 1," *Midland, MI*, 1983, November.

45 Kinetico, Inc., "Comparisons Between Countercurrent and Concurrent Regeneration," *Newbury, OH*, 1990.

46 Battelle, "Selective Ion Exchange (SIE) Unit Test Results," *Columbus, OH*, 1991, August.

47 Hutchins, Terry, "Control Those Solutions," *Products Finishing*, 1991, September.

48 Berg, Jack H., "Maintaining Your Filtration System," *Products Finishing*, 1990, November.

49 Kunces, David, "How to Treat Spent Electroless Nickel Baths," *Products Finishing*, 1990, August.

50 Steward, F.A. and McLay, W.J., "Recovery/ Regeneration of Process Baths, Part IV," *Plating and Surface Finishing*, 1985, November.

51 Van der Meerakker, J., "Recovering EDTA from Old Electroless Copper Solutions," *Plating and Surface Finishing*, 1991, April.

52 Merlin Enterprises, "Merlin," *Huntington Beach, CA*, 1990.

53 Synder, Donald, "Trivalent: The Second Decade," *Products Finishing*, 1988, March.

54 Goshart, Earl, "The Hazards of Chromium Compounds," *Metal Finishing*, 1988, September.

55 Guffie, Robert, "Hard Chromium Plating," *Products Finishing*, 1990, October.

56 Detrisac, M. Arthur, "Treatable Cleaners," *Metal Finishing*, 1991, September.

57 Scislonski, Stan, "Emulsion and Solvent Cleaners," *Metal Finishing*, 1990, May.

58 Smith, George H., "Escape to Aqueous Cleaning," *Metal Finishing*, 1991, September.

59 Science Applications International Corp., "Evaluations of Cleaners for Solvent Substitution, Air Force Plant No. 6, EPA Contract No. 68-C8-0061," *EPA Risk Reduction Engineering Laboratory, Cincinnati, OH*, 1991, January.

60 Polhamus, Robert L., "Precision Cleaning of Metal Parts Without Solvents," *Metal Finishing*, 1991, September.

61 Quitmeyer, Jo Ann, "Aqueous Cleaners Challenge Chlorinated Solvents," *Pollution Engineering*, 1991, December.

62 Heydanek, Dennis, "The Protocol and the Clean Air Act: Ultimate Phasing Out of 1,1,1-Trichloroethane," P*lating and Surface Finishing*, 1991, January.

63 Kolene Corporation, "Salt Bath Cleaning Technology," *Detroit, MI*, 1990.

64 Tally Cleaning Systems, "Turbo Spray Washer," *Attleboro Falls, MA*, 1990.

65 Porter, Mark C. (Ed.), "Handbook of Industrial Membrane Technology," *Noyes Publications, Inc., Park Ridge, NJ*, 1990.

66 Patterson, James W. and Passino, Robert (Ed.), "Metals Specification, Separation, and Recovery," *Lewis Publishing, Chelsea, MI*, 1990.

67 Martin, Thomas H., "Fine Tuning Waste Treatment," *Products Finishing*, 1991, August.

68 Wedl, David J. and Faulk, Richard J., "Cyanide Destruction in Plating Sludges by Hot Alkaline Chlorination," *Metal Finishing*, 1991, November.

69 Langan, John, "Alternative for Tank Chromium Using Brush Plating," *AESF Surfin Conference*, 1992, June.

70 Hammond, Aubret J., "Ion Vapor Deposition (IVD) of Aluminum Coatings," *McClellan AFB*, 1987, November.

71 Mattox, Donald M., "Physical Vapor Deposition (IVD) Process," *Plating and Surface Finishing*, 1992, January.

72 Bibben, John W., "Chrome-Free Deoxidizing," *Products Finishing*, 1991, November.

73 Ko, C.H., et.al., "A Comparison of Cadmium Electroplate and Some Alternatives," *Plating and Surface Finishing*, 1991, October.

74 Jeanmenne, Robert A., "EN for Hard Chromium," *Products Finishing*, 1990, January.

75 Hinton, Bruce R.W., "Corrosion Prevention and Chromates: The End of an Era?," *Metal Finishing*, 1991, October.

76 Spessard, John E., "The IVADIZER as a Replacement for Cadmium Electroplating at U.S. Army Installations," *Anniston Army Depot, Alabama*, 1990.

77 Bradley, Robert E., "Resin-Sealed Anodized Coatings," *Plating and Surface Finishing*, 1988, June.

78 Cameron, Rick, "Replacement of Cyanide Containing Electroplating Solutions Within the DOE

Weapons Complex," *Pollution Prevention Advisor (ISSN 0156-4586), Department of Energy*, 1992, June.

79 Werner, Douglas B. and Mertens, James A., "Replacing 1,1,1-Trichloroethane: Consider Other Chlorinated Solvents," *Plating and Surface Finishing*, 1991, November.

80 Mohler, J.B., "Alkaline Cleaning for Electroplating," *Metal Finishing*, 1984, September.

81 Fuchs, F. John, "Ultrasonic Cleaning," *Metal Finishing*, 1984, January.

82 Batutis, Edward F., "Keep Your Cleaners Clean," *Products Finishing*, 1989, October.

83 Jansen, George and Teavoort, Jan, "New Concept for Alkaline Cleaning," *Metal Finishing*, 1985, April.

84 Yazdani, Azita, "Source Reduction of Chlorinated Solvents In the Electronics Industry," *Plating and Surface Finishing*, 1990.

85 Ehrler, A.J., "Closed-Loop Absorption for Solvent Recovery," *Metal Finishing*, 1987, November.

86 Lawson, Carl and Noble, William, "Aqueous Systems for Cleaning Strip," *Metal Finishing*, 1987, May.

87 Titch, Archer, "Aqueous Ultrasonic Cleaning Systems," *Metal Finishing*, 1988, April.

88 Berg, Jack, "Filtered Thoughts: Start at the Cleaning Line," *Metal Finishing*, 1986, November.

89 Brassard, Jeff, "Decorative Chrome Electroplating - A Process in Transition," *Metal Finishing*, 1988, June.

90 Norris, J.C., "Brush Plating - Part I," *Metal Finishing*, 1988, July.

91 Rubenstein, Marvin and Rosen, Al, "Electrochemical Metallizing for Tin and Tin Alloy Applications," *Plating and Surface Finishing*, 1988, August.

92 Graves, Beverly, "One Step Ahead," *Products Finishing*, 1990, August.

93 Spearot, Rebecca M., "The Environmental Audit: A Proactive Approach to Environmental Management," *Metal Finishing*, 1991, November.

94 Veit, Peter, "Waste Treatment - Evaluating Black Boxes, Miracle Machines and Useful Tools," *Metal Finishing*, 1987, November.

95 Schroeden, John S., "The Key to Waste Management," *Products Finishing*, 1991, August.

96 Nathan, George, "Ion Exchange of Metal from Waste Water of Complex Sulfide Hydrometallurgy," *Hydrometallurgy, 23 (Netherlands)*, 1990.

97 Rosenblum, John and Naser,Mazen J., "Heavy Metals Waste Minimization: Practice and Pitfalls," *Plating and Surface Finishing*, 1991, April.

98 Baker Brothers/Systems, "At A Glance," *Clearwater, FL*, 1992.

99 ELTECH Systems Corporation, "RETEC Applications," *Sugarland, TX*, 1990.

100 ELTECH Systems Corp., "RETEC - Heavy Metal Recovery Systems for the Metal Finishing Industry," *Sugarland, TX*, 1989.

101 ELTECH Systems Corp., "RETEC Technical Data," *Fairport, OH*, 1989, August.

102 ELTECH Systems Corp., "RETEC Cells - Technical Specifications," *Sugarland, TX*, 1990.

103 Wiaux, J.P., "Industrial Electrolysis Applications in the Recovery and Recycling of Metals," *ELTECH Systems Corp., Geneva, Switzerland*, 1990.

104 Wiaux, J.P., "The Recovery and Recycling of Metal Utilizing Electrolytic Techniques," *ELTECH System Corp., Geneva, Switzerland*, 1990.

105 Wiaux, J.P. and Dietrich, J.J., "Effective Electrolytic Recovery of Heavy Metals from Dilute Effluents," *ELTECH Systems Corp., Geneva, Switzerland*, 1990.

106 Baker Brothers Systems, "HMT Electrowinning Metal Recovery Systems," *Stoughton, Mass.*, 1990.

107 Chan, Michael T., "Economic Analysis for Retrofit of a HMT Recovery Unit in an Existing Chemical

Precipitation System," *Baker Brothers Systems, Stoughton, Mass.*, 1987, September.

108 Chan, Michael T., "Economic Analysis for Installation of a New Waste Treatment System in a Small Printed Circuit Board Facility," *Baker Brothers Systems, Stoughton, Mass.*, 1987, September.

109 Memtek, "Ion Exchange - Electrolytic Recovery Wastewater Treatment Systems," *Billerica, MA*, 1989.

110 Jaynor, Inc., "Jaynor's Gold Digger," *Falls Church, VA*, 1990.

111 Precious Metals Processing Consultants, "The Gold Bug," *Palisades Park NJ*, 1990.

112 Manchester Corp., "Manchester Electrolytic Recovery System," *Harvard, MA*, 1990.

113 Eco-Tec, "Recon Electrowinning System, Bulletin No. ET-5-86-SM," *Pickering, Ontario*, 1990.

114 Hulbert, Gary et. al., "Printed Circuit Board Manufacturer Installs Customized Waste Reduction System," *Plating and Surface Finishing*, 1989, April.

115 Toxic Recovery Systems International, "TRSI Data," *Elk Grove, CA*, 1989, October.

116 Covofinish Co., Inc., "Covofinish Data," *North Scituate, RI*, 1988, December.

117 Hradil, Edward, "Polarized Electrode Recovery of Metals Used in PCB Manufacturing," *Electronics*, 1985, June.

118 Bailey, Dan et. al., "High-Mass-Transfer Electrolytic Recovery: A Case Study," *Plating and Surface Finishing*, 1988, April.

119 Precious Metals Processing Consultants, Inc., "IONNET," *Palisades Parks, NJ*, 1990.

120 Dejak, Michael, "Ion Exchange + Electrowinning = Recovery at Hewlett Packard," *Plating and Surface Finishing*, 1988, April.

121 "Case History: Wastewater Treatment," *Printed Circuit Fabrication, Vol. 7, No. 5*, 1984, May.

122 "Case Study: Beyond Compliance," *Printed Circuit Fabrication, Vol. 11, No. 8*, 1988, August.

123 Wiaux, J.P. and Nguyen, T., "Anodic Reactions and the Recycling of Metals Utilizing Electrolysis," *Metal Finishing*, 1990, November.

124 Wiaux, J.P. and Nguyen, T., "Recovered Value from Electroplating Industry Waste," *Metal Finishing*, 1990, June.

125 Shaulys, Joseph, "Cadmium Compliance Achieved with Electrowinning," *Plating and Surface Finishing*, 1991, September.

126 Scholler, M., et. al., "Fluidized Bed Pellet Reactor to Recover Metals or Anions," *Metal Finishing*, 1991, November.

127 Beninati, C.A. and McLay, W.J., "Electrolytic Metal Recovery Comes of Age," *Plating and Surface Finishing*, 1983, March.

128 Bailey, Dan and Chan, Mike, "Electrolytic Recovery = Theory, Application, Advantages," *8th AESF/EPA Conference*, 1987, February.

129 Druce, Roy, "Electrolytic Recovery of Metal Finishing Waste Water," *HAZTECH Canada Conference and Exhibition*, 1987, May.

130 BEWT Recovery Technologies, Inc., "BEWT Data," *Whitby, Ontario*, 1989, July.

131 "Direct Silver Recovery Brings Immediate Savings," *Finishing Industries*, 1981, March.

132 "Chemelec Leads to Major Savings in Silver at Kaynar (UK) Ltd.," *Products Finishing*, 1983, December.

133 "Restmor Solves Plating Effluent Problems by Installing Chemelec Unit," *Products Finishing*, 1983, February.

134 Lancy International, Inc., "Lancy Data," *Warrendale, PA*, 1989, October.

135 Stewart, F.A., "Waste minimization Alternative Recovery Technologies," *Metal Finishing Guidebook and Directory*, 1989.

136 Trionetics, Inc., "Trionetics Data," *Twinsburg, OH*, 1992, June.

137 Phillips, C.T., et. al., "Solids Detoxification - Metals Recovery," *Enviroscience, Inc.*, 1990.

138 Capaccio, Robert S. and Gingras, Robert K., "Wastewater Treatment for Electroless Plating," *Products Finishing*, 1989, August.

139 Tharp, James W., "Waste Treatment of Chromium," *Products Finishing*, 1989, March.

140 Crane, Al, "Recovery of Metals in Circuit Board and Metal Plating Manufacturing," *General Dynamics Pomona Division*, 1990.

141 Humphreys, Paul G., "New Line Plates Non-Cyanide Cadmium," *Products Finishing*, 1989, May.

142 Sricharoenchaikit, Paul, "Ion Exchange Treatment for Electroless Copper - EDTA Rinse Water," *Plating and Surface Finishing*, 1989, December.

143 Jones, Evan O. and Stewart, Terri L., "Recycling and Treatment of Metal-Bearing Spent Acids," *Pacific Northwest Laboratories*, 1990.

144 Johannes, Richard D., et. al., "Electroplating/ Metal Finishing Wastewater Treatment: Practical Design Guidelines," *43rd Purdue Industrial Waste Conference*, 1989.

145 Peters, Robert W. and Bennett, Gary F., "The Simultaneous Removal of Oil and Heavy Metals from Industrial Wastewaters Using Hydroxide or Sulfide Precipitation Coupled With Air Flotation," *Hazardous Waste & Hazardous Materials, Vol. 6, No. 4*, 1989.

146 Ying, Wei-chi, et. al., "Removal of Hypophosphite and Phosphite from Electroless Nickel Plating Baths," *43rd Purdue Industrial Waste Conference Proceedings*, 1989.

147 Brooks, Clyde S., "Treatment and Metal Recovery for Electroless Plating Wastes," *43rd Perdue Industrial Waste Conference Proceedings*, 1989.

148 Edwards, Marc and Benjamin, Mark M., "Adsorptive Filtration Using Coated Sand: A New Approach from Treatment of Metal-Bearing Wastes," *Journal WPCF, Vol. 61, No. 9*, 1989, September.

149 Desher, Drew, et. al., "Meeting Water Quality Standards With Advanced Treatment Processes at an Aerospace facility," *62nd Annual WPCF Conference*, 1989.

150 Kane, James E., "The Use of Innovative Treatment Technologies to Upgrade Performance of an Existing Pretreatment System," *43rd Purdue industrial Waste Conference*, 1989.

151 "Purifying Plating - Water Discharges," *American Machinist*, 1989, January.

152 Semmens, Michael J. and Chang, Yu-yuen, "Selective Cyanide Recovery From Wastewater Containing Metal Cyanide Complexes," *43rd Purdue Industrial Waste Conference*, 1989.

153 Foecke, Terry, "Source Reduction Opportunities in the Plating Industry," *Minnesota Technical Assistance Program (MNTAP)*, 1990.

154 Tsai, Eric Chai-Ei and Nixon, Roy, "Simple Techniques for Source Reduction of Wastes from Metal Plating Operations," *Hazardous Waste & Hazardous Materials, Vol. 6, No. 1*, 1989.

155 Holford, Ray, "Mechanical Plating and Mechanical Galvanizing Update," *Metal Finishing*, 1989, July.

156 Scheren, G., et. al., "A Unique Nontoxic Precipitating Agent for Pretreatment of Heavy Metals in Wastewater," *62nd Annual WPCF Conference*, 1989.

157 Rich, Ronald R. and von Kuster, Thomas Jr., "Recovery of Rinse Water and Plating Bath from Process Rinses Using Advanced Reverse Osmosis," *Water Technologies, Inc.*, 1990.

158 Durr, Larry, "Solvent Recovery," *Metal Finishing*, 1986, June.

159 Janikowski, Stuart K. and Pryfogle, Peter A., "Evaluation of Non-cyanide Metal Strippers," *44th Purdue Industrial Waste Conference*, 1990.

160 Westra, Mark A. and Rose, Bryan L., "Oil and Grease Removal from a Concentrated Source in the Metal Finishing Industry," *44th Purdue industrial Waste Conference*, 1990.

161 Manzione, Mark A., et. al., "Meeting Stringent Metals Removal Requirements with Iron

Adsorption/Coprecipitation," *44th Purdue industrial Waste Conference*, 1990.

162 Cheremisinoff, Paul N. (Ed.), "Encyclopedia of Environmental Control Technology, Vol. 4, Hazardous Waste Containment and Treatment," *NTIS PB91-191163*, 1990.

163 Hinatsu, J.T. and Foulkes, F.R., "Electrochemical Kinetic Parameters for the Cathodic Deposition of Copper from Dilute Acid Sulfate Solutions," *The Canadian Journal of Chemical Engineering, Vol. 69*, 1991, April.

164 Stucki, S., et. al., "Electrochemical Waste Water Treatment Using High Overvoltage Anodes, Part II: Anode Performance and Applications," *Journal of Applied Electrochemistry, Vol. 21*, 1991.

165 Stucki, S., et. al., "Electrochemical Waste Water Treatment Using High Overvoltage Anodes, Part I: Physical and Electrochemical Properties of Sn O2 Anodes," *Journal of Applied Electrochemistry, Vol. 21*, 1991.

166 Kirsch, F.W. and Looby, G.P., "Waste Minimization Assessment for Multilayered Printed Circuit Board Manufacturing," *U.S. EPA Risk Reduction Engineering Laboratory, EPA/600/M-91/021*, 1991, July.

167 Dalton, Phillip J., "Water Conservation and Waste Minimization at Hudgins Plating," *Plating and Surface Finishing*, 1991, August.

168 Hillier, Andrew and Walton, Clifford W., "Modeling Electroplating Rinse Systems Using Equation-Solving Software," *Plating and Surface Finishing*, 1991, November.

169 Stankovic, V.D. and Stankovic, S., "An Investigation of the Spouted Bed Electrode Cell for the Electrowinning of Metal from Dilute Solutions," *Journal of Applied Electrochemistry, Vol. 21*, 1991.

170 Abda, M., et. al., "Removal of Chromium from Aqueous Solutions by Treatment with Fibrous Carbon Electrodes: Column Effects," *Journal of Applied Electrochemistry, Vol. 21*, 1991.

171 Arumugam, Sam. V. and Bhagat, S.K., "Dewatering Characteristics of Nickel Hydroxide Sludges with Preformed Ferric Hydroxide Sludge Solids," *None*, 1990.

172 Pletcher, D. et. al., "Reticulated Vitreous Carbon Cathodes for Metal Ion Removal from Process Streams, Part I: Mass Transport Studies," *Journal of Applied Electrochemistry, Vol. 21*, 1991, February.

173 Huang, Y.L., "Min-Cyanide: An Expert System for Cyanide Waste Minimization in Electroplating Plants," *Environmental Progress (Vol. 10, No. 2)*, 1991, May.

174 Reisser, Arye and Lewin, Daniel R., "Computerized Batch Process for Treating Electroplating Wastes," *Environmental progress, Vol. 10, No. 3*, 1991, August.

175 Bartolozzi, M., "Cadmium Recovery in Batch and Flow Reactors from Solution at Different pH," *Resources, Conservation and Recycling, Vol. 6*, 1991.

176 Stefanowicz, Tadeusz, "Tin Recovery from an Electroplating Sludge," *Resources, Conservation and Recycling, Vol. 6*, 1991.

177 Chiu, Shen-Yann, et. al., "Waste Minimization Study for a Printed Circuit Board Manufacturing Facility in Taiwan," *Hazardous Waste & Hazardous Materials, Vol. 8, No. 2*, 1991.

178 Sorensen, Frode, "A Process -Based Method for Substitution of Hazardous Chemicals and its Application to Metal Degreasing," *Hazardous Waste & Hazardous Materials, Vol. 8, No. 1*, 1991.

179 Roy, A., et. al., "Solidification/Stabilization of a Heavy Metal Sludge by a Portland Cement/Fly Ash Binding Mixture," *Hazardous Waste & Hazardous Materials, Vol. 8, No. 1*, 1991.

180 Merrill, Douglas T. and McLearn, Mary E., "Managing Solid Residues from a Treatment Process for Priority Metals," *45th Purdue Industrial Waste Conference*, 1991.

181 Ayyaswami, Arul, et. al., "Design, Construction and Operation of Industrial Wastewater Treatment Plant," *45th Purdue Industrial Waste Conference*, 1991.

182 Brooks, Clyde S., "Selective Precipitation of Mixed Metal Hydroxides," *45th Purdue Industrial Waste Conference*, 1991.

183 Jawed, Mohammud and Tare, Vinod, "Application of Starch Xanthates for Cadmium Removal: A Comparative Evaluation," *Journal of Applied Polymer Science, Vol. 42*, 1991.

184 "Recovery/Recycling in Plating," *Products Finishing*, 1992, December.

185 Gallerani, Peter and McCarvill, Rick, "Waste Minimization and Pollution Prevention at Pratt and Whitney Aircraft," *Plating and Surface Finishing*, 1991, March.

186 Hassan, Sardar Q., et. al., "Treatment Technology for Aqueous Metal and Cyanide Bearing Hazardous Wastes (F007)," *Air Waste Management Association, Vol. 41*, 1991.

187 Erten, Mujde, et. al., "Evaluating Toxicity of Speciated Metals from Metal Discharging Industries," *45th Purdue Industrial Waste Conference*, 1991.

188 Sabatka, Win and Moison, Robert L., "Controlling Degreasing Emissions," *Products Finishing*, 1992, December.

189 Dickens, Paul, "Waste Elimination at Memc Electronic Materials, Inc.," *45th Purdue Industrial Waste Conference*, 1991.

190 Kaszubowski, Kenneth E., "Development of a Treatment Scheme for Spent Cleaning Solutions Using Statistical Process Control Concepts," *45th Purdue Industrial Waste Conference*, 1991.

191 Baldwin, Phil C., "A Solidification System for Metal Finishing Wastes," *Metal Finishing*, 1992, November.

192 Bailey, Dan and Howard, Tim, "Acid Recovery with Diffusion Dialysis," *Metal Finishing*, 1992, November.

193 Wilk, Lisa F. and Capaccio, Robert S., "Application of Ion Exchange Technology in Pollution Prevention," *Metal Finishing*, 1992, November.

194 Dhonau, Roger, "CWA Reauthorization," *Metal Finishing*, 1992, November.

195 Santiago, Ivonne, "Removal of Hexavalent Chromium from Water Using Tailored Zeolites," *Metal Finishing*, 1992, November.

196 Fabro, Mario, "Vacuum Distillation for the Recovery of Rinse Water and Concentrated Solutions in the Electroplating Process," *Metal Finishing*, 1992, November.

197 Mandich, N.V. and Krulik, G.A., "Substitution of Nonhazardous for Hazardous Process Chemicals in the Printed Circuit Industry," *Metal Finishing*, 1992, November.

198 Walton, Clifford W. and Loos, Kevin J., "Options for Minimizing Metal Finishing Waste," *Plating and Surface Finishing*, 1992, November.

199 Hall, Mitchell S. , et. al., "Atmospheric Releases of Hexavalent Chromium from Hard Chromium Plating Operations," *Plating and Surface Finishing*, 1992, November.

200 Martin, Thomas H., "Chasing Those Elusive Cyanide Ions," *Plating and Surface Finishing*, 1992, November.

201 Pacek, Michael R., "Clean Water for Better Rinsing," *Products Finishing*, 1992, March.

202 Lucas, David, "Land Ban," *Products Finishing*, 1988, March.

203 Graves, Beverly, "Industrial Toxics Project: The 33/50 Program," *Products Finishing*, 1992, June.

204 Warrender, Bob, "Efficient Filtration," *Products Finishing*, 1992, March.

205 "Chromium Plating," *Products Finishing*, 1988, March.

206 Altmayer, Frank, "Zero Discharge," *Plating and Surface Finishing*, 1992, January.

207 Gotthelf, Beth and Spearot, Rebecca, "You Learn You're Operating Without a Necessary Permit. What Do You Do?," *Plating and Surface Finishing*, 1992, April.

208 Vidmat, Kevin P., "Three Dimensions of Waste Minimization and Project Interaction," *Plating and Surface Finishing*, 1992, April.

209 Zavodjanick, John, "Aerospace Manufacturer's Program Focuses on Replacing Vapor Degreasers," *Plating and Surface Finishing*, 1992, April.

210 "Emerging technologies: The Impact on Surface Finishing," *Plating and Surface Finishing*, 1992, April.

211 Graham, Glen and Patry, Michael, "Hazardous Waste Minimization at Tinker AFB," *Plating and Surface Finishing*, 1992, August.

212 Temple, Stephen D., "Impingement: The Key to Effective Aqueous Cleaning," *Metal Finishing*, 1992, August.

213 Altmayer, Frank, "The Clean Air Act and Chromium Emissions," *Plating and Surface Finishing*, 1992, July.

214 Whitall, Kevin L., "Low Emission Solvent Cleaning," *Metal Finishing*, 1992, August.

215 "Shop Problems," *Metal Finishing*, 1992, March.

216 Henderson, Jeff, "Reducing Disposal Costs," *Products Finishing*, 1992, August.

217 Mabbett, Arthur N., "Waste Treatment Optimization," *Products Finishing*, 1992, August.

218 Marcus, Steve, "Closed-Loop Vibratory Finishing," *Products Finishing*, 1992, August.

219 Semmens, M. J., et. al., "A Gas Membrane - Ion Exchange Process for Cyanide Recovery," *Metal Finishing*, 1987, November.

220 Lawrence Livermore National Laboratory, "Waste minimization Plan Building 141 Processing Facility Electronic Manufacturing Group," *NTIS DE90-010205*, 1990, February.

221 Maltby, Peter, "The Efficient Use of Aqueous Cleaning for Precision Components," *Metal Finishing*, 1992, August.

222 Hosea, J. Michael, et. al., "Metal Recovery by Ion Exchange - Seven Crucial Issues," *AEDSF SUR/FIN Conference*, 1990.

223 Greiner, Timothy J., "Closed-Looped Metal Finishing Processes," *AESF SUR/FIN Conference*, 1990.

224 Osley, Paul L., "Nickel Recovery from Wastewater by Solvent Extraction," *AESF SUR/FIN Conference*, 1990, July.

225 Daniali, S., "Solidification/Stabilization of Heavy Metals in Latex Modified Portland Cement Matrices," *Journal of Hazardous Materials, Vol. 24*, 1990.

226 Ohtake, H., et. al., "Reduction of Toxic Chromate in an Industrial Effluent by Use of a Chromate-Reducing Strain of Entrobacter Cloacae," *Environmental Technology, Vol. II*, 1990.

227 Wiaux, J.P., "The Recovery and Recycling of Metal Utilizing Electrolytic Techniques," *Recycling*, 1990.

228 Chohji, Tetsuji, et. al., "Removal Efficiency of Heavy Metals From Laboratory Wastewater Containing Ligands by an Aluminum Hydroxide Co-Precipitation System," *Environmental Technologies, Vol. II*, 1990.

229 Davis, Mark W. and Sandy, Tom, "Zero Sludge/Zero Discharge Pretreatment Systems for the Metal Finishing and Plating Industries," *44th Purdue Industrial Waste Conference*, 1990.

230 Carpenter, Charles, et. al., "Sodium Sulfide/Ferrous Sulfate Metals Treatment for Hazardous Waste Minimization," *44th Purdue Industrial Waste Conference*, 1990.

231 Eyal, Aharon M., et. al., "New Process for Recovery of Zinc Sulfate and Sulfuric Acid from Zinc Electrowinning Bleed Solutions," *Solvent Extraction and Ion Exchange, Vol. 8, No. 2*, 1990.

232 Ferrante, Louise M., "Waste Minimization is a Serious Business," *The National Environmental Journal*, 1992, April.

233 Cheremisinoff, Paul N., "Hazard Analysis for Waste Reduction," *The National Environmental Journal*, 1992, April.

234 Li-Yang and McCoy, Benjamin J., "Waste Minimization for Printed Circuit Board Manufacturing," *Hazardous Waste & Hazardous Materials, Vol. 7, No. 3*, 1990.

235 Ficker, Charles, et. al., "Non-Conventional Metal Precipitation at a Printed Circuit Board Manu-

facturer: A Case Study," *45th Purdue Industrial Waste Conference*, 1991.

236 Dharmavaram, Seshhasayi, et. al., "Automated Economic Analysis Model for Hazardous Waste Minimization," *Journal of Air Waste Management Association, Vol. 40*, 1990.

237 Gemmell, Alison S. and Higgins, Thomas E., "Waste Minimization Assessments," *AICHE 1990 Summer National Meeting*, 1990, August.

238 Walton, Clifford W. and Poppe, Gregg L., "Applying Microcomputers to the Analysis of Waste Treatment and Recovery Processes," *Plating and Surface Finishing*, 1990, June.

239 Tarabocchia, John and Karg, Al, "Printed Circuit Board Waste Treatment Removes Metals, Reduces Costs," *Plating and Surface Finishing*, 1990, February.

240 Ortego, J. Dale, "Spectroscopic and Leaching Studies of Solidified Toxic Metals," *Elsevier Science Publishers (Netherlands)*, 1990.

241 Swalheim, D.A., "Recovery and Reuse of Chemicals in Plating Rinses," *American Electroplaters and Surface Finishers Society*, 1988.

242 Schwartz, Morton and Swalheim, Donald A., "Electrochemistry for the Surface Finisher," *American Electroplaters and Surface Finishers Society*, 1992.

243 "Treatment of Cyanide-and Chromium-Bearing Wastes," *American Electroplaters and Surface Finishers Society*, 1991.

244 Yazdani, Azita, "Regulatory Pressures on Chlorinated Solvents: How Can Users Cope?," *13th AESF/EPA Conference on Environmental Control for the Surface Finishing Industry*, 1992.

245 Willis, Dennis G., "Decorative Plater Eliminates Vapor Degreasing," *13th AESF/EPA Conference on Environmental Control for the Surface Finishing Industry*, 1992.

246 Zavodjancik, John, "Pratt & Whitney's Vapor Degreaser Replacement Program," *13th AESF/EPA Conference on Environmental Control for the Surface Finishing Industry*, 1992.

247 DeWitt, James W., "Pollution Prevention: Developing Trends in Connecticut," *13th AESF/EPA Conference on Environmental Control for the Surface Finishing Industry*, 1992.

248 Dini, J.W. and Steffani, C.P., "Waste Minimization in a Non-Production Oriented Metal Finishing Operation," *13th AESF/EPA Conference on Environmental Control for the Surface Finishing Industry*, 1992.

249 Vidmar, Kevin P., "Three Dimensions of Waste Minimization, and Project Integration at Stanley Fastening Systems," *13th AESF/EPA Conference on Environmental Control for the Surface Finishing Industry*, 1992.

250 Koop, Steven W., "Wastewater Treatment Methods in the Metal Finishing Industry and the Recycling Factor," *13th AESF/EPA Conference on Environmental Control for the Surface Finishing Industry*, 1992.

251 Hanewald, R.H., et. al., "Recycling Metal Bearing Wastes Through Pyrometallurgical Technology," *13th AESF/EPA Conference on Environmental Control for the Surface Finishing Industry*, 1992.

252 Reinhard, Fred P., "The Recovery of Used Acid Utilizing Diffusion Dialysis," *13th AESF/EPA Conference on Environmental Control for the Surface Finishing Industry*, 1992.

253 Lindstedt, John, et. al., "Detection Level Limitations for Hexavalent Chromium," *13th AESF/EPA Conference on Environmental Control for the Surface Finishing Industry*, 1992.

254 Bleeks, Thomas W. and Davidson, Tamara S., "An Alternative to Cyanide Copper," *13th AESF/EPA Conference on Environmental Control for the Surface Finishing Industry*, 1992.

255 Mayer, Ernest, "Heavy Metals Removal with the DuPont/Oberlin Microfiltration Technology (SITE)," *13th AESF/EPA Conference on Environmental Control for the Surface Finishing Industry*, 1992.

256 Beemster, Bernard J., "Two Emerging Technologies for On-Line and In-Line Monitoring of Metals in Electroplating Wastes," *13th AESF/EPA Conference on Environmental Control for the Surface Finishing Industry*, 1992.

257 Van de Walle, Richard, et. al., "Use of Granular Magnesium-Based Adsorbent for Heavy Metal Removal and Acid Neutralization," *13th AESF/EPA Conference on Environmental Control for the Surface Finishing Industry*, 1992.

258 Von Kuster, Thomas and Malmberg, Joel, "Closed-Loop Rinse Recovery, Using Advanced Reverse Osmosis, Following Copper Cyanide Plating," *13th AESF/EPA Conference on Environmental Control for the Surface Finishing Industry*, 1992.

259 Zisson, Peter S. and McNeel, Thomas E., "Considerations on the Use of Dithiocarbamates for the Removal of Metals from Water," *13th AESF/EPA Conference on Environmental Control for the Surface Finishing Industry*, 1992.

260 Potamis, Gerald, "EPA Region I Effluent Toxicity Regulations," *12th AESF/EPA Conference on Environmental Control for the Surface Finishing Industry*, 1991.

261 Westfall, Brian A., "Technical Assistance for Waste Minimization," *12th AESF/EPA Conference on Environmental Control for the Surface Finishing Industry*, 1991.

262 Westfall, Brian A., "Toxicity Characteristics Leacheate Procedure or TCLP: An Overview of Its Effect on Waste Generators and Its Methodology," *12th AESF/EPA Conference on Environmental Control for the Surface Finishing Industry*, 1991.

263 Reinke, Daniel, et. al., "Comprehensive Waste Reduction at Von Duprin, Inc.," *12th AESF/EPA Conference on Environmental Control for the Surface Finishing Industry*, 1991.

264 Rosenblum, John, "Heavy Metals Waste Minimization: Practice and Pitfalls," *12th AESF/EPA Conference on Environmental Control for the Surface Finishing Industry*, 1991.

265 Gallerani, Peter, "Waste Minimization and Pollution Prevention," *12th AESF/EPA Conference on Environmental Control for the Surface Finishing Industry*, 1991.

266 Roditi, Solomon, "Implementation of Waste Minimization Program: A Case Study," *12th AESF/EPA Conference on Environmental Control for the Surface Finishing Industry*, 1991.

267 Backus, Scott, et. al., "Toxicity Reduction at Har-Conn Chrome Company," *12th AESF/EPA Conference on Environmental Control for the Surface Finishing Industry*, 1991.

268 Wood, David A., "The Use of Simple Material Balances to Solve Problems in a Circuit Board Manufacturer's Waste Water," *12th AESF/EPA Conference on Environmental Control for the Surface Finishing Industry*, 1991.

269 Goudie, William and Vitale, Americus C., "Membranes for Waste Minimization and Recycling," *12th AESF/EPA Conference on Environmental Control for the Surface Finishing Industry*, 1991.

270 Abbott, James and Stevens, Bruce W., "Waste Minimization in Decorative Chromium Plating Baths," *12th AESF/EPA Conference on Environmental Control for the Surface Finishing Industry*, 1991.

271 Yazdani, Azita, "Source Reduction of Chlorinated Solvents in Electronics Industry," *12th AESF/EPA Conference on Environmental Control for the Surface Finishing Industry*, 1991.

272 Vidmar, Kevin P., "Elimination of 1,1,1 Trichloroethane Through Critical Evaluation and Aqueous Cleaning," *12th AESF/EPA Conference on Environmental Control for the Surface Finishing Industry*, 1991.

273 Hovarth, Gene, "In-Tank Filtration/Solution Purification: A New Concept in Waste Minimization," *12th AESF/EPA Conference on Environmental Control for the Surface Finishing Industry*, 1991.

274 Hillier, Andrew C. and Walton, Clifford, "A Demonstration of the Application of an Equation Solving Program for Modeling Rinse and Recovery Systems," *12th AESF/EPA Conference on Environmental Control for the Surface Finishing Industry*, 1991.

275 Kirman, Lyal and Kovach, Jerome, "Recycling Anodizing Rinse Water Using Ion Exchange, A Case Study," *12th AESF/EPA Conference on Environmental Control for the Surface Finishing Industry*, 1991.

276 Fitzgerald, Nancy M., et. al., "Removal of Toxic Metals from Metal Finishing Waste Streams," *12th*

AESF/EPA Conference on Environmental Control for the Surface Finishing Industry, 1991.

277 Lewis, Tom III, "Recovery of Chromium for Waste Streams," *12th AESF/EPA Conference on Environmental Control for the Surface Finishing Industry*, 1991.

278 Siefert, Kristine S. and Lampert, Kerstin E., "ORP for Chemical Dosage Control in Metal Precipitation," *12th AESF/EPA Conference on Environmental Control for the Surface Finishing Industry*, 1991.

279 DeWitt, James W., "Water Quality-Based Permitting Strategies in Connecticut," *11th AESF/EPA Conference on Environmental Control for the Surface Finishing Industry*, 1990.

280 Curtis, Michael D., "Implementation of Toxicity Regulations in the State of Connecticut and the impact on Connecticut Electroplaters," *11th AESF/EPA Conference on Environmental Control for the Surface Finishing Industry*, 1990.

281 Hemmelggarn, David J. and Gullett, Lisa, "Biomonitoring - - One Plater's Experience," *11th AESF/EPA Conference on Environmental Control for the Surface Finishing Industry*, 1990.

282 Dhonau, Roger and Malinowski, John, "Effluent Toxicity Reduction: Issues & Strategies for Metal Finishers," *11th AESF/EPA Conference on Environmental Control for the Surface Finishing Industry*, 1990.

283 Flannery, Kelly A. and Zawadzki, Stanley, "Practical Applications of Stabilization," *11th AESF/EPA Conference on Environmental Control for the Surface Finishing Industry*, 1990.

284 Philipp, C.T. and Bennett, G.F., "Recycle Options for F006 Sludges," *11th AESF/EPA Conference on Environmental Control for the Surface Finishing Industry*, 1990.

285 Murphy, Kevin P., "Replacing CFC-113 Solvents in the Electroplating Industry," *11th AESF/EPA Conference on Environmental Control for the Surface Finishing Industry*, 1990.

286 Richmond, John, "Prolonging the Life of Electroless Nickel Plating Baths by Phosphite Removal," *11th AESF/EPA Conference on Environmental Control for the Surface Finishing Industry*, 1990.

287 Harnsberger, Steve, "A Fluoride Resistant Titania Ceramic Electrode for Metal Recovery and Pollution Control," *11th AESF/EPA Conference on Environmental Control for the Surface Finishing Industry*, 1990.

288 Carpenter, Charles J., "Overview of the Air Force Engineering and Service Center's Hazardous Waste Minimization Program," *11th AESF/EPA Conference on Environmental Control for the Surface Finishing Industry*, 1990.

289 Stuart, Larry G. and Greenberg, Lawrence A., "Optimized Wastewater Treatment for a Printed Circuit Board Facility," *11th AESF/EPA Conference on Environmental Control for the Surface Finishing Industry*, 1990.

290 Lee, Robert F., "A Circuit Board Manufacturer's Wastewater Treatment System, A Semi-Closed Loop Case Study," *11th AESF/EPA Conference on Environmental Control for the Surface Finishing Industry*, 1990.

291 Holmes, V.L. et. al., "The Substitution of IVD Aluminum for Cadmium," *Air Force Engineering and Service Center, Tyndall Air Force Base*, 1989, August.

292 Kate, Emily, "Solving Air Emissions Problems from Metal Cleaning Operations," *The National Environmental Journal*, 1992, August.

293 "Inspecting Your Degreaser," *Products Finishing*, 1992, December.

294 Warheit, Kevin E., "Zero Discharge - Fact or Fiction. A Waste Minimization Case Study," *Kinetico*, 1991, June.

295 Grosse, Douglas W., "Innovative Practices for Treating Waste Streams Containing Heavy Metals," *U.S. EPA Risk Reduction Engineering Laboratory, EPA/600/J-91/081*, 1991.

296 Rajcevic, Harold P., "A Hydrometallurgical Process for the Treatment of Industrial Wastes," *Plating and Surface Finishing*, 1990, July.

297 Kirsch, F. William and Looby, Gwen P., "Waste Minimization Assessment for a Bumper Refinish-

ing Plant," *U.S. EPA Risk Reduction Engineering Laboratory, EPA/600/M-991/020*, 1991, July.

298 Harries, Robert C., et. al., "Using a Waste Audit Approach to Determine Waste Management Alternatives at a Printed Circuit Board Manufacturing Plant," *43rd Purdue Industrial Waste Conference*, 1989.

299 Viet, Peter L., "The Evaporator - A Great Tool, But No Free Lunch," *Metal Finishing*, 1989, November.

300 Hartley, Howard S., "The Evolution of Evaporative Recovery," *Plating and Surface Finishing*, 1982, January.

301 U.S. EPA Risk Reduction Engineering Laboratory, "Guides to Pollution Prevention, The Fabricated Metal Products Industry," *EPA/625/7-90/006*, 1990, July.

302 U.S. EPA Risk Reduction Engineering Laboratory, "Facility Pollution Prevention Guide," *EPA/600/R-92/083*, 1992, May.

303 ICF Technology, Inc., "New York State Waste Reduction Guidance Manual," *New York State Department of Environmental Conservation*, 1989, March.

304 Science Applications International Corporation, "Pollution Prevention Opportunity Assessment: Implementation of Brush Plating, Tinker AFB," *EPA Contract No. 68-C8-0062*, 1992, November.

305 Cushnie, George C., "Summary Report: Control and Treatment Technology for the Metal Finishing Industry," *U.S. EPA, EPA 625/8-82-008*, 1982, August.

306 Durney, Lawrence J. (Ed.), "Electroplating Engineering Handbook," *Van Nostrand, Reinhold, Fourth Edition, New York*, 1984.

307 Lindstedt, John S., "Metal Loading Study, Metal Finishing Industry 1989 to 1991, Findings and Conclusions," *14th AESF/EPA Conference on Environmental Control for the Surface Finishing Industry* , 1993, January.

308 Sibik, Larry K., "The Clean Air Act Compliance Options for Electroplating and Surface Finishers," *14th AESF/EPA Conference on Environmental Control for the Surface Finishing Industry* , 1993, January.

309 Collins, Michael A., "Storm Water Management for the Electroplating and Surface Finishing Industry," *14th AESF/EPA Conference on Environmental Control for the Surface Finishing Industry,* 1993, January.

310 Boehm, Arthur J., "Metal Finishing Ventilation Design," *14th AESF/EPA Conference on Environmental Control for the Surface Finishing Industry* , 1993, January.

311 Schwering, Hans, et. al., "Crossflow Microfiltration for Extending the Service Life of Aqueous Alkali Degreasing Solutions," *14th AESF/EPA Conference on Environmental Control for the Surface Finishing Industry* , 1993, January.

312 Moebius, A., "Maintenance of Electroplating and Pickling Baths - The Key for the Minimization of Waste," *14th AESF/EPA Conference on Environmental Control for the Surface Finishing Industry* , 1993, January.

313 Peterson, Dave, "A Design of Experiments to Optimize Copper Cleaning and Achieve Waste Minimization," *14th AESF/EPA Conference on Environmental Control for the Surface Finishing Industry* , 1993, January.

314 Zavodjancik, John, "A Methodology to Minimize Rinsewater Use at Pratt & Whitney," *14th AESF/EPA Conference on Environmental Control for the Surface Finishing Industry* , 1993, January.

315 Marino, Frank A., "Alternative Cleaning Technology (ACT) A Corporate Approach for the Aerospace Industry," *14th AESF/EPA Conference on Environmental Control for the Surface Finishing Industry,* 1993, January.

316 Ebasco Environmental, "Hazardous Waste Minimization Study, Air Force Plant No. 6," *Lockheed Aeronautical Systems Company, Marietta, GA*, 1992, May.

317 Walton, Clifford W., "Software Review: Plato's Process Planner," *Plating and Surface Finishing*, 1992, December.

318 Drake, David and Gallerani, Peter, "Wastewater Management for the Metal Finishing Industry in

the 21st Century," *14th AESF/EPA Conference on Environmental Control for the Surface Finishing Industry* , 1993, January.

319 Nerkowski, Susan and Garriepy, David, "A Systematic Method To Address Solvent Use in Handwiping Operations," *14th AESF/EPA Conference on Environmental Control for the Surface Finishing Industry* , 1993, January.

320 Peterson, Dave, "Elimination of Chlorinated Solvents at Modine Manufacturing: An Integrated Approach," *14th AESF/EPA Conference on Environmental Control for the Surface Finishing Industry* , 1993, January.

321 Thom, James E. and Lankford, Lance, "Reducing Use of Ozone-Depleting Substances and Chlorinated Solvents at McClellan AFB," *14th AESF/EPA Conference on Environmental Control for the Surface Finishing Industry* , 1993, January.

322 Marino, Frank A., "A New Process Alternative for Replacing Ozone-Depleting Solvent Cleaners - Advanced Vapor Degreasing," *14th AESF/EPA Conference on Environmental Control for the Surface Finishing Industry*, 1993, January.

323 Jensen, Allan H., et. al., "Pollution- Free Plating Machines for Use in Workshops," *14th AESF/EPA Conference on Environmental Control for the Surface Finishing Industry* , 1993, January.

324 Wentler, Lloyd and Perrone, Stephen, "General Electric Closes the Loop at Their North Carolina Facility," *14th AESF/EPA Conference on Environmental Control for the Surface Finishing Industry* , 1993, January.

325 Williams, Bill, et. al., "Case Study: Achieving Stringent Toxicity Limits at a Job Plating Shop with Waste Minimization and End-Of-Pipe Treatment," *14th AESF/EPA Conference on Environmental Control for the Surface Finishing Industry* , 1993, January.

326 Jenkins, Robert W. and Sacco, Robert M., "Waste Minimization Program Implemented at Hamilton Standard Through Optimization of Wastewater Treatment System," *14th AESF/EPA Conference on Environmental Control for the Surface Finishing Industry* , 1993, January.

327 Lindstedt, John and Wagner, Ernestine, "Electroplating Effluent Polishing Using Tertiary Ion Exchange: A Case Study," *14th AESF/EPA Conference on Environmental Control for the Surface Finishing Industry* , 1993, January.

328 Tragellis, Stratton G., "Sludge Recycling Versus Waste Minimization," *14th AESF/EPA Conference on Environmental Control for the Surface Finishing Industry* , 1993, January.

329 Zapisek, Stephen, "Recycling Effluent of Conventional Waste-Water Precipitation Treatment Via Ion Exchange and Evaporation," *14th AESF/EPA Conference on Environmental Control for the Surface Finishing Industry* , 1993, January.

330 Brooks, Clyde S., "Metal Recovery from Electroless Plating Wastes," *Metal Finishing*, 1989, May.

331 Humpal, Grey J. and Chiesa, Robert, "Selection, Design and Operation of Inclined Plate Settlers," *Metal Finishing*, 1990, November.

332 Renard, D.E., "Metal Recovery for Leached Plating Sludge," *Plating and Surface Finishing*, 1987, October.

333 Spinna, Robert J., "AES Research at Work," *Plating and Surface Finishing*, 1981, July.

334 Arniella, Elio F. and Blythe, Leslie J., "Solidifying Traps Hazardous Wastes," *Chemical Engineering*, 1990, February.

335 Grace, Christen, "Glass Manufacturing Can Be a Conduit for Waste Management," *Hazmat World*, 1989, November.

336 Deuschle, Andreas, "Diffusion Dialysis - An Economical Technology for Recovery of Acids From Pickling Processes," *14th AESF/EPA Conference on Environmental Control for the Surface Finishing Industry* , 1993, January.

337 Hepworth, M.T. and Becksted, L.W., "Chromium Recovery from Electroplating Sludge," *Plating and Surface Finishing*, 1989, November.

338 American Electroplaters and Surface Finishers Society, "AESF Shop Guide, 9th Edition," *Orlando, FL*, 1992.

339 Gallerani, Peter, "Summary Report: Minimization of Metal Finishing Wastes at Pratt & Whitney," *Integrated Technologies*, 1993, February.

340 Wood, William G. (Coordinator), "The New Metals Handbook, Vol. 5. Surface Cleaning, Finishing, and Coating," *American Society for Metals*, 1990, May.

341 Graham, A. Kenneth, "Electroplating Engineering Handbook, Third Edition," *Van Nostrand Reinhold Co.*, 1971.

342 Reinhard, Fred P., "Utilization and Practical Application of Ion Exchange - A Critical Review," *14th AESF/EPA Conference on Environmental Control for the Surface Finishing Industry* , 1993, January.

343 Baldwin, Phil C., "A Disposal System for Metal Finishing Sludges," *Metal Finishing*, 1989, June.

344 McCarthy, James A., "Characterization of the Leachability of Electroplating Wastewater - Treatment Sludges," *Plating and Surface Finishing*, 1979, November.

345 "Federal Register, Vol. 51, No. 56," *p. 10168*, 1986, 24 March.

346 Zaki, Nabil, "Risk reduction in Chrome Emissions," *Metal Finishing*, 1990, June.

347 DeWitt, James W., "The F006 Land Ban - Its Impact on the Metal Finisher," *Plating and Surface Finishing*, 1989, April.

348 Palmer, S.A.K., et.al., "Metal/Cyanide Containing Waste Treatment Technologies," *Noyes Data Corporation, Park Ridge, NJ*, 1988.

349 Brown, Craig J., "Metal Recovery by Ion Exchange and Electrowinning," *4th International Forum on Electrolysis in the Chemical Industry*, 1990.

350 Lowenheim, Frederick A., "Electroplating Fundamentals of Surface Finishing," *McGraw-Hill Book Company, New York, NY*, 1978.

351 Ontario Research, "Factors Which Affect the Success of Metal/Chemical Recovery Installations in Metal Finishing Industry," *American Electroplaters' Society, Canadian Branch*, 1987, June.

352 Maintino, Phil, "Waste Recovery Using the Rotating Cathode," *Circuit Tree Magazine*, Unknown.

353 Grey, Jeri, et. al., "Atmospheric Evaporative Recovery on a Nickel Plating Operation," *Waste Reduction - Pollution Prevention: Progress and Projects within North Carolina*, 1988, March.

354 King, Bill, "SAMSCO," *SAMSCO, Inc.*, 1992, January.

355 Choate, Cliff, "Atmospheric Evaporators," *Metal Finishing*, 1990, March.

356 Ryder, George, "Make Pollution Prevention Pay," *Metal Finishing*, 1987, July.

357 Yates, William, "Leveraging Recovery," *Products Finishing*, 198, February.

358 Yates, Bill, "Atmospheric Evaporators," *Plating and Surface Finishing*, 1986, April.

359 Yates, Bill, "Natural Recovery," *Finishers' Management*, 1988, March.

360 Fischer, Craig, "Technotreat Vendor Literature," *The Technotreat Corporation (OK)*, 1989, April.

361 Sawyer, Clair N. and McCarty, Perry L., "Chemistry for Sanitary Engineers," *McGray-Hill Book Company*, 1967.

362 Kemmer, Frank N. (Ed.), "The Nalco Water Handbook," *McGraw-Hill Book Company*, 1979.

363 Munns, Kevin, "Western Forge Maintains High Quality With Purification System," *Finishers' Management*, 1990.

364 Letherdale, Joe, "Acid Purification for Tube Mill," *TPQ*, 1992.

365 Brown, Craig J., "Productivity Improvements Through Recovery of Pickle Liquors with APU Process," *Iron and Steel Engineer*, 1990, January.

366 "Anodizing: Better Control Means Consistent Color," *Modern Metals*, 1990, August.

367 Dubpernell, George, "A History of Chromium Plating," *Plating and Surface Finishing*, 1984, June.

368 Knill, Edmund C. and Chessin, Hyman, "Purification of Hexavalent Chromium Plating Baths," *Plating and Surface Finishing*, 1986, August.

369 Knoll, Edmund C. and Chessin, Hyman, "Contamination of Hexavalent Chromium Plating Baths," *Plating and Surface Finishing*, 1986, August.

370 Cushnie, George C., "Removal of Metal Cations from Chromium Plating Solutions," *10th AESF/EPA Conference on Environmental Control for the Surface Finishing Industry* , 1989, January.

371 Mandich, N.V., "Removal of Contaminants in Chrome Plating Solutions via Porous Pot Method," *Proceedings AESF Hard Chromium Workshop*, 1992, January.

372 Klink, Kevin, et. al., "Prototype Evaluation of Electrodialysis Process," *McClellan AFB (Contract F04699-90-D0035, Delivery Order 5029)*, 1991, December.

373 "Handbook of Evaporation Technology," *None*, 1990.

374 Crabbe, David and McBride, Richard, "The World Energy Book," *Nichols Publishing Co., NY*, 1978.

375 "Evaporator Handbook, Third Edition," *AVP Group*, Not Given.

376 CENTEC Corporation, "Control Technology for the Metal Finishing Industry - Evaporators," *U.S. EPA 625/8-79-002*, 1979.

377 Swenson Corporation, "Swenson Evaporators," *Not Given*, Not Given.

388 "Finishing Facts: Ion Exchange Helps," *Plating and Surface Finishing*, 1983, May.

389 Altmayer, Frank, "Introducing the COPS," *Plating and Surface Finishing*, 1983, March.

390 Cushnie, George, "Personal Communication: Max Smith (Pentol) and George Cushnie CAI Engineering," 1991, December.

391 Centec Corporation, "Environmental Regulations and Technology," *USEPA 625/10-85/001*, 1985, October.

392 Centec Corporation, "Economics of Wastewater Treatment Alternatives for the Electroplating Industry," *USEPA 625/5-79-016*, 1979, May.

393 Centec Corporation, "Reducing Water Polution Control Costs in the Electroplating Industry," *USEPA 625/5-85/016*, 1985, May.

394 Centec Corporation, "In-Plant Changes," *USEPA 625/8-82-008*, 1982, August.

395 Centec Corporation, "Sludge Handling, Dewatering, and Disposal Alternatives for the Metal Finishing Industry," *USEPA*, 1982, October.

396 Centec Corporation, "Control and Treatment Technology for the Meal Finishing Industry, Ion Exchange," *None*, 1981, June.

397 Vaughan, Daniel, "Closed-Loop Processing of Chromic Acid Solutions," *10th AESF/EPA Conference on Environmental Control for the Surface Finishing Industry*, 1989, January.

398 Backus, Scott, "Plato's Process Planner," *Finisher's Management*, 1993, May.

399 Rockstraw, David A., "Removal of Heavy Metals from Water Elelctrodialysis in the Presence of Acids and Salts," *10th AESF/EPA Conference on Environmental Control for the Surface Finishing Industry*, 1989, January.

400 David, Thomas A., "Application of Ion Exchange Membranes to Industrial Separation," 1990.

401 Osmonics, "Pure Water Handbook," *Osmonics*, 1991, October.

402 CAI Engineering, "Evaluation of the Chrome Plating System at the U.S. Mint, Philadelphia," *United States Mint P.O. No. MT-90-222*, 1990, September.

403 Koch Engineering Company, "Mist Eliminators," *Wichita, KS*, 1990.

404 Powers, William E. "Air Pollution Control Equipment for Selected Plating Shop Precosses Based on Emission Test Results," *Naval Energy and Environmental Support Activity*, 1985 (est.).

405 Venturini, Peter, D., "Hexavalent Chromium: Data Control Measure for Chrome Platers," *Air Resource Board, Sacramento, CA*, 1987, March.

406 "Finishing Facts," *Plating and Surface Finishing*, 1988, June.

407 Peer Consultants, "Determination of the Efficiency of a Meshpad Mist Eliminator," *USEPA Emissions Measurement Branch (EPA Contract No. 68-02-4346)*, 1989, August.

408 Peer Consultants, "Detemination of the Efficiency of a Mesh-Pad Mist Eliminator," *USEPA Emisions Measurement Branch (EPA Contract No. 68-02-4346)*, 1989, February.

409 Cushnie, George, "Removal of Metals from Wastewater, Neutralization and Precipitation," *Noyes Publications, Park Ridge, NJ*, 1984.

410 Hoffman, Robert O., "Filter Presses for the Electroplater," *Plating and Surface Finishing*, 1983, June.

411 Turgeon, Irwin, "Filter Press Maintenance," *Metal Finishing*, 1987, August.

412 Hall, Don and Uhrich, Kevin, "Simultaneous Removal of Several Metals-A Case Study," *Andco Environmental Processes, Inc.*, 1990.

413 Bowers, Alan R., et. al., "Iron Process for Treatment of Cr (VI) Wastewaters," *Metal Finishing*, 1986, November.

414 Casaday, Joseph E., "Membrane Filtration System Cleans Up at UniFirst Plants," *Textile Rental Magazine*, 1992, February.

415 USEPA, "Guide to Clean Technology-Alternative Metal Finishes (draft)," 1992, July.

416 USEPA, "Guide to Clean Technology-Alternatives to Chlorinated Solvents for Cleaning and Degreasing," 1992, July.

417 USEPA, "Guides to Pollution Prevention-The Printed Circuit Board Manufacturing Industry," *Risk Reduction Engineering Laboratory*, 1990, June.

418 Patry, Michael and Graham, Glen, "Hazardous Waste Minimization Report," *Oklahoma City Air Logistics Center*, 1991.

419 Norgaad, Gregor E., "Centralized Waste Treatment Facility Opens in Twin Cities Area," *Plating and Surface Finishing*, 1988, November.

420 Odle, Robert R., et. al., "The Recycling of Hazardous Metal Plating Wastes," *The Journal of the Minerals, Metals & Materials Safety*, 1991, June.

421 Murphy, Michael (ed), "Metal Finising Guidebook and Directory Issue '93," *Metal Finishing*, 1993, January.

422 HPD Incorporated, "Evaporators," *Napperville, IL*, 1990.

423 Perry, Robert H. and Green, Don, "Perry's Chemical Engineers' Handbook, Sixth Edition," *McGraw-Hill Inc.*, 1969.

424 Dewitt, James E., "F006 Land Ban—Its Impact on the Metal Finisher," *10th AESF/EPA Conference on Environmental Control for the Surface Finishing Industry*, 1989, January.

425 Cashen, Tim, "An Integrated System for Assuring Safe Disposal of Elelctroplating Sludge Waste," *10th AESF/EPA Conference on Environmental Control for the Surface Finishing Industry*, 1989, January.

426 Munster, Graf, "The Application of Various GFS Solidification Methods for Wastes from Metal Treatment Industries," *10th AESF/EPA Conference on Environmental Control for the Surface Finishing Industry*, 1989, January.

427 Butlin, John, "European Developments in Hazardous Waste Management," *10th AESF/EPA Conference on Environmental Control for the Surface Finishing Industry*, 1989, January.

428 Zaki, Nabil, "Complying with Air Quality Standards with New Trivalent Chromium Plating Technology," *10th AESF/EPA Conference on Environmental Control for the Surface Finishing Industry*, 1989, January.

429 Miles, Thomas, "An Ultra-High Efficiency Control Device," *10th AESF/EPA Conference on Environmental Control for the Surface Finishing Industry*, 1989, January.

430 Camp, William O. and Meeker, Nicholas, "Discharge Compliance for PWB Manufacturing

Using Microfiltration Membranes," *10th AESF/EPA Conference on Environmental Control for the Surface Finishing Industry*, 1989, January.

431 Martin, Hollis L., "Wastewater Filtration Enhancement," *10th AESF/EPA Conference on Environmental Control for the Surface Finishing Industry*, 1989, January.

432 Susa, Thomas J. and Tata, Richard A., "A Case Study of an Electrodialysis Reversal Syste, for the Recovery of Nickel Salts," *10th AESF/EPA Conference on Environmental Control for the Surface Finishing Industry*, 1989, January.

433 Eason, John, "The Use of Canister Ion-Exchange Technology in a Zero Discharge Facility," *10th AESF/EPA Conference on Environmental Control for the Surface Finishing Industry*, 1989, January.

434 Ryan, James et. al., "Case Study of a Minimum Discharge Heavy Metal Waste Reduction System at Aeroscientific Corporation," *10th AESF/EPA Conference on Environmental Control for the Surface Finishing Industry*, 1989, January.

435 Weber, Walter J., "Physicochemical Processes," *Wiley-Interscience*, 1972.

436 "Environmentally Speaking...the Progress Continues," *Pratt & Whitney*, 1991.

437 Warheit, Kevin E., "A Sorption Filter System Helps Manufacturers Meet Strict Effluent Limits," *9th AESF/EPA Conference on Environmental Control for the Surface Finishing Industry* , 1988, January.

438 Scamehorn, John F. and Christian, Sherril D., "Removal of Dissolved Metals from Water Using Micellar-Enhanced Ultrafiltration," *9th AESF/EPA Conference on Environmental Control for the Surface Finishing Industry* , 1988, January.

439 Couture, Stephen D. et. al., "Meeting Water Quality Based Discharge Standards With Advanced Wastewater Treatment Technology," *9th AESF/EPA Conference on Environmental Control for the Surface Finishing Industry* , 1988, January.

440 Gopalratnam, Venbakm C. et. al., "Effect of Collectors and Their Dosages on the Removal of Metals from Industrial Wastewaters Using Precipitation/Air Flotation," *9th AESF/EPA Conference on Environmental Control for the Surface Finishing Industry* , 1988, January.

441 Uhrich, Kevin, D. and Hall, Don, "Simultaneous Removal of Several Metals A Case Study," *9th AESF/EPA Conference on Environmental Control for the Surface Finishing Industry* , 1988, January.

442 Pajunen, Paul, "Copper and Nickel Removal in Printed Circuit Board Processing by Ion Exchange and Electroforming," *9th AESF/EPA Conference on Environmental Control for the Surface Finishing Industry* , 1988, January.

443 Grotelueschen, Ralph, D., "An Innovative Wastewaer Treatment Process that Simultaneously Improves Effluent Quality and Reduces Heavy-Metals EP Toxicity," *9th AESF/EPA Conference on Environmental Control for the Surface Finishing Industry* , 1988, January.

444 Boyce, A.R. and Kavanaugh, D.J., "Electrolytic Regeneration of Chromic/Sulfuric Acid Etchant," *9th AESF/EPA Conference on Environmental Control for the Surface Finishing Industry* , 1988, January.

445 Pardus, Michael J., "Cost Effective Compliance with the Pretreatment Standards: A case Study in Waste Minimization," *9th AESF/EPA Conference on Environmental Control for the Surface Finishing Industry* , 1988, January.

446 Reinhard, Fred P., "Ion Exchange as a Tool for Water and Metal Recycling," *9th AESF/EPA Conference on Environmental Control for the Surface Finishing Industry* , 1988, January.

447 Bailey, Daniel and Chan, Mike, "High-Mass-Transer (HMT) Electrochemical Process: Integration of Metal Recovery and Regulatory Compliance (A Case Study)," *9th AESF/EPA Conference on Environmental Control for the Surface Finishing Industry* , 1988, January.

448 Tison, Richard P., "Electrochemical Metals Recovery from Dilute Solutions Using Ion-Exchange Materials," *9th AESF/EPA Conference on Environmental Control for the Surface Finishing Industry,* 1988, January.

449 Forrestal, Brian, "Factors Which Affect the Success of Metal/Chemical Recovery Installations in

the Metal Finishing Industry-Survey Results," *9th AESF/EPA Conference on Environmental Control for the Surface Finishing Industry* , 1988, January.

450 McCoy, Daniel, "Toxicology of Metals in Plating and Surface Finishing," *9th AESF/EPA Conference on Environmental Control for the Surface Finishing Industry* , 1988, January.

451 Glover, Connie, "New Directions for Hazardous Waste Testing," *9th AESF/EPA Conference on Environmental Control for the Surface Finishing Industry* , 1988, January.

452 Raifman, Stuart M., "Optimizing Filtration Systems," *Products Finishing*, 1993, May.

453 Nudo, Lori, "Heirs to the Throne," *Pollution Engineering*, 1993, June.

454 Rosendahl, John, "Environmentally Friendly Cleaning Solutions: The Lingo," *Products Finishing*, 1993, Feb..

455 Poll, Jerry, "Chromates, Recycling and Zinc Sludge," *Products Finishing*, 1993, Feb..

456 Bar, Rob, "Surpassing Chromium Emission Standards," *Products Finishing*, 1993,July.

457 Graves, Beverly, "Switching to Trivalent Chromium Plating," *Products Finishing*, 1993, July.

458 Bassett, Bill, "Carbonates and Alkaline Zinc Plating," *Products Finishing*, 1993, July.

459 Monopoli, A. V., "Removing Dissolved Inorganics from Industrial Wastewater," *The National Environmental Journal*, 1993, Aug..

460 Huang, Chein-Ho and Chen, A.C., "Influences of a Wetter and Other Plating Metals On Removal of Dissolved Nickel Metal," *Plating and Surface Finishing*, 1993, March.

461 Travis, Glen and Boster, Cynthia, "Elimninating 1,1,1-Trichlor Vapor Degreasing In Aerospace Repair Applications," *Plating and Surface Finishing*, 1993, March.

462 Horvath, Gene and Isham, Steve, "In-Tank Filtration for the Anodizing Industry," *Plating and Surface Finishing*, 1993, March.

463 Ehmann, I., "Electrolytic Effluent Treatment for Acid or Cyanide Zinc Plating Baths," *Metal Finishing*, 1993, May.

464 Zhou, C.D. and Chin, D.T., "Copper Recovery and Cyanide Destruction with a Plating Barrel Cathode and a Packed-Bed Anode," *Plating and Surface Finishing*, 1993, June.

465 Altmayer, Frank, "Improving the Operation of Cyanide Destruct Systems," *Plating and Surface Finishing*, 1988, April.

466 Coulter, Kenneth, "Centralized Treatment of Metal Finishing Wastes," *Plating and Surface Finishing*, 1979.

467 Lancy, L.E. and Steward, F. A., "Disposal of Metal-Finishing Sludges- The Segregated Landfill Concept," *Plating and Surface Finishing*, 1978, December.

468 Innes, W.P. and Toller, W.H. and Tomasello, D., "Electrolytic Regeneration Of Chromic Acid Etchants," *Plating and Surface Finishing*, 1978, November.

469 Cadotte, A. and Golomb, A. and Monteith, H., "Contaminant Build-up in the Plating Bath Resulting from Closed Loop Operation," *Plating and Surface Finishing*, 1978, October.

470 Cadotte, A.P. and Monteith, H. D. and Golomb, A., "Contaminant Build-up in the Plating Bath Resulting from Closed Loop Operation - Nickel Plating," *Plating and Surface Finishing*, 1979, January.

471 Hall, E.P. and Lizdas, D.J. and Auerbach, E.E., "Recovery Techniques in Electroplating," *Plating and Surface Finishing*, 1979, February.

472 "Recovery Pays," *Plating and Surface Finishing*, 1979, February.

473 Spatz, Dean, "A Case History of Reverse Osmosis Used for Nickel Recovery in Bumper Recycling," *Plating and Surface Finishing*, 1979, July.

474 Wang, Victor and Merchant, Abid N., "Metal-Cleaning Alternatives for the 1990s," *Metal Finishing*, 1993, April.

475 Baxley, Sylvia L. and Safranek, William H. and Isham, Steven B., "EPA/AES Conferees Dissect

Pollution Control," *Plating and Surface Finishing*, 1979, March.

476 Bishara, J.I. and Brannan, J.R. and Horvath, R.J., "Recovery of Hexavalent Chromium from Plating Rinse Waters," *13th AESF/EPA Conference on Environmental Control for the Surface Finishing Industry* , 1992.

477 Karrs, Stanley R. and McConahy, Joan L., "A Comparison Of Filtration Techniques to Meet Tight Effluent Metal Limits," *13th AESF/EPA Conference on Environmental Control for the Surface Finishing Industry* , 1992.

478 Asada, Katsutoshi and Gerdes, Lisa and Kawahara, Takuo, "Electrodialysis of Effluents From Treatment of Metallic Surfaces," *12th AESF/EPA Conference on Environmental Control for the Surface Finishing Industry*, 1992.

479 Hosea, J. Michael and Nelis, Patrick M. and Mayne, Michael d. and Greene, Michael, "Metal Recovery by Ion Exchange - Seven Crucial Issues," *13th AESF/EPA Conference on Environmental Control for the Surface Finishing Industry* , 1992.

480 Mayr, M. and Blatt, W. and Busse, B. and Heinke, H., "Electrolytic Systems for Applications in Fluoride-Containing Electrolytes," *Fourth International Forum On Electrolysis in the Chemical Industry*, 1990, November.

481 Schwering, Hans and Golisch, Peter and Kemp, Alan, "Crossflow Microfiltration for Extending the Service Life of Aqueous Alkali Degreasing Solutions," *Plating and Surface Finishing*, 1993, April.

482 Altmayer, Frank, "Comments/Suggestions for Interim Reports," *NCMS Steering Committee*, 1993, September.

483 Weismantel, Guy E., "Sludge Dewatering Technology," *Pollution Engineering*, 1993, April.

484 Williams, Bill; Schaffman, Paul; Backus, Scott; Steir, George, "Case Study - Achieviing Stringent Toxicity Limits At a Job Plating Shop With Waste Minimization and End-of-Pipe Treatment," *Plating and Surface Finishing*, 1993, April.

485 Langan, James P., "The Montreal Protocol"," *Plating and Surface Finishing*, 1993, April.

486 Kula, Darrell, "The Cost Benefit of Metal Recovery in the Plating Shop Instead of Waste Treatment While Attaining Compliance," *IPC Fall Meeting, Avionics Division*, 1983, September.

487 Altmayer, Frank, "Complying With Cadmium Standards," *Plating and Surface Finishing*, 1993, April.

488 Durney, Larry, "PF Finishing Clinic," *Products Finishing*, 1990, May.

489 Waitz, William H., "Ion Exchange in Heavy Metals Removal and Recovery," *Rohm and Haas Co.*, 1979.

490 Unknown, "Compliance At Four-D Metal Finishing," *Finishers' Management*, 1984, October.

491 Pardus, M.J. and Regan, R.W., "Hazardous Waste Minimization at a Brass Wire Mill," *Plating and Surface Finishing*, 1989,April.

492 Lindstedt, John and Doyle, Michael, "Silver Recovery with Ion Exchange and Electrowinning," *AESF./EPA*, 1993, January.

493 Kirman, Lyle, "Learning to Live with TDS Limits," *Vendor Supplied Paper*, unknown date.

494 Hosea, Michael, "Sludgeless Pollution Control," *Products Finishing*, unknown date.

495 Cadwallader, J.F., and McCauley, J.L., "An Energy-Efficient System for Closed-Loop Treatment of Electroplating Wastewater," *Plating and Surface Finishing*, 1982, January.

496 None, "New Soldering Process REquires Less Energy and Reduces Waste," *Waste Minimization and Recycling Report, Government Institutes, Inc.*, 1993, February.

497 Kornosky, Ronald, "Finishers' Think Tank," *Plating and Surface Finishing*, unknown date.

498 Unknown, "Enfinity, Electroless Nickel Process," *Stapleton Technologies, Inc.*, 1993, February.

499 Unknown, "Kinetico Nickel Recovery System," *Kinetico Engineered Systems Division*, unknown date.

500 Hirsch, Stanley, "Deionization and its Application for Electroplating," *Metal Finishing*, 1988, June.

501 Capaccio, Robert S. and Gingras, Robert K., "Wastewater Treatment for Electroless Nickel Plating," *Products Finishing*, 1989, August.

502 Fleet, Bernard: Small, Colin E.; and Kowalski, Mark D., "Technologies, Strategies and Economics of Managing the Electronic Industry's Toxic Wastes," *AESF/EPA*, unknown date.

503 Whitmore, Pilar, "Closed Loop in Two Steps (A Practical Approach to Zero Discharge of Wastewater)," *Surface Finish 91*, 1991.

504 Whittle, Larry; Ross, Claire E.; and Vachon, Derek, "A Study of the Effectiveness of Cyanide Destruction Using Free Residual Chlorine Measurement to Control the Alkaline Chlorination Process," *AESF/EPA*, unknown date.

505 Munns, Kevin, "Recycling chemicals on the Anodizing Line - Cost Savings and Quality Improvements," *AESF/EPA*, unknown date.

506 McCullough, Matthew, "Evaluating the Economics and Effectiveness of Source Reduction Options in Metal Finishing," *AESF/EPA*, unknown date.

507 Moebius, Andreas, "Electrochemical Processing for the Minimization of Wastes in the Electroplating Industry - A Critical Review," *AESI/EPA*, unknown date.

508 Horner, Gregg Daniel; and Duffey, Joseph G., "Electrochemical Removal of Heavy Metals from Wastewaters," *AESF Conference*, 1983, June.

509 Markovac, V. and Heller, H.C., "Engineering Aspects of Electrodialysis for Nickel Plating Rinsewater," *Plating and Surface Finishing*, 1982, January.

510 Roach, Edward T., "Evaluation of Donnan Dialysis for Treatment of Nickel Plating Rinsewater," *USAEPA*, 1985, August.

511 Louchart, Gary W., "Howard Plating Cleans Up its Act with Magnesium Hydroxide," *Plating and Surface Finishing*, 1987, November.

512 Scholler, M. and van Dijk, J.C., "Recovery of Heavy Metals by Crystallization," *Metal Finishing*, 1987, November.

513 Brown, Craig, "Recovery of Metals from Effluents by High Efficiency Air Agitation Electrowinning," *AESF, 73nd Annual Technical Conference*, 1986, June.

514 DuPont, "Axrel Cleaning Agents," *Wilmington, DE*, 1992, November.

515 Smiley, B.C. et.al., "A Membrane System for Aemi-Aqueous Defluxing," *DuPont Company*, 1992.

516 Meagher, Michael and Gallerani, Peter, "Zero Discharge Permitting - From Swamp to Quagmire," *AESF/EPA Conference*, 1994, January.

517 U.S. EPA, "Guidance Manual for Electroplating and Metal Finishing," *U.S. EPA, Washington, DC*, 1984, February.

518 Lott, John, "Comments/Suggestions forNCMS Interim Reports," *DuPont Electronics*, 1993, November.

519 U.S. Department of Energy, "Environmental Consciousness: A Strategic Competitiveness Issue for the Electronics and Computer Industry," *DOE Study*, 1993, March.

520 Bormann, Gerhard, "Membranelektrolyse zur Instandhaltung von Hartchrombadern," *Galvanotecnik, D-7968 Saulgau*, 1992.

521 USEPA, "Proposed Standards for Chromium Emissions from Hard and Decorative Chromium Electroplating and Chromium Anodizing Tanks; Proposed Rule," *40 CRF Part 63*, December 16, 1993.

522 Reinhard, Fred, "Advanced Reverse Osmosis-A Practical Tecnology for Water and Metal Recycling," *15th AESF/EPA Conference on Environmental Control for the Surface Finishing Industry*, 1994, January.

523 Cushnie, George C., "Electroplating Wastewater Pollution Control," *Noyes Publications, Park Ridge, NJ*, 1984.